CIGR Handbook of Agricultural Engineering

Volume III
Plant Production Engineering

Edited by CIGR—The International Commission of Agricultural Engineering

Volume Editor:
Bill A. Stout
Texas A&M University, USA

Co-Editor:
Bernard Cheze
Ministry of Agriculture, Fisheries and Food, France

Published by the American Society of Agricultural Engineers

LCCN 98-93767 ISBN 1-892769-02-6

For Information, contact:

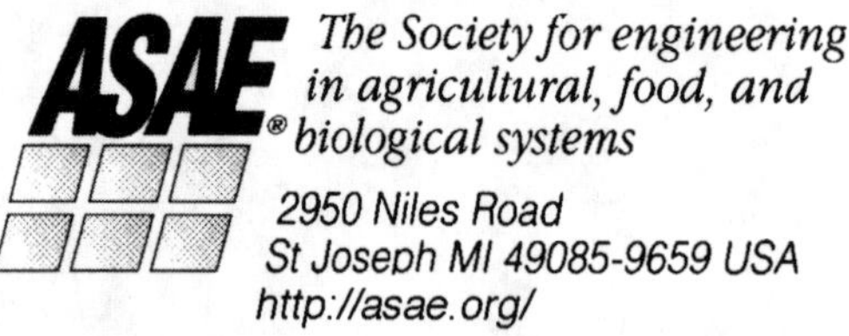

2950 Niles Road
St Joseph MI 49085-9659 USA
http://asae.org/

Manufactured in the United States of America

Editors and Authors

Volume Editor

B. A. Stout
Agricultural Engineering Department, Texas A&M University, College Station, Texas 77843-2117, USA

Co-Editor

B. Cheze
MAPA/DEPSE 5, 78 Rue de Varenne, 75349 Paris SP 07, France

Authors

P. F. J. Abeels
Professor Em., Univ. Catholique de Louvain, Faculte des Sciences Agronomiques, Department of Environmental Sciences and Land Management, Agricultural and Forest Engineering, Place Croix du Sud 2, Box 2, B 1348 Louvain La Neuve, Belgium

H. Auernhammer
Institut für Landtechnik, der TU München, Am Staudengarten 2, 85350 Freising-Weihenstephan

P. Balsari
Universitá degli Studi di Torino, Dipartimento di Economia e Ingegneria Agraria, Forestale e Ambientale, Grugliasco, Italy

J. F. Billot
TMAN CEMAGREF, Parc de Tourvoie, 92160 Antony, France

D. Blary
Mécanisation agricole, CIRAD-CA Programme GEC, BP 5035 34090 Montpellier Cedex 1, France

E. H. Bourarach
Institute of Agronomy and Veterinary Medicine, Hassan II, Department of Agricultural Engineering, B.P. 6202 Rabat Instituts, Rabat, Morocco

A. G. Cavalchini
Università Degli Studi, Istituto di Ingegneria Agraria, 20133 Milano—Via G. Celoria, 2, Italy

W. Chancellor
Biological and Agricultural Engineering Department, University of California Davis, Davis, California 95616-5294

B. Cheze
MAPA/DEPSE 5, 78 Rue de Varenne, 75349 Paris SP 07, France

B. Fritz
Agricultural Engineering Department, Texas A&M University, College Station, Texas 77843-2117, USA

C. E. Goering
Agricultural Engineering Department, University of Illinois, Urbana, Illinois 61801, USA

R. Hahn
Standards Coordinator, American Society of Agricultural Engineering, 2950 Niles Road, St. Joseph, MI 49085-9659, USA

M. Havard
CIRAD-SAR, BP 5035, 73, Rue J.F. Breton, 34090 Montpellier, Cedex 1, France

H. J. Heege
Christian-Albrechts-Universität Kiel, Institut für Landwirtschaftliche Verfahrenstechnik, Max-Eyth-Strasse 6, 24118 Kiel, Germany

R. O. Hegg
Agricultural and Biological Engineering Department, Clemson University, McAdams Hall, Box 340357, Clemson, South Carolina 29634-0357, USA

J. W. Hofstee
Department of Agricultural Engineering and Physics, Wageningen Agricultural University, Agrotechnion Bomenweg 4, 6703 HD Wageningen, The Netherlands

D. R. Hunt
University of Illinois, Agricultural Engineering Department, University of Illinois, Urbana, Illinois 61801, USA

H. D. Kutzbach
Inst. F. Agrartechnik -440-, Universitaet Hohenheim, Garbenstr. 9, D-70599 Stuttgart, Germany

P. S. Lammers
Institut für Landtechnik, Universität Bonn, Postanchrift: 53115 Bonn, Nussallee 5, Germany

A. Lara Lopez
Consul for Science and Technology, State of Guanajuato, Mineral de Valencia No. 20, Marfil, Guauajuato, Mexico

E. Manfredi
Universita degli studi di Bologna, Dipartimento di Economia e Inegneria Agrarie, Via Zamboni, 1, 40126 Bologna, Italy

R. Oberti
Universitá degli Studi di Milano, Istituto di Ingegneria Agraria, Milano, Italy

E. U. Odigboh
University of Nigeria, Faculty of Engineering, Nsukka, Enugu State, Nigeria

J. Ortiz-Canavate
E.T.S.I.AGRONOMOS, Dpt. Ing. Rural, Universidad Politécnica de Madrid, Ciudad Universitaria s/n, 28040—Madrid, Spain

C. B. Parnell
Department of Agricultural Engineering, Texas A&M University, College Station, Texas 77843-2117, USA

R. Peters
KTBL-Versuchsstation Dethlingen, Diethlingen 14, D-29633 Munster, Germany

R. Pirot
Mécanisation agricole, CIRAD-CA Programme GEC, BP 5035 34090 Montpellier Cedex 1, France

G. Quick
Department of Agricultural and Biosystems Engineering, Iowa State University, Davidson Hall, Ames, Iowa 50011-3080, USA

K. T. Renius
Institut fur Landmaschinen, Technical University of Munich, Arcisstr. 21, 80333 Munchen, Germany

A. G. Rijk
Asian Development Bank, P.O. Box 789, Manila 0980, Philippines

M. Ruiz-Altisent
E.T.S.I.AGRONOMOS, Dpt. Ing. Rural, Universidad Politécnica de Madrid, Ciudad Universitaria s/n, 28040—Madrid, Spain

J. Sakai
Kyushu University, Agricultural Machinery Laboratory, Department of Agricultural Engineering, Hakozaki. Higashi-ku. Fukuoka, 812 Japan

B. Scheufler
Amazonen Werke H. Dreyer, Postfach 51, D-49202 Hasbergen-Gaste, Germany

J. K. Schueller
Department of Mechanical Engineering, University of Florida, P.O. Box 116300, Gainesville, FL 32611, USA

B. Shaw
Agricultural Engineering Department, Texas A&M University, College Station, Texas 77843-2117, USA

L. Speelman
Department of Agricultural Engineering and Physics, Wageningen Agricultural University, Agrotechnion Bomenweg 4, 6703 HD Wageningen, The Netherlands

T. Tanaguchi
Obihiro Univ. Of Agr. & Vet. Med., Department of Agro-Environmental Science, Inada-ohou Obihiro Hokkaido 0808555, Japan

H. J. Tantau
Institute of Horticultural and Agricultural Engineering, University of Hannover, Herrenhaeuser Str. 2, D-30419 Hannover, Germany

A. A. Wanders
IMAG-DLO, Mansholtlaan 10-12, P.O. Box 43, NL-6700 AA, Wageningen, The Netherlands

G. H. Weise
Institut für Landtechnik der Justus-Liebig-Universität Giessen, Braugasse 7, D-35390 Giessen, Germany

R. Wilkinson
212 Farrall Hall, Department of Agricultural Engineering, Michigan State University, East Lansing, Michigan 48824, USA

Editorial Board

Contents

Foreword xxv
Preface xxvii

1 Machines for Crop Production 1
1.1 Power Sources 1

Human-Powered Tools and Machines 1
1.1.1 Technical Characteristics of Human Power 1
Introduction 1
Power Production and Consumption by Humans 1
Human Work Output 3
Some Compensating Attributes of Human Labor 3
Importance of Human-Powered Agricultural Tools/Machines in the LDC's 4
1.1.2 Human-Powered Tools and Machines for Field Operations 5
Definitions 5
Classification by Field Operations 5
Hand-Tools for Land Preparation 5
Manual Planting Tools and Machines 6
Manual Weeding Tools and Machines 8
Manual Harvesting Tools and Machines 11
Economics of Human-Powered Tools/Machines for Field Operations 13
1.1.3 Human-Powered Tools and Machines for Post-Harvest Operations 15
Definitions and Some General Remarks 15
Some Common Tools for Crop Processing by Peasant Farmers 15
Some Common Human-Powered Processing Machines 17
1.1.4 The Sociology and Future of Hand Tool Technology in LDC's 19
References 21

Animals 22
1.1.5 Efficient Use of Energy Potential by a Draft Animal Power Unit 22
1.1.6 Control of Animal's Energy Potential 24
Environmental Conditions 24
Choice of Animals 25
Use of Animals 26
Livestock Management 28
1.1.7 Farm Equipment for Transport, Tillage, Secondary Cultivation and Sowing 29
Equipment for Transport 29

Equipment for Tillage 31
Sowing 35
Secondary Tillage Operations 37
1.1.8 Use of Draft Animals 38
1.1.9 Conclusion 39
References 40

Engines 41
1.1.10 Types of Engines Used in Agriculture 41
1.1.11 Fuel Injection in CI Engines 43
1.1.12 Engine Performance Parameters 43
1.1.13 Turbocharging and Intercooling 49
1.1.14 Engine Performance Maps 51
1.1.15 Engine Utilization 52
1.1.16 Engine Design Goals 53
1.1.17 Engine Selection 53

Tractors: Two-Wheel Tractors for Wet Land Farming 54
1.1.18 The Role for Small Scale Farms 54
1.1.19 Differences of Farming Principles Between Upland Fields and Paddy Fields for Tractors 55
Plow Pan Layer 55
Depth of Plowing 56
Flatness and Size of a Field Lot 56
Planting Systems and the Principle of a Transplanting System in Paddy Cultivation 56
1.1.20 Types and Durability of Two-Wheel Tractors 57
Types of Two-Wheel Tractors 57
Durability Classification of Two-Wheel Tractors 58
1.1.21 Principles of Mechanisms and Mechanics 59
Engines on Two-Wheel Tractors 59
Engine-base Assembly 61
Handle Assembly 62
Power Transmission Mechanisms 62
Front and Rear Hitches and Hitch-pins 64
Wheels and New Wheel Dynamics 65
Two-Wheel Tractors with the Plow and Plowing 70
Two-Wheel Tractors with Rotary Tillers 77
References 93

Tractors: Two-Wheel Tractors for Dry Land Farming 95
1.1.22 Description and Types of Two-Wheel Tractor Designs 95
1.1.23 Production and Concentration of Two-Wheel Tractors 101
1.1.24 Mechanics of Two-Wheel Tractors 103
1.1.25 Field Operations with Two-Wheel Tractors 105

1.1.26a Management of Two-Wheel Tractors 110
1.1.26b Trends in the Development of Two-Wheel Tractors 114
References 114

Tractor: Two Axle Tractors 115
1.1.27 History, Trends, Concepts 115
Tractor: Definition, History and Trends 115
Tractor Concepts 115
1.1.28 Tractor Mechanics 118
Single-Wheel and Tire Mechanics 118
Soil Compaction under Tires 120
Mechanics of Two-Axle Tractors Pulling Implements or Trailers 122
Mechanics of Hillside Operation and Overturning Stability 125
1.1.29 Chassis Design 127
Traction Tires: Requirements, Design, Specifications 127
Chassis Concepts, Four-wheel Drive 129
Brakes 131
Steering 134
Track Width and Wheel-to-hub Fixing 136
1.1.30 Diesel Engines and Fuel Tank 137
Development Trends 137
Motorization Concept 138
Diesel Engine Installation 139
Practical Fuel Consumption and Fuel Tank Size 139
1.1.31 Transmissions 139
Introduction 139
Requirements 140
Fundamentals of Speeds and Torque Loads 141
Rear Power Takeoff (Rear PTO) 143
Symbols for Transmission Maps 144
Stepped Ratio Transmissions 145
Continuously Variable Transmissions (CVTs) 148
Master Clutch and Shift Elements 151
1.1.32 Working Place 153
Introduction: Role of Comfort, Health and Safety 153
Work Load on Machine Operators 153
Technical Aids for the Operator: Survey and Related Standards 154
Operator's Seating Accommodation and Access 155
Tractor Ride Dynamics (Vibrations) 157
Tractor Noise Control 161
Tractor Safety 163
1.1.33 Implement Control and Hydraulics 165
Beginnings of Implement Control by Hydrostatic Hitches 165

Concept and Dimensions of the Three-point Hitch 166
Kinematics and Forces of the Three-point Hitch 168
Control Strategies for Three-point Hitches 169
Fluid Power Systems: Symbols and Vocabulary 172
Hydraulic Circuits: Basic Systems 174
References 176

1.2 Tillage Machinery 184
1.2.1 General Importance of Tillage Operations 184
Appropriate Tillage According to Soil Conditions 185
Socio-Economical Aspects of Tillage 187
1.2.2 Soil Engaging Components 188
Basic Elements and Materials of Tillage Tools 188
Drawn Implements 190
PTO-Driven Implements 193
1.2.3 Tillage Systems 193
Conventional Tillage System 193
No-Tillage System 210
Tillage Systems for Special Conditions 211
References 215

1.3 Seeders and Planters 217
1.3.1 Introduction 217
1.3.2 Seeders 217
Seed-Spacing and Seeding-Depth 217
Precision Drilling 218
Bulk Seeding 226
1.3.3 Planters 235
Potato Planters 236
Transplanters 238
References 239

1.4 Fertilizer Distributors 240
1.4.1 Introduction 240
1.4.2 Fertilizer Distributor Types 241
Spinning Disc Spreaders 241
Oscillating Spout Spreaders 248
Boom Spreaders 252
Aerial Spreaders 255
Liquid Fertilizer Spreaders 256
1.4.3 Particle Trajectories 258
1.4.4 Spread Pattern Analysis 261
1.4.5 Developments 266
Multiple Hoppers 266

		Site Specific Spreading	267
		Electronics	267
	References		268
1.5	Pest Control Equipment		269
	1.5.1	Pest Control Methods	269
		Non-Chemical Methods	269
		Chemical Control Methods	269
	1.5.2	Non-Chemical Techniques	269
		Mechanical Weeding	269
		Thermal Treatments	271
		Vacuum	272
		Biological Pest-control Equipment	272
	1.5.3	Chemical Control Methods	273
		Formulations	273
		Characteristics of Droplets	275
		Methods of Application	276
		Meteorological Constraints	280
	1.5.4	Spraying Equipment	281
		Hand Sprayers	281
		Power Equipment	283
		Air-assisted Spraying	284
		Aerial Application	285
	1.5.5	Sprayer Components	286
		Pumps	286
		Sprayer Tanks	290
		Agitation	290
		Strainers and Screens	292
		Plumbing and Controls	293
		Nozzles	295
	1.5.6	Choice of Equipment	300
		Work Rate	300
		Fans and Capacity	301
		Crop Spraying	301
		Orchard Spraying	302
	1.5.7	Spraying Techniques	302
		Drift reduction	302
		Chemical Mixing and Disposal of Excess Pesticide	303
		Cleaning Equipment	303
		Closed-Transfer Systems	303
		Ergonomics and Operator Safety	304
	1.5.8	Sprayer Calibration	304
		Sprayer Ground Speed	304

Timed-flow Method for Calibrating Boom Sprayers 304
Calibration Flow Check 305
1.5.9 Granular Applicators 306
Metering Devices 306
Calibrating and Using Granular Applicators 307
Application Rate 307
1.5.10 Future Trends 308
Developing Countries 308
Developed Countries 308

1.6 Harvesters and Threshers 311

Grain 311
1.6.1 Functional Components of Combine Harvesters 311
1.6.2 Threshing and Separation 312
Tangential Threshing Unit 312
Straw Walker 314
Rotary Separators 315
Rotary Combines 315
Comparison of Tangential and Axial Threshing Units 317
Separation Theory 317
Cleaning 319
Cleaning Theory 322
1.6.3 Combine Harvester Performance 323
Work Quality 323
Crop Properties and Harvest Conditions 325
Combine Type and Design 326
Engine Power 326
Harvest Management 327
Combine Testing 327
1.6.4 Information and Control Systems 328
Cabs and Controls (Operator Compartment) 328
Information Systems 328
Control Systems 330
1.6.5 Combine Attachments and Variants 331
Header 332
Chopper 333
1.6.6 Rice Harvesting 333
Rice Combines: Why a Rice Combine? 333
Custom-Made Rice Combines 334
Combine Categories 335
Red and Green Dominate 335
Stripper Fronts Gaining on Rice Fields 336

Traction and Flotation Assistance for Combines in Rice Fields 337
A Question of Teeth 339
Rice Combines in Asia 340
Chinese Combines 340
Combines in Southeast Asia 340
Summary 341
1.6.7 Power Threshers as Precursors of Mechanization 341
Other Countries in Asia 343
Chinese Threshers 343
IRRI Axial-Flow Thresher Developments 343
Throw-in Threshers That Chop the Straw for Stockfeed 344
A Quantitative Assessment of Power Threshers 345
Summary 346
References 346

Forage Crops 348
1.6.8 Foreword 348
1.6.9 Meadow-Type Forages 349
Harvest, Treatment and Storage Methods 350
Machines and Equipment 357
1.6.10 Forage Cereals 374
The Methods 374
Machines and Equipment 376

Root Crops 381
1.6.11 Sugar Beet Harvesting 381
Main Stages of Mechanical Harvesting 381
1.6.12 Potatoes 397
Lifting 397
Sieving 400
Haulm, Clod, and Stone Separation 400
Hoppers/Discharge Elevators 402
Single-row, Two-row and Four-row Harvesters 403
Self-propelled Harvesters 404
Two-Stage Harvesting 404
Damage to Potato Tubers 407
Bibliography 408

Fruits and Vegetables 408
1.6.13 Introduction 408
1.6.14 Harvesting Functions 409
1.6.15 Principles and Devices for the Detachment 410
Low-height Herbaceous Structures (Vegetables, Strawberries, etc.) 410

	Bushy Structures (Small Fruits, Wine Grapes)	416
	Tree Structures (Fruits, Nuts)	419
1.6.16	Complementary Operations	425
1.6.17	Mechanical Aids to Manual Harvest	426
1.6.18	Electronic Sensors and Robotic Harvesting	428
	List of Symbols	429
	References	430
	Tropical Crops	431
1.6.19	Introduction	431
1.6.20	Sugar Cane	431
	Whole-stick Harvesting	432
	Self-propelled Full Stick Harvesters	433
	Harvest of Chopped Cane	437
1.6.21	Cotton	438
	Defoliation	438
	Desiccation	438
	Actual Harvest	439
1.6.22	Groundnuts	441
	Lifting	441
	Uprooting	442
	Drying on the Soil	444
	Threshing	444
1.6.23	Tropical Tubers	446
	Cassava	446
	Yam	449
1.6.24	Millet	450
1.6.25	Coffee	452
	References	453
1.7	Transportation	455
1.7.1	Introduction	455
1.7.2	Powered Farm Vehicles for Use in the Field	455
	Monowheelers	455
	Tri-wheelers	456
	Four-wheel Carriers	457
	Multiwheel Carriers	457
	Crawler-type Carriers	458
1.7.3	Motor Trucks (Used for Local and Long-distance Transport)	460
	Subcompact Trucks	460
	Farm Trucks	461
1.7.4	Trailers	463
	Trailers for Use with Walking Tractors	463
	Trailers for Use with Four-wheel Tractors	463

Trailers with Hydraulic Tippers 464
Grain Trailers 464
Trailers Used for Transporting Combine Harvesters and Heavy Equipment 465
1.7.5 Loaders 465
Hoist and Truck Loaders 465
Universal Elevators 465
Tractor-mounted Loaders 465
Self-propelled Loaders 466
Forklifts 466
1.7.6 Monorails 468
References 468

1.8 Specific Equipment Used for Cultivation Inside Greenhouses 469
1.8.1 Introduction 469
1.8.2 Seeders and Seeding Lines 469
Requirements 470
Principles of Operation 470
1.8.3 Transplanters 471
1.8.4 Transportation 472
1.8.5 Benches 472
1.8.6 Irrigation Systems 472
Overhead Irrigation 473
Drip-and-Trickle Irrigation 473
Subirrigation 474
Soil-less Culture 475
1.8.7 Artificial Lighting 476
1.8.8 Greenhouse Climate Control 477

1.9 Forest Engineering 480
1.9.1 Harvesting in General 480
Tree Harvesting 480
Processing Techniques 480
Handlings and Maneuvers 481
1.9.2 Felling 481
Felling Process 481
Felling Tools 482
Mechanical Tree Fellers 484
Ergonomics 485
1.9.3 Delimbing 485
1.9.4 Additional Interventions 486
1.9.5 Topping, Bucking 486
1.9.6 Debarking 486
1.9.7 Wood Comminution 487

1.9.8 Harvesters 487
1.9.9 Operation 487
1.9.10 Security and Safety 487
1.9.11 Timber Transportation 488
Log Extraction 488
Skidding 488
Winch Skidding 490
Grapple Skidding 491
Forwarding 492
Locomotion 492
1.9.12 Special Logging 494
Semiaerial Systems 494
Aerial or Off-the-Ground Systems 497
Water Transportation 498
1.9.13 Forest-Stand Maintenance 498
Special Comments 498
Cleaning and Thinning of Young Stands 498
Pruning 498
Fertilization 498
Fire Protection 498
Soil Restoration 499
1.9.14 Forest-Stand Establishment 499
General Comments 499
Site Preparation 500
Planting 500
1.9.15 Forest Roads 501
1.9.16 Forest Regulations 502
Ordering Forest Machines 502
Machines and Environment 502
Best Management Practices 502
Techniques and Future 502
References 502

1.10 Standardization 503
1.10.1 Standardization on Workplace Health and Safety in the European Union 517
1.10.2 Relationships Between European and International Standardization 519
1.10.3 E.U. Standards for Environmental Protection 519
1.10.4 Other Directives for Machinery 520
1.10.5 National Standards Development 520
1.10.6 Standards Searches 520
References 520

2 Mechanizations Systems 521
2.1 Systems Engineering, Operations Research, and Management Science 521
2.1.1 Optimization 522
Linear Programming 522
2.1.2 Time and Project Management 525
CPM 525
PERT 527
2.1.3 Queuing Theory 528
Single-server M/M/1 528
Multiple-Server Systems 530
Queuing-Theory Example 531
2.1.4 Simulation 532
Random Numbers and Random-Number Generation 533
Uniform Distribution 533
Exponential Distribution 534
Triangular Distribution 535
References 536

2.2 Agricultural Mechanization Strategy 536
2.2.1 Definitions 536
2.2.2 Introduction 537
2.2.3 Need for Mechanization and Productivity-enhancing Technology 538
2.2.4 Adoption Process for Mechanization and Labor Productivity–Enhancing Technology 539
2.2.5 Basic Guidelines and Principles for Strategy Formulation 541
2.2.6 The AMS Formulation Process 542
Who Should Formulate the Strategy? 543
How to Proceed with Formulation? 543
Composition of the Strategy Formulation Team 544
Outline of the Strategy Document 545
Implementation of the Strategy 546
2.2.7 Some Frequently Raised Issues 546
2.2.8 Key Policy Instruments for Formulation of an AMS 547
Subsidies 548
Credit for Agricultural Machinery 548
Taxes and Duties 549
Private, Cooperative, or Government Ownership 550
Input and Output Prices 550
Public Investments 550
References 553

2.3 Transfer of Technology 554
2.3.1 Introduction 554
2.3.2 TT in Industrialized Countries 554
2.3.3 TT in Less Industrialized Countries 556
Prerequisites and Constraints Connected with TT in the Rural Areas 557
Steps and Structure of TT 558
The Role of Farmers 558
The Role of Farm Machinery Industries for Local Manufacturing 558
Facilities for TT 560
Examples of TT Successfully Carried out in Developing Countries 561
The Role of Organizations and Associations 562
The Role of Farm-Machinery Institutes 562
The Role of Education and Training 563
The Role of the Government 563
The Protection of Intellectual Property and the Risk of a Possible Misuse of Know-how 564
References 564

2.4 Field Machinery Management 565
2.4.1 Field Operations 565
2.4.2 Field Patterns 567
2.4.3 Calibration 570
2.4.4 Loss Determination 572
2.4.5 Field Adjustments 573
2.4.6 Repair and Maintenance 573

2.5 Cost Analysis 574
2.5.1 Field Capacity of Machines 574
2.5.2 Costs of Operation 576
2.5.3 Machinery Selection 578
2.5.4 Replacement Policies 582
References 584

3 Trends for the Future 585
3.1 Sustainable and Environmental Engineering 585
3.1.1 Definition and Background 585
3.1.2 Policy 589
3.1.3 Social, Economic, and Regional Differences 589
India 589
Chile 590
Philippines 590

3.1.4 Components 590
Soil 591
Water 591
Air 593
Nutrients 593
Plant Base 594
Integrating Plant and Animal Systems 594
Energy 595
3.1.5 Summary 597
References 598

3.2 Precision Farming 598
3.2.1 Introduction 598
3.2.2 Positioning in Precision Farming 600
Satellite Navigation Systems 600
Other Location Systems 602
3.2.3 Concepts of Precision Farming Systems and Required System Elements 602
Map-Based Systems 603
Real-Time Systems 603
Real-Time Systems with Maps 604
3.2.4 Yield Mapping 604
Grain Combine Harvester Systems 604
Other Continuous Crops 605
Non-Continuous Yield 607
Data Storage and Mapping 607
3.2.5 Soil and Weed Mapping 608
Soil Sampling 608
Weed Mapping 609
Remote Sensing 610
3.2.6 Control of Field Operations 610
Requirements and System Components 611
Fertilization 611
Pesticide Application 612
Other Controller Operations 612
3.2.7 Information Management 613
BUS Systems on Mobile Equipment 613
Data Transfer to and from Farm Management 614
Data Management and Geographic Information Systems 614
Decision-support Systems 615
References 616

Index 617

Foreword

This handbook has been edited and published as a contribution to world agriculture at present as well as for the coming century. More than half of the world's population is engaged in agriculture to meet total world food demand. In developed countries, the economic weight of agriculture has been decreasing. However, a global view indicates that agriculture is still the largest industry and will remain so in the coming century.

Agriculture is one of the few industries that creates resources continuously from nature in a sustainable way because it creates organic matter and its derivatives by utilizing solar energy and other material cycles in nature. Continuity or sustainability is the very basis for securing global prosperity over many generations—the common objective of humankind.

Agricultural engineering has been applying scientific principles for the optimal conversion of natural resources into agricultural land, machinery, structure, processes, and systems for the benefit of man. Machinery, for example, multiplies the tiny power (about 0.07 kW) of a farmer into the 70 kW power of a tractor which makes possible the production of food several hundred times more than what a farmen can produce manually. Processing technology reduces food loss and adds much more nutritional values to agricultural products than they originally had.

The role of agricultural engineering is increasing with the dawning of a new century. Agriculture will have to supply not only food, but also other materials such as bio-fuels, organic feedstocks for secondary industries of destruction, and even medical ingredients. Furthermore, new agricultural technology is also expected to help *reduce* environmental destruction.

This handbook is designed to cover the major fields of agricultural engineering such as soil and water, machinery and its management, farm structures and processing agricultural, as well as other emerging fields. Information on technology for rural planning and farming systems, aquaculture, environmental technology for plant and animal production, energy and biomass engineering is also incorporated in this handbook. These emerging technologies will play more and more important roles in the future as both traditional and new technologies are used to supply food for an increasing world population and to manage decreasing fossil resources. Agricultural technologies are especially important in developing regions of the world where the demand for food and feedstocks will need boosting in parallel with the population growth and the rise of living standards.

It is not easy to cover all of the important topics in agricultural engineering in a limited number of pages. We regretfully had to drop some topics during the planning and editorial processes. There will be other requests from the readers in due course. We would like to make a continuous effort to improve the contents of the handbook and, in the near future, to issue the next edition.

This handbook will be useful to many agricultural engineers and students as well as to those who are working in relevant fields. It is my sincere desire that this handbook will be used worldwide to promote agricultural production and related industrial activities.

Osamu Kitani
Editor-in-Chief

Acknowledgments

At the World Congress in Milan, the CIGR Handbook project was formally started under the initiative of Prof. Giuseppe Pellizzi, the President of CIGR at that time. Deep gratitude is expressed for his strong initiative to promote this project.

To the members of the Editorial Board, co-editors, and to all the authors of the handbook, my sincerest thanks for the great endeavors and contributions to this handbook.

To support the CIGR Handbook project, the following organizations have made generous donations. Without their support, this handbook would not have been edited and published.

Iseki & Co., Ltd.
Japan Tabacco Incorporation
The Kajima Foundation
Kubota Corporation
Nihon Kaken Co., Ltd.
Satake Mfg. Corporation
The Tokyo Electric Power Co., Inc.
Yanmar Agricultural Equipment Co., Ltd.

Last but not least, sincere gratitude is expressed to the publisher, ASAE; especially to Mrs. Donna M. Hull, Director of Publication, and Ms. Sandy Nalepa for their great effort in publishing and distributing this handbook.

Osamu Kitani
CIGR President of 1997–98

Preface

Effective crop production requires machines—hand tools, animal-drawn implements and engine-powered equipment. This volume presents the fundamentals of various agricultural tools and machines and describes the types most commonly used for various operations. The scope of crop production is defined rather broadly to include greenhouse production and forestry in addition to field crops. Important peripheral topics also are covered, such as transport machines, machine systems, and technology transfer.

Since machines for crop production represent a substantial capital investment for individual farmers, principles and guidelines are given for proper selection and machine management to achieve the greatest return. On a broader scale, policies and strategies are given for effective national or regional mechanization programs. Transfer of technology from industrialized to developing countries is also discussed.

Standards are crucial in the design, testing, marketing and use of agricultural machines. Globalization of the agricultural machinery manufacturing industry requires that machines built by one manufacturer operate effectively with power units built by another. Also, safety issues require standards to assure protection of the operator and the general public.

The future is always hard to predict, but one thing is certain: If humans are to survive and thrive on planet Earth, agricultural practices must be sustainable over the long term. Machines and associated farming practices can have a profound impact—both positively and negatively—on soil erosion, precise chemical application, air quality and other environmental aspects. Precision farming techniques are designed to vary the fertilizer and chemical application rates in accordance with the crop needs and thereby save money and help maintain the environment.

Many individuals and agencies have contributed to this handbook. The various chapters were written by 41 individuals—all of whom are experts in their particular area of specialization. These authors are from 12 countries and represent many languages other than English. Although every effort was made to standardize the format of each chapter, it is hoped the reader will overlook minor variations in format and terminology resulting from the broad authorship.

It is not possible to acknowledge individually the hundreds of authors of the references cited, although their work contributed significantly to this volume.

The chapter manuscripts were reviewed by two world renowned experts in the field of agricultural mechanization. Their questions and comments were considered by the chapter authors and resulted in substantial improvement in the manuscripts. In addition, all the chapter manuscripts were reviewed by Ms. Lynette James, Department of Agricultural Communications at Texas A&M University, a very capable editor who helped standardize the format and make the volume more readable.

A project of this type would have been impossible without many competent and dedicated research assistants, typists, reviewers and other helpers. The editors extend a sincere thanks to everyone who contributed to this volume.

B. A. Stout, Editor Volume III
B. Cheze, Co-Editor Volume III

1 Machines for Crop Production

1.1. Power Sources

Human-Powered Tools and Machines

E.U. Odigboh

1.1.1. Technical Characteristics of Human Power

Introduction

To *mechanize* means to use machines to accomplish tasks or operations. A machine may be as simple as a wedge or an inclined plane, or as complex as an airplane. Agricultural mechanization, therefore, is the use of any machine to accomplish a task or operation involved in agricultural production. It is clear from this definition that agriculture anywhere has always been mechanized, employing a combination of three main sources of power: human, animal and mechanical/engine, giving rise to three broad levels of agricultural mechanization technology classified as *hand-tool technology* (HTT), *draft-animal technology* (DAT) and *mechanical-power* or *engine-power technology* (EPT).

Hand-tool technology is the most basic level of agricultural mechanization, where a human being is the power source, using simple tools and implements such as hoes, machetes, sickles, wooden diggers, etc. A farmer using hand-tool technology can cultivate only about one hectare of land. He cannot do more than that because of certain scientifically established facts.

Power Production and Consumption by Humans

As a source of power, the human being operates essentially like a heat engine, with built-in overload controls or regulators. Chemical energy input in the form of food is converted into energy output, some of which is useful for doing work. On the average, a healthy person in temperate climates consumes energy at a sustainable rate of only about 300 W, while in tropical climates, as a result of heat stress the rate is reduced to only about 250 W. Many tasks for agricultural production can be performed only at higher rates of energy consumption, however, as shown in Table 1.1. Some actual manual work rates for certain field operations are presented in Table 1.2.

The fact that many primary agricultural production operations demand higher rates of energy than the maximum sustainable rate of energy consumption by humans necessitates

Table 1.1. Human Power Consumption for Various Farming Activities

Activity	Gross power consumed (Watts)
Clearing bush and scrub	400–600
Felling trees	600
Hoeing	300–500
Ridging, deep digging	400–1000
Planting	200–300
Plowing with animal draft	350–550
Driving single axle tractors	350–650
Driving 4-wheel tractor	150–350
Driving car on farm	150

Source: mainly from Dumin and Passmore, 1967, Energy, work and leisure. Heineman as given by Inns (1992).

Table 1.2. Some Field Operation Rates by Farmers Using Hand-Tools

Operation	Average manual work rate (man days/ha)*	
Land clearing	32.6	(20.1–47.8)
Ridging for cassava	43.8	(29.7–64.5)
Mound making for yams	57.8	(35–93)
Cassava planting	28.3	
Yam planting	17.3	
Weeding root crops	36.7	(22.3–77.6)
Weeding general	40.0	
Cassava harvesting	28.5	
Yam harvesting	32.0	

* Range values in parenthesis [1].

rest periods in manual work. The rest period required can be estimated using the formula [2],

$$T_r = 60(1 - 250/P)$$

where,

T_r = required rest period in min/h of work, and
P = actual rate of energy consumption in watts.

Using the formula, it follows that the manual ridging operation which demands 400–1000 W (Table 1.1) requires rest periods of between 22.5 and 45 minutes per hour of work. Note that at the 1000 W rate of energy consumption, the farmer can work only for 15 minutes, and must rest for 45 minutes, per hour of work. It must be noted here also that an appropriate rest period, as estimated using the above formula, is a physiological necessity inherent in manual work.

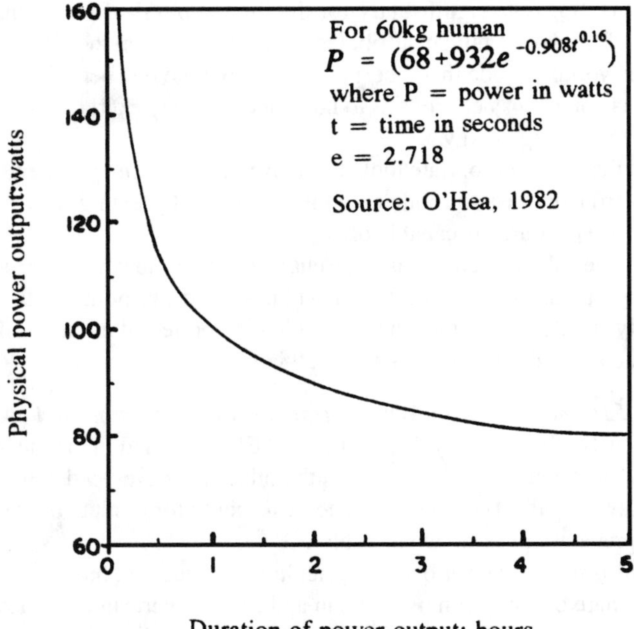

Figure 1.1. Sustainable Physical or Power Output by Humans (Inns, p. 2).

Human Work Output

Only about 25 percent of the energy consumed when handling relatively easy tasks such as pedaling, pushing or pulling is converted to actual human work output. Under more difficult work conditions, the efficiency of converting consumed energy to physical work may be as low as 5 percent or less. This means that, at the maximum continuous energy consumption rate of 0.30 kW and conversion efficiency of 25 percent, the physical power output is approximately 0.075 kW sustained for an 8–10 hour work day. Naturally, higher rates can be maintained for shorter periods only, as shown in Fig.1.1 [2].

Some Compensating Attributes of Human Labor

The discussion thus far and the facts given in Tables 1.1 and 1.2 make it abundantly clear that power is the major limitation to increasing the area cultivated by the hand-tool farmer. It should be noted that the problem is not necessarily with the tools used, especially for primary production operations, since efforts made to redesign them have yielded no significant improvements [3, 4]. The toil, drudgery, and severe power constraint on timely field operations, which limit production and earning capacity, are the inherent characteristics of peasant farmers using hand-tool technology; change the technology and you change the farmer's status [5].

Still, the peasant farmer and his hoe and machete are efficient companions in crop production at the subsistence level where he operates. This is so because of certain human attributes that compensate significantly for the limited physical power that the farmer can generate. The relevant human attributes are exhibited when the farmer:

- Adopts a working mode that incorporates appropriate rest periods.
- Makes instantaneous decisions as to how much force to exert to accomplish a task, thereby conserving energy.
- Chooses the most appropriate tools for a given production unit operation.
- Changes from one task to another readily and rationally, exhibiting a versatility that no other power source is capable of.

In spite of the inherent compensating characteristics, however, the power needed to operate any human powered tool or machine should not be more than the farmer can potentially supply; the farmer should employ the preferred modes of human power application such as pedaling or simulated walking.

Importance of Human-Powered Agricultural Tools/Machines in the LDC's

All three levels of technology, HTT, DAT and EPT, are used in the mechanization of agriculture in most countries of Africa and the other less developed countries (LDC's) of the tropical world. But HTT predominates, especially for production field operations such as land preparation, as shown in Table 1.3.

Table 1.4 also shows that for overall agricultural production, human power accounts for the lion's share of work in most African and Latin American countries. It has been suggested that a power-use intensity of 0.4 kW/ha is required for effective levels of agricultural mechanization. While that figure may well be controversial, the facts and figures presented in Tables 1.3 and 1.4, and especially those in Table 1.5, show that the power-use intensity in Africa is so low that it should be of serious concern to all. Considering the natural limitations of human powered tools and machines, their predominance in the agriculture of developing countries is an important factor to address when dealing with overall economic development of those countries.

Table 1.3. Sources of Power for Various Primary Land Preparation Operations in Various Countries

% of Total Land Cultivated		Draught	Engine
Country	Human	animal	(Mech.)
Nigeria	86	4	10
Botswana	20	40	40
Zimbabwe	15	30	55
Tanzania	80	14	6
Kenya	84	12	4
Ethiopia	10	80	10
Zambia	55	15	30
Swaziland	15	35	50
Uganda	70	20	10
China	22	26	52
India	18	21	61

Source: [6].

Table 1.4. Sources of Power for Overall Agricultural Production in Latin America and Africa (% Share)

Source of Power	Latin America	Africa	Nigeria
Human power	59	89	90
Animal power	19	10	8
Engine power	22	1	2

Source: [7, 8].

Table 1.5. Engine Power Available for Agriculture in Different Countries and Continents

Country/Continent	W/ha	(Hp/acre)
USA	783	(0.430)
Europe	694	(0.340)
Latin America	201	(0.110)
China	142	(0.080)
Africa	37	(0.020)
Nigeria	18	(0.008)

Source: Adapted from [9].

1.1.2. Human-Powered Tools and Machines for Field Operations

Definitions

The description of a machine in the introduction to Section 1.1, which grouped a wedge together with an airplane, may be valid only at a certain level of conceptualization. But in a more formal sense, a machine is a device or mechanical contrivance consisting of *two or more relatively constrained components* which is energized by a power source to transmit and/or modify force and motion to accomplish some desired kind of work. In contrast, a tool is a human powered instrument or implement *usually without parts that move relative to one another*, like a hoe, a dibber, or the like, used to facilitate mechanical manual operations.

Classification by Field Operations

Field operations are tasks performed in the field at different phases of crop production. The major operations include land preparation, planting, weeding, and harvesting. Based on these operations, the tools/machines used are classified into: land preparation tools/machines; planting tools/machines; weeding/cultivation tools/machines; and harvesting tools/machines.

Hand-Tools for Land Preparation

Hoes

Naturally, soil preparation is usually the first task in crop production, undertaken to achieve a variety of basic interrelated objectives such as seedbed preparation, weed control, soil and water conservation, soil compaction amelioration, etc. In peasant agriculture, soil or land preparation to achieve a combination of these objectives usually involves tilling with a hoe, and constitutes the most significant characteristic of the hand-tool (mechanization) technology.

Curiously, no manually operated machine for land preparation is commonly available. The hoe is the most popular and most versatile tool used in developing countries of the world, where peasant farmers account for close to 90 percent of the area under cultivation. The hoe is the tool used almost exclusively in land preparation of peasant agriculture, for combined primary and secondary tillage, and for land-forming operations such as ridging, bedding, mounding, bunding, ditching, etc. Hoes for land preparation come in different sizes, weights and peculiar shapes, having evolved over the years to suit widely varying crops and conditions of soil, farming culture, farmers' physiques and temperaments. Described generally as long-handled implements with thin, flat blades set transversely, common technical features of hoes include long handles and heavy heads carrying the cutting blades or shares. Handles vary a great deal in length, shape and curvature. Blades also vary a great deal in shape, size and curvature, leading to an intriguingly varied world of hoes, as illustrated by the small sample given in Fig. 1.2. Wide-bladed hoes are used for digging, ridging and mounding under normal soil conditions; narrow-bladed ones are used for hard soil conditions; while tined hoes, which are not very common, are used for stony conditions.

Machetes/Spades

Other hand tools used to complement the hoe in land preparation under peasant agriculture include machetes, axes, spades, forks, and rakes, which also vary in sizes and shapes, as illustrated in Fig. 1.2. Next to the hoe, the machete is one of the most important tools in peasant agriculture, where it is indispensable in land clearing and a host of other crop production operations.

Manual Planting Tools and Machines

Hoes

The hand hoe of appropriate size and shape is the most versatile tool used by the peasant farmer in planting cereals, root crops and other crops. The farmer with the hand hoe can use his judgment and experience to place the seeds or planting materials at optimum depths and appropriate spacings within and between rows, and provide just the right firming pressure to achieve good yields. Hoes used for planting, while varying greatly according to diverse cultural preferences, usually are lighter and smaller than those for primary tillage or ridging, mounding, bedding or ditching operations, because less energy is demanded (see Table 1.1) and closer attention required.

Manual Planters

Unlike the case for land preparation, there are many hand-operated machines available for planting and sowing, often with improved results in terms of uniformity of plant spacing and row configuration. The manual planters may be as simple as dibbers, which are pointed instruments made of steel or wood tipped with steel, used to place seeds in the ground. Or they may be as sophisticated as the various types of jab planters or pushed

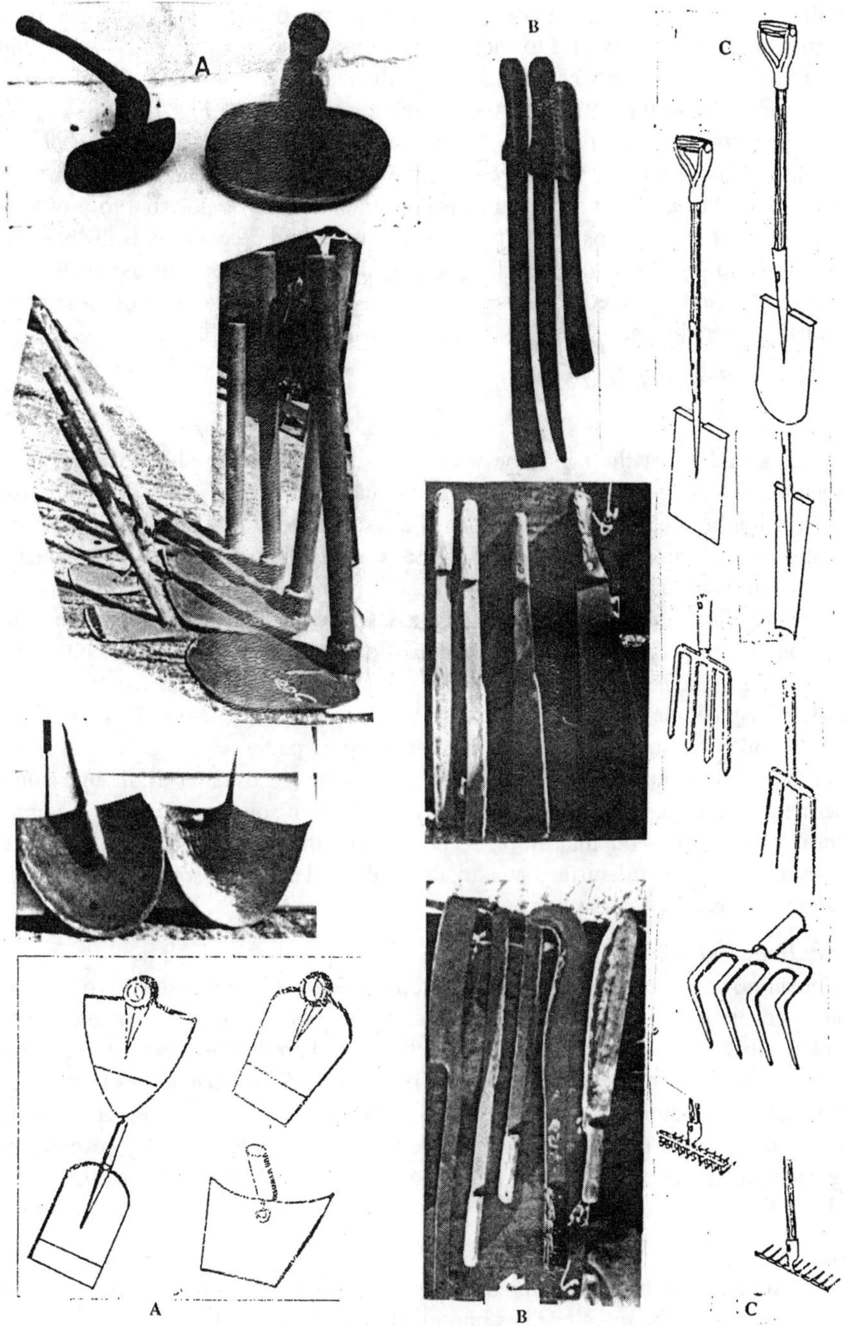

Figure 1.2. A Variety of Hand Tools for Land Preparation. A - Hoes; B - Machetes; C - Shovels, spades, forks and rakes.

or pulled seed drills with more complex seed metering devices. In this case, different metering mechanisms give rise to such planter types as seed-roller, fluted-roller, slide-roller and chain- and- sprocket driven seed drills/planters. Illustrations of some of the major types of manual planting tools and machines are given in Fig. 1.3.

It is important to state here that the more sophisticated pushed or pulled planters, which usually are equipped with seed coulters or other furrow openers, do require well-prepared seed beds, which a typical peasant farmer usually is not able to provide. In fact, a peasant farmer, whose only or major means of land preparation is the hand hoe, is not likely to prepare enough land area to make the ownership or use of the more sophisticated hand-operated planters economical. As a result, adoption of these pushed or pulled planters by peasant HTT farmers is very limited indeed.

Manual Weeding Tools and Machines

Hoes

What is said about the hand hoe with respect to planting applies to weeding and cultivation. Generally speaking, in peasant agriculture, the heavy work of land preparation using big hoes is handled by the men while subsequent field operations, especially weeding, are undertaken by women and children, using the smaller and lighter hand hoes that come in three major types: digging hoes, chopping hoes and pushing/pulling hoes. Most peasant farmers own only the digging hoe type, which they use for different tillage operations, often with designs that are peculiar to certain traditional communities, such as the design called *ikeagwu-agadi* (literally meaning "exhaustion free for the aged") by Igbo-speaking people of Nigeria, which is a very popular hoe (see Fig. 1.4) for weeding under all soil conditions, soil topography and cropping patterns.

By implication, the chopping hoes, used to chop the weeds and soil, though suitable under hard or friable soil conditions and all conditions of soil topography and cropping pattern, are much less popular. Still less popular are the pushing/pulling hoes, used to cut weeds under the soil surface but suitable only under friable soil conditions. Some examples of weeding hoes are given in Fig. 1.4.

Rotary Hoes and Wheeled Cultivators

Human-powered rotary hoes for weeding do exist but are mainly used for row cropped paddy rice or upland crops in friable soils. Also, many designs of human-powered wheeled cultivators, with different kinds of weeding shares (tines, hoes, etc.) are available but are suitable only for row crops in friable soils. Some examples of rotary hoes and wheeled cultivators are given in Fig. 1.4. Naturally, use or ownership of these more sophisticated human-powered weeders is very much restricted, thereby severely limiting their impact on the activities of peasant or small-holder farmers of the tropical world.

Slashers

For completeness, human-powered slashers, most commonly in the form of machetes or cutlasses, should be mentioned as important human-powered weeding tools used by peasant farmers. Slashers are used to cut down above-ground parts of weeds and are especially useful in controlling weeds in plantations or perennial crops.

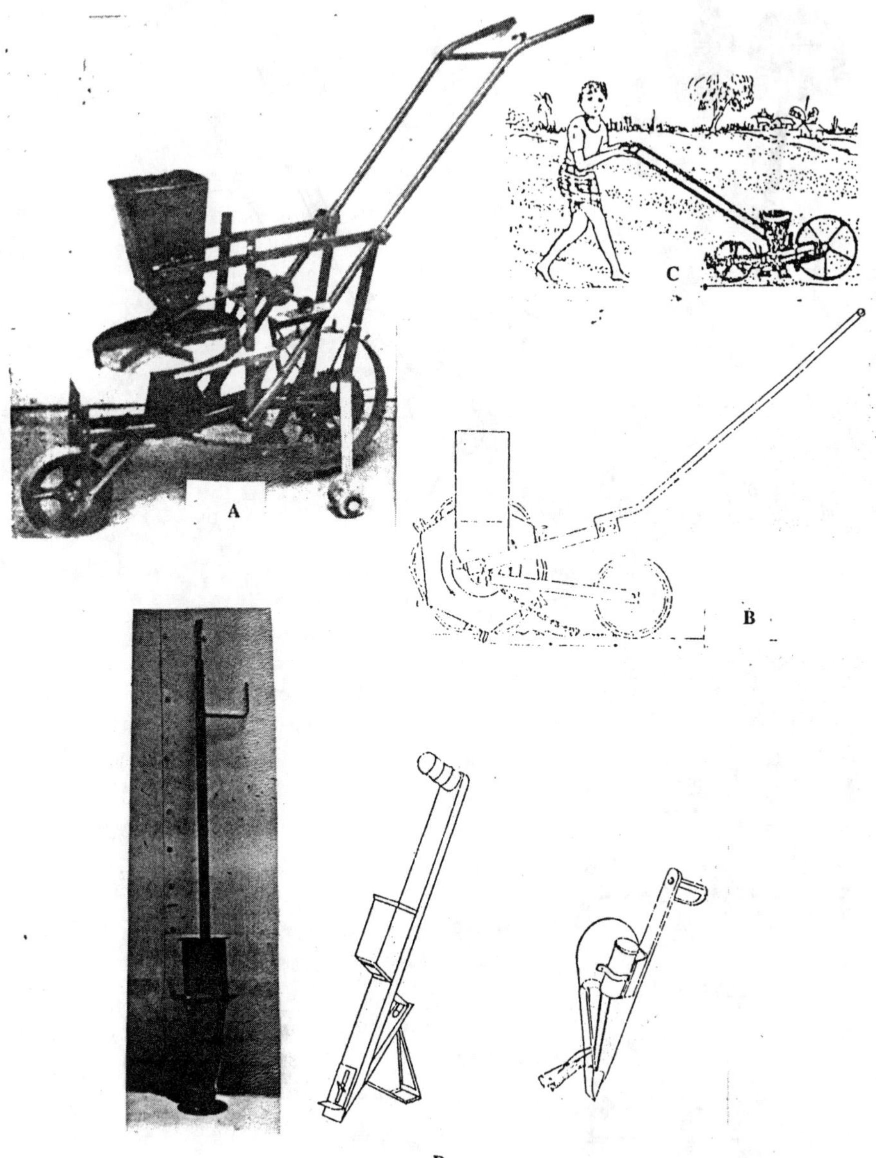

Figure 1.3. Some Examples of Manual Planting Machines. A - Hand-pushed centrifugal grain/fertilizer broadcaster; B - Hand-pushed rotary injection planter; C - Hand-pushed seed drill D - A variety of jab planter.

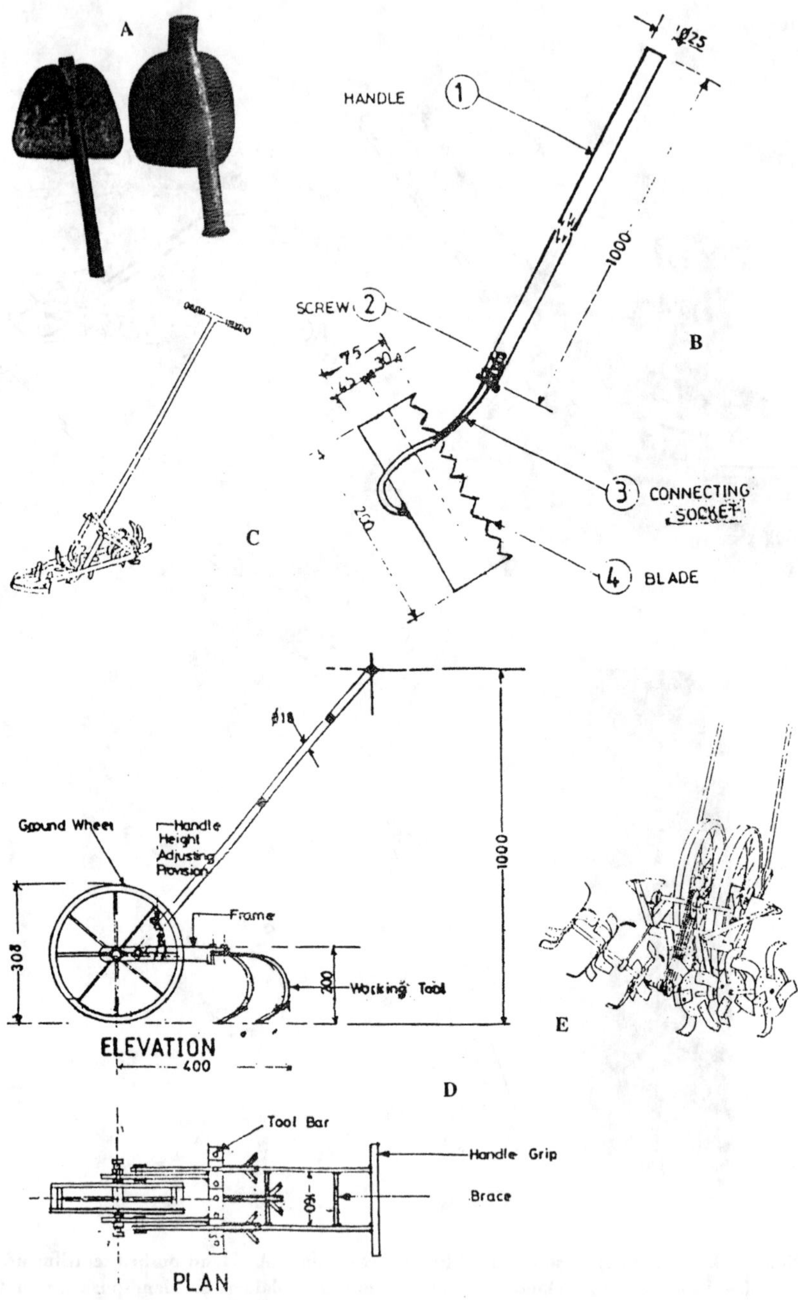

Figure 1.4. Some Examples of Weeding Tools and Machines. A - Weeding hoes (Ikeagwu-agadi); B - Improved weeding hand hoe; C - Hand-pushed rice weeder; D - Wheeled hand-pushed weeder; E - Hand-pushed ridge-profile weeder.

Manual Harvesting Tools and Machines

From discussions thus far, it is evident that tools and machines for field operations used by peasant farmers have retained their pristine forms and sizes as developed by their ancestors centuries ago. This is particularly true of harvesting operations for which the hoe, various diggers, machetes and knives, sickles and scythes, persist as the major tools available to peasant farmers of the developing countries of the world. A few manual harvesting machines have been developed here and there, but they cannot compete favorably with the manual harvesting tools in terms of cost and efficiency.

Hoes

If the hoe is thought ubiquitous in peasant agriculture, well so it is. It is the principal tool used by small holder farmers to harvest root and tuber crops (yams, cocoyams, potatoes, corn, cassava, etc.) as well as all crops that develop underground, such as groundnuts. Of course, the type of hoe used depends on the crop, the topography (flat, beds, ridges or mounds) and the soil type or condition (hard or friable, plastic or muddy). Happily, in most cases, a suitable hoe is always available.

Diggers and Lifters

A variety of simple tools consisting of long-handles with sharpened or speared digging tips made wholly of wood, wooden with steel tips, or made wholly of steel, form a second group of tools known as diggers, which are used to harvest root and tuber crops, especially yams. Often they are used together with hoes to deal with roots and tubers that develop at considerable depths in the ground. Sometimes, shovels and forks, where available, are used in place of wooden diggers. There are also a number of designs of hand tools called lifters, used for root crops, especially cassava, as illustrated in Fig 1.5.

Machetes and Knives

For harvesting cereals (millet, corn, rice, sorghum) peasant farmers use various types of machetes or knives developed over the centuries to cut the plant stalk or grain heads, in a once-over operation or selectively, with the special advantage that shattering losses are minimized. Another advantage is that inclusion of unnecessary vegetation is drastically reduced, making for lower transport costs and safer storage. The main disadvantage is the inherently high labor requirements, which can be considerably higher than those for sickles, especially for heavy crops. Special knives have been developed for some crops, such as sugar cane and oil-palm.

Sickles

A sickle is a peculiarly shaped knife, very popular in the harvesting of cereals, consisting of a curved metal blade with a sharpened edge and a wooden handle fitted on to a shank known as the tang. The length of the blade and its curvature, as well as the shape of the handle and its angle of attachment, varies a good deal from one culture to the other, as illustrated in the sample presented in Fig. 1.5. Typical specifications of a sickle also are shown.

Scythes

A scythe is a variant of a sickle, composed of a long, curving blade with a sharp edge, made fast at one end to a long, bent shaft with a handle forming a unit called the snath.

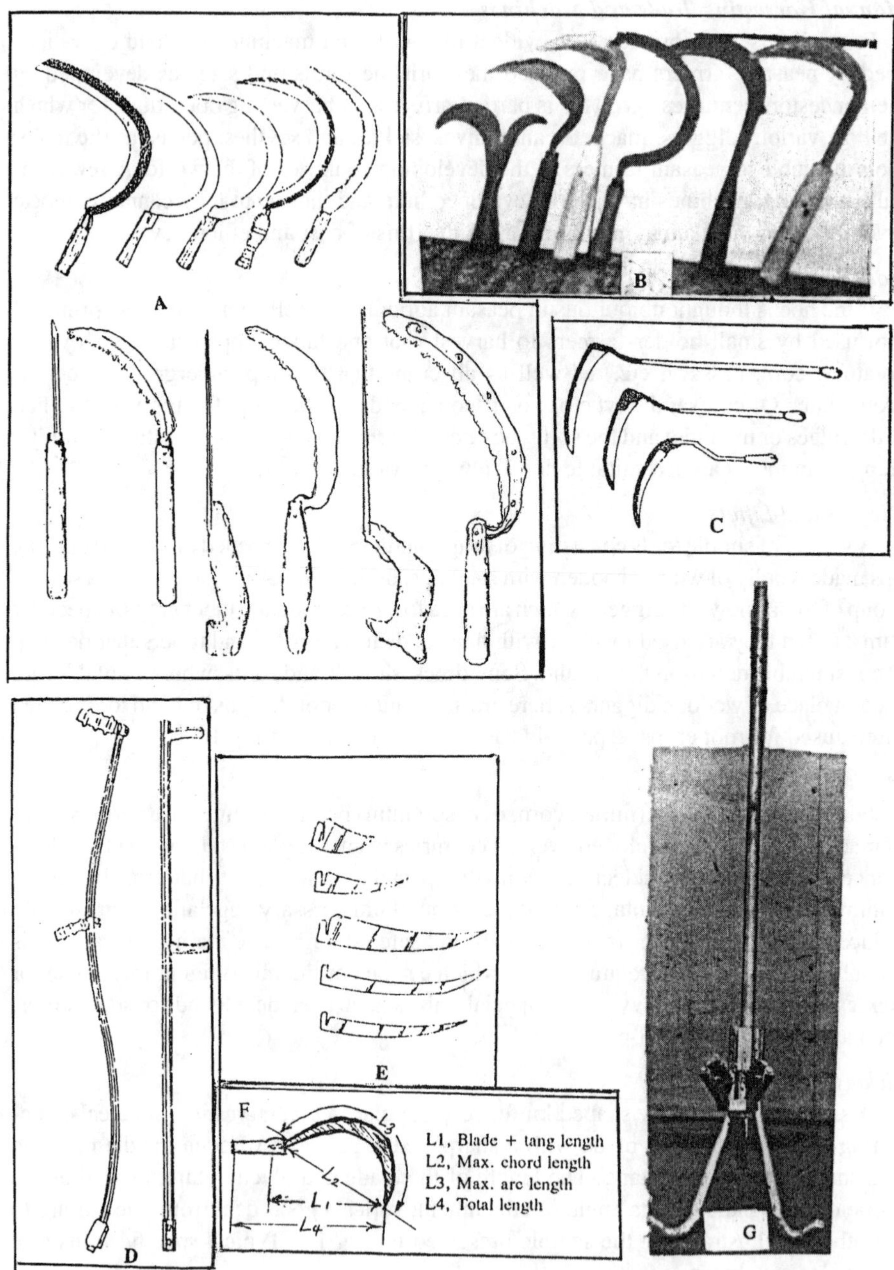

Figure 1.5. Some Examples of Harvesting Tools other than Hoes and Machetes. A - Different traditional sickle shapes; B - Some Nigerian sickles; C - Various harvesting hooks; D - Scythe handles; E - Different scythe blades; F - Sickle dimensions; G - Cassava lifter.

The curved lengths of scythes vary quite considerably, with the shorter bladed ones being more suitable for difficult terrains such as hill slopes. The lengths and curvatures of the snaths also vary a good deal as they are designed to permit users of various statures to operate with both hands and outstretched arms. Specifications for scythes are similar to those for sickles. Samples of scythe blades and snaths are shown in Fig. 1.5.

Harvesting Hooks

The cereal or grass harvesting tool called the reaping hook was developed as a sort of hybrid or cross between the sickle and the scythe. Reaping or harvesting hooks, like sickles and scythes, come in varying sizes and traditional shapes as illustrated in Fig. 1.5. They have short handles and very sharp blades, so that a user crouches to cut the free standing crop without supporting or holding it. Preference for the harvesting hooks appears to be a cultural trait.

Economics of Human-Powered Tools/Machines for Field Operations

The discussions so far have shown that human-powered tools/machines for field operations are quite limited, consisting essentially of hoes, machetes, knives, hooks and diggers, virtually preserved in their pristine shapes and sizes by the peasant farmers who inherited them from their great ancestors of many centuries ago. The overriding characteristics of these implements are their relatively low energy demand, low labor productivity, low technology, low output and inherently high laboriousness and tedium, all of which have direct and indirect relevance to the economics of hand-tool technology (HTT), over and above the purely capital cost considerations.

The low-technology characteristic is important because it implies that these implements can be, and generally, are fabricated by the farmers themselves, as well as by local artisans and blacksmiths, so that their supply is largely in response to their demand. Although local manufacture in itself is a good thing and should be encouraged for various economic reasons, it must be noted that, given the intrinsically low-volume, the lack of formal scientific basis, and the absence of quality control in the production process, the products generally are of low quality.

In addition to artisans and blacksmiths, some urban-based companies in some developing countries manufacture the hand tools of interest, as shown in Table 1.6. But such companies often have to contend with many local problems, such as: lack of the right kind of steel locally; inimical or non-conducive manufacturing environment; poor technological support institutions, infrastructure and superstructure; stiff or unfair competition with directly imported alternative goods; and unrealistic excises and sometimes discouraging duties on imported raw materials. Additionally, the companies face the serious problem of low or uncertain volume of sales arising from the fact that their customers, the peasant farmers, who are usually among the poorest of the poor in their country, have limited capacity to invest in new implements, even when the implements they use are too old and or worn out and require replacement. As a result of these severe local problems, the few existing companies lack the requisite incentives to invest in the improvement of the products or expansion of their operations, all with predictable consequences. This means that needs for most of these implements are usually satisfied by importation from the developed countries, such as Austria, Belgium, Germany, Spain, the United Kingdom, and the United States.

Table 1.6. Some Formal Companies in Some Developing Countries that Manufacture Some Human Powered Implements for Production Field Operations

Country	Company	H	MK	RH	WH	S	R	F	MT	P	SK	SY	GH	PL	CL
Brazil	Azotupy Ind. Met. Ltd	*				*									
Bangladesh	Zahed Metal Ind.	*													
Cameroon	Tropic	*					*	*		*			*		
Chile	Famae	*				*									
China	China Imp/Exp Corp	*													
India	Kumaon Nursery		*					*							
	Kumar Ind.		*			*		*	*		*				
	Yantra Vidyalaya	*				*	*								
	Bharat Ind. Corp	*													
	Cossul & Co. Ret. Ltd	*													
Kenya	Datini Mercantile Ltd						*	*							
	Oyani Rural Centre	*													
Malawi	Chillington Agric Ltd	*	*								*				
Nepal	Agric.Tools Fac. Ltd	*					*		*		*				
Niger	A.F.M.A.										*				
Nigeria	Crocodile Match. Ltd		*			*									
	W.Nig. Tech Ibadan													*	
	Zartech, Ibadan														*
Peru	Fahena S.A					*	*	*							
	Herramieutas S.A	*				*	*	*	*	*	*				
	Herrandina						*				*				
Phillipines	Agric. Mech. Dev. Pro	*													
Sri Lanka	Kanthi Ind.							*							
	Agric. Impl. Factory	*													
Tanzania	Sarvudaya Kandy	*													
	Ubongo Farm Imp/	*					*								
	Zana Za Kilimo Ltd	*							*						
Uganda	Chillington Tool Ltd	*													
Zimbabwe	Bulawayo Steel Prod.								*						
	Garba Ind. (Pvt) Ltd	*						*							
	Temper Tools" Ltd		*				*	*			*				
	Tool making Eng'g.														

Note: H - hoes; M/K - machetes/knives; RH - rotary hoes; WH - wheeled hoes; S - spades/shovels; R - rakes; F - forks; MT - mattocks; P - picks/axes; SK - sickles; SY - scythes; GH - grass hooks; PL - jab/roller planters; CL - cassava lifters.

It is clear that there are many factors that combine with capital cost considerations to make the economics of hand-tool technology in developing countries quite intriguing. With small and usually irregularly shaped plots planted with a mixture of crops, a cropping system that tolerates the feasible use only of primitive hand-tools that may be old and worn out, leading to poorly cleared land, poorly tilled soils and poor planting in irregularly formed rows, the outcome is inevitably a poor harvest. Under these circumstances, with low cost advantage and low productivity, the economics of peasant agriculture using HTT, exhibits the vicious cycle of poverty-begetting-poverty.

1.1.3. Human-Powered Tools and Machines for Post-Harvest Operations

Definitions and Some General Remarks

Post-harvest operations refer to those activities undertaken to transport, process, transform, preserve or store harvested agricultural products in order to enhance their economic value by increasing their nutritional value and availability over time and space and therefore, their price or market value. The relevant activities include such unit operations as threshing, cleaning, sizing, shelling, peeling, grating, cutting, slicing, chipping, grinding, milling, comminuting, cooking, drying, pasteurizing, fermentating, handling and transporting. Although some authorities treat threshing together with harvesting (of cereals especially), at the level of operation of small-holder farmers it is considered more appropriate to treat threshing as a post-harvest operation because the two activities are quite separate.

Tools and machines retain the definitions given in the introduction. Therefore, we are concerned here with tools used, and machines powered by human beings to perform the indicated post-harvest unit operations. Naturally, small-holder or peasant farmers form the majority of those who use the human-powered tools and machines, especially the tools for post-harvest operations. Some post-harvest unit operations for certain crops, such as peeling of cassava or extraction of melon (*egusi*) seeds from the pod, can be performed using only hand tools because viable machines for them, human-powered or otherwise, simply do not exist. In such cases, there is really no choice but to live with the inherent tedium and low efficiency.

But because of the nature of the majority of post-harvest operations involved, small-to-medium-scale commercial farmers, and even non-farmers, also use the human-powered tools and especially, machines, to some advantage. Nevertheless, the fact of the situation is that, given the natural limitations of human-powered machines (tedium, low power, low capacity, and low efficiency), most human-powered machines for post-harvest unit operations have motorized counterparts, powered by electric motors or internal combustion engines. Of course, motorized machines are preferred by small-scale commercial farmers, or by non-farmers who use them mostly to serve peasant farmers, who naturally prefer the custom services to the tedium of manual operation.

Nevertheless, there are literally hundreds of food products processed by peasant farmers using hand tools and human-powered machines not found in the literature. Many of them are specific to certain localities outside of which they may not be known even in the same country. As already stated, manual processing of these products is time-consuming and tedious; the conditions prevalent at this level of operation generally are unsanitary and inherently unhygienic, with little attention paid to quality control, making the wholesomeness and quality of the products, perforce, variable and uncertain.

Some Common Tools for Crop Processing by Peasant Farmers

Knives

Many post-harvest unit operations for processing of agricultural products by peasant farmers are performed with the group of tools classified as knives. For example, cutting, peeling, slicing, chipping and many size-reduction operations are carried out

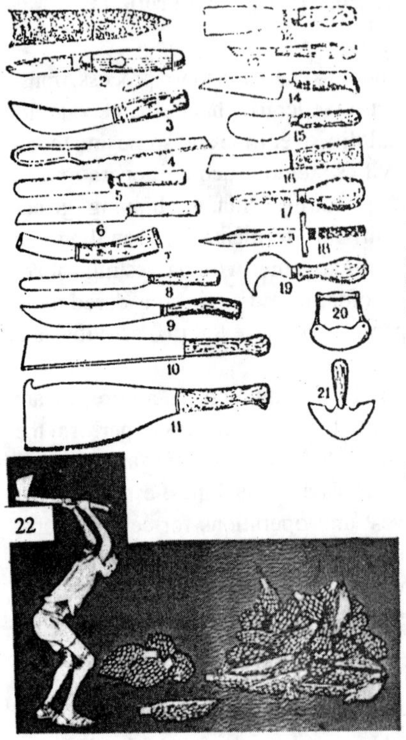

1 - Indian jasper blade; 2 - Pocket knife;
3 - Skinning knife; 4 - Bread knife; 5 - Table knife;
6 - Gold beater's knife; 7 - Farrier's knife;
8 - Palette knife; 9 - Carving knife; 10 - Corn knife;
11 - Cane knife; 12 - Putty knife;
13 - Wood carver's knife; 14 - Felt knife;
15 - Paper hanger's knife; 16 - Hacking knife;
17 - Oyster knife; 18 - Hunting knife;
19 - Oil cloth knife; 20 & 21 - Chopping knife;
22 - Axe for quatering oil palm fruit bunches.

Figure 1.6. World of Knives and Cutters for Traditional Crop Processing.

using knives, which come in numerous sizes and shapes, as shown in Fig. 1.6. Knives are as indispensable and ubiquitous in post-harvest crop processing operations by peasant farmers as hoes and machetes are in their field operations for primary agricultural production.

Mortar and Pestle Systems

Next to knives, mortars and pestles are the next most commonly used tools in peasant crop processing operations. They are used for threshing, grinding, milling, size reduction and all unit operations that can be performed by pounding or rubbing. A mortar is a strong vessel, usually of wood in this context, in which materials are pounded or rubbed with a pestle. The popularity of the mortar-and-pestle system derives from the fact that the peasant farmers are able to produce the tools for themselves. Mortars are produced from trunks and pestles from appropriately sized branches of hard or high-density wood of trees such as the iroko, oil-bean trees, etc. Naturally, mortars and pestles vary a great deal in size, shape and weight, depending on the intended applications, as illustrated in Fig. 1.7. A mortar may be large enough to admit six to eight persons, standing, with long pestles to pound its contents together/simultaneously, to macerate parboiled palm fruits in peasant palm-oil processing operations, or, in the preparation of yam *foo-foo* or *fufu*

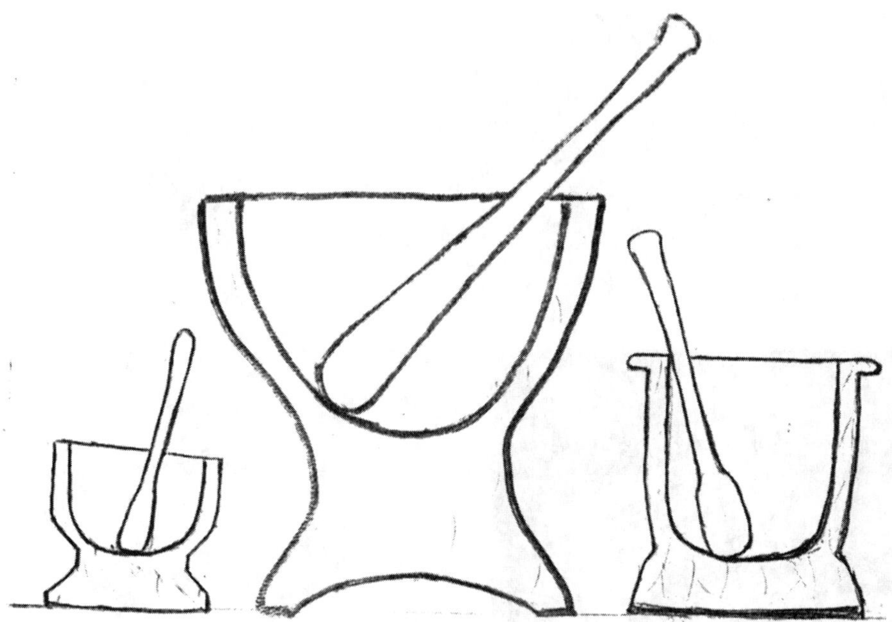

Figure 1.7. Mortar and Pestle Systems.

(Sierra Leone), for communal work-team feeding. Or the mortar/pestle system may be just large enough to grind a few grains of pepper for a small pot of soup for a bachelor.

Miscellaneous Processing Tools

There are, of course, many processing unit operations by peasant farmers that cannot be carried out with knives or mortar/pestle systems but require unique tools such as: shellers for maize, groundnuts or such prodded crops; graters for cassava and other roots and tubers; sifters for meals and flours; and winnowers for grain. Some of these tools are illustrated in Fig. 1.8 without further comments, since they generally are very simple in form and construction.

Some Common Human-Powered Processing Machines

Manual Machines That Can Replace the Knife

For a limited number of unit operations for which the peasant farmer uses a knife, there exist some human-powered machines or equipment. Such machines that can replace the knife in traditional or peasant processing operations include roller root cutters, cassava graters, vegetable cutters/slicers, etc. Some of the machines are illustrated in Fig. 1.9. On the whole, such machines do not make a significant impact on the post-harvest processing operations of peasant farmers.

Manual Machines That Replace the Mortar/Pestle System

It is perhaps correct to state that for most mortar/pestle systems in peasant or traditional operations, there are viable human-powered machines as alternatives. The machines are

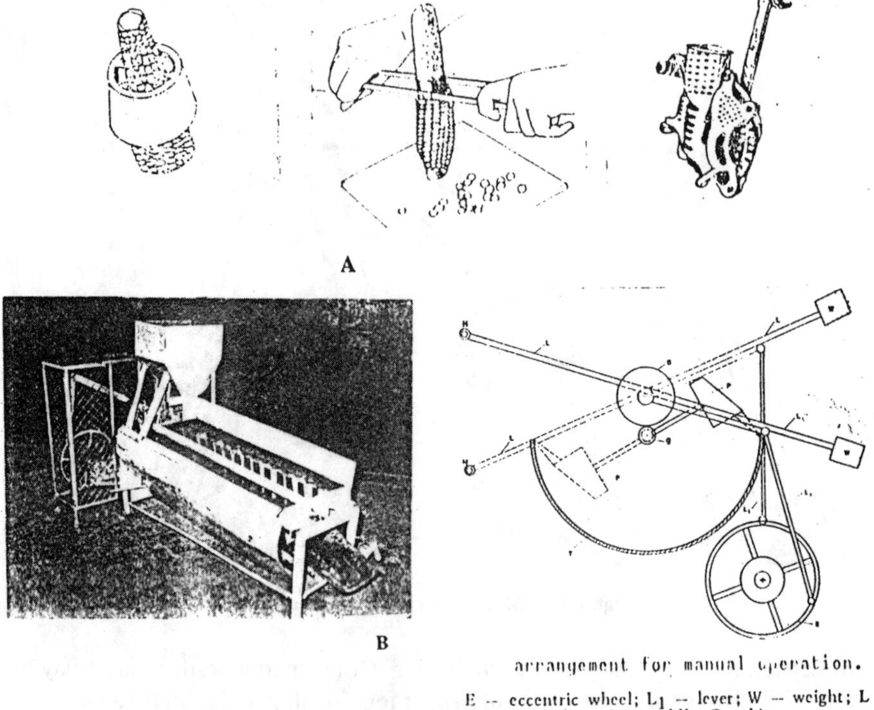

Figure 1.8. Some Miscellaneous Manual Processing Tools and Machines. A - Different types of hand tools for shelling corn; B - Gari frying machine with manual operation arrangement; C - Manual Bitter-leaf processing machine.

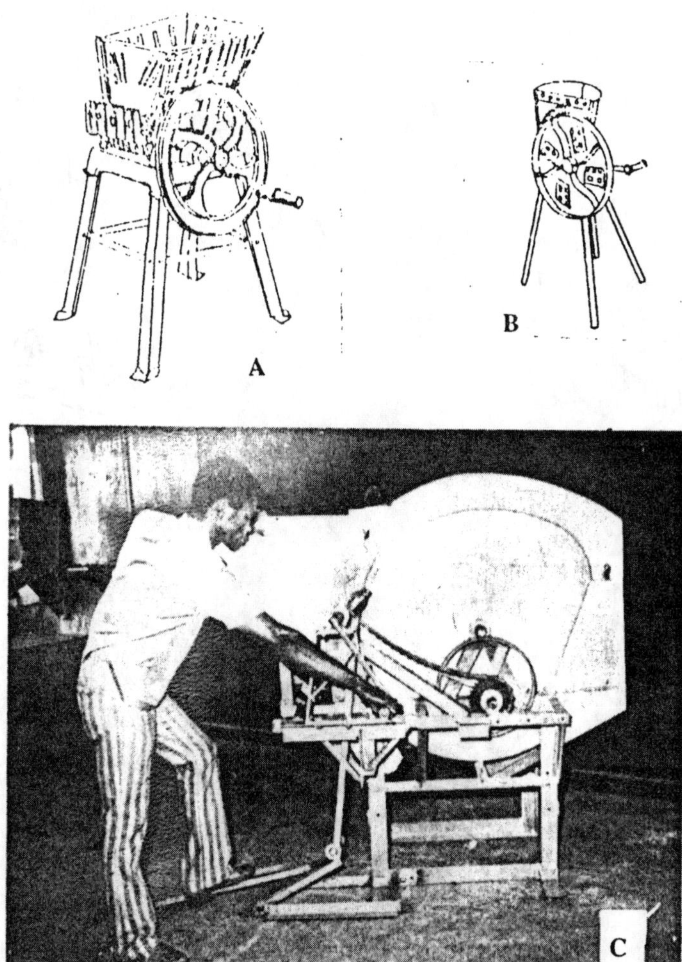

Figure 1.9. Some Manual Processing Machines That Can Replace the Knife. A - Roller root cutter; B - Disk vegetable cutter/slicer; C - Pedal-assisted manual cassava grate.

either hand- or foot-operated, and commonly include grinders, mills, hullers, decorticators, shellers, crushers, presses and winnowers/separators, as illustrated in Fig. 1.10. Their impact on peasant farmers' operations is quite significant, especially in terms of the custom services that small-scale non-farmer operators of these machines provide, as mentioned earlier.

1.1.4. The Sociology and Future of Hand Tool Technology in LDC's

In Section 1.1.1, the predominance of HTT in the agriculture of LDC's was identified as an important factor in the overall economic development of the countries. The sociology

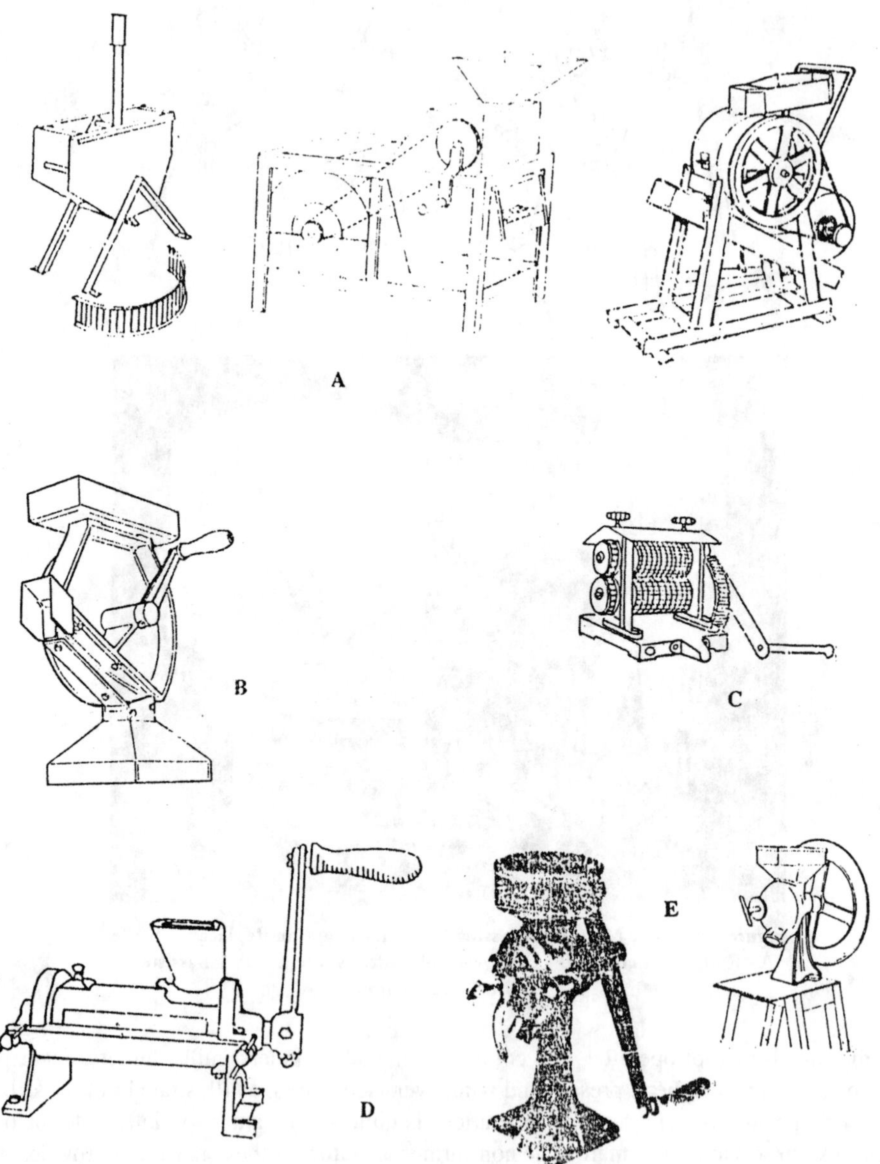

Figure 1.10. Some Manual Machines That Replace the Mortar and Pestle. A - Groundnut shellers (hand-operated, hand-cranked and foot-operated types; B - Disk coffee puler; C - Hand operated crusher; D - Rice huller; E - General purpose hand grist mills.

of HTT thus becomes an important subject to consider in discussing its present or its future in LDC's.

The man with the hoe, a title credited to Edwin Markham [10], still remains an apt description of the peasant farmer in developing countries of Africa, even on the verge of the 21st century. In spite of decades of enormous expenditures and investments in agriculture, the peasant farmer using HTT remains an indigent serf, regarded by today's youths as a dreadful anachronism. With HTT the prevalent technology, the agricultural industry in those countries has degenerated into a world for and of losers [11], populated by aged and aging peasants, whose precarious hand-to-mouth existence seems to stand like a huge signboard screaming to the youths: KEEP AWAY!

As such, HTT curiously is no longer sustainable in many of the LDC's because, frozen as it is in its pristine stages of hoe-and-machete or mule-and-oxen technologies, it leads to the prevalence of mass poverty, which the young people of the LDC's abhor and roundly reject [12].

It ought to be appreciated by all actors involved that the continuing policy of near exclusive advocation or promotion of HTT in the LDC's is becoming counterproductive. Therefore, it has become necessary and urgent to reduce the undue emphasis on HTT as a conceptual and psychological point, and to begin to change the undignified image of peasant farmers, thereby make farming more attractive to the youths. An official policy to deemphasize HTT and promote or encourage higher levels of engine power mechanization technology (EPMT) is necessary to rejuvenate the agricultural industry and foster some hope of a better future for the industry in the LDC's of Africa.

References

1. Anazodo, U. G. N. (1976). A field study and analysis of problems of and prospects for mechanization of family farms in Anambra and Imo States of Nigeria: A Research Report. Department of Agricultural Engineering, University of Nigeria, Nsukka.
2. Inns, F. (1992). Field Power. In Tools for Agriculture, Fourth Edition Intermediate Technology Publications Ltd., London.
3. Odigboh, E. U. (1991). Continuing Controversies on Tillage Mechanization in Nigeria. *J. Agric. Sci. Technol.* 1(1): 41–49.
4. Makanjuola, G. A., Abimbola, T. O. and Anazodo, U. G. N. (1991). Agricultural Mechanization Policies and Strategies in Nigeria. In Mrema, G. C. (Ed.) Agricultural Mechanization Policies and Strategies in Africa: Case Studies from Commonwealth African Countries. Commonwealth Secretariat, London.
5. Odigboh, E. U. (1996). Small-Medium-Scale Farmer Oriented Mechanization Strategy for Energizing Nigerian Agriculture. *Proceedings of the 1996 National Engineering Conference of Nigerian Society of Engineers*, pp. 96–115.
6. Mrema, G. C. and Mrema, M. Y. (1993). Draught animal technology and agricultural mechanization in Africa: its potential role and constraints. *Network for Agricultural Mechanization in Africa (NAMA) Newsletter* 1(2): 12–33.
7. Comsec (1990). Report of expert consultation on agricultural mechanization in Commonwealth Africa. Commonwealth Secretariat, Malborough House, Pall Mall London SW1Y 511X: 83.

8. Anazodo, U. G. N., Opara, L., and Abimbola, T. O. (1989). Perspective plan for agricultural development in Nigeria (1989–2004). Agricultural Mechanization Study Report, Fed. Agric. Coordinating Unit (FACU), Ibadan.
9. Anazodo, U. G. N., Abimbola, T. O. and Dairo, J. A. (1987). Agricultural machinery use in Nigeria: the experience of a decade (1975–1985). *Proc. Nigerian Soc. of Agric. Engineers (NSAE)* 11: 406–429.
10. Gunkel, W. W. (1963). Nigerian Agriculture. Paper No. NA 63–107, prepared for distribution at North Atlantic Section Meeting of American Society of Agricultural Engineers. University of Maine, Orono, Maine, August 25–28, 1963.
11. Wainan, S. (1990). "Major Agricultural Reforms Needed", Experts say. *African Farmer* 3: 34.
12. Odigboh, E. U. and Onwualu, A. P. (1994). Mechanization of Agriculture in Nigeria: A Critical Appraisal. A Commissioned feature article. *Journal of Agricultural Technology, JAT* 2(2): 1–58.
13. FAO (1988). Agricultural Mechanization in Development: Guidelines for Strategy Formulation. FAO Agricultural Services Bulletin 45, FAO, Rome.

Animals

M. Havard and A. Wanders

Animal power accounts for about 20 percent of agricultural mechanization in developing countries; human power accounts for 70 percent, mechanical power 10 percent. The world number of draft animals has been estimated to about 400 billion head, mainly in Asia. Cattle, buffaloes, horses, donkeys and mules are the main draft animals. Camels, elephants, llamas and yaks also may be used. The factor limiting the use of animals for work is their reduced energy potential, which is determined by characteristics and working ability of the species.

1.1.5. Efficient Use of Energy Potential by a Draft Animal Power Unit

Several parameters such as energy, power, draft force, etc. are used to define the work performed by animals. Energy is provided to animals from feeds previously metabolized into fats and carbohydrates, and then assimilated in muscles. The energy output required for such processing is not well known.

A draft animal power (DAP) unit (i.e., animal plus equipment), which exerts a tractive force, can be compared with a system consisting of a resistant part (equipment) and a power unit (animals). The energy accumulated by animals is partially released in a mechanical form when pulling equipment or carrying a load.

Work W is the energy required to move an object with a draft force F through a distance L: $W = FxL$. The power P of a DAP unit is the work performed per unit of time T: $P = W/T$. The example given hereafter refers to pulling a plow (Fig. 1.11).

Draft resistance *Fr* is a downward diagonal. On a well-adjusted plow that the operator can easily hold straightward, *Fr* consists of three forces:

- Force exerted by soil on the working parts *Fs*
- Implement weight P which in fact does not affect the resistance effort significantly

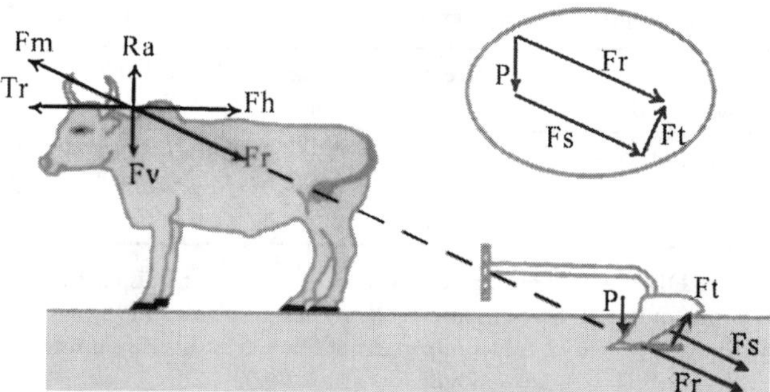

Tr = tractive force (pull, opposed by Fh)
Fm = motive force (opposed by Fr)
Fh = horizontal resistance to draft
Fv = vertical load
Ra = animal reaction (opposed by Fv)
P = weight of implement
Fr = result of resistance forces
Ft = soil resistance of plow heel
Fs = soil resistance to forward movement

Figure 1.11. Diagram of tractive forces.

- Soil reaction at level of the implement bottom part *Ft*. This is an almost vertical component which counterbalances the plow weight to a large extent.

The motive force *Fm* exerted by animal(s) is opposed by the resistance force; it makes it possible to perform work. The direction of these forces is defined as the line of draft. In our example, the draft chain between the implements and animals is a flexible link coinciding in position with the line of draft. *Fm* can be analyzed in terms of two axes–a horizontal component corresponding to the effort exerted by the animal to move the implement (i.e., tractive force *Tr*) and a vertical component corresponding to additional load exerted on the animal's forward members *Fv*. This can result in additional fatigue.

The sustained effort corresponding to a working time of about 2–3 hours per day, at a normal speed (0.6–0.8 m/sec), determines the animal team's capability. According to animals and working conditions, the next-to-optimum effort is between 9% and 12% of the live weight for oxen and buffaloes, between 12% and 16% for donkeys and camels (Table 1.7, Fig. 1.12)[12, 14, 18]. Nevertheless this is difficult and time-consuming to determine. Knocking against obstructions results in a maximum instantaneous effort. This equals the animal live weight in value and determines strength constraints for the materials to be used in farm implements.

Use of the animal energy available to perform sustained work varies in efficiency according to local working conditions (types of cultivation works, implements and harnessing systems used). For given work task, energy or net output (energy requirement to energy expenditure ratio, W/DE) can be estimated as a continuous effort at 26%–29% for donkeys, 24%–27% for horses, and 16%–20% for zebus [18].

Table 1.7. Efforts and energy expenditures for some draft animal species*

Animals	Live Weight (LW) kg	Force pc of LW	Speed m/s	Work Time h	DE (MJ)
Donkey	100–150	10–16	0.7–0.9	3.0–6.0	5–8
Horse	200–300	10–16	0.9–1.0	4.5–6.5	16–24
Zebu	300–450	9–15	0.6–0.8	4.5–6.5	24–40

Where DE, energy expenditure at work = DE horizontal displacement + DE pull output + De transport + DE vertical displacement [6].

* In India [15], similar forces in percent of live weight have been recorded with camels (15–16 pc) and buffaloes (10–12 pc).

Source: [18].

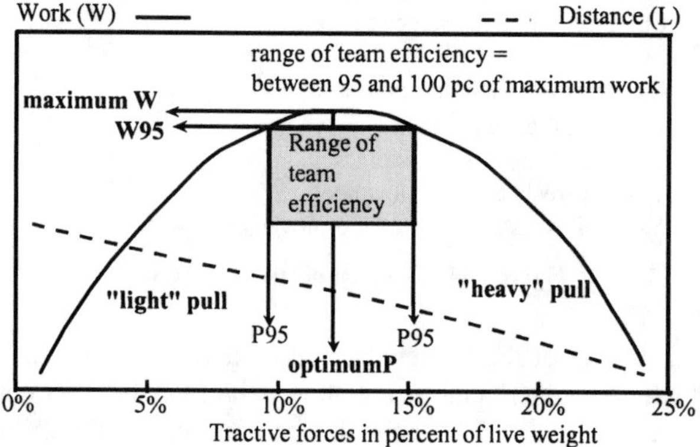

Figure 1.12. Trend of curves of the distance travelled and pull output in terms of tractive force.

Selecting the most suitable parameter for characterizing team capability is difficult. Available power and energy vary with the mean pull required per day. Capability depends on the direction of the line of draft and ground surface conditions. Efficiency decreases from 1 in straight-line work to 0.8 in circular work (capstan). Tillage in dry soils causes high frequencies and a wide range of variations in forces. This results in pronounced vibrations and greatly affects the animal's comfort. Working in loose soils limits the animal's draft capability because of the sinking effect.

1.1.6. Control of Animal's Energy Potential

In-depth knowledge of the factors influencing the work achieved by animals is required. Some of these can be controlled by farmers, others cannot (Fig. 1.13).

Environmental Conditions

Environmental factors (soil and climate) that define working conditions are uneasy to control. Farmers can improve such conditions but to a small extent. They may prefer to work their animals at the beginning or end of the day, when heat is acceptable. This is profitable in terms of animal capabilities and endurance.

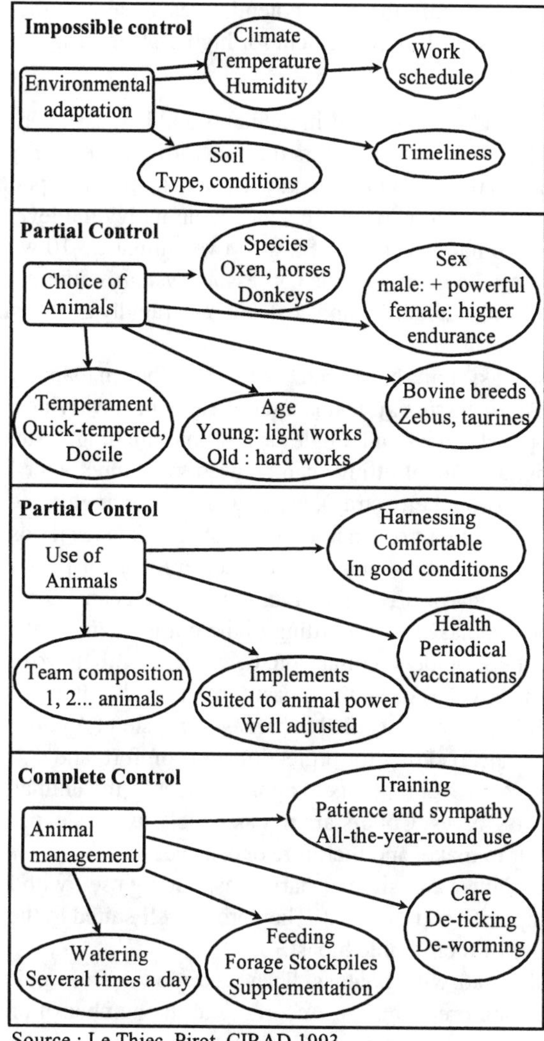

Source : Le Thiec, Pirot, CIRAD 1993.

Figure 1.13. Factors influencing the capabilities of animal teams.

Choice of Animals

The characteristics of animals (breed, species, sex, age, temperament) determine their working abilities. Farmers cannot control these. Their only room for maneuver is in the choice they can make between the various species locally available or affordable. Heavy and slow species (e.g., elephants, buffaloes, bovine crosses) must be preferred for hard work. Light and fast species (e.g., donkeys, horses, camels) are particularly suited for light work such as sowing, weeding and transport. Whatever the species, local breeds are preferable. In terms of draft, the cows potentialities are almost the same as those of oxen. Nevertheless, cows are less convenient for high draft forces.

Use of Animals

Team composition and the choice of implements and harnessing systems depend on the farmer's decisions. They are key factors for transforming the energy accumulated by animals into mechanical power. With current yokes and harnessing, pooling two animals or more in a team results in a reduced efficiency at an individual level. If the available power is 1 with one animal, it is only 1.85 with two animals, 3.10 with four, and 3.80 with six [12]. Choosing the more suitable harnessing system, equipment and number of animals depends on local availability and cost, but may rapidly enable significant energy gains.

The main harnessing systems have been largely described in many manuals and books [3, 4, 6, 9, 10, 16]. A *harnessing system* is a set of elements involving a harness, driving fittings (steering ropes, bridles) and single or multiple hitching systems (abreast or in tandem). For carting, additional fittings can be used to assume other functions such as the cart balance (back strap, belly strap), braking, and reversing (breeching strap).

The harness is the main part of a harnessing system. It makes it possible to optimize the energy potential provided by an animal to exert a force for pack transport, pulling a cart or a farm implement, or driving an animal-powered gear. There are various types of harnesses that can be classified according to the point where they apply work to the animal, e.g., before the shoulders (collar), on the withers (withers yoke), just behind the horns on the neck (neck yoke) or on the breast (breast band or breast strap) (Fig. 1.14).

The *collar* generally is the most suitable harnessing system. A collar includes a frame for fitting on the animals, padding for protection and comfort, and a device for hitching an implement. There are several points to apply a collar to the animal, which results in a better distribution of forces. Collars are not as widely used as expected because they are relatively difficult to make, and therefore expensive.

Breast bands are lighter and simpler harnesses, widely used with horses and mules because of their simplicity and low cost. They prove well-suited to the conformation of such animals because of their ample breast.

Yokes are mainly used with bovines. They take power from points higher on the animals than collars and breast bands. According to the number of animals harnessed, yokes can be single with a single animal, double with a pair of animals, or sometimes triple for training a young animal between two older ones.

Head yokes described as forehead yokes are tied in front of the horns, and are rather uncommon. They were known in Spain and largely popularized in Switzerland and Germany. Head yokes described as *neck yokes* are tied just behind the horns. They were widely used in Europe before the introduction of power-driven equipment. Mainly used with humpless cattle (*Bos taurus*) with strong necks and horns, their form varies from the simplest uncarved wooden pole to yokes shaped into more or less pronounced bows. Padding is required between the yoke and the animal's neck. Incorrectly shaped or fitted neck yokes, with excessively loose or thin securing ropes, provoke injury, horn wearing and sawing. This results in reduced power from the draft animals harnessed.

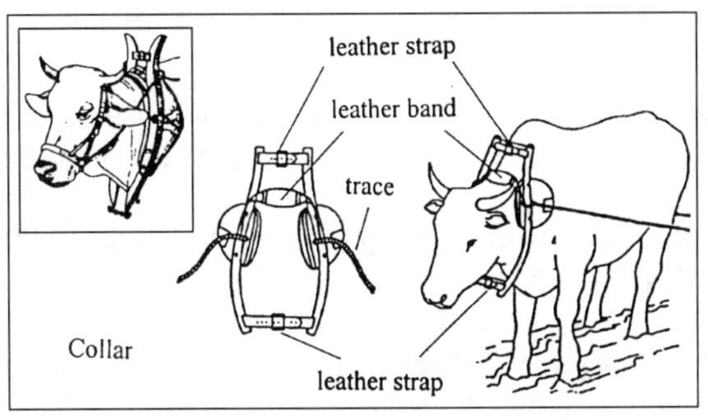

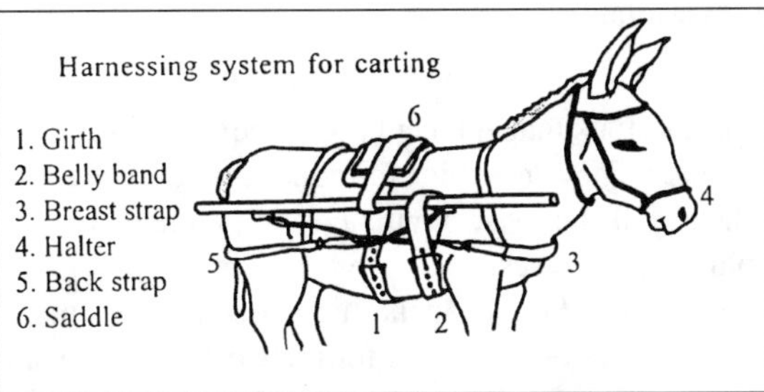

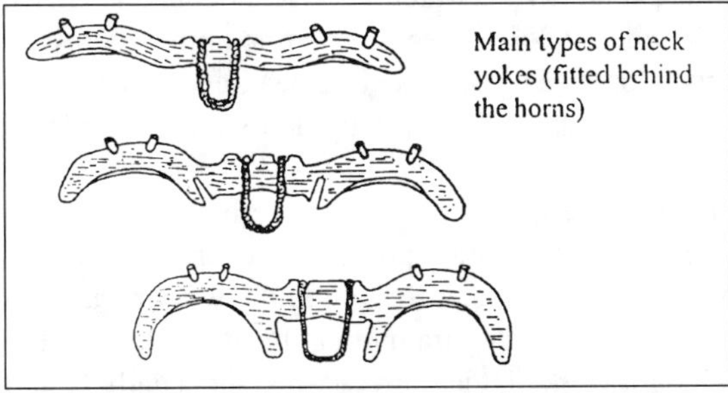

Main types of neck yokes (fitted behind the horns)

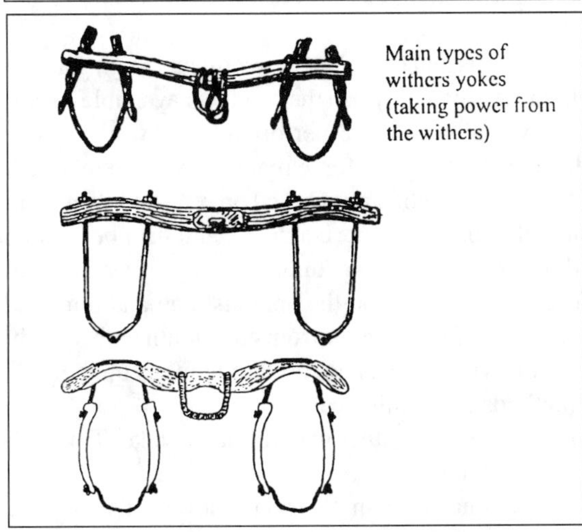

Main types of withers yokes (taking power from the withers)

Figure 1.14. Different types of harnessing systems.

In Burkina Faso [9] and India [15], the use of improved yokes has made it possible to achieve respective energy gains of 15% and 15%–25% as compared with traditional yokes.

Withers yokes apply on the withers, in front and over the shoulders. They are naturally suited to hump cattle (*Bos indicus*) such as zebus. They can also be used with taurines as N'Damas. Withers yokes are predominant in Africa.

Whatever the type of yoke, lowering the attachment point also requires lowering the applying points towards the points of the shoulders. This reduces the slope of the line of draft (an angle of 15° is suitable).

Livestock Management

The components of livestock management (i.e., feeding, training, care, and watering) depend entirely on the user. They determine the conditions and steadiness of the working animal, resulting in the actual energy accumulated and available for working with respect to the energy potential of the species.

Feeding must cover the nutritional expenditures that allow animals to maintain their weight. To produce work, milk or meat, and for their maintenance, animals require energy, proteins, minerals, vitamins and water. The energy contained in feeds is called gross energy, 25%–55% of which is lost in feces. The remaining energy is called digestible energy. This undergoes losses as gas, urine and heat, and leads to the net energy used by animals for their maintenance, for working and sometimes also for other production.

Stock watering is of prime importance to counterbalance the water losses induced through the respiration and transpiration processes occurring when an effort is exerted. Water quantities must be determined according to the work required and climatic conditions. They can easily double. The minimum is 15 liters of water for an ox weighting 300 kg, performing a light work, in the rainy season, and a daily feed ration of DM (dry matter) 4–5 kg [3].

Timeliness in vaccinations and other health treatments (e.g., parasite control) are also beneficial, although farmers can only use the products available locally.

The training level will affect animal performance at work. Docile animals, which prove easy to handle and steady in the force required, will be more efficient. A training program consists of a succession of stages based on repeated commands and constraints to one or several animals, so as to achieve a docile and firm behavior at work. Training will preferably be done before the agricultural campaign, when animals are about three years old. Duration depends widely on the animals' age and temperament, and on the trainer's skill as well. Training time ranges from one month with already familiar animals to more than two months with animals not used to human company.

There are three methods for training cattle:

1. Hitching two young bulls together with the same yoke. This is the most common, but the more difficult, method.
2. Training a young animal with an older one for one or two weeks before they are yoked together. The advantage is reduced stress for the young animal because of the example offered by the older one.

3. Yoking three animals together, with the young animal between two already well-trained animals. This is certainly the most rapid training method with vicious, nervous and restive cattle. It is widely employed in southeast Africa.

The future user must train draft animals personally, or at least take an active part in the training. The first stage is a mutual familiarization. The trainer must show patience, calm, sympathy and firmness. Further stages involve putting the yoke on the animal and making it accept the yoke, and training the animal to walk and pull a light load, exert a tractive force, pull farm implements and work alone.

1.1.7. Farm Equipment for Transport, Tillage, Secondary Cultivation and Sowing

This chapter has voluntarily been limited to the most common mechanized operations such as transport, tillage, secondary cultivation and sowing. Nevertheless, animal power also can be used for water-lifting or harvesting (e.g., groundnut lifting). For further details on the mechanization of such operations, see the many manuals available [1–4, 6, 13].

Equipment for Transport

In rural and urban areas, the daily transport of people and goods is a highly important activity. Various means are used: riding, pack and cart transport (two-wheeled carts or four-wheeled trailers), and sledges.

Riding donkeys, horses and camels is common throughout the Mediterranean and sub-Saharan areas. Riding donkeys without any harnessing system is predominant because of their high hardiness. Oxen also may be employed, as in Mauritania and Chad. Camels and horses are more prestigious. They are harnessed with specific saddles locally manufactured.

Carrying loads on the back of animals is very common throughout all the tropical areas. Loads vary between 80 and 100 kg in weight for donkeys, to 300 kg for dromedaries. Three methods can be applied: bulky loads are directly placed over the animal's back and held in place with ropes; rigid materials (stones, firewood, water containers) are transported on pack saddles with a wooden frame fitting on protective paddings; and in-bulk products are placed in symmetrical pannier baskets over the animal's back (Fig. 1.15).

In certain areas of southern and eastern Africa and Madagascar, artisan-made wooden sledges are used. Their main advantage is that they are cheap, being easy to make and maintain. They are narrower than carts, have a low center of gravity, and can be used on steep, wet, or unbearing ground. Nevertheless they cannot be recommended on tracks because the tractive force required is higher than for a cart, and repeated passages accelerate erosion. That is why they have been banned in some hilly countries such as Lesotho and Zimbabwe. Using a sledge, a pair of oxen can carry a load of about 200 kg, at 0.8 m per second, over several kms.

In the tropics, animal-drawn carts are the most widely used equipment for transport. In rural areas, farmers, artisans and traders employ carts for domestic needs (water and firewood), agriculture (seeds, fertilizers, manure, harvest), trade and social purposes.

Figure 1.15. Using horses and donkeys for carrying a load or carting. *Source:* [11].

The load capacity of a cart (the load a cart can carry across country without distortion or breaking) is 500 kg with a donkey and 1,000 kg with a pair of oxen. The tractive force *Tr* required to move a cart is the product of the total weight *P* of the load (load capacity + dead weight), by the rolling coefficient *K* (which varies with the soil surface state) and the slope *i* (slope coefficient in percent), that is: $Tr = Px(K + i)$, where Tr

and P are in kg. A braking system is required in hilly areas. Carts may be fitted with wooden or steel wheels. But wheels fitted with pneumatic tires–similar to those on light power-driven vehicles–are becoming increasingly common.

In the past, wheels with wooden spokes and metal rims were used on all carts. Made today in Northern Africa, Egypt and Madagascar, this type of wheel is rather uncommon in other countries.

Metal wheels are being abandoned in many areas. They have been introduced for transport purposes in the sub-Saharan wet areas (having high ground clearance, well-suited to negotiate ruts and holes), and are still used in the sandy areas where spiky bushes can cause problems of punctures with pneumatic tires.

Two-wheeled carts with pneumatic tires are the best suited to satisfy the varied needs and situations. Tired wheels require a minimum tractive force as their rolling coefficient is optimum–above 35% on dry track and up to 64% on muddy ground.

Tip-carts with tip-up boxes facilitate the unloading of materials such as earth or manure but are little used because of their high manufacturing cost.

Four-wheeled carts (trailers) can be found in southern African countries, in urban areas and near cities (Bamako, Cairo). They also are used on some plantations. They allow heavy loads up to 3 tons because the animals only support a very small part of the load weight.

Equipment for Tillage

Under Dry Conditions

Under dry conditions tillage objectives are as follows:

- To loosen the soil for creating conditions conducive to in-depth aeration and water movement, which promotes root growth, facilitates rain water infiltration, and achieves convenient conditions for germination.
- To control weeds by uprooting, burying, or even promoting weed emergence for later destruction.
- To plow in fertilizers and soil improvers and mix them with soil.

Choosing a tillage method depends on the type of soil, ground moisture, and on the cultivation system applied. For cultivation in flat ground, tined implements are suitable on dry sandy soil, while plowing is essential on wet, silty soils. For bedding or ridge cultivation, the use of a plow or a ridger is essential.

There are four main categories of operational sequences to be performed prior to planting:

- Tine tillage on dry soil (subsoiling) which can be done before surface tillage and plowing operations. Currently tested in tropical areas.
- Tine tillage on the row.
- Surface tillage on dry or wet soil, in a single or several passes, with tined or rotary implements.
- Plowing on wet soil with or without subsequent cultivations.

Subsoiling under dry conditions consists of breaking up the surface layer of soil and penetrating to an average depth of 7–10 cm with a rigid tine (*Coutrier*) developed by

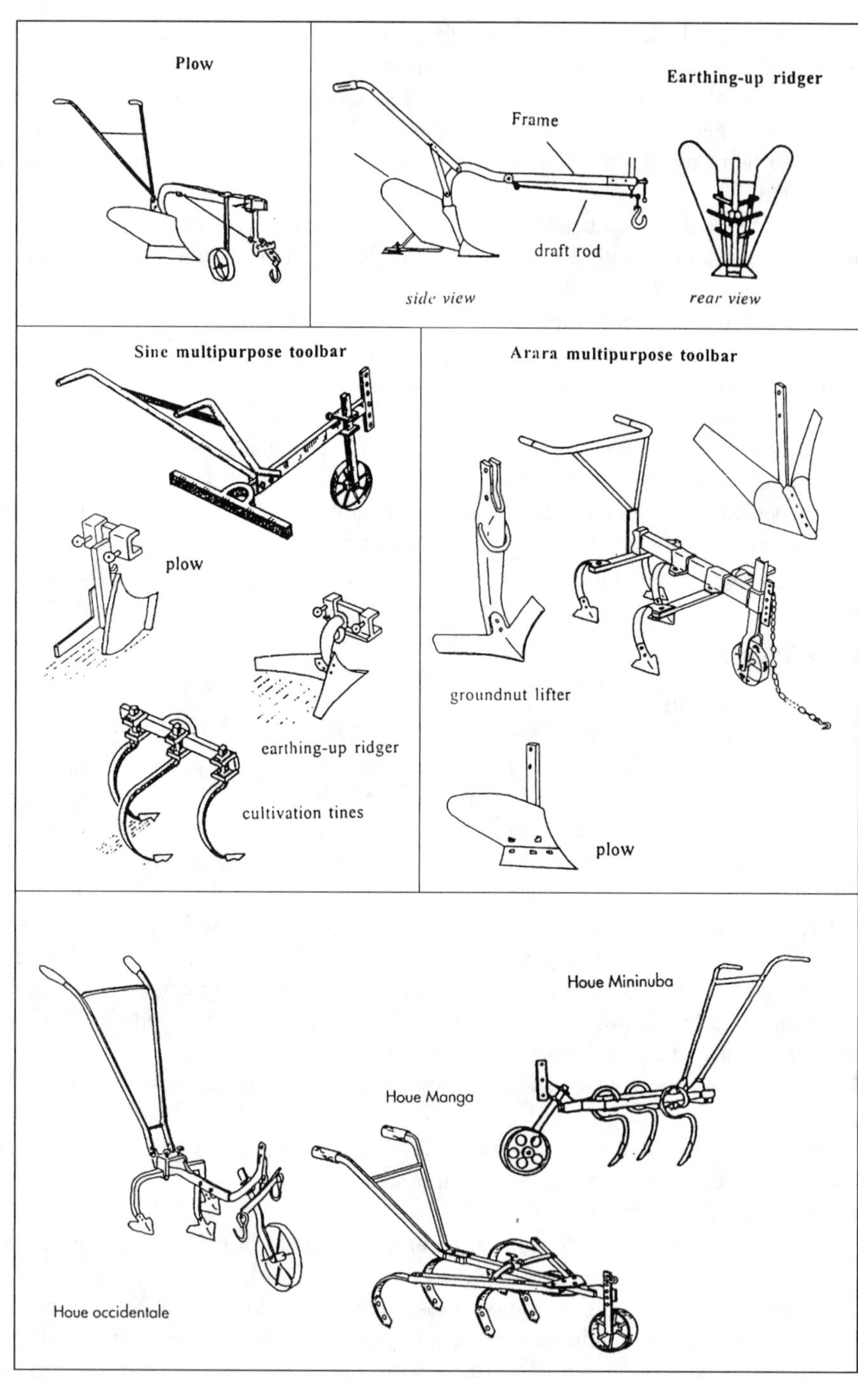

Figure 1.16. Some tillage implements used in sub-Saharan Africa.

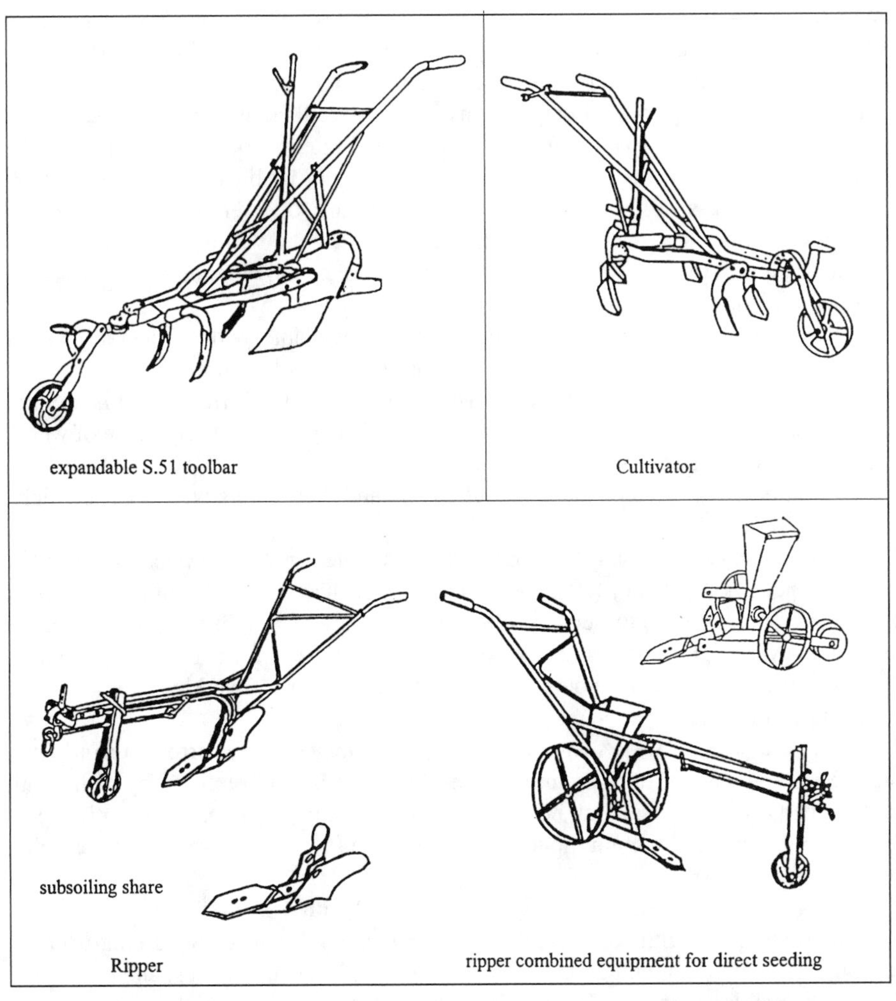

Figure 1.16. (Continued)

Centre d'Etudes et d'Expérimentation du Machinisme Agricole Tropical (CEEMAT) as to favor infiltration of the water of the first rains. This technique is particularly suitable for semi-arid areas where *soil sealing* and the crusting of silty soils prevent the water of the first rains from infiltrating, and prevent traditional tools with cutting edges from penetrating. The draft force required for the operation can be supplied by a yoke of oxen. This results in a cloddy profile with a pass every 50 cm (10–12 hours per ha) to a depth between 8 cm in crusting soils, and 20 cm in sandy soils. On the other hand, the technique makes it possible to stagger the operations of land preparation before sowing.

Ripping [1] consists of working on the row with a subsoiling tine mounted on a plow frame or a toolbar (Fig. 1.16). Compared with plowing, this technique results in valuable time savings because of highly reduced working times and also because it makes it possible to work earlier at the beginning of the rainy season. Combined with a seeder, such equipment may be used for direct seeding in light soils.

The main advantage of soil preparations by surface tillage without soil inversion is that they make early sowing possible because of the limited working time required. The implements used loosen the soil up to 8–10 cm in depth by breaking it up into small clods without excessive fine soil. As they do not turn the soil over, they do not bury the organic matter, but mix the surface layer and so destroy the weeds. They can be used prior to seeding for direct soil preparation or to complete plowing, or after seeding for weeding operations.

Three main types of surface tillage implements are available: with a point (ard plows); with tines (cultivators and harrows); of the rotary type (disk harrows).

Ards, dating back to antiquity, are still widely used in North Africa, Ethiopia, Central America and some regions in Asia. In ancient times, they were entirely made of wood. Today, steel is used for some working parts.

Tined implements include harrows, cultivators and their derivatives (e.g., scarifiers, weeders, hoes).

Harrows consist of a steel or wooden frame mounted with rigid tines. Rarely used among African farmers, harrows are used to complete the work made with a plow. They crush the clods, level the loosened surface layer and provoke light compression on the disturbed layer. Light harrows (25–30 kg) comprise 20 tines for a 1m working width. Working times are between 8 hours and 15 hours per hectare at a working depth between 3 cm and 7 cm.

While harrows have wide shares, cultivator shares are generally narrow. According to the species and number of draft animals used, they can be differentiated by their width (determined by the type and number of tines) and their working depth. Cultivation tools are commonly mounted on multipurpose toolbars used for tilling, weeding and ridging purposes (Fig. 1.16).

Rolling implements (disk harrows) are rarely used with draft animals.

Soil preparations with soil inversion (plowing, ridging) consist of turning over the topsoil, between 10 cm and 25 cm in depth, plowing in plant residues, and destroying the weeds. Soil inversion results in an irregular and slumpy soil, which must be further prepared carefully with other implements such as harrows or cultivators to obtain a convenient seedbed.

Plowing generally is done with a simple moldboard plow, and ridging with a ridger or ridging plow (Fig. 1.16). There are various designs of plows in service. They mainly differ by the species and number of draft animals used. A donkey can pull a plow between 25 kg and 30 kg in weight with shares between 15–18 cm in size. A pair of oxen of 600–800 kg can pull a plow weighting more than 35 kg, with shares between 25–30 cm in size.

Working times are about 20–40 hours per hectare.

In certain regions, ridging substitutes for plowing. Ridging consists of combining two inverted furrow slices to make a backfurrow, or ridge, and an open furrow. The center of the backfurrow is not loosened and this can later interfere with root growth. The ridging time with a simple plow or a ridging plow is about 10–15 hours per hectare.

Under Irrigated Conditions

In all irrigated areas, rice cultivation prevails because rice offers good tolerance to flooding. Soil preparation of rice fields depends on the water control level (zero control, partial control, or full control), and also on the planting method applied (dry sowing, wet sowing, wet transplanting). For dry sowing, the tillage methods described earlier in Section 1.1.7 also apply. For wet sowing or transplanting with water control, soil preparation generally consists of a plowing operation with subsequent cultivations using harrows or other implements on flooded rice fields to obtain a smooth and level mud layer for efficient destruction of weeds.

Where clay soils prevail, plowing with a moldboard plow–very difficult with draft animals under dry conditions–is done under wet conditions but on dried swamps. For plowing in flooded fields, light plows of Asian design (described as reversible turn-wrest type) can be used. They have a triangular share and a slatted moldboard with slats that can move separately. Their working width is between 20 and 30 cm (Fig. 1.17).

The final seedbed preparation can be done in dry soil with harrows. Nevertheless the draft effort required is high in clay soils found in rice cultivation. Working times are about 12–15 hours per hectare.

Puddling on flooded soil requires specific equipment. Comb and rotary harrows are used widely in Asia but little in Africa (Fig. 1.17). In heavy and sticky soils, rotary harrows can be particularly effective because their working parts are rotating. Their working time is about 8–10 hours per hectare.

Sowing

Sowing takes place in the same period as soil preparation, or just after. Conversely to hand sowing, mechanical sowing makes it possible to achieve a good regularity of seed distribution. The objective is to place seeds in optimal conditions to use natural resources (light, water, minerals), and achieve a uniform distribution of seeds.

Sowing density and pattern depend on the varieties and species concerned and on environmental conditions.

Sowing effectiveness varies with the conditions of seedbed preparation (importance of tillage), and use of sowing implements (depth adjustment, row and seed spacing, appropriate choice of the feed system).

Using animal-drawn seeders may be impossible, for instance, in ridge cultivation because of the lack of appropriate equipment. It proves also poorly suitable for mixed cropping (different seed sizes and sowing densities). Among the various sowing techniques applied with cultivation in flat ground (the most common method of soil preparation), some can be performed with animal-drawn implements as dibbling and row seeding (or drilling).

Broadcasting is a manual or power-mechanized operation.

Dibbling, i.e., single or several seeds dropped together and buried in a same pocket, may be done manually or with animal-drawn single seeders.

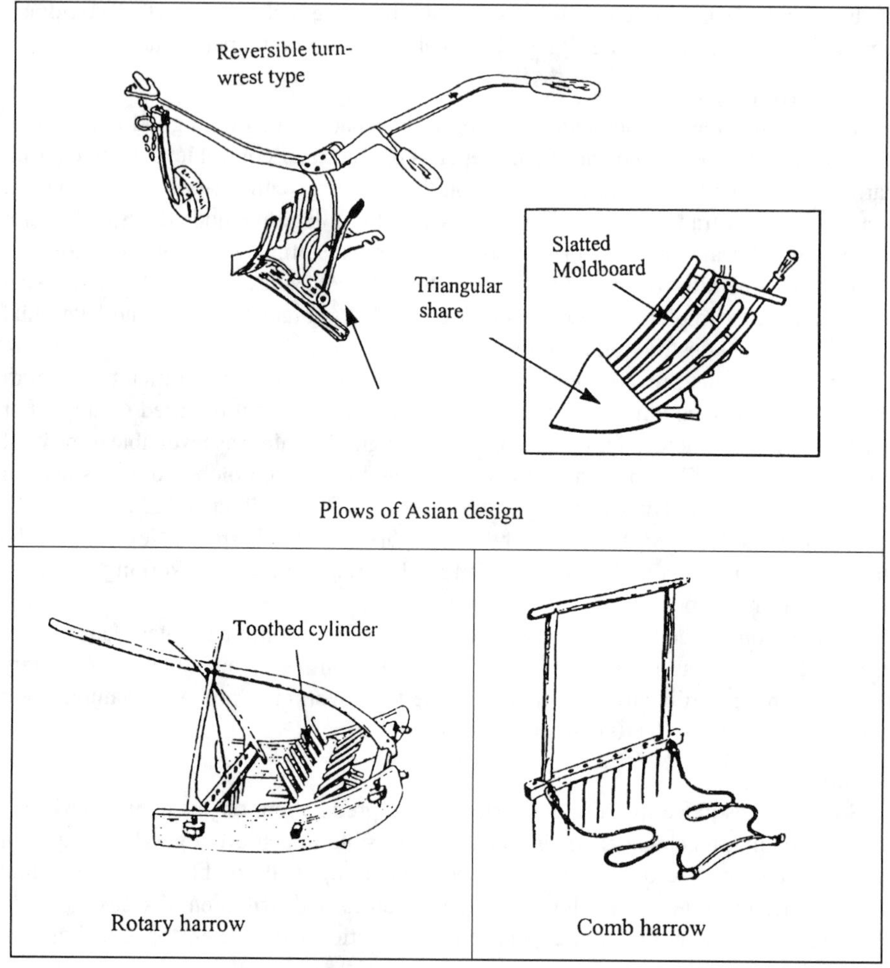

Figure 1.17. Some implements for tilling on flooded soils.

Drilling can be carried out with multiple-row seeders (3 to 9 rows). With such equipment, it is possible to control row spacing, seed rate and covering. These seeders are of little use in sub-Saharan Africa because of their high cost and also because soil preparation and seeds are poor in quality. For a seeder with a 1–1.5 m working width, pulled by a pair of oxen, the working time is between 3 hours and 5 hours per hectare.

Precision seeding assumes evenness in seed or pocket spacing in the row. In sub-Saharan Africa, one-row seeders pulled by a single animal are the most widely used implements for dibbling grain crops (maize, millet, sorghum), and also with single seeders for groundnut and cowpea (Fig. 1.18). Light, simple, robust and low in cost, these seeders require good working conditions: conveniently loosened soil, without weeds or

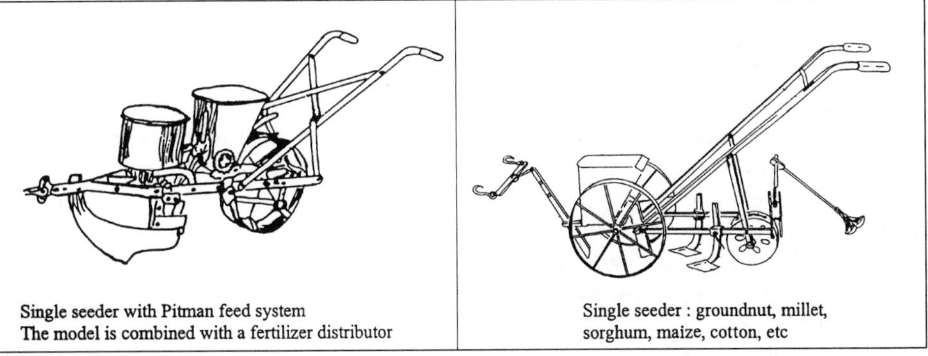

Single seeder with Pitman feed system
The model is combined with a fertilizer distributor

Single seeder : groundnut, millet, sorghum, maize, cotton, etc

Figure 1.18. Animal-drawn sowing equipment.

plant residues; well adjusted equipment, and appropriate feed system; clean, sorted and graded seeds; and a 0.8–1.1 m per second working speed. Pulled by a donkey or a horse, their working time is between 4 hours and 6 hours per hectare. With several draft animals, two or more seeders can be combined.

Direct seeding consists of dropping the seeds into a furrow open with a coulter or a disk, without previous seedbed preparation (Fig. 1.16). Although rarely used in animal traction, the technique is increasingly appreciated in light soils by cotton companies in southern African countries (e.g., Zambia, Zimbabwe, Tanzania), and in Brazil as well. Because good weed control is required, herbicides are used in high rainfall areas.

Secondary Tillage Operations

The less suitable the soil preparation and sowing, and the later the weeding operations, the faster weeds invade crops.

To delay the emergence of weeds and limit their spread, farmers must control weeds as soon as possible at the various stages when effective weed control means can apply:

- At seedbed preparation, prior to seeding, by uprooting and burying weeds (see Section 1.1.7).
- At sowing, by using seed densities resulting in plant populations hindering the growth of weeds (see Section 1.1.7).
- At young plant level, by destroying weeds.

Using animal-drawn equipment for weeding row crops is an easy and economic method available to many farmers.

During growth of the sown crop, two main techniques can be applied: harrowing at an early date (early weeding) and interrow weeding and cultivation.

Early weeding makes it possible to limit the development of the young weeds significantly and makes them easy to remove. Spring tine harrows are used to perform this shallow cultivation (1–2 cm deep).

Secondary tillage operations must be repeated to loosen the part of the topsoil that promotes root development, and to control weeds. Interrow weeding, which consists of breaking up the topsoil crust (3–5 cm), favors infiltration of rain water and limits surface evaporation. Weeding operations consist of cutting weeds at a small depth with cutting working parts. In fact, both operations are often combined in a single implement. Several passes are required.

Under dry conditions, mechanized secondary tillage operations make it essential that regular interrow spacings are wide enough (more than 0.5 m) for the passage of an animal. The work performed, most frequently limited to a single interrow, corresponds to a working width between 0.3 m and 0.8 m suitable for the draft force of a single animal in light soils, or for a pair of animals in heavier soils. Interrow cultivators or weeding tines attached to multipurpose toolbars are used (Fig. 1.16). The types of tines and shares depend on the operation to be done and soil conditions:

- In light soils, straight rigid tines with wide shares are used for weeding purposes, with narrower shares for hoeing.
- In clay soils, spring tines with duckfoot shares, assuming high penetration, are preferred.
- In sandy clay and gravelly soils, spring loaded tines with narrow shares are recommended for hoeing operations, with wider shares for weeding.

Interrow working times are between 10 hours and 15 hours per hectare, according to animals and crops.

Additional cultivations such as earthing-up, which may be applied to strengthen the bottom of plants with soil, also have a weeding effect on interrows as they cover weeds to some extent.

1.1.8. Use of Draft Animals

Generally speaking, oxen worked in pairs are preferred for tillage; horses and mules for seeding and secondary cultivations; and donkeys for transport purposes [13]. There are particular examples according to environmental conditions, history and traditional use of animal traction throughout the world.

Until the beginning of this century, European agriculture, army and transport services used animal power mainly with horses and mules. Since that period, the use of animal power in Europe has greatly decreased and even disappeared, except, for example in certain western countries, such as in Portugal, and for logging in Scandinavia.

Animal traction has been used in Asia for thousands of years. Today, it remains a vital component of a farmer's life. Draft animals—usually an animal team controlled by one person—are conveniently maintained and trained. Female working animals are common. Water buffaloes and cattle are worked mainly in pairs for tilling paddy fields, rice threshing and transport. Elephant numbers are decreasing. They mainly assist with logging. In India, donkeys, horses, mules, camels and yaks are used for pack transport, but rarely for farm cultivation [15].

Animal traction was introduced into Latin America several centuries ago by colonists. In the tropical zone, cattle are worked in pairs, usually controlled by one person, for plowing and transportation. In the highland and plateau areas of certain countries (e.g., Mexico, Chile, Argentina, Brazil), horses, donkeys and mules are preferred to oxen because they walk faster and prove better suited to pack and cart transport.

In North Africa, cattle, donkeys, horses and camels are still used widely for water-raising, pack transport and carting. As power mechanization has developed, animal traction for farming and transport has strongly decreased in plain areas, but not to the same extent in highland areas. In the Nile valley, water buffaloes also are used for plowing.

Animal traction in sub-Saharan Africa, excluding Ethiopia, was introduced for farming at the end of the past century. The number of work animals is low (estimated to 12 million head including 6 million for Ethiopia) compared with the size of the continent. They are widely used in semiarid and subhumid areas. In cotton and groundnut zones of western Africa and some southern African countries (e.g., Zambia, Zimbabwe), 30 percent to 90 percent of farmers use animal traction [5]. This technology also can be found in some humid areas, but is almost nonexistent in vast countries such as the Democratic Republic of Congo. In most cases, two or three persons are required to work with a team, which shows the low training level of animals. These are often excessively weak because they are underfed. Donkeys and horses, lower in number than cattle, are used in semiarid areas for light work (i.e., sowing and weeding) and also for cart and pack transports. Cattle principally are worked in pairs, mainly for plowing and ridging, and to a lesser extent for sowing, weeding and transport purposes.

1.1.9. Conclusion

Animal power available for work is seldom used with maximum efficiency. Possible improvements must take into account the whole system, including the user, the animal, the harnessing system, the equipment, and all the other factors that determine animal performance [17]. Long-term improvements must focus on improved management (feeding, care, training) of animals by farmers, and on the selection of animals better suited to work. In the short term, animal performance can be improved significantly through the selection of appropriate and diversified equipment and harnessing systems [17].

There are fewer research studies on the efficient use of animal power in agriculture than on farm equipment and stock-breeding aspects. In sub-Saharan Africa, studies have focused on farm equipment design and testing and on socio-economic aspects of mechanization. In Asia, research mainly has concerned the design and production of equipment and the economic aspect of using animal-drawn equipment versus tractors. In India, the effect of work on cattle and water buffaloes' physiology and stress has been studied.

For the last 10 years, studies on animal power have beneffited from new methods induced by the development of electronic and computerized instruments that can be used for in-field measurements on working animals [7, 8, 14, 18].

These studies are being conducted by research centers and universities:

- Europe: Centre for Tropical Veterinary Medicine (CTVM), UK; Hohenheim University, Stuttgart, Germany; Agricultural University, Wageningen, The Netherlands; Centre de Coopération Internationale en Recherche Agronomique pour le Développement (CIRAD), France; Silsoe Research Institute (SRI), UK.
- Africa: International Livestock Centre for Africa (ILCA); International Crop Research Institute for the Semi-arid Tropics (ICRISAT);

- Asia: International Rice Research Institute (IRRI); Central Institute of Agricultural Engineering (CIAE), India.

These methods must lead to a better knowledge of the work efficiency of animals. According to Vall [18], such knowledge will make it possible to limit the empirical part involved when choosing a DAP unit by optimizing animal-implement appropriateness. Precise data are required on the physical capacity of the species (draft force, speed, time extent, energy expenditure), and draft force variations over time.

References

1. Anonymous Animal Traction Training Manual. Department of Agriculture, Lusaka, Zambia.
2. Anonymous Animal Traction Training Manual. Institute od Agricultural Engineering, Harare, Zimbabwe.
3. CIRAD, 1996. Agriculture africaine et traction animale. Montpellier, CIRAD, France. p. 355.
4. FAO, 1994. Draft animal power manual. FAO, Rome, Italy.
5. Havard, M., Le Thiec, G., Vall, E., 1996. Les tendances majeures d'utilisation de la traction animale dans les pays francophones d'Afrique Sub-saharienne. Communication au deuxième symposium international de l'Association Africaine pour la Recherche sur les Systèmes de Production, la Vulgarisation et la formation. Ouagadougou, Burkina, Aug. 21–23 p. 12.
6. Munzinger, P., 1982. Animal Traction in Africa. GTZ, Eschborn, Germany, p. 522.
7. Kemp, D. C., 1987. La culture attelée. Travaux récents et activités en cours. Revue mondiale de Zootechnie, 63, pp. 7–14.
8. Lawrence, P. R., Stibbards, R. J., 1990. The energy costs of walking, carrying and pulling loads on flat surfaces by brahman cattle and swamp buffalo. Animal Production, 50, pp. 29–39.
9. Le Thiec, G., 1991. Les harnachements pour bovins. De l'Europe à l'Afrique. Montpellier, CIRAD, France, p. 41.
10. Ministère de la coopération, CEEMAT, 1971. Manuel de culture avec traction animale. Paris, France, La documentation française, collection Techniques rurales en Afrique.
11. Poitrineau, E., 1988. Manuel d'enseignement à l'usage des vulgarisateurs au service des paysans. Aubervilliers, Groupe de recherche et de réalisations pour le développement rural dans le tiers monde (GRDR), France. pp. 80, 119, 121.
12. Ringelman, M., 1908. Génie Rural appliqué aux colonies. Paris, Librairie Maritime et Coloniale, p. 693.
13. Schmitz, H., Sommer, M., Walter, S., 1991. Animal Traction in Rainfed Agriculture in Africa and South America. GTZ, Eschborn, Germany, p. 310.
14. Slingerland, M. A., 1989. Selection of animal for work in sub saharan Africa: Research at the ICRISAT Sahelian Centre. pp. 203–210. In: Hoffmann, D., Nari, J., Petheram R. J. (editors). Draft Animals in Rural Development. ACIAR Proceedings 27, p. 347. Canberra, Australia.

15. Srivastava, N. S. L., 1989. Research on Draft Animal Power in India, p. 53–60. In: Hoffmann, D., Nari, J., Petheram, R. J. (editors). Draft Animals in Rural Development. ACIAR Proceedings 27, p. 347. Canberra, Australia.
16. Starkey, P. H., 1989. Harnessing and implements for animal traction. GTZ, Eschborn, Germany, p. 245.
17. Upadhyay, R. C., 1989. Performance limiting factors in draft animals: Can they be manipulated to improve output? P. 166–175. In: Hoffmann, D., Nari, J., Petheram, R. J. (editors). Draft Animals in Rural Development. ACIAR Proceedings 27, p. 347. Canberra, Australia.
18. Vall, E., 1996. Capacités de travail, comportement à l'effort et réponses physiologiques du zébu, de I'âne et du cheval au Nord-Cameroun. Thèse de doctorat, Ecole nationale supérieure agronomique, Montpellier, France, p. 418.

Engines

C. Goering

Engines convert fuel chemical energy into mechanical power. Combustion of the fuel causes one or more reciprocating pistons, each contained in a cylinder, to turn a crankshaft. The crankshaft delivers the power via a flywheel and an output shaft to the desired load. A cross-sectional view of an engine is shown in Fig. 1.19.

1.1.10. Types of Engines Used in Agriculture

There are three broad classification systems for engines. Engines are classified as two-cycle or four-cycle, depending on whether the combustion cycle is carried out in two or four piston strokes. Also, engines are classified as spark ignited (SI) if the air/fuel mixture is ignited by a spark, or compression-ignited (CI) if ignition occurs by compression of the air/fuel mixture. Because of its higher compression ratios and leaner combustion, a CI engine can deliver up to 40 percent better fuel economy than a similarly loaded SI engine. Finally, engines are classified as air or water cooled. Small, single-cylinder SI engines are air cooled. With only a few exceptions, large, multi-cylinder engines are water cooled.

Typical two-cycle SI engines use the crankcase as an air pump to compress the air/fuel mixture as the piston moves toward crank dead center (CDC), also called bottom dead center (BDC), and then deliver it to the combustion chamber via ports in the cylinder wall as the piston moves toward head dead center (HDC), also called top dead center (TDC). Exhaust gases exit the combustion chamber via other ports in the cylinder wall, so these engines have no valves. Use of the crankcase as an air pump interferes with its use as an oil sump, so the engine lubricating oil is typically mixed with the fuel. Because they have a high power-to-mass ratio, two-cycle SI engines are used for special applications, such as powering chain saws, small lawnmowers or boats. Their more difficult starting, erratic idling and poorer fuel economy (compared to four-cycle SI engines) limits their more widespread use.

Two-cycle CI engines use a mechanically-driven blower to pressurize air, which enters the combustion chambers via slots in the cylinder walls. As a piston approaches HDC, fuel is injected to mix and burn with the air. Approximately half way through the piston

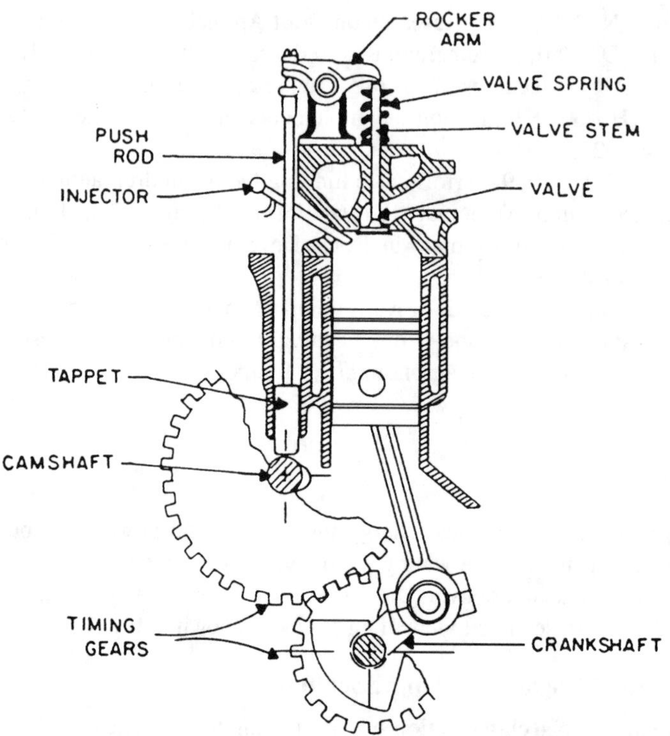

Figure 1.19. Front cross section of a four-cycle CI engine.

downstroke, a conventional poppet valve in the cylinder head opens to begin expelling exhaust gases. These engines run as smoothly and efficiently as four-cycle CI engines, and the crankcase can be used as a lubricating oil sump. The added cost of the blower is the main disadvantage.

Only four-cycle engines will be considered in the remainder of this handbook, since they far outnumber two-cycle engines. For both SI and CI engines, the four strokes include intake, compression, power and exhaust. Valves are provided for admitting air to CI engines, or air/fuel mixture to SI engines, and for expelling exhaust gases. Typical valve timing, in relation to the four piston strokes, is illustrated in Fig. 1.20. Note that two revolutions of the crankshaft are thus required for each engine cycle and the timing gears (Fig. 1.19) are arranged to allow two revolutions of the crankshaft for each revolution of the camshaft.

CI combustion chamber designs fall into two categories: direct-injection (DI) and indirect injection (IDI). Figure 1.21 illustrates the DI design and one example of an IDI design. The IDI design is more fuel-tolerant, that is, it will accommodate fuels with a wider range of viscosities and cetane ratings than the DI design, but the latter has about 8% to 10% better fuel economy. Thus, the trend is toward the use of the DI design in newer CI engines.

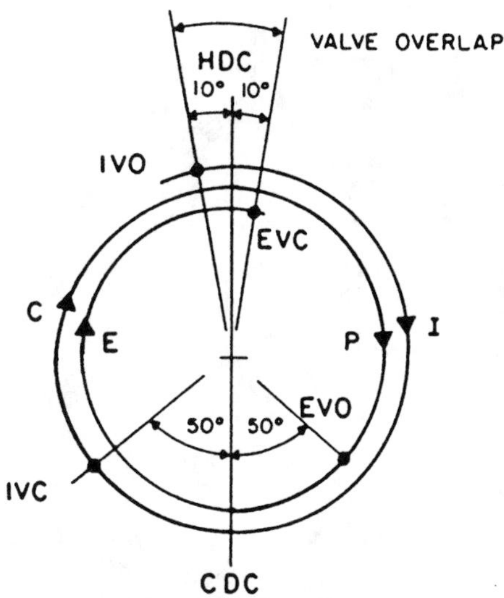

Figure 1.20. Typical valve timing spiral of a four-cycle CI engine.

1.1.11. Fuel Injection in CI Engines

Fuel injection is a critical process in CI engines. Present technology includes the use of an injector pump to deliver individual fuel injections via high-strength lines to injectors that inject fuel directly into the engine cylinders. However, the current trend is toward use of electronic injection; one example system is shown in Fig. 1.22. A high-pressure pump delivers fuel to a manifold (common rail). An electronic control unit controls the timing and duration of injections by opening solenoid valves. Injection timing and duration can be tailored to meet various objectives, including maximizing power or fuel economy or minimizing exhaust emissions.

1.1.12. Engine Performance Parameters

Liquid fuels are a highly concentrated form of chemical energy storage. Combustion of these fuels releases fuel equivalent power into the combustion chambers, i.e.:

$$P_{fe} = \frac{H_f \dot{m}_f}{3600} \tag{1.1}$$

where:

P_{fe} = Fuel equivalent power, kW
H_f = Heating value of fuel, kJ/kg (see Table 1.8)
$\dot{m}_f$ = Rate of fuel consumption, kg/h

Combustion cannot occur unless air and fuel are delivered to a combustion chamber in the proper proportion. The required air/fuel ratio can be calculated as

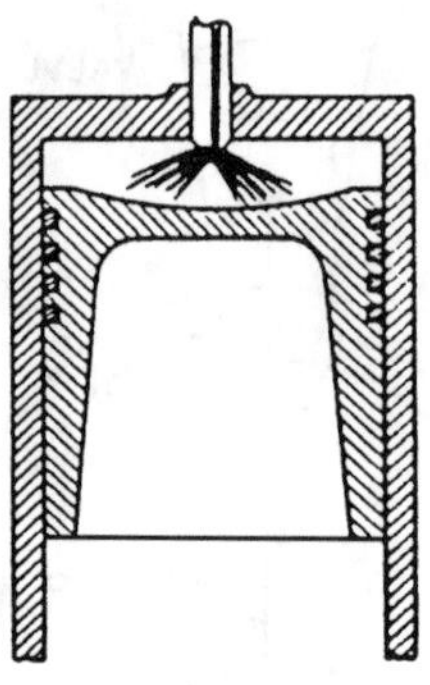

(a) Direct Injection

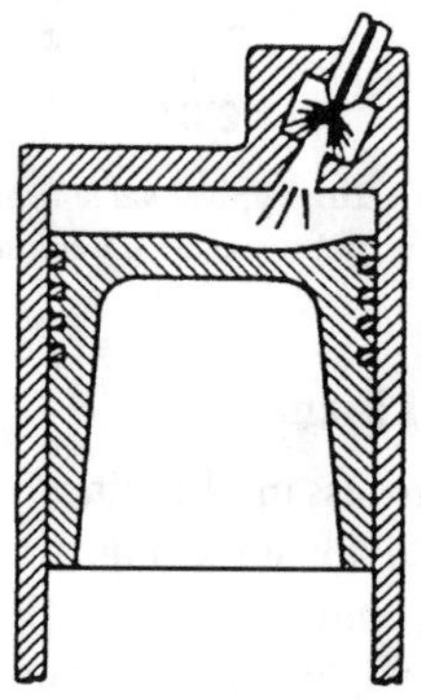

(b) InDirect Injection

Figure 1.21. Illustration of DI and IDI combustion chambers of CI engines.

follows:

$$A/F = \frac{137.3\left(x + \frac{y}{4} - \frac{z}{2}\right)}{\phi(12x + y + 16z)} \tag{1.2}$$

where:

A/F = Mass of air required per unit mass of fuel
x, y, z = number of carbon, hydrogen and oxygen atoms, respectively, in the fuel molecule (see Table 1.8)
ϕ = Equivalence ratio

Table 1.8. Comparison of Properties of Several Fuels

Fuel	Density kg/l	Higher Heating Value kJ/kg	Research Octane Number	Cetane Number	Boiling Range °C	Formula	Stoich Air-Fuel Ratio
Propane	0.580	50,300	111		–44	C_3H_8	15.5
Gasoline	0.735	47,600	93		30–230	C_8H_{18}	15.2
No. 1 Diesel	0.832	45,700		40*	160–260	$C_{16}H_{34}$	15.0
No. 2 Diesel	0.834	45,500		40*	200–370	$C_{16}H_{34}$	15.0
Methyl Soyate	0.885	38,379		51*		$C_{19}H_{36}O_2$	12.5
Methanol	0.792	22,700	110		65	CH_4O	6.49
Ethanol	0.785	29,700	110		78	C_2H_6O	8.95

* Minimum cetane rating for diesel fuel; typical cetane number for methyl soyate.

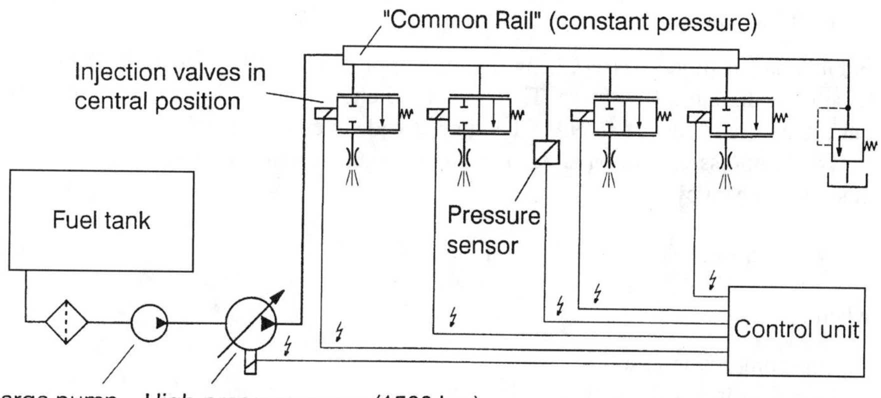

Figure 1.22. An electronically-controlled, diesel injection system (Courtesy of Karl Renius, Technische Universitat München).

The fuel equivalence ratio indicates whether the air/fuel mixture is lean ($\phi < 1$), stoichiometric, i.e., chemically-correct ($\phi = 1$) or rich ($\phi > 1$). The heterogeneous combustion in CI engines limits their ϕ to approximately 0.7 or less. From Eq. 1.2 with $\phi < 0.7$, the A/F for burning diesel fuel in CI engines is greater than 20; thus the amount of power that can be released in an engine is limited by the engine's air-handling capacity. SI engines use a homogeneous air/fuel mixture and must operate close to stoichiometric, i.e., with $0.87 < \phi < 1.25$.

Displacement is an indication of engine size. The displacement of a single cylinder is:

$$V_1 = \frac{\pi(\text{bore})^2}{4}\,\text{stroke} \tag{1.3}$$

where:

V_1 = displacement of a single cylinder, cm^3
bore = cylinder diameter, cm
stroke = stroke length, cm

For a multicylinder engine with n cylinders, the engine displacement is simply n times V_1. Engine displacement is given in cm^3 or in liters. The air handling capacity of an engine is given by:

$$\dot{m}_a = 0.03 V_e N_e \rho_a \eta_v \tag{1.4}$$

where:

$\dot{m}_a$ = Air handling capacity, kg/h
V_e = Engine displacement, L
N_e = Engine speed, rev/min
ρ_a = Density of ambient air, approximately 1.19 kg/m^3
η_v = Air delivery ratio

For naturally aspirated (NA) CI engines, η_v is approximately 0.85, but turbocharging can raise η_v as high as 2.0 or more. The η_v of an SI engine is greatly reduced by throttling, but can approach that of an NA, CI engine at wide open throttle (WOT).

The compression ratio of an engine, an important design parameter influencing combustion, is given by:

$$r = \frac{V_1 + V_c}{V_c} \tag{1.5}$$

where:

r = compression ratio
V_c = clearance volume, i.e., cylinder volume with piston at HDC, cm^3

The heterogeneous combustion in CI engines results in an energy release pattern as illustrated in Fig. 1.23. Fuel is injected late in the compression stroke, then begins to evaporate and mix with air during an ignition delay. The delay is short (less than a few ms at typical engine speeds) and thus mixing is incomplete. Thermal energy used to evaporate the fuel during the ignition delay causes an apparent negative energy release initially. When ignition occurs, all of the air/fuel mixture that has been prepared during the ignition delay burns suddenly, resulting in a triangular-shaped spike of premixed burning. Fuel vapor and air then diffuse toward each other in the zones in which premixed burning has occurred and combustion continues as diffusion-limited burning. The combination of premixed and diffusion combustion produce the total energy-release which drives the piston during the power stroke.

Premixed combustion is efficient in converting fuel energy and, except for production of oxides of nitrogen (NOx), is also clean combustion. However, the rapid energy release in premixed combustion produces engine knock and high engine stress. Slower diffusion burning is quieter and less stressful on the engine but also is less efficient and produces

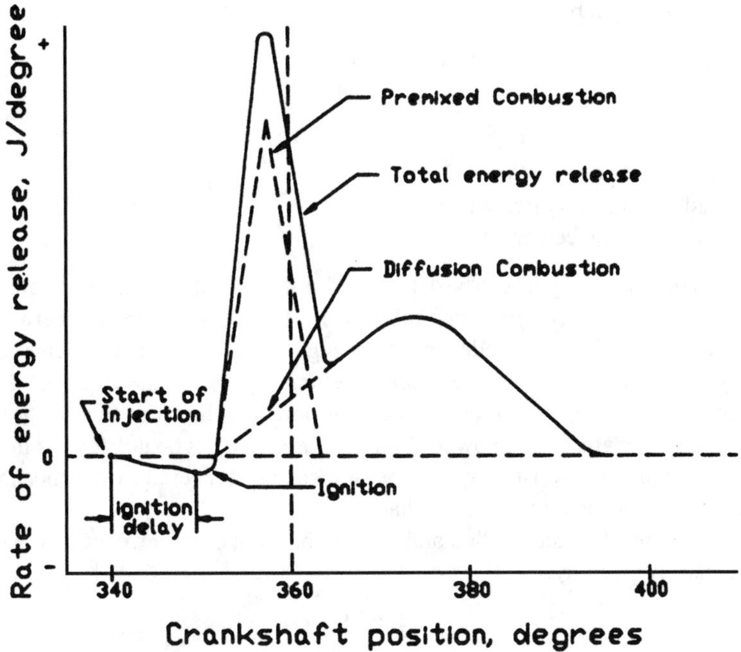

Figure 1.23. Rate of energy release from fuels in a CI engine.

exhaust smoke and carbon monoxide (CO) emissions. The relative proportion of diffusion burning is increased by increasing the compression ratio of the engine, by adjusting the start of injection closer to HDC, by using fuels of higher cetane rating, or by increasing the load on the engine. Thus, combustion in a CI engine is influenced by its design, adjustment and the mode of operation.

Governmental regulation of exhaust emissions from CI engines is becoming common. Strategies for reducing emissions include turbocharging, using injection pressures above 100 Mpa, electronically controlling fuel injection, and using oxidizing catalysts to reduce CO and HC (unburned hydrocarbons) emissions. Improved fuel economy helps reduce carbon dioxide (CO_2) emissions.

In CI engines, r must be high enough to cause the fuel to self-ignite. Typically, $14 < r < 22$ for CI engines. Combustion in an SI engine is similar to the premixed combustion in a CI engine, i.e., the air and fuel are well mixed. Ideally, a flame front is initiated by an electrical spark and sweeps uniformly across the combustion chamber. However, autoignition and knock may occur, depending upon the engine compression ratio and the octane rating of the fuel. The high octane of alcohols (Table 1.8) permit an r up to 12 : 1 without knock, but the limiting r is lower for petroleum fuels.

Most of the fuel equivalent power is lost to heat loss and to friction before it reaches the flywheel to do useful work. The indicated power (the mechanical power developed at the pistons) is given by:

$$P_i = \frac{p_{\text{ime}} V_e N_e}{120000} \tag{1.6}$$

where:

P_i = Indicated power, kW
p_{ime} = Indicated mean effective pressure, kPa
V_e = Engine displacement, L

The p_{ime} of a prototype engine can be determined as the area of its cyclic pressure-volume diagram, divided by V_1. The p_{ime} for a fully-loaded CI engine is typically between 1000 and 1700 kPa, depending on the design of the engine, including the turbocharger if one is used on the engine. The indicated thermal efficiency, η_{it}, is the fraction of the P_{fe} that is converted into P_i. Thermodynamic considerations limit η_{it} to 0.45 to 0.55 for well-designed CI engines. The power difference, $P_{fe} - P_i$, is heat that exits the engine primarily through the exhaust valves; some of it can be recovered by a turbocharger or, if a heat source is needed, by a heat exchanger.

The power at the flywheel (called brake power because of its early measurement with a prony brake) is given by:

$$P_b = \frac{2\pi T_b N_e}{60000} \tag{1.7}$$

where:

P_b = Brake power, kW
T_b = Brake torque, N.m

The difference between the indicated power and brake power is, by definition, friction power, P_f. Included in P_f is the power used in engine friction and in running such accessories as the oil pump, water pump, cooling fan, and even an air conditioner compressor. The mechanical efficiency, η_m, is the fraction of P_i that is converted into P_b. Typically, η_m is close to 0.85 in a fully-loaded CI engine with normal accessories. The overall, or brake thermal efficiency, η_{bt}, is the fraction of the P_{fe} that is converted to P_b. It can be shown that $\eta_{bt} = \eta_{it} * \eta_m$.

An equation similar to Eq. 1.6 is often used for brake rather than indicated power; the corresponding pressure is brake mean effective pressure, p_{bme}. There is no direct measurement of p_{bme}; rather, Eq. 1.6 is used to calculate p_{bme} from known values of P_b, V_e and N_e. The p_{bme} is also called specific torque because it is proportional to the brake torque that can be achieved per unit of engine displacement, i.e.:

$$p_{\text{bme}} = 4\pi \frac{T_b}{V_e} \tag{1.8}$$

Figure 1.24 shows typical p_{bme} ranges for several diesel engines. The pressures are given in bar, where 1 bar = 100 kPa.

A friction mean effective pressure, p_{fme}, also can be calculated using Eq. 1.6 with known values of P_f, V_e and N_e. The p_{fme} increases with the square of the engine speed.

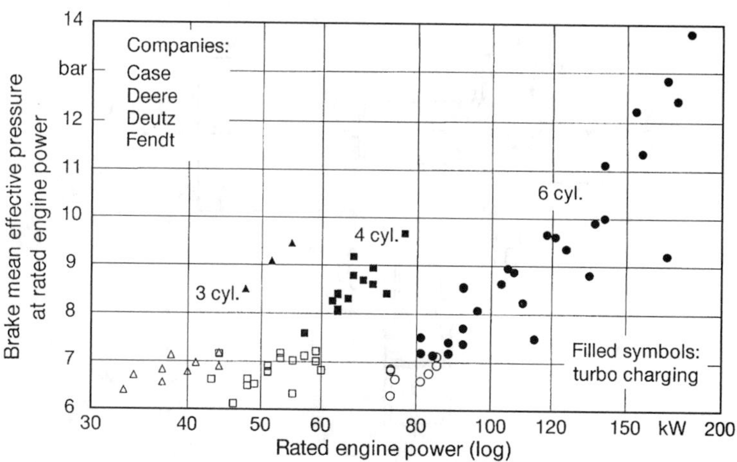

Figure 1.24. Brake mean effective pressures of some diesel engines. (Courtesy of Karl Renius, Technische Universitat München).

Consequently, p_{fme} and P_f decline and thus engine efficiency increases at reduced engine speeds.

The term, specific fuel consumption (SFC), indicates the amount of fuel that must be burned in an engine to accomplish a specific amount of work. SFC is calculated using the following equation:

$$XSFC = \frac{\dot{m}_f}{P_x} \tag{1.9}$$

where:

XSFC = Specific fuel consumption, kg/kW.h
$\dot{m}_f$ = Fuel consumption, kg/h
P_x = Power, kW
x = B for brake power, I for indicated, etc.

It is necessary to specify which power measurement was used in calculating SFC, i.e., BSFC is derived from brake power, ISFC from indicated power, etc. SFC is the inverse of efficiency, i.e., a low SFC value implies high efficiency and vice versa. A BSFC of 0.20 kg/kW.h is an optimistic goal for a well-designed CI engine running at full load on No. 2 diesel fuel.

1.1.13. Turbocharging and Intercooling

Turbochargers greatly increase the power output and versatility of engines. The concept of a turbocharger is illustrated in Fig. 1.25; engine exhaust gases drive a turbine which, in turn, drives a compressor. The compressor compresses ambient air to increase air delivery to the engine. Since an engine's power capability is limited by its air handling capacity, turbochargers permit greatly increased power output from a given engine. The

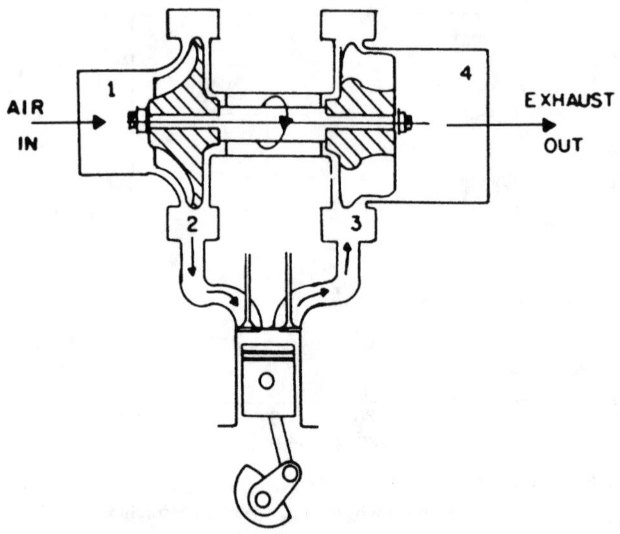

Figure 1.25. The concept of a turbocharger.

pressure ratio across the compressor is defined as:

$$p_{rc} = \frac{p_1 + \text{boost}}{p_1} \tag{1.10}$$

where:

p_{rc} = Pressure ratio across the compressor
p_1 = Inlet air pressure, approximately equal to barometric pressure
boost = increase in pressure provided by the compressor.

Compressing the air increases its temperature; the temperature ratio across the compressor is:

$$\Theta_{rc} = 1 + \frac{p_{rc}^{0.286} - 1}{\eta_c} \tag{1.11}$$

where:

$\Theta_{rc} = \Theta_2/\Theta_1$ = Ratio of absolute temperatures (°K) across the compressor.
η_c = Compressor efficiency (decimal).

Typical efficiencies of a well-designed compressor on a fully-loaded CI engine are 0.6 to 0.7. The air delivery ratio (see Eq. 1.4) of a CI engine running at full load is approximately equal to:

$$\eta_v = \frac{p_{rc}}{\Theta_{rc}} \tag{1.12}$$

For example, with $p_{rc} = 2.1$ and $\eta_c = 0.65$, $\Theta_{rc} = 1.36$ and $\eta_v = 1.54$, assuming ambient air temperature at 20°C (293°K), air would enter the engine at 126°C. To reduce such high temperatures, an intercooler (a heat exchanger which absorbs heat from the compressed air and rejects it to a secondary fluid) can be used. The secondary fluid can be ambient air (air-to-air intercooler) or the engine coolant (air-to-water intercooler). An intercooler can produce only a small improvement in air delivery ratio, but does help protect the engine from excessive temperatures.

Engine manufacturers may use turbocharging and intercooling to obtain a family of engines from one basic engine block and cylinder head. The smallest (i.e., lowest powered) engine in the family is naturally aspirated (NA). The largest is highly turbocharged (TC) and intercooled (IC). The greatly increased air delivery with TC and IC permits greatly increased fuel delivery to boost the power output. Intermediate engine sizes are achieved by using smaller fueling increases. The entire family of engines requires only one design and manufacturing effort and only one parts inventory instead of one for each engine size. However, all of the engines in the family must be made strong enough to handle the power output of the largest engine in the family.

1.1.14. Engine Performance Maps

An engine performance map is useful both in engine selection and in choosing desirable operating points for an engine. Figure 1.26 illustrates an engine map for a DI, CI engine equipped with a speed governor (most agricultural engines have speed governors). Speed, torque and power values are given as a percentage of those at governor's maximum. The curve extending from point A through Point B to Point C is a curve of

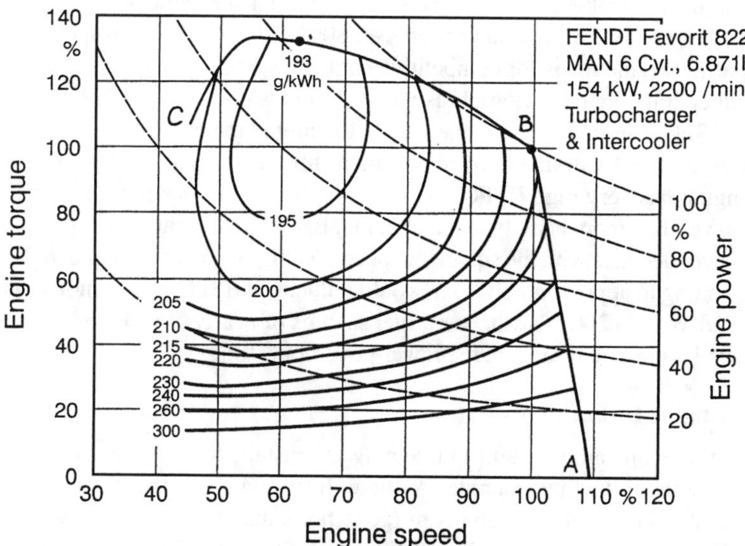

Figure 1.26. Performance map of a diesel engine (Courtesy of Karl Renius, Technische Universitat München).

maximum engine torque versus engine speed. Each of the solid contours show torque-speed combinations for the constant BSFC shown on the curve. Each of the dashed curves show torque-speed combinations for the constant brake power output shown on the curve.

Point A on the map is the high idle point, so called because the speed is high and there is no load on the engine. Between Points A and B, the engine is operating under governor control; the governor is regulating the fuel delivery to accommodate the load with little change in speed. Point B is called the governor's maximum point; if more load is applied to the engine, the governor is unable to increase fuel delivery to accommodate the increased load and the speed falls sharply. High governor regulation indicates the governor allows little change in speed between Points A and B and vice versa. The curve shown between Points A and B is for one setting of the governor speed-control lever. By adjusting this lever, the curve between A and B can essentially be slid left or right, i.e., the governor can be set to control at a different speed range than the one shown on the map.

Between Points B and C, the engine is operating under load control; the speed is controlled only by the torque load on the engine. Points between B and C are also called the *lugging* range. The increase in torque from Point B to the highest point on the torque curve is called torque reserve. Torque reserve is also expressed as a percentage of the torque at governor's maximum. High torque reserve allows an engine to accommodate a temporary overload without stalling. Continued operation in the lugging range is undesirable because the governor is unable to control the engine speed. Engines with torque reserve as high as 50 percent are available.

If the torque rise essentially offsets the speed reduction as the engine enters load control, the engine is capable of *constant power* operation, a desirable feature. In Fig. 1.26, the engine has nearly constant power between 80% and 100% of rated speed.

Comparing engine maps for competitive engines is a useful way to select an engine for an application. The map shows brake power outputs that are available and the corresponding BSFC values. The lower the BSFC, the more efficient the engine. A map also can be used to select a desirable operating point for a given engine. Note that a lightly loaded engine has very high BSFC; most fuel equivalent power is being consumed as friction power. In a more heavily loaded engine, BSFC is lower at lower speeds because friction power declines with the square of speed. On Fig. 1.26, for example, reducing the high idle setting to approximately 95% would reduce the BSFC at governor's maximum from 210 g/kW.hr to 200 g/kW.hr, for a fuel savings of nearly 5%. However, the engine would have less torque reserve at the lower governor setting.

1.1.15. Engine Utilization

The engine map can be useful in planning the installation and use of an engine, as illustrated by the following example: Assume, from Fig. 1.26, that the engine is to be operated at 2, 200 rev/min and 668 N.m (governor's maximum), for a power output of 154 kW. The estimated BSFC is 215 g/kW.hr; multiplying the BSFC by the power gives a predicted fuel consumption of 33.1 kg/hr. Using a fuel density of 0.834 kg/l (Table 1.8) and assuming at least 6 hours between refills, a fuel tank capacity of at least 238 l would be required. Next, assume the fuel is to be No. 2 diesel fuel and an equivalence ratio of 0.65 is desired; then, from Eq. 1.2, the A/F ratio would be 23 and the required air delivery would be 762 kg/hr. The inverse of the A/F ratio, the F/A ratio, would be 0.0435. Since, on a mass basis, the exhaust flow is the sum of the air and fuel flows, the exhaust flow would be $762 * (1 + 0.0435) = 795$ kg/hr. Selecting the air cleaner and inlet pipe diameter requires the volumetric air flow rate. The following two equations can be used to estimate densities of the inlet air and exhaust gases, respectively:

$$\rho_a = \frac{p}{0.287\Theta} \tag{1.13}$$

$$\rho_{ex} = \frac{p}{0.277\Theta} \tag{1.14}$$

where:

ρ_a, ρ_{ex} = density of inlet air or engine exhaust, kg/m^3
p = gas pressure, kPa
Θ = gas temperature, °K

For either the inlet air or the exhaust gas, the pressure will be close to one atmosphere, i.e., 100 kPa. Assuming an ambient temperature of 20°C (293°K) and a maximum exhaust temperature of 600°C (873°K), the densities would be 1.19 kg/m^3 for the inlet air and 0.414 kg/m^3 for the exhaust. Corresponding volumetric flow rates are 10.7 m^3/min for the inlet air and 32.0 m^3/min for the exhaust. The air cleaner and inlet piping must be sized to accommodate 10.7 m^3/min of flow without excessive pressure loss; the muffler and exhaust piping must be sized to accommodate 32.0 m^3/min without excessive pressure loss. A pressure loss of 5 Pa across either the inlet or the exhaust would be acceptable.

Finally, the engine mounting structure and the drive shaft must be able to accommodate the peak torque. Although the mean torque (from Fig. 1.26) was 668 N.m, shock loading could produce peak torques several times higher.

1.1.16. Engine Design Goals

Several design goals are aimed at improving the economy of diesel engines. One goal is to reduce BSFC below 0.2 kg/kW.hr. Another is to reduce the engine mass/power ratio, i.e., the kg/kW. A third is to reduce engine maintenance and repair costs, and a fourth is to extend the engine B10 life to 6, 000 hours. The B10 life is the number of hours at which no more than 10 percent of the engines in a given group have failed.

1.1.17. Engine Selection

Selection of an engine for a specific application involves many factors. CI engines are a good choice in situations where the maintenance and repair infrastructure is limited. SI engines require more frequent tuneups; CI engines can operate for long periods with minimal maintenance beyond regular oil changes and maintenance of fuel cleanliness. Also, IDI-type engines tolerate wider variations in fuel viscosity and cetane ratings than the DI type. Selection of air-cooled CI engines eliminates the maintenance required for liquid cooling systems.

Engines vary as to the amount of torque reserve available; the amount of torque reserve needed varies with the engine application. Engines used to drive steady loads, such as those imposed by water pumping, require minimal torque reserve. Conversely, high torque reserve is important in engines used in agricultural tractors, since loads on such tractors can vary widely within a given field. Good fuel economy is obtained by operating the engine in governor control but loaded close to governor's maximum. Under those conditions, high torque reserve allows the engine to accommodate torque overloads of short duration without stalling or requiring shifting of gears. Thus, high torque reserve can improve both productivity and fuel efficiency.

Tractors: Two-Wheel Tractors for Wet Land Farming

J. Sakai

1.1.18. The Role for Small Scale Farms

As a rule, a country is expected to have all industries, *primary*, *secondary* and *tertiary industries* developed in good harmony. In such a case, the percentage of the population employed in agriculture as a part of the primary industry to the total employed population in all industries in any country has a tendency to decrease (Fig. 1.27a) because of the increase in secondary and tertiary industries with national economic development. Figure 1.27b shows how the decreasing number of people engaged in the primary industry has supported the development of the other industries.

There are many agricultural countries which mainly have small-scale family farms with small fields. When those are mechanized with large-scale machines, it is difficult for farmers to own them due to high prices, and, as history shows, they are invited or forced to participate in group farming or in contract farming. In such cases, farmers must depend on the drivers in charge of each machine for the accomplishment of their important farmwork, and will come only to watch drivers' work, while they are originally independent, eager to work and harvest by themselves.

National modernization should be supported by the promotion of stable farming attained by means of rational utilization of large-, medium- and small-scale farm machinery so as to promote farmers' technical ability and productivity.

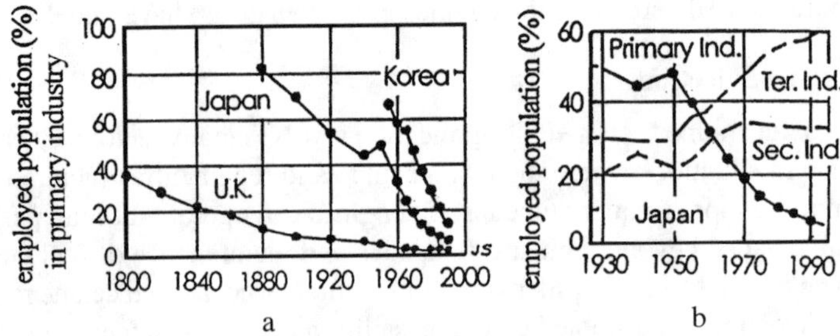

Figure 1.27. [1].

Two-wheel tractors and their equipment can be owned by them due to lower prices, and can be driven by any family member as they are simple and easy to handle.

Therefore, small-scale farmers can shift from animal-powered farming to engine-powered farming. Although the labor productivity of two-wheel tractors is lower than that of four-wheel tractors, it will be higher than that of animal-powered farming, and the farmer can enjoy speedy and timely completion of farmwork with less labor. Farmers also may be assured that almost all farm work done by animal power can be carried out by two-wheel tractors, still making use of their technical knowledge of conventional animal-powered farming. Thus, two-wheel-tractors are useful mainly to small-scale farms.

1.1.19. Differences of Farming Principles Between Upland Fields and Paddy Fields for Tractors

Two-wheel tractors are clearly separated into two types. One type is for upland fields, and the other is for paddy fields as well as upland fields.

The perfect understanding of different farming principles will provide the basic idea to attain successful development and appropriate use of two-wheel tractor-implement systems to small scale farms (Table 1.9).

Plow Pan Layer

It is common sense in upland farming that the formation of the *plow pan* under the topsoil, disturbed layer, should be avoided. Upland farmers like and maintain *plow pan-free fields*.

On the contrary, it is essential for Asian paddy farming to form and maintain the optimum plow pan layer, because paddy farming countries with *abundant rainfall* and

Table 1.9. Different principles between upland field farming and paddy field farming (Sakai)

	Euro-American Upland Field Farming (Wheat, Corn, Vegetables, etc.)	Asian Paddy Field Farming (Paddy Rice, Wheat, Vegetables, etc.)
Annual rainfall	300–600 mm/year	1500–3000 mm/year max. 4500 mm/year
Depth of tillage	20–30 cm: The deeper, the better. (trial of minimum tillage)	10–15 cm: The shallower, the easier to work. depth of tilth after puddling: 15–20 cm
Level & flatness of a field-lot	No need	Must be strictly horizontal and flat. recommended to be (by Japan Ministry of Ag.) ± 2.5 cm lot-level
Levees	No need	Absolutely needed
Size of a field lot	The larger, the better.	The smaller, the easier to make it level & flat. traditionally 0.1–0.3 ha recommended to be (by Japan Ministry of Ag.) 50 m × 20 m ⇒100 m × 20–30 m ⇒ ≤ 1 ha
Plow pan	Must not be formed. When formed, break it. Expectation of better root growth. Farmers like fields without plow pan.	Must be formed and maintained, in order to prevent excessive percolation of irrigation water. Farmers dislike deep and leaking paddy fields.
Planting system	Mainly sowing seeds after tillage	Mainly transplanting seedlings after puddling

rich *ground water* have different structures of underground soil layers from those in upland farming countries. In paddy regions of high groundwater levels (water tables) through the year, the plow pan layer has very important functions:

① In usual paddy fields, farmers, animals or machines should be able to stand and work well on the plow pan layer, which prevents the fields from becoming too deep. If not, tractor farming is difficult.

② The plow pan layer is expected to protect irrigation water from excess *percolation* into subsoil and groundwater. In paddy countries in general, groundwater is moving to lower places such as a spring, river, or directly to the sea. Percolation means the leaking loss of irrigation water with nutrients of fertilizer, resulting in a decrease in not only water-use efficiency, but also crop yield.

Depth of Plowing

In modern upland farming, it is common to maintain the cultivated topsoil layer of at least 20 cm, and if possible, 30 cm, depth of plowing.

In common paddy field farming, however, conventional depth of plowing has been only 10 cm to less than 15 cm since the age of animal-power farming. Plowing is done in the early rainy season after a half year's dry season. Then, the tight topsoil will become a soft topsoil layer with irrigation water after puddling to keep 15–20 cm thickness.

Paddy farmers dislike deep fields and fields with irrigation-water leakage. Therefore, careless adoption of modern mechanization, especially of inadequate steel lug-wheels and too powerful plowing systems in the shallow topsoil of animal-powered paddy farming will have a serious possibility of scraping, harming year by year, and at last destroying the conventional plow pan structure and culture within one decade.

Flatness and Size of a Field Lot

The surface of paddy fields should be perfectly horizontal and flat due to the nature of water surface, while upland fields are not necessarily flat and level. Asian ancestors made as many small paddy fields as they could in their history. A smaller lot more easily allows a level and flat topsoil layer.

Planting Systems and the Principle of a Transplanting System in Paddy Cultivation

Farmers in most paddy rice farming countries have traditionally adopted not the system of sowing seeds (*direct sowing system*), but the *system of puddling & transplanting young-plants*.

In the paddy rice farming countries, if farmers continue to apply a direct sowing system without *puddling*, they will recognize the following three problems within several years after starting direct sowing system without puddling. (In the countries which have a tight subsoil layer with very little or no groundwater, the field will be free from those problems.)

① A decrease in yield due to increasing percolation of irrigation water because of many root holes in subsoil. It is important that puddling plugs the root holes.

② A decrease in yield due to active growth of weeds and insects because of the difficulty in controlling them in the direct-sowing system without puddling. As a result, the amount and frequency of *herbicide* and *insecticide* use will be increased.

In the transplantation system, however, weeds and insects are cut and buried under the topsoil layer through primary tillage and puddling. The rice seedlings after transplantation are more vigorous than the weeds and insects, which now decompose to be manure, called *green manure*.

③ Difficulty for farmers to maintain flatness and level of the field surface without puddling.

1.1.20. Types and Durability of Two-Wheel Tractors

Two-wheel tractors have a variety of names: single axle tractor, hand tractor, walking tractor, walk-behind tractor, etc. Fig. 1.28 shows some of them.

The two-wheel tractor can accomplish many kinds of farm work with various types of implements attached to the tractor as shown in Fig. 1.29. Those implements are called *attachments*. The two-wheel tractor with a tillage implement is called a *power tiller*.

Types of Two-Wheel Tractors

The machines are classified by size as follows.

Mini-Tiller Type (2–3 PS, 1.5–2.2 kW)

This is the smallest type of two-wheel tractors (Fig. 1.28). Many kinds of rotor blades are installed on the drive axles instead of wheels, and a drag-stake is attached to the rear hitch of the machine. This unit is called a *motor tiller* or wheelless cultivator, and used for only domestic gardening and not for professional farming.

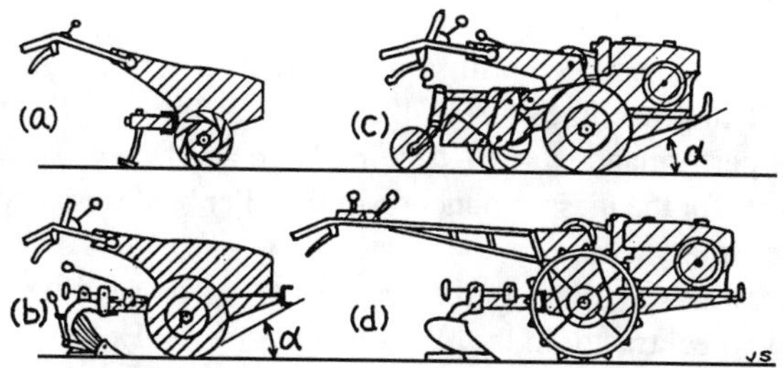

Figure 1.28. Types of two-wheel tractors (Sakai).

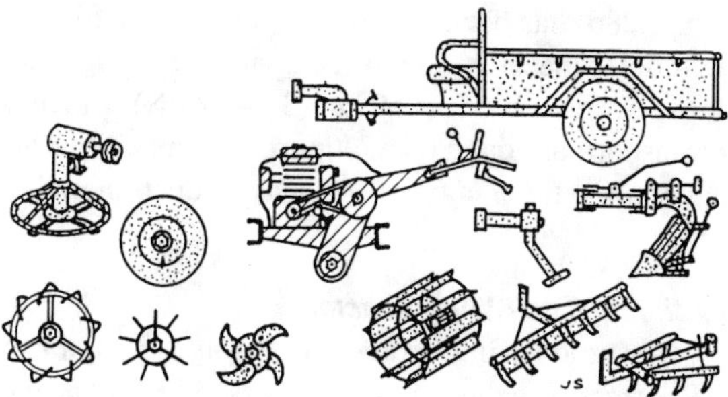

Figure 1.29. Traction type and attachments (Sakai).

Traction Type (4–6 PS, 2.9–4.4 kW)

This machine is used for plowing with a plow and for transportation with a trailer, not for rotary tillage (Figs. 1.28b and 1.29). By changing the attachments, this type has better multipurpose performance than the other types to carry out all farm work formerly done by animal power.

The dry weight (weight without oil and fuel) of the machine with standard tires and without any attachment or additional weight is only about 100–140 kgf (1.0–1.4 kN). The machine has just enough power to pull a single plow instead of a draft-animal and to realize good multipurpose performance, but it should not be excessively heavy, which requires high-level R&D knowhow. As a rule, a forced-air-cooled engine is mounted in order to make it small and light. It is said that as many as 50 kinds of attachments for one traction-type machine should be prepared to meet the needs of broad-range marketing. The IRRI tractor in the 1970s was a trial model of this type.

Dual Type (5–7 PS, 3.7–5.2 kW)

This is an intermediate size which falls between the traction type and the drive type. Plowing and narrow rotary tillage can be done. Although its tillage performance is lower than that of the drive type, its multipurpose performance is better.

Drive Type (7–14 PS, 5.2–10.3 kW)

This machine cultivates soil by transmitting the engine power mechanically to the tillage implement coupled immediately behind the two-wheel tractor as shown in Fig. 1.28c, as a specialized machine for tillage, plowing and harrowing attained by only one-pass driving.

The two-wheel tractor coupled with a rotary tiller is called a *rotary power tiller*. The total weight of the machine is 300–400 kgf (2.9–3.9 kN). So, the machine has poor multipurpose performance due to its being large and heavy. The establishment of manufacturing plants for the mass-production of the tiller blades is very essential for the smooth diffusion of rotary tillage.

There were other kinds of tillage implements, such as a crank-tiller, a gyro-tiller, etc., coupled to the two-wheel tractor in the 1950s and the 1960s. However, all of these have disappeared from the market.

Thai Type (8–12 PS, 5.9–8.8 kW)

This machine (Fig. 1.28d) has been locally developed in Thailand, and has a simple structure with a water-cooled diesel engine and comparatively longer handles. The weight of the machine with cage wheels is 350–450 kgf (3.4–4.4 kN), which achieves powerful plowing and trailer transport, although its multipurpose capabilities are limited because it is far heavier than common traction types. Annual production in the 1990s was about 70,000 to 80,000 units.

Durability Classification of Two-Wheel Tractors

Two-wheel tractors are grouped into professional *farm-use tractors* called agricultural tractors, and *hobby-use tractors* called garden tractors, mini-tillers, etc.

According to the design and test know-how in the company, the necessary durability, the life, of the machine has to be indicated as the total necessary driving hours under a full load condition in a durability test.

It is said that required durability of a hobby-use machine may be equivalent to 15–25 hours' driving every year, and total driving hours over the life of a machine may be less than only 150 hours in northern developed countries. This is based on the idea of a common office worker using the machine less than two hours every weekend during the four weeks or so in each season, spring and autumn, which amounts to several to 10 years' life on average.

The durability level of a professional tractor for a small-scale farmer who holds about one hectare of double cropped land, for example, is estimated to be equivalent to 200–250 hours' operation every year under a full load condition, for several years' life. If the tractor is used for contract operation, it is expected to withstand at least 500–600 hours' driving every year. This level of use may be estimated with the expectation of 8–10 hours' operation every day for one month in each of two farming seasons per year. Then, the required total hours of the durability test may be at least 2,000 hours, which would necessitate almost one year's test operation as follows:

The machine should be started every workday morning by the test engineer, and should have 8 hours' operation a day, including daily cooling after work, and a periodic inspection of one to two weeks for all machine elements and parts every 200–300 hours of test driving.

1.1.21. Principles of Mechanisms and Mechanics

The two-wheel tractor consists of five components: (Fig. 1.30)

① *Engine*
② *Engine-base assembly* with a front hitch and a stand
③ *Transmission gear-case assembly* with a master clutch and a rear hitch
④ *Handle assembly* with several control levers
⑤ *Farm-wheels*

Engines on Two-Wheel Tractors

An industrial engine is mounted on the tractor. A forced-air-cooled *gasoline engine* with a single cylinder using either a two- or four-stroke cycle is mounted on the machine so as to make it light. A water-cooled, single-cylinder *diesel engine* is mounted on heavy two-wheel tractors. Figure 1.31 shows each of them and an example of actual *engine*

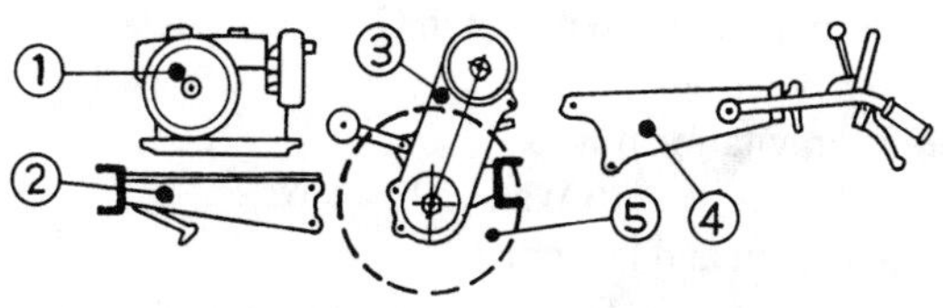

Figure 1.30. Main components.

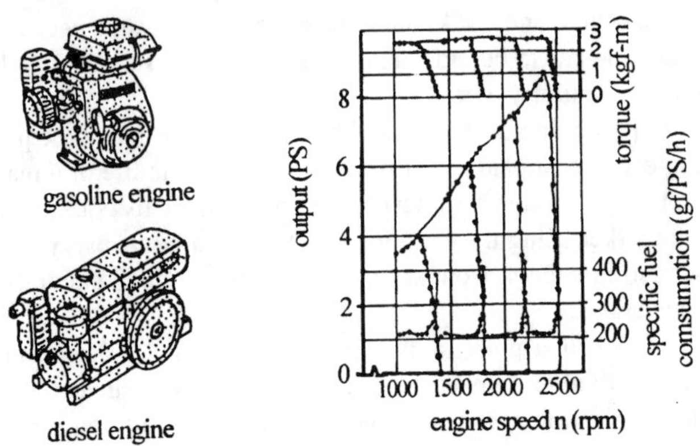

Figure 1.31. Engines and actual performance curves (Sakai).

performance curves measured in the engine test room of an R&D company. Engine performance curves in catalogues are modified by the company in a much simplified expression of these basic data. (There are a few models of small and light *air-cooled diesel engines*. Those engines, however, need a high level of engineering technology to develop and a limited number of countries produce them.)

Output Shaft

The rotation direction of the output shaft of the engine should be counter-clockwise from the farmer facing it, according to the industrial standard.

Power Output of the Engine

The expression of engine output is recommended to be in SI units. However in general, conventional units such as horsepower, kilogram-meter, etc. are still in popular use, especially in manufacturing companies. The expression of engine output power varies in Watt units slightly from country to country. They are HP, PS and kW as follows:

1 HP: horsepower in the U.K. system of units:

550 ft·lbf/s = 76.0402 ··· kgf·m/s

∵ 1 lb (English system of units) represents the gravitational force acting on a mass of 0.453592 ··· kgf,

ft (English system of units) = yard/3 = 0.914399 ··· /3 = 0.304799 ··· m

W = HP · G = 745.6996 ··· ≒ 746 Watt ≒ 0.75 kW

∵ G = 9.80665 ··· m/sec^2 as an international standard.

1 PS: Pferdestärke, German for horsepower in Germany DIN, Japan-JIS, etc.: called *metric horsepower*

75 kgf·m/s (f means gravitational force)

W = PS · G = 735.4987 ··· ≒ 736 Watt ≒ 0.74 kW

Thus, "PS ≒ HP" in design and test practice.

Moreover, the measuring method for the power of the engine under a prevailing industrial or engineering standard differs depending on each country. For example, the SAE standard in the United States calls for the measurement of two kinds of engine output, *gross power rating* of a basic gasoline engine which is measured without equipment and accessories such as a muffler and an air cleaner, etc., and *net power rating* of a fully equipped one. (Refer to SAE J1995, J1349.) Both test data are corrected to what would be expected under standard atmospheric pressure and temperature conditions. Some companies indicate the gross power rating in the catalogue.

DIN in Germany and JIS in Japan (JIS B 8017, 8018) call for measurements to be made with such equipment in place, and the data are corrected to German or Japanese standard atmospheric conditions. Their companies indicate the horsepower rating of the fully equipped engine in their catalogues.

Moreover, the *determination manner* of a catalogued horsepower obtained from many basicdata (refer to Fig. 1.31) may be different from one manufacturer to another.

Therefore, in an actual case, the engines of the same catalogue-horsepower, PS or HP, from different countries produce considerably different power ratings, more or less 20%, in actual farming work.

The measured values of the horsepower of small industrial diesel engines also have practical differences.

Force Calculation of Engine Output

The driving force P (kgf or N), produced at the effective radius, r (m), of the output-pulley and driven by the output Ne (PS ≒ HP) of an engine, is obtained by the following equations as illustrated in Fig. 1.32, showing a historically original expression of one horsepower as an example:

$$P = 60 \cdot 75Ne/(2\pi rn) \quad [kgf] \tag{1.15}$$

$$\therefore P = 716.2Ne/(nr) \quad [kgf] \tag{1.16}$$

$$P = 7023.5Ne/(nr) \quad [N] \tag{1.17}$$

Engine-base Assembly

The engine-base is tightly bolted and supported by a folding-type front stand located under the forwardmost section of the base. It should be safe and convenient for the farmer if he or she can operate the front stand while holding and pressing down the tractor handles.

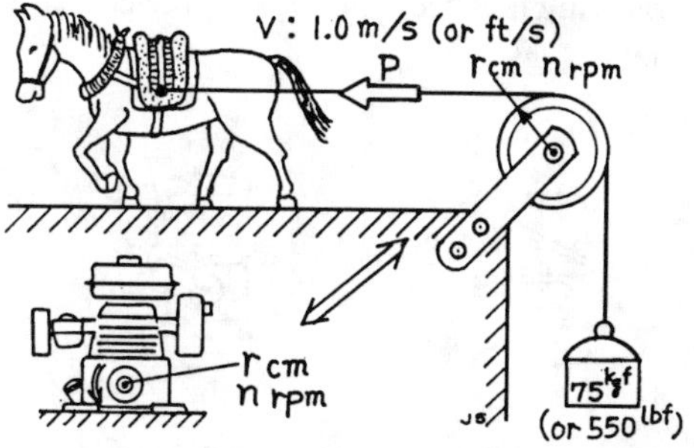

Figure 1.32. One horsepower expression (Sakai).

When the two-wheel tractor, coupled with standard tillage equipment, is put on a flat surface, the *elevation angle*, α, at the front of the engine base (Fig. 1.28) should be as large as possible, more than 30°, and if possible 40°, in order to avoid causing damage to the levees around the paddy fields or to the plants on upland fields.

Handle Assembly

The handle assembly has several levers to operate the master clutch, parking brake, gearshift mechanism, steering clutches, and engine governor, etc. and each of these levers must be installed in its proper location.

The handles should be optimum in height and width to give a comfortable operating posture to the farmer in both operations of transportation and fieldwork.

Power Transmission Mechanisms

The transmission mechanism consists of a master clutch on the input shaft, a PTO shaft, shiftable transmission mechanisms, a parking brake and a final reduction drive using gears or a chain and sprocket mechanism. The final drive includes a set of steering clutches and drive axles, which are usually hexagon shafts.

The Size of Power Transmission Mechanisms

The size of the whole mechanism is designed with the following ideas: for a given level of power transmitted, all components are subjected to torque and forces that are inversely proportional to their rotational speed (refer to Eqs. 1.16 and 1.17). In order to make their structure smaller and economical, the first half of the transmission mechanism, including the multiratio gear mechanism, should operate with small reduction ratios at high rotating speeds, but within the range of noiseless rotation speeds.

However, additional ideas are needed to distribute the total reduction ratio to all the gears in an appropriate way, because having too great a reduction ratio for the final drive may cause a problem of reduced ground clearance for the tractor, due to excessive diameters of the final gears.

Master Clutches

The master clutch may be categorized as the following types:

① *Belt clutch:* Idler tension type, engine tension type as shown in Fig. 1.33.

② *Disk clutch:* Single, double or multiple disk type (all disk clutches are usually of a dry type).

③ *Cone clutch:* Large clutch capacity, easy to produce, but big and heavy due to cast-iron components.

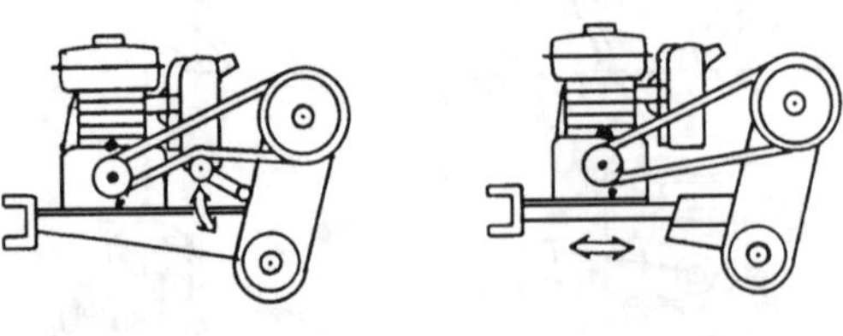

Figure 1.33. Tension belt clutches.

④ *Centrifugal clutch:* Easy to drive. In design, it is necessary to select carefully the optimum rotation speed and torque capacity at the start of clutch engagement.

Torque Capacity of the Master Clutch

The torque capacity of the master clutch has to be determined under the loading conditions. The torque capacity of the clutch for long life may be as follows:

maximum engine-torque × 1.5–2: for traction work
maximum engine-torque × 2–4: for rotary tillage

Multiratio Gears and Travel Speeds

The farmer requests the following two ranges of travel speeds: (Table 1.10)
human walking speeds for farm work
transportation speeds in trailing

Table 1.10. Travel speeds of two-wheel tractors

	(cm/s)	(km/h)
Rotary tillage	25–50	0.9–1.8
Miscellaneous field work*	50–70	1.8–2.5
Plowing	70–120	2.5–4.3
Transportation**		15 or 25 or 30

* Puddling, inter-row cultivation, seeding, mowing, etc.
** Nominally traffic law may determine legal speeds. Actual max. speeds may be set by local customs.

Two-wheel tractors, mainly of a traction type, frequently have a multispeed gearing system for general farm work and a range-shift gearing system as follows:

Farm work shift: 2–4 forward and 1 reverse gear ratios
Range shift: A farmwork range and a transport range

Two-wheel tractors of the drive type frequently have no range-shift for transportation, because of the difficulty of detaching the tillage implement.

Steering Mechanisms

There are four types of steering mechanisms as follows:

① *Loose pin-hole type of wheel hub:* The pin-hole of each wheel hub is tangentially elongated, while a round pin is attached to the axle to transmit torque to the wheel hub. If the driver pushes the tractor handle strongly to the left or right, both wheels can tolerate slight rotation differences to allow the turn. This is applied to simple hobby tractors.

② *Dog type:* This is the most common type, with one clutch for each wheel. When engine power is more than 7–8 PS (5.2–5.9 kW), some farmers may have difficulty in operating the clutch lever because of a high level of torque acting on the dog-clutch. Therefore, the clutches are recommended to be installed on the upper shaft of higher rotational speed. In order to avoid this problem, there are a few alternate mechanisms such as gear clutches, planetary gear clutches, etc. (Fig. 1.34).

③ *Planetary gear type:* The steering clutch-lever of this mechanism is easily operated. This mechanism can reduce the total number of transmission shafts, and give a larger reduction ratio to the final drive.

④ *Differential gear type:* This type is useful for easy operation of the steering drive for trailer transport. A differential-lock mechanism is not always necessary. The brake mechanism with a waterproof structure should be installed to drive wheels. This type has the disadvantage of high cost and is rarely used for two-wheel tractors.

Front and Rear Hitches and Hitch-pins

The two-wheel tractor has a hitching mechanism at the rear and sometimes at the front of the tractor as shown in Fig. 1.29. They are usually of the same dimensions.

Figure 1.35 shows their main hitch and pin dimensions of the JIS B 9209 and TIS 781-2531. Figure 1.36 shows types of the hitch for swinging functions of attachments (refer to p. 75). JIS recommends rolled steel for general structures, such as SS41, as their materials.

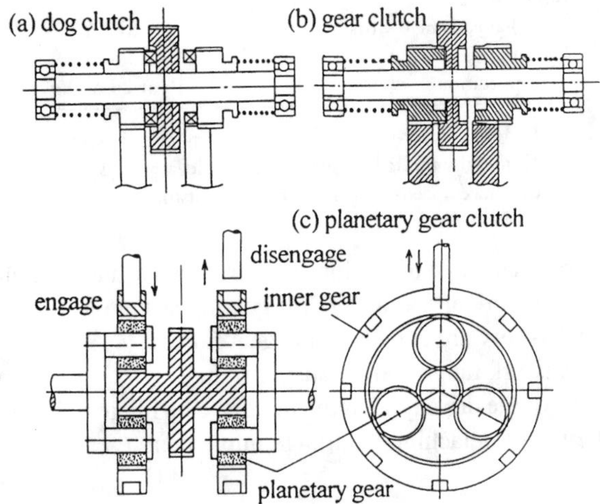

Figure 1.34. Steering clutches.

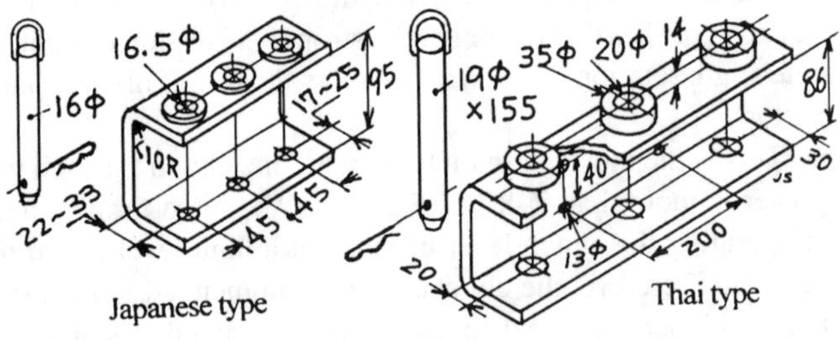

Figure 1.35. Hitches and pins.

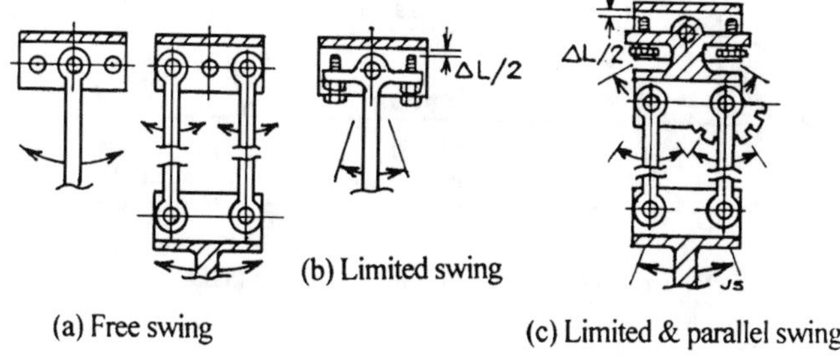

Figure 1.36. Types of swinging functions (Sakai).

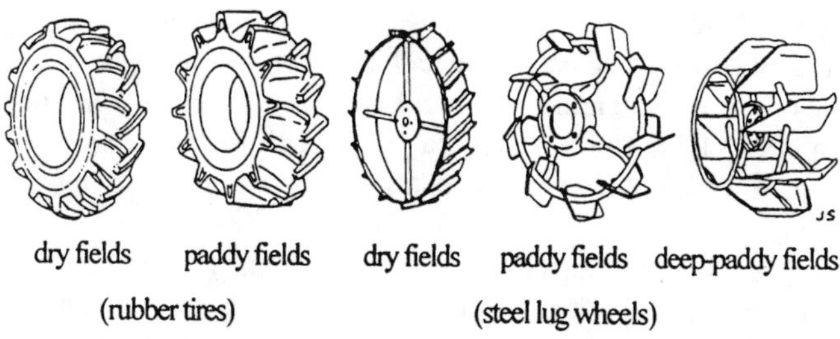

Figure 1.37. Rubber tired wheels and steel wheels.

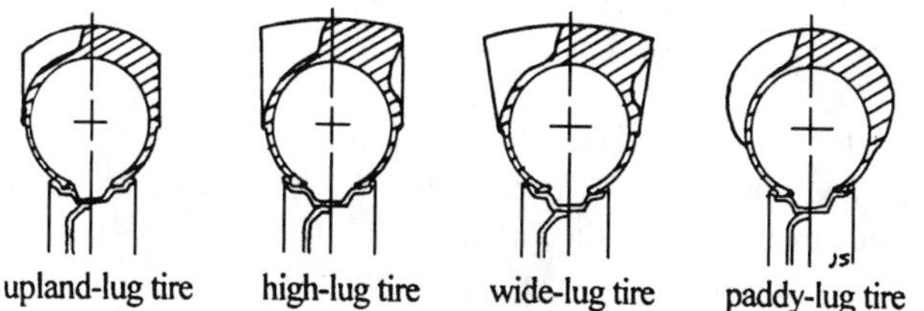

Figure 1.38. Cross sections of rubber tires.

Wheels and New Wheel Dynamics

In this section, minimizing the conventional principles of wheel structures, only new principles of wheels for two-wheel tractors on paddy soil will be explained. In order to avoid disorder, the process of finding and setting terminology will be summarized herein.

Types of Wheels and Lugs

They are grouped into two kinds, *rubber tired wheels* or *steel wheels*, and *upland-use wheels* or *paddy-use wheels*. Those with unique lugs are called *lugged wheels* (Fig. 1.37).

a. Rubber-Tired Wheels. Paddy-use tires are called *high-lug tires*, *wide-lug tires* and *paddy-lug tires*, developed in the 1960s in Japan (Fig. 1.38), of which the lug size is about two times higher, the *lug pitch* is longer and the *lug width* is less than those of the

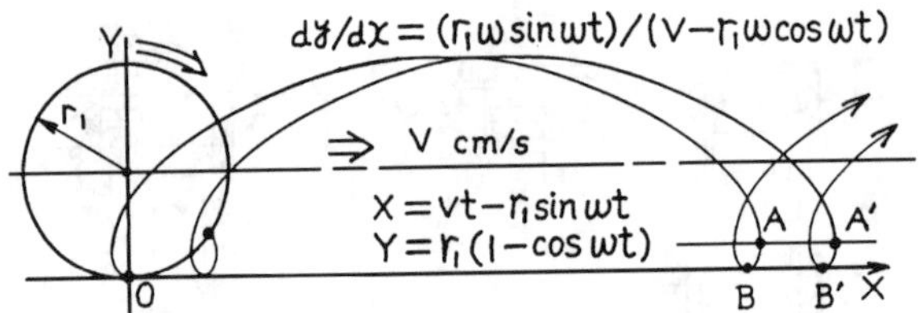

Figure 1.39. Minimum number of lugs (Sakai).

upland-use tire. The *overall width* of the total lug pattern is wider than the *tire section* (refer to the terminology in ASAE S296.4 DEC95).

b. Steel Wheels. These are classified into two types. In general, the upland wheel has many lugs on its plate-rim. The paddy wheel has fewer and larger lugs on its pipe-rim than the upland one. The usual number of lugs of a paddy wheel for two-wheel tractors is 6–12 as shown in Fig. 1.37. The reason is that the wider lug spacing and smaller number of lugs on a pipe rim are effective in preventing the wheel from trapping adhesive soil clods between lugs.

To minimize the number of lugs N_L in planning design is one of the important tasks for the design engineer. This is calculable by the following radian equations, with the consideration that the lugs should be given a maximum *lug spacing* to move only in the downward and backward directions (from A to B in Fig. 1.39) in the soft paddy soil [2]:

$$N_L \geq 2\pi / \cos^{-1}\{30v/(\pi n r_1)\} \tag{1.18}$$

or

$$N_L \geq 2\pi / \cos^{-1}(1 - S) \tag{1.19}$$

where v : expected travel speed of the tractor (cm/s)
n : rotational speed of the wheel (rpm)
r_1 : outside radius of the wheel (cm)
S : expected travel reduction. The practical value is 0.10–0.20 in paddy fields in general.

Loading Pressure and Mobility

If the machine has better mobility than that of the farmer, the farmer cannot follow the machine. Approximate human foot-pressure is 0.4 kgf/cm^2 (39 kPa). In order to have machine mobility similar to human mobility and to maintain the plow pan surface, the loading pressure of the lug is recommended to be 0.2–0.3 kgf/cm^2 (20–30 kPa).

The Shape of a Model Lug and Classifications

Figure 1.40 shows a new expression of model lugs from the design point of view. Namely, a basic lug shape is formed with five points (A, B, C, D and E). Lug geometry is described by the following four straight or curved surfaces:

① AB surface: called a *lug side, trailing*.
② BC surface: called a *lug face*.
③ CD surface: called a *lug side, leading*.
④ DE surface: called an *undertread face*.

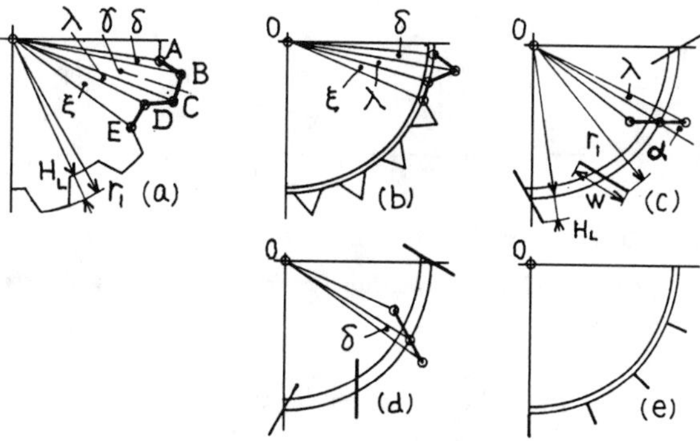

Figure 1.40. Model lug-shapes (Kishimoto).

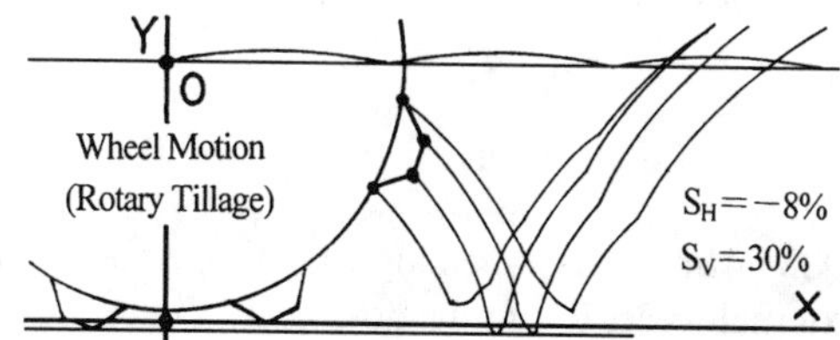

Figure 1.41. An example of rigid-lug motions (Kishimoto).

The Motion of Lugs and the Lift Reduction

Lug motion is conventionally described with the idea of the *travel reduction* S_H in the X-axis direction only. A new reduction, S_V, in the Y-axis direction was proposed by T. Kishimoto, Japan [3]. This value, S_V%, has been named *Jyosho-teika-ritsu* (上昇低下率) in Japanese, and *lift reduction* in English, and has been defined as follows:

$$S_V \equiv 100(H_{V0} - H_V)/H_{V0} \qquad (1.20)$$

where H_{V0} : max. lift of the axle on a rigid level surface.
H_V : actual lift due to sinking of the lug into soft soil.

The lift reduction of zero% means lug motion on the rigid surface, and that of 100% means smooth straight motion of the wheel center on the soft soil. Figure 1.41 shows the motion equations with travel and lift reductions, S_H and S_V, of the rigid lug make it possible to simulate rigid-lug motions on/in soft soil as well as on a rigid surface.

External Forces and Lift Resistance

The conventional theory shows that there are two kinds of basic external forces acting on the towed wheel as follows: (Fig. 1.42)

① *Soil reaction force* R_N in the upward direction, to the *dynamic load* W_N in the downward direction as an action force. Their values should be equal: $R_N = W_N$.

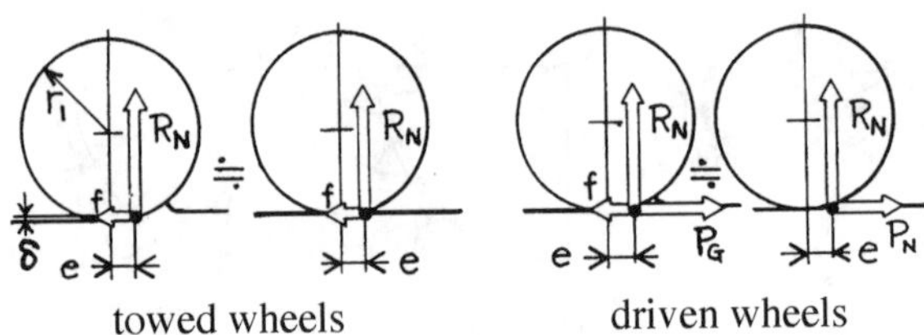

Figure 1.42. Conventional expression of forces.

② *Motion resistance* f in the backward direction.

$$f \equiv \mu R_N \quad (\mu \text{ is a } \textit{motion resistance ratio}) \tag{1.21}$$

There is a *gross traction* P_G in the direction of travel on a driven wheel, and a *net traction* P_N is:

$$P_G - f \equiv P_N \quad (\text{because if } P_N \Rightarrow \text{zero}, P_G \Rightarrow f) \tag{1.22}$$

It was noticed in 1988 by Sakai that when lug wheels operate on compressed adhesive soil, a considerable external force, R_L, in the downward direction, resisting the upward motion of the lug at the bottom of trochoidal motion, is generated on the lug. This was reported in 1991 [5] after getting the experimental proof measured with two kinds of new sensors [4, 6]. This force has been named *Jyosho-teikoh-ryoku* (上昇抵抗力) in Japanese, *Sangseung-jeohang-ryeok* in Korean, and *lift resistance*. [5, 7]:

The principal factor contributing to lift resistance is a downward resistance, A_P, acting on the *lug face* on the soil surface when the lug starts to move up. This force A_P has been called *contra-retractive adhesion* [5] or *perpendicular adhesion* [7].

There are other factors affecting the lift resistance on the leading or the trailing lug side. An experiment showed these were small, representing several to 20% of the total lift resistance at high levels of travel reduction [11].

Conventional slow-speed wheel dynamics shows that the sum of dynamic loads, ΣW_N, acting on all the supporting soil surface on average is the same as the total weight W_T of the car. There was, however, a new finding in 1993 by J. S. Choe, [7, 9, 10] that on the plow pan surface, the actual dynamic load W_G acting on the lug to the soil is greater than the *dynamic load*, W_N. Namely, the average sum of the actual dynamic loads, ΣW_G, acting from all the wheels to the supporting soil surface of a tractor is greater than the total weight W_T of the tractor. $\Sigma W_G \geq W_T$ Fig. 1.43 shows the principle of how a 60 kgf person produces 80 kgf load and soil reaction on his left foot [7].

This phenomenon has been called a *load transfer phenomenon* [8]. So as to harmonize with the conventional definitions and data of the dynamic load, motion resistance ratio, traction efficiency, etc., the *lift resistance ratio* C_{LR} has been defined as follows [9, 10]:

$$C_{LR} \equiv R_L/R_N \qquad \therefore\ C_{LR} \equiv R_L/(R_G - R_L) \tag{1.23}$$

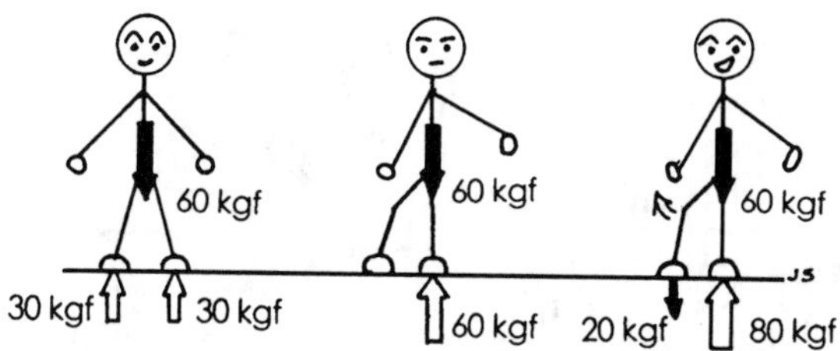

Figure 1.43. Actual forces on the human foot (Sakai).

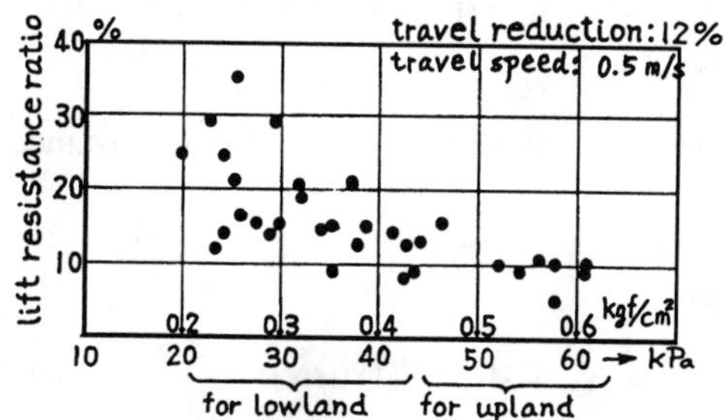

Figure 1.44. Lift resistance ratios (Choe and Kishimoto).

R_N : soil reaction force without the lift resistance. W_N, has been named the *net dynamic load* [8]. R_N can be called a *net dynamic-load reaction.*
R_G : (= W_G), *gross dynamic-load reaction* with the lift resistance:

$$R_G = |R_N| + |R_L| \tag{1.24}$$

Experimental data [11] showed that the lift resistance ratio C_{LR} for light clay soil is in the range of 0.05–0.35 (Fig. 1.44), and proved that the higher the travel speed of the lug, the larger the value of the ratio becomes [7].

Acting Locations of the Lift Resistance

The acting location of lift resistance changes delicately. However, reasonable approximation can be made similar to the method used for a soil reaction force. The equation to calculate the distance e of R (= R_N) with a motion resistance f ($\equiv \mu R$) on the wheel is: (Fig. 1.42)

$$e \risingdotseq \mu r_1 \tag{1.25}$$

$$\because \ Re = f(r_1 - \delta) \qquad \delta \to \text{ very small}$$

where μ : motion resistance ratio. $\mu \equiv f/R = f/R_N$
δ : height of the point of action of the motion resistance from the compressed soil surface: approximately less than a half of wheel sinking depth or tire deformation.

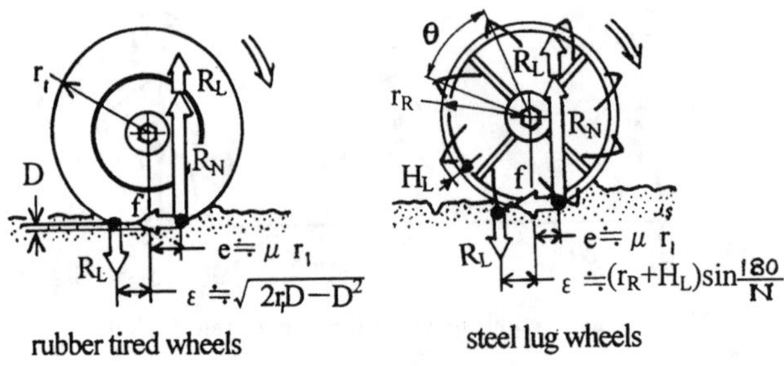

Figure 1.45. New expression of external forces (Sakai).

Practical equations for the location distance ε of the lift resistance might be: (Fig. 1.45) [9]:

① Rubber tire wheel $\varepsilon \fallingdotseq (2r_1D - D^2)^{1/2}$ $\quad \because r_1^2 - \varepsilon^2 = (r_1 - D)^2$ (1.26)

② Rigid lug wheel

$$\varepsilon \fallingdotseq (r_R + H_L)\sin(180/N) \qquad \because \theta = 360/N \tag{1.27}$$

where r_R : radius of the rigid wheel-rim (cm)
D : deformation of tire in radius direction (cm)
N : number of lugs
θ : lug pitch angle (°)

These principles for tractor dynamics on paddy soil are available to those on dry soil by substituting zero for R_L.

Two-Wheel Tractors with the Plow and Plowing

In order to be accepted satisfactorily by farmers, plowing performance of two-wheel tractors should be definitely better than that of the existing animal draft plowing with *native plows* to which they are accustomed. Agricultural engineers need to master the native plow and plowing technology in order to develop such a locally-made tiller.

Plows attached behind two-wheel tractors may be classified into two types. They are originally a European type and a Japanese type (Fig. 1.46).

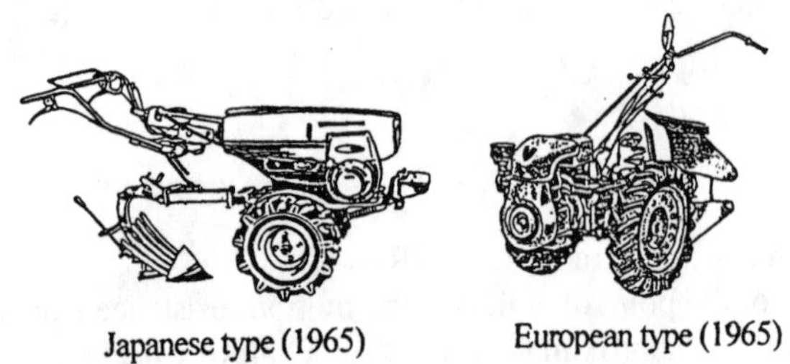

Figure 1.46. The two-wheel tractor with a plow [16].

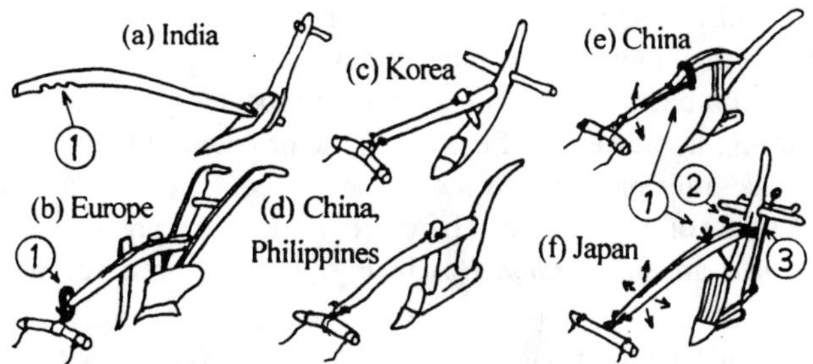

Figure 1.47. Animal draft native plows [12, 13, 16, etc.].

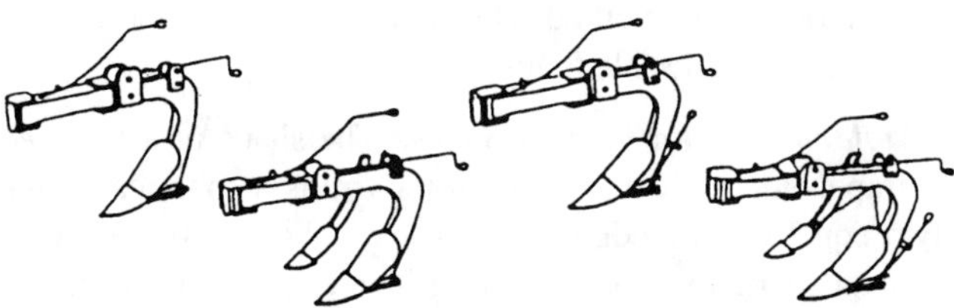

Figure 1.48. Japanese plows for two-wheel tractors (Sakai).

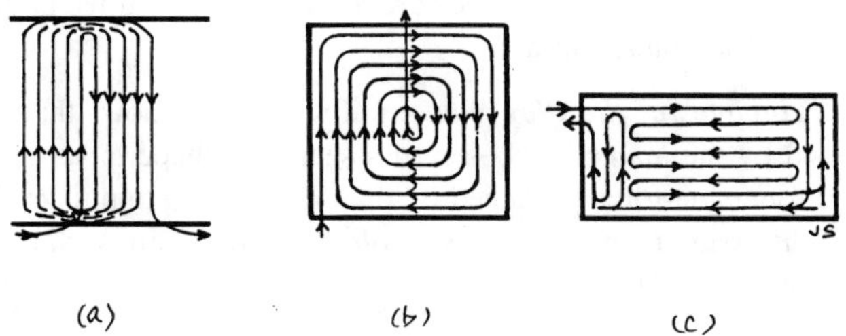

Figure 1.49. Plowing methods on fields.

Differences between European Plows and Asian Plows

Figure 1.47 shows some examples of animal-draft native plows. Figure 1.48 shows illustrations of Japanese plows for the two-wheel tractor.

Historically, European plows have been developed basically for upland farming, while Asian plows have been developed for paddy farming as follows.

a. Plowing Technology on Small Fields. There are two kinds of basic plowing methods and their modifications in large farm fields (Fig. 1.49). One is the *return plowing method* (a) and the other is the *continuously-circuitous plowing method* (b). These plowing methods are carefully followed by the farmer with consideration given to the gathering and casting of the furrow slices.

However, in small fields, especially in paddy fields, these plowing methods are not so practical. The method in which the plow is raised on the pass inside the levee causes the headland to be compacted by driving the tractor repeatedly, and continuous plowing involves busy turning in the center of the field. The problem of both methods is poorly leveled field surface after plowing, so the farmer is required to do additional work in puddling to make the soil surface flat. Therefore, if there is a practical reversible plow, a *continuous return plowing method* ((c) in Fig. 1.49) is preferred in narrow fields [13].

In museums in Asian countries, there are ancient plows whose shares and moldboards of symmetrical shape are set at the center front tip of the plow handle bottom (refer to (c) and (d) in Fig. 1.47). It is supposed that they were used as *two-way plows* by changing the *tilt angle* of the plow beam-handle to the left or right, and their small fields were plowed with a continuous return method. Those *Asian native plows* were originally very simple but convenient on small scale fields.

b. Plowshares, Moldboards and Furrow Slices. The shape of a *European plowshare* is like a narrow trapezoid, while an *Asian plowshare* is a spherical triangle or a half-oval, and has two equal cutting edges. As shown in Fig. 1.50, the moldboard of the European plow has a shape to impart a strong inverting action to furrow slices. Usually surface residues, stubble and weeds are buried well under the disturbed topsoil layer.

The Asian plow, however, does not need to overturn furrow slices so much because, in paddy farming, plant residues and weeds are mixed into and buried under the topsoil through puddling before transplanting.

c. Supporting Principles of the Plow Bottom. Figure 1.50 shows that the European plow is supported by the *landside* whose cross-sectional shape is vertical in order to counteract the strong side-force produced by the plow, and to minimize the formation of a plow pan. The Asian native plow has a *sole*, a wide bottom surface, to form and maintain a plow pan (Table 1.9).

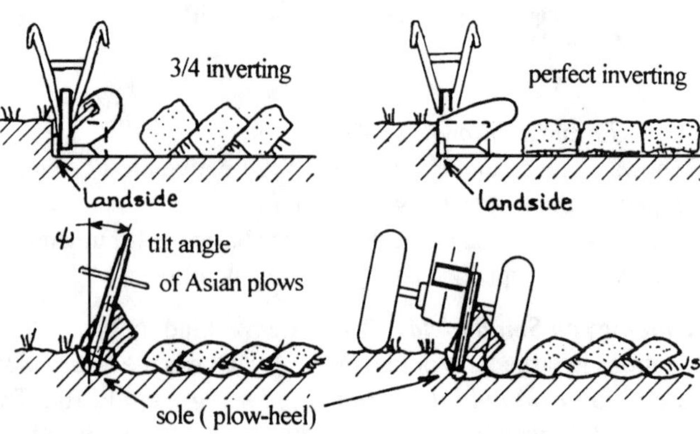

Figure 1.50. Plows, furrows and machines.

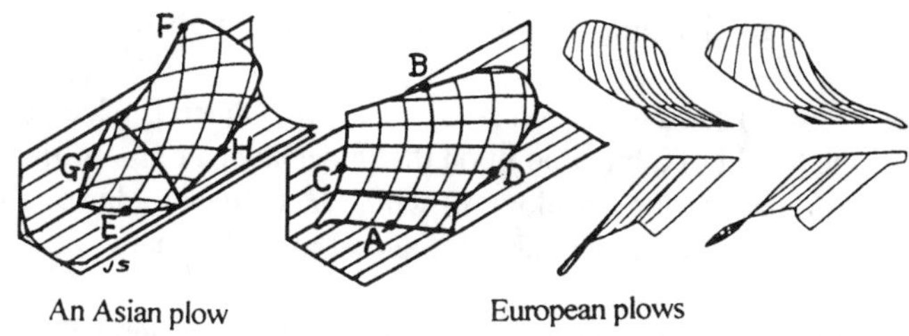

Figure 1.51. Meaning of profilograph expression (Sakai).

d. Operation Principles for the Beam-Handle. As shown in Fig. 1.50, it is an basic operation principle of the *European moldboard plow* that the plow-beam, or *plow-tail*, has to be kept vertical. This principle is simple for the farmer. However, this might pose a technical problems for the two-wheel tractor, because it is usually inclined during plowing. In Asia, the native animal plows have been supported not only perpendicularly but at the optimum angle ψ of 15°–30° to the left or right so as to give desirable throwing and inverting effects to the furrow slice. The angle ψ is called *rishin-kaku* (犂身角) in Japanese, or *tilt angle* [16, 19].

e. Specific Resistance. Specific resistance, kgf/cm^2, (N/cm^2) of the plow changes depending on the soil, the shape of the plow and plowing speeds. Figure 1.51 shows comparison between a common European plow and a Japanese plow using a *profilograph* expression.

With the European plow, the furrow slice is cut with the share, which may receive strong side-force as well as backward cutting resistance from the soil. Then, the furrow slice receives gradual lifting-force along curve AB. However, if a *knife-* or *disk-coulter* is not installed at the front of the *plow point*, the furrow slice receives a sudden, strong side-force along the moldboard of an almost straight line CD to attain a better inverting function to the furrow slice. The European plow has a long landside to support such a strong side force generated by the plow. Some European plows have a curved CD line.

The Asian plowshare has two equal cutting edges with comparatively small rake angles, if the share has a rational curved surface. The furrow is cut and gradually lifted and pushed along the curves EF and GH without a special inverting force. The overturning of the furrow slice is assisted only by the upper part of the twisted moldboard or the *flexible mold-fork*, called *jiyu-hera* (自由へら) in Japanese (1948, S. Takakita, Patent No. 184859, Japan).

In general, the specific resistance of modern Asian plows is much smaller than that of European plows.

f. Controlling the Depth, Width and Inverting Direction of Furrow Slices. Farmers need to adjust plowing depth and width. Adjusting mechanisms of the plow were developed in the age of animal power. In order to adjust plowing depth, the *draft-angle* of the *draft-line* had to be changed. The depth of European plows (b in Fig. 1.47) of fixed beam structure was adjusted by changing the hitchhole on the beam tip to which the yoke was connected.

The Japanese plow (f) had a *depth control screw* ① (S. Ohtsu, 1902, patent No. 5317, Japan) to adjust the height of a hitch point on the movable beam tip. In Jiangsu, China, in 1983, there was a native wooden plow (e) which had a depth adjusting mechanism ① by which the beam could be bent about a lateral axis. The name of its inventor is unknown.

In order to adjust tillage width, the European plow (b) had the lateral plate with several holes or a sliding hitch on the fixed beam tip, to which the yoke was connected. The Japanese plow had a *width control function* ② (1902, S. Ohtsu, Patent No. 5317, Japan) to provide side-swing movement for the beam.

There was another development of a practical reversible function of the plow. The Japanese plow had a *turn-wrest-lever* ③ behind the plow beam handle to change the direction in which the furrow slice was inverted. (1901, G. Matsuyama, Patent No. 4975, Japan [14]) ((f) in Fig. 1.47). Its use involved turning about a longitudinal axis and twisting the plowshare and the moldboard to the right or left, called *soh-yoh-suki* (双用犁) in Japanese *and Japanese turn-wrest plow* in English.

Although there were some kinds of reversible or turn-wrest plows (Fig. 1.52) in Europe [12, 13, etc.], most reversible plows for European two-wheel tractors in the 1960s had two bottoms (Fig. 1.46) [16].

Steel Lug-Wheels for Plowing and Tillage Depth

As a rule, plowing starts at the beginning of the rainy season when the soil moisture content is low following a half year's dry season. Rational steel wheels are effective in providing smooth plowing. Figure 1.53 shows the relation of wheels to plowing. The *outer-lug* and *inner-lug* must have rational shapes fit for the soil surface.

There are several design equations for the plowing lug-wheels. [16, 17] Rational outside radius r_1 of the wheels for the maximum depth H_M of plowing is obtained by the following equation:

$$r_1 \fallingdotseq [L(H_M + D_{SL} + D_{SH})/2\{L^2 - (H_M + D_{SL} - D_{SH})^2\}^{1/2}] + r_C + H_C \qquad (1.28)$$

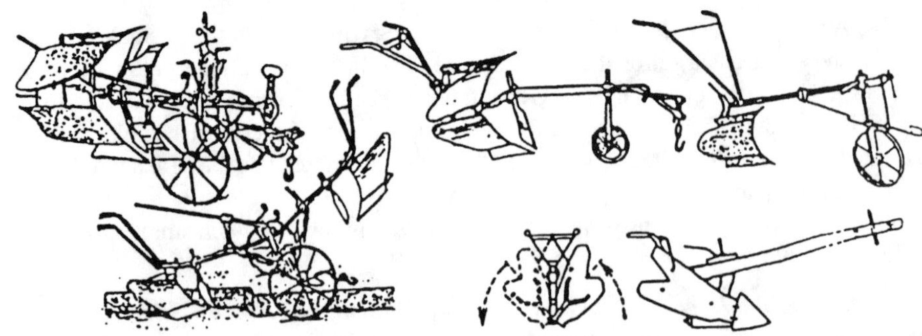

Figure 1.52. European reversible or turn-wrest native plows.

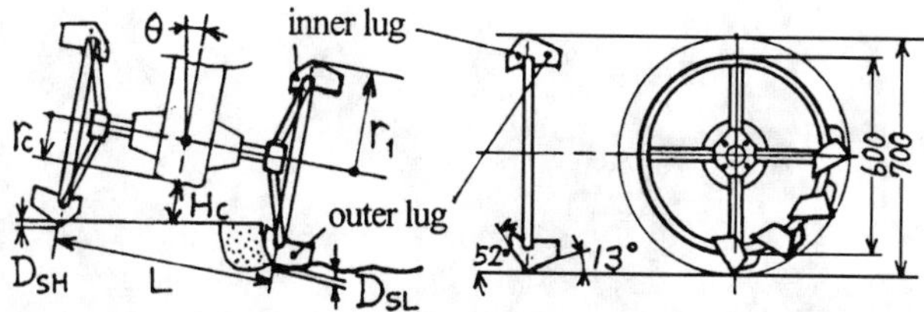

Figure 1.53. Plowing wheels (Surin).

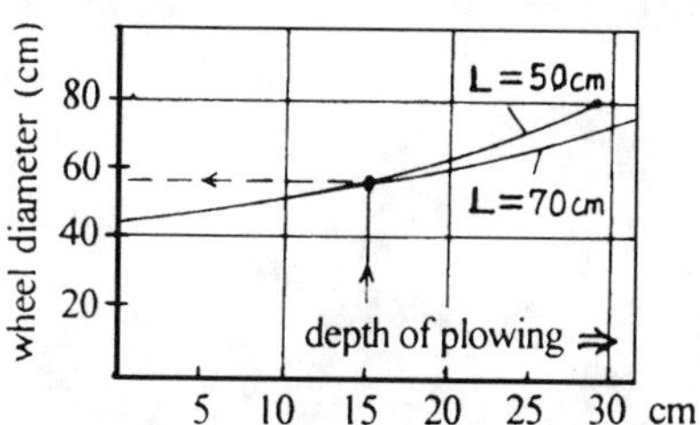

Figure 1.54. Diameter of plowing wheels (Surin).

where L : actual width of wheel treads (cm)
H_M : maximum depth of plowing (cm)
D_{SL} : sinking depth of the wheel on plow pan (cm)
D_{SH} : sinking depth of the wheel on topsoil (cm)
r_C : bottom radius of the transmission gear case (cm)
H_C : expected ground clearance between center gear case bottom and topsoil surface in design (cm)

Figure 1.54 shows that the wheel diameter of 50–60 cm is normal for the two-wheel tractor, if the plowing depth is 12–15 cm. Two-wheel tractors in some developing countries are equipped with large cage wheels of 70–85 cm diameter for plowing in order to enable operation on deep fields having 20–30 cm depth of the topsoil layer with high levels of irrigation water.

Operation Principles of Skillful Plowing

The two-wheel tractor with the optimum setting of rational wheels and a reasonable plow will attain smooth plowing by almost hands-free operation, as shown in Fig. 1.55. In order to attain such a smooth plowing, the following principles should be adopted: [16]

a. Setting the Center of Gravity. The gravity center of the whole machine should be set above the wheel center.

b. Deviating Angle of the Plowing Tractor. The direction of the tractor has to be set so as to deviate a certain angle η from the plowing direction to the unplowed side as shown

Figure 1.55. Hands-free plowing with skillful technologies under good field conditions (1995, Surin). ***Remarks:*** **In practice, the farmer slightly puts his hands on the handle to steer.**

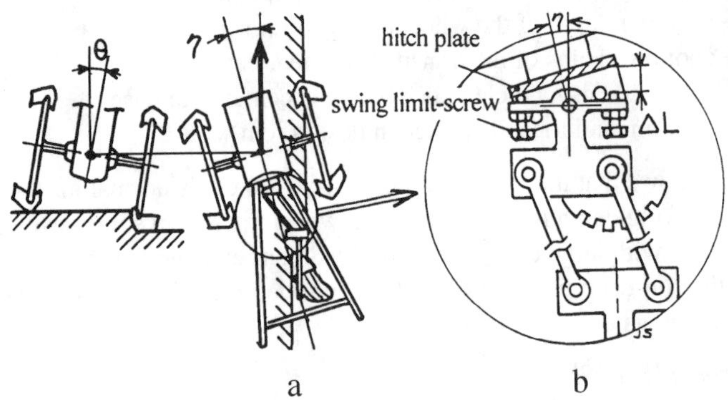

Figure 1.56. Deviating angle (Surin).

in Fig. 1.56. The angle η has been named *deviating angle*. The optimum deviation angle, 4°–6° in general, of the tractor can be set with the plow tipped *swing-limit screws* to the hitch-plate (Fig. 1.36, 1.56b) by giving optimum *stabilizing clearance* ΔL between the limit screw-tip and the hitch-plate. Then, the tractor will travel smoothly along the *furrow-wall*.

Plowing technologies with economical machines of 4–6 PS (refer to Fig. 1.29, *traction type*) were enthusiastically accepted by small scale farmers all over Japan between about

Table 1.11. rpm of the tillage shaft

rpm	Tillage for	Travel Speed (cm/s)
150–200	Sandy or wet soft soil, puddling	50–70
200–300	Common soil or sticky soil	30–50
300–400	Very sticky or dry hard soil	20–30

Remarks: The axles of motor tillers rotate at 40–70 rpm.

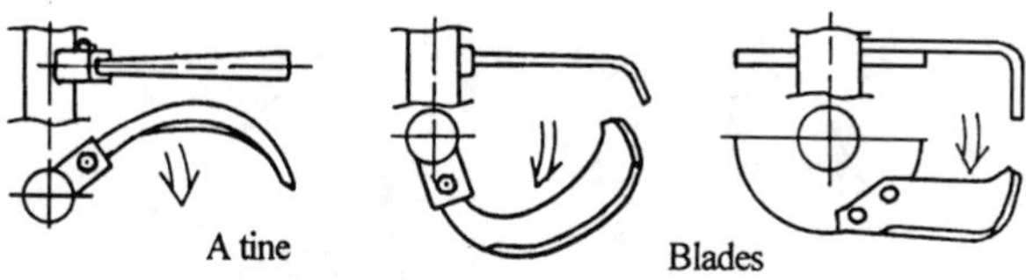

Figure 1.57. Tillage tines and blades.

1955 and 1970. Those farmers had finished shifting from animal-powered farming to two-wheel tractor farming within only 15 years.

Two-Wheel Tractors with Rotary Tillers

The rotary tiller coupled behind the two-wheel tractor (refer to Fig. 1.28c) achieves rotary tillage with many kinds of tines or knife-blades (Fig. 1.57) installed on a rotating lateral axle driven by the tractor engine. There were many kinds of tiller tines, called *pick-tines* or a *hook-tines* whose tillage resistance was much smaller than that of knife-blades. However, they disappeared due to their tendency to hook plant residues.

The blade is called a *rotary tiller blade*, tiller blade, tillage blade, rotor blade, etc. The lateral axle is called a *rotary tiller shaft*, rotary tiller axle, or simply tiller shaft, tillage shaft, etc. Their rotational speeds are usually in the range of 150–400 rpm as shown in Table 1.11.

Rotational speeds higher than 400 rpm are not practical. Such a high-speed rotation causes power loss due to violent breakage and high-speed spread of soil clods backward.

A speed range of 50–70 cm/s is useful for interrow cultivation between plant rows (with or without a furrow opener) on vegetable fields or orchards. A driver's seat supported by a mono-wheel is attachable behind the tiller instead of the rear wheel, and steering the machine is done by controlling the mono-wheel with the driver's feet.

The *tillage pitch* P of the blade is calculated by:

$$P = 60v/(nZ) \tag{1.29}$$

where v : machine travel speed (cm/s)
n : rotational speed of the tillage axle (rpm)
Z : number of blades in one rotational plane

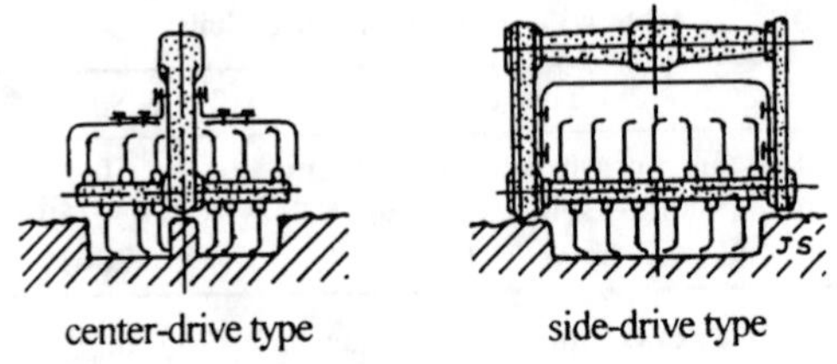

Figure 1.58. Types of power transmission systems.

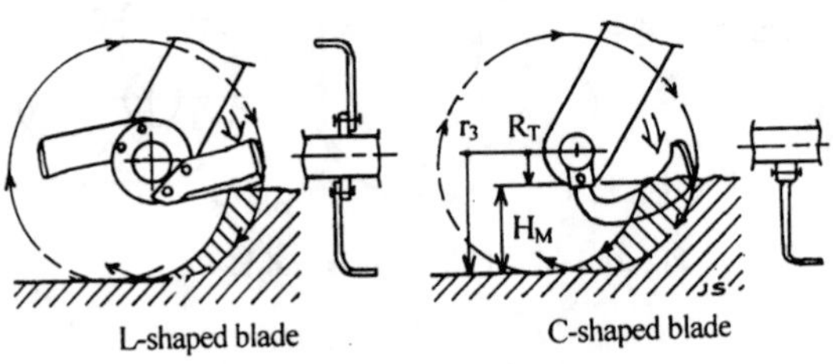

Figure 1.59. L-shaped blades and C-shaped blades.

Actual bite length P_A, however, will be much shorter than the calculated pitch P depending on the shearing by its neighboring blade tips as shown in Fig. 1.64.

The Power Transmission and Tillage Depth

There are two types of transmission mechanisms, a *center-drive type* and a *side-drive type* (Fig. 1.58). The center-drive type is adopted for use only on less than 8 PS (5.9 kW) small tractors. The maximum depth of tillage, H_M, which must be specified in the catalogue, is determined by the following equation: (Figs. 1.58 and 1.59) [20]

$$H_M = r_3 - R_T \tag{1.30}$$

where r_3 : rotation radius of the rotary tiller blades
R_T : bottom radius of the final transmission case

Types and Characteristics of Rotary Tiller Blades

Blades are classified into two categories: a European-type called an *L-shaped blade* and a Japanese-type called a *C-shaped blade* (Fig. 1.59).

The L-shaped blade was invented in 1922 by A. C. Howard in Holland for cultivating upland fields in 1922. However, common L-shaped blades were difficult to use for paddy farming, because weeds and plant residues were difficult to cut on soft paddy soil, though they could be cut easily on upland hard soil. Moreover, rice stalks have greater strength than that of wheat stalks, so the blades easily hook them, and they make the rotor turn into a drum of straws.

The C-shaped blade was invented by K. Honda and R. Satoh in 1940 (patent No. Shou-15-15990, Japan) under such paddy field conditions. The blade has a bent and curved shape of changing thickness from the holder portion to the blade tip. Although these

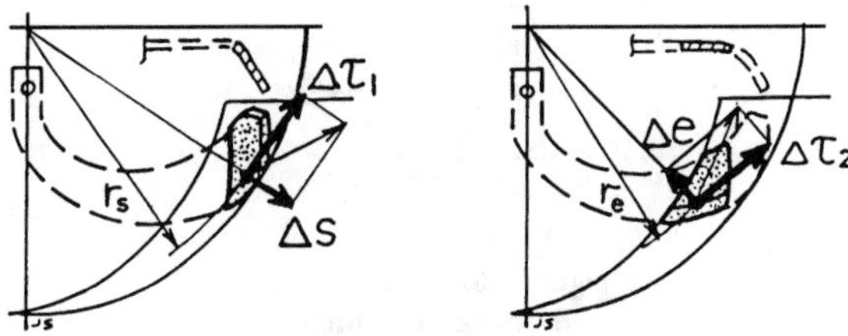

Figure 1.60. Radial suction forces Δs on scoop surface, and centripetal force Δe on straight blade [25].

structures require several press-forming processes on a production line [50], the blade is free from a hooking problem with plant residues, if its *edge-curve* is rational (Refer to Figs. 1.75 and 1.76).

External Forces Acting on the Rotary Tiller

Two kinds of external forces from the soil act on the rotary tiller of a two-wheel tractor as follows:

① *Tillage resistance forces* acting on many tiller blades.

② *Reaction forces of dynamic loads* acting on the rear stabilizer wheel which controls the depth of tillage.

a. Tillage Resistance Forces Acting on Blades. L-shaped blades were studied in the 1950s [21] and 1960s [22], etc., which will be explained later. For C-shaped blades, their forces have been measured with strain gauges since the 1950s also [20, 23, 24, 25] as follows:

① The *total tillage resistance force* T_3 consists of four kinds of external forces acting on the blade in the soil slice as shown in Fig. 1.60.

② A *radial suction force*, Δ**s**, and a *torque resistance force*, $\Delta\tau_1$, act on the scoop surface, and a *centripetal force*, Δ**e**, and *torque resistance force*, $\Delta\tau_2$, act on the straight blade portion. These forces are to be the resultant forces S and T_1, E and T_2, respectively with the following definitions: (Figs. 1.61 and 1.62)

$$S \equiv \Sigma\Delta s \tag{1.31}$$

$$E \equiv \Sigma\Delta e \tag{1.32}$$

$$\tau \equiv T_1 + T_2 \equiv \Sigma(\Delta\tau_1 + \Delta\tau_2) \quad \text{and} \tag{1.33}$$

$$T_3 \equiv S + E + \tau \tag{1.34}$$

T_3 is called a *total tillage resistance force* of rotary tillage, S is a *total radial suction force*, E is a *total centripetal force* and τ is a *total torque resistance force.*

b. The Total Torque Resistance Force and the Total Tillage Resistance Force Acting on the Rotary Tiller. The total torque resistance force τ acting on the tiller shaft and regression functions to describe the torque pattern on each C-shaped blade were obtained with experimental proof that the *total torque resistance curve* acting on the

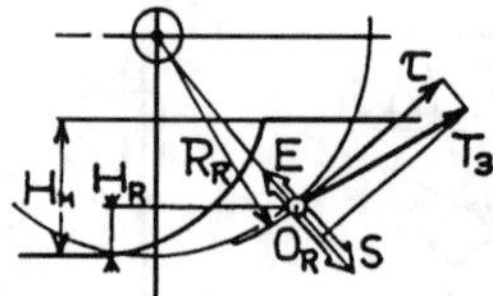

Figure 1.61. Total tillage resistance T_3 which consists of E, S and τ [25].

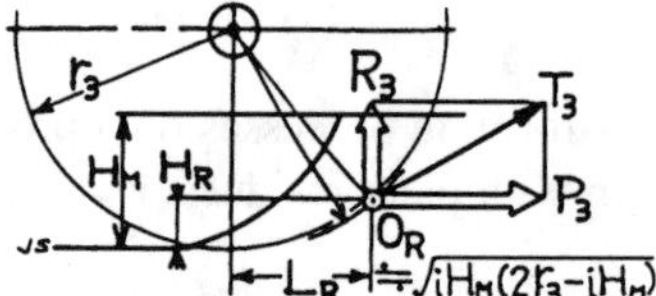

Figure 1.62. T_3 and location of the virtual point of action, O_R [25].

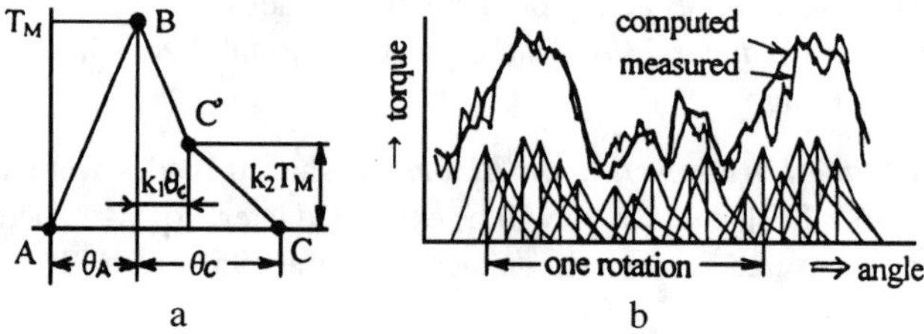

Figure 1.63. [24].

shaft computed by these functions were practically similar to the curve measured in the experiment as follows: [24]

① In the case of adhesive hard soil, a quadrilateral model of ABC′C (Fig. 1.63a) is better than a triangular model for computing the torque resistance curve of a rotary tiller axle as shown in Fig. 1.63b.

② The torque curve of the blade depends on the shape and size of each soil slice. These are represented by each *area* S (cm^2) of the *cutting pattern* on the field surface as shown in Fig. 1.64. The cutting patterns are classified practically into six types, I–VI, and each area is calculated with *actual bite length* P_A. The directions of the shear lines generated by the neighboring blade tips as projected on the soil surface were practically 45° to the back from the lateral face of the blade except the pattern VI of 90° from the same face of the blade. Figure 1.65 shows examples of the torque curve for each cutting pattern.

③ The maximum torque T_M (kgf·m) and the torque curve of the quadrilateral model (Fig. 1.63a) for the C-shaped blade can be obtained by the following multiple regression functions whose parameters are P_A (cm), S (cm^2), rotational angles θ_A

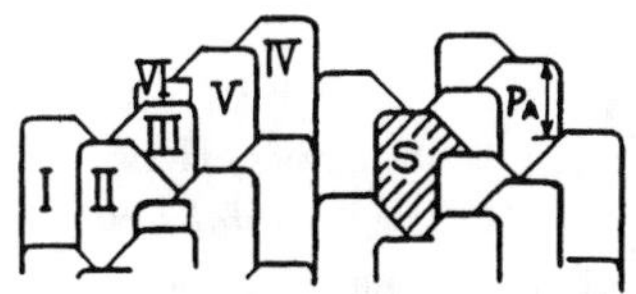

Figure 1.64. Cutting patterns on the soil surface [24].

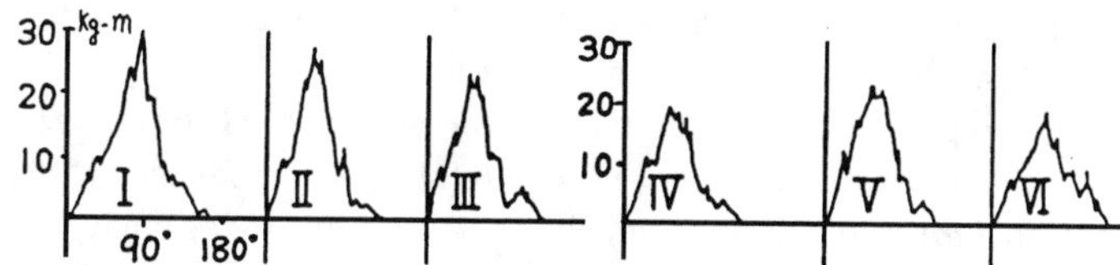

Figure 1.65. Examples of torque curves measured [24].

and θ_C (°), and the coefficients k_1 and k_2 for the point C′:

$$T_M : 2.17P_A + 0.226S \quad \text{for the pattern I–V} \tag{1.35a}$$

$$2.17P_A + 0.226S + 7.1 \quad \text{for the pattern VI} \tag{1.35b}$$

$$T_M(N - m) : 21.3P_A + 2.21S \quad \text{for I–V} \tag{1.36a}$$

$$21.3P_A + 2.21S + 70 \quad \text{for VI} \tag{1.36b}$$

$$\theta_A : 3.2P_A + 22 \tag{1.37}$$

$$\theta_C : 85°(\text{I}), 62°(\text{II–V}), 72°(\text{VI})$$

$$k_1 : 0.36 \qquad k_2 : 0.47$$

These regression functions are indispensable to the CAD of the blade location arrangement on the tiller shaft.

c. The Virtual Point of Action for the Total Tillage Resistance Force and Its Direction. There is a resultant force, the *total tillage resistance force* T_3, acting on the rotary tiller. It had been thought from the 1950s to 1970s by many scientists in the world that the total rotary tillage resistance might be located inside the soil being sliced, which means inside the peripheral circle of the blade tip.

In 1977, Sakai reported [23] for the C-shaped blade that the point of action, O_R, of the total tillage resistance T_3 is located outside the soil being sliced, at the *intersection* of the *horizontal line* of 1/3–1/2 height from the bottom of the tillage depth H_M and *the circle* with a longer radius by 1–4% than the circle drawn by the blade rotation radius r_3 (Fig. 1.62). The point of action, O_R, is called *zen-koun-teikou-kasou-sayouten* (全耕うん抵抗の仮想作用点) in Japanese, the *virtual point of action for the total rotary tillage resistance*. Its location height H_R and the radius R_R of the virtual point of action are expressed with the *radius coefficient*, C_R, and *height coefficient*, i, as follows:

$$R_R \equiv C_R r_3 \qquad C_R = 1.01\text{–}1.04 \qquad \therefore R_R \fallingdotseq 1.02r_3 \tag{1.38}$$

$$H_R \equiv iH_M \qquad i = 0.3\text{–}0.5 \qquad \therefore H_R \fallingdotseq 0.4H_M \tag{1.39}$$

The direction of the total tillage resistance T_3 fluctuates, although it is generally oriented upward and forward. As shown in Figs. 1.61 and 1.62, the direction is also determined by the sum of three kinds of vectors. They are the *total radial suction force*, S, the *total torque resistance force*, τ, and the *total centripetal force*, E, acting on all the blades in the soil. For C-shaped blades, the direction of the total tillage resistance, T_3, is in the range of 25°–35° or about 30° above the horizontal at usual depth of tillage [23, 25].

d. Tillage Thrust Force and Tillage Lifting Force. The total tillage resistance force T_3 is divided into a forward reaction force P_3 as a horizontal component force, and an upward reaction force R_3 as a vertical component force.

The forward-acting force P_3 has been named *kou-shin-ryoku* (耕進力) and the upward-acting force R_3 has been named *hane-a-ge-ryoku* (跳上げ力) in Japanese [20]. Namely, P_3 is a *tillage thrust force* and simply called a forward thrust or a tillage thrust, and R_3 is a *tillage lifting force* and simply called an upward force or a lifting force. Maximum or mean values (kgf) of these forces, R_3, P_3 and T_3, which are expected to be produced by the rotary power tiller from the design point of view, are obtained from the engine output-power Ne (PS) of the tractor as follows: [20]

$$T_3 \fallingdotseq 71620 Ne\eta/(C_R r_3 n_3) \quad \text{(kgf)} \tag{1.40a}$$

$$R_3 \fallingdotseq 71620 Ne\eta C_L/(C_R r_3 n_3) \quad \text{(kgf)} \tag{1.40b}$$

$$P_3 \fallingdotseq 71620 Ne\eta C_T/(C_R r_3 n_3) \quad \text{(kgf)} \tag{1.40c}$$

(The constants, 71620 in the equations, might be $71620 \times 9.8 \times 0.736 \fallingdotseq 516600$ for T_3, R_3 and P_3 (N), and Ne (kW) calculations).

In which

η: power transmission efficiency from the engine to the rotary tiller axle. Its efficiency is 2–5% higher than that of the *open-loop system* without the *feedback power* (refer to Fig. 1.67). For design and tests in the company, the value of efficiency 5% higher than that of the open-loop system is used as a technical guideline.

C_R: *radius coefficient* of the virtual point of action for the total tillage resistance. $C_R = 1.01–1.04$

1.00 is a practical approximation for C-shaped blades as a design guideline (refer to Eq. 1.38).

C_L: *coefficient of the tillage lifting force* [20], R_3/T_3, 0.7–1.0 is used as a technical guideline.

1.0 is used for estimating the approximate peak value of the tillage lifting force, while 0.7 is used for an average value of this force as a design rule of thumb.

C_T: *coefficient of tillage thrust force* [20], P_3/T_3, 1.0–1.4 is used as a technical guideline.

1.4 is for the approximate peak value of the tillage thrust force, while 1.0 is for an average value of this force.

Relative location of the virtual point of action for the total rotary tillage resistance is calculated practically by the following equations illustrated in Fig. 1.62:

$$L_R^2 \fallingdotseq r_3^2 - (r_3 - iH_M)^2$$
$$\therefore\ L_R \fallingdotseq \{iH_M(2r_3 - iH_M)\}^{1/2} \qquad (1.41)$$

where L_R : horizontal distance from the blade rotation center to the virtual point of action (cm)
i : height coefficient of the virtual point of action 0.4 is used for a tiller equipped with C-shaped blades, (refer to Eq. 1.39)
H_M : maximum depth of tillage (cm)
r_3 : rotation radius of the blades (cm)

e. Forces Acting on the Rear Stabilizer Wheel. The rear wheel is called a *gauge wheel* also. This wheel supports the rotary tiller and is used for the adjustment of tillage depth by changing its relative height to the tiller. The dynamic principle of adjusting the tillage depth is to adjust the contact load of the rear wheel on the soil. The dynamic load on the rear wheel is in the range of zero to several tens kgf. If not, the driver cannot pull up the handlebars.

Feedback Power and Transmission Efficiency

The rotary-tilling tractor shows *negative values of travel reduction* during tillage work. There were such case studies in the 1950s and 1960s. These phenomena for a four-wheel tractor were analyzed [30] and the following theorems were proved [31, 32]: A part of the tillage power transmitted from the engine to the rotary tiller is returned back to the tractor during tillage. This power has been named *kangen-douryoku* (還元動力) in Japanese and *return power* in English. Part of the return power is consumed by the traveling resistance of the whole machine, and the surplus of the return power is transmitted as a *feedback power* (ﾌｨｰﾄﾞﾊﾞｯｸ動力) into the power transmission gears of the tractor through its drive wheels and to the rotary tiller axle. A rotary tilling tractor contains a *closed-loop* for the power flow.

Those principles were applied to analyze the two-wheel tractor [33], and proved again [34] as follows: As shown in Fig. 1.66, the tractor wheel torque becomes a negative value during tillage work. The drive wheels of the rotary power tiller are not driven by the engine but driven by the *tillage thrust force* on the rotary tiller, and act not as driving wheels but as braking wheels in order to keep a constant traveling speed. Figure 1.67 shows a block diagram of the power flow in a rotary power tiller to calculate the *power transmission efficiency* η_R *of the closed loop* from the engine to the tiller shaft.

These return and feedback power phenomena have an important effect, equivalent to a 2–5% increase of power transmission efficiency from the engine to the rotary tiller, in the case of a two-wheel tractor [34]. These principles are useful to find new technical know-how in design and test aspects to develop a rational rotary tilling tractor.

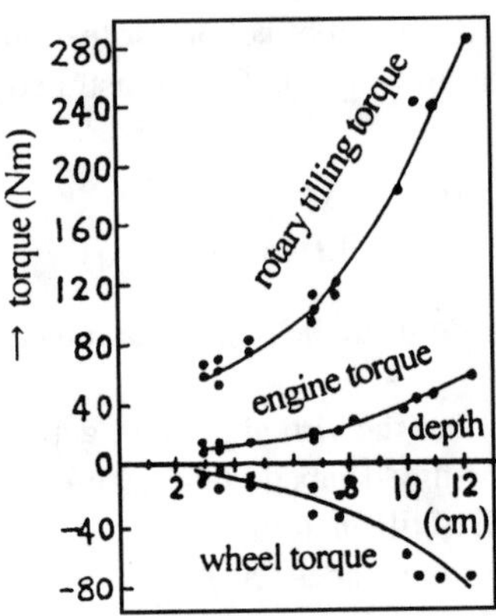

Figure 1.66. Test data of the feedback power (Zou).

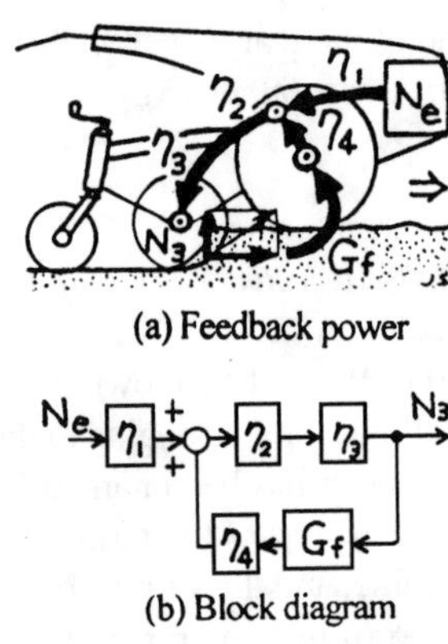

Figure 1.67. Power flow (Sakai).

Motion and Design Principles of C-shaped Blades and Tiller Shafts

a. Motion Equations and the Relief Angle. The practical locus curve equations for the tiller blade, with the plow pan surface as the X-axis, are [20] (Fig. 1.68)

$$x = vt - R\sin\omega t \tag{1.42a}$$

$$y = R(1 - \cos\omega t) \tag{1.42b}$$

where $\omega : 2\pi n/60 = \pi n/30$

v : machine travel speed (cm/s)

R : the rotation radius of the rotary tiller blade (cm)

n : rotational speed of the rotary tiller blade (rpm)

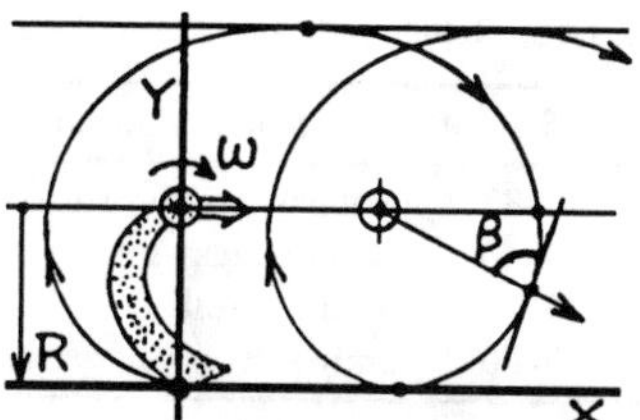

Figure 1.68. The blade-tip motions and its angle relation.

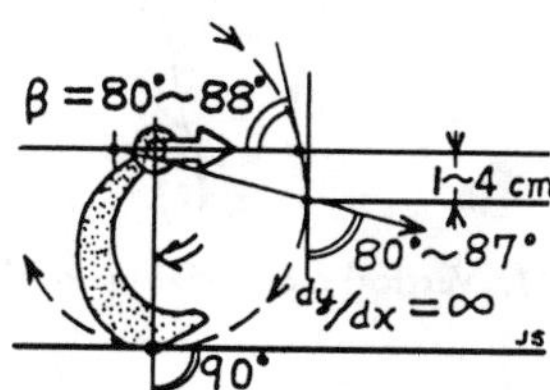

Figure 1.69. Actual angles of blade motions.

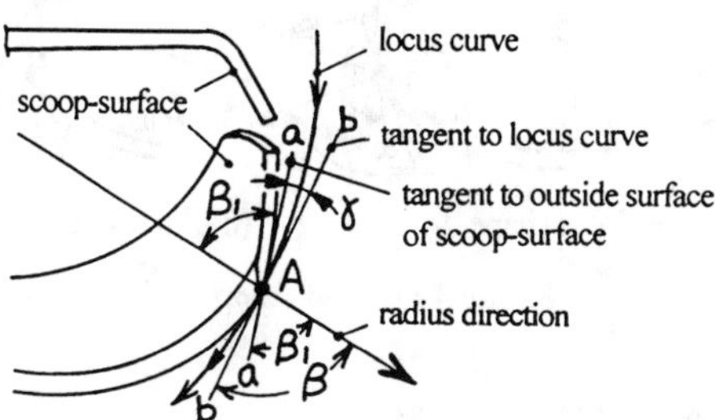

Figure 1.70. Scoop surface and angle relations.

It is important to understand the operation characteristics of the blade through the relative angle, β, (Figs. 1.68–1.70) between the motion direction of any point on the locus curve and line from the center of the rotary tiller axle to that point. The angle β (degrees) at tillage depth H (cm) is calculated by the following equation: [20]

$$\beta = \cos^{-1}[(30v/R)[H(2R/H)/\{(30v)^2 - 60n\pi v(R - H) + (Rn\pi)^2\}]^{1/2}] \quad (1.43)$$

b. Tip Blade and the Scoop Surface. The inside surface of the tip blade is called *sukui-men* (掬い面) in Japanese and *scoop surface* in English (Fig. 1.70).

When the tip blade is cutting into the soil and prior to its reaching the maximum depth of tillage, the outside surface of the scoop surface must not interfere with the face of the soil cut by the tip blade, and hence a rational suction function should be given to the tip blade. Therefore, the whole outside surface of the scoop part should have a smaller

Table 1.12. Relief angle and scoop angle β_1

γ°	$\beta_1{}^\circ$	Soil Condition
3–5	80–83	Very hard or very sticky soil
5–10	70–80	Hard or sticky soil
10–20	60–70	Common soil
20–30	50–60	Sandy soft soil or wet soft soil

Remarks: v = 25–50 cm/s and n = 200–300 rpm.

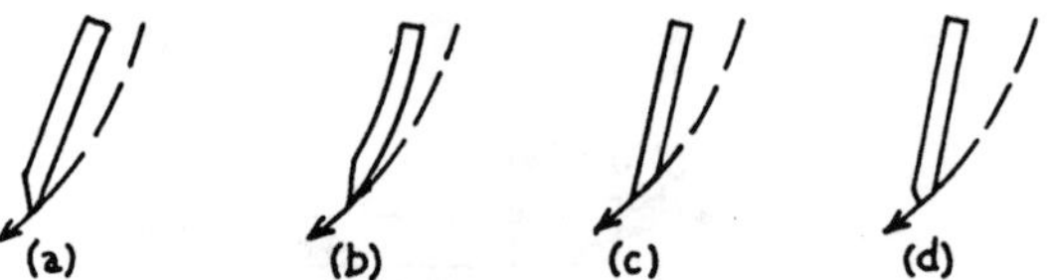

Figure 1.71. Vertical cross sections of tip blade.

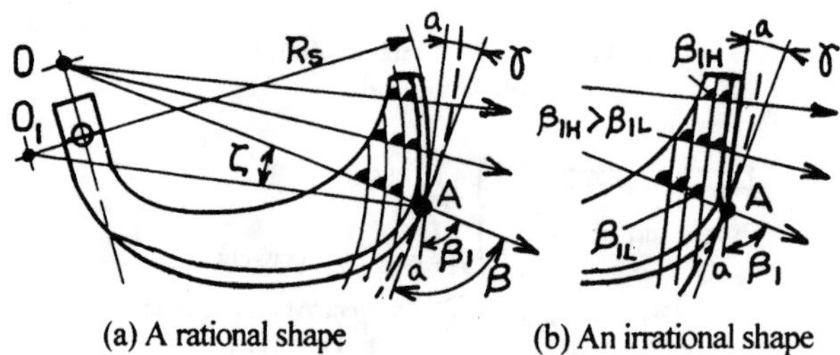

Figure 1.72. Rational shape of scoop surface (Sakai).

angle1 β_1 than β as follows: [20]

$$\beta_1 \equiv \beta - \gamma \qquad \therefore \gamma = \beta - \beta_1 \tag{1.44}$$

where (*Special notice* : $\gamma \neq 90^\circ - \beta_1$)

γ : the *relief angle* between the outside surface, a-a, of the scoop part and its motion direction, b-b (degrees) as shown in Fig. 1.70.

Figure 1.71 shows the shapes of vertical cross-sections of various tip blades. The vertical cross-section of the scoop surface has an equal thickness and only type (b) can have the smallest γ as a design guideline. The angle β_1 has been named *hai-kaku* (背角) in Japanese [20] and *scoop angle* [4]. Table 1.12 shows these angles and soil conditions.

Figure 1.72 shows angle relations of the scoop surface [20]. Their relations should be $\beta_{1H} \geq \beta_{1L}$. Thus, as shown in Figs. 1.72 and 1.73, the curvature radius R_S of the scoop surface and the radius R of the tip point A should be $R_S \leq R$, and the center of the curvature ought to be on the line AO_1 [28]. Usually R_S is 9–15 cm under Asian paddy

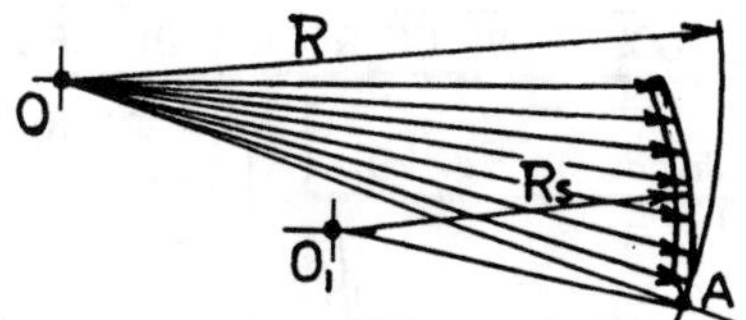

Figure 1.73. Design of scoop surface.

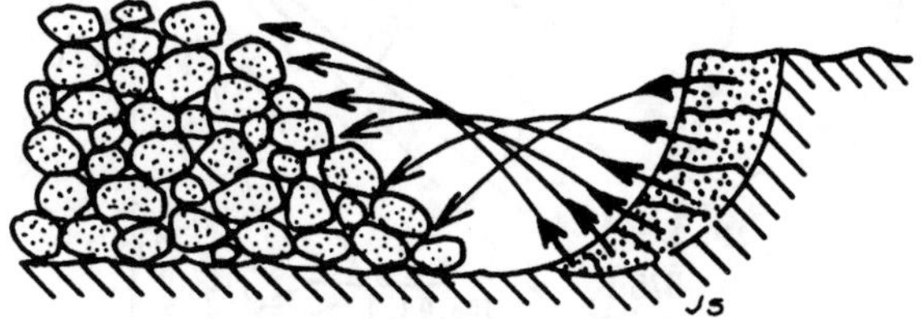

Figure 1.74. Ideal motion of soil clods.

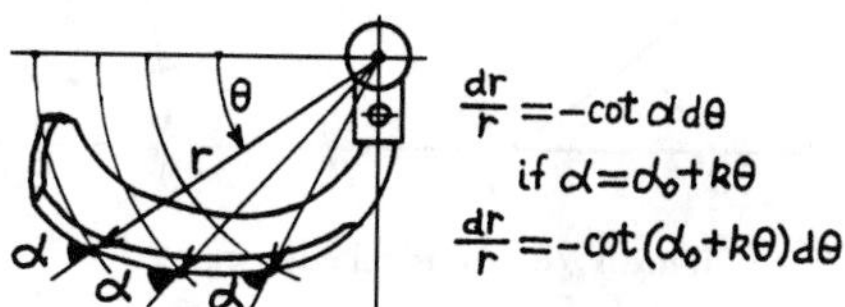

Figure 1.75. Edge-curve and edge curve angle.

field conditions. The angle, ζ, $\angle OAO_1$ is:

$$\zeta = 90^\circ - \beta_1 \quad \text{at the blade tip point A} \tag{1.45}$$

Figure 1.74 shows ideal motions of soil clods.

c. Straight Blade and the Edge-curve Angle. The angle α in Fig. 1.75 is one of the important design parameters, which has been named *hai-raku-kaku* (排絡角) in Japanese, and *edge-curve angle* in English. A design principle for α is as follows: [26, 27]

The edge-curve angle smaller than 55°–57.5° is apt to cause trash entwining trouble at the tip of the blade, and the angle smaller than 65°–67.5° is also apt to have trouble at the blade holder portion. The following equation is available to obtain a rational edge curve as shown in Fig. 1.76.

$$r = r_0 \sin^{1/k} \alpha_0 \sin^{-1/k}(\alpha_0 + k\theta) d\,\theta \tag{1.46}$$

in which
- r and θ : radius (mm) and angle (°) in polar coordinates
- r_0 and α_0 : radius and edge curve angle at the tip of the edge curve in polar coordinate, when $\theta = 0^\circ$
- k : the *increasing ratio* of the edge curve angle. 1/18: Increase of 10°(57.5° ⇒ 67.5°) in the range of 180°

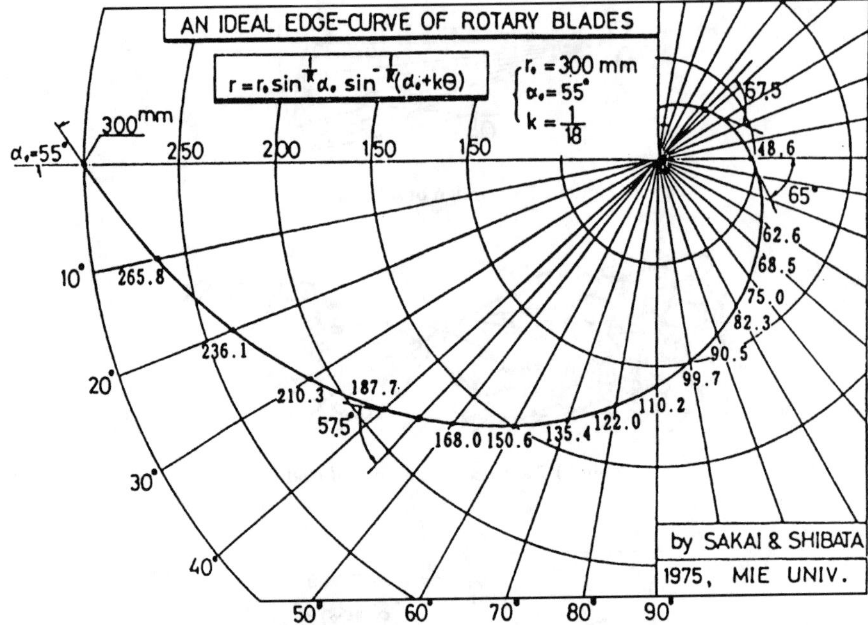

Figure 1.76. An ideal edge-curve.

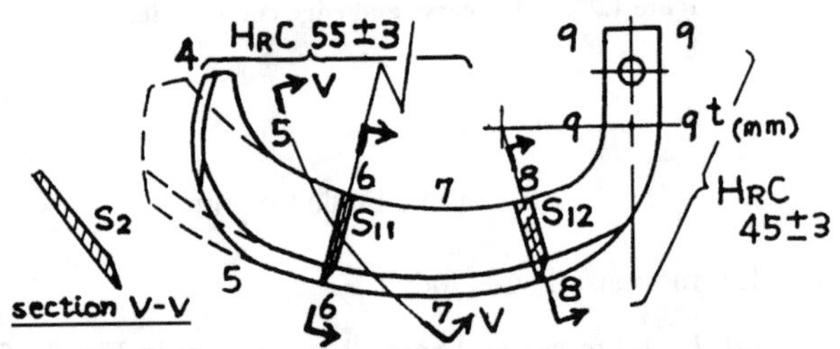

Figure 1.77. The straight blade and cross sections.

d. Shapes of Cross-sections and the Blade Edge. Figure 1.77 shows an example of design expressions for cross sections, S_{11} and S_{12}, of the straight blade. One of the technical keys is to consider the shapes of not only the cross-section S_{11} and S_{12} in the radius direction but the *actually operating cross-section* S_2 along the trochoidal locus curve V-V. Figure 1.78 shows their various shapes and cutting forms in the soil. The ε of the wedge-typed blades, (a) and (b), shows that the soil ought to be compressed to the untilled, immovable side, and gives strong friction resistance. The thickness of C-shaped blades changes from the holding portion to the blade-tip. Therefore, as shown in Fig. 1.77, the cross-section S_{11} of a single-edged rectangular shape forms an *actually operating cross-section* S_2 of an inverted wedge type which has a relief angle as shown by (d) in Fig. 1.78 [35, 36].

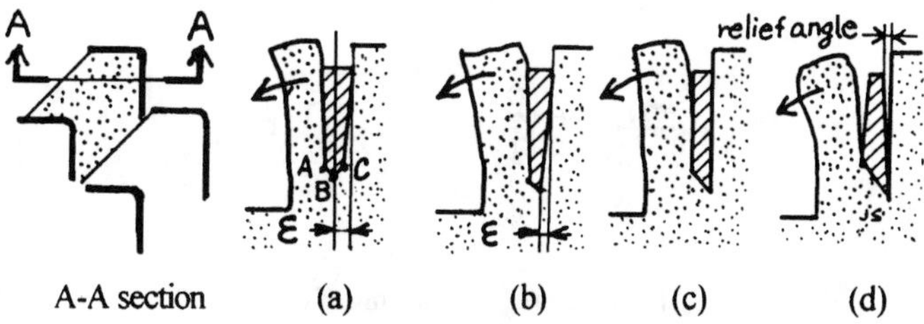

Figure 1.78. Cutting operation in A-A sections (Sakai).

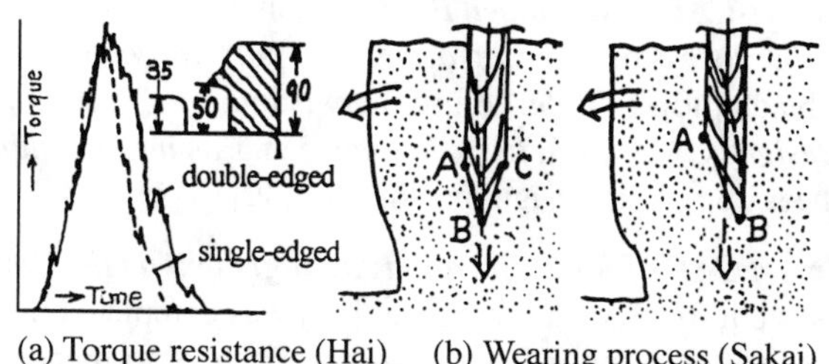

(a) Torque resistance (Hai) (b) Wearing process (Sakai)

Figure 1.79. Wearing process of the blade edge.

Tillage resistance of C-shaped blades of a single edge and rectangular cross-sections decreases at least 10%–30% from that of common wedge-type blades, and the blade tends to keep its original shape in the wearing process as shown in Fig. 1.79, and moreover, its superiority tends to increase [36, 37].

e. Dimensions of the Blade Holder Portion. In the past there have been many types and sizes. Nowadays, however, only one type (Fig. 1.80) with three sizes of blade radius R, 225, 235 and 245 mm, is used for two-wheel tractors, as shown in Table 1.13 (a part of JIS B9210).

f. Materials, Heat Treatment and Hardness. Carbon steel S58C, spring steel SUP-6 or SUP-9 are recommended by JIS (B9210). SUP-6 is mostly used. (SK-5 of car leaf springs is not recommended.) Two kinds of hardness have been adopted since the 1960s as follows (Fig. 1.77) [44]:

Tip & straight blades: H_RC 52–58 to be resistant to wear
Fixed part & bent shank: H_RC 42–48 to be tough

Table 1.13. Dimensions of holder portion (mm)

R	A	B	F_1	G	E	D
225	25	9	25	≥23	50/55/58	10.5 ± 0.2
235	25	9	25	≥23	58	10.5 ± 0.2
245	25/27	10	25	≥27	50/55/65	10.5/12.5 ± 0.2

Remarks: Slicing width of the tip blade is to be 35–40 mm. Thus, holder interval is 45–50 mm in general.

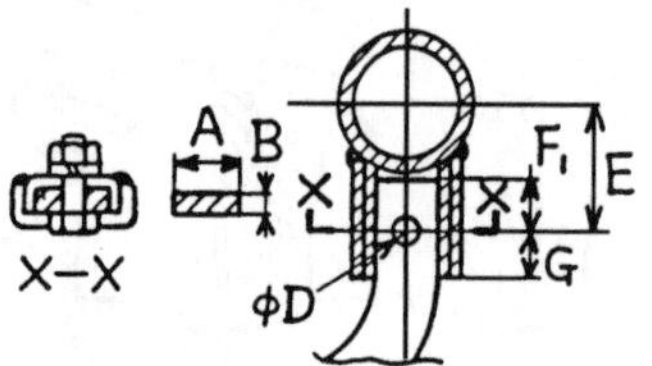

Figure 1.80. Dimensions of the blade holder portion.

g. Design and Expert-CAD of C-shaped Blades. It may take 10–20 days to achieve by hand all the drawing process of *planning design* and *part design* along with a *flowchart* for the C-shaped blade [40]. In order to use the wisdom of skilled engineers in *expert computer aided design* (CAD) [39], *fuzzy inference* and its *membership functions* have been utilized to the scoop-angle design by Noguchi [40].

h. Design and Expert CAD of Blade Location Arrangement on the Rotary Tiller Shaft. An Expert CAD system for blade arrangement [41] was completed in 1990 [42, 43] so as to make a database of only non-inferior solutions. Each performance parameter of the tiller shaft is improved by the system. In particular, the required average tillage power decreases several percent from that of existing machines.

Dynamic Principles and Pertinent Design Guidelines of the Rotary Power Tiller for Stable Tillage

Hereafter, explanation will be practically done with gravitational units.

a. Balancing Conditions at Rotary Tillage Work. Figure 1.81 shows all external forces acting on the rotary power tiller that must satisfy the following three balancing conditions as a free body on the earth:

Vertical balance:

$$W_T + R_{L2} + R_{L4} = R_{N2} + R_{L2} + R_3 + R_{N4} + R_{L4} \quad \text{or}$$
$$R_{N2} = W_T - R_3 - R_{N4} \tag{1.47}$$

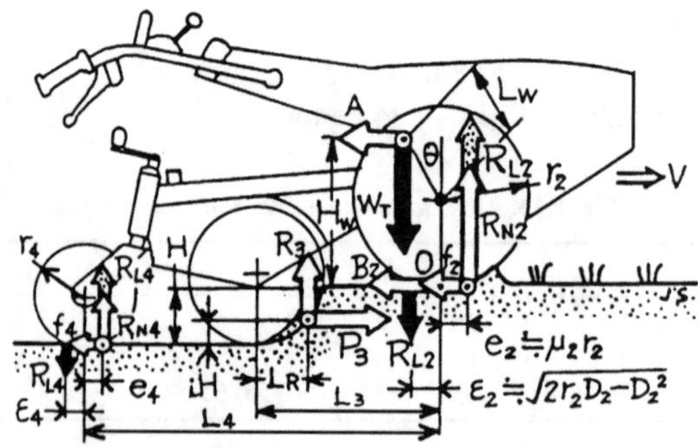

Figure 1.81. External forces on a rotary power tiller.

Horizontal balance:

$$P_3 = f_2 + B_2 + f_4 \tag{1.48}$$

Moment balance about the point O under the wheel center:

$$\Sigma(\text{counterclockwise moments}) = \Sigma(\text{clockwise moments}):$$

$$W_T L_W \sin\theta + (R_{N2} + R_{L2})e_2 + R_{L2}\varepsilon_2 + AH_W + P_3(H - iH) + R_{L4}(L_4 - \varepsilon_4)$$
$$= R_3(L_3 - L_R) + (R_{N4} + R_{L4})(L_4 - e_4) + f_4 H \tag{1.49}$$

where W_T : total weight of the whole machine at work (kgf)
A : *acceleration resistance* (kgf)

$$A = m\,dv/dt \tag{1.50}$$

R_{N2} and R_{N4} : reaction of *net dynamic load* acting on wheels without the lift resistance (kgf) (refer to Fig. 1.42)
R_{L2} and R_{L4} : *lift resistance* (kgf) (refer to Fig. 1.45)
R_3 : *tillage lifting force* (kgf) (refer to Eq. (1.40b))

$$R_3 = 71620 Ne\eta C_L/(C_R r_3 n_3)$$

P_3 : *tillage thrust force* (kgf) (refer to Eq. (1.40c))

$$P_3 = 71620 Ne\eta C_T/(C_R r_3 n_3)$$

f_2 and f_4 : *motion resistance* (kgf) (Eq. (1.21))

$$f_2 = \mu_2 R_{N2} \quad \text{and} \quad f_4 = \mu_4 R_{N4} \tag{1.51}$$

B_2 : *frictional reaction* of the lug wheels on the soil (kgf). This force has not been measured well. However, practical estimation of its value may be obtained in a similar way to the coefficient of friction as follows:

$$B_2 \fallingdotseq C_B R_{N2} \quad \text{and} \quad C_B \fallingdotseq 0.8 \tag{1.52}$$

L_W : distance between the center of gravity and the wheel center (cm)
H_W : height of the center of gravity of the machine from the field surface (cm)

$$H_W \fallingdotseq r_2 + L_W \cos\theta \tag{1.53}$$

e_2 and e_4 : horizontal distance between R_{N2} or R_{N4} and the wheel center (cm) (refer to Eq. (1.25))

θ : location angle of the center of gravity from the vertical line on the wheel center (degrees)

$$\theta \fallingdotseq \tan^{-1} H/L_3 \qquad \therefore \ \sin\theta \fallingdotseq H/\left(L_2^2H^2\right)^{1/2} \tag{1.54}$$

ε_2 & ε_4 : horizontal distance between R_{L2} & R_{L4} and the wheel center (refer to Eq. (1.26, 1.27)) (cm)

H : depth of tillage (cm)

iH : height of the *virtual point of action*, O_R, for the total tillage resistance (refer to Eq. (1.44)) (cm)

$$(H - iH) \fallingdotseq 0.6H \tag{1.55}$$

L_3 : horizontal distance between the wheel center and the tillage shaft center. The shorter, the better (cm)

L_R : horizontal distance from the virtual point of action, O_R, for the total tillage resistance to the tillage shaft center. (refer to Eq. (1.41))

b. Design Principles for Optimum Weight, Engine Power and Other Main Specifications. Two of the worst causes of accidents during tillage are a sudden *upward jumping motion* of the rotary tiller due to a large *tillage lifting force*, and a *forward dashing motion* of the machine due to a large *tillage thrust force* on hard soil.

In order to achieve stable tillage without engine power shortage, the following two design equations, under strict soil conditions with a lifted rear wheel, are useful to calculate and adjust main specifications and dimensions of the rotary power tiller [55]. Then, the forces, R_{L2}, R_{L4}, R_{N4}, f_4 and A are zero:

$$W_T \le \{71620Ne\eta/(C_R r_3 n_3)\}[C_L + \{C_T/(C_B + \mu)\}] \tag{1.56a}$$

$$W_T \le [71620Ne\eta\{C_L(L_3 - (0.4H(2r_3 - 0.4H))^{1/2} + \mu_2 r_2) - 0.6HC_T\}]/\left[C_R r_3 n_3\left\{L_w H/\left(L_3^2 + H^2\right)^{1/2}\right\} + \mu_2 r_2\right] \tag{1.56b}$$

in which the following strict conditions of dry hard soil are applied to calculate as R&D design know-how:

Ne : not catalogue horsepower of the engine, but actual maximum output (PS $\fallingdotseq$ HP, refer to Fig. 1.31)

η : power transmission efficiency (the value of openloop system +5%) (refer to Fig. 1.67)

C_R : *radius coefficient*, 1.01–1.04: 1.0 as a strict value (refer to p. 82)

C_L : *coefficient of tillage lifting force*, 0.7–1.0: 1.0 as a strict value (refer to p. 82)

C_T : *coefficient of tillage thrust force*, 1.0–1.4: 1.4 as a strict value (refer to p. 82)

C_B : *coefficient of frictional brake*, 0.8 (Eq. (1.52))

μ : *motion resistance ratio* on upland soil: 0.05

When the weight of a rotary power tiller is more than the calculated weight, the engine tends to be overloaded during tillage on the hard soil. In such a case, *engineers must not easily exchange the engine for a strong heavy one. Machine weight must be decreased instead.*

References

1. A Charted Survey of Japan, 1967–1994/95, Tokyo: Kokusei-sha.
2. Sakai, J., Surin, P., Kishimoto, T. 1987. A Study on Design Theories of Iron Wheels for Plowing, AMA 18(4): 11–18, Shin-Norinsha, Tokyo.
3. Sakai, J., Kishimoto, T., Surin P. 1988. Basic Studies on Design Theories of Agricultural Wheels, (Part1), J. JSAM 50(6): 11–18.
4. Sakai, J., Zhou, W., et al. 1990: An Equipment for Measuring Forces Acting on the Agricultural Wheel Lug, Res. Bul. JSAM Kyushu Section. (39): 6–10.
5. Sakai, J. Kishimoto, T. Inoue, E., et al. 1991. A Proposal of Lift Resistance on a Wheel Dynamics, Proc. Int'l Agric. Mechanization Conf. 2: 116–133 Beijing.
6. Kishimoto, T., Taniguchi, T., Sakai, J. et al. 1991. Development of Devices for Measuring External Forces Acting on Agric. Lug-Wheels, (Part 2). Res. Bul. Obihiro Univ. 279–287.
7. Sakai, J., Choe, J. S., Kishimoto, T., Yoon, Y. D. 1993. A Proposal of A New Model of Wheel & Tractor Dynamics that includes Lift Resistance, Proc. Inter. Conf. Agric. Machinery and Process Eng'g, 1176–1185 Seoul: KSAM.
8. Kishimoto, T., et al. 1993. Effect of Lift Resistance on Dynamic Load acting on a Circular wheel, 1166–1175 ibid.
9. Sakai, J., Yoon, Y. D., Choe, J. S., Chung, C. J. 1993. Tractor Design for Rotary Tillage Considering Lift Resistance, J. KSAM 18(4): 344–350.
10. Sakai, J., Choe, J. S., Inoue, E. 1994. Lift Resistance of Wheels and Design Theories of Wheel Lugs (Part 1), J. JSAM 56(2): 3– 10.
11. Kishimoto, T. Choe, J. S., Sakai, J. 1997. Lift Resistance Acting on Wheel Lugs and Lift Resistance Ratio, Res. Bull. Obihiro Univ. 20(2): 127–132.
12. Hirobe, T. 1913. Nouguron, (Agricultural tools), Tokyo: Seibi-dou.
13. Mori, S. 1937. Rito Rikou-ho (Japanese Conventional Plows and Plowing Methods), Nihon-Hyouronsha, Tokyo.
14. Kishida, Y. 1954. Matsuyama Genzo Ou Hyoden, Shin-Norinsha Co., LTd., Tokyo.
15. Hopkin, H. 1969. Farm Implements for Arid and Tropical Regions, FAO, Rome, 5961.
16. Sakai, J., Surin P., Kishimoto, T. 1987. Study on Basic Knowledge of Plowing Science for Asian Lowland Farming, AMA 18(1): 11–21, 18(2): 11–17 Shin-norinsha Tokyo.
17. Sakai, J., Surin P., Kishimoto, T. 1987. Engineering Design Theories of Hand-Tractor Plows, AMA 18(3): 11–19 ibid.
18. Surin P., Sakai, J., Kishimoto, T. 1988. A Study on Engineering Design Theories of Hand-Tractor Plows II, AMA 19(2): 9–19 ibid.

19. Sakai, J. 1990. Principles of Walking Tractor Plowing and Design Theories of Japanese Plows, J. JSAM Kyushu Section. (30): 38–46.
20. Sakai., J. 1960. A Theoretical Approach to the Mechanism and Performance of the Hand Tractor with a Rotary Tiller, together with Practical Application, Ph. D. Thesis, Kyushu Univ. 160 pages, published in 1962, Shin-norinsha Tokyo.
21. Söhne, W. 1957. Einfluss von Form und Anordnung der Werkzeuge auf die Antribsmomente von Ackerfräsen, Grundlg. D. Landtechn 9: 69–87.
22. Bernacky, H., 1962. Theory of the Rotary Tiller, Inst. Of Mech. And Elect. Of Agric. In Warsaw, Bul. (2): 9–64, of which English translation is in MTML, USAID, USA.
23. Sakai, J., Salas Sr., C. G. 1977. Graphical Studies on Resultant Force of Rotary Tillage Resistance (Part 1), Bul. Fac. Ag. Mie Univ. 54: 223–258.
24. Shibata, Y., Sakai, J. 1977. Study on Rotary Tillage Resistance of a Japanese C-Shaped Blade, J. JSAM 39(4): 447–457.
25. Sakai, J. 1979. Eng'g Characteristics of Rotary Tillage Resistance of Japanese Rotary Tillers with Tractors, Proc. 8th Conf. ISTRO, 2: 415–420, Session 14, Bundesrepublik.
26. Sakai, J. Shibata, Y., Taguchi, T. 1976. Design Theory of Edge-Curves for Rotary Blades of Trzactors, J. JSAM 38(2): 183–190.
27. Sakai, J. 1977. Some Design Know-how of Edge-Curve Angle of Rotary Blades for Paddy Rice Cultivation, AMA 8(2): 49–57, Shin-norinsha Co., Ltd, Shin-norinsha Tokyo.
28. Sakai, J., Shibata, Y. 1977. Design Theories on Scoop-Surface of Rotary Blades for Tractors, J. JSAM 39(1): 11–20.
29. Sakai, J. 1978. Designing Process and Theories of Rotary Blades for Better Rotary Tillage (Part 1) (Part 2), JARQ 12(2): 86–93, 12(4): 198–204., Tropical Ag. Research Center, Japan Ministry of Ag., For. and Fish.
30. Sakai, J. Shibata, Y. 1977. Studies on Feedback Power and Power Transmission Systems of Closed-loop for Rotary-Tilling Tractors, J. JSAM 39(3): 287–297.
31. Shibata, Y. Sakai, J. 1978. Dynamic Characteristics of Tractor mounted with Rotary Tillers (1), J. JSAM 40(3): 345–353.
32. Shibata, Y., Sakai, J. 1979. Ibid (II) & (III), 41(1): 37–42 & 41(2): 207–214, Japan.
33. Sakai, J., Zou, C. 1987. Ibid, (Part 2), J. JSAM 49(1, 2): 109–116, and (Part 3), J. JSAM 49(3): 189–195.
34. Sakai, J., Zou, C. 1988. Ibid (Part 4), J. JSAM 50(1): 19–26.
35. Sakai, J. 1979. Studies on Design Theories of Single-edged and Double-edged Rotary Blades for Tractors. Bull. Fac. Ag. Mie Univ. 58: 129–135.
36. Sakai, J., Hai L. V. 1982. The Reduced Tillage Energy of Japanese Rotary Blades, Proc. 9th Conferance of ISTRO 639–644, Osijek, Croatia.
37. Hai, L. V., Sakai, J. 1983. Studies on the Tillage Characteristics of Single and Double Edged Blades for Japanese Rotary Blades, (Part 1). J. JSAM 45(1): 49–54.
38. Hai, L. V. 1983. Basic Research on Characteristics and Design Theories of Rotary Tiller Blades, Ph. D thesis, Kyushu Univ.

39. Hai, S. 1993. Design Theories and Production Technology of Japanese Rotary Tiller Blades, textbook, 81 pages, Farm Mach. Section, Tsukuba Int'l Agric. Training Center, JICA.
40. Mizota, T., Noguchi, R. et al. 1991. Optimum Design of the Agricultural Rotary Tiller Blade by Fuzzy Inference, (1). J. J S Design Eng. 26(10): 511–516, 26(11): 548–550.
41. Sakai, J., Shibata, Y. 1978. Design Theory of Rotary Blades' Arrangement for Tractors, J. JSAM 40(1): 29–40.
42. Sakai, J., Chen, P. 1990 Studies on Optimum Design Theories of a Rotary Shaft and The Expert CAD System (Part 2). J. JSAM 52(4): 45–52.
43. Chen, P., Sakai, J., Noguchi, R., 1991. Studies on Optimum Design Theories of a Rotary Shaft and The Expert CAD System, (Part 3) (Part 4). J. JSAM 53(2): 53–61, 53(3): 35–45.
44. Sakai, J., Hai L. V. 1966. Agricultural Engineering of Rotary Tilling Tractors, Lecture Text No. 11–1, 2, 1983 edition, Tsukuba Int'l Agric. Training Center, JICA.

Tractors: Two-Wheel Tractors for Dry Land Farming

A. Lara-Lopez & W. J. Chancellor

1.1.22. Description and Types of Two-Wheel Tractor Designs

Two-wheel tractors are sources of power designed to perform most field operations. Due to the size of such tractors, they become an economic alternative for scale farming. In addition, two-wheel tractors are much more productive than animal traction and they require less time for attendance and preparation, giving the individual farmer more independence and contact with modern technology. Also, due to the simple design, local manufacture of two-wheel tractors has been implemented in several countries successfully, increasing the employment opportunities.

Two-wheel tractors may be classified according to the engine power in three categories. Table 1.14 shows the main physical characteristics for each category.

Two-wheel tractors of Category I are normally dedicated to garden work, transportation or light interrow cultivation. Many of such tractors can incorporate a rotary tiller on the axle, with a working width of 500–800 mm, Fig. 1.82. Due to their low weight and small clearance, these machines are not suitable for plowing dry and hard soils. However they may perform a number of field operations, such as lifting water, forage cutting, transport and cultivation.

Two-wheel tractors of Category II may be used for the same field operations as those of category I. In this case tiller width is of 1 m. It is possible to cultivate and plow soft soils. Some designs incorporate wing hubs with pawls to assist turning. Many tractors of this category are equipped with a rotary tiller driven from a PTO, Fig. 1.83.

Two-wheel tractors of Category III cover the same applications as those of categories I and II, but in addition some designs are sufficiently heavy to plow in dry-land agriculture. For silty clay, the required draft for a 250 mm mouldboard plow working at 200 mm depth with a speed of 3 km/h is approximately 3.7 kN. This value is similar to the

Table 1.14. Main characteristics for two-wheel tractors[2]

Category	Overall Dimensions L × W × H mm	Track Width mm	Clearance mm	Mass kg	Maximum Traction kN	Speed km/h	Power kW	Engine
I	1500 × 410 × 1000	315	150	45–60	0.3–0.5	1.55	2.7–4	2 or 4 cycle gasoline or diesel
II	830–530 × 1800 × 1230	400–700	200	75–148	0.6–1.2	1–12.6	5–7	4 cycle gasoline or diesel
III	1900 × 560 × 800	400–750	200	175–465	1.37–3.7	1–16.3	8–10.2	4 cycle gasoline or diesel
	2680 × 960 × 1250		546[1]	5.47[1]	3.91[1]			

[1] Parameters for a high clearance two-wheeled tractor of special design. Lara-López et al. 1982.
[2] *Source:* catalogues from manufacturers from Europe and Asia and publications on special designs.

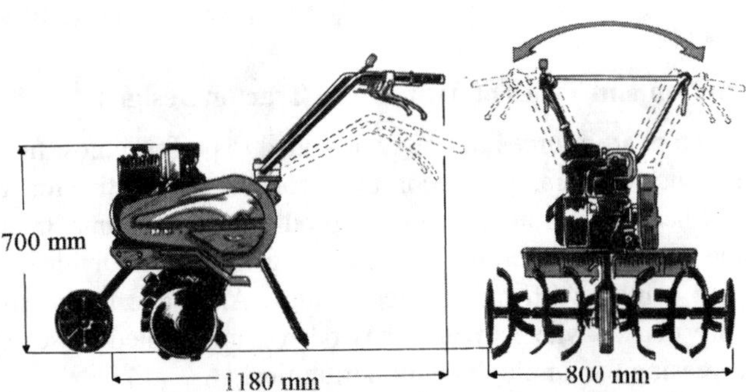

Figure 1.82. Two-wheel tractor of Category I. 2.5 kW (3.5 hp) with tiller on the axle. Mass of 50 kg. Meccanica Banassi.

maximum draft obtained from some of the heaviest commercial models of two-wheel tractors, although for dry soils with crop residues the draft requirements may be higher (Fig. 1.84).

Two-wheel tractors designed for wet land may be lighter than those designed for dry land, where high traction capability is more necessary than flotation. Wet-land tractors are equipped with steel wheels with lugs to provide traction. Satisfactory results in dry-land agriculture were obtained with a power-to-mass ratio of 0.014 kW/kg. Heavy-duty two-wheel tractors designed for wet land conditions show ratios of 0.022 kW/kg.

Row crops such as corn, sorghum and vegetables are cultivated several times requiring high clearance tractors for multirow cultivation. Regions such as Latin America, where such crops are basic foods, require such type of equipment. However, the production of high-clearance tractors is rather small. Figure 1.85 shows a two-wheel tractor with axle

Figure 1.83. Two-wheel tractor of Category II. 8.8 kW (12 hp) with tiller on seperate axle, four stroke gasoline engine or diesel engine. Mass of 175 kg. Goldoni.

Figure 1.84. Two-wheel tractor of Category III. 8 kW (11 hp) heavy duty machine air-cooled or water cooled diesel engine or water cooled gasoline engine. Mass of 376 kg. Mitsubishi.

Figure 1.85. A Regular two-wheel tractor with axle extensions for incorporation of high clearance feature.

extensions for providing high clearance. Figure 1.86 shows the walking and riding version of a high clearance tractor designed for dry-soil agriculture and local manufacture. [1] and [2].

Many tractors of Category I have a rigid axle and steering is done by pulling handles sideways causing tines on the tiller to slide with respect to the ground. Other small and intermediate tillers are equipped with towing hubs with pawls to assist turning. Some of the two-wheel tractors manufactured in industrial nations are equipped with a differential gear with locking device and independent brakes. Some designs for local manufacture incorporate steering clutches, with brakes for heavy tractors. The tractor of Fig. 1.86 steers by controlling the tension on two independent belts. Two-wheel tractors incorporate articulated sulkies to increase productivity. In this case the chassis angle is controlled. Modern designs incorporate driven wheels for the sulky increasing, tractive capability.

Engine power is engaged by tensioning v-belts, in some designs. This is done with idler pulleys, which in some cases are supported by springs and in other cases with an over-center linkage. This device has the advantage that the idler pulley is fixed in a position independently of the variation of the torque of the engine that may take place during one cycle. Another option for tensioning belts used in some garden equipment is by sliding the engine along guides. Some designs use this principle for loosening belts and at the same time producing contact between pulleys of special design to obtain

Figure 1.86. A high clearance two-wheel tractor designed for local manufacture, developed at the University of California and at the University of Guanajuato, Mexico. a) Walking version. b) Riding version with sulky and simple hidraulic lifting device.

reverse movement of the tractor. An economic multiratio belt drive consists of changing belts to pulleys of different diameter. Most of the two-wheel tractors manufactured in industrialized countries incorporate friction clutches and transfer power from the engine to a gear box with several gear ratios, as shown for a small tiller in Fig. 1.87. Many designs for local manufacturing incorporate roller chain drives or a combination of roller chain drives and gear drives. Power for a separate rotary tiller axle is obtained from one shaft of the gear box and an engaging device is incorporated. Other power take-offs are incorporated to drive stationary equipment from the gear box or directly from the engine.

Figure 1.86. (Continued)

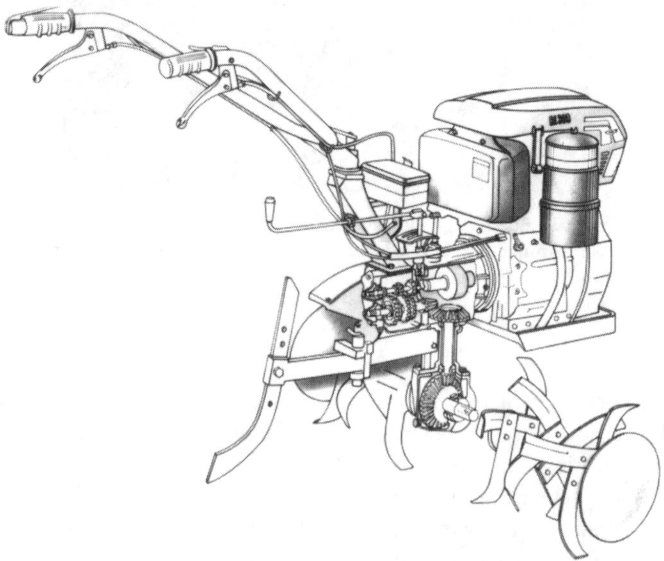

Figure 1.87. Mechanical transmission of Category II two-wheel tractor based on gears.

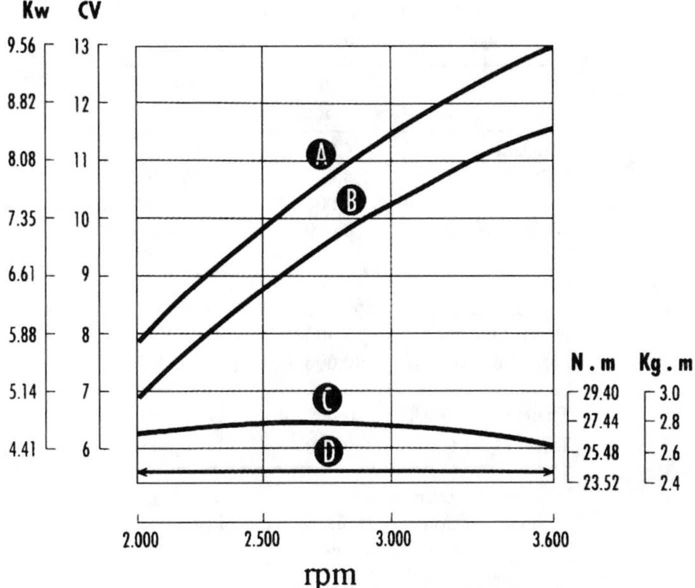

Figure 1.88. Characteristic curves for typical engines use on two-wheel tractors. A and B are power curves, and C and D are the respective torque curves.

Two and four-cycle gasoline and four-cycle diesel engines are used for two-wheel tractors of Category I. For categories II and III, four-cycle diesel and gasoline engines are used. Diesel engines of new design incorporate an air cooling system and operation speeds up to 3600 rpm. This type of diesel engine is competing with traditional four-cycle gasoline engines and with lower-speed water-cooled diesel engines. Figure 1.88 shows typical characteristic curves for diesel and gasoline engines.

1.1.23. Production and Concentration of Two-Wheel Tractors

Production and demand data for two-wheel tractors are affected by the scheme of categorization used by the organization compiling the figures. Data presented here will include parallel figures for both two-wheel and four-wheel tractors (Tables 1.15 and 1.16).

The concentration of two-wheel tractors has been of particular significance in certain countries of Asia, especially those in which lowland rice is a major crop. Data for some of these countries are presented in Tables 1.17 and 1.18.

The numbers of two- and four-wheel tractors in a given country tend to increase slowly in introductory phases of agricultural mechanization. If mechanization is found to be economically advantageous, and thus continues, the numbers increase rapidly. As mechanization approaches completion tractor numbers become nearly constant and then begin to decrease gradually as more powerful units replace older, lower-power machines, as is seen in Table 1.19 to be the case for Japan.

Table 1.15. Average annual demand for agricultural tractors in 1993–1995

Country	Power Tillers (Two-Wheel)	Tractors (Four-Wheel)
Bangladesh	4,300	700*
India	5,400	131,700
Indonesia	80,000	600*
Isl. Rep. of Iran	10,000	15,000
Nepal	100*	800*
Pakistan	1,700*	31,100*
P. R. of China	1,100,000	84,000
Philippines	2,800	400
Rep. of Korea	90,000	8,400
Sri Lanka	8,600*	2,000*
Thailand	76,000	6,200
Total (RNAM)	1,379,200	280,900

* Computed estimates.
Source: Regional Network for Agricultural Machinery National Institutes.

Table 1.16. Annual production (in thousands) of tractors greater than 7.5 kW and of garden tractors less than 7.5 kW

	North America*		P. R. of China		Japan		Europe	
Year	< 7.5 kW	>7.5 kW	<7.5 kW	>7.5 kW	<7.5 kW	>7.5 kW**	<7.5 kW	>7.5 kW
1984	107	766	689	38	362	217	101	696
1987	122	941	1106	37	276	196	103	561
1990	117	1134	1101	39	269	174	111	582
1993	117	1359	961	38	225	155	68	658

* Total shipments.
** Includes crawler tractors.
Source: United Nations, Department for Economic and Social Information and Policy Analysis, 1993. Industrial Commodity Statistics Yearbook of Production and Consumption Statistics.

Table 1.17. Numbers and concentration of tractors greater than and less than 14.7 kW in the Peoples Republic of China

	Tractors Less Than 14.7 kW				Tractors Greater Than 14.7 kW		
Year	Number	(Two-Wheel)	kW/ha	Total Power (GW)	Number	kW/ha	Total Power (GW)
1970			0.0038			0.031	
1980	1,875,247	1,875,247	0.16	16.19	743,560	0.24	23.74
1988	5,958,000	4,528,080	0.555	53.19	870,187	0.30	28.96

Sources: Chinese Academy of Agricultural Mechanization Sciences Newsletter, Vol. 1, No. 1/2. July, 1990.

Table 1.18. Concentration of two-wheel and four-wheel tractors (tractors/1000 ha of arable land) in selected Asian countries, 1979–1980

Country	Two-Wheel Tractors	Four-Wheel Tractors
P. R. of China	73	3
India	0.1	2.3
Indonesia	0.2	0.1
Japan	504	269
Rep. of Korea	114	1
Nepal	0.2	1.1
Pakistan	—	3.8
Philippines	9	3
Sri Lanka	10	18
Thailand	14	3

Source: Adulavidhaya, Kampol and Bart Duff, 1983. The growth and impact of small farm mechanization in Asia, In: Workshop Papers: The Consequences of Small Rice Farm Mechanization in Thailand, November 10–11, 1983, Bangkok Thailand (available from the International Rice Research Institute, Manila Philippines).

Table 1.19. Numbers (in thousands) of two-wheel and four-wheel tractors in Japan

	Two-Wheel		Four-Wheel	
Year	Number on Farms	Annual Production	Number on Farms	Annual Production
1950	13			
1955	89	63		
1960	764	305		
1965	2190	136	32	
1970	3144	350	219	42
1975	3279	303	647	207
1980	2771	322	1471	202
1985	2579	277	1853	209
1990	2185	273	2142	115
1992	1786	265	2002	144

Source: Farm Machinery Statistic (1969, 1978, 1982 and 1993 editions). Farm Machinery Industrial Research Corporation, Shin-Norinsha Co., Ltd. Tokyo, Japan.

1.1.24. Mechanics of Two-Wheel Tractors

The net draft on a tire is the difference between the traction force and rolling resistance. Both forces depend on the normal force on the tire due to contact with soil. Plowing a dry field with a two-wheel tractor normally requires addition of ballast to the tractor wheels.

Proper location of the center of mass relative to the axle and lateral location of tillage implements are critical for proper operation of two-wheel tractors. Location of the center of gravity on the axle or on a vertical line crossing the axle facilitate lifting the implement and produces good tractive efficiency of the tires.

During plowing, one tire is rolling on the bottom of the furrow and the other rolls on the unplowed land. The normal force on the tire rolling on bottom is greater than that rolling on the land. Figure 1.89 shows forces on a two-wheel tractor during plowing. The hitch point of the plow must be located slightly closer to the tire running on the borrow bottom to stabilize tractor attitude. If the plow is not properly located, a correcting torque about

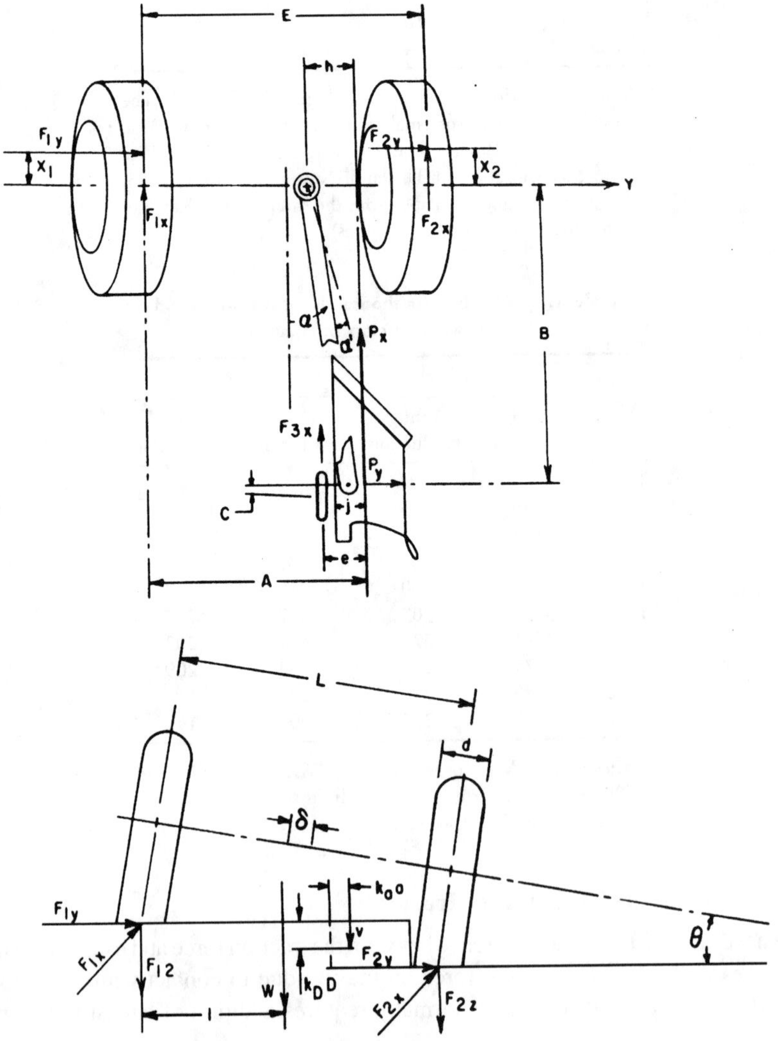

Figure 1.89. Forces on two-wheel tractor during plowing.

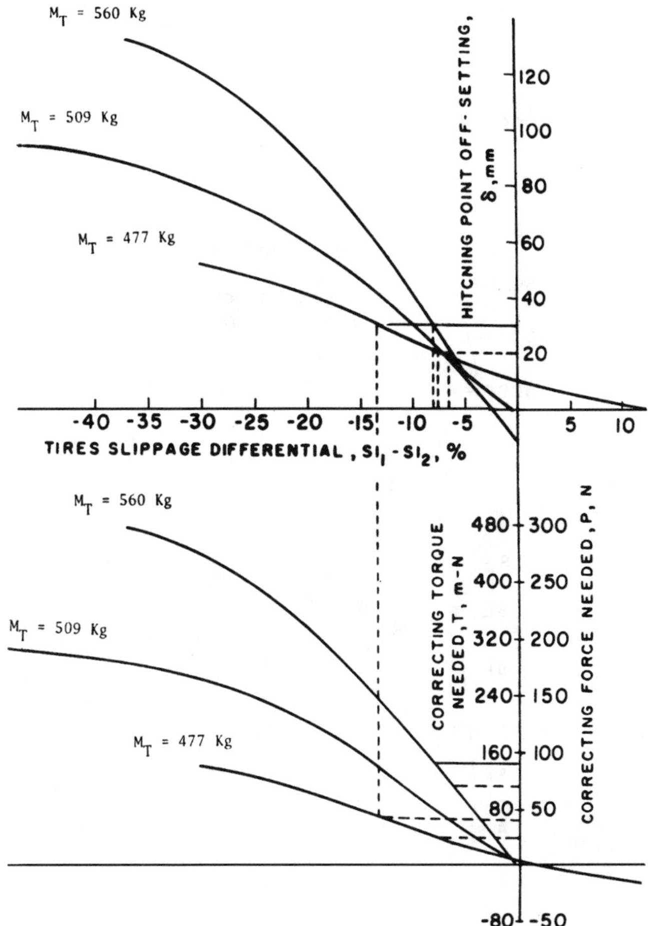

Figure 1.90. Effect of the hitch-point off-setting on slippage differential correcting torque and operators force on controls, for tractors of several weights.

the vertical axis of the tractor must be exerted by the operator pushing handles laterally. Figure 1.90 shows the effect of the hitch-point offset and correcting force required by the operator to maintain the two-wheel tractor plowing in a straight line [3, 4]. According to these figures, it is possible to locate the plow in a position that would not require lateral correction by the operator. In practice some correction is needed due to the changes in soil conditions in the same field. Soil considered in simulation is a black vertisol typical of the central region of Mexico. From the same analysis, the tractive performance of a Category III two-wheel tractor is presented in Fig. 1.91.

1.1.25. Field Operations with Two-Wheel Tractors

Evaluation of field performance of a tractor requires the definition of some basic concepts related to the function of the machine. One of such concepts is the field capacity

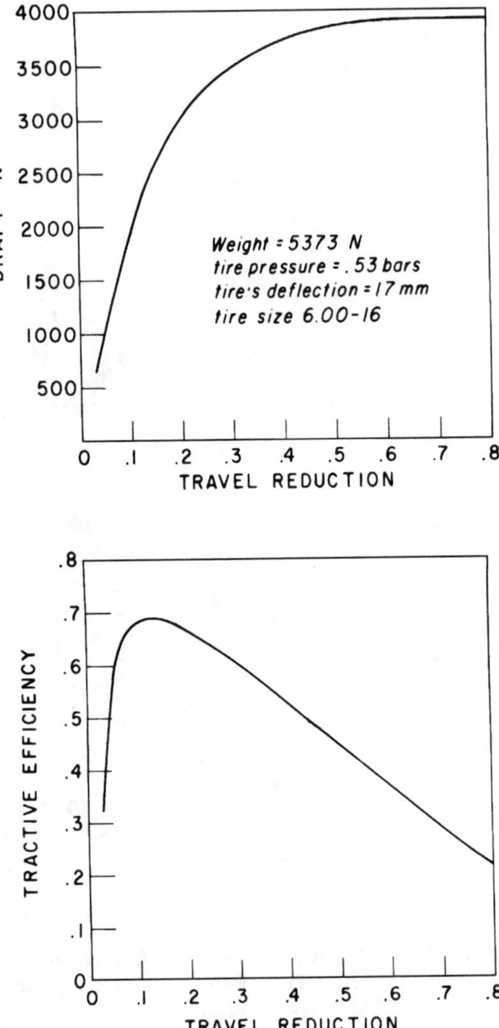

Figure 1.91. Traction performance at various levels of travel reduction (Slip) draft force (upper diagram) and tractive efficiency (lower diagram).

of the tractor, which is obtained by dividing the total area worked by the period of time spent from the beginning of the first round to the end of the last round.

Field efficiency is calculated for each test as the ratio of the area that the tractor completed, to the area that the tractor would work operating at the average speed of the test and using the nominal width of work, as shown by the following equation.

$$FE = \frac{100Aw}{VW\Delta T} \tag{1.57}$$

Where:

FE = Field efficiency, %
Aw = Worked area during the test
V = Average speed, m/s
ΔT = Recording period of time, s
W = Nominal working width, m

Travel reduction (slippage) of tractor tires during field test is obtained by the following equation.

$$S = \frac{L_f - L_\ell}{L_f} 100 \tag{1.58}$$

Where:

S = Travel reduction or slippage
L_f = Distance traveled by the tractor, free of draft load on hard ground (concrete) in ten revolutions of the drive wheels
L_ℓ = Distance traveled by the tractor, under normal working conditions, in ten revolutions of the drive wheels

Slippage is related to the efficiency of a tractor in doing effective work. The term tractive efficiency, defined as the ratio of the draw-bar power (applied to an implement) to the axle power, is used to evaluate the performance of the tractor tires. Tractive efficiency may be estimated by the following equation, if the normal force on tires is known [4].

$$TE = (1 - S)\left(1 - \frac{\frac{1.2}{Cn} + 0.04}{0.75(1 - e^{0.3CnS})}\right) \tag{1.59}$$

Where:

TE = Tractive efficiency
S = Slippage in decimal form (not percent)

$$Cn = \frac{(CIbd)}{W} \tag{1.60}$$

where:

CI = Soil cone index
b = Unloaded tire section width
d = Unloaded overall tire diameter
W = Dynamic normal force on tire

Table 1.20 shows values related to the field performance of the high clearance tractor of Fig. 1.86 in dry land agriculture [2].

In addition to tillage operations there are also implements for forage harvesting that may be attached to two-wheel tractors (Fig. 1.92). Grain harvesting, particularly in Asia, is sometimes performed using the two-wheel tractor PTO to drive reapers.

Table 1.20. Average values recorded and calculated during normal field operation

Field operation	a / b	Average Speed km/h	Effective Width mm	Working Depth mm	Travel Reduction %	Field Capacity ha/h	Field Efficiency %	Draft kN	Tractive Efficiency %	Fuel Consumption L/ha
Chisel harrowing	a	3.10	1600	100	19.00	0.476	96.00	2.52	68.00	7.18
Plowing	b	4.16	1600	100	20.00	0.60	92.00	—	66.00	7.96
	a	3.23	306	172	14.07	0.0774	74.50	1.86	68.50	54.78[d]
	a	3.37	250	200	22.70	0.0705	83.70	—	—	31.46
Disk harrowing	b	3.33	1133	100	26.00	0.33	83.08	3.30	62.00	9.85
Planting	a	3.40	1600		7.00	0.36	66.00		70.20	
Primary row cultivation	a	2.95	1600	60	19.00	0.46	99.00	2.30	70.30	7.56
Secondary row cultivation	a	3.13	1600	190	20.30	0.459	96.85	2.51	66.70	7.25
Spraying	b	4.37	7.14			2.25	92.00[c]			

[a] Walking option.
[b] Riding option.
[c] Time spent in filling the tank is not considered.
[d] Engine adapted for kerosene.

Figure 1.92. Forage harvesting with two-wheel tractors.

Figure 1.93. Undriven Trailer.

Two-wheel tractors are widely used for local transportation towing trailers. Brakes are incorporated in the tractor and the trailers. Trailers may be undriven as in Fig. 1.93 or driven. Driven trailers are more popular in Europe.

1.1.26a. Management of Two-Wheel Tractors

One of the distinguishing features of two-wheel tractors is that they are available in small sizes which makes them adaptable to farmer owner-operator type of management on small farms. The following equations describe the economics of the farmer owner-operator management system:

$$\text{Total annual cost} = Z = PAK + BLW + LW(C + LD)/P \tag{1.61}$$

Then by differentiating Z with respect to P and setting the result equal to zero:

$$\text{Optimum tractor power (kW)} = P^* = [LW(C + LD)/AK]^{1/2} \tag{1.62}$$

Introduction of P^* into Eq. 1.19 gives the optimized total cost per unit work done,

$$(Z/LW)^* = B + 2[AKD/W) + (AKC/LW)]^{1/2} \tag{1.63}$$

in which:

A = Ratio of annual fixed costs to new machine price (0.2 m, Table 1.21)
B = Fuel, repair and wear costs per rated kW·h (0.47 kg of unhusked rice per kW·h in Table 1.21)

Table 1.21. Costs of tractor-powered operations on various sized farms*

	Farm Size (ha)						
	1	2	4	8	16	32	64
Optimum tractor size (kW)	1.2	1.7	2.9	5.2	9.6	18.9	36.2
Annual fixed costs (kg**/ha)	375	287	244	219	202	194	189
Annual labor cost (kg/ha)	279	176	105	59	32	17	9
Annual cash and capital cost (kg/ha)	745	565	446	373	330	307	294
Annual timeliness loss costs (hg/ha)	93	118	140	158	171	178	182
Total annual cost (kg/ha)	838	683	586	531	501	485	476
Annual hours of tractor operation	186	235	280	316	341	363	367
Cash plus capital costs per unit work (kg/kW·h)	3.63	2.75	2.17	1.82	1.60	1.50	1.43

* For the case of one optimally sized tractor per farm.
** All costs are given as kg of unhusked rice equivalent.

C = Operator wage per hour (1.5 kg/h in Table 1.21)
D = Average timeliness penalty per unit area for one working hour of delay (0.5 kg/ha·h in Table 1.21)
K = Tractor price per rated kW (1667 kg/kW in Table 1.21)
L = Land area serviced annually (farm size, ha)
P = Power rating of tractor (kW)
T = Annual hours of operation $= LW/P$
W = Tractor work required per hectare (205 rated kW·h in Table 1.21)
$*$ = An optimized level of the variable designated

Equation 1.63 shows that the cost per unit of tractor work has two components not affected by farm size, and has a third component inversely proportional to farm size. This last component, (AKC/LW), is responsible for machinery being of reduced financial feasibility for small-scale farmers using the farmer owner-operator mode of machinery management. This characteristic is illustrated in Table 1.21 in which it can be seen that even with optimum tractor size selection, small farms with small tractors have higher power inventories per hectare, are operated fewer hours annually, and thus incur higher interest charges per hectare-year, than is the case for large tractors on large farms. Furthermore, despite the higher power levels per hectare, the operator hours required per hectare are greater for small tractors on small farms than for large tractors on large farms. Both these above-mentioned factors cause the cash plus capital costs per unit tractor work done to be higher for small tractors on small farms than for large tractors on large farms. This effect is mitigated somewhat by the fact that per-hectare timeliness losses are less for small farms with small tractors than for large farms with large tractors. This advantage is not sufficient under typical economic conditions, however, to offset the economic disadvantages associated with small farm size and corresponding small tractor size.

After the matter of selecting the optimum sized tractor has been taken care of, there are three categories of methods for further improving the economic feasibility of tractor use on the small farms most commonly associated with two-wheel tractors:

1. Using equipment that offers low initial cost per unit of performance.
2. Multipurpose use of the tractor or tractor engine.
3. Multifarm use through custom-hire operations or group use.

The purchase of second-hand tractors is one route by which low-cost per unit-performance equipment can be acquired. This, however, may involve increased repair costs and reduced machine life. A second source of such equipment is the use of equipment designed to reduce cost by minimizing convenience and appearance features.

An example of such design is the Thai two-wheel tractor which initially had no differential gear or steering clutches and relied on long handles to allow the operator to effect turns. This tractor had only one speed forward and used an operator-controllable drive-belt tension idler pulley in the place of a clutch. Simple design permitted local manufacturers to further reduce cost. This type of design in 1983 had purchase costs of 1040 kg of unhusked rice per rated engine kW in Thailand and 1273 kg/kW in the Philippines. This was approximately 64 percent of the costs of similarly powered units from Japan sold in Thailand and the Philippines, and 49 percent of the cost of large four-wheel tractors sold in those countries (Table 1.22).

The economies of multipurpose use of a two-wheel tractor or its engine can be encouraged by making available a large range of attachment types to allow the tractor to serve a broad range of on-farm functions, and thus offset costs associated with accomplishing these functions by other means. This is particularly important when the tractor is used to replace draft animals. One of the attachments most frequently found to be economically advantageous is the trailer, which permits not only on-farm transport as well as farm-to-market transport, but which also allows use of the tractor for commercial road transport. The engine of a two-wheel tractor frequently carries a value greater than half

Table 1.22. Prices of tractors in selected Asian countries (kg unhusked rice/kW) in 1980

Country	Two-Wheel Tractors	Four-Wheel Tractors
Japan	374	303
Rep. of Korea	539	—
P. R. of China	701	564
Thailand	1051	2154
Philippines	1271	2590
Sri Lanka	1541	2316
Malaysia	1823	2391
Bangladesh	2356	—
Indonesia	2808	2635
Nepal	3444	1394
Burma	5318	—

Source: Adulavidhaya, Kampol and Bart Duff, 1983. The growth and impact of small farm mechanization in Asia. In: Workshop Papers: The Consequences of Small Rice Farm Mechanization in Thailand, Nov. 10–11, 1983, Bangkok, Thailand (available from the International Rice Research Institute, Manila, Philippines).

of that of the tractor. The engine, while mounted on the tractor, has use for powering stationary pumps and crop processing machines. In some cases the engine can be removed from the tractor and used to power boats and locally manufactured transport vehicles in the off-season.

Multifarm use of tractors is most commonly associated with large four-wheel tractors. However, the same advantages of reduced capital investment per hectare served also apply to two-wheel tractors. The generally lower speeds and the need to change field type wheel equipment for road transport of two-wheel tractors as compared with four-wheel tractors has resulted in multifarm use of two-wheel tractors being confined to small service areas. For this reason multifarm use of two-wheel tractors depends on means to secure a large portion of the area to be served within a short distance of the base at which the tractor is located. Systems for such management fall in the following categories:

1. Joint ownership.
2. Group farming.
3. Custom hire services in conjunction with a farmers' organization.

The above three management systems all require highly developed programs of agreements and understandings among all the persons involved if the systems are to be effective. There are costs associated with establishing, carrying out and maintaining these programs of agreement. However, the economic potential of multifarm use can result in costs per unit of tractor work which may be as low as half of those associated with the farmer owner-operator system of management.

A study in Malaysia of two-wheel power tillers (with attached rotary tillers) and four-wheel tractors using mainly rotary tillers for custom-hire service cultivation of irrigated rice land found the economic parameters given in Table 1.23.

Table 1.23. Economic parameters for two-wheel and four-wheel tractors performing rotary tillage of rice land

Parameter	Two-Wheel Tractors	Four-Wheel Tractors
Power (kW)	6.9	43.3
Initial price (kg)*	11,402	52,522
Tractor power price (kg/kW)	1,652	906
Annual hours of operation	395	1,042
Field work (h)	358	894
Travel time (h)	37	148
Field area served annually (ha)	24.7	180.5
Cost of tractor work (kg/ha)	221	210
Annual fixed costs/(price)	0.199	0.143
(Fuel + repair + wear) costs/energy delivered (kg/kW·h)	0.829	0.264
Operator wage (kg/h)	3.8	3.7
% downtime in working season	16.2	14.5
Annual repair costs/(price)	0.0703	0.0687
Energy to service 1ha (kW·h)	214.5	100
Diesel fuel use per hour (liters)	1.13	5.11
(Repair costs)/(fuel costs)	2.85	1.07

* Economic values are given in terms of kg of unhusked rice equivalent, which in Malaysia in 1969 reflected a price of $ (US) 92.70 per 1000 kg.

Source: Chancellor, W., 1970. Survey of tractor contractor operations in Thailand and Malaysia, agricultural Engineering Dept., University of California, Davis, California, USA. p. 179.

1.1.26b. Trends in the Development of Two-Wheel Tractors

During recent years two-wheel tractors have been more adapted to sulkies to reduce operator fatigue and increase tractor productivity with an increment in powers. This tendency occurs specially in Europe. New special applications have been developed for two-wheel tractors such as drilling holes, cutting wood and moving snow. Also there has been observed a tendency to substitute for gasoline engines, air-cooled diesel engines of similar speed, for better reduced consumption.

References

1. Lara-López, A., Chancellor, W. J., Kepner, R. A., and Kaminaka, M. S. 1982. A Two-Wheeled Tractor for Manufacture in México. TRANSACTIONS of the ASAE. 25(5): 1189–1194, 1203.
2. Lara-López, A. 1996. Field Evaluation of a High Clearance, Two-Wheeled Tractor Designed for Local Manufacture in Mexico. AMA. 27(1): 59–62.
3. Lara-López, A. 1980. Design and Development of a Two-Wheeled Tractor for Production by Small Scale. Manufacturers in Mexico. Doctorate Dissertation. University of California, Davis.
4. Lara-López, A. 1996. Mechanics of two-wheeled tractors. Proceedings of the 2nd meeting of the Mexican Society of Mechanical Engineers. Guanajuato, Mexico (Spanish).
5. ASAE Standard D497. Agricultural Machinery Management Data. ASAE Standards 1990. pp. 285–291.
6. Cervantes-Sánchez, J. and Lara-López, A. 1988. Simulation of the Performance of a Two-Wheeled Tractor with a Riding Sulky, Proceedings of the 14th Meeting of the National Academy of Engineering of Mexico (Spanish). pp. 171–175.
7. Chancellor, W. J. 1969. Selecting the Optimum Size Tractor. Transactions of the ASAE. 12(4): 411–414, 418.
8. Club of Bologna. 1991. Development and Present Situation of Agricultural Mechanization. The Case of 25 Countries. 3rd Meeting of the Club, XXII EIMA. Bologna-November. p. XV.1.
9. Colín-Venegas, J. and Lara-López, A. 1987. Adaptation of a Row Planter to a High-Clearance Two-Wheeled Tractor. Proceedings of the 13th Meeting of the National Academy of Engineering of Mexico. (Spanish), pp. 466–470.
10. Lozano-Lucero, G. P. and Lara-López, A. 1982. Row Cultivation with a High Clearance Two-Wheeled Tractor. Proceedings of the 8th Meeting of the National Academy of Engineering of Mexico (Spanish), pp. 107–111.
11. Valdivia-Hernández, R. and Lara-López, A. Sulky for a High Clearance Two-Wheeled Tractor. Proceedings of the X Meeting of National Academy of Engineering of Mexico (Spanish). pp. 165–169.

Tractors: Two Axle Tractors

K. T. Renius

1.1.27. History, Trends, Concepts

Tractor: Definition, History and Trends

The view of tractor history depends on the definition of the *agricultural tractor.*

The agricultural tractor is a vehicle for offroad and onroad operation, being able to carry, guide, pull and drive implements or machines – moving or stationary – and to pull trailers (see also ISO 3339).

Using this definition, the famous steam vehicles of Fowler of Great Britain, developed around 1860 and pulling plows by cable systems, may be regarded as forerunners. (Table 1.24). The first tractors could be seen in the *steam traction engines*, of the USA around 1880 in small quantities. Later on, steam engines were replaced by internal combustion engines (1890–1910). At first, costs of those tractors were very high – one reason for the German company Stock to develop in 1907 a self-propelled plow, which could also pull trailers and drive threshers. This principle was successful in Europe until about 1925. The 1920s were clearly dominated by the famous *Fordson*, USA (1917), which achieved a worldwide market share above 50% with only one model for several years. At that time, the tractor was used mainly for plowing and threshing. Diesel engines, pneumatic tires and the PTO introduction became the highlights of the 1930s. After World War II, remarkable technical development was typical for the industrialized countries, with an increasing number of tractor functions. The rear three-point hitch, with at least draft control, was generally introduced and, in many countries, the mean rated engine power of tractors rose dramatically. In some areas such as Europe and Japan, the additional front drive for standard tractors changed its role from optional to standard equipment. The same happened for quiet cabs.

The most recent trend is characterized by an increasing role of information technology (electronics) in connection with improved power-train concepts (power shifts, continuously variable drives), highly sophisticated hydraulics (load sensing, proportional valves), and a continually increasing level of driver comfort [13].

Manufacturing systems had by the 1990s changed dramatically from mass production of a few models to production of individual tractors on customer demand.

Tractor Concepts

A simplified "family tree" is shown in Fig. 1.94. Track laying tractors have had, and still have a very limited importance because of their high initial costs, limited speeds and tendency to damage crops when turning. Their main advantage can be seen in a considerably better tractive efficiency and higher pullweight ratios compared to wheel tractors, while the often stressed advantages of reduced soil compaction generally are not as important due to the footprint effect of the conducting road wheels of the chain or the band (peak contact pressure can be double nominal).

Regarding wheel tractors, the *walking tractor* is often the typical first mechanization step – as happened in several far eastern countries, such as Japan, China and Korea.

Table 1.24. Milestones of tractor history [1–12]

1860	Predecessor: Steam cable plowing machines, Fowler, UK. Promoted by Max Eyth (1836–1906), Germany
1878	Steam traction engine, Case, USA
1890	Steam crawler tractor (B. Holt), USA
1889	First gasoline tractors, USA ("at least one company" [7])
1892	Froelich gasoline tractor, USA
1904	"Ivel" tricycle tractor, UK
1904	"Caterpillar" crawler tractor (B. Holt), USA
1908	Self propelled plow (gasoline), R. Stock, Germany
1914	"Waterloo Boy" (later J. Deere), USA
1917	"Fordson," block chassis, first mass production, most successful tractor ever built, USA
1918	Power take off (PTO), International Harvester (IH), USA
1920	Nebraska-Test No. 1 (Waterloo Boy), USA
1921	"Lanz-Bulldog," 1 cyl./2-cycle, Germany
1922	First tractor with Diesel engine, Benz-Sendling, Germany
1925	Patent application of H. Ferguson, No. 253 566 (issued 1926): 3-point hitch, draft control, UK
1927	First PTO standard, ASAE, USA
1932	Low pressure pneumatic tires for tractors by Goodyear/Allis Chalmers, USA (Nebraska-Test 1934)
1938	Direct injection (DI) Diesel engines for tractors, system MAN, Germany
1948	"Unimog" (25 PS, 50 km/h) by Boehringer. Since 1951 Daimler-Benz, Germany
1948	Air cooled Diesel engines, Eicher, Germany (1950 all Deutz tractors air cooled)
1950	Front end loader. Ferguson, Hanomag, Baas, Germany
1950	Fluid coupling for tractor gearbox by Voith for Allgaier-Porsche, 1968 for Fendt, Germany
1954	Hydrostatic NIAE research tractor, Silsoe UK
1954	"Torque Amplifier" (Hi–Lo power shift), IH, USA
1954	Power steering for large tractors, J. Deere, USA
1959	Kubota and Shibaura produce the first small Japanese two axle-tractors
1959	Pendulum test for new tractors, Sweden
1960	Constant pressure hydraulics with variable displacement pump, J. Deere, USA
1963	Full "Power Shift" (8 forw./4 rev.), J. Deere, USA
1965	"Center input" for driven front axles, Same, Italy
1972	First quiet cab Nebraska-tested, J. Deere, USA
1973	Load-sensing hydraulics, Allis Chalmers, USA
1973	Deutz "Intrac" and Daimler Benz "MB-trac," Germany
1978	Electronic 3-point hitch control, Bosch/Deutz, Germany (electronic force sensor 1982)
1980	40 km/h standard tractors, Fendt and Schlüter, Germany
1982	50 deg inner steering angle and integrated steering cylinder for driven front axles, ZF, Germany and Deutz/Sige, Germany/Italy
1986	Front axle overdrive for sharp turns of standard tractors. Kubota "Double Speed Turn," Japan
1987	Break through of load-sensing hydraulics, beginning by Case-IH "Magnum," USA
1987	Passive soft cab suspension, Renault, France
1988	"Munich Research Tractor" (frame, suspended engine, flat bonnet, CVT, very low bystander noise level)
1991	Flat bonnet tractors by Deutz, Germany
1992	Frame chassis tractors with suspended engine by J. Deere, Germany and USA
1993	50 km/h top speed for standard tractors and hydropneumatic front axle suspension, Fendt, Germany
1996	Continuously variable hydrostatic power split gearbox for standard tractors, Fendt, Germany

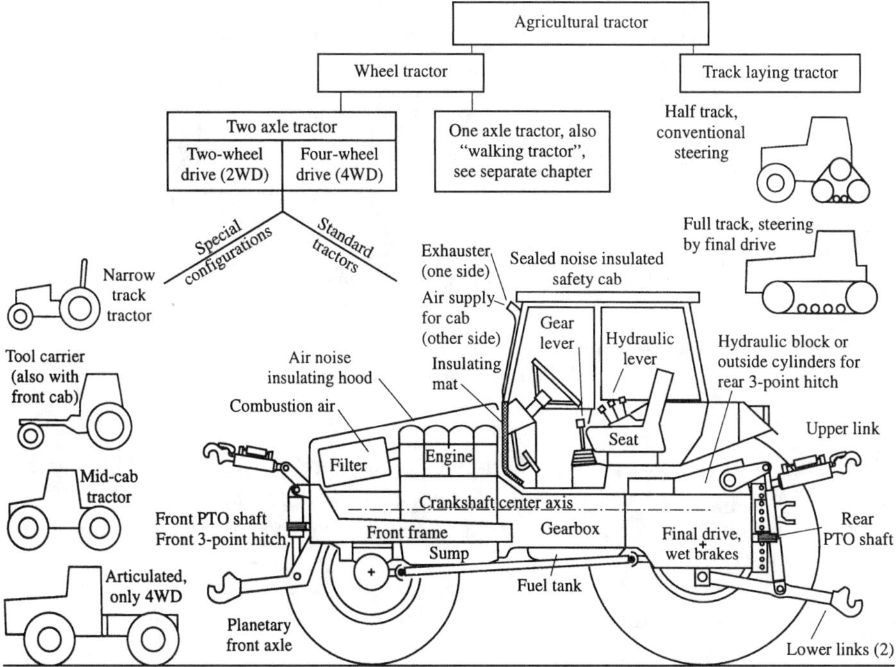

Figure 1.94. Tractor concepts.

Initial costs (per kW) are much lower than for two-axle tractors. Comfort, however, is also lower due to the high forces required by the operator and due to the walking mode.

In a worldwide view, two-axle tractors clearly are of greatest economic importance compared to the others. Four-wheel drive can be superior to two-wheel drive, mainly for soft soil conditions and for front end loader applications. It may be noted that the term *four-wheel drive* is used in most countries to express a concept with four driven wheels. In some countries, such as the USA, the definition has been restricted for a long time to equal diameter tires, addressing mainly the large articulated *four-wheeler*, which often uses unit size *dual tires*.

Two-axle tractors can be designed in many different basic concepts. The special configurations represent, however, only a limited market penetration, while the so called *standard tractor* can be regarded as the most important concept. Figure 1.94 shows a typical model standard tractor, popular in high developed countries in the late 1990s. A four-cylinder turbocharged diesel engine with direct injection delivers the power (i.e., 90 kW) for the drive train, the power take off (PTO), and the auxiliary consumers, such as hydraulic pumps, air compressors, air conditioning compressors, electric generators, and others. Four-wheel drive is the dominating version in Central Europe (the same in Japan, but the mean rated engine power there is much lower). Other modifications relate to the main objectives of application (i.e., more for row crops or more for tillage – more for upland or more for paddy fields).

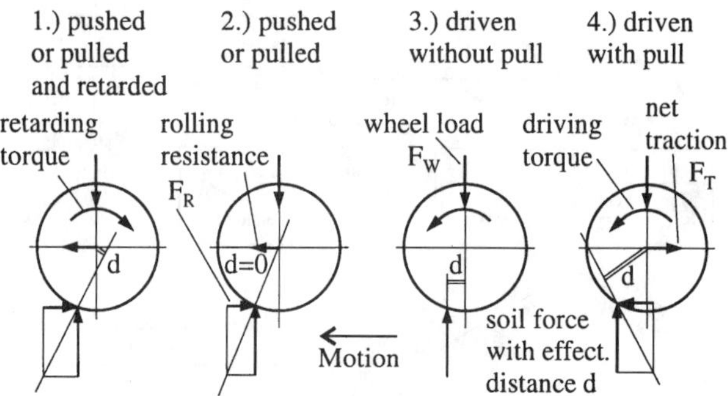

Figure 1.95. Four cases of wheel-soil interaction, rigid wheel (simplified) with forces and torques.

The number of required tractors for a given agriculturally used area (AUA) can be calculated from the required density of rated engine power per ha of AUA.

Guideline: Small farms >3 kW/ha, mid farms (around 100 ha) 2 kW/ha, big farms ≤1.5 kW/ha [13].

1.1.28. Tractor Mechanics

Single-Wheel and Tire Mechanics

Four typical cases can be defined, as shown in Fig. 1.95 [14].

1.) Represents a wheel with a retarding torque generated by a brake or by the engine drag.
2.) Represents the front wheel of a 2WD tractor.
3.) Represents the wheel of a one-axle tractor or of a 4WD tractor driving without any pull.
4.) Represents the standard configuration of a pulling wheel creating traction.

Tire *slip* is a result of soil deformations (including shear effects) and tire deformations [14]. Slip can be positive (*travel reduction*) or negative.

Definition:

$$\text{slip i} = \frac{\text{travel without slip minus true travel}}{\text{travel without slip}} \tag{1.64}$$

Unfortunately, there is a basic problem with the definition of *zero slip*, which is a question of the effective radius [14]. The author recommends the following procedure: Measure the travel distances for a given number of wheel revolutions for the cases 2.) and 3.) of Fig. 1.95, and calculate from the mean distance the effective tire radius. A simplified method uses only case 3.), which can, however, only be recommended for firm soils (the use on concrete can overrepresent lug effects).

Due to Fig. 1.96, *gross traction* = *net traction* plus *rolling resistance*:

$$F_U = F_T + F_R \tag{1.65}$$

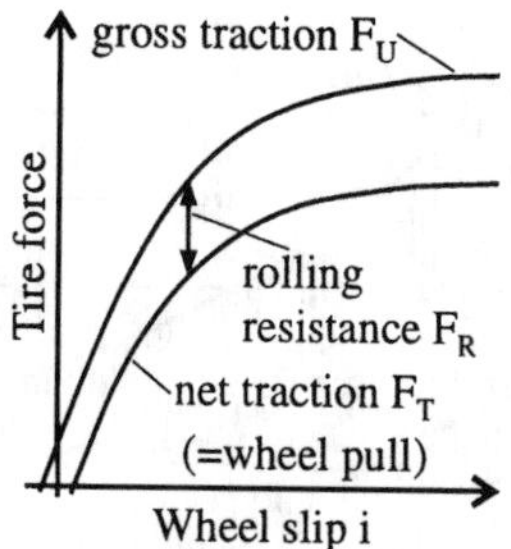

Figure 1.96. Gross traction, net traction and rolling resistance for tires.

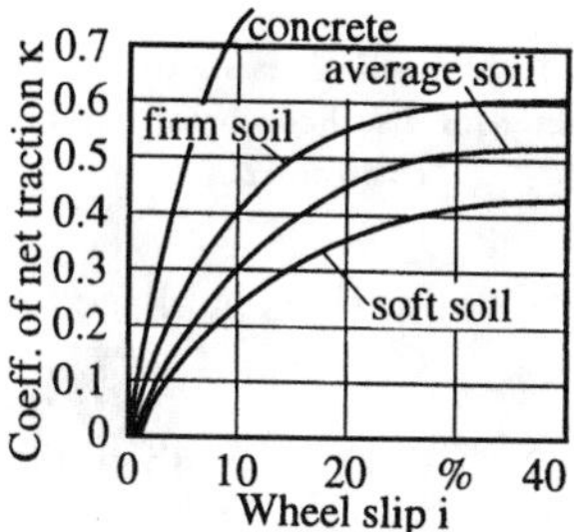

Figure 1.97. Net traction of general purpose tractor tires, diameters >1 m, radials.

The gross traction is a force that cannot be measured directly but can be calculated from the driving torque and the effective radius (circumferential wheel force). Handling of tire mechanics can be simplified by dimensionless coefficients, using the tire load F_W:

$$\text{Net traction coefficient } \kappa = F_T/F_W \tag{1.66}$$

A rough approach for κ is presented by Fig. 1.97. Mathematical models often use the expression $a \cdot (1 - e^{-b \cdot i})$, where e is the base of nat. log.; a and b are functions of the soil conditions, the tire specifications and the inflation pressure [15, 16].

Empirical tire performance data can be taken from [17]. Soil conditions can be evaluated by measuring the penetration forces of standardized *cone penetrometers*, [18, 19].

Rolling resistance also can be expressed dimensionless:

$$\text{Rolling resistance coefficient } \rho = F_R/F_W \tag{1.67}$$

Rough values for typical slips and tire diameters >1 m:

0.02	concrete, high inflation pressure
0.06	firm soil, uncultivated
0.10	average soil
0.1–0.3	soft soil, cultivated

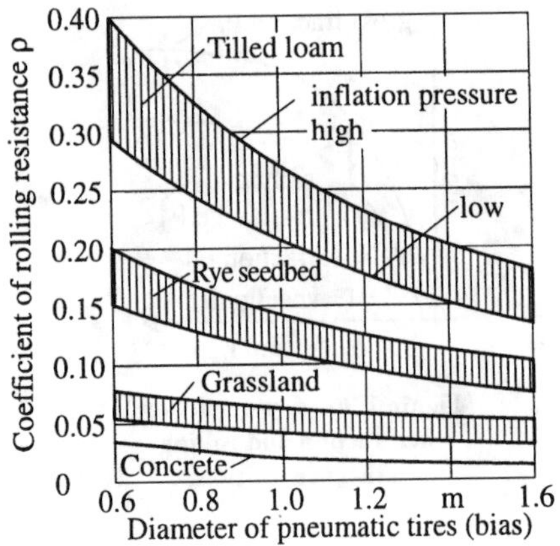

Figure 1.98. Rolling resistance versus tire diameter for 4 soil conditions and zero traction [20].

For average soil and diameters >1 m, various published measurements can be summarized by the rough model:

$$\rho \approx 0.070 + 0.002 \cdot \text{i (with slip i in \%)} \tag{1.68}$$

Influence of tire diameter is shown by Fig. 1.98.

Traction efficiency of a driven wheel is

$$\eta_T = \frac{\text{net traction power}}{\text{input power}} = \frac{\kappa}{\kappa + \rho}(1 - \text{i}) \tag{1.69}$$

where the term $\kappa/(\kappa + \rho)$ considers the rolling resistance and (1 – i) the slip. The author sees no preference which one of these two terms should be taken first to calculate losses – two definitions of losses seem to be possible [21].

The maximum of η_T typically is between 0.6 (soft soil) and 0.8 (firm soil) at slip values of about 15% and 8%. Onroad efficiencies peaker at 0.90 to 0.94 at slip values of about 6%–4% (see experience from *Nebraska Tests* [9, 22] and from *OECD Tests*).

Soil Compaction under Tires

Optimal soil structure can be regarded as an important factor of plant production. Two conditions are important:

Optimal porosity

Optimal size distribution of the pores

Definition of *porosity* n:

$$\text{n} = \frac{\text{pore volume (gas + water)}}{\text{total volume}} \tag{1.70}$$

Optimal values figure between about 42% (light soils) and 48% (heavy soils) in connection with adequate pore size distributions. Soil compaction can effect lower values.

Table 1.25. Soil compaction

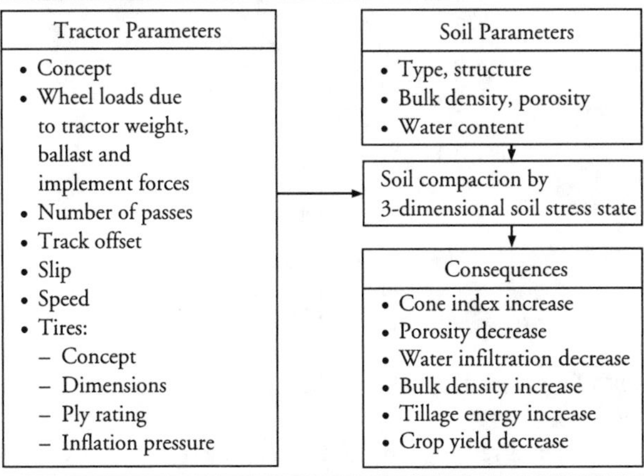

Tractor Parameters
• Concept
• Wheel loads due to tractor weight, ballast and implement forces
• Number of passes
• Track offset
• Slip
• Speed
• Tires:
– Concept
– Dimensions
– Ply rating
– Inflation pressure
Soil Parameters
• Type, structure
• Bulk density, porosity
• Water content
Soil compaction by 3-dimensional soil stress state
Consequences
• Cone index increase
• Porosity decrease
• Water infiltration decrease
• Bulk density increase
• Tillage energy increase
• Crop yield decrease

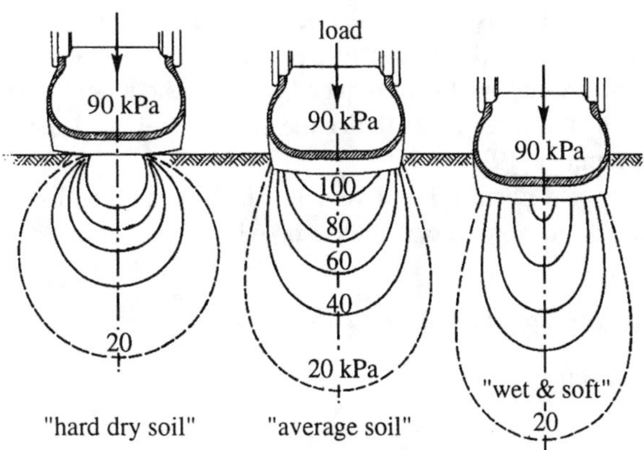

Figure 1.99. "Söhne's pressure bulbs" (calculated main normal stress) under a tractor tire for different soil conditions. Tire size 13.6–28, load 10 kN [24].

Tractor and soil parameters result in a three-dimensional soil stress state which induces compaction with some typical consequences, Table 1.25.

Models of three-dimensional stress distributions in soils under tires considering the elastic-plastic deformations were presented first by Söhne [23, 24], who developed his *pressure bulbs* on the basis of earlier models of Boussinesq and Fröhlich (foundations of buildings), Fig. 1.99.

The following general rules for controlled soil compaction can be based on these fundamentals:

1. Compaction in the upper layers is mainly a function of the tire contact pressure, while compaction in the lower layers is mainly a function of the tire load.

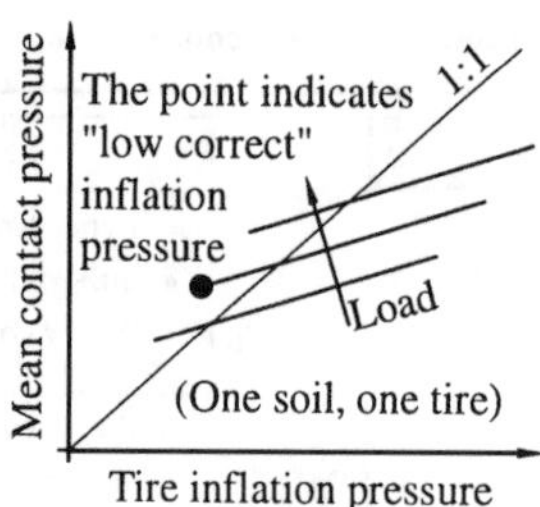

Figure 1.100. Mean contact pressure versus inflation pressure [27].

2. Compaction by a given stress state is in most practical cases higher for wet soils than for dry soils.
3. Low contact pressures require large tires in connection with low inflation pressures. In the case of *low correct inflation* [25] (lowest value to carry the load), mean tire-soil contact pressure is typically little above inflation pressure, Fig. 1.100. The difference is mainly related to the carcass stiffness. The soil strength can be evaluated by cone penetrometer measurements [18]. Lugs can create extra compaction [26].

A replacement of tires by tracks for reduced soil compaction often is less successful than expected due to the peak loads under the road wheels and the multipass effect according to their number [19].

Tire dimensions for a given offroad vehicle can be evaluated by relating the load to the vertical tire projection (tire diameter d, width b) as proposed by several researchers.

$$\textit{Contact pressure index}\ [8] = \frac{\text{tire load}}{\text{bd}} \tag{1.71}$$

Recommended values for typical traction tires [29]:

$$\leq 25 \text{ kPa (soft soils)}, \leq 35 \text{ kPa (standard)}.$$

In many countries, agricultural tractors also must offer good onroad performance. A wide band of inflation pressures should be available to meet both the demands offroad and onroad use, Table 1.26. Inflation pressure remote control can help to make best use of modern tire concepts which offer a large bandwidth of specified inflation pressures [28].

Mechanics of Two-Axle Tractors Pulling Implements or Trailers

Two-Wheel Drive

The standard case can be seen in a rear-wheel drive tractor, pulling an implement or trailer on a horizontal terrain without acceleration, Fig. 1.101.

The tractor is regarded as a plane, rigid body with summarized front (f) and rear (r) tire forces. Tire mechanics of the front axle see case 2.) in Fig. 1.95, page 118 and those of the rear axle see case 4.) respectively.

The term $\mathrm{Fp} \cdot \cos\alpha$ can be called *productive implement pull*, as this force has the same direction as the implement speed v, thus creating the implement power or *drawbar*

Table 1.26. Favorable tire inflation pressures [21]

	Inflation Pressure	
x Means: Favorable	High	Low
• Initial costs per unit load capacity	x	
• Required space per unit load cpacity	x	
• Tire weight per unit load capacity	x	
• Rolling resistance on the road	x	
• Soil contact pressure		x
• Offroad performance		x
• Riding comfort		x

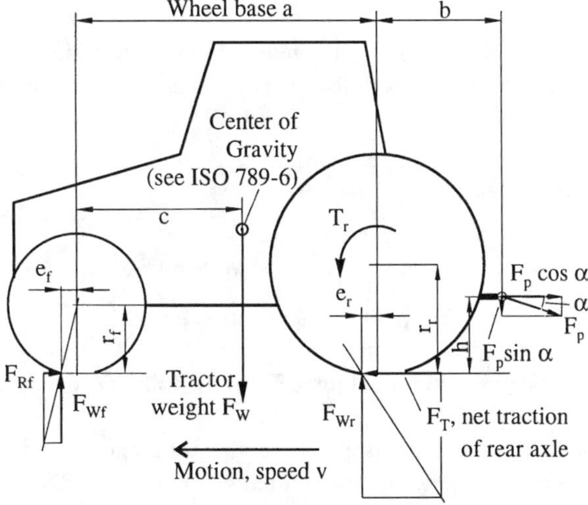

Figure 1.101. Forces on a pulling rear-wheel drive tractor on horizontal terrain at constant speed.

power P_{net}:

$$P_{net} = F_p \cdot \cos\alpha \cdot v \tag{1.72}$$

Horizontal equilibrium is easy to express

$$F_p \cdot \cos\alpha = F_T - F_{Rf} \tag{1.73}$$

Calculation of F_T and F_{Rf} needs, however the vertical axle loads F_{Wf} and F_{Wr}, as

$$F_T = \kappa_r \cdot F_{Wr} \quad \text{and} \quad F_{Rf} = \rho_f \cdot F_{Wf}. \tag{1.74}$$

κ and ρ see page 119. It may be noted that $\rho = e/r$

$$F_{Wr} = \frac{F_W(c + e_f) + F_p \cdot h \cdot \cos\alpha + F_p(a + b + e_f)\sin\alpha}{a + e_f - e_r} \tag{1.75}$$

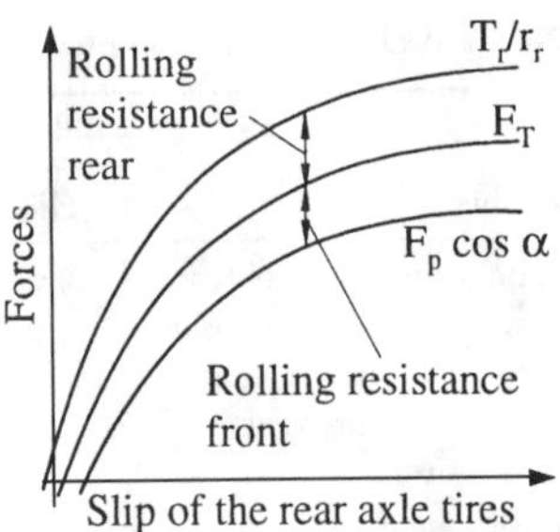

Figure 1.102. Influence of rolling resistance on productive implement pull [29].

It is important that $F_p \cdot \sin\alpha$ increases the rear axle load, thus increasing also the rear axle net traction (this was one reason for the famous draft control system, invented in the 1920s by Harry Ferguson).

$$F_{Wf} = F_W + F_p \cdot \sin\alpha - F_{Wr} \tag{1.76}$$

Rear axle torque

$$T_r = F_T \cdot r_r + F_{Wr} \cdot e_r = (\kappa_r + \rho_r)\, F_{Wr} \cdot r_r \tag{1.77}$$

The symbols r_f and r_r indicate the points of application for the horizontal tire forces (their vertical distance from the tire centers).

If we use the virtual circumferential force T_r/r_r, the rolling resistance can be demonstrated by Fig. 1.102.

The diagram indicates that there is a considerable difference between the productive implement pull $F_p \cdot \cos\alpha$ and the circumferential rear axle force T_r/r_r. A certain portion of the precious rear axle net traction is needed for pushing the front axle.

Four-Wheel Drive

In the case of four-wheel drive (other conditions the same), productive implement pull is equal to the sum of front and rear net traction forces:

$$F_p \cdot \cos\alpha = \kappa_f \cdot F_{Wf} + \kappa_r \cdot F_{Wr} \tag{1.77a}$$

The so-called *multipass effect* results in

$$\kappa_r > \kappa_f \quad \text{and} \quad \rho_f > \rho_r. \tag{1.78}$$

Measurements carried out by Holm [30] have been used to derive the following rough model [29]:

$$\kappa_f + \rho_f \approx \kappa_r + \rho_r \tag{1.79}$$

Example for "average" field conditions, wheel diameters >1 m, 15% slip and good balanced tire loads:

$$0.39 + 0.10 = 0.43 + 0.06$$

Good balanced tire loads means that the ratio between the actual tire load and the nominal tire load (for actual speed) is about the same for all tires, and that this ratio is not too much below 1.

The rule $\kappa_f + \rho_f \approx \kappa_r + \rho_r$ simplifies the calculation of axle torques:

$$\frac{\text{axle torque front}}{\text{axle torque rear}} \approx \frac{F_{Wf} \cdot r_f}{F_{Wr} \cdot r_r} \tag{1.80}$$

Full drawbar power requires a certain minimum speed due to traction limits. In a high percentage of practical offroad operations, this minimum speed can be found between 4 km/h and 8 km/h for common tractor concepts, common weight-to-power ratios and common implement (trailer) use. The larger the tractor, the higher this speed, due to decreasing weight-to-power ratios. The opposite trend can be observed if soil conditions change from soft/wet to firm/dry.

After a "run-in" procedure, a radial tire can produce on a cleaned concrete surface $(\kappa + \rho)$ values around 1.0.

Typical minimum speeds on concrete figure therefore between 3 km/h and 5 km/h, influenced again by the weight-to-power ratio but also by the ballast mode (high ballast leads to low minimum speeds. See Nebraska and OECD Test results). If full power is applied at speeds below 3.5 km/h (on concrete) for a longer time (days), the transmission can be damaged as it is overloaded compared to common field operations.

Mechanics of Hillside Operation and Overturning Stability

Critical Slopes

Slopes can limit the use of tractors and other agricultural machinery. The main criteria for the maximum possible slopes are:

- Sufficient contact between the tires and the surface (side forces, drift angles, reduced traction)
- Overturning (static and dynamic, all directions)

Maximum slopes for dynamic tractor operation are much lower than for static conditions. Combinations with implements or trailers can reduce them again. Therefore, *overturning* is one of the most frequent and important accidents.

Figure 1.103 demonstrates the critical slopes for all the travel directions of a typical European standard tractor without implements.

The slope limits can be regarded to be typical for all similar standard tractor models within a range of about 2000 kg to 5000 kg net weight. The most critical slope angles can be found for driving 90° uphill and along a *contour line* (0° and 180°), while all downhill directions are less critical; sliding will often be more dangerous in these cases than overturning.

Implements and trailers can considerably reduce the described limits, dramatically in the case of front end loading and pulling a trailer uphill.

Calculation of the critical slope for 90° uphill is easy, as the equation of axle load distribution only has to be directed to the particular case of front axle load = zero. Nevertheless, this type of overturning is very dangerous, as the driver has no chance of a successful reaction.

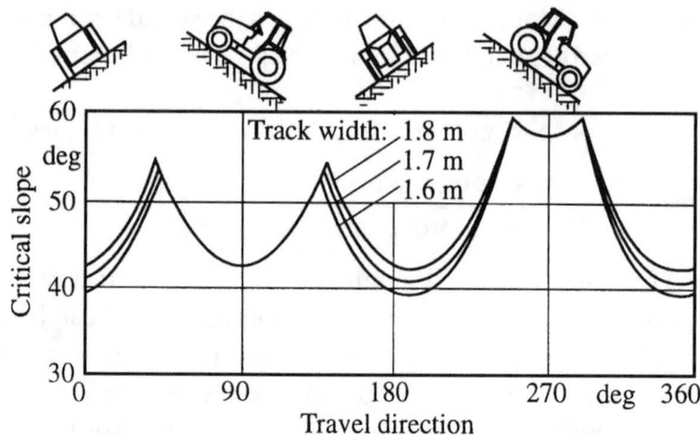

Figure 1.103. Critical slopes for static overturning without implements. Tractor Deutz Fahr DX 4.70 (66 kW), 1987. (Applied software of H. Schwanghart [31]).

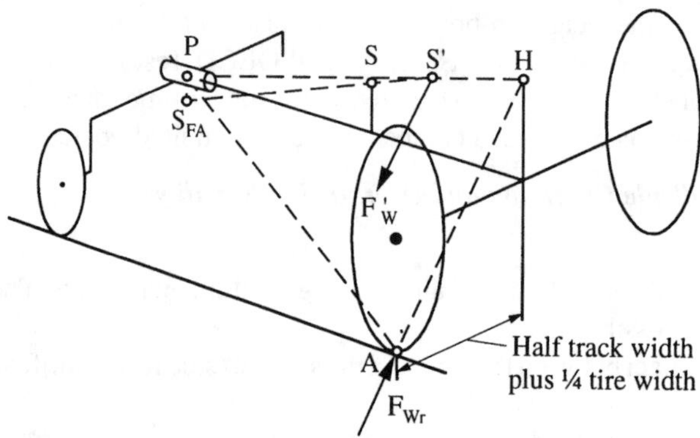

Figure 1.104. Critical state for static sideways overturning [29].

Mechanics of Sideways Overturning

A calculation of the critical slope for the standard case of sideways overturning (without implements), Fig. 1.104, has to consider the pivot position P, the gravity center of the front axle S_{FA} and that of the tractor S, in addition to common tractor specifications.

In the first stage of sideways overturning, the front axle remains in contact with the surface. Therefore, a third gravity center S′ has to be calculated (tractor minus front axle) using S and S_{FA} with the connected masses. A straight line through P and S′ delivers the auxiliary point H. Instability is on hand if the force directions of F'_W (tractor weight minus front axle weight) or of the rear axle load F_{Wr} are within the plane AHP. A first approximation (used for Fig. 1.103 above) uses the simplification of a chassis with rigid front axle, taking S instead of S′ and F_W instead of F'_W.

Static slope limits are, however, not valid for most practical tractor operations, as dynamic effects such as vertical acceleration due to uneven surfaces reduce the slope limits considerably. As a rough approximation, the *slope limits* for dynamic operation should be 50%–60% lower than the static values (smaller tractors more critical than bigger ones due to suspension effects of the tires). This means that a hilly field should not have slopes above about 20° (36%), if a standard tractor is chosen for mechanization. For this reason all lubrication systems and hydraulic circuits of a standard tractor should be able to handle a permanent tractor tilt angle of 20°. In the case of slopes above 20° (i.e., grass land), special tractors have been developed with higher track width and lower position of gravity center mainly due to the use of low section height tires (small diameter, large width). These vehicles can handle slopes up to 25°–30° but their operation requires four-wheel steering to compensate the high drift angles of the tires (side forces) when driving along a contour line.

In some countries, health and safety authorities require that agricultural tractors must be able to stop the overturning rotation after the first contact of the safety structure with the terrain surface on a slope of 40%. The problem of this rule is that a tractor with a good general stability against overturning tends to fail in this test.

1.1.29. Chassis Design

Traction Tires: Requirements, Design, Specifications

The first replacements of rigid steel wheels by pneumatic tires took place for onroad operation in Europe in the late 1920s. The first commercial traction tires for offroad operation were introduced 1932 in the USA by Goodyear/Allis Chalmers and in 1934 in Germany by Continental/Lanz. The first Nebraska Test of a tractor using pneumatic tires was the No. 223 (Allis Chalmers WC tractor, 1934). The change to pneumatic tires was one of the most effective tractor innovations, and thus introduced within a few years of their development.

The main duties of a traction tire for an agricultural tractor are listed in Table 1.27 [29, 32].

These requirements should be met in connection with adequate initial costs.

The maximum nominal vertical tire loads of a tractor have an important influence on its *payload*, as the payload is usually equal to the sum of nominal tire loads minus tractor net weight. High payloads have become popular as they enable heavy mounted implements, large ballast weights and productive containers for seeds and chemicals.

Table 1.27. Demands on traction tires for tractors

- Carrying loads
  - Vertical: weight, ballast and transferred implement forces
  - Longitudinal: pull and braking
  - Lateral: side forces due to slopes or centrifugal forces
- Self-cleaning (to remove the soil between the lugs)
- Low soil compaction and surface damage
- Elasticity and damping (to reduce vehicle vibrations)
- To serve as fluid ballast container
- Low wear, long life, sensitivity against sharp obstacles

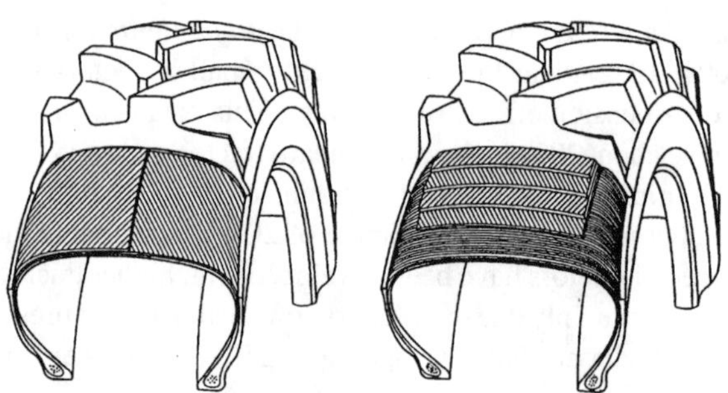

Figure 1.105. Bias ply and radial ply tires, [33].

A *power index* has been introduced to evaluate the dimensions of powered tires: Engine power is related to the sum of the projection areas (width b, diameter d) of all powered tires $\Sigma(b \times d)$ [29]:

$$\frac{\text{Rate eng. power}}{\Sigma(\text{b} \times \text{d})\text{powered}} \leq \frac{1}{4} \cdots \frac{1}{\underline{3}} \cdots \frac{1}{2} \frac{\text{kW}}{\text{dm}^2} \tag{1.81}$$

High values are typical for high-speed cultivation (tractors with low weight-to-power ratios) or for high importance of PTO operations (orchard tractors, vineyard tractors, implement carrier).

Two basic design concepts have been developed for tires: *bias ply* and *radial ply tires*, Fig. 1.105.

Radial ply tires are superior to bias in many criteria (Table 1.27) with however little lower damping and higher sensitivity against sharp obstacles sideways. Typical for radials is a little larger contact area and a better internal torque transfer due to the belt concept in connection with thin and very flexible side walls. Traction coefficients and traction efficiencies are typically little higher for radials than for bias plies for the same slip [34], but a good bias ply tire at least can achieve the performance of a poor designed radial tire [35].

The most important factor for dimensioning tires for tractors (and for most other vehicles) is the nominal vertical load, which is standardized along with standardized dimensions, Table 1.28.

Table 1.28. ISO standards for tractor tires

ISO 4223-1	(1989, AMD 1992): Definitions for pneum. tires
ISO/DIS 4251-1	(1994): Tires (PR rated) and rims: designation, dimensions
ISO/DIS 4251-2	(1994): Load ratings for bias
ISO 4251-3	(1994): Rims
ISO 4251-4	(1992): Tire classification code
ISO/DIS 7867-1	(1994): Tires & rims: combinations
ISO/DIS 7867-2	(1994): Tires & rims: loads, speeds
ISO 8664	(1992): Radials, load-index, speed symbol

Due to ISO, a bias ply tire 16.9 – 38 8 PR ("diagonal construction") is specified as follows (example):

16.9 = nominal tire width (≈ width in inch)
38 = nominal rim diameter (diam. in inch)
8 PR = *Ply rating* (indicates strength of carcass)

Basic tire loads (BTL) for reference inflation pressures are listed in ISO standards for 30 km/h reference speed.

Lower maximum speeds allow higher nominal loads, for bias (ISO/DIS 4251-2): 25 km/h (+7%), 20 km/h (+20%), 10 km/h (+40% with 30 kPa increased inflation pressure).

According to ISO, a radial ply tire is specified as follows (example, same tire size as above):

16.9 R 38 141 A6

R = radial, 141 = *load index* (2575 kg at 160 kPa), A6 = *speed symbol* (30 km/h reference speed, A8 would indicate 40 km/h reference speed).

Low section height tires are specified by the ratio of section width and height, for example 70%:

520/70 R 38 145 A6

These tires offer, at about the same diameter, greater width (and thus more contact area) than the standard versions. Low section height tires have become popular as an increase of width has only minor influences on the vehicle speeds and transmission torques.

Chassis Concepts, Four-wheel Drive

Chassis Configurations

The following aspects have to be considered:

- Standard tractor or special concept
- Four-wheel drive or two-wheel drive
- Steering concept
- Block or frame chassis

The well-known *block chassis design* uses the engine and transmission housing instead of a frame, Fig. 1.106.

It was first applied for mass tractor production by Henry Ford with his famous Fordson in 1917. The block principle became the most important rule for standard tractor design, saving initial costs and allowing the drive line elements to be perfectly encapsulated and well-lubricated. There were some exceptions where frames were used, for instance, small Japanese tractors and German garden tractors or special vehicles as the Unimog and MB-Trac. Many tractors used half frame configurations to reinforce chassis durability around the engine and simplify front attachments.

German research works carried out at the Technische Universität München [36, 37] finally resulted in the *Munich Research Tractor* (1988) and pointed out some important

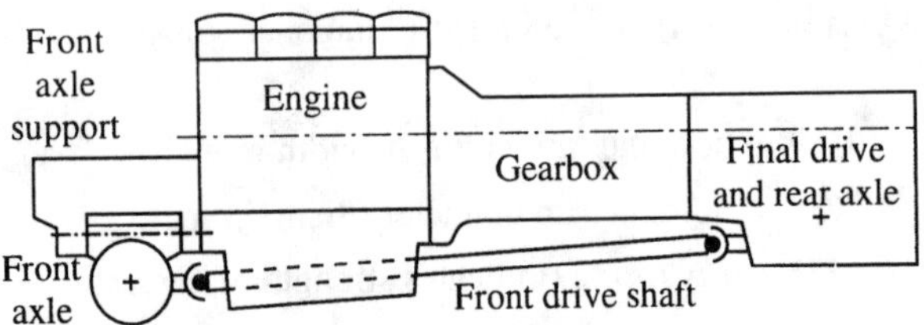

Figure 1.106. Block chassis concept for a standard tractor.

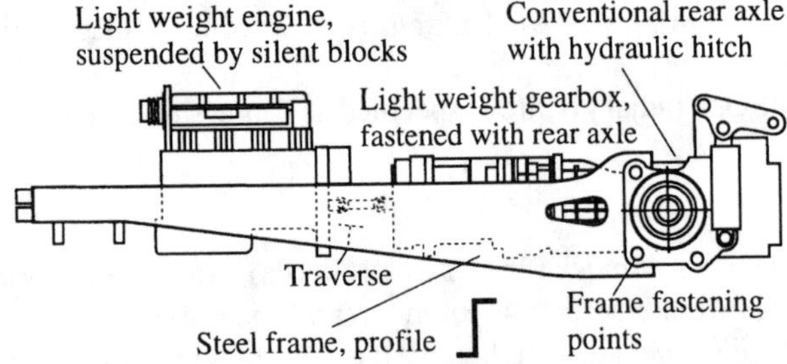

Figure 1.107. Frame and driveline configuration of the new J. Deere tractor line "6000" (1992).

arguments in favor of the *frame chassis* in comparison to the block chassis:

- Higher potential for bystander noise reductions
- Improved basis for front-end loader and front hitch
- Higher component flexibility
- Simplified maintenance and repair

In late 1992, John Deere presented newly designed tractor lines ("6000" & "7000") using frames [13], Fig. 1.107.

A change from block to frame requires high investment costs, as almost every component needs a new design.

Four-Wheel Drive versus Two-Wheel Drive

After many historical attempts [38], a breakthrough came relatively late starting in Western Europe in the 1970s [38], resulting in a considerable change of the standard tractor running gear in the 1980s as an option became the preferred concept. The same happened in Japan. Four-wheel drive offers advantages for both the designer and the farmer. Advantages for the farmer [38–42]:

- Increased pull and drawbar power. The softer the soil, the higher the benefits
- Improved tractive efficiency offroad, typically 15% resulting in fuel savings of the same level
- Improved productivity for front-end loading
- Improved maneuverability on slopes and under other difficult conditions, also in the forest
- Improved braking performance

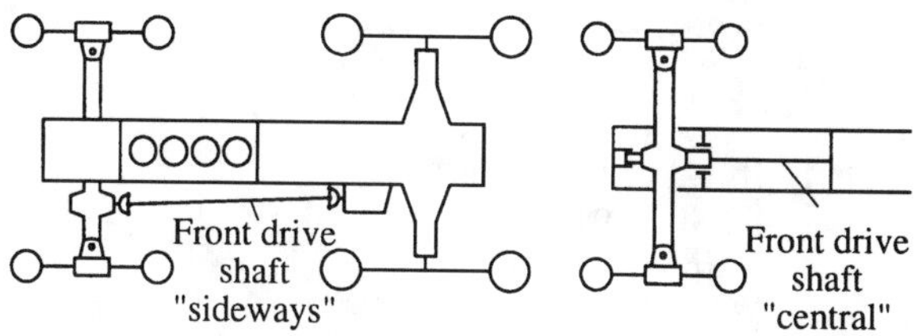

Figure 1.108. Basic front axle drive shaft concepts [21].

The first listed benefit has often been underevaluated, when the additional pull force was derived directly from the additional vertical axle load. The true gain in traction is considerably higher due to the multipass effect and the direct compensation of the front axle rolling resistance [40]. Typical example: 20% additional vertical load results in 30%–35% more total pull.

Advantages for the designer:

- More freedom for adequate tire volumes related to tractor size, power and type of mission
- Possibility of a low cost front axle brake on the front drive shaft (important for higher speeds)
- Reduced torque loads at the rear axle

The disadvantages concern mainly the additional initial costs of about 16%–20% [41, 42] and the little higher losses for onroad operation at higher speeds.

Two drive line concepts are used for driven front axles of standard tractors, Fig. 1.108. In each case, the power typically is diverted from the gearbox output shaft, controlled by a wet clutch. This leads to a fixed ratio between rear and front axle speed. The *sideways shaft concept* can be preferred, if the four-wheel drive has the status of an option. The *central shaft concept* is cheaper and offers more space for other components such as the fuel tank. Maximum possible steering angles are usually not influenced (except in specific cases). An offset in height at the *central* concept (Fig. 1.108) needs a universal joint but favors the maximum steering angle.

Brakes

Requirements

Brakes are necessary to retard and to fix the tractor for onroad driving and for hillside operations including trailers [43]. Onroad performance is in the foreground; single wheel braking, however, is required as a typical offroad specification.

There are usually three groups of functions for the tractor brakes:

- Service braking (main function)
- Emergency braking (if the service brakes fail)
- Park braking (only for stop position)

Emergency and park braking are often combined and typically operated by a hand lever.

Tractor retardation can be demonstrated by the braking distance for a given initial vehicle speed, Fig. 1.109. Brakes at the rear axle have low initial costs, and are usually

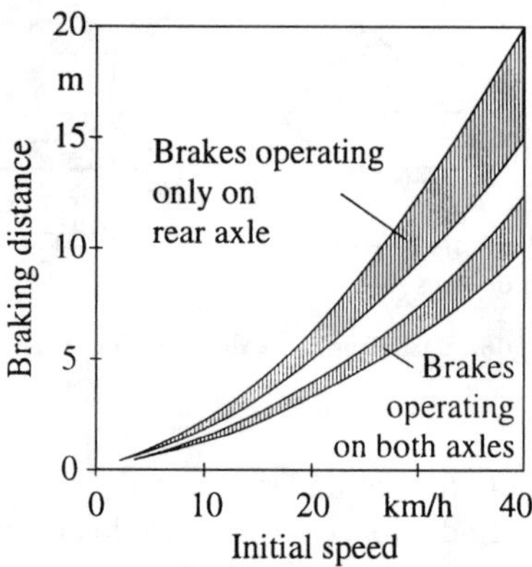

Figure 1.109. Guide line for braking distances on dry concrete for maximum tractor weight and maximum axle load [44].

sufficient for top speeds up to about 32 km/h (20 mph). Higher top speeds such as 40 or 50 km/h require service brakes operating on both axles. These became popular first in Europe. Front-axle brakes at the front wheels are expensive, costs can although be saved if the front axle braking function is combined with the front axle drive. Basic demands for service brakes are listed in Table 1.29 [29].

Basic requirements and test procedures are standardized by ISO 5697-1 and -2 (see most recent versions). Many countries have traffic or health and safety regulations (see, for example EC regulations). Measured brake performance characteristics can be picked up from OECD test reports.

Brake Concepts

There are four typical concepts in use for tractors, Table 1.30. *Wet disc brakes* seem to be the best solution for the rear axle, and are increasing in popularity. Drag losses can be a problem for higher speeds, but there are some design rules to keep them low [45].

Table 1.29. Demands for service brakes

- Max. pedal force 80–120 N related to 1 m/s^2 retardation for the tractor with full ballast
- One-side braking possible to reduce the turning radius, but with reliable mode change for onroad breaking
- Able to handle the generated heat
- Resistant against dust, water and mud
- If possible, maintenance-free
- No vibrations, no screech
- Low drag torque losses
- Low initial costs, low repair costs, long life

Table 1.30. Brake concepts for tractors, after [21]

Concept	Pro's	Con's
Drum brake, dry	Lowest first costs. Easy to repair, very low drag losses	High scattering of braking torques, sensitive to dust, water and mud
Caliper brake, dry	Able to handle very high sliding speeds, lining easy to replace, very low drag losses	Wear by dust and mud, wear at lining and disc, high temperatures (neighbored parts)
Disc brake, dry	Longer life than caliper brakes, low drag losses	High initial costs, sensitive to mud (if not capsulated), lining often difficult to replace
Disc brake, wet	Very low wear (life time up to tractor life), resistant against dust, water, mud, zero maintenance	Certain drag losses (sometimes high), oil must be specified for this brake type, special know how

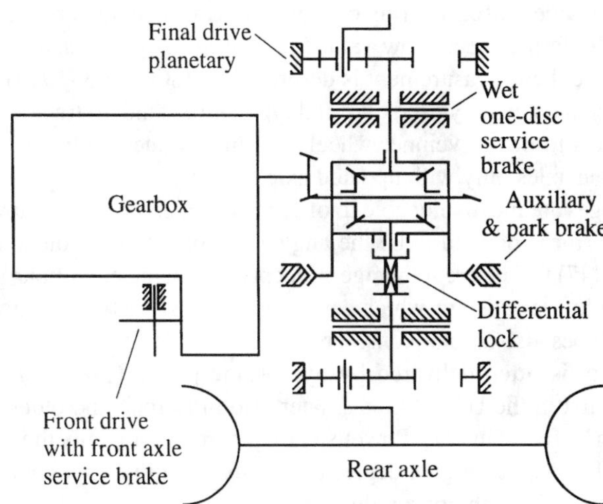

Figure 1.110. Typical brake positions within a tractor drive.

Initial costs of tractor brakes are not only influenced by their concept but also by their position: high rpm positions (on counter shafts or pinion shafts) lead to smaller dimensions than positions directly at the wheels.

Brake Position within the Drive Line

A typical configuration is demonstrated by Fig. 1.110 for a European standard tractor. The dry *caliper brake* on the front drive shaft can be called the cheapest design for a front

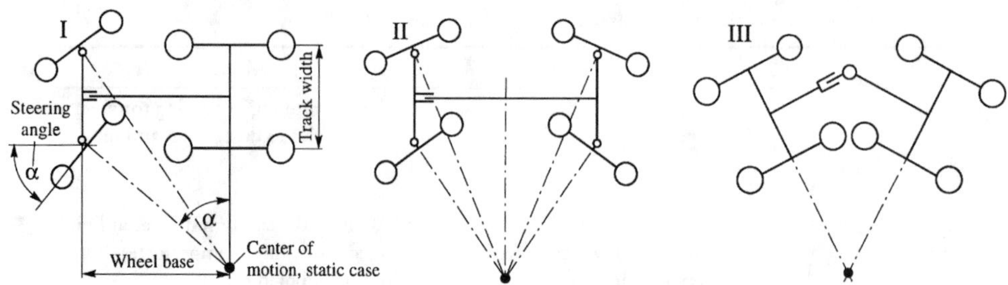

Figure 1.111. Steering concepts: I front-axle steering ("Ackerman steering"), II front and rear-axle steering, III steering by articulated frame; center of motion defined for static conditions.

axle brake. Another economic concept uses the rear axle brakes for all-wheel braking: When the driver operates the brake pedal, the front drive is automatically engaged (wet clutch) connecting both axles and thus offering excellent retardation [44, 46].

Steering

Concepts

The three most widely used steering concepts for two-axle tractors are demonstrated by Fig. 1.111. The farmer needs low values for the *turning and clearance diameters.* Their definition and their measurement is described by ISO 789-3 (1993).

Concept I is the most widely used one, and typical for standard tractors. The *steering angle* always refers to the curveinner wheel. Required values figure 50°–60° which is achieved for driven axles only with updated axle designs.

Steering linkages are mostly not able to offer the exact geometry as shown. Therefore a steering angle error is defined to be the angle error of the curve-outer wheel. Values can be read from [47] for different linkage concepts. The static conditions as shown only are valid for very low speeds, for which centrifugal forces and tire drift angles must not be considered. Slopes also are not considered.

Steering energy is often delivered by hydrostatic power (*power steering*), which became popular first in the United States, later also in Europe and other areas. Design fundamentals can be found in [47]. Power steering increases not only the comfort for the driver but also offers a much higher degree of design flexibility (pipes and hoses instead of linkages) also favoring cab noise reduction.

Concept II is used for special vehicles with four equal tires such as hillside multipurpose tractors. It typically offers three steering modes: front steering (on the road), front and rear opposite and front and rear parallel (to compensate, for example, hillside drifts). The full use of the steering angle is often not possible due to the required clearance for the lower links of the rear three-point hitch. An advantage can be seen in the *track-in-track driving* (multipass, no transmission wind-ups).

Concept III is typical for big four-wheel-drive tractors, offering again track-in-track driving. Main disadvantages relate to the high level of required hydrostatic power (losses) [47] and the poor ability for high speeds on the road. Concept III is sometimes also applied for vineyard and orchard tractors.

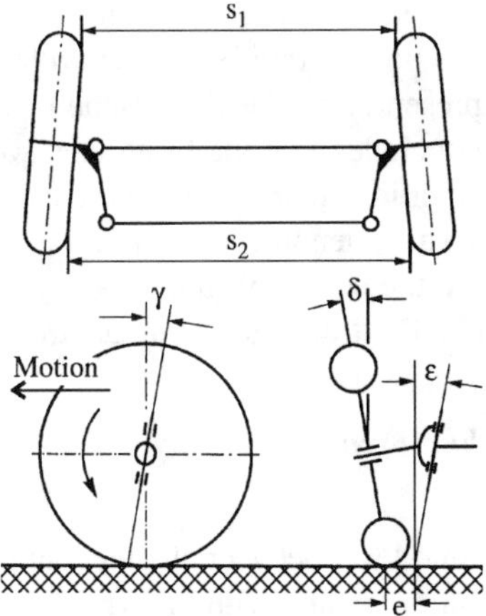

Figure 1.112. Ackerman steering, geometry.

Details of the "Ackerman" Steering

There are some typical specifications for the three-dimensional geometry, Fig. 1.112.

$s_2 - s_1$ Toe-in 3 ··· 5 mm

γ Caster angle 0 ··· 5 (12) deg

δ Camber angle 2 ··· 4 deg

ε Kingpin inclination 3 ··· 8 deg

e Kingpin offset 0 ··· 100 mm

Unusual large *caster angles* have been applied by John Deere to achieve higher maximum steering angles for a given tire-chassis clearance situation (see also the famous Citroen 2CV passenger car). Kingpin offset values are not only influenced by the basic steering geometry, but also by the rim design. If the rims are changed or turned or modified to adjust the track width, kingpin offset can be influenced considerably.

Practical Operation of Tractors with "Ackerman" Steering

The steering torque (usually defined for the kingpin axis) consists of three portions:

- Main portion, to overcome the tire friction. (calculation see [47])
- Smaller portion for lifting the tractor front part (due to the kingpin inclination)
- Small portion to overcome friction in the kingpin bearings, in the linkage, etc.

Maximum *hand forces* at the steering wheel are specified by road traffic regulations in many countries. In the case of power steering, the farmer requires that the system is just able to perform the maximum steering angle on dry concrete with maximum axle load even if the tractor is not moving (important for front end loading). This demand is usually stronger than common traffic regulations.

In the case of four-wheel drive, the steering behavior is substantially influenced by the power train concept. If both axles are connected by a fixed-ratio gearing (which is the most popular concept for standard tractors), the front axle input speed is in sharp turns too low and therefore creates braking effects instead of traction. A first approach to solve this problem was presented by Kubota in 1986, with a front axle overdrive for sharp turns [48]. A four-wheel drive power train prototype with an infinitely variable front axle speed control was first presented by the Technische Universität München [49, 50]. Research with this concept concluded that the front axle overspeeding demand during turns is lower for dynamic conditions than for the static case (i.e., 20% instead of 30% for 50° inner steering angle). This phenomenon can be explained by the drifting angles of all tractor tires which move the center of motion away from the rear axle axis to a position a little more forward. Drift angles of tractor tires have been investigated by several researchers [51–54].

Track Width and Wheel-to-hub Fixing

Track Width

This is the distance at ground level between the median planes of the wheels on the same axle, with the tractor stationary and with the wheels in position for traveling in a straight line. Track width is important mainly for operating the tractor in row crops [55] and to enable *multipass effect* for offroad tractor-trailer operations.

There are three typical concepts of track width adjustment:

- Infinitely variable, often power assisted
- Adjustment stepwise achieved by several modes of *rim* assembling (itself and to hub)
- Two basic track widths by turning the rims

ISO 4004 (1983) recommends the following basic track widths:

1500 mm ± 25 mm
1800 mm ± 25 mm
2000 mm ± 25 mm

Further popular track widths for standard tractors are: 730 mm (vine yard), 1000 mm (orchard), 1250, 1350, 1600, 2200 and 2400 mm. Problems with small values typically relate to limited horizontal clearance of the tires, mainly at the front axle for high steering angles. Problems with high track widths can arise due to traffic regulations (which limit the tractor width, e.g. 3 m). It also has to be considered that the axle body and chassis stresses increase with the track width. Multipurpose tractors in high developed countries with about 60 kW–100 kW rated engine power typically offer a track width range of about 1500 mm to 2000 mm (1500 mm only with very narrow tires due to given cab width of usually about 1000 mm). The 60″ (1524 mm) track width is popular in North America for row crop tractors.

Wheel-to-hub Fixing

This is a critical area regarding durability. Problems sometimes occur if a tractor is overloaded or if tires are used with diameters above the released limits.

ISO 5711 (1995) clearly favors the *flat attachment* with separated centering (in comparison with integrated centering by spherical or conical studs/nuts, as usually applied for passenger cars). ISO 5711 incorporates an informative annex for typical measures of spherical or conical studs/-nuts.

Figure 1.113 and Table 1.31 give some important dimensions as recommended by ISO 5711.

Table 1.31. Flat attachment type of wheel-to-hub fixing, dimensions as standardized by ISO 5711 (1995)

Number of equally spaced stud holes	Wheel					Hub	
	D_1 nom.	D_2 $0+1$	D_3 $0+0.5$	D_4 min.	Stud Ø*	D_5 $0-0.2$	D_6 $0-5$
4	100	15	61	145	12	60.8	140
5	140	17	96	185	14	95.8	180
6	205	21	161	255	18	160.8	250
8	203.2	21	152.4	257	18	152.2	252
	275	24	221	325	20	220.8	320
10	335	26	281	390	22	280.8	385
12	425	26	371	470	22	370.8	465

* For information only.
Note: All dimensions in mm.

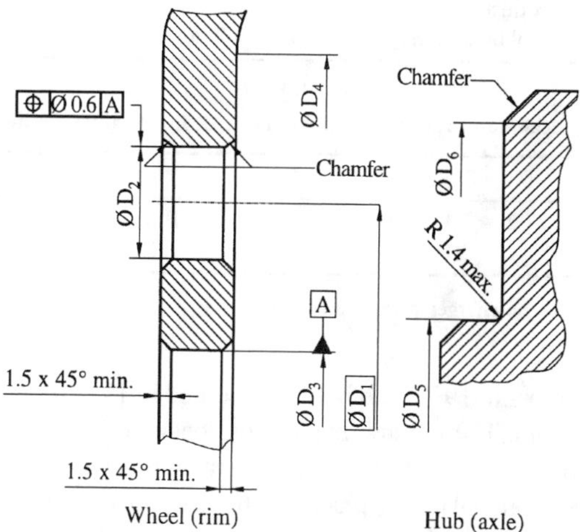

Figure 1.113. Flat attachment type of wheel-to-hub fixing, measures as standardized by ISO 5711 (1995).

Thickness of the wheel plate (rim) is not standardized and can be modified to meet the durability needs. High values and adequate steel quality can be recommended for critical cases.

1.1.30. Diesel Engines and Fuel Tank

This chapter covers only Diesel engine applications for two-axle tractors. For general fundamentals of engine design see chapter "Engines" (page 41 ff) and references [7, 11]. Net engine power measurement is standardized by ISO 2288.

Development Trends

The Diesel engine was introduced in Europe earlier than in the United States [56] because of higher fuel prices in Europe. The German Benz-Sendling inaugurated the

Table 1.32. Development trends of DI diesel engines in high developed countries

- Average rated power increased considerably
- 4-cylinder engine dominates since the seventies
- Turbo charging became very important: enables higher power density, lower emissions [57] and improved performance for high altitudes
- Rated engine speed was first increasing (to produce more power) but is since the seventies stagnating (at 2000 to 2500 rpm) in favor of fuel economy and noise limitation
- Specific fuel consumption (optimum) <200 g/kWh
- High heating performance for the tractor cab using the engine waste heat
- High torque reserves (definition see below) became popular, also their use to form a "constant power" speed range
- Maintenance has been simplified continuously
- Longer engine life became popular; rule for professional tractors: 6000 h with 90% probability under realistic loads
- Market share of air cooled engines decreased
- Electronic engine control and extreme high injection pressures (>100 MPa) have been introduced since the nineties.
- Emission control became very important (regulations)

Table 1.33. Typical motorization pattern for a tractor line (tractor families I, II, III)

Family/eng. cyl.	I/3			II/4				III/6			
Tractor model	I.1	I.2	I.3	II.1	II.2	II.3	II.4	III.1	III.2	III.3	III.4
Air charging*	–	–	T	–	T	T	TI	T	T	TI	TI

* T: Turbo charger, TI: Turbo charger with intercooler.

new engine era for tractors in 1922, and MAN presented in 1938 the first direct-injection Diesel engine for tractors. The remarkable success became possible with precise high-pressure fuel injection pumps (Bosch). The general replacement of indirect injection (IDI) with direct injection (DI) took place for tractors in Europe after World War II, improving the fuel economy. The price per kW output is higher for Diesel than for petrol engines due to the expensive injection pump and high cylinder pressures. Some typical development trends of the recent decades in high developed countries are listed by Table 1.32.

The *torque reserve* (also *torque backup*) is usually defined for dropped engine speed as the torque increase from the rated engine torque to the maximum torque related to the rated torque. High values favor driving comfort but increase the engine cost per kW rated power.

The future requirements of low exhaust emissions (regulations) together with low fuel consumption continue to favor the use of air charging, mostly by turbo chargers. If the charging pressure exceeds values of about 100 kPa (1 bar), intercooling is recommended to reduce the heat load on the engine and to increase the air mass flow.

Motorization Concept

A complete tractor line is usually formed by at least three families using engines with three, four and six cylinders [21], Table 1.33 (sometimes further family IV/6).

Table 1.34. Deutz "five-point-cycle"

Point No.	Torque % "Rated"	Speed % "Rated"	Power % "Rated"	Time Portion %
1	88	95	83.6	31
2	48	85	40.8	18
3	40	53	21.2	19
4	15	100	15.0	20
5	0	40	0	12

A unique *displacement* of 1.2 l/cyl offers, for example, rated engine power values from about 40 kW (tractor model I.1) to about 220 kW (tractor model III.4), assuming a rated engine speed of about 2300–2500 rpm.

Diesel Engine Installation

Installation requires many procedures; guidelines have been presented by [58]. The conventional block chassis concept uses the engine as a part of the chassis. If the engine is installed into a frame by iso mounts, there is much more freedom and design flexibility (including the use of truck or passenger car engines), and also a higher potential for noise reduction [13, 37].

Practical Fuel Consumption and Fuel Tank Size

It is well known that the tractor engine is never working all the time at rated engine power. John Deere refers to an average of about 55% of maximum power used on a year-round basis [59]. The author's experience points out that this value is a realistic choice only for those tractors that are mainly used for heavy tillage. For tractors with a wider-spread spectrum of operations, the *Deutz five-point-cycle* [21] may be more realistic, Table 1.34. It results in an average of 40.2% rated engine power. Fuel consumption maps of updated Diesel engines deliver upon this cycle an average resulting fuel consumption of about 100–110 g (0.12–0.13 l) per installed rated engine kW. Utility tractors are often loaded even below this cycle level.

A statistical evaluation of fuel tank sizes [60] leads to the following recommendations (Diesel fuel tank capacity in liter per installed kW rated engine power):

- Light duty (i.e., family I in Europe) ~1.5 l/kW
- Medium duty (i.e., family II in Europe) ~2.0 l/kW
- Heavy duty (i.e., family III in Europe) ~2.5 l/kW

1.1.31. Transmissions

Introduction

The tractor transmission (also called *transaxle*) is usually defined as demonstrated by Fig. 1.114: a combination of the vehicle speed change gear box, the rear axle with brakes, the power take off (PTO) and – if required – arrangements for the front axle drive and for the drive of auxiliary units (mainly hydraulic pumps).

The transmission represents about 25%–30% of the total tractor initial costs, much more than the engine cost. Some review papers on tractor transmissions have been presented by [61 to 72].

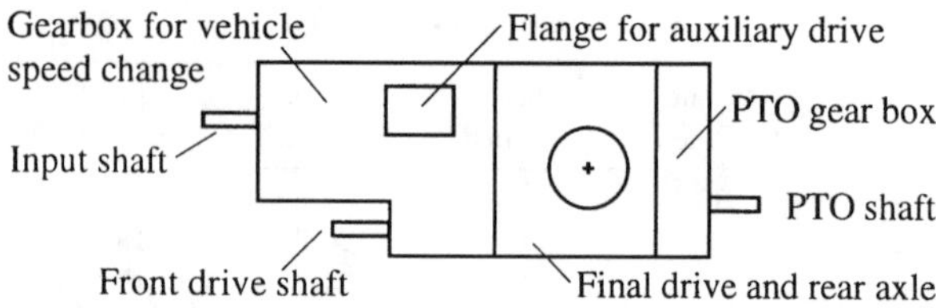

Figure 1.114. Tractor transmission: typical structure.

Requirements

The main requirement as related to the gearbox of the vehicle speed change: It should be able to use the offered engine performance characteristics (torque, power, specific fuel consumption) as economically as possible for given operational conditions.

A worldwide view shows that the present bandwidth of tractor transmission requirements is wider than ever before. Three main parameters can be identified for every market:

1. Technology level
2. Power level
3. Political conditions

Table 1.35 addresses some typical technology levels.

A low-tech level forbids high-tech transmissions. The higher the engine power, the more can the economic balance lead to a higher technology level (see also the historical development in Europe).

Level I may be typical for an initial mechanization phase, the high top speed of 20–25 km/h considers the importance of transports in many developing countries. Level II may be typical for the booming phase of a mechanization process (see, for example, India). Level III and IV represent requirements for markets such as Europe, North America, Australia and Japan. They are explained more in detail by Table 1.36.

The development within the next decade will open another level V addressing continuously variable vehicle drives (see page 148 ff).

Table 1.35. Typical requirements for stepped ratio transmissions of standard tractors by technology levels – worldwide view

	Nomin. Speeds, km/h				
Level	Forward	Reverse	No. of Speeds Forw./Rev.	Shift	PTO Speeds rpm
I	2–20(25)	3–8	6/2 to 8/2	SG, CS	540
II	2–30	3–10	8/4 to 12/4	CS, SS	540/(1000)
III	(0.5)2–30(40)	3–15	12/4 to 16/8	SS, HL	540/1000
IV	(0.3)2–40(50)	2–20	16/12 to 36/36 (or more)	SS, PPS, FPS,	540/1000 (750/1250)

Note: SG Sliding gear, CS Collar shift, SS Synchro shift, HL HiLo power shift, PPS Partial power shift (3 or more speeds), FPS Full power shift, ( ) options.

Table 1.36. Basic specifications for stepped ratio transmissions of standard tractors (levels III/IV, Table 1.35)

1. Speeds forward
   - Band-width of nominal speeds according to market level; Minimum 2 to 30 km/h, maximum (0.3) 2 to 40 (50) km/h
   - Ratio of neighbored speeds: $\varphi = 1.15 \cdots 1.20$ for main working range ($4 \cdots 12$ km/h) Higher ratios below 4 km/h, little higher above 12 km/h
2. Speeds reverse
   - Band-width of nominal speeds min (2) $3 \cdots 15$ (20) km/h
   - Ratio of neighbored speeds: $\varphi = 1.20 \cdots 1.40$ (Low values in case of reverse harvesting with PTO implements)
   - Reverse speeds should have corresponding forward speeds (reverse same speed or little faster)
3. Speed shift
   - Shiftability: Safe, fast, comfortable, precise
     Low number of levers, low shift forces, adequate lever positions and shift travels, shift pattern easy to understand
   - Onroad: "One-lever-shift" should be possible when starting with a gear of $\leq$10 km/h rated speed
   - Simplified shift forward-reverse ("shuttle")
4. Pedals: Low pedal forces (clutch, brakes)
5. Rear power take off (PTO), see also Table 1.37 below
   - PTO type 1 ISO 500, 540 rpm (at about 90% rated engine speed, which is not standardized)
   - PTO type 2 ISO 500, 1000 rpm (at 95 to 100% rated engine speed, which is not standardized) shiftable from type 1 shaft or with easy shaft change; Some markets require 4 speeds (i.e., 540/750/1000/1250 rpm, 750 and 1250 as "economy" versions for 540 and 1000 rpm)
6. Rear axle
   - Adequate clearance (track width see "chassis design")
   - Differential lock, easy to disconnect
   - Brakes see chapter "chassis design"
7. Durability, repairs, maintenance, total efficiency
   - Transmission life: 6000 to 12000 h (probability 90%), according to market and power levels (high values for large tractors)
   - Repairs only for typical wear out parts such as dry clutches and dry brakes (trend: wet lifetime clutches and brakes)
   - Maintenance demands as low as possible (i.e., filter replacement on request, indicated by pressure drop sensor)
   - Total transmission efficiency (input shaft to wheels) at least 85% at full load for the main working range (4–12 km/h) [65]

Fundamentals of Speeds and Torque Loads

Speed Concepts and Life Calculation

Lifetime portions for the tractor forward speeds can be derived from speed collectives which can vary considerably from market to market, Fig. 1.115.

The diagram indicates that about 2/3 of the tractor life is related to nominal speeds between 4 km/h and 12 km/h (*main working range*). The calculation of nominal transmission speeds for a given number x, a given minimum speed V_{min} and a given top speed V_{max} can be based in a first approach on a geometrical sequence:

$$C = (V_{max}/V_{min})^{1/(x-1)} \tag{1.82}$$

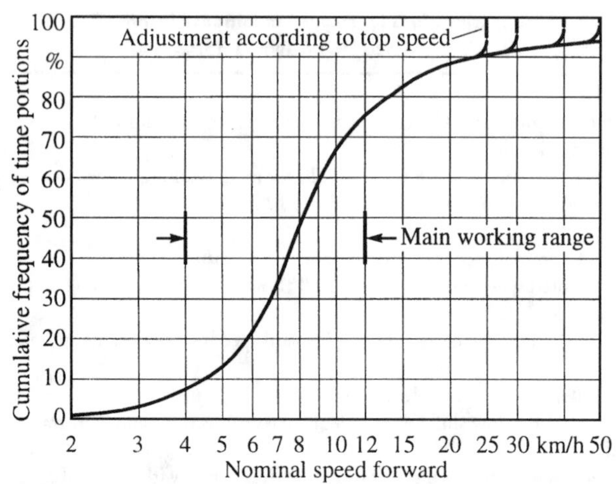

Figure 1.115. Distribution of nominal forward speeds for mid Europe and tractors > 40 kW [21].

C is called the *uniform step ratio* (ratio of neighbored speeds). Example: $x = 24$, $V_{min} = 0.5$ km/h, $V_{max} = 38$ km/h, $C = (38/0.5)^{1/23} = 1.20719$.

This rough approach can be refined by a concentration of speeds within the 4 km/h–12 km/h range (i.e., $C = 1.16 \cdots 1.18$) to the debit of little higher step ratios outside (mainly 0.5 km/h–4 km/h).

Figure 1.115 can be used to find lifetime targets for the different tractor speeds. Example: Calculate life for a 7 km/h speed, if the next higher one would figure 8 km/h. Solution: Life (7 km/h) $= 48\% - 34\% = 14\%$.

Transmission Loads and Fatigue Life Calculation

If the transmission speeds, the rated engine power, the engine full load torque characteristic and the transmission full load efficiency (85%–88% for the main working range) are given, a log-log *axle performance diagram* can be plotted, Fig. 1.116 [73]. Traction limits (also for Nebraska or OECD test planning) can be introduced.

A simplified fatigue calculation for the gear wheels of the forward speeds can be made as follows:

- *Gearbox input load*: According to rated eng. torque
- *Traction limit*: According to $\kappa + \rho = 0.6$ with typical tire loads

Traction limit is used for all those speeds in which engine torque cannot be transmitted. Based on these loads, safety factors for fatigue limit for bending should be 1.2 to 2.0 and for pitting 0.6 to 1.5 depending on the total number of gear wheel revolutions (life target). If the above mentioned loads are also applied for the calculation of antifriction bearings, the calculated lives should be below real-world life targets. These empirical rules are sometimes confusing. A clarified background can be achieved by the *random load fatigue theory* based on *load collectives* [74], which have been applied meanwhile very successfully [75, 76].

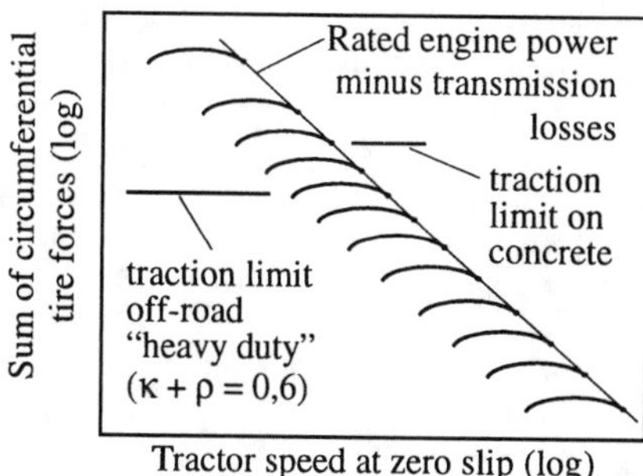

Figure 1.116. Axle performance diagram for a given transmission.

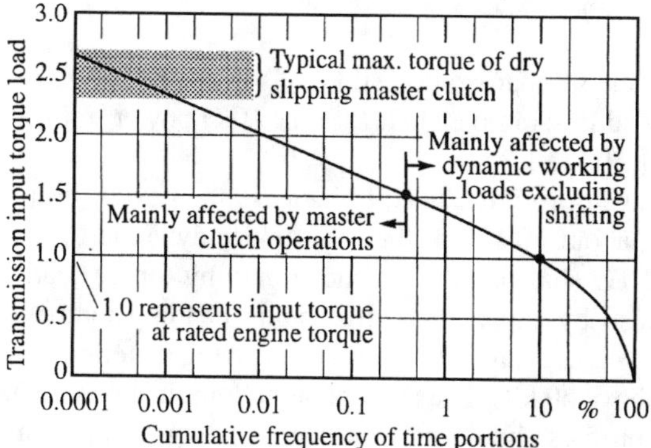

Figure 1.117. Standard load collective "Renius" for the input shaft of stepped ratio tractor transmissions with dry master clutch. Engine torque reserve was originally defined as about 15% rated torque [74], but little higher torque reserves seem to have only minor influences.

As the most important example, Fig. 1.117 shows the load collective of a common stepped-ratio transmission (with dry master clutch) at the gearbox input shaft [74]. The author thinks that it covers 90–95% of all practical tractor applications in agriculture, and that local conditions have only minor influences. Local influences are, however, more important for the torque load collectives at the traction wheels [74].

Application of the collectives for calculation and lab test is published in [75] and [76], a shortcut on optimal gear wheel dimensioning also in [63]. New input torque measurements of a continuously variable transmission confirms Fig. 1.117, with the amendment that automatic power control leads to a little higher load level in the right diagram section [77].

Rear Power Takeoff (Rear PTO)

The rear power takeoff (PTO) is very popular worldwide and there is a highly developed standardization which has its roots in the famous first ASAE standard of 1927.

Table 1.37. Basic Specifications of rear PTO, as standardized by ISO 500 (status 1997)

PTO Type		1	2	3
Direction of rotation		Clockwise viewed from behind		
Nominal speed, rpm		540	— 1000 —	
Nom. diameter, mm		35	35	45
Number of splines		6	21	20
Spline profile		Straight	— Involute —	
Shaft location	Height* min, mm	450	550	650
	max, mm	675**	775**	875**
	Offset, mm	— max. ± 50 horiz. —		
Clearance and shielding		See details in ISO 500		

* Above surface.
** Upper region recommended.

Today's worldwide standardization (ISO 500 1979 and 1991) defines three types of rear PTOs, mainly addressing the shaft (not the PTO power train). Some basic specifications are listed by Table 1.37.

Type 2 and 3 have been introduced to increase the power limit of type 1, which is, for a good design, about 60 kW at the shaft under dynamic loads (not standardized). Durability of the PTO shaft is not influenced only by torque loads (with usually high amplitudes), but also by superimposed dynamic bending moments (as a reaction of universal joints). In case of uniform loads (i.e., driving water pumps inline), the fatigue limit is therefore above 60 kW. The most popular combination can be seen in type 1 and 2. Different shaft profiles should prevent overspeeding of 540 rpm implements. Most farmers prefer however, both speeds shiftable from one type 1 shaft. Early doubts about safety, expressed by health and safety authorities, have not been confirmed.

Regarding common PTO power train designs, there are three basic concepts, Fig. 1.118.

The PTO shaft speed is proportional to the engine speed for the concepts I and III and proportional to the *zero slip* vehicle travel speed for concept II. Concept I was the typical initial one, easy and cheap but with the disadvantage that clutch operations stop both the tractor and the PTO. This was prevented by the development of *independent* concepts III, which became popular after World War II. Concept II has limited importance. It can, for example, be used to power the axles of trailers. The trailer transmission ratio must be properly adjusted to the tractor PTO power train ratio to secure synchronized operation. A rotational shaft speed level of 5–10 revolutions per m tractor travel can be recommended.

Today, most tractors offer concept III (some with mode b from II as option). As economy version is more typical of smaller tractors, while the modern version with life shaft and hydraulically operated multidisc clutch is more typical for updated bigger tractors (see also chapter "Stepped ratio transmissions," page 145).

Symbols for Transmission Maps

Uniform symbols were introduced by the author for vehicle transmissions in 1968 [61] for a better representation and understanding of transmission concepts, Fig. 1.119. Symbols for fluid power elements see "Implement Control and Hydraulics," page 165.

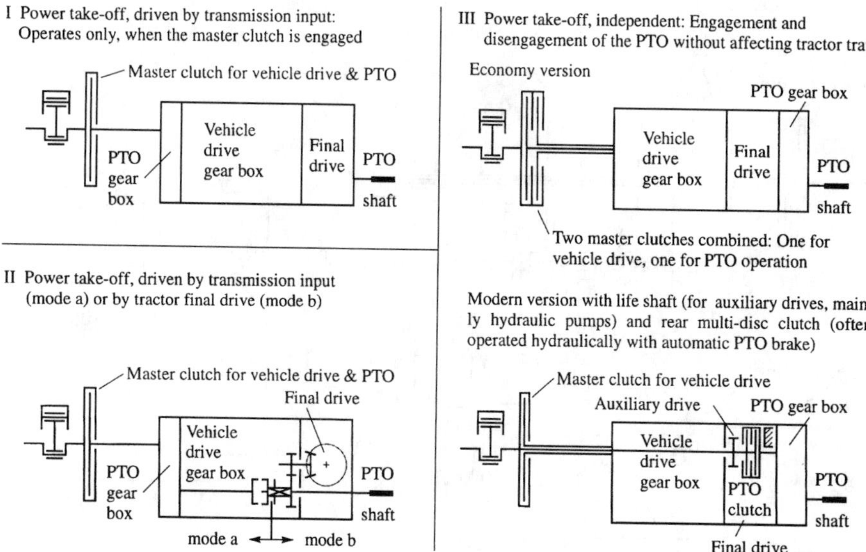

Figure 1.118. Concepts for rear power takeoff (rear PTO).

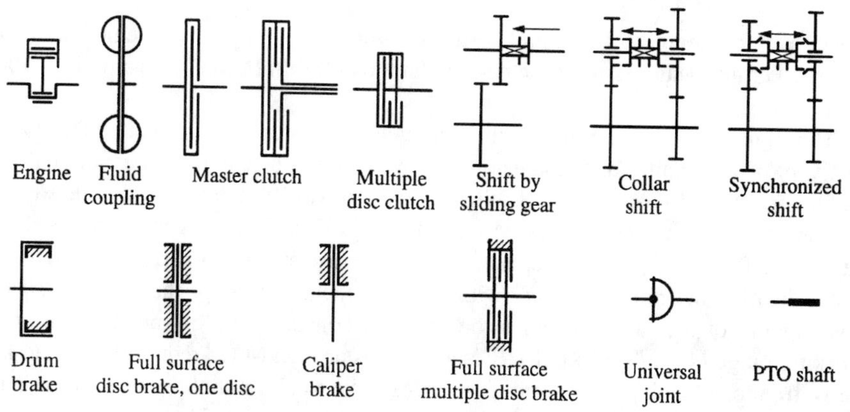

Figure 1.119. Symbols for transmission maps.

Stepped Ratio Transmissions

The high numbers of demanded forward and reverse speeds can be submitted only by using the *range principle*: 12/4 speeds (forw./rev.) are, for example, achieved by combining four *basic speeds* 1, 2, 3, 4 with three ranges L, M, H, and one reverse range R. The main reason for this principle is to save gearwheels (reducing initial cost, weight, gearbox length, efficiency, oil volume etc.). Guidelines for gearbox evaluations addressing the number of gearwheels in relation to the number of speeds have been presented by [63].

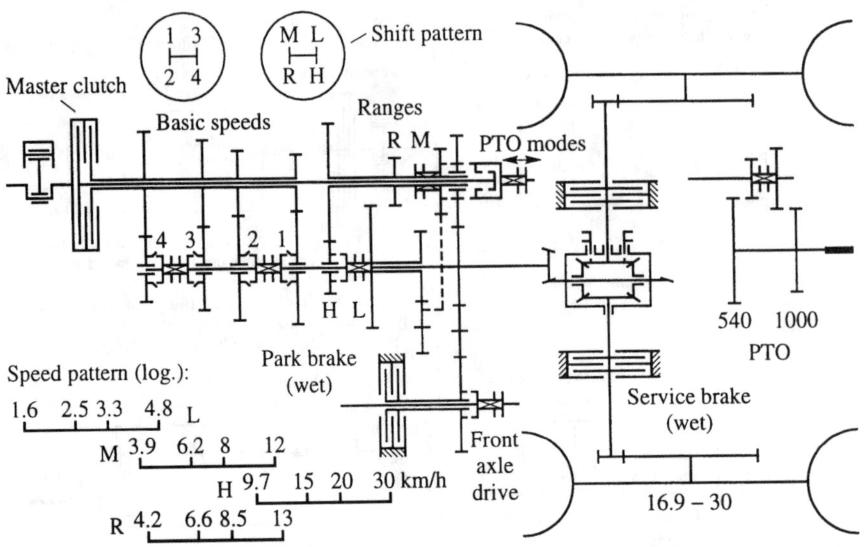

Figure 1.120. Tractor transmission 12/4, Massey Ferguson 1977 (44/50/55 kW); speed pattern relates to MF 274 S.

As a first example for a complete transmission map, Fig. 1.120 demonstrates the Massey Ferguson successor of the earlier (famous) "Multi Power" transmission (1961) [21].

The engine drives the combination of dry vehicle master clutch and dry PTO clutch – this way of combining dry clutches has proved to be outstandingly economical. The gearbox contains four basic speeds, three ranges forward and one reverse. The working speeds are synchronized, ranges M and H uses collar shift, ranges L and R sliding gear shift. The total speed concept can be presented best using logarithmic speed scale and indicating the ranges by bars with tines on them indicating the working speeds. Overlapping of the ranges enables the required concentration of speeds in the main working range (4 km/h–12 km/h). R is a little faster than M (as required for head land turns, front end loading etc.) – shift forward-reverse can be straightly done by the range shift lever without moving the shift lever for the basic speeds. Their *shift pattern* is in accordance with popular shift patterns for passenger cars. The rear axle is equipped with a differential which can be locked, with wet multiple disc service brakes and spur gear final drives. Spur gears offer high axle clearance and outstanding flexibility for the brakes [63], but planetary final drives have the advantage of lower costs and higher maximum possible speed reduction ratios. Therefore, other transmission versions have been equipped with planetaries (in the upper power class). The PTO concept offers the "two speeds from one shaft concept" using a double sliding gear. The driver also can choose between two basic PTO modes: "independent" and "vehicle speed dependent." Almost all the shift elements represent two functions by two shift directions (i.e., speed 1 and 2, or range M an L, or PTO speed 540 and 1000) – this is a proved principle to save costs.

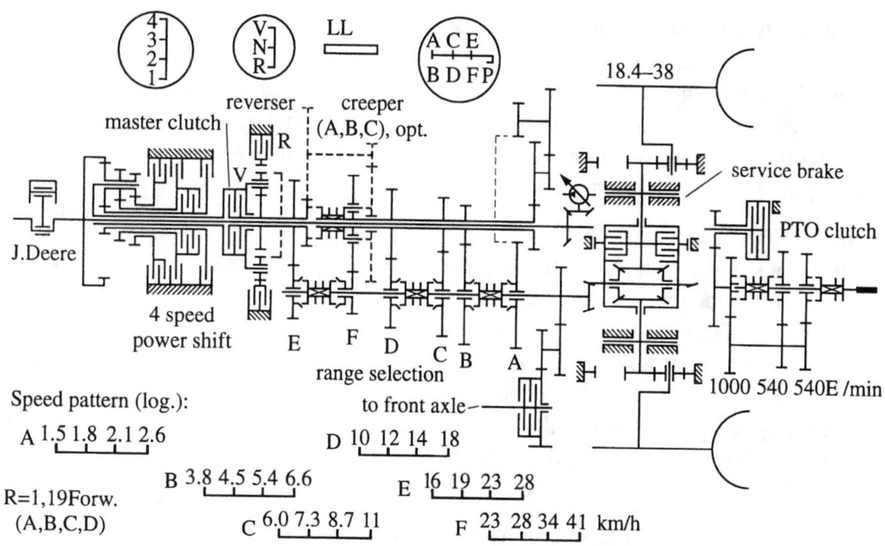

Figure 1.121. John Deere "PowrQuad" 36/28 transmission with power shifted four basic speeds and reverser for "6000" tractor series. Similar concept is used for the upper "7000" series. Optional creeper speeds from 0.15 km/h.

The second example of a tractor transmission is shown in Fig. 1.121. It represents a popular concept of John Deere, which came out together with the new "6000" series in 1992 (initially 55–73 kW) [71].

This transmission offers a large number of options according to the worldwide sales of the tractor: From the simple 12/4 SynchroPlus to the fully equipped 36/28 PowrQuad. The diesel engine directly drives a planetary gear set for four *power-shifted speeds* (4th speed direct), followed then by the wet master clutch and the power shifted reverser. Its design (outer meshing planetary gear engaging with inner meshing planetary gear) is a recommended principle for automatic vehicle transmissions. Following this in the power flow is the synchronized range selection A to F and the optional, slow *creeper range* (by further reduction of A, B, C).

Speeds are plotted for rated engine speed in logarithmic scale. The overlaps in ranges D, E and F simplifies range changing when hauling heavy loads. The transmission is operated by three shift levers (four with creeper). The rear axle works with an electro-hydraulically operated multidisc *differential lock.* The PTO at the rear offers an extra 540 rpm economy drive (reduced engine speed), a typical European demand since its introduction by Fendt in 1980. The rear-sited PTO clutch allows the autonomous drive of the variable axial piston pump for the newly introduced load-sensing hydraulics (replacing the former constant pressure system).

All western tractor companies have offered in the 1990s concepts with similar specifications, which first were presented by CASE with its Maxxum tractor line in 1989 [71].

Continuously Variable Transmissions (CVTs)

They are the dream of every progressive tractor engineer for a long time, and many attempts have been made to introduce them. Some physical principles are evaluated for tractors by Table 1.38. Forecasts for the future favor "hydrostatic direct" for small tractors (see Japan), "mechanical" chain variators for mid-sized tractors, and "hydrostatic with power split" for large tractors [71].

Required Efficiencies

Most CVT developments for tractors had efficiency problems. Market requirements depend on the tractor concept and size. Small garden tractors became popular with hydrostatic CVTs in spite of low efficiencies [78]. Considerably higher efficiencies are necessary for agricultural multipurpose standard tractors due to heavy pull operations. Figure 1.122 shows a full load efficiency target, as proposed by [13].

According to unpublished studies of the author, this target is only slightly below the efficiencies of well-designed, full power shift tractor transmissions.

Table 1.38. Evaluation of CVT principles for tractors

Physical CVT Principles	Pro's	Con's	Comments
Hydrostatic direct	Components available, excellent shuttle comfort	Low efficiency, high noise level	IH 1967 [62] limited success, popular for small tractors
Hydrostatic with power split	High efficiency	High initial cost, some concepts with complicated range shift	Fendt "Vario" 1996, many other developments
Hydrodynamic	Low initial cost, very low noise level, low wear	Low efficiency, ratio not stiff	Certain improvements possible by movable vanes
Mechanical by traction/friction	High efficiency, low noise level	No start from zero, hydrostatic control	Becoming important for car transmissions
Electrical	Low noise level, excellent shuttle comfort	Adequate concepts are very expensive	Long term improvements expected by fuel cells

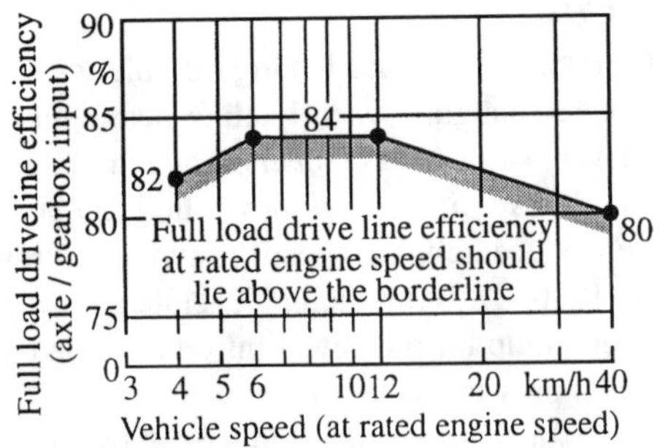

Figure 1.122. Full load efficiency target for tractor drive CVTs [13] (small tractors excepted).

I II III

Figure 1.123. Hydrostatic CVTs simplified.

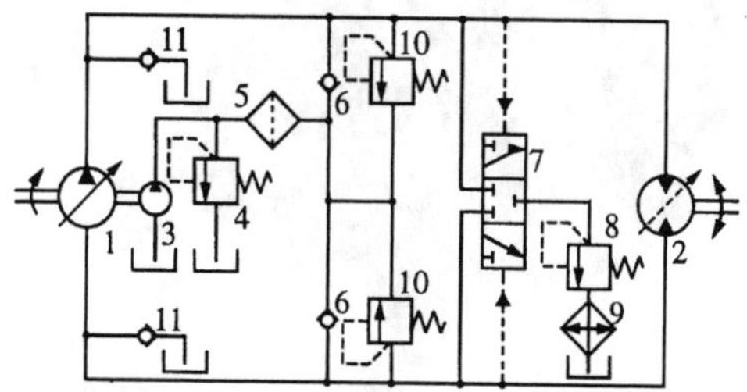

Figure 1.124. Hydrostatic CVT "direct," typical circuit. 1 pump, 2 motor, 3 charge pump, 4 charge pressure safety valve, 5 filter, 6 charge check valve, 7 flushing output valve, 8 charge pressure relieve valve, 9 heat exchanger, 10 high pressure safety valve, 11 emergency suction valve.

Hydrostatic CVTs "Direct"

A hydrostatic CVT is formed by the combination of at least one hydrostatic pump and at least one motor. At least one unit must have a continuously *variable displacement*, Fig. 1.123 (Symbols ISO 1219-1).

The most popular is concept I, which was also used for the famous NIAE research prototype 1954 [21].

Concept II offers better efficiency characteristics while III is not used for vehicles due to problems with zero output speeds and reversing. An example circuit is shown by Fig. 1.124 (Symbols ISO 1219-1).

The charge pump (3) is always feeding the low pressure pipe by check valve (6) to replace system leakages and to enable cooling by flushing the surplus flow via the flushing valve (7). This is controlled automatically by the high pressure side. High pressure safety valves (10) sometimes are replaced by a limitation of the maximum pressure via a pump displacement redirection (less heat generation, but more expensive).

Mechanical CVTs

Among a broad variety of mechanically working CVTs, high potential is seen for the *chain variator* principle. Two typical concepts of chains are known: The push type [79] and the pull type [80]. The pull type chain CVT has been used for the Munich Research Tractor and for a Schlüter prototype [69, 81], Fig. 1.125.

Full load efficiencies of these units (including fluid power control) can peak clearly above 90% if clamping forces are well controlled and if the hydrostatic control system is energy-efficient [82].

Figure 1.125. Mechanical CVT of the Munich Research Tractor with pull type chain (P.I.V.).

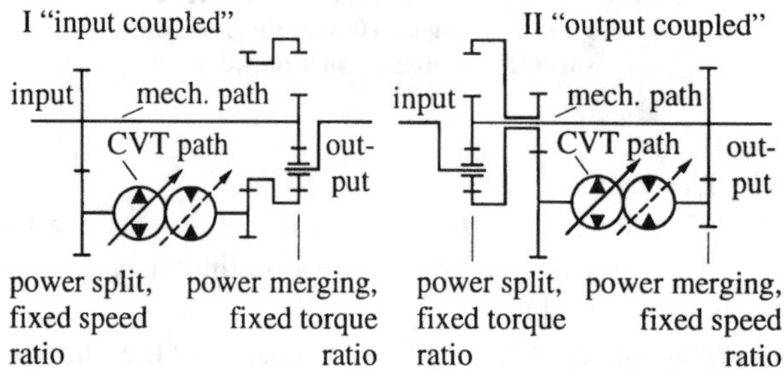

Figure 1.126. Two basic concepts of continuously variable transmissions using the power split principle.

Power Split CVTs

The power is split into two parallel paths: one mechanical with fixed ratio and the other by the CVT. Both energy flows are then again merged. The objective is mostly to increase total efficiency above that of a CVT "direct" unit [64], Fig. 1.126.

Fendt/Germany introduced in 1996 the first commercial standard tractor with a *hydrostatic power split CVT*, concept II [83], based on a Fendt patent [84]. Outstanding high efficiencies (covering the target of Fig. 1.122) have been achieved not only due to the power split, but also due to new "high angle" bent axis pumps/motors with spherical pistons (with rings), developed by Fendt with assistance of Sauer Sundstrand. The outstanding efficiency potential of such pumps was demonstrated in 1970 [85]. A CVT concept I was introduced by Class/Germany (1997) in small numbers [86], and prototypes are also known from other companies (Steyr, ZF et al.). There were several confidential cooperations in the field of CVTs between companies and Universities (Bochum, Munich etc.) to support the various new developments.

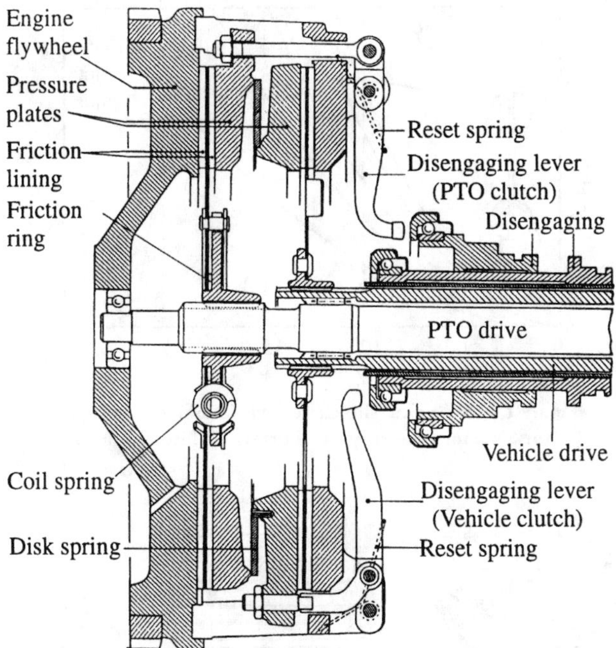

Figure 1.127. Typical design of a dry master clutch package for vehicle drive gearbox and PTO [21].

Master Clutch and Shift Elements

The most popular type of master clutch has two functions (Fig. 1.127): Disengaging and engaging smoothly the power train for the gearbox and the PTO. Linings mostly work dry using organic or *cerametallic* materials (friction coefficients 0.2 to 0.5 [64], service factor ≈2.5 for drive, ≈2.0 for PTO). The dry-wet discussion leads to similar pros and cons as for disc brakes (see chapter "brakes," page 131 ff).

Additional *fluid coupling* increases comfort and decreases powertrain vibration (mainly of interest for PTO). The disadvantage is the energy loss due to slip and ventilation (mean value about 2%–3%). Typical slips are plotted by Fig. 1.128. Practical slip evaluation can be achieved by using engine load cycles (see chapter "diesel engines and fuel tank," page 137 ff). Fendt first used fluid couplings to pick up torques from slip and speed (1993) for improved power shift control.

Figure 1.129 summarizes the three most important principles for manual gear shift within the gearbox and the PTO. Function of *synchronizer* with automatic blocker is as follows:

Move of shift collar creates a certain friction torque resulting in a limited circumferential offset (indexed position) between body splines and blocker splines. If speeds are synchronized, torque disappears, and shift can be completed [64].

There are many different types in use. In case of high friction loadings (specific heat loads), cooling by transmission oil pump can be highly recommended.

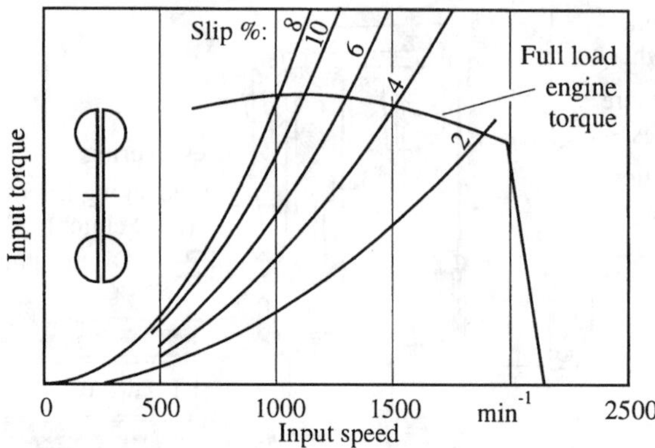

Figure 1.128. Typical slip characteristics of a fluid coupling for a tractor transmission (courtesy F. Görner, Fendt).

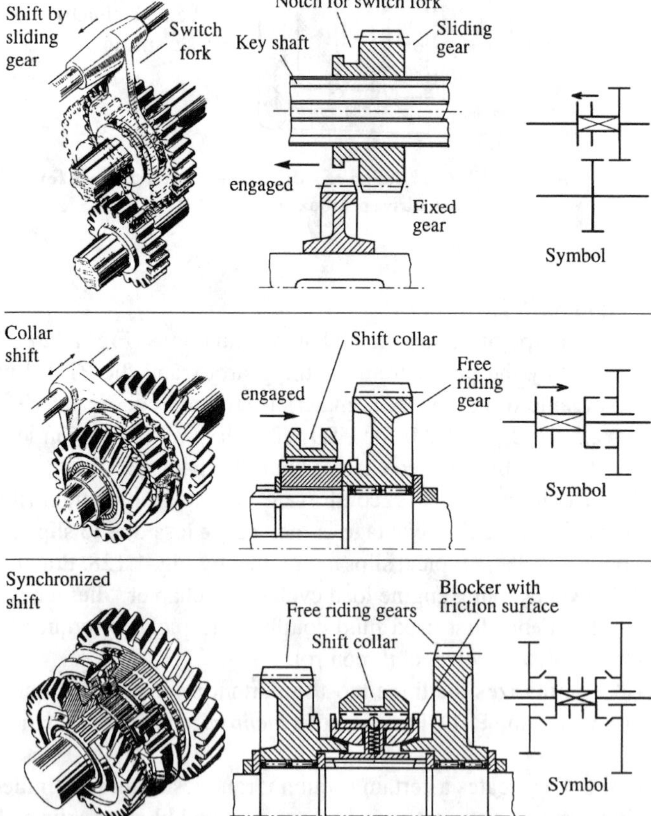

Figure 1.129. Three important elements for manual gear shift [21] (left three-dimensional views courtesy Zahnradfabrik Friedrichshafen).

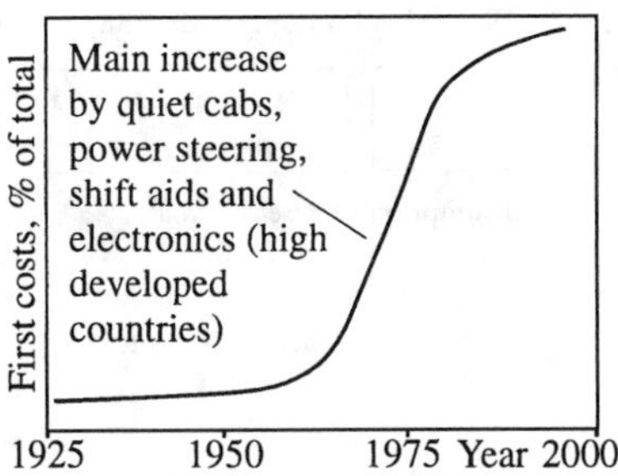

Figure 1.130. Typical increase (trend) for initial costs as related to comfort, health and safety (high developed countries).

1.1.32. Working Place

Introduction: Role of Comfort, Health and Safety

The scientific investigation of man-machine relations is still a young discipline that has been developed mainly since the 1950s [87 to 94]. Old tractors are characterized by a low level of technical aids – expenses for the driver (seat, levers, steering wheel, instrumentation) were only some percent of total tractor value, Fig. 1.130. These expenses increased dramatically in the 1970s and 1980s in the highly developed countries; developing countries follow with delay [95]. Health and safety authorities have initiated bigger steps (see history of safety frames and quiet cabs), promoting typical prejudices such as "not competitive in initial costs," "will not pay," "more parts more problems," etc. It seems to be an interesting message for developing countries [95] that the speed of technical developments for comfort, health and safety was mostly underestimated (see introduction of quiet cabs), and that legal requirements are often surpassed after a certain period of time by market requirements (see noise level inside of cabs). Initial costs of good cabs generally are more critical for smaller tractors than for bigger ones, typically from 15% (small tractors) to 10% of total (very big tractors).

Work Load on Machine Operators

The loads are usually classified into four sections (see Table 1.39) each with some characteristic criteria. The total workload results from all these particular points and is balanced by the operator's working potential and his working willingness. Physical and psychological stress creates *human fatigue*, which will usually reduce human performance. The risks of accidents also can increase, but the risk level and the consequences can be controlled by adequate *design for safety.* (see, for example safety frames versus driver injury or death, in case of overturning).

Stresses which exceed for longer periods of time individual stress limits can damage health. A typical example can be seen in intervertebral disc problems due to vibration overloads, or defective hearing as a consequence of inadequate high noise levels.

Table 1.39. Work load on operators ([88] modified)

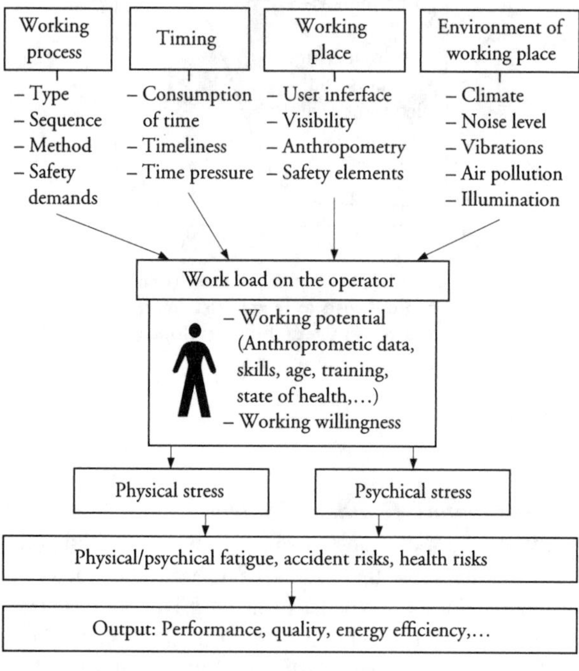

Millions of people worldwide may suffer under these health problems because they did not recognize the cumulative damaging effect early enough. The goals for the driver can be identified by general criteria such as productivity, quality of work output, energy efficiency, machinery wear, maintenance costs, comfort, health, safety, and others. Some of these criteria are difficult to measure; the operator can, however, set targets for particular parameters. Productivity can be, for example, kept high by maximum engine power, and energy efficiency by limited wheel slip or *shift up throttle down*.

The target setting and output control can be understood as a *closed loop control circuit*, Fig. 1.131. An important element of this representation is the feedback of the result to the operator, so that he is able to compare output with the task and correct the working process, if necessary. A further reduction of the workload on the driver can be achieved, if the machine itself controls the result by "automatic" operation.

A famous example for this strategy is represented by the automatic draft control of the three-point hitch, invented by H. Ferguson during the 1920s. Progress with electronics has initiated since the 1980s several new approaches.

Technical Aids for the Operator: Survey and Related Standards

If a certain working process and timing is given, workload and stress on the operator can be reduced considerably by technical aids, Table 1.40.

It is a well known experience that these aids increase tractor initial costs considerably, but nevertheless there is a strong trend towards them – in the highly developed countries mainly above about 50 kW rated engine power.

Table 1.40. Typical workloads on the operator and technical aids to reduce them

Work Load Categories	Technical Aids for the Operator and Related International Standards (ISO)
	Working Process
• Set tasks	– Instrumentation, board computer (ISO 11783, 11787, 11788)
• Compare tasks with output	– Visibility aids (ISO 5721)
	– Sensors, instrumentation (ISO 11786)
	– Board computer, GPS (ISO 11788)
	– Display, signals, symbols (ISO 3767-1)
• Correct process	– Well positioned levers, pedals, buttons (ISO 3789 part 1–4, 8935, 15077)
	– Adjustable steering wheel, adjustable seat etc. (ISO 3462, 4252, 4253, 5353)
	– Adequate resistance forces of control elements (ISO/TR 3778, ISO 15077)
	– Symbols (ISO 3767-1) and movement directions (ISO 15077) for control elements
	– Shift aids such as synchronizers, power shift, automatic shift
	– Continuously variable actuators (CVT for vehicle speed, remote control by proportional valves, . . .)
	Environment of Working Place
• Climate	– Cab with blower, tinted glass, shadow wings, air conditioning, heating system (ISO 7730, 14269-1 and -2, ISO/TR 8953)
• Noise	– Cab with noise insulation (ISO 5128, 5131)
	– Power train components noise insulated (ISO 362, 3744, 10844)
• Dust	– Closed cab with pressurized air (ISO 3737, 14269-5)
	– Air input through filter (ISO 14269-4)
• Vibrations	– Passive reduction of accelerations by springs and dampers for seats, cabs, axles, mounted implements (ISO 2631, 5007, 5008, 8608)
	– Active driver position control
• Illumination	– Working lamps for night
	– Illumination of instrumentation

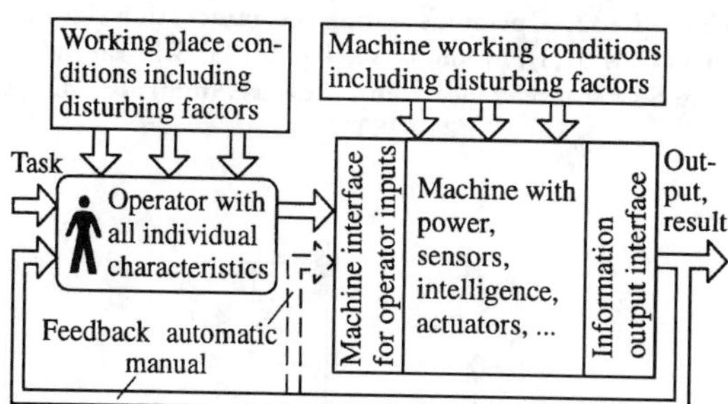

Figure 1.131. Closed loop control circuit of an operator – machine system.

Operator's Seating Accommodation and Access

For many years the standard ISO 4253 (1977) was used to recommend dimensions for operator's *seating position* in relation to steering wheel and pedals. The seat position was defined by the so-called *seat reference point* (SRP). The newer standard ISO 4253 (1993) has been modified replacing the SRP by the *seat index point* (SIP, see ISO 5353), Fig. 1.132 and Fig. 1.133.

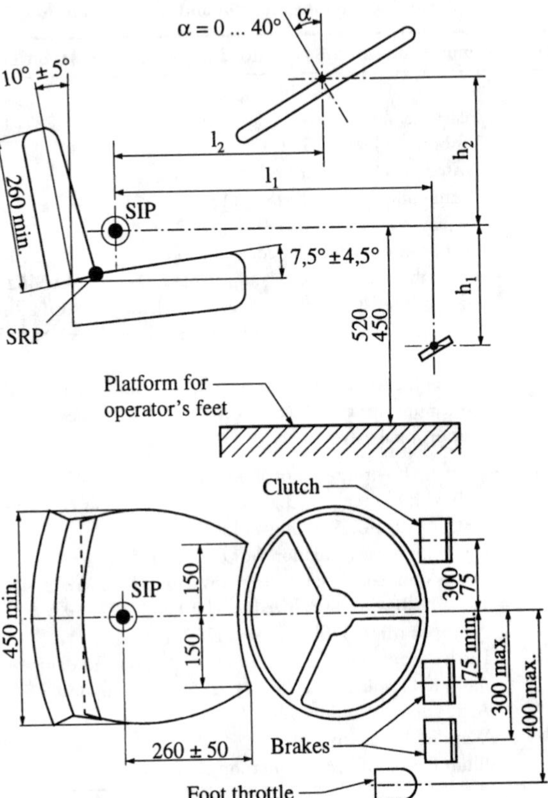

Figure 1.132. Operator's seating accommodation side and plan view (l_1, l_2, h_1 and h_2 see Fig. 1.135) SIP seat index point; SRP seat reference point. Measures (mm) according to ISO 4253 (1993).

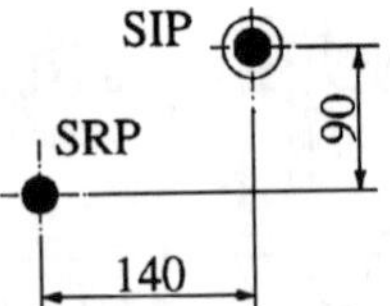

Figure 1.133. Offset measures between SRP and SIP (status used 1997 according to draft amendment ISO 3462: 1980/DAM1, still under discussion).

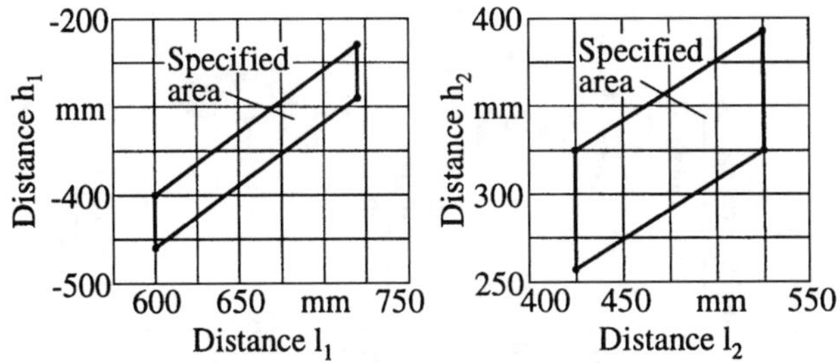

Figure 1.134. Pedal and steering wheel positions relative to SIP (ISO 5253: 1993).

Recommended relations between l_1 and h_1 and between l_2 and h_2 can be read from Fig. 1.134.

Optimum locations for control elements are specified by ISO/DIS 15077 (1997). They cannot, however be used completely due to space conflicts with access/exit requirements. This is one reason to concentrate gearshift and hydraulic shift levers on the right in the cab, other controls around the steering wheel. (*Control forces* and displacements for controls, see Table 1.41.) There are sometimes problems with the pedal force of nonassisted brakes in case of high tractor ballast (or payloads) in connection with high retardation. State of the art comfort level for braking pedal forces can be reviewed best by using actual OECD test reports indicating the relation of total tractor weight, mean retardation and pedal forces (see also chapter "brakes," page 131 ff).

Minimum internal clearance dimensions are specified by Fig. 1.135, based on ISO 4252. The lateral clearance of twice 450 mm = 900 mm leads to a minimum outside *cab width* of about 1000 mm, limiting the lowest possible *track width* for a given tire (i.e., about 1500 mm track width for 16.9″ nominal tire width).

ISO 4252 identifies also *access/exit doorway dimensions*, Fig. 1.136, and requires at least three emergency exits.

Tractor Ride Dynamics (Vibrations)

All terrain surfaces and all roads are more or less uneven, influencing considerably the *riding comfort*, and often limiting top speed. Unevenness can be measured easily and is usually represented by the so-called *spectral unevenness density* versus the *unevenness wave length* L or the *circular displacement frequency* $2\pi/L$ [96–102] (see also ISO 8608 (1995)). Correlation of the unevenness of two parallel field tracks (tractor tires left and right) is important only for very large wavelengths, but unimportant for short waves [102].

Vibration load on the driver is measured and evaluated by accelerations. In principle, six components have to be considered: three directions x, y, z of rectilinear vibrations and three angular (or rotational) vibrations related to the three axes x, y and z. In the case of sitting or standing operator, z-axis is defined vertical and x, y horizontal (see, for example, ISO 2631). Rectilinear vibrations are the most important ones for tractors. They are measured by picking up the accelerations and by calculating *root-mean-square values* (RMS values).

Table 1.41. Control forces and displacements of selected actuators, extracted from ISO/DIS 15077 (1997)

	Displacement				Actuating Force			
			Maximum				Normal	
							Normal Operation	
Control	Unit	Minimum	Frequent Operation	Infrequent Operation	Unit	Minimum	Frequent	Infrequent
Finger operated								
– Push buttom	mm	2.0	6.0	8.0	N	2.5	10.0	13.0
– Slide/toggle Rocker	mm	3.0	25.0	32.0	N	2.5	10.0	13.0
– Selector switch-stop to stop	deg	10.0	30.0	40.0	Nm	0.15	0.60	0.75
Hand operated								
– Steering wheel	deg	—see actual regulations/standards—			N	—see actual regulations/standards—[1]		
– Push-pull handle	mm	15.0	45.0	60.0	N	12.0	48.0	60.0
– Lever (longitud.)	mm	100.0	300.0	400.0	N	20.0	80.0	100.0
– Lever (transverse)	mm	60.0	190.0	250.0	N	15.0	60.0	75.0
Foot operated								
– Push button	mm	15.0	45.0	60.0	N	20.0	60.0	90.0
– Pedal, light duty	mm	20.0	60.0	80.0	N	12.0	48.0	60.0
Leg/foot operated								
– Pedal, heavy duty with back support	mm	30.0	90.0	120.0	N	30.0	300.0	375.0
– Brake pedal	mm	—see actual regulations/standards—			N	—see actual regulations/standards—[2]		

[1] Max. steering forces: 250 N (non assisted, ISO 789-11), 400 N (assisted, case of oil pressure supply fail, ISO 10998).

[2] Max. foot pedal breaking force: ISO/DIS 15077 specifies 600 N (excluding small drivers), which sometimes has been measured in OECD tests for maximum ballasted tractors and high retardations.

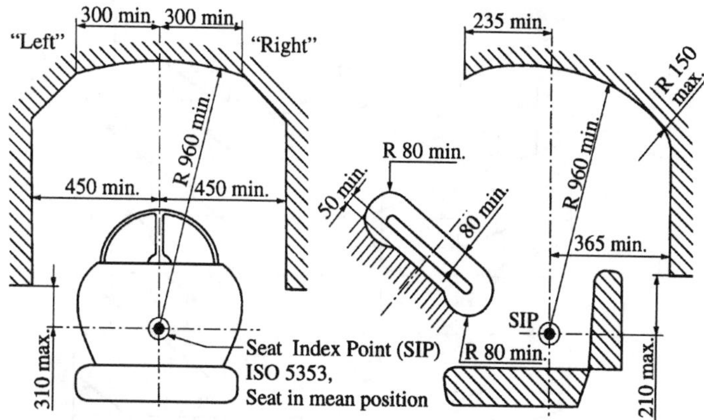

Figure 1.135. Minimum internal clearance dimensions for tractors with one operator position and at least 1150 mm track width (ISO 4252: 1992).

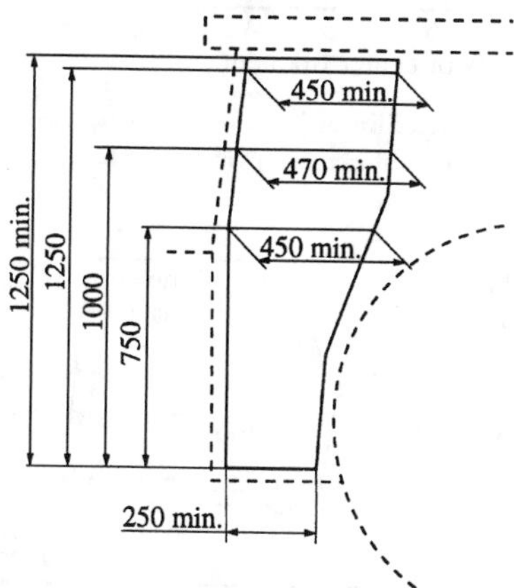

Figure 1.136. Access doorway dimensions (ISO 4252: 1992) step dimensions see ISO 4254-1.

ISO 2631 defines three criteria for vibration evaluation

1. The preservation of working efficiency: *Fatigue-decreased proficiency boundary*
2. The preservation of health or safety: *Exposure limit*
3. The preservation of comfort: Reduced *comfort boundary*

Limits for the first criterion are plotted by Fig. 1.137 according to ISO 2631. The diagram demonstrates that the human body reacts most sensitively to accelerations within a frequency band of 4 Hz to 8 Hz. Limits for the two other criteria can be calculated from the same diagram by multiplying the acceleration values (factor 2 for *exposure limits* and 1/3, 15 for *reduced comfort boundary*).

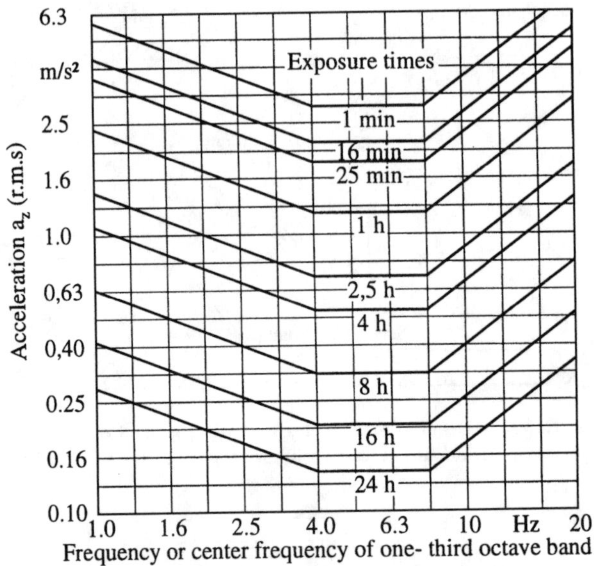

Figure 1.137. "Rectilinear" vibration limits, z-axis, ISO 2631 for "fatigue-decreased proficiency boundary."

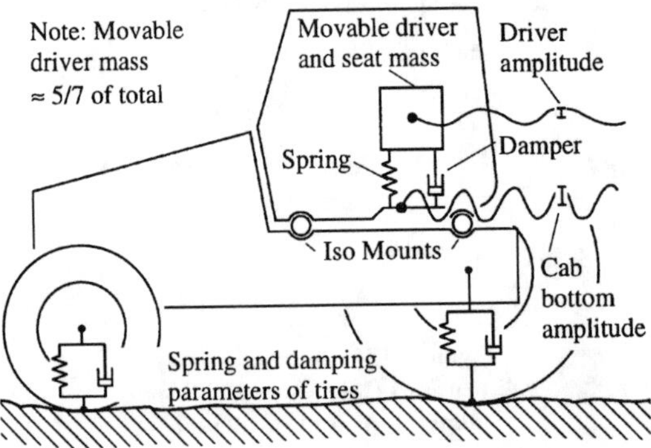

Figure 1.138. Simplified plane tractor model for vertical vibrations (no spring/dampers for axles, cab and others).

Regarding influence of design parameters on vibration load of the driver, the *seat* is usually the most important component [103], Fig. 1.138. Tractor tires have certain insulation effects [96], *axle suspensions* have increasing market shares – same for *cab suspensions* and other principles. Cab iso mounts influence mainly noise insulation.

Insulation effect of the seat can be measured by the *transfer function*, Fig. 1.139. The diagram indicates that vibrations (amplitudes) can be reduced only if their frequency is at least $\sqrt{2}$ times higher than the *natural circular frequency* of the system *driver-seat*,

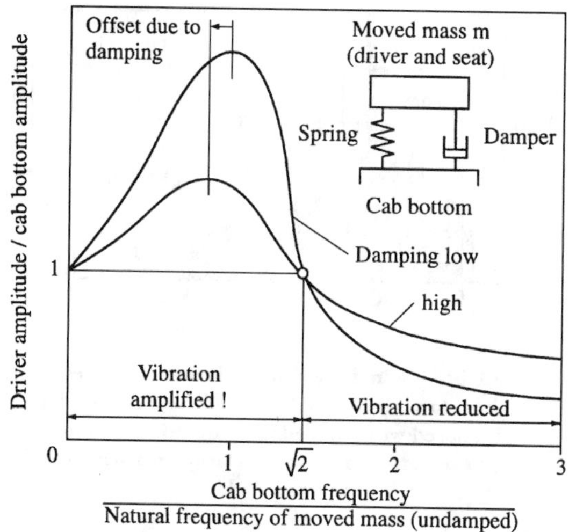

Figure 1.139. Transfer function for vertical vibrations from the cab bottom to the operator, expressed by amplitude ratio.

which can be calculated as follows:

$$\omega_o = (c/m)^{1/2} \qquad \text{(without damping)} \tag{1.83}$$

$$\omega'_o = \omega_o(1 - D^2)^{1/2} \qquad \text{(with damping)} \tag{1.84}$$

c spring rate, m movable mass, D damping ratio.

The consequence of these laws is that ω'_o of the driver-seat-system should be as low as possible. Good values figure below 1.5 Hz [88] but there are limits due to minimum possible spring rates and seasickness potentiality. D should be 0.2 to 0.4 [88].

Tractor Noise Control

Tractor noise problems can be classified in two categories:

- Noise at the operator's ear (on the tractor)
- Standby noise (at a certain distance)

The first category is clearly the more important one, as the driver is exposed to the noise all the time operating the tractor, while people in the neighborhood of the tractor or on connected implements are exposed only for limited periods of time [104].

The physical analysis of noise leads to the definition: noise = pressure vibrations (air, gas, liquids) or stress vibrations (solid materials). The operator is usually exposed to noise in the form of air pressure vibrations at certain frequencies. The air pressure alternates around the mean pressure (atmospheric pressure). For this reason RMS values are used to describe sound pressures, expressed in dB. This is, however, not a pressure unit, but a logarithmic ratio of pressures. The lowest pressure level that can be noticed by the human ear at 1000 Hz, is defined as the general basis $p_o = 2\ 10^{-5}$ Pa. If a measured

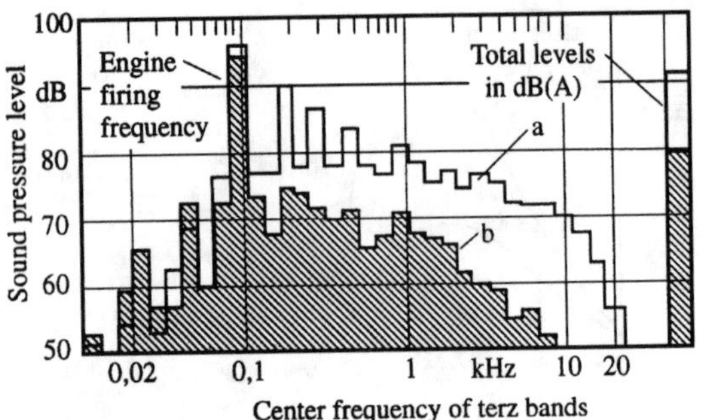

Figure 1.140. Sound pressure levels by frequencies and weighted total dB(A) levels at the drivers ear at rated engine speed without load. a "conventional" (block concept simulated), b engine in elastic suspensions and encapsulated.

pressure (RMS) is called p, the *noise pressure level* is calculated as follows

$$L = 20 \log p/p_o \tag{1.85}$$

The result is dimensionless, but is identified by the added term dB. Noise pressure levels can be measured by special adjusted microphones.

A *noise* is usually a combination of various frequencies with particular noise pressure levels each. Figure 1.140 shows a noise analysis (Fourier analysis) carried out for the *Munich Research Tractor* [37].

As the human sensitivity against noise load depends on the frequency, noise pressure levels are multiplied by weighting factors, which are defined by ISO 2204. The most used procedure is the A weighting leading to dB(A) levels.

Regulations such as 77/311 EWG are guided by the principle that the highest possible noise level at the driver's ear should not exceed 90 dB(A) for a maximum of 8 hours (future limits will probably be stronger). Noise load on a tractor with cab depends not only on the acoustic design quality but also on operational circumstances:

- Engine rotational speed
- Engine load
- Cab doors/windows open or closed

Therefore it can be recommended to measure noise levels over a longer period of time and to represent them by *noise load spectra* [36, 105] as shown in Fig. 1.141.

These spectra allow, for example, the calculation of *equivalent noise levels* for a complete working day. They tell much more about realistic noise loads than maximum values. The curve for plowing indicates, for example that there have been very uniform working conditions regarding all three factors mentioned above. As maximum values are, however, easy to measure, they are picked up at OECD and Nebraska test procedures (see related test codes and ISO 5131).

Design for low noise levels require high skills in acoustics, Table 1.42.

Table 1.42. Design for low noise level at the operators working place, recommendations

- Operator enclosure (cab) well sealed, keeps air noise outside
- Cab on soft elastic suspensions, keeps body noise outside
- All linkages, pipes, tubes, electric wires and other elements between chassis and cab designed for low body noise transfer properties (one reason for power steering: tubes transport less noise than solid linkages)
- Heavy cab glazing in elastic mountings
- Damping mats outside the cab in front (to damp engine noise) and under bottom (to damp transmission and hydraulic noise)
- Frame chassis: Components in elastic suspensions to interrupt body noise
- Very stiff housing walls of noise emitting components to prevent noise amplification by resonances
- Optimized muffler for engine exhaust gas and air intake
- Silencer and damper to compensate ununiformity of hydraulic pump(s)
- Optimized (stiff) sheet metal elements, damping mats
- Components with low noise emission, helical gears

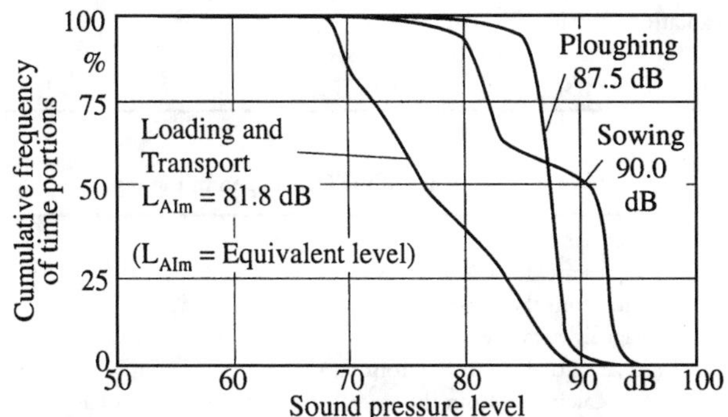

Figure 1.141. Noise load spectra (drivers ear) within a tractor cab (first generation) for different operations. Equivalent levels formed with filter "A" and recording dynamics impulse [105].

Tractor Safety

Design for safety became popular first in the highly developed countries in the 1950s and 1960s. The objective was a reduction of tractor accidents and their consequences. Many suggestions were made by health and safety insurance companies and legal authorities. It was recognized early by scientists in Europe and the United States that there was a high percentage of fatal tractor accidents related to overturning, and that these consequences could be mitigated considerably by adequate safety structures [106–113]. There were in 1969 in Germany, for example, 1448 fatal accidents in agriculture, 318 of them related to tractors and 181 of these (57%) caused by overturning [114].

The first phase of fighting against fatal overturning accidents was characterized by developing a test procedure, Table 1.43. The most investigated problem was to define realistic test energies. The pendulum test as specified by Table 1.44 was important for the time period between about 1960 and 1990. Since then there has been a clear trend towards a complete static test. The test is passed when a given *clearance zone* (as specified in the standards, i.e., ISO 3463) is not violated. The *rollover protective structures* (ROPS)

Table 1.43. Brief review on the introduction of roll-over protective structures (ROPS) and test procedures

1954	Early studies of overturning in Sweden by H. A. Moberg
1957	First proposal for a pendulum test [109]
1959	Roll-over Protective Structures obligatory for newly registered tractors in Sweden
1966	Tests and research in USA [106, 110, 111, 117]
1967	ASAE Standard R 306 (USA) [106]
1967	Pendulum test becomes optional part of OECD test
1967	Legislation for safety cabs in the UK [107]
1970	OECD test code, revised version
1970	ROPS obligatory for new registered tractors in Germany
1970	NIAE research on rearward overtun energy
1974	Revised OECD test code (rearward energy)
1976	Basic expertise for EEC commission [118]
1977	Regulation 77/536/EEC (dynamic test)
1977	ROPS obligatory for total German tractor fleet [31, 119]
1979	Regulation 79/622/EEC (static test)
1981	ISO 3463 (dynamic test) ISO 5700 (static test)
1983	Static OECD test (becoming popular)

Table 1.44. Sequence of the OECD pendulum test

1. Impact from rear
2. Vertical static load "rear"
3. Impact from the front
4. Impact sideways
5. Vertical static load in the front part
Energies: calculated from the tractor net mass (linear influence) and wheelbase (squared), represented by the adjusted height of a 2000 kg mass pendulum.

must be designed for controlled absorption of a certain percentage of the rollover energy, mainly by plastic deformation. Solid profiles therefore can be favorable even for bending (opposite to the rules of elastic bending). (For more design recommendations, see [106, 119].) Meanwhile, it can be said that *safety frames* (today often cab-integrated) could dramatically reduce *overturning fatalities* as demonstrated for Sweden by Fig. 1.142 and for West Germany by Fig. 1.143.

The reduction of fatalities was in West Germany more than 90%. Test energies seem to be well settled: They cover about 94% of various overturning accidents according to a British report [116]. A more recent study for Iowa indicates that from 1988 to 1992 55% of all fatal tractor accidents were related to overturns, as many older tractors had not yet been equipped with ROPS [117]. The authors warmly recommend the use of ROPS and of *seat belts* for ROPS equipped tractors (ISO 3776).

After this progress, safety level of tractors was increased also in other areas by many regulations. One example for such a safety framework is the *EU directive on machine safety*, effective 1\1\93, the so-called *Machinery Directive.* It defines work safety specifications to be met by manufacturers before a machine is sold. The basic

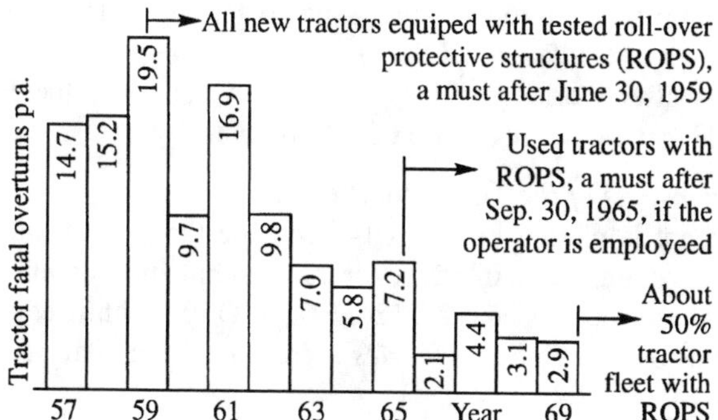

Figure 1.142. Tractor fatal overturns in Sweden, 1957–69, per 100000 tractors [112].

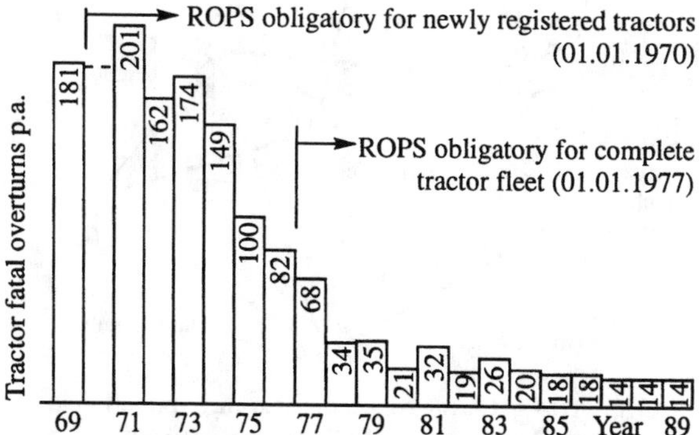

Figure 1.143. Total tractor fatal overturns in West Germany, 1969–89 [114, 115, completed].

safety requirements are defined by the European Commission, but they are completed in detail by EN standards developed by the European Standardization Committee (CEN) [120].

1.1.33. Implement Control and Hydraulics

Beginnings of Implement Control by Hydrostatic Hitches

The beginnings are connected with the name of Harry Ferguson, who developed in the early 1920s the idea of the closed-loop *draft control* for rear-attached implements, by a *three-point hitch* [121]. At that time Ferguson was mainly engaged in plow design, where he recognized the poor technical level of tractor-plow combination [122]. His new ideas initiated one of the most important innovations in tractor design, first applied to a Fordson F in the 1920s (prototypes), later to british tractors (Black Tractor 1933, David Brown 1936) in small quantities, and since late 1939 commercially to Fordson tractors (the "Handshake Agreement" [122]). The design was characterized by a coil spring upper link force sensor and a one-piston pump working directly on the lift cylinder and controlled by a pump intake throttle valve (scheme see Fig. 1.151 and [123–125]). A general breakthrough of the *Ferguson System* as a common principle of tractor design took place directly after World War II. By the end of the 1960s there were very few tractors worldwide that did not make use of a draft control.

Concept and Dimensions of the Three-point Hitch

The basic concept is plotted in Fig. 1.144. As the three-point hitch has been for many decades the most important interface for tractor-implement combinations, a strong standardization has been developed [123, 126, 127]. ISO 730-1 (third edition 1994) classifies the dimensions by four *hitch categories* according to nominal tractor PTO power, Figs. 1.145 and 1.146 and Table 1.45.

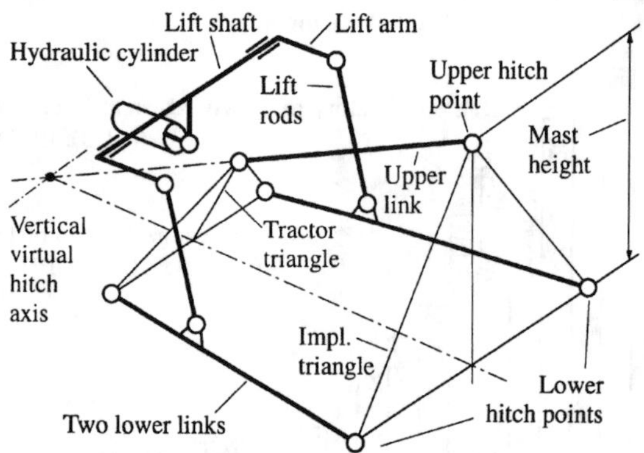

Figure 1.144. Classic concept of rear three-point hitch (modern version: cylinder(s) also outside).

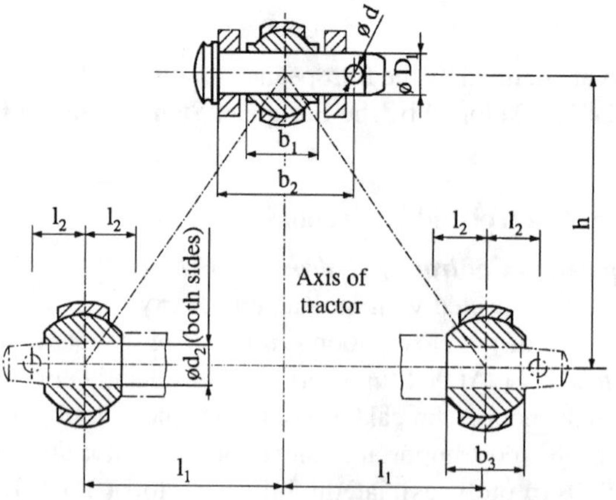

Figure 1.145. Dimensions of tractor rear hitch points (ISO 730-1: 1994). Front hitch almost the same (ISO 8759-1).

Table 1.45. Important tractor rear hitch dimensions (ISO 730-1: 1994)

	Hitech Categories with PTO Power in kW at Rated Eng. Speed:	1	2	3	4L	4H
		Up to 48	Up to 92	80 to 185	—150 to 350—	
		Upper Hitch Point				
D_1	Diameter of hitch pin	$19^{\ 0}_{-0.08}$	$25.5^{\ 0}_{-0.13}$	$31.75^{\ 0}_{-0.2}$	$45^{\ 0}_{-0.8}$	$45^{\ 0}_{-0.8}$
b_1	Width of ball	44 max.	51 max.	51 max.	64 max.	64 max.
b_2	Linch pin hole distance	76 min.	93 min.	102 min.	140 min.	140 min.
d	Diameter for linch pin hole	12 min.	12 min.	12 min.	17.5 min.	17.5 min.
		Lower Hitch Point				
d_2	Diameter of hitch pin hole	$22.4^{+0.25}_{\ 0}$	$28.7^{+0.3}_{\ 0}$	$37.4^{+0.35}_{\ 0}$	$51^{+0.5}_{\ 0}$	$51^{+0.5}_{\ 0}$
b_3	Width of ball	$35^{\ 0}_{-0.2}$	$45^{\ 0}_{-0.2}$	$45^{\ 0}_{-0.2}$	$57.5^{\ 0}_{-0.5}$	$57.5^{\ 0}_{-0.5}$
l_1	Lateral distance from lower hitch point to centerline of tractor	359	435	505	610 or 612	610 or 612
l_2	Lateral movement of lower hitch point	100 min.	125 min.	125 min.	130 min.	130 min.
L	Distance from PTO to lower hitch points	500 to 575	550 to 625	575 to 675	575 to 675	610 to 670
h	Mast height	460 ± 1.5	610 ± 1.5	685 ± 1.5	685 ± 1.5	1000 ± 1.5

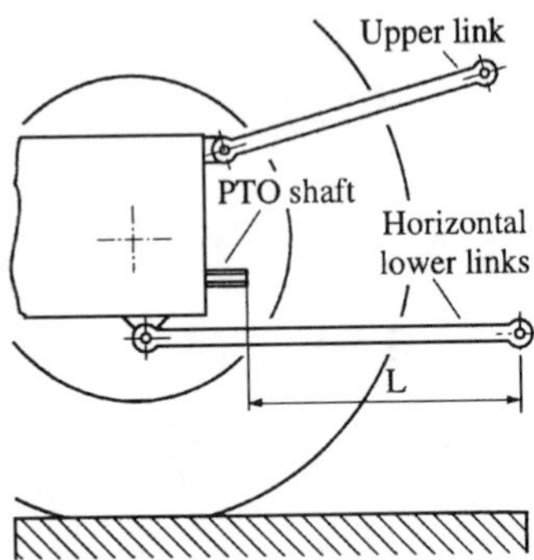

Figure 1.146. Distance from PTO to rear hitch lower link points (ISO 730-1: 1994).

Kinematics and Forces of the Three-point Hitch

Kinematics

The three-point linkage is a three-dimensional system. Plane views, Fig. 1.147, lead to four-bar linkages with *virtual hitch points* and *convergence distances* (ISO 730-1: 1994). ISO 730-1 recommends following values for horizontal convergence distances:

Cat. 1: 1700 to 2400 mm
Cat. 2: 1800 to 2400 mm
Cat. 3: 1900 to 2700 mm
Cat. 4: 1900 to 2800 mm

If a certain implement (weight) shall be carried, the necessary torque at the lift shaft can be determined graphically as demonstrated by Fig. 1.148. Theoretical *lifting force* maps have been developed by [129], practical values see OECD Test Reports.

Forces in Case of Floating Implement.

Floating position is characterized by "no pressure in the cylinders" (connected with oil sump). Implement is controlled only by the equilibrium of implement forces and the virtual hitch point, as demonstrated for a mounted plow, Fig. 1.149. The total plow force results from soil resistance, plow weight and supporting *heel force* (with friction). The line of draft must cross the *virtual hitch point*, vertical load transfer from implement to tractor and from front to rear axle is not very important. Kinematics of front three-point hitch have been analyzed by [132].

Forces in Case of Hydraulic Implement Control

Implement is mainly carried by the tractor hitch in this case, demonstrated by Fig 1.150. Soil resistance and plow weight are the same as for the above explained floating position, but the heel force is reduced considerably, influencing the *line of draft* as shown. This principle (as invented by Harry Ferguson) has three important advantages:

1. Improved implement control by hydraulics (draft control, position control and others)

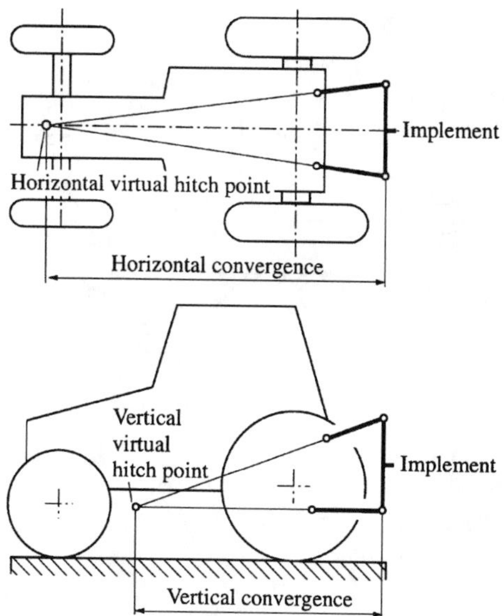

Figure 1.147. Horizontal and vertical convergence (ISO 730-1: 1994).

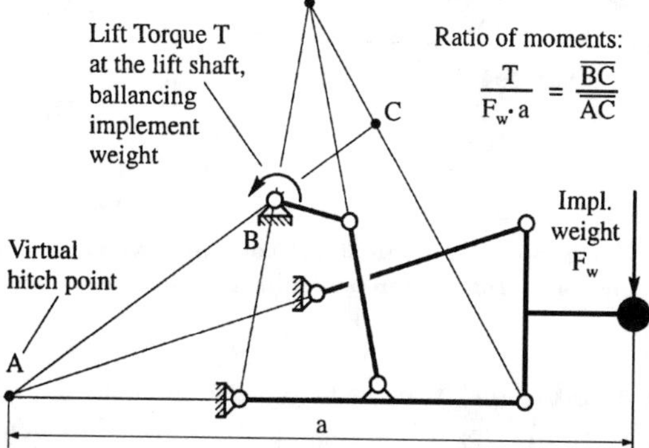

Figure 1.148. Graphical determination of moments for tractor rear three-point hitch [128].

2. Improved traction by transfer of vertical implement forces to the tractor chassis
3. Reduced friction losses due to reduced heel forces or rolling resistance of supporting wheels.

Control Strategies for Three-point Hitches

The various systems are addressed by Table 1.46. They relate mainly to rear hitches (control strategies for front hitches have not yet been developed as well). *Floating* and

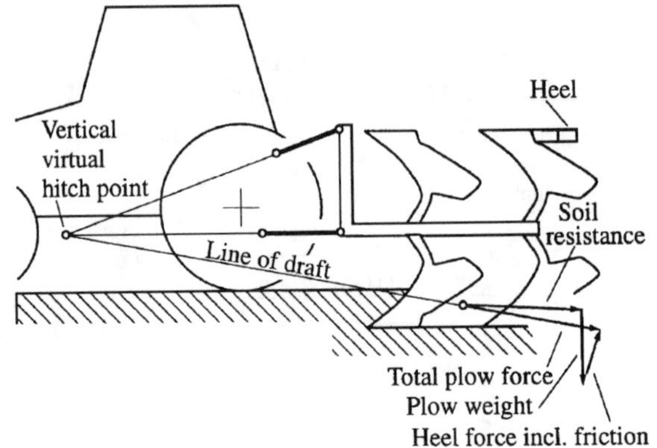

Figure 1.149. Three-point linkage operating in floating position; forces simplified (more in detail see [130, 131]).

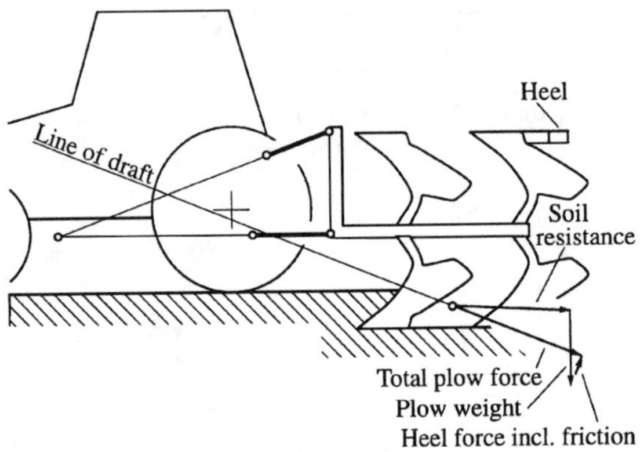

Figure 1.150. Three-point linkage operating with hydraulic implement control; forces simplified (more in detail see [130, 131]).

force control by upper link were the first strategies. An early stage of the *Ferguson System* is represented by Fig. 1.151. Upper link force is transformed to a clockwise movement of the control linkage operating the control valve. This opens the pump intake port, creating a certain flow to the lift cylinder, lifting in the same time the implement and finally reducing the upper link force (approaching the task set by the control lever).

Figure 1.152 shows a common system that combines strategies 3 and 4. They can be run both separately or in a *mixed mode.* Compared with 2 and 3 the mix reduces depth variations in case of soil density variations and improves dynamic control stability (problem of force control [133]).

Development of control strategies and systems is influenced considerably by sensor and processor development. Figure 1.153 demonstrates some important principles of forces sensing. Principles M1 and M2 are typical for low-cost *upper link sensing* systems (Ferguson). M3 is very economic, as it uses the lower link fastenings at the same time

Table 1.46. **Basic tractor hitch control strategies [29]**

No	Control Strategy	Typical Applications, Comments
1	"Free floating implement" (valve in "sinking position")	Set-in procedure of plows, all implements having supporting wheels (such as power tiller, rotary mower)
2	"Force control by upper link" (original Ferguson-System)	Mounted implements (smaller tractors) for tillage
3	"Force control by lower links"	Mounted and semi-mounted implements for tillage (bigger tractors)
4	"Position control"	Mounted implements (i.e., for distribution of chemicals and seeds)
5	"Mix of force and position control"	Tillage implements (in relation to 1, 2, 3: reduced depth variations)
6	"Transmission torque control"	Tillage implements
7	"Hydraulic pressure control"	Load transfer from trailer or implement wheels or other supporting elements to tractor wheels
8	"Wheel slip control"	Tillage implements
9	"Control strategies" working with implement sensors	Mounted implements, i.e., tillage implements with depth sensor
10	"Control strategies including external cylinders"	Semi-mounted tillage implement with actuated rear gage wheel

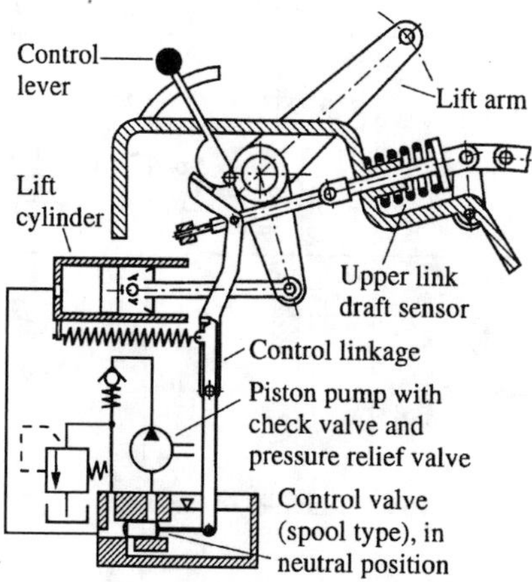

Figure 1.151. Early configuration of the "Ferguson System" with upper link force sensing [123, 124 modified].

as a force sensor. (J. Deere 1965). H2 has been used by Case for many years and E3 represents the electronic *force sensor bolt* first introduced by Bosch in 1982. Electronic data processing (Bosch/Deutz 1978) has had a dramatic increase since the mid 1980s. The first analog working generations have been replaced more and more by digital systems offering higher flexibility.

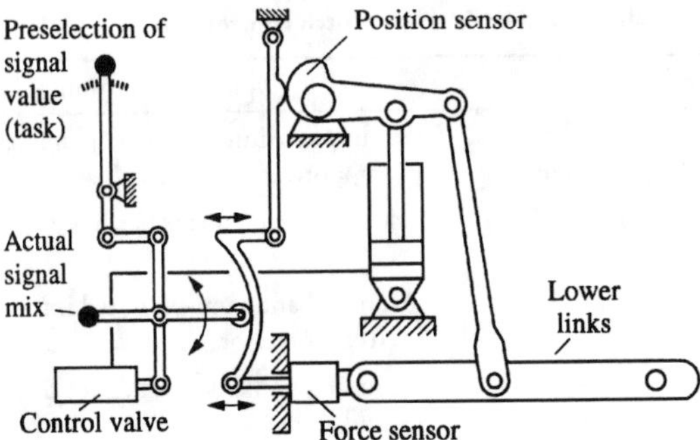

Figure 1.152. Hitch control system with lower link force control and position control – also combined (mixed); very popular system (mechanically or electronically) [29].

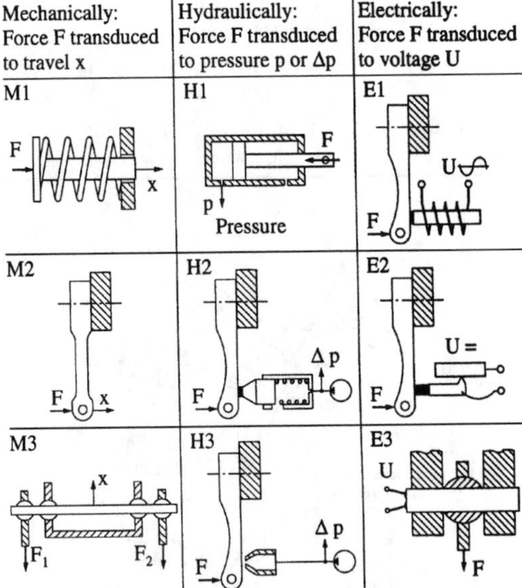

Figure 1.153. Principles of force sensors [125].

Fluid Power Systems: Symbols and Vocabulary

Communication in hydraulics can be simplified considerably by graphic *symbols*, as in other disciplines such as electrics, electronics, software engineering, transmissions etc.

The first international standard ISO 1219 (1976) was replaced by ISO 1219-1 in 1991 introducing some modifications (i.e., concerning pressure control valves and hydraulic cylinders). An extract has been formed by Table 1.47. ISO 1219-1 (1991), focusing more on general design rules than ISO 1219 (1976) and containing, for example, standardized relations of typical dimensions (circles, squares, rectangles, triangles etc.).

Table 1.47. Extract of ISO 1219-1 (1991) "Fluid Power Systems and Components - Graphic Symbols and Circuit diagrams. Part 1: Graphic Symbols"

Pumps and Motors

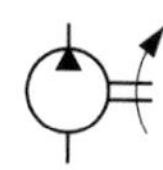
Hydraulic pump with one direction of flow, fixed displacement and one direction of rotation

Hydraulic pump with two directions of flow, variable displacement and one direction of rotation

Hydraulic motor with one direction of flow, fixed displacement and one direction of rotation

Hydraulic variable speed drive unit, pump and motor with variable displacement, two directions of output rotation

Hydraulic Cylinders

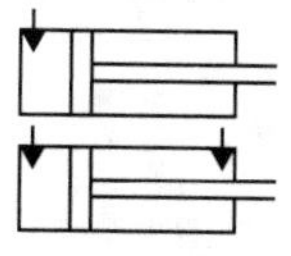
Single acting hydraulic cylinder

Double acting hydraulic cylinder

Lines

Working line

Control line

Flexible line

Enclosure for integrated functions

Diverse Components

Reservoir

Filter

Cooler (left) Heater (right)

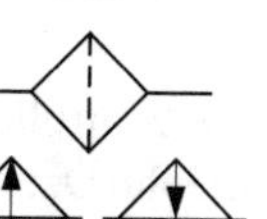
Accumulator

Directional Control Valves

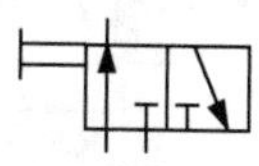
Directional valve 3/2: Three ports and two distinct positions, muscular control

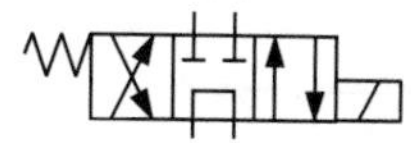
Directional valve 4/3: Four ports, three distinct positions. Actuated by electrical solenoid and return spring

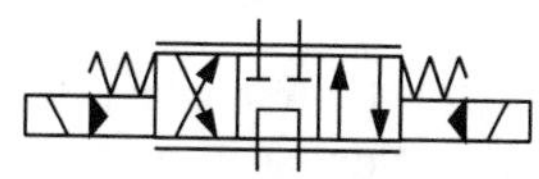
Directional valve 4/3: Four ports, three distinct positions and infinite intermidiate positions (throttle effects). Electric input control signals, internal operation by internal pilote pressure, centering spring

Pressure Control Valves

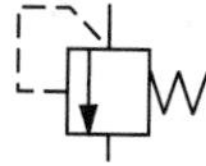
Pressure relief valve. Character of control line indicates internal control function: Input pressure ballanced by spring load

Flow Control Valves

Adjustable restriction valve

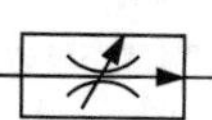
Flow control valve with two ports: Flow is about constant independant of pressure levels

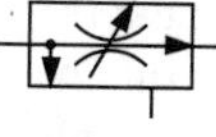
Flow control valve with three ports: Flow is about constant independant of pressure levels. Bypass for surplus flow

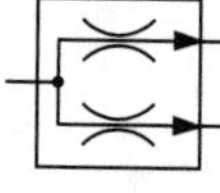
Flow divider valve: two output flows of a fixed ratio being widely independant of pressure levels

Non-return Valves and Quick Coupling

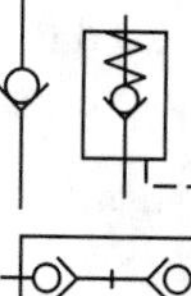
Non-return valves (check-valve), right: valve can be opened by a pilot control pressure

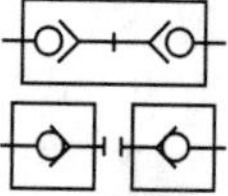
Quick-release coupling with mechnically opened non-return valves. Above: coupled, below: disconnected

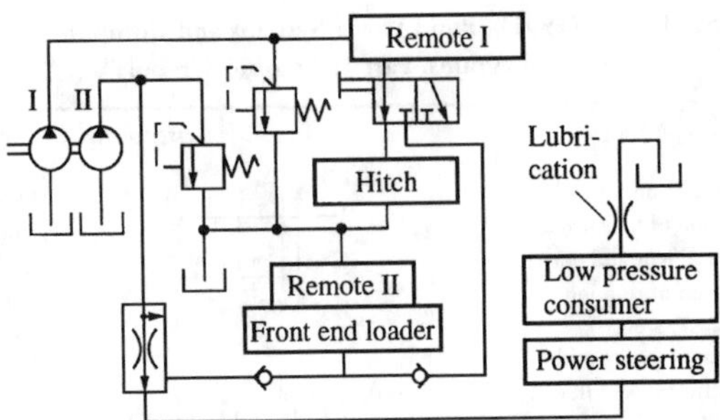

Figure 1.154. Hydraulic circuit with fixed displacement pumps and use of flow control valve bypass output [134, Fendt].

A second part ISO 1219-2 (1996) contains typical circuit diagrams by examples.

The ISO 1219 standardization has been well accepted worldwide (in Germany, for example, by taking it over as national standard DIN-ISO 1219).

A separate vocabulary (English – French – German) has been developed by ISO 5598.

Hydraulic Circuits: Basic Systems

Most existing circuits of tractor hydraulics can be classified into five basic systems, which are explained below. The terms *open center* and *closed center* address the neutral position of the service valves.

Systems with Fixed Displacement Pump(s) and "Open Center" Control ("PF-open Center")

This is the most used system worldwide, commonly working with one or two gear pumps. Output flow is approximately constant, if speed at the input shaft is constant (therefore also called *constant flow system*). Figure 1.54 demonstrates a typical European concept. A first gearpump I provides a flow for one remote line and the hitch, a second gearpump II for power steering (priority by flow control valve) – the bypass flow is used for front-end loader and remote II. As the front end loader requires high flow rates, bypass and pump I flow can be summarized by the 3/2 valve. The "PF-open-center" creates the lowest initial cost; disadvantages are: full flow circulating all the time, high losses in case of low required flow rates at high pressures, low flows when engine is idling.

Constant Pressure Systems with Fixed Displacement Pump and Closed Center ("PF-constant Pressure")

A fixed displacement pump is used to charge accumulators by a pressure-controlled charge valve (unsteady). This principle has been used previously by Ford and John Deere.

Constant Pressure Systems with Variable Displacement Pump and Closed Center ("PV-constant pressure")

This type of circuit was introduced for mass-produced tractors by John Deere in 1960 (3010, 4010) [135]. The pump is only delivering the required flow, Fig. 1.155.

Adjustment of system pressure to load induced pressure needs *flow controller* (losses), but all actuators can be operated without reciprocal influence.

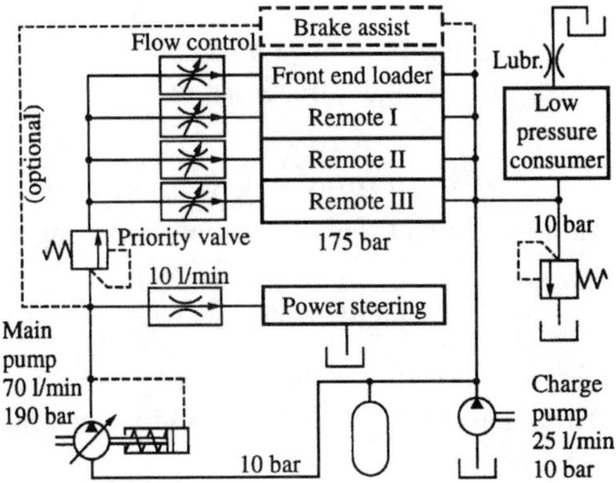

Figure 1.155. Hydraulic circuit with variable displacement pump and constant pressure control [134, J. Deere 1985].

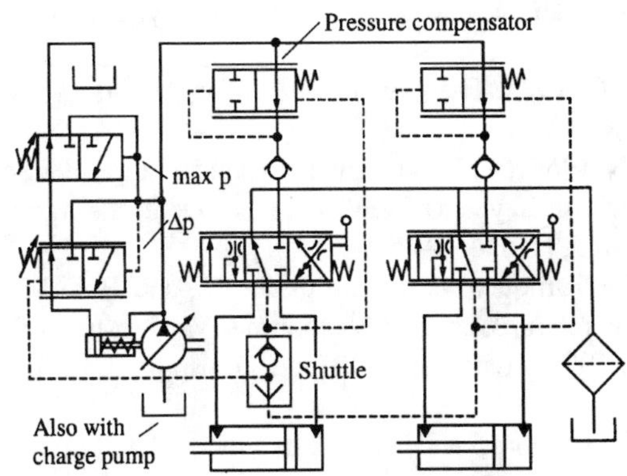

Figure 1.156. Load sensing hydraulic circuit [141].

Flow on Demand Systems with Variable Displacement Pump and Closed Center ("PV-load-sensing")

Circuits of this type are usually called *load-sensing* (LS) and were introduced first to agricultural tractors by Allis Chalmers (USA) in 1973 [136]. Their potential of energy saving by *on demand flow* and pressure is well confirmed [134, 137–141]. The second objective of LS-systems is the excellent ability for *proportional flow control valves*, Fig. 1.156. A variable displacement pump is controlled by the pressure drop of the service valve; the pump thus delivers for any valve setting just that flow rate necessary to keep the value Δp constant. Therefore the flow through the valve is directly proportional to the valve opening area (assuming constant orifice flow coefficient). When the valve is closed, the pump automatically moves to a "near zero flow" position, reducing at the same time its pressure to Δp (i.e., 20 bar). If the working pressure exceeds a given limit, the pump also reduces flow output, saving a pressure relief valve and its energy losses (heat).

In the case of more than one actuator (with the task of no reciprocal influence), the highest actuator pressure is used for the pump control (shuttle). In order to have the same pressure drop Δp also at the other directional valve, pressure compensator valves are added.

Flow on Demand-systems with Fixed Displacement Pumps ("PF-load Sensing")

They use pressure drop controller to ensure the constant Δp at the service valve [141].

References

1. –,–. 1993. Significant contributions and contributors to the mechanization of agriculture and construction 1893–1993. EMI Equipment Manufacturers Institute. Chicago, IL.
2. Worthington, W. H. 1967. The engineers history of the farm tractor. Impl. & Tractor 82(2) through (11).
3. Fletcher, L. J. 1922. Factors Influencing Tractor Development. Agric. Engineering 3(11): 179–182.
4. Holt, P. E. 1925. The development of the track type tractor. Agric. Engineering 6(4): 76–79.
5. Kranich, F. N. G. 1925. The power take off for tractors. Agric. Engineering 6(9): 204–208.
6. Boltinski, W. N. 1967. Der Traktorenbau in den 50 Jahren der Sowjetmacht (Tractor development within 50 years of sowjet power). Deutsche Agrartechnik 17(8): 366–367, (9): 442–443, (10): 454–456.
7. Liljedahl, J. B., Turnquist, P. K., Smith, D. W., and Hoki M. 1989. Tractors and their power units. Fourth edition. Melbourne: Van Nostrand Reinhold.
8. Freitag, D. R. 1979. History for Offroad Transport. J. of Terramechanics 16(2): 49–68.
9. Wendel, C. 1985. Nebraska Tractor Tests Since 1920. Sarasota/Florida: Crestline Publishing.
10. Williams, M. 1991. Tractors since 1889. Ipswich: Farming Press.
11. Goering, C. E. 1992. Engine and Tractor Power. 3rd edition. St. Joseph, Michigan: American Society of Agricultural Engineers (ASAE).
12. Göhlich, H. 1984. Development of tractors and other agricultural vehicles. J. Agric. Engng. Research 29(1): 3–16.
13. Renius, K. T. 1994. Trends in Tractor Design with Particular Reference to Europe. J. Agric. Engng. Research 57(1): 3–22.
14. Söhne, W. 1986. Final lectures on terramechanics at the Technische Universität München.
15. Wismer, R. D., and Luth H. J. 1972. Offroad traction prediction for wheeled vehicles. ASAE paper 72–619.
16. Upadhyaya, S. K., Wulfsohn, D., and Jubbal, G. 1989. Traction prediction equations for radial ply tires. J. of Terra-mechanics 26(2): 149–175.
17. Dwyer, M. S., Comely, D. R., and Everden, D. W. 1974. Handbook of Agricultural Tyre Performance. NIAE Silsoe.

18. Bekker, M. G. 1969. Introduction to Terrain-Vehicle Systems. Ann Arbor: The University of Michigan Press.
19. Wong, J. Y. 1989. Terramechanics and Offroad Vehicles. Amsterdam, Oxford, New York, Tokyo: Elsevier.
20. McKibben, E. G., and Davidson, J. B. 1949. Transport Wheels for Agricultural Machines, Part IV. Agric. Engineering 21(2): 57–58.
21. Renius, K. T. 1987. Traktoren. 2nd edition. München: BLV Verlag.
22. Leviticus, L. I., and Reyes, J. F. 1985. Tractor Performance on Concrete. Transactions ASAE 28(5): 1425–1429.
23. Söhne, W. 1953. Druckverteilungen im Boden und Bodenverformung unter Schlepperreifen. (Soil pressure distribution and soil deformation under tractor tires). Grundlagen der Landtechnik 3(5): 49–63.
24. Söhne, W. 1958. Fundamentals of Pressure Distribution and Soil Compaction Under Tractor Tires. Agric. Engineering 39(5): 276–291.
25. Lanças, K. P., Santos Filho, A. G., and Upadhyaya, S. K. 1996. Energy and time saving using "low/correct" inflation pressure in radial ply tires. Proc. Int. Conf. Agric. Engng., AgEng '96 Madrid, 23.09.–26.09.96, Report No. 96-A-021.
26. Hammel, K. 1994. Soil stress distribution under lugged tires. Soil & Tillage Research 32(2/3): 163–181.
27. Schwanghart, H. 1994. Effect of Reduced Tire Inflation Pressure on Agricultural Tires. Proc. 6th European ISTVS Conference 28.–30.9.1994, Vienna, 1: 277–294.
28. Rempfer, M., and Schwanghart, H. 1996. Economic and ecological benefits of central tire inflation systems in Agricultural Vehicles. Proc. Int. Conf. Agric. Engng. AgEng '96 Madrid, 23.09.–26.09.96, Report No. 96-A-068.
29. Renius, K. T. 1996. Traktoren und Erdbaumaschinen. (Tractors and Earth Moving Machinery). Lectures held at the Technische Universität München (since 1982).
30. Holm, I. C. 1969. Multi-pass Behaviour of Pneumatic Tyres. J. of Terramechanics 6(3): 47–71 (see also Ph. D. thesis Technische Universität München 1971).
31. Schwanghart, H. 1982. Umsturzverhalten von Traktoren ... (Tractor overturn mechanics...). Forschungsbericht Agrartechnik MEG No. 73. Institut für Landmaschinen, Technische Universität München.
32. Ellis, R. W. 1977. Agricultural Tire Design. Requirements and Selection Considerations. ASAE Lecture Series (3). St. Joseph, MI, American Society of Agricultural Engineers.
33. Steinkampf, H. 1974. Ermittlung von Reifenkennlinien und Gerätezugleistungen für Ackerschlepper. (Determination of tire performance characteristics and drawbar power for tractors). Ph. D. thesis Techn. Univ. Braunschweig/Germany. Landbauforschung Völkenrode, Sonderheft 27(1975).
34. Wulfsohn, D., Upadhyaya, S. K., and Chancellor, W. J. 1988. Tractive Characteristics of Radial Ply and Bias Ply Tyres in a Californian Soil. J. of Terramechanics 25(2): 111–134.

35. Steinkampf, H. 1981. Ergebnisse aus Reifenvergleichsversuchen. (Tire performance comparisons). Landtechnik 36(10): 479–482.
36. Bacher, R. 1981. Möglichkeiten zur Lärmreduzierung an Ackerschleppern. (Potential of noise reduction for tractors). Ph. D. thesis Technische Universität München.
37. Kirste, T. 1989. Entwicklung eines 30 kW-Forschungstraktors als Studie für lärmarme Gesamtkonzepte. (Development of a 30 kW research tractor as a study for low-noise concepts). Fortschritt-Berichte VDI Series 14(43). Düsseldorf: VDI-Verlag.
38. Renius, K. T. 1989. Gedanken zur frühen und neueren Entwicklung des Allradantriebes bei Traktoren. (Considerations on previous and recent developments of tractor fourwheel drives). Landtechnik 44(10): 420–425.
39. Sonnen, F. J. 1969. Drawbar Performance of High-Powered Farm Tractors with Rear-Wheel and Four-Wheel Drive. J. of Terramechanics 6(1): 7–21.
40. Renius, K. T. 1979. European four-wheel drive: are technical advantages profitable? ASAE paper 79–1555.
41. Rackman, D. H. 1985. Four-wheel drive tractor – a review. J. of Agric. Engng. Research 31(3): 185–201.
42. Paul, M., and Wilks, E. 1989. Driven Front Axles for Agricultural Tractors. ASAE Lecture Series (14). St. Joseph, MI: American Society of Agricultural Engineers.
43. Dwyer, M. J. 1970. The braking performance of tractor-trailer combinations. J. of Agric. Engng. Research 15(2): 148–162.
44. Simuttis, A. F. 1985. Über das Bremsverhalten landwirtschaftlicher Traktoren im Geschwindigkeitsbereich bis 40 km/h. (Braking performance of agricultural tractors up to 40 km/h. top speed). Fortschritt-Berichte VDI Series 12(56). Düsseldorf: VDI-Verlag.
45. Reiter, H. 1990. Verluste und Wirkungsgrade bei Traktorgetrieben (Losses and efficiencies of tractor transmissions). Fortschritt-Berichte VDI Series 14(46). Düsseldorf: VDI-Verlag.
46. Product information on Steyr tractors. 1995. Steyr Landmaschinentechnik GmbH, Austria.
47. Wittren, R. A. 1975. Power Steering for Agricultural Tractors. ASAE Lecture Series (1). St. Joseph, MI: American Society of Agricultural Engineers.
48. Fukui, T. 1984. Fahrzeug mit Allradantrieb. (Vehicle with four-wheel drive) Application for patent by DE 3 408 991 F. R. G. (12.03.1984 by Kubota). See also SAE paper 901 630 (1990).
49. Grad, K. 1994. Improvements of FWD-Tractor Performance by Electronic Controls. AgEng '94 International Conference 29.08.–01.09.1994 Milan. Paper 94-D-003.
50. Grad, K. 1997. Zur Steuerung und Regelung des Allradantriebes bei Traktoren. (Control of the four-wheel drive of agricultural tractors). Fortschritt-Berichte VDI Series 14(82). Düsseldorf: VDI-Verlag.
51. Schwanghart, H. 1968. Lateral Forces on Steered Tyres in Loose Soil. J. of Terramechanics 5(1): 9–29.

52. Crolla, D. A., and El-Razaz, A. S. A. 1987. A review of the combined lateral and longitudinal force generation of tyres on deformable surfaces. J. of Terramechanics 24(3): 199–225.
53. Kutzbach, H. D., and Barrelmeyer, T. H. 1996. Measurement and prediction of lateral and longitudinal tyre forces of agricultural tyres. AgEng '96. International Conference Madrid 23.–26.09.1996. Paper 96A–097.
54. Barrelmayer, T. 1996. Untersuchung der Kräfte an gelenkten und angetriebenen Ackerschlepperrädern bei Gelände- und Straßenfahrt. (Forces of steered and powered tractor tires offroad and onroad). Fortschritt-Berichte VDI Series 14(79). Düsseldorf: VDI-Verlag.
55. Parsons, S. D., Griffith, D. R., and Doster D. H. 1984. Equipment wheel spacing for ridge-planted crops. Agric. Engineering 65(8): 10–14.
56. Kiene, W. 1968. Die Entwicklung des Dieselmotors im Ackerschlepper. (Development of the Diesel engine for agricultural tractors) Landtechnik 23(10): 302–318.
57. Ahlers, C. T. J., and Ihnen, M. H. 1992. Diesel Engine Technology for low Emissions for Agricultural and Industrial Equipment. ASAE Lecture Series (17). St. Joseph, MI: American Society of Agricultural Engineers.
58. Sprick, W. L., and Becker, T. H. 1985. The Application and Installation of Diesel Engines in Agricultural Equipment. ASAE Lecture Series (11). St. Joseph, MI: American Society of Agricultural Engineers.
59. Machinery Management. Fourth Edition 1992. Moline, IL: Deere & Company.
60. Renius, K. T., and Pfab, H. 1990. Traktoren 1989/90. (Agricultural Tractors 1989/90). ATZ Automobiltechnische Zeitschrift 92(6): 334–346.
61. Renius, K. T. 1968. Grundkonzeptionen der Stufengetriebe moderner Ackerschlepper. (Concepts of stepped ratio transmissions of modern tractors) Grundlagen der Landtechnik 18(3): 97–106.
62. Renius, K. T. 1969. Stufenlose Drehzahl-Drehmoment-Wandler in Ackerschleppergetrieben. (Tractor transmissions with continuously variable speed shift). Grundlagen der Landtechnik 19(4): 109–118.
63. Renius, K. T. 1976. European Tractor Transmission Design Concepts. ASAE paper 76-1526.
64. Browning, E. P. 1978. Design of Agricultural Tractor Transmission Elements. ASAE Lecture Series (4). St. Joseph, MI: American Society of Agricultural Engineers.
65. Renius, K. T. 1984. Neuere Getriebeentwicklungen bei Ackerschleppern. (Recent developments of tractor transmissions) Grundlagen der Landtechnik 34(3): 132–142.
66. Renius, K. T. 1991. Anforderungen und Entwicklungstendenzen bei Traktorgetrieben. (Requirements and development trends for tractor transmissions). In: VDI-Berichte 878: 41–56. Düsseldorf: VDI-Verlag.
67. Renius, K. T. 1992. Developments in Tractor transmissions. The Agricultural Engineer 47(2): 44–48.

68. Müller, F., and Sailer H. 1993. New Future-Oriented Tractor Transmissions for ZF. SAE paper 932 420.
69. Tinker, D. B. 1993. Integration of Tractor Engine, Transmission and Implement Depth Controls: Part I. Transmissions. J. Agric. Engng. Research 54(1): 1–27.
70. Scarlett, A. J. 1993. Integration of Tractor Engine, Transmission and Implement Depth Controls: Part II. Control Systems. J. Agric. Engng. Research 54(2): 89–112.
71. Renius, K. T. 1994. Tractor Transmission Developments. Proc. XII CIGR World Congress on Agric. Engng. Mailand 29.08.–01.09.94. 2: 1167–1174.
72. Renius, K. T. et al., 1988/89/90/91/92/93/95/96/97/98/99. Motoren und Getriebe bei Traktoren – Tractor engines and transmissions. Annual review in: Jahrbuch Agrartechnik – Yearbook Agric. Engng. (from 1991 bilingual german/engl.). Münster: Landwirtschaftsverlag. (Issue "1994" doesn't exist due to modified counting).
73. Wendeborn, J. O. 1966. Unter Last und stufenlos schaltbare Fahrantriebe für Schlepper. (Powershift and continuously variable tractor transmissions). Grundlagen der Landtechnik 16(2): 51–59.
74. Renius, K. T. 1976. Last- und Fahrgeschwindigkeitskollektive als Dimensionierungsgrundlagen für die Fahrgetriebe von Ackerschleppern. (Load and speed collectives for dimensioning of tractor travel drives). Fortschritt-Berichte VDI Series 1(49). Düsseldorf: VDI-Verlag.
75. Renius, K. T. 1977. Betriebsfestigkeitsberechnungen von Maschinenelementen in Ackerschleppern mit Hilfe von Lastkollektiven. (Fatigue calculation of random loaded tractor machine elements applying load collectives) Konstruktion 29(3): 85–93.
76. Renius, K. T. 1979. Betriebsfestigkeitsberechnung und Laborerprobung von Zahnrädern in Ackerschleppergetrieben. (Fatigue calculation and lab test of random loaded gear wheels in tractor transmissions). In: VDI-Berichte 332: 225–234. Düsseldorf: VDI-Verlag.
77. Vahlensieck, B. 1996. A Power Controlled Continuously Variable Transmission for Tractors: Effects on Load Spectra. Proc. Int. Conf. Agric. Engng. AgEng '96 Madrid, 23.09.–26.09.96, Report No. 96 A-025.
78. Betz, M. 1991. Hydrostatic Transmission Application Considerations for Commercial Turf Maintenance Vehicles. ASAE Lecture Series (16). St. Joseph, MI: American Society of Agricultural Engineers.
79. Fushimi, Y., Fujii T., and Kanehara, S. 1996. A Numerical Approach to Analyze the Power Transmitting Mechanisms of a Metal Pushing V-Belt Type CVT. SAE paper 960720.
80. Schönnenbeck, G., and Heinrich, J. 1991. Continuously Variable Chain Drives for High Torque Application. Paper presented at ISME Conference "Motion and Power Transissions." Nov. 1991. Hiroshima, Japan.
81. Sauer, G., and Renius, K. T. 1991. Traktorantriebe mit stufenlosen Kettenwandlern. (Tractor drives with continuously variable chain variators). Landtechnik 46(7/8): 322–328.

82. Sauer, G. 1996. Grundlagen und Betriebsverhalten eines Zugketten–Umschlingungsgetriebes. (Fundamentals and operational characteristics of a pull type chain variator). Fortschritt-Berichte VDI Series 12(293). Düsseldorf: VDI-Verlag.
83. Dziuba, F., and Honzek, R. 1997. Neues stufenloses leistungsverzweigtes Traktorgetriebe. (A new power-split tractor transmission). Agrartechnische Forschung 3(1): 19–27.
84. Marshall, H. (Fendt), 1973. German Patent 2335629 (Application July 13, 1973).
85. Hoffmann, D. 1970. Verlustleistungen und Wirkungsgrade eines Hydro-Axialkolbengetriebes mit sphärischen Kolben. (Power losses and efficiencies of a hydrostatic transmission using bent axis units with spherical pistons). Landtechnische Forschung 18(3): 65–69.
86. Publications of Jarchow, F., Emamdjomeh, N., and Fredriksen, N., (since 1989).
87. Matthews, J., and Talamo, J. D. C. 1970. A study of tractor noise control. Conference on agricultural and allied tractors. London, Oct. 6–7, paper No. 16.
88. Göhlich, H. 1987. Mensch und Maschine (Operator – machinery relations). Lehrbuch der Agrartechnik (5). Hamburg/Berlin: Verlag P. Parey.
89. –,–. 1972. Nebraska Test No. 1110 – John Deere 4430 Quad-Range Diesel (Low noise safety cab).
90. Sellon, R. N. 1976. Design of operator enclosures for agricultural equipment. ASAE Lecture Series (2). St. Joseph, MI: American Society of Agricultural Engineers.
91. Purcell, W. F. H. 1980. The human factor in farm and industrial equipment design. ASAE Lecture Series (6). St. Joseph, MI: America Society of Agricultural Engineers.
92. Batel, W. et al., 1976. Senken der Belastungen am Arbeitsplatz auf fahrenden Arbeitsmaschinen durch technische Maßnahmen. (Reduction of operator loads at the working place in mobile machinery by technical means). Several articles within Grundlagen der Landtechnik 26(2): 33–69.
93. Schmittke, H. (Editor). 1981. Lehrbuch der Ergonomie. (Handbook of ergonomics). München, Wien: Carl Hanser Verlag.
94. Dreyfuss, H. 1967. The measure of man. New York: Whitney Library of Design.
95. Yadar, R., and Tewan, V. K. 1997. Cab for Indian Tractors: A case study. AMA 28(2): 27–29.
96. Göhlich, H. 1991. Tractor Ride Dynamics. Chapter 3 in "Progress in Agricultural Physics and Engineering" (edited by J. Matthews). Wallingford: C. A. B. International.
97. Lines, J. A. 1987. Ride vibrations of agricultural tractors: transfer functions between the ground and the tractor body. J. Agric. Engng. Research 37(1): 81–91.
98. Wehage, R. A. 1987. Vehicle dynamics. (Overview of 13 papers presented on the 9th ISTVS Conf. Barcelona, Spain, 1987). J. of Terramechanics 24(4): 295–312.
99. Xu, Da-Ming, and Yong, R. N. 1993. Autocorrelation model of road roughness. J. of Terramechanics 30(4): 259–274.

100. Prasad, N., Tewari, V. K., and Yadav, R. 1995. Tractor ride vibration – a review. J. of Terramechanics 32(4): 205–219.
101. Wendeborn, J. O. 1966. The irregularities of farm roads and fields as sources of farm vehicle vibration. J. of Terramechanics 13(3): 9–40.
102. Omiya, K. 1986. Characteristics of farm field profiles as sources of tractor vibration. J. of Terramechanics 23(1): 23–36.
103. Stikeleather, L. F. 1981. Operator seats for agricultural equipment. ASAE Lecture Series (7). St. Joseph, MI: American Society of Agricultural Engineers.
104. Meyer, R. E., Schwab, C. V., and Bern, C. J. 1993. Tractor Noise Exposure Levels for Bean-bar Riders. Transactions ASAE 36(4): 1049–1056.
105. Witte, E. 1979. Stand und Entwicklung der Lärmbelastung von Schlepper- und Mähdrescherfahrern. (Review of operator noise loads on tractors and combines) Grundlagen der Landtechnik 29(3): 92–99.
106. Teaford, W. J. 1993. Roll-over Protective Structures (ROPS) for Wheeled Agricultural Tractors. ASAE Lecture Series (18). St. Joseph, MI: American Society of Agricultural Engineers.
107. Hunter, A. G. M. 1991. Stability of Agricultural Machinery on Slopes. Chapater 4 in "Progress in Agricultural Physics and Engineering" (edited by J. Matthews). Wallingford: C. A. B. International.
108. Manby, T. C. D. 1963. Safety aspects of tractor cabs and their testing. Journal and Proceedings Institution of Agric. Engineers 19(3): 53–68.
109. Moberg, H. A. 1964. Tractor safety cabs. Test methods and experiences gained during ordinary farm work in Sweden. Report of the National Swedish Test Institute for Agricultural Machinery, Upsala.
110. Bucher, D. H. 1967. A protective canopy for the farm tractor. Agric. Engineering 48(9): 496–499 and 506.
111. Willsey, F. R., and Liljedahl, J. B. 1969. A study of tractor overturning accidents. ASAE paper 69–639.
112. Nordström, O. 1970. Safety and comfort test program. Swedish approach. ASAE paper 70–105.
113. Manby, T. C. D. 1970. Energy absorbed by protective frames for drivers when tractor overturns rearward. Departmental Note DN/TMPD/25/70. NIAE, Silsoe.
114. Heidt, H. 1982. Die Arbeitsunfälle mit Landmaschinen und ihre Verhütung. (Working accidents with agricultural machinery and how to avoid them). Grundlagen der Landtechnik 32(3): 78–85.
115. Heidt, H. 1984. Schwerpunkte landwirtschaftlicher Arbeitsunfälle und Möglichkeiten zu ihrer Verhütung. (Important working accidents in agriculture and the potential of avoiding them) Landtechnik 39(1): 35–40.
116. Chisholm, C. J., and Seward, P. C. 1975. The Correlation between Damage to Rollover Protective Structures in Tractor Overturning Accidents and in Standard Tests. Departmental Note DN/E/674/4600. NIAE, Silsoe.
117. Lehtola, C. J., Marley, S. J., and Melvin, S. W. 1994. A Study of Five Years of Tractor-related Fatalities in Iowa. Applied Engng. in Agriculture 10(5): 627–632.

118. Boyer, F., Chisholm, C. J., and Schwanghart, H. 1976. Umsturzschutzvorrichtungen für landwirtschaftliche oder forstwirtschaftliche Zugmaschinen auf Rädern. (Rollover protective structures for agricultural and forestry wheel tractors). Expertise for EEC commission XI/1036/76-D, Brussels.
119. Schwanghart, H. 1975. Festigkeitsbetrachtung von Schlepper–Umsturzsicherheitsrahmen, besonders für die Nachrüstung. (Loads and deflections of roll-over protective structures, mainly for after-sales equipment) Landtechnik 30(10): 441–445.
120. Lüttel, L. 1995. Product-related implementation of the new EU safety regulations. Yearbook Agricultural Engineering 7, chapter 1.3: 30–35. Münster (Germany): Landwirtschaftsverlag.
121. Ferguson, H. G. 1925. Apparatus for Coupling Agricultural Implements to Tractors and Automatically Regulating the Depth of Work. British Patent No. 253.566 (applied Feb. 12, 1925, accepted June 14, 1926).
122. Booth, C. E., and Condie, A. T. 1986. The new Ferguson album. Carlton/UK: Allan T. Condie Publications.
123. Morling, R. W. 1979. Agricultural tractor hitches – analysis of design requirements. ASAE Lecture Series (5). St. Joseph, MI: American Society of Agricultural Engineers.
124. Seifert, A. 1951. Ölhydraulische Kraftheber für den Ackerschlepper. (Hydraulic hitches for tractors) Grundlagen der Landtechnik (1): 45–60.
125. Hesse, H. 1982. Signalverarbeitung in Pflugregelsystemen. (Signal processing in tractor hitch controls) Grundlagen der Landtechnik 32(2): 54–59.
126. Flerlage, B. 1956. Normung der Dreipunktaufhängung am Schlepper. (Standardization of three-point tractor hitch). Grundlagen der Landtechnik (7): 89–106.
127. Erwin, R. L. 1960. The Three-Point Hitch Standard. ASAE paper 60–138.
128. Hain, K. 1953. Das Übersetzungsverhältnis in periodischen Getrieben von Landmaschinen. (input-output ratios of periodical linkages for agricultural machinery) Landtechnische Forschung 3(4): 97–108.
129. Have, H., and Kofoed, S. S. 1972. Die Hubkraft-Kennlinien eines Dreipunkt-Systems. (Lifting force characteristics of a three-point linkage system). Grundlagen der Landtechnik 22(1): 16–20.
130. Skalweit, H. 1951. Kräfte zwischen Schlepper und Arbeitsgerät. (Forces between tractor and implement). Grundlagen der Landtechnik (1): 25–36.
131. Cowell, P. A., and Len, S. C. 1967. Field Performance of Tractor Draft Control Systems. J. Agric. Engng. Research 12(3): 205–221.
132. Thompstone, R. G., and Cowell, P. A. 1990. The influence of front linkage geometry on tractor-implement interaction. J. Agric. Engng. Research 45(3): 175–186.
133. Pfab, H. 1995. Grundlagen zur Auslegung des geregelten Krafthebers bei Traktoren. (Fundamentals for the design of closed loop controlled tractor hitches). Fortschritt-Berichte VDI, Series 14(70). Düsseldorf: VDI-Verlag.
134. Garbers, H., and Harms, H. H. 1980. Überlegungen zu künftigen Hydrauliksystemen in Ackerschleppern. (Considerations on future hydraulic systems for tractors). Grundlagen der Landtechnik 30(6): 199–205.

135. Fletcher, E. H. 1961. Closed Center Hydraulic System with a Variable-Displacement Pump. ASAE paper 61–644. See also Agric. Engng. Vol. 6, 1963(1): 18–21.
136. Khatti, R. 1973. Load sensitive hydraulic system for Allis-Chalmers models 7030 and 7050 agricultural tractors. SAE paper 730 860. Warrendale, PA, USA: Society of Automotive Engineers.
137. Harms, H.-H. 1980. Energieeinsparung durch Systemwahl in der Mobilhydraulik. (Energy savings by system analysis for mobile machinery hydraulics) VDI-Z. 112(22): 1006–1010.
138. Jarboe, H. J. 1983. Agricultural load-sensing hydraulic systems. ASAE Lecture Series (9). St. Joseph, MI: American Society of Agricultural Engineers.
139. Garbers, H. 1986. Belastungsgrößen und Wirkungsgrade im Schlepperhydrauliksystem. (Load spectra and efficiencies of tractor hydraulics) Fortschritt-Berichte VDI, Series 14(30). Düsseldorf: VDI-Verlag.
140. Matthies, H. J. 1995. Einführung in die Ölhydraulik. (Introduction to oil hydraulics). 3rd edition. Stuttgart: Teubner Verlag.
141. Friedrichsen, W., and van Hamme., T. 1986. Load-Sensing in der Mobilhydraulik. (Load-sensing for mobile machinery hydraulics). ölhydraulik und pneumatik 30(12): 916–919.

1.2. Tillage Machinery

G. Weise and E. H. Bourarach

Tillage means the preparation of the growth zone in the soil (about 10 to 90 cm of the top layer of soil) for plant development. As large areas of the surface of the earth are subject to tillage, man has tried to ease the cumbersome and time-critical work of tillage and developed machines which allow in most places of the world to perform this task with ease and efficiency.

1.2.1. General Importance of Tillage Operations

The task of tillage is to prepare soils for productive use. Usually tillage is limited to the arable layer of soil, which contains organic matter and where plant life actually can occur. Tillage has to be performed to clear virgin soils of plants and animals for agricultural use. Furthermore, it must be performed to bring the seedlings into the soil and procure for them a good environment for further development.

Another objective of tillage is to control weeds and animals living in the soil, such as mice or slugs. This is, compared to the use of chemical means, an energy and time-consuming way to control pests. Another important point is surface leveling because most operations in mechanized agriculture depend on level surfaces. Irregularities in the soil niveau may be caused by traffic on the soil, harvesting or climatic effects. Together with this goes the need to distribute clods and porosity according to plant need (Fig. 1.157).

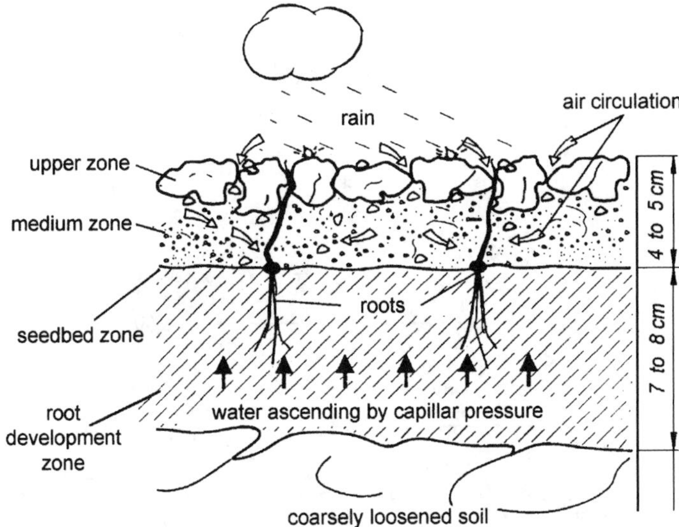

Figure 1.157. Stratified seedbed (from [1]).

The seeds should be covered by small clods for protection while around the seeds, fine soil should prevail. Under the seeds, porosity must not be too high, while smaller and larger clods should give structure to the soil. Producing this distribution of smaller and larger clods (stratified seedbed) is one of the main objectives of primary tillage. Producing fine soils for the environment of the seedling and the structure of the seedbed is the main objective of secondary tillage and seedbed preparation. Warming up the soil and bringing air to deeper layers stimulates life in the soil. At the same time, loosening makes it easier for plant roots to penetrate into deeper soil layers. An optimum porosity will also facilitate the infiltration of air and water for the plant roots, and the ascention of water from deeper soil layers during dry periods. Loosening the subsoil may be necessary to break up a hardpan, which can be created by trafficking and smearing the bottom of the tillage zone as it happens with plowing or which may develop naturally as in sodopol soils. Finally, it can be necessary to undertake soil improvements such as bringing down organic matter into the sterile subsoil or bringing up sand/clay subsoil into arable layers containing too much sand/clay in their texture.

Appropriate Tillage According to Soil Conditions

An important characteristic of agricultural soil is its texture. It is usual to divide the smallest mineral particles forming the soil matrix into the three diameter classes of sand (particles between 2 mm and 0.05 mm), silt (particles between 0.05 mm and 0.002 mm) and clay (particles smaller than 0.002 mm) [2]. Gravel and cobbles (over 2 mm) appear in agricultural soils but are usually unwanted because they make tillage hazardous and keep little organic matter. Sand-sized and larger particles can be fractionated by sieving. Silt or clay particles must be estimated by hydrometer or pipette analysis. Soils are classified into several types, according to the distribution of the three particle size classes. The

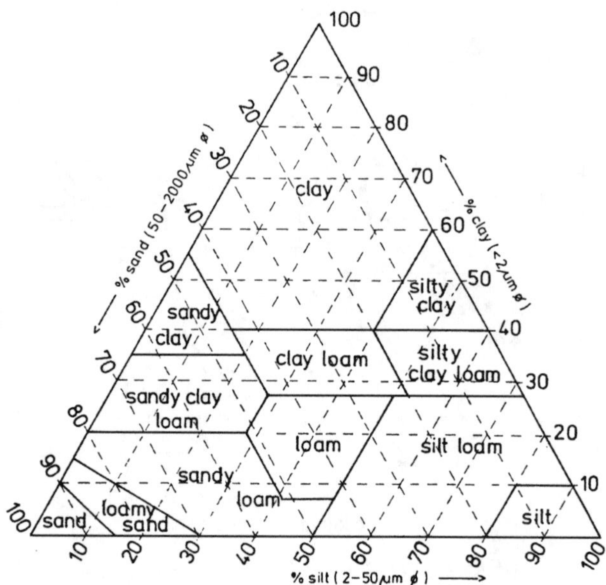

Figure 1.158. Soil classification according to sand, silt and clay content (textural triangle) (from [2]).

percentages of the three particle classes usually are drawn in a textural triangle where a soil type can be determined (Fig. 1.158).

Soil type has great influence on its workability. Tillage on *sands and sandy soils* (light soils) is easy at all moisture contents. But water storage capacity is low while infiltration rate and water conductivity are high. This can be an advantage for irrigated soils as there is only little danger of salt residues. Due to high water conductivity and easy warming as a result of a low heat capacity, organic matter is reduced fast and the content of nutrients and humus is comparatively low. Yields on these soils are usually lower than on other soil types. Melioration by incorporating clay or silt into the sand is possible.

Silty and loamy soils (medium soils) are usually the most advantageous agricultural soils. Their ability to hold water is still high enough for plant growth, and they show sufficiently large pores for good aeration. Their nutrition content is comparatively high, and the limits of their workability are much wider than with clay soils. They are usually soils where the highest yields can be achieved.

Because of its influence on soil properties, clay takes in a greater area of the textural triangle than sand. *Clay soils* (heavy soils) are difficult to cultivate as the soil becomes very hard at low moisture contents, rendering tillage operations almost impossible. At high moisture contents, plastification will make crumbling almost impossible and produce very high draft forces for tillage implements together with severe smearing at the soil-tool interfaces. Thus, these soils can be worked only in a limited range of moisture contents. Water conductivity is low so clay soils can be wet in a humid climate. Porosity is high, but most pores are fine pores that do not allow sufficient aeration. The ability to hold water is high, so drying and swelling can lead to cracks and to a kind of auto tillage of such soils. These soils can be meliorated by incorporating sand from deep soil layers or by deep loosening.

Soil particles agglutinate to soil aggregates. They form the actual soil matrix with the system of fine, medium and large pores. The medium-sized pores are most important for water storage functions, while large pores are important for conducting air to the plant roots. Fine pores are of less importance to plant development as contained water is inaccessible to plants. Large- and medium-sized pores can be produced by shrinking and swelling in clay soils, by earthworms and other animals in the soil and by tillage operations. Thus, achieving optimum porosity must always be one of the main objectives of tillage. An optimum tilth is usually obtained when crumbs of a diameter smaller than 50 mm are obtained and a sufficient amount of stable aggregates remains to create porosity and as a protection against erosion. Especially small aggregates show a small stability against water impact and wind erosion. When particles are destroyed by water, silting-up and encrustation occurs, which can obstruct plant growth. Such crusts must be broken; in order to avoid silting-up and encrustation, minimum tillage, soil cover and mulch should be applied. Generally in humid climates, one should keep soils in a more loose condition in order to avoid silting up and encrustation and to increase aeration. Under dry climatic conditions, a denser soil condition is advantageous since loosening will reduce the water storage capacity of the soil and make the water supply from deeper soil layers more difficult.

Socio-Economical Aspects of Tillage

Research carried out in the United States showed that, as the amount of tillage decreases (from conventional tillage to reduced tillage or no-till), the size and the number of machines decreases and costs for machinery and labor decrease, too [3]. Major costs are affected by machinery and herbicides. For corn production, the results are summarized below:

- Ridge-till and no-till are the most profitable systems in different soil types.
- In all cases mulch-till systems (fall chisel, disk or field cultivator) are more profitable than fall plow but not as good as ridge-till and no-till.
- In poorly drained soil, ridge-till is more profitable than no-till. Slopes higher than 4% favour no-till systems.

Time requirements can be reduced by 65% and 43% when replacing conventional tillage with zero tillage and reduced tillage, respectively. At the same time, energy requirements can be reduced from more than 90 kW h/ha for conventional tillage to about 60 and 10 kW · h/ha, respectively, for reduced and zero tillage system.

The adoption of soil conservation tillage systems suffers from many constraints, particularly in developing areas:

- Lack of financial means (difficult procedures to obtain credits and subsidies, high interest rates).
- Climatic conditions impose more risks with the new tillage systems.
- Low technical level of farmers due to lack of research and extension.

In developed countries, energy and labor saving and the protection of soil and water from pollution and degradation are the major aspects connected with tillage. The search for adapted systems assumes that the user's technical level and the farmer's financial funds are taken into account. A simulation based on data of an investigation carried out in Morocco shows that the cost of direct seeding is less than for other tillage systems if the annual planted surface is larger than 60 ha.

There is an urgent need to introduce sustainable agriculture in fragile ecosystems, particularly in the developing countries where low-input farming increases the environmental degradation and perpetuates low yields and low revenue (Lal [4], Nicou [5], Merzouk [6], Oussible [7], Papendick and Cambell [8]).

1.2.2. Soil Engaging Components

During tillage, different forces are applied to the soil, such as shearing, compression, deformation, impact, vibration and acceleration. The resulting products take effect at the soil aggregate level: cutting, crumbling, inversion, mixing, sorting, leveling and shaping a particular section profile of the soil. The action of any given implement is the combination of two or more of those effects.

Basic Elements and Materials of Tillage Tools

The main active part of almost any tool is the edge. Other parameters such as the shape, the speed of the tool action, the depth and the forward speed in addition to the soil physical properties have a great effect on the resulting soil structure. The most important form for tillage tools is the wedge. When entering the soil the force applied to the tool is multiplied before acting on the soil, thus causing either soil failure under dry conditions (crumbling) or plastic deformation in wet soils. Usually soil failure is the desired effect while plastic deformation is an unwanted effect unless applied for puddling in paddy field tillage.

For soil-loosening, usually only half a wedge is used in order to avoid friction forces on a second soil tool contact interface. Thus, the normal form of a soil-loosening implement will be the raked blade. It can be shown that the soil before such a raked blade fails in a way that a curved slipline starts at the tip of the tool and goes from there to the soil surface (Fig. 1.159). This pattern is valid for comparatively wide tools as side effects are neglected. For small tools, side effects must be taken into account. The application of this theory has been described by various researchers [9, 10, 11]. It is usually suitable for elements that break the soil without turning it, such as cultivators, sweeps and other tines or (bulldozer and) excavator blades. If the tool is comparatively small, the soil will start from a certain depth to flow around the tool instead of breaking before it, which will cause compaction around the tool instead of loosening. This depth is the so called *critical depth* and the failure pattern of the soil is shown in Fig. 1.159. For plows, there are empirical formulae which were basically developed by Goriachkin and which state a quadratic relationship between the plowing resistance and the working speed.

Secondary tillage, whether performed with drawn implements or with animated tools, is always worked on pre-loosened soils where individual clods must be further reduced to smaller aggregates. Possible ways of achieving this aim are shown in Fig. 1.160. Tillage machines that use drawn tines are performing this mainly by cutting, breaking, milling and to a smaller extent by beating and throwing soil aggregates. Their effect usually becomes stronger with speed. At the same time, tools with drawn tines are able to separate the different size classes of the soil. Implements with driven tools usually break

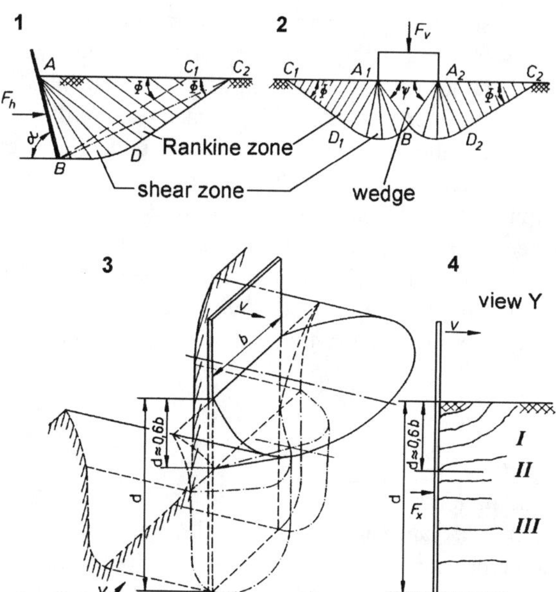

Figure 1.159. Failure patterns of soil before a tillage tool: 1: Failure according to the model of passive earth pressure; 2: Failure according to the model of bearing capacity; 3: Failure before a narrow tine; passive earth pressure failure above critical depth and bearing capacity failure below critical depth; 4: Failure surfaces in the soil: passive earth pressure failure zone (I), intermediate zone (II), and bearing capacity failure zone (III) (from [11]).

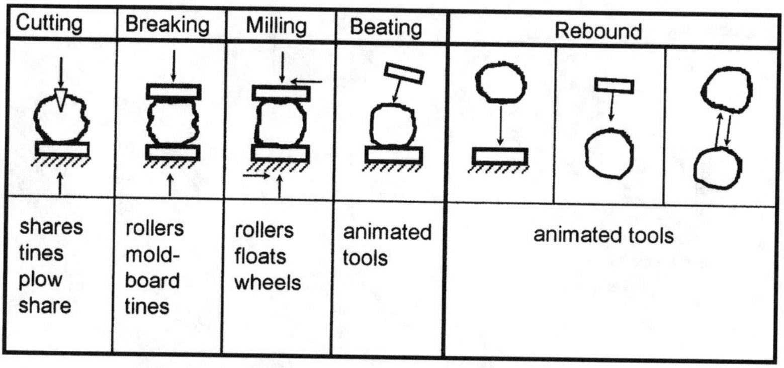

Cutting	Breaking	Milling	Beating	Rebound
shares tines plow share	rollers mold- board tines	rollers floats wheels	animated tools	animated tools

Figure 1.160. Mechanisms for clod breaking (from [11]).

clods by beating them and throwing them against some surface from where they will rebound. Driven tillage tools usually do not separate soil into size classes, but their size reduction effect is rather larger than that of drawn implements. So far, a complete theoretical prediction model for secondary tillage machines is still missing.

Following frequent use, the tool edge of tillage tools undergoes wear, which results in a parabolic edge profile, depending on the wearing material, the thickness of the tool and the conditions of use (Richardson [12, 13]). Richardson [12] demonstrated that wear resistance increases as the hardness of the wearing material exceeds about 80% of the hardest particles of the soil. DIN 11100 recommended at least 500 HV. Presently, the use of a protecting layer of carbide or other hard materials on the nozzle of the share can be found. In terms of wear, the lifetime of a disk can be doubled by this measure.

Drawn Implements

Share and Moldboard

The share and the moldboard are the main engaging parts of the moldboard plow (Fig. 1.161). The share is a plane part with trapezoidal shape. It cuts the soil horizontally and lifts it. Common types are regular, winged plane, bar-point and share with mounted or welded point. The regular share conserves a good cut but is recommended on stone-free soils. The winged plane, share is used on heavy soil with a moderate amount of stones. The bar-point share can be used in extreme conditions (hard and stony soils). The use of the share with mounted point is somewhere between the last two types.

The function of the moldboard is to lift the soil cut by the share and to let it undergo an action of torsion and inversion. The intensity of this action depends on the type of the moldboard. At present three types are mostly used: helicocylindrical (universal), digger (cylindrical), and helical (semidigger) (Fig. 1.162).

The standard (universal) type is by far the most widely used one because of its adaptation to almost all situations and soil types, except in the most extreme conditions.

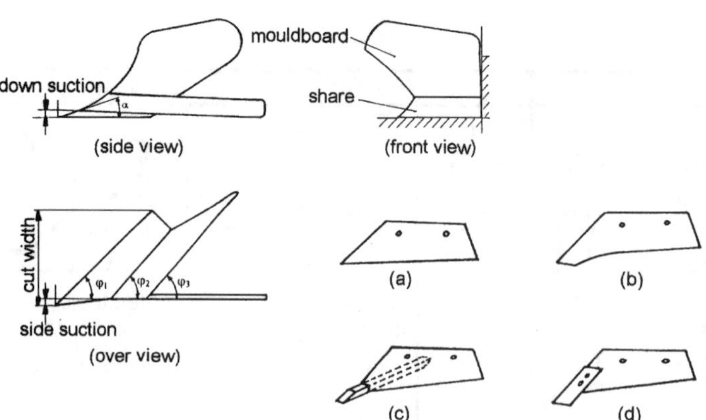

Figure 1.161. Moldboard plow bottom with main parameters and share forms (a) regular, (b) winged plane, (c) bar point, (d) mounted or welded point.

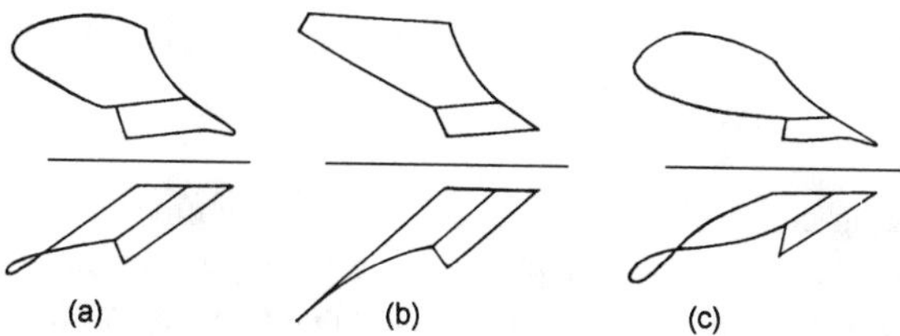

Figure 1.162. Moldboard forms (a) helico-cylindrical (universal), (b) digger (cylindrical), (c) helical (semidigger).

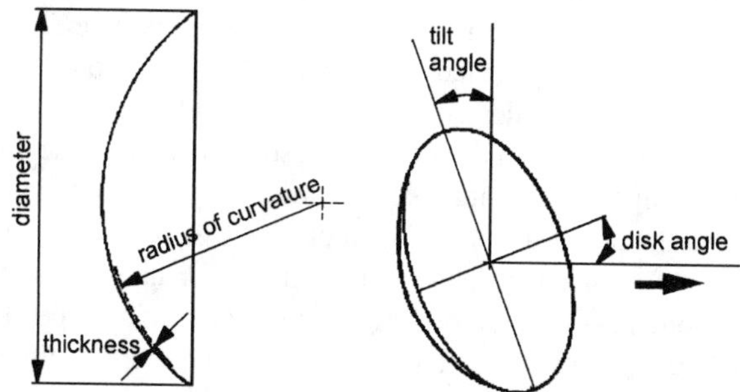

Figure 1.163. Dimensions and working angles at disks for tillage.

The cylindrical moldboard is very intense in its action (*bulldozer action*). On the other hand, the semidigger type is well suitable for rapid and returning action but limited on the depth.

Disks

Disks used for plowing and harrowing are made out of a portion of a sphere, thus creating the disk shape. Their common dimensions are 410–1270 mm in diameter and a thickness of 4.5–9 mm and a curvature of about 2.5 times the diameter for the plow and 1.2 times the diameter for harrows (Fig. 1.163). The disk will wear externally, showing a parabolic profile with external bevel. To facilitate the penetration into hard soil, the disk could be made with indentation. However, moist residues can limit the penetration.

Disks can be used for disk harrows or disk plows. For harrow disks, the tilt angle is zero. The disk angle can be adjusted from 15° to 25° (Fig. 1.163). In the case of a disk plow, the tilt angle can be adjustable from 10° to 25°, the disk angle can be either fixed or adjustable (40° to 50°).

The disk action on the soil is somewhat similar to that of the cylindrical moldboard type. The double obliquity placement of the disk causes its rotation once it is moved forward. This movement facilitates the cutting of the soil and its lateral displacement as a result of soil-metal friction and the soil aggregate acceleration. The working depth depends on both disk angle and the weight per body. There are two types of harrow disks: medium harrows with less than 80 kg/disk and heavy ones having more than 120 kg/disk.

Chisel-Type Tools

The soil-engaging components of a chisel-type are symmetrical, plane or concave. They till the soil profile to a certain depth, leaving plant residues near the surface. They can be used to loosen the soil, break up compacted layers, slightly bury organic residues, without soil inversion, and in certain cases, control weeds and surface crusting. There are different share types (Fig. 1.164). The rake angle varies between 20° and 60°. The chisel share and the diamond-pointed share are strong and commonly used. The regular tine is also very used, but it is less strong and does less burying. The twisted share and duckfoot share are used in stubble tillage because of their mixing effect. The sweep tine is popular for weed control and surface crust breaking. They also are used in situations where water conservation is needed, in fallow fields, and to maintain row crops. The winged tine is used in conservation tillage and if a short machine is desired as one single tool, it can loosen a comparatively large band of soil. The wings only loosen soil without mixing the layers so the structure of soil is kept intact. If mixing is desired, a straight or a chisel share must be used instead of the chisel. For harrows and cultivators the soil-engaging components are made of rigid tine or spring-tine type (Fig. 1.164). The second type has a better performance and is sometimes combined with a rotary harrow. The spring tines sort and level aggregates. Sometimes they are reinforced with a flat or a helical spring. The tool is a chisel-type share, simple or double tine or duckfoot share.

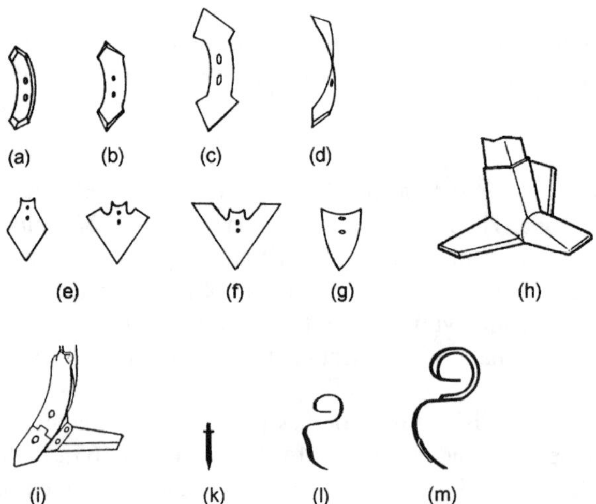

Figure 1.164. Different chisel type tools (a) straight, (b) chisel share, (c) diamond pointed, (d) twisted share, (e) duckfoot shares, (f) sweep, (g) stubble share; winged shares: (h) non mixing, (i) mixing; Harrow and cultivator tines: (k) rigid, (l) spring tine, (m) reinforced tine.

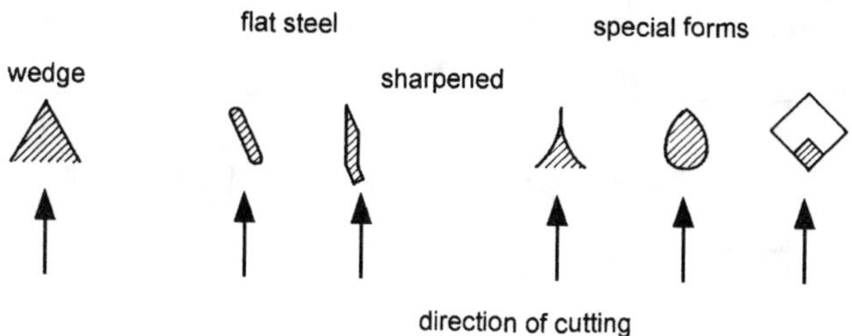

Figure 1.165. Cutting edges of tines for PTO-driven implements.

PTO-Driven Implements

Contrary to drawn implements, PTO-driven implements act intensively on the soil, rather independently of the forward speed of the tractor. This allows the rapid achievement of the desired soil structure. However, this forced action can lead to soil structure damage, including very fine aggregates in dry crumbling soil, smear or compaction under the worked soil layer in wet conditions and the destruction of fauna which is beneficent to the soil. Weise [14] investigated the crumbling effect of different types of tools for PTO-driven implements (Fig. 1.165): blade with or without a sharpened edge (twisted, curved or L-shaped blade); tines with a corner and special tines with a complex shape. In weak-structured soil, drop-shaped tines and blades without a sharpened edge are recommended. Twisted and curved blades are suitable for stubble tillage. Special and L-shaped blades are used more and more for primary tillage but still used for secondary tillage under hard soils conditions.

1.2.3. Tillage Systems

Tillage machines typically are grouped together in tillage systems, which can be of various levels of intensity. One tillage system comprises all the machines necessary for primary and secondary tillage and enable to perform seeding or planting. A general idea of the energy requirement of various tillage machines is given in Table 1.48.

Plows and chisel plows have the highest energy requirements. Reducing the intensity of tillage will considerably reduce the energy required per m tillage width. In addition, soil type and moisture are important for the actual energy requirement and must be taken into account. The numbers given are only relevant for the tractor power requirement caused by the tillage implement. The tractor itself needs further power to overcome rolling and slope resistance. The values given are valid for four-wheel drive tractors. In the case of a two-wheel drive tractor, it is necessary to add between 2–5 kW to the total energy requirement to account for the rolling resistance of the front axle. The operations which can be performed by various tillage systems are shown in Table 1.49.

Conventional Tillage System

The reference system for tillage is still the *conventional tillage system*, which is based on a high intensity of soil engagement and inversion of the soil. Its characteristic

Table 1.48. Specific tractor power (per m working width) requirements for different tillage operations (from [15])

Tillage tool	Working Speed m/s	Working Dept. cm	Soil Condition Light KW/m	Medium KW/m	Heavy KW/m
Plow	1.5–2.0	20–30	20–35	30–60	60–120
Cultivator	1.5–2.0	15–25	12–25	20–45	35–80
Stubble plow	2.0–2.5	10–15	12–25	20–38	40–75
Disk harow	2.0–2.5	7–10	5–10	10–20	20–40
Rotary spade harrow	2.0–3.0	8	8–13	12–22	22–40
Curry-comb	2.25	3	2–4	4–6	6–12
Rotary tiller	1.5–2.0	8–10(15)	15–22	20–35	35–60
Rotary harrow	1.5–2.0	8–10	10–18	15–30	30–50
Rotovator	1.5–2.0	8–10	6–12	10–30	25–50
Mulching rotor	1.5–2.5	3–10	value of machine used		
Combined tillage machinery	addition of the power requirement of the single elements				
Direct drill	2.5–4.5	2–7	9–15	12–30	20–45

implement is the moldboard plow. As plowing produces comparatively large aggregates and a rough soil surface, it must be followed by secondary tillage in order to prepare a seedbed. The main advantage of the system is to produce controlled soil conditions, a thorough loosening and a field clear of plant residues. Organic matter and weed seeds or shattered seeds are buried. But the soil is left bare for a comparatively long time, so it may be subject to wind and water erosion. Furthermore, tractors running in the furrow and the plow bodies can create a plowpan compaction.

Primary Tillage Machinery

Working depth for primary tillage varies between 15 cm and 80 cm depending of soil and climate. Various machinery exist for this purpose, although the plow is still the most important one.

Plows. The actual plow is derived from the chisel plow, which is completed by a moldboard. An alternative to the moldboard plow is the disk plow, which is widely used in arid climates as it is less susceptible to obstacles in the soil and to rough conditions. Many plow frames are still square-framed while more modern types rely on one central tubular beam. The most important parameters for adjusting plows are the working depth and the front furrow width. All plow adjusting systems rely on the fact that plow bodies will travel only in directions parallel to their bottom and their landside. This self-lining effect is the principle used for the adjustment of all plows.

Moldboard Plows. One form of a moldboard plow which is still in use is the trailed plow (Fig. 1.166). It consists of a carrier frame running on wheels bearing the actual plow bodies. Depth control is performed by varying the position of the carrier wheels while the front furrow width is adjusted by the position where the plow is hitched to the tractor. Trailed plows are normally used as one-way plows only.

Table 1.49. Actions of various tillage implements (changed from xx)

Tillage tool	Action									
	Loosening Coarsely	Loosening Finely	Loosening Shallow	Loosening Deep	Inverting	Mixing	Crumbling	Leveling	Compacting	Weed Control
Plow										
moldboard	X			X	X		(X)			X
disks	X			X	X		(X)			X
rotary	X	X		X	X	X	X	(X)		X
Spading machine	X			X	(X)	(X)	(X)			X
Deep loosener	X			X						
Cultivators										
Chisel	X			X		(X)				
Winged share	X		X	X		X	(X)			X
Spring tine		X	X			X	X	X		X
Toothed harrow		X	X			X	X	X		X
Tined rotors							X	(X)	X	
Disk harrow		X	X			(X)	X			X
Rotary spade harrow		X	X			(X)	X			X
Furrow press									X	
Roller							(X)	(X)	X	
Float							(X)	X	X	
Rotary cultivator		X	X			X	X	X		X
Power harrow		X	X			X	X	X		X
Dyna drive		X	X			(X)	X	(X)		

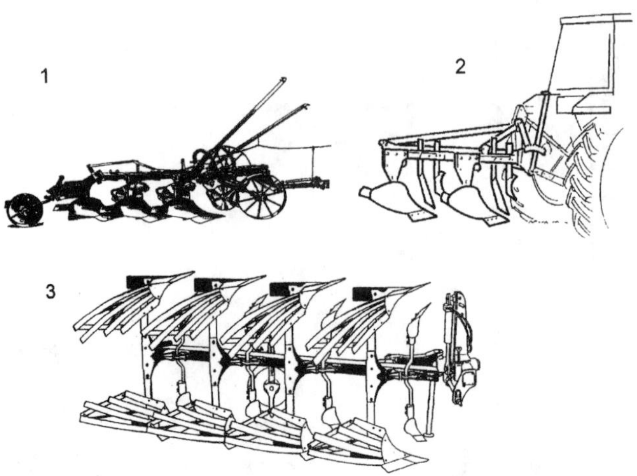

Figure 1.166. (1) Trailed plow, (2) and tractor mounted one way plow, (3) and tractor mounted reversible plow (from [1]).

The most important plow variety is the tractor-mounted moldboard plow. It can be used as a one-way plow and as a reversible plow. One-way plows (Fig. 1.166) usually are less complicated, cheaper and less heavy than reversible plows. But it is necessary to divide the fields into beds and join these beds properly, which can become time-consuming on smaller fields. Working widths of the shares may range from about 25–55 cm.

The most important variety of tractor-mounted plows is the reversible moldboard plow (Fig. 1.166). It allows for going up and down the field in one furrow, thus increasing productivity considerably, especially on small and medium-sized plots (0.5–5 ha). The three-point hitch is equipped with a hydraulic turning mechanism which allows turning the plow 180°. These plows usually are equipped with advanced lining systems in order to get optimum performance from tractor and implement. They allow the independent adjustment of the front furrow and to position the virtual linkage point of the hitch couplers of the tractor, in order to minimize lateral forces on the tractor or to reduce draft. One of these systems is shown in Fig. 1.167. By a mechanism linkage with four joints the position of the lower links of the tractor can be changed while the plow rests in place with reference to the tractor.

If the number of shares (working width per share about 45 cm) exceeds 5, usually semimounted plows are used. The number of shares in this system usually increases up to 7, which gives working widths of 3–3.5 m. For even greater number of shares (up to 16) plows with two-wheel undercarriage are built (Fig. 1.168).

For semimounted plows, depth adjustment is performed by changing the vertical position of the three-point hitch of the tractor and by adjusting the vertical position of the carrier wheels. Semimounted plows need large headlands as they have to be driven like a very long single-axle trailer.

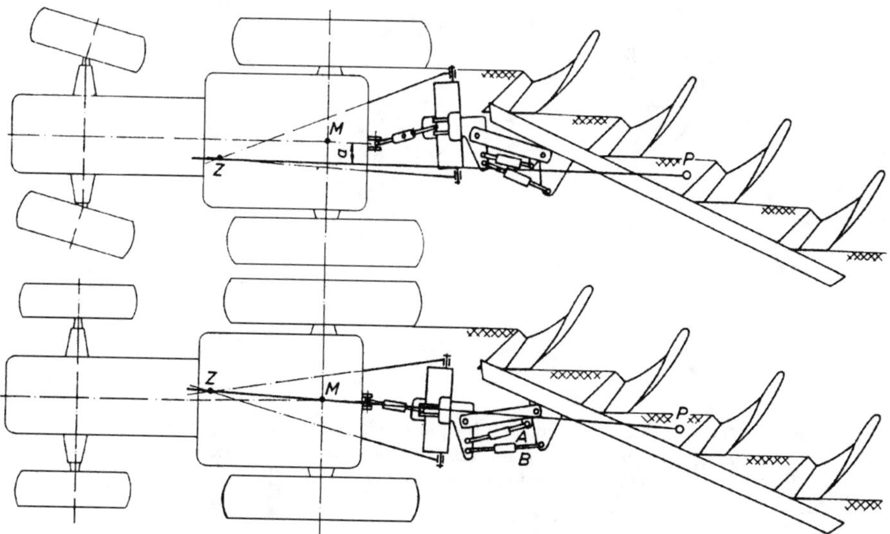

Figure 1.167. Lining system (Lemken) for tractor mounted plows.

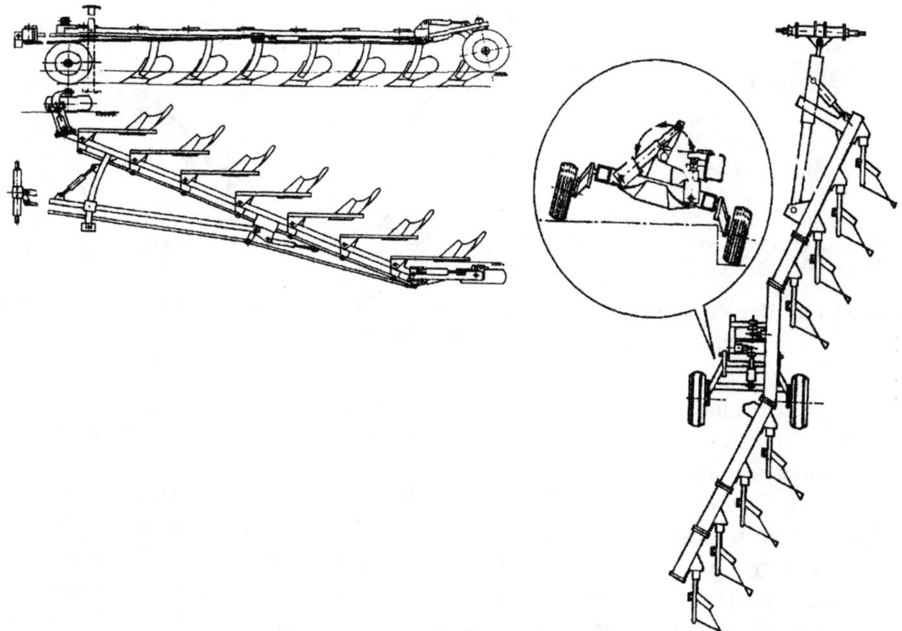

Figure 1.168. Semimounted plows: one way plow with carrier wheel at the end (left) and reversible plow with intermediate undercarriage (right) (from [1] and [11]).

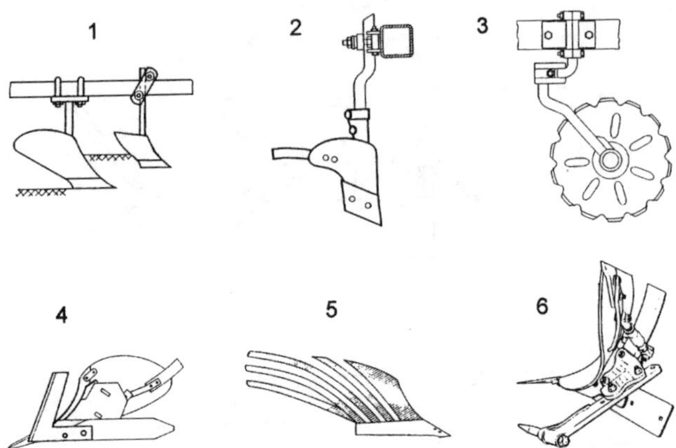

Figure 1.169. Special tools for plows: (1) skimmer, (2) manure burying coulter, (3) disk coulter, (4) coulter attached to the share, (5) strip moldboard, (6) underground spike.

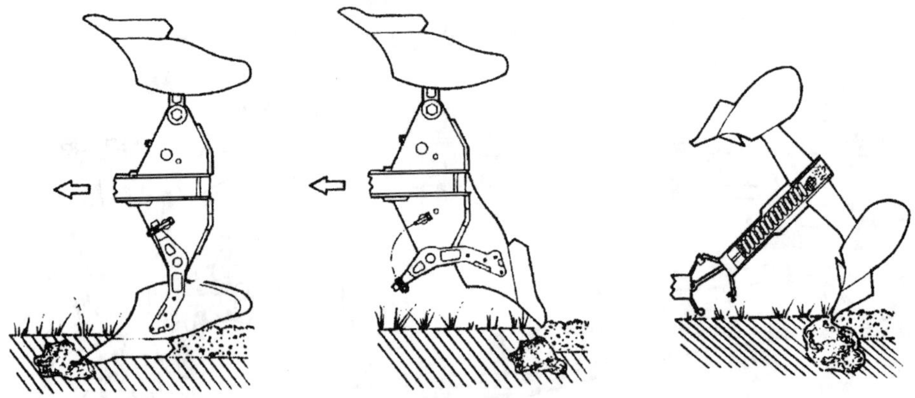

Figure 1.170. Overload protection systems; predetermined breaking point (left/center) and non-stop system with spring (right) (from [1]).

To facilitate plow work there is quite a number of additional tools (Fig. 1.169).

The most important ones are skimmers and manure-burying coulters, which throw organic matter, manure, or fertilizer into the open furrow, thus considerably easing incorporation work. Coulters (straight or disks) cut, in a defined way, the earth beam, which makes it turn more easily and helps to make straight rectangular furrows, thus facilitating driving along the furrow wall. Finally, there can be various tools working under the plow body, such as a horizontal spike used to break plow pans. In sticky soils, strip moldboards can help to reduce draft forces since they have a smaller surface subject to adhesion. For heavy or stony conditions, various kinds of overload protections (Fig. 1.170) can be installed.

The simplest overload protections are predetermined breaking points, such as bolts, which will let, when breaking, the body swing away from the obstacle. More advanced

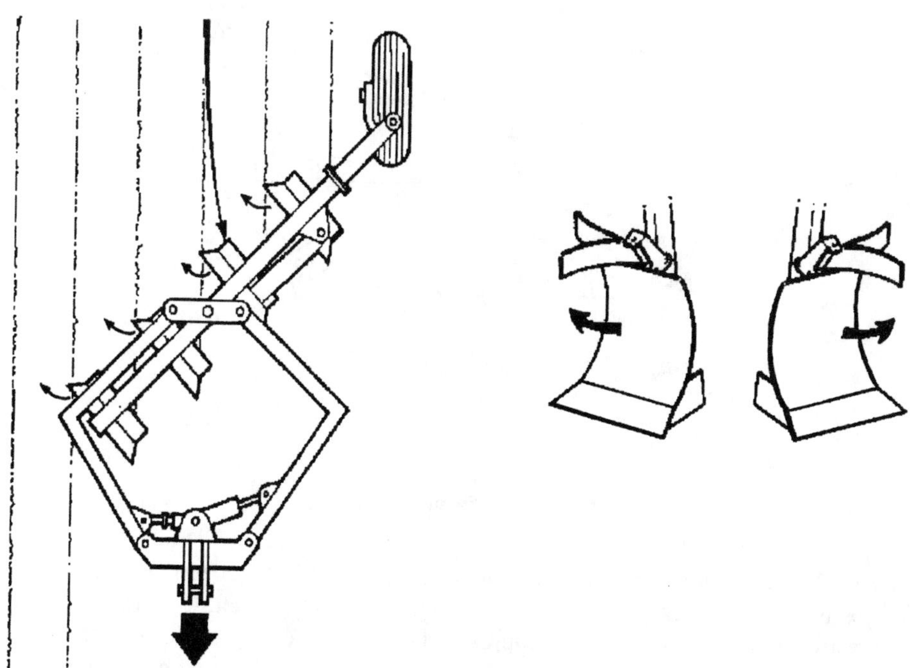

Figure 1.171. Swing plow (left) and symmetrical plow body (right) (from[1]).

systems allow the body to avoid the obstacle and then return into their original position by means of a spring mechanism or a hydraulic system.

An economical way to produce a reversible plow with the number of plow bodies required for a one-way plow is the swing plow (Fig. 1.171) [16]. This plow uses a symmetrical part of a cylinder as a share. The main beam can be turned around a central joint, thus creating a plow which turns the soil to both sides. This plow system can be cheaper and less heavy than a comparable reversible plow and allows rather wide furrows. But as the shares do not create a geometrically exact turning of the soil beam, the soil must crumble while sliding up the moldboard. If the soil does not crumble, such as heavy wet plastic soil, big blocks of soil and an uneven soil surface will occur.

Disk Plows. An alternative to the moldboard plow is the disk plow (Fig. 1.172). These plows do not need an overload protection as the disks can roll over obstacles. As the disks have no landside or slip heel, all vertical and horizontal forces perpendicular to the direction of travel must be taken up by the tractor or by guiding wheels.

The quality of inverting and crumbling the soil is lower than with the moldboard plow, and great amounts of organic residue can be a problem. Working at shallow depths over a longer period can result in heavy plow pans. Finally, due to the shape of the soil-engaging implement, the bottom of the furrow is not level. Nonetheless, due to its simplicity and robustness, the disk plow plays a major role in areas with difficult soil conditions where only limited amounts of drawbar power are available.

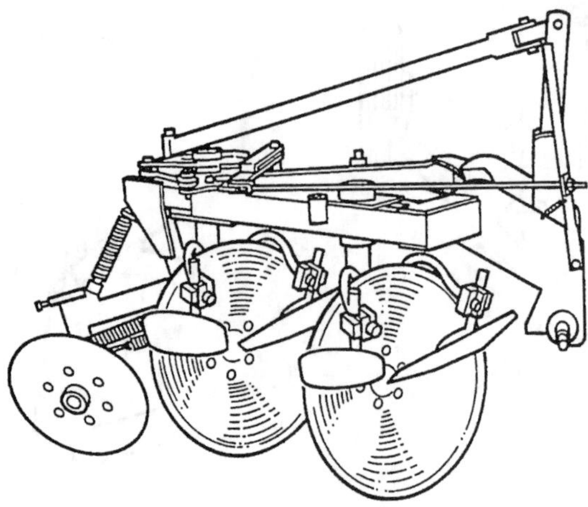

Figure 1.172. Disc plow (from [1]).

Driven Primary Tillage Implements. The advantage of PTO-driven primary tillage machines is that the energy of the tractor is transmitted directly, eliminating the rather ineffective power transmission via slipping wheels and allowing to use less ballasted tractors. The disadvantage of these machines is that they usually have high energy requirements, can be complicated, and are subject to wear. This means more maintenance work and they do sometimes crumble soil too much so that there may be an increased danger of encrustation and erosion.

Rotary Cultivator. The best known and most universal driven tillage machine is the rotary cultivator (Fig. 1.173), whose use ranges from primary and secondary tillage in conventional systems to minimum tillage and use in paddy field rice cropping. This machine performs primary and secondary tillage in one pass, even on difficult soils, while on sandy soils with little structure and internal cohesion, the use of the rotary cultivator can be problematic. An advantage of the machine is that organic matter is disintegrated and mixed into the whole worked area of soil. A problem can be that soil over the whole worked area is cut into very small clods, which can cause encrustation and erosion and, under wet conditions, smear at the bottom of the worked area may occur. A possibility to lessen these problems and reduce energy consumption is to reduce the rotor speed, but then it may happen that the seedbed is not sufficiently fine. The rotary harrow can be used for minimum tillage, too. Certain machines perform seeding either by using a seeder bar right behind the rotor so that the seedlings are put into the soil at working depth or on the plane created by the roller and covered by that portion of soil which is going over the roller (Fig. 1.173). Thus, only the seedbed zone or a slightly deeper soil area is loosened.

Rotary plow. It has been tried several times to combine the plow with rotating implements. This principle is realized in the powered disk plow. (Fig. 1.174). It is actually a disk harrow using comparatively large disks which are rotated by the tractor PTO in order to ease cutting of the soil and reduce draft forces. The disks are indented or can be

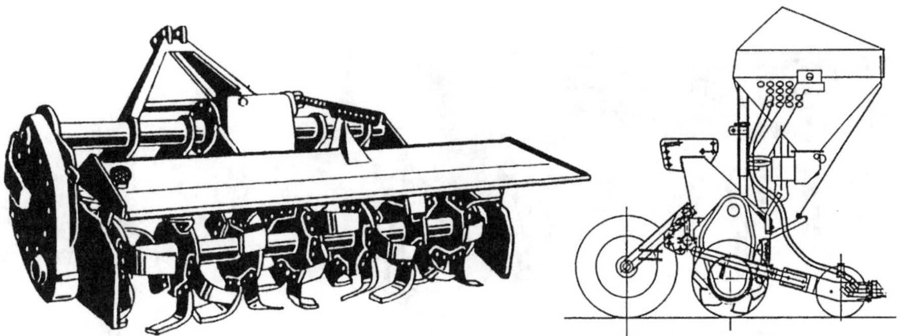

Figure 1.173. Rotary cultivator (left) and version for minimum tillage (right).

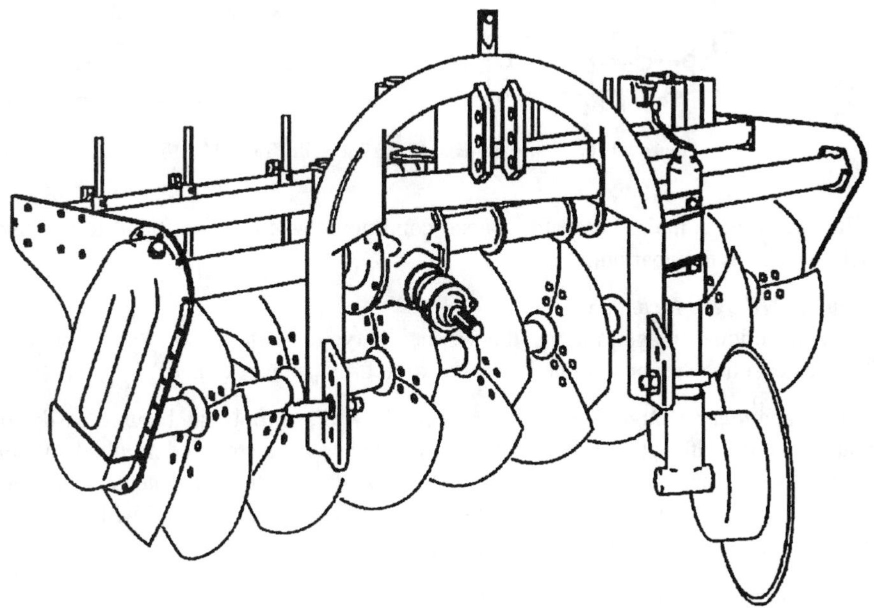

Figure 1.174. Driven disk plow (from [17]).

equipped with additional tines, which perform further crumbling of the turning soil. By driving the disks, the formation of a hardpan can be avoided. As the machine needs both PTO and drawbar power without creating a seedbed, this is a machine only for especially difficult and wet soil conditions.

Spading Machine. Another driven tillage machine is the spading machine (Fig. 1.75). It imitates the manual work of digging with a spade. The advantage is that this machine forms almost no hardpan as the path of the tools in the soil is almost never parallel to the soil surface. But the machine is rather complicated and comparatively energy consuming.

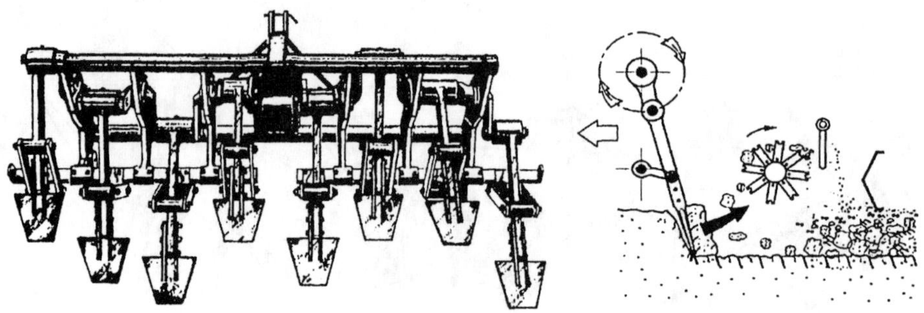

Figure 1.175. Spading machine (left) and working principle (right) (from [1]).

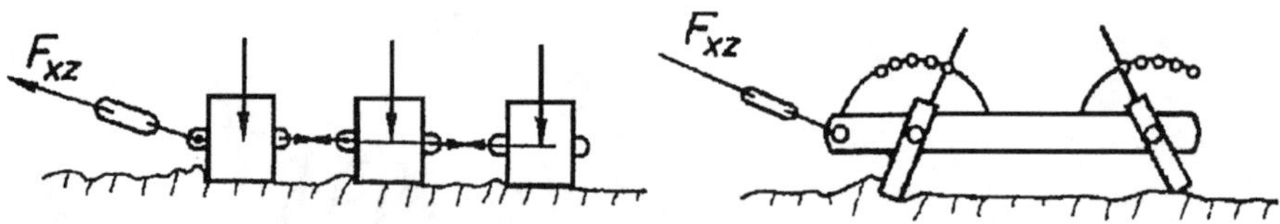

Figure 1.176. Floats; adjustable version right (from [11]).

So this machine is mainly used for special conditions, where it is difficult to transmit tractor power as drawbar pull.

Secondary Tillage Machinery

In conventional tillage systems, the seedbed is typically prepared in a second step. This creates a fine seedbed on the loosened soil left by primary tillage.

Drawn Secondary Tillage Tools. Drawn secondary tillage tools can be used easily for great working widths at high working speeds. They create a stratified seedbed and are comparatively simple and cheap. Their great advantage is that the intensity of work usually increases with working speed. On more difficult soils, together with mulch or under wet conditions, the use of drawn secondary tillage tools is limited.

Floats. Floats (Fig. 1.176) are amongst the most simple secondary tillage tools.

They are pulled over the soil surface without engaging implements in the soil. Their effect is mainly leveling, and they are used to grade ridges or ruts from previous crops. At the same time, they have a certain crumbling and mixing effect.

Harrows. Harrows are the most important machines for seedbed preparation. They are still used in low input agriculture to bury the seedlings spread by hand or with disk broadcasters. In a similar way, they serve to incorporate fertilizer. They have to perform crumbling, loosening, mixing, leveling and to a certain extent stratifying the seedbed. At the same time, they are used to control weeds after the emergence of the actual crop. A further task of harrows is stubble cleaning; that is, cutting, loosening and mixing of the root zone in order to induce germination of fallen out grains and prepare the field for primary tillage.

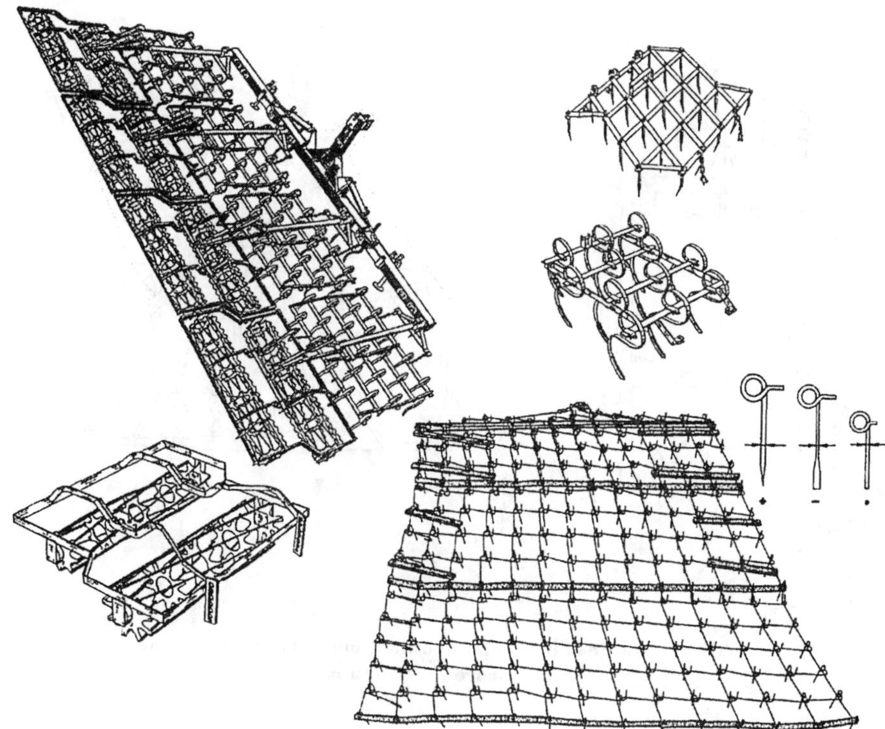

Figure 1.177. Harrow, harrow elements with rigid and spring tines, roller with cutter bars and chain harrow (right below) with tines (from [11]).

The conventional harrow consists of tines grouped together in a frame and several of these frames grouped together for the complete harrow; various tines of different intensity (rigid or spring) and additional rollers for better crumbling and good leveling are available (Fig. 1.177). These elements are grouped in carrier frames for seedbed preparation. The tillage intensity is comparatively high. A harrow with a smaller working intensity is the chain harrow (Fig. 1.177), which adapts well to any soil profile or ridges, and which is used especially for weed control and for leveling dams and ridges.

A machine with a very great importance is the disk harrow. It usually consists of several sets of disks on a horizontal axis, which are angled towards the direction of travel with an adjustable angle (Fig. 1.178). This allows to control the intensity of the soil engagement of this machine. In order to avoid lateral forces and to avoid soil transport to one side, there usually are two (symmetrically) angled sets of disks grouped one after the other. In order to achieve central force application, disks of one row are angled in two opposite direction. Due to its scraping effect, the disk harrow is well-suited for stubble cleaning. The comparatively strong inverting effect allows tillage with disk harrows alone as its effect comes close to that of a disk plow. At the same time, it can also be used for secondary tillage to break clods, create a seedbed and level the soil after plowing.

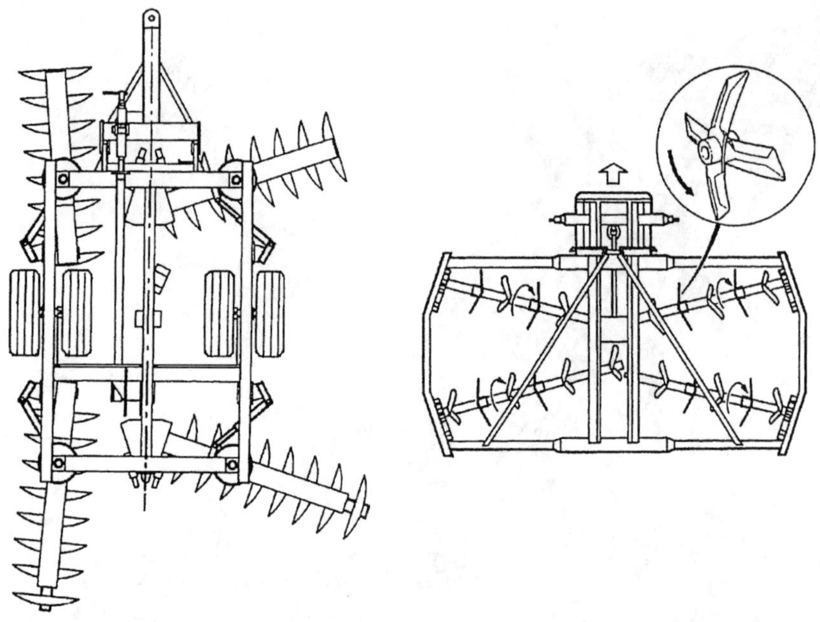

Figure 1.178. Disk harrow (half in transport position and half in work position) (left) and rotary spade harrow (right) (from [1]).

A problem of this device can be that it may create a smear zone; furthermore, it is not pulled into the soil by the form of its implements, which means that it must be fairly heavy in order to sufficiently penetrate into the soil.

Rotating Secondary Tillage Tools. Another variety of drawn secondary tillage machines includes those with rotating implements. The most important principle is that implements such as cutter bars, spikes, or teeth are attached to a wheel or roller, and are thus penetrated into the soil. Bars usually cut the soil while tines or spikes have a effect on the penetrated soil and tend to break it, so soil is mainly crumbled. Besides this, the spikes do mix soil and residues. This is especially true if the tines are combined with a disk (e.g., the rotary spade harrow Fig. 1.178). Simpler machines are varieties of spiked rotors (Fig. 1.179), which just roll over the soil surface. To increase the levering and crumbling effect of such rotors, a machine with two rotors was designed, the Dyna-drive (Fig. 1.179). It consists of two rotors, which are joined by a chain drive. The gear ratio from the first to the second rotor is 1 : 2 up to 1 : 3. Thus, the first rotor drives the second rotor. The effect of the first rotor mainly is to break and loosen the soil, and it works best if it finds some resistance breaking up settled soil, while the second rotor performs crumbling and mixing.

Seedbed Combinations. In order to profit from the special capabilities of drawn implements, they can be combined to achieve a seedbed in the desired form. An example is given in Fig. 1.180. This machine first performs a leveling action by a skimming bar

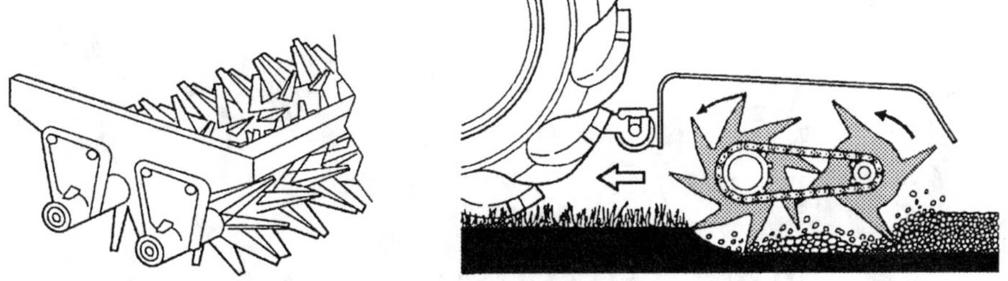

Figure 1.179. Spiked rotor (left) and Dyna-drive (right) (from [1]).

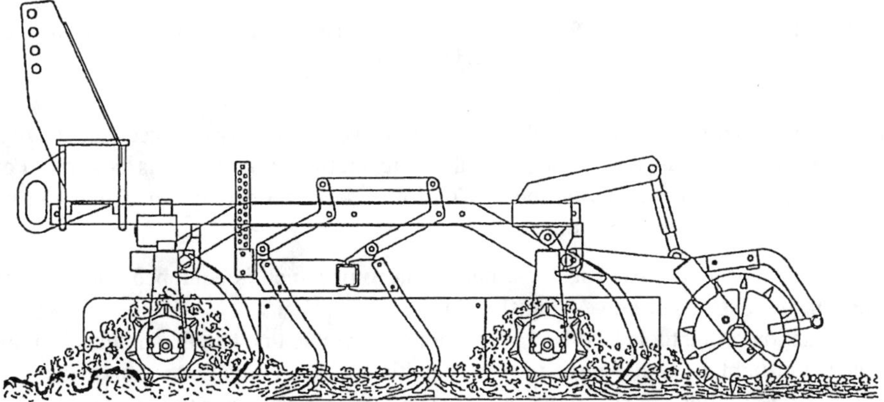

Figure 1.180. Combined seedbed preparation devices.

together with some crumbling by the first roller, which is equipped with cutter bars. This action is followed by a light cultivator, which breaks the surface and loosens the soil. These implements are followed by a second crumbling and leveling unit. The last tool is a cross kill roller for recompaction.

Such a unit can create a rather specific seedbed for most crops, depending on the depth adjustment of the single implements, the number and kind of implements, and the working speed. Working widths of 4 m and more are possible, thus creating very powerful machines for high performance. Especially on lighter soils and under favorable conditions (correct moisture content for tillage), these machines form a rather effective and economical way for seedbed preparation. When several implements are combined, the limitations of the soil conditions are less severe as the machine acts like several passes of a single implement.

Driven Tillage Tools. The implements of driven tillage tools are actuated via the PTO of the tractor. The advantage of these implements is that a rather large amount of energy can be directed to the soil even at low working speeds. Heavy soils especially can be crumbled at most conditions unless they are in a plastic state. At the same time, these machines are rather compact, which makes the handling on the field and on the road easier, and

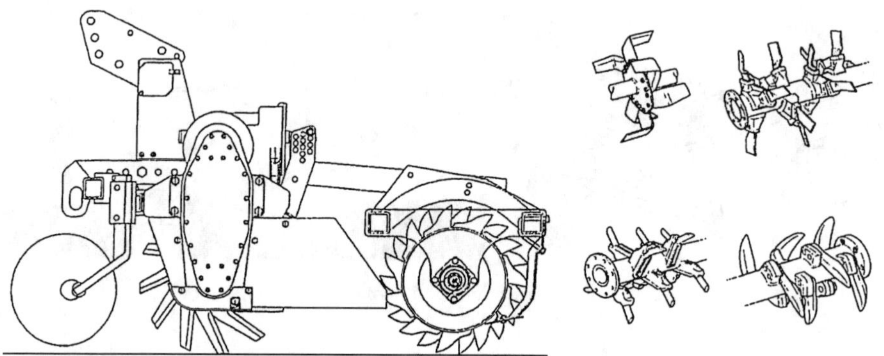

Figure 1.181. Rotovator with different rotors (from [1]) (rotary cultivator, angled tine, spike, and blade) and furrow press.

allows combination with other implements. This can considerably increase productivity, as even under difficult conditions a seedbed preparation in one pass is possible. For further increase of productivity, most PTO-driven secondary tillage machines can be combined with a drilling machine.

Rotovators. Rotovators consist of a horizontal rotor which is equipped with various kinds of tines, spikes or knives (Fig. 1.181).

The action of the tools is to beat or cut along the whole tine length, and thus is more aimed at crumbling clods, while the rotary cultivator cuts typical bites out of the soil. The mixing abilities are still rather good. Using tines or spikes causes a considerable reduction in the power requirement compared to a rotary tiller [14]. This means it can be used with some success in settled soils for stubble cleaning or mulching, and it is a very important machinery for reduced tillage. When used for secondary tillage alone, it causes good crumbling but the leveling effect may be inferior to other implements. For recompaction, further crumbling and as a means of depth control, rotovators usually are equipped with a bearing roller. If needed, rotors can be changed easily.

Power Harrows. A rather important PTO-driven machine for secondary tillage especially for heavy soils is the rotary harrow (Fig. 1.182).

It consists of a number of rotating disks, which usually are equipped with two or three tines that are angled towards the soil. Thus, they mainly beat and crush clods on the soil surface and press them into the soil. During the operation of the rotary harrow, a wall of earth builds up before the machine but due to the horizontal movement of the tools, almost no mixing occurs. This earthwall creates a level plane surface after the passage of the machine. Compared to the rotovator, the energy consumption is higher. This is partly due to the necessary drawbar pull and partly to the less effective tool operation. The machine is not suitable for plant residues as the great number of rotating axles may wind up plant residues, thus causing blockage of the machine. If the machine should be used in unloosened soil, it is necessary to angle the tines away from the direction of travel in order to achieve a rake angle that breaks up soil towards the surface. Rotary harrows need a roller to bear the weight of the machine, for depth control and for recompaction.

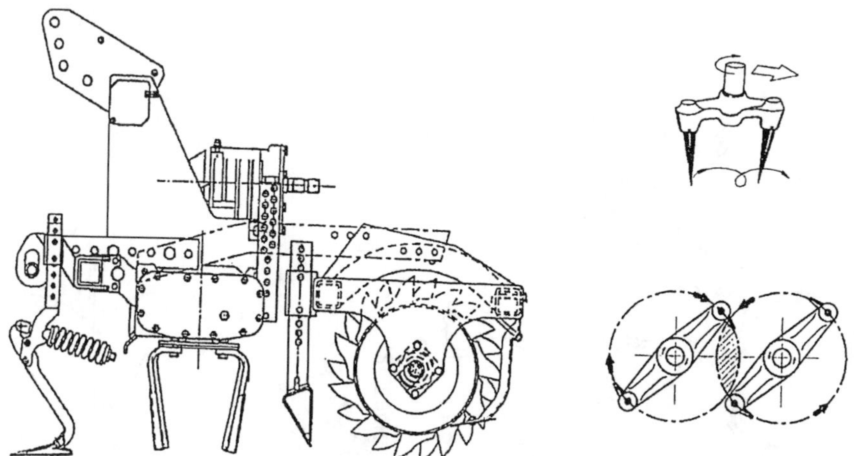

Figure 1.182. Rotary harrow with furrow press (left) and tool and interaction of neighbouring tools (right) (from [1]).

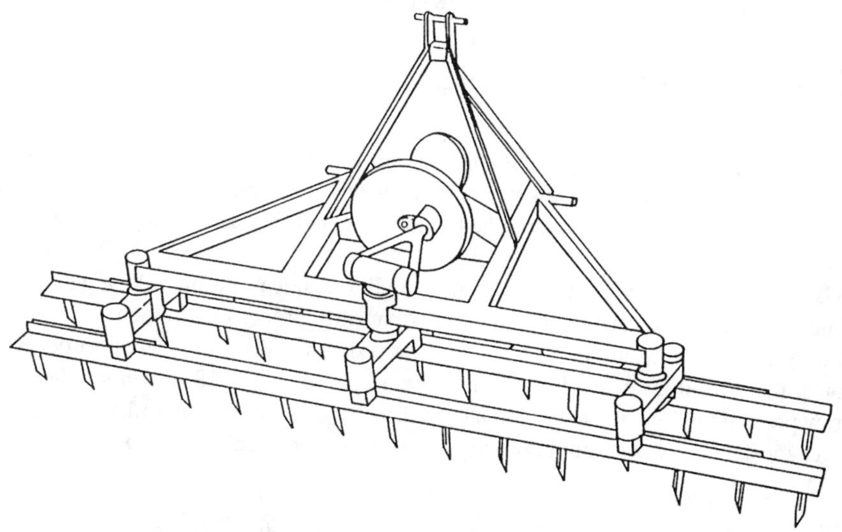

Figure 1.183. Reciprocating harrow (from [18]).

A further variety of PTO-driven harrows is the reciprocating harrow (Fig. 1.183). It has a similar effect as the rotary harrow, but the impact of the machine is smaller than that of a rotary harrow. It usually consists of two beams which bear the tools and which oscillate in opposite directions.

Dam Forming. For special purposes, it can be necessary to create dams or ridges (e.g., potatoes). Usually they are formed with specially designed drawn implements (Fig. 1.184). At the same time, there are specially designed rotary cultivators for this

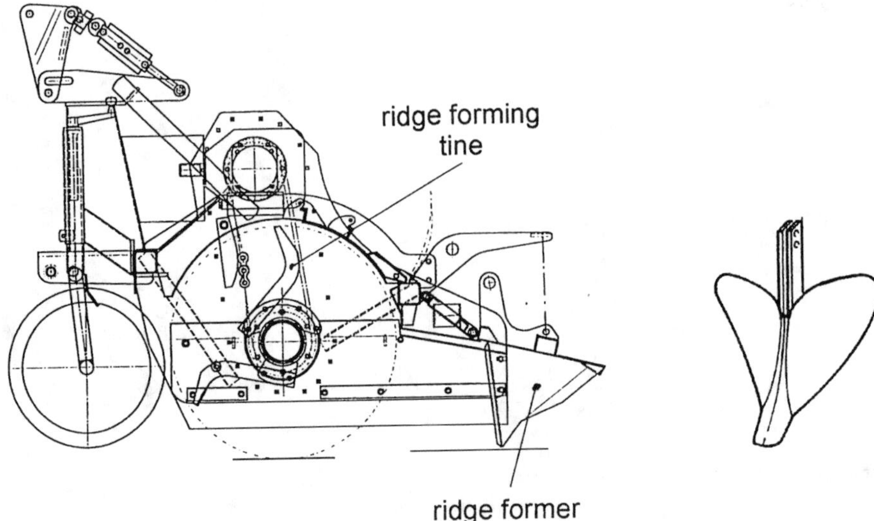

Figure 1.184. Rotary harrow for ridging (left) and ridger (right) (from [1]).

purpose, creating the dam in one pass together with the necessary crumbling and secondary tillage (Fig. 1.184).

Rollers. Rollers serve mainly for recompacting, leveling and crumbling. At the same time, their task is to create a better contact between seeds and soil. Furthermore, they can be used as a means of spreading the tractor weight evenly across the whole width of the vehicle.

Usually, rollers should be used under dry conditions since wet soil will clog them easily unless scrapers are used to clean them permanently. According to their form, rollers can only compact the surface or act at a certain depth if they are shaped to penetrate through the topsoil. Many different forms are available (Fig. 1.185). Among them, the furrow press should be noted, which is used directly after plowing in order to give back to over-loosened soils optimum porosity and to make them trafficable again.

Conservation Tillage Systems

The conservation tillage system is defined as a system that "provides means of profitable crop production while minimizing soil erosion caused by wind and/or water... moisture, energy, labor and even equipment conservation are additional benefits" (Midwest Plan Service [3]); similar advantages were shown by other researchers (Tebruegge [19]). The ASAE defines the following conservation systems:

- **Mulch tillage:** Tillage or preparation of the soil in such a way that plant residues or other mulching materials are specifically left on or near the surface.
- **Reduced tillage:** A system in which the primary tillage operation is performed in conjunction with special planting procedures in order to reduce or eliminate secondary tillage operations.

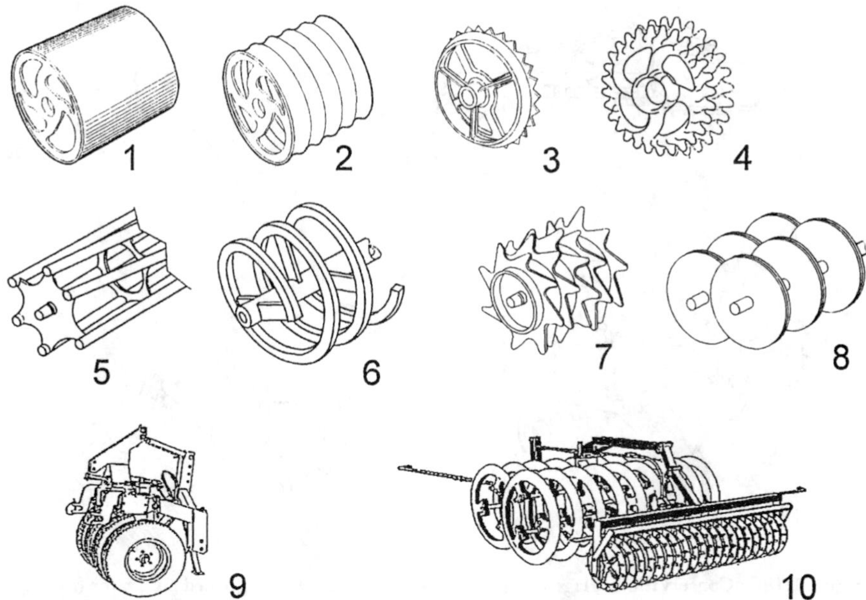

Figure 1.185. Various types of rollers: (1) plain roller, (2) grooved roller, (3) Cambridge roller, (4) cross kill roller, (5) cage roller, (6) spiral roller, (7) toothed packer, (8) disk roller, (9) tyre furrow press, (10) conventional furrow press with cross kill roller (from [1]).

- **Strip tillage:** A system in which only isolated bands of soil are tilled.
- **No-tillage planting:** A procedure whereby a planting is made directly in an essentially unprepared seedbed.

Conservation tillage systems try to disturb the soil as little as possible to conserve its natural structure, leave the maximum vegetal residue next to the soil surface, and/or try to build a rough surface. Thus, the soil will be protected against wind and runoff erosion, and water evaporation can be limited, especially in arid areas. Typical machines for reduced tillage are chisels, wing-tine cultivators, and, to a certain extent, animated rotating implements or disk harrows. Chisels and other forms of cultivators loosen the soil without turning it, leaving residue at the surface. According to the tool used, they may mix residues to a certain extent into the top layer or they may leave the soil unmixed. If crumbling is not sufficiently fine, seedbed preparation may be necessary. Furthermore, it is possible to reduce the tillage depth (3–10 cm), which will reduce the amount of moved soil and thus, considerably, energy requirements. To reduce the amount of work for tillage, it is possible to combine several steps of tillage in one machine (Fig. 1.186).

Residue management and the use of appropriate tillage methods will improve infiltration rates and consequently decrease the runoff. In Nebraska, soil loss by erosion in silt loam soil grown with oats under a slope of 10% were reduced respectively by 16%, 24% and 80% for the reduced tillage (blade plow-disk-harrow); disking system

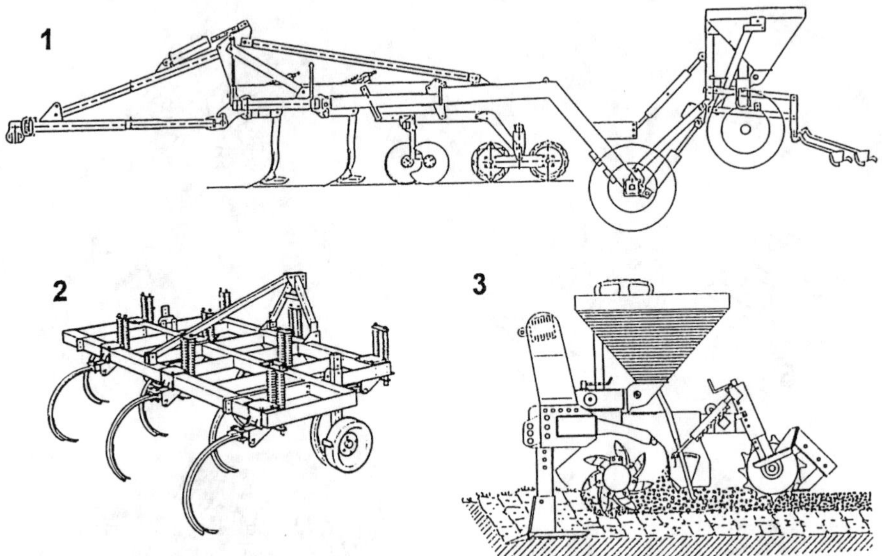

Figure 1.186. Conservation tillage equipment: (1) semimounted wing-tined cultivator (mixing type) combined with distributing disks, crumbling roller and drilling machine, (2) chisel plow (from [1]), (3) wing-tined cultivator combined with tine rotor, roller and drilling machine (from [14]).

(disk-disk-harrow) and the no-till compared to the conventional system using the moldboard plow, disk and harrow (MidWest Plan Service, [3]). When considering water balance in rainfall agriculture, water losses are due essentially to the runoff, evaporation and transpiration. The plant residues that are found on the surface protect the soil from the sunbeams and reduce water losses due to evaporation (Willcocks, [20]; Al-Darby et al., [21]). Willcocks [22] has shown in the semiarid area of Botswana that the soil temperature can reach 70°C when the soil is without cover and plant residues can reduce the temperature by 10%. However, the use of this method in arid areas is restricted by the residue amount, which is small and generally vested upon livestock.

In semiarid areas, systems that let more surface residues are more efficient in water use. The necessity to adjust the growing season with the rain period has been shown repeatedly (Bourarach [23]). In order to achieve this goal, there are two solutions: soil tillage as soon as possible after the first precipitation, or summer work on dry soil. As to energy saving, Dycker and Bourarach [24] have shown that in a loamy-sandy soil, fall cereal planting in direct seedling requires 3.2 l/ha of fuel, while the sequences using a heavy teeth cultivator (chisel) followed by a light disk harrow, heavy disk harrow, and rotary cultivator respectively consume 20.3, 17.6, and 19.9 l/ha of fuel.

No-Tillage System

In no-tillage systems, planting is made directly in an essentially unprepared seedbed, even if some tillage is done to create a small initial seedbed for the seedling. Sowing and

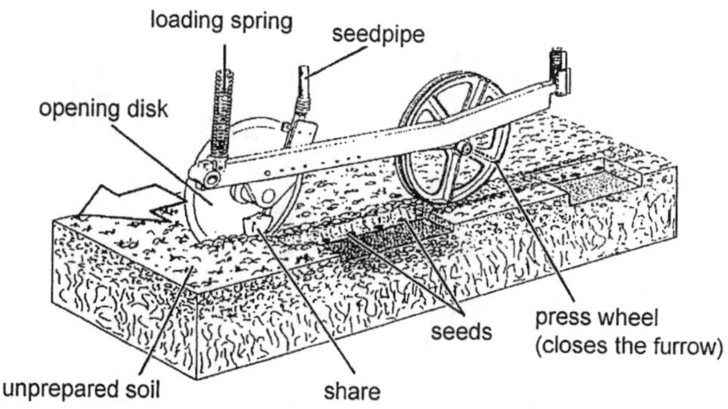

Figure 1.187. Principle of a no-till machine (from [1]).

fertilizing equipment in no-till systems must be able to handle residue, penetrate hard soil, regulate the working depth, cover seeds and give sufficient soil-seed contact. For these reasons, the no-till seeding and planting equipment must be heavier and stronger than conventional equipment. In some cases, conventional equipment can be converted to a no-till machine, although specific machines are advisable (Fig. 1.187).

The basic action which must be performed by a no-till machine consists of residue and soil cutting, furrow opening, seed and fertilizer placement, and covering. Rear, front or side wheels are often used to regulate the working depth and to ameliorate seed-soil contact and to limit water losses. Weed and pest control has to be done chemically. Disk coulters are commonly used for cutting crop residue (Fig. 1.188(a)). Fluted and rippled disks give good soil disturbance, but require more weight and slower speed than smooth disks. In extreme soil and residue conditions, the use of a powered disk is required. There are different furrow openers (Fig. 1.188(b)). The single and double disk are more adapted to soft or moist soils or for large amounts of residue. Contrarily to disks, chisel and hoe openers push the residue aside and disturb the soil more. But they are more sensitive to plaguing in loose residue. In hard soil, point openers perform better than hoe openers because they disturb the soil less and produce smaller aggregates. The runner opener is adapted more for row-crop planters. Wide flat press wheels are not recommended for no-till because they are sensitive to soil roughness and wet conditions. The sweep opener is more suitable for the air seeder, which needs usually some stubble cleaning for successful work. When the press wheels are not sufficient to cover the seeds, special devices are used (Fig. 1.188(c)). A pair of angled wheels is efficient in closing seed furrow but it needs more clearance between the rows than the other press wheel types (Fig. 1.188(d)). Nowadays, several direct drills are combined with fertilizer and pesticide applicators. To avoid germination damage, fertilizers generally are applied besides or below the seed rows.

Tillage Systems for Special Conditions

For virgin soils and some unusual conditions, it is necessary to use special tillage machinery.

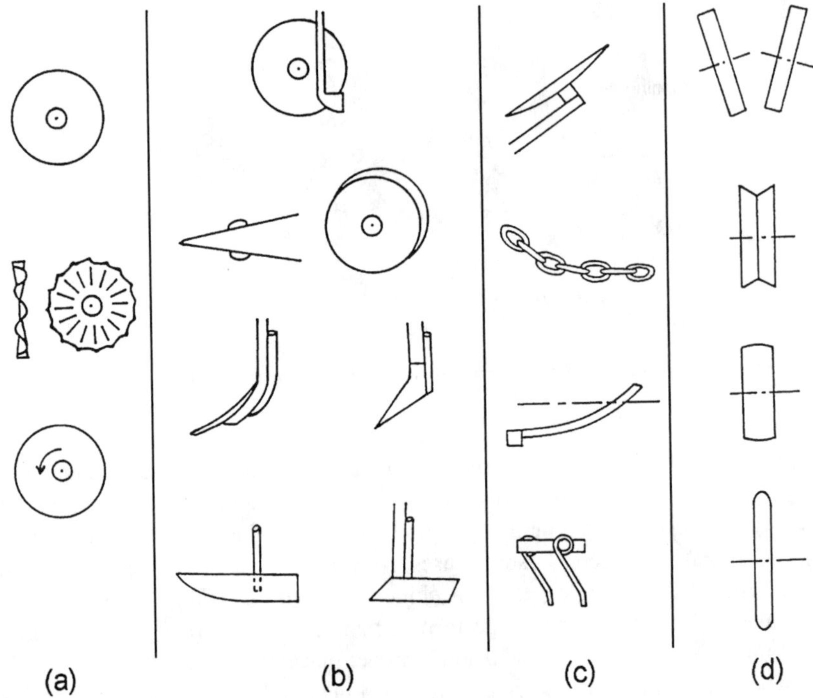

Figure 1.188. Main devices of no-till equipment for (a) residue cutting (b) furrow opening and seed placement (c) seed covering and (d) pressing the furrow.

Land Reclamation

In order to make virgin soils arable, it is necessary to clear them from all other plants and roots. This is best performed by plowing. At the same time, it may be necessary to bring organic matter into deeper layers of soil to get a larger arable layer and to remove stones. Some fen soils can be mixed with the sandy layers below, bringing up some mineral material in order to achieve a more stable arable soil. For this purpose, special deep plows have been developed (Fig. 1.189) which allow for plowing as deep as 2 m.

As the drawbar forces of large shares are very high, it is not possible to reach great working widths. For drainage purposes, there are specially designed ditch plows allowing to create ditches in one pass if the use of an excavator is not possible. Figure 1.190 shows an example of such a machine which is used to drain intraffickable fields. It consists of a cable-drawn ditch plow, which makes drainage ditches.

Another way to create drainage is the mole drain plow (Fig. 1.191), which uses the effect of critical depth [26] to stabilize the drainage channel. The plow consists of a share and an expander, creating the drain channel. Around the channel, cracks form in the soil, which conduct the water into the drain channel. It is usually not necessary to lay a pipe in these channels, although this type of plow can be used for pipe laying, too.

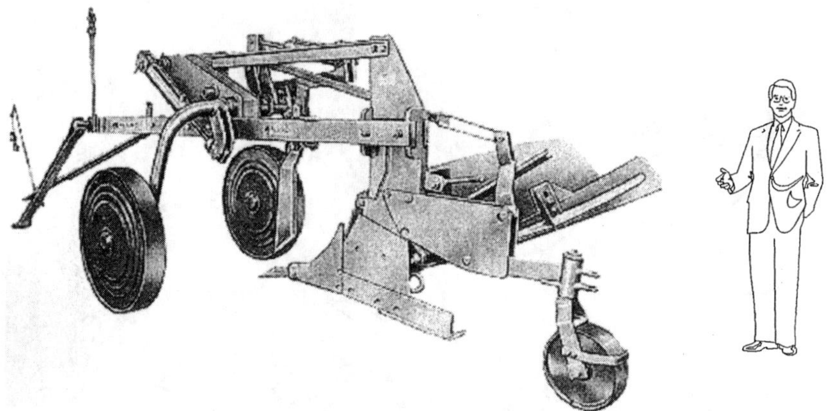

Figure 1.189. Deep plow.

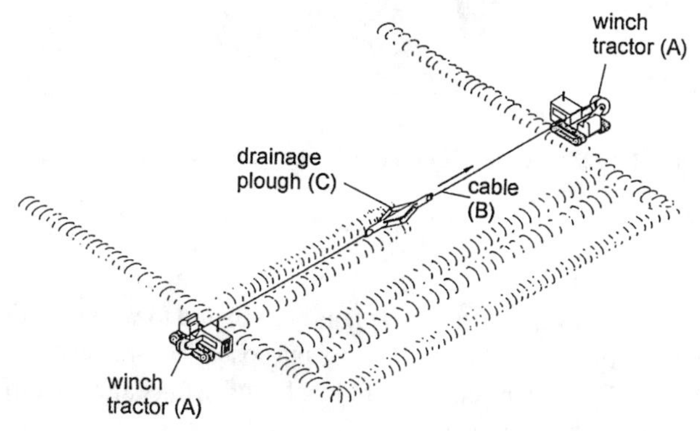

Figure 1.190. Cable drawn drainage plow (from [25]).

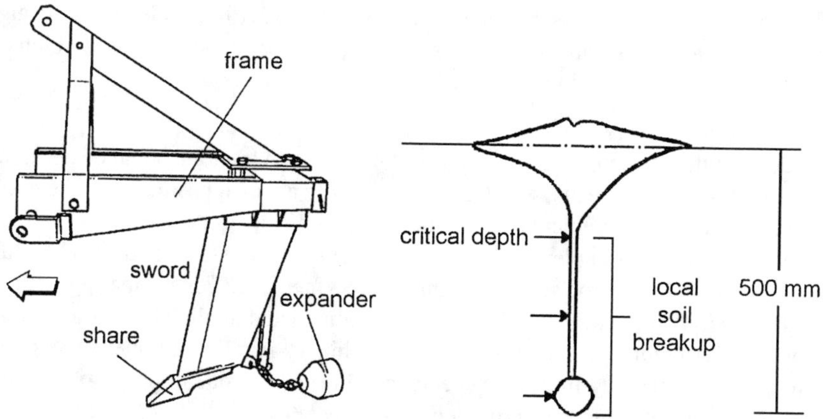

Figure 1.191. Mole plow and mole channel (from [1] and [26]).

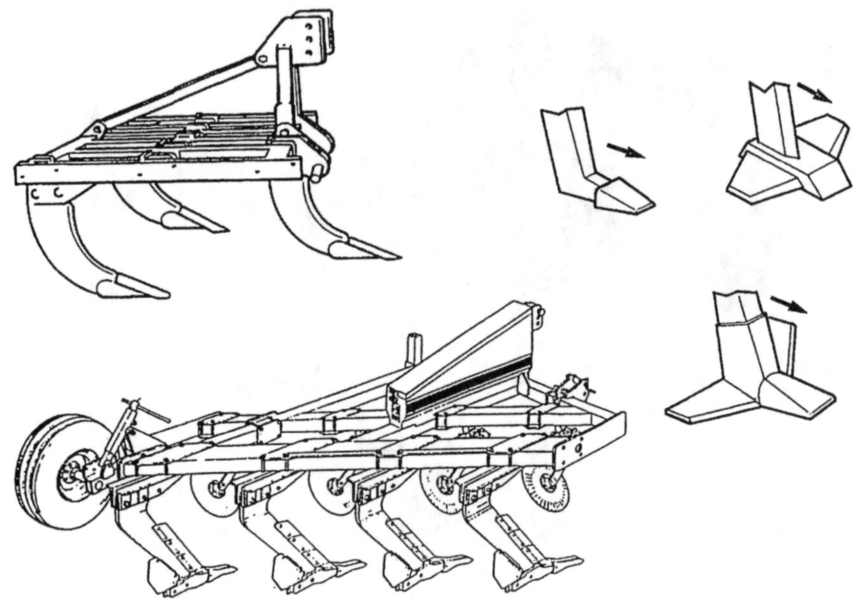

Figure 1.192. Deep loosener, share forms (from [1]) and para plow.

Deep Loosening

Hardpans can form a severe obstacle to plant growth and to water conductivity. Thus, it can become necessary to break them. For this purpose, deep-loosening equipment has been developed. Usually devices consist of chisels and wings attached to a vertical leg in order to increase critical depth (Fig. 1.192).

As the necessary draft forces can become very high, the number of implements is limited. Some models have driven oscillating shares in order to reduce draft requirement. Another variety of subsoil loosener is the para plow (Fig. 1.192). Its leg is not straight but angled. This allows the attachment to the leg of a further adjustable wing, which can be used to control the loosening intensity.

Cable-Drawn Implements

In order to cultivate soils which are not trafficable with tractors, it can be necessary to use cable-drawn implements, which are moved using a winch positioned at the end of the field that pulls the implement over the field (Fig. 1.193).

Applications for such implements in very wet soils or very steep fields can be found in mountainous countries or special cultures like vineyards. For normal agriculture, there is the desired side effect that the soil is not compacted at all. This will result in a very good soil structure. But the use of such machinery is rather cumbersome so that their application is actually limited to special conditions, and due to progress in tractor technology, the use of cable-drawn implements is decreasing.

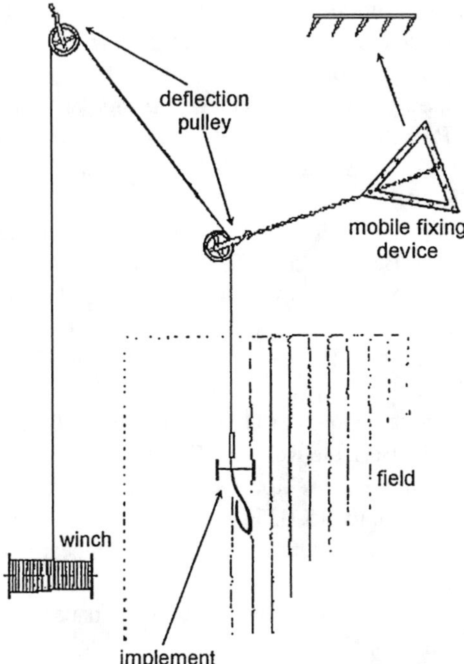

Figure 1.193. Installation for cable drawn implements (from [27]).

References

1. Cédra, C., 1991. Les Matériels de travail du sol, semis et plantation. Collection Formagri Vol. 3. CEMAGREF Antony: Lavoisier.
2. Scheffer, F., and Schachtschabel, P., 1992. Lehrbuch der Bodenkunde. Stuttgart: Ferdinand Enke Verlag.
3. Midwest Plan Service, 1992. Conservation tillage systems and management. MWPS 45. Iowa State University.
4. Lal, R., 1991. Tillage and agricultural sustainability. Soil and Tillage Research, 20: 133–146.
5. Nicou, R., 1974. Contribution a l'étude et l'amélioration de la porosité des sols sableux et sablo-argileux en zone tropicale sèche Ouest Africaine. Conséquences agronomiques. Agro. Trop. 29(11): 1100–1126.
6. Merzouk, A., 1985. Relative erodability of nine selected Moroccan soils as related to their physical, chemical and mineralogical properties. PhD. Thesis University of Minnesota, St. Paul.
7. Oussible, M., 1993. Tassement du sol agricole: production et productivité. Communication présentée au journée AMSOL. Rabat.

8. Papendick, R. I., and Campbell, G. S., 1988. Concepts and management strategies for water conservation in dryland farming. Proc. Conf. on Dryland Farming. Amarillo/Bushland. Texas, pp. 119–127.
9. McKeyes, E., 1985. Soil Cutting and Tillage. Amsterdam: Elsevier.
10. Hettiaratchi, D. R. P., Witney, B. D., and Reece, A. R., 1966. The Calculation of Passive Pressure in Two-Dimensional Soil Failure. Journal of Agricultural Engineering Research 11(2): 89–107.
11. Soucek, R., and Pippig, G., 1990. Maschinen und Geräte für Bodenbearbeitung, Düngung und Aussaat. Berlin: Verlag Technik GmbH.
12. Richardson, R. C. D., 1967. The wear of metals by hard abrasives. Wear 10: 291–309.
13. Richardson, R. C. D., 1968. The wear of metals by soft abrasives. Wear 11: 245–275.
14. Weise, G., 1996. Möglichkeiten zur Optimierung einer Bodenbearbeitungskombination. VDI Fortschritt-Berichte Reihe 14(76). Düsseldorf: VDI Verlag.
15. Fröba, N., 1995. Benötigte Traktormotornennleistung bei landwirtschaftlichen Arbeiten. KTBL Arbeitsblatt No. 0255 in Landtechnik 50(5): 277–282.
xx. Estler, M., and Knittel, H., 1996. Praktische Bodenbearbeitur Frankfurt: DLG-Verlag.
16. Weise, G., and Grosse, K., 1993. Schwenkpflüge genauer betrachtet. Landtechnik 48(1/2): 32–35.
17. Japanese Society of Agricultural Machinery, 1994. Introduction to Japanese Agricultural Machinery (I)-Tractors and Implements-11.
18. Eichhorn, H., 1985. Landtechnik. 6th edition Stuttgart: Ulmer.
19. Tebrügge, F., 1994. Beurteilung von Bodenbearbeitungssystemen unter den Aspekten von Bodenschutz und Ökonomie. in Tebrügge, F. et al. Beurteilung von Bodenbearbeitungssystemen hinsichtlich ihrer Arbeitseffekte und deren langfristige Auswirkungen auf den Boden. Giessen: Wissenschaftlicher Fachverlag Dr. Fleck, pp. 5–16.
20. Willcocks, T. J., 1981. Tillage of clod-forming sandy loam soils in semi-arid climate of Botswana. Soil & Tillage Res. (1): 323–350.
21. Al-Darby, A. M., and Lowery, B., 1984. Conservation tillage: comparaison of methods. Ag. Eng. 65(10).
22. Willcocks, T. J., 1980. Dryland farming research scheme. Botswana. Final scientific report of phase II. Vol. 1.
23. Bourarach, E-H., 1989. Mécanisation du travail du sol en céréaliculture pluviale: performances techniques et aspects économiques dans une région semi-arides au Maroc. Thèse ès-Sc. Ag. IAV Hassan II.
24. Dycker, J., and Bourarach, E. H., 1992: Energy requirements and performances of different soil tillage systems in the Gharb and Zaer region. Hommes Terre et Eaux 22(87). ANAFID. Rabat.
25. Shaw, L. N., 1995. Cable Drawn Drainage Plough for Land Reclamation in Unconsolidated Phosphate Clay Sediments. Journal of Agricultural Engineering Research 62: 215–219.

26. Spoor, G., and Godwin R. J., 1978. An Experimental Investigation into the Deep Loosening of Soil by Rigid Tines. Journal of Agricultural Engineering Research 23: 243–258.
27. Hefti, and Brugg, J., 1955. Untersuchungen über das Zugkraft- und Transportproblem im Hangbetrieb. Alpenwirtschaftliche Monatsblätter (reprint) 5(6/7).
28. McKeyes, E., 1989. Agricultural Engineering Soil Mechanics. Amsterdam: Elsevier.

1.3. Seeders and Planters

H. Heege and J. F. Billot

1.3.1. Introduction

Seeders and planters are essential for the reproduction of crops. Their function is metering and placing in the soil of seeds or plants or of parts thereof. Thus, seeders are used for generative reproduction, whereas planters aid in vegetative propagation. However, in some parts of the world, the term *planter* is used for a seeder. Both seeders and planters are either used as solo machines or in combinations with preceding soil cultivating machines. In some cases, combinations with fertilizing equipment also are common.

1.3.2. Seeders

Generally, seeders comprise one or several hoppers, which contain the seeds of metering parts, and of equipment for seed placement in the soil. The metering either aims at equidistant spacing of the seeds or is restricted to feeding a stream of seeds into a conveying tube. The former case can be defined as precision seeding, whereas for the latter case, the term *bulk seeding* might be appropriate. Since in most cases the seeds are placed in a row, the definitions of precision drilling as well as bulk drilling make sense.

Precision drilling is used mainly for rather widely spaced crops, such as corn, beans, sugar beets and sunflower. With closely spaced crops, precision drilling is too expensive and therefore bulk drilling common (Fig. 1.194).

Seed-Spacing and Seeding-Depth

The mean area per seed with drilled crops is the product of row-spacing and the average seed distance within the row.

Row-spacing varies widely with crops, climatic conditions and production technology. Precision-drilling of crops occurs mainly with row-spacings between 25 cm and 90 cm. Within this span narrow row spacings are used for peas and beans, medium row spacings for beets and sunflower, and the widest spacings for corn. Since in humid areas the water supply allows generally for higher seed and crop densities, row spacing in these areas is closer than in dry regions. Furthermore, mechanical weed control and harvesting methods often require minimum row spacings.

Seed spacing within the row always is a fraction of row spacing, though under ceteris paribus conditions, equal spacings within and between the rows would promote high yields. However, realizing a given seed area by narrower row spacings associated with wider seed distances in the row increases the investment for precision seeding as well as

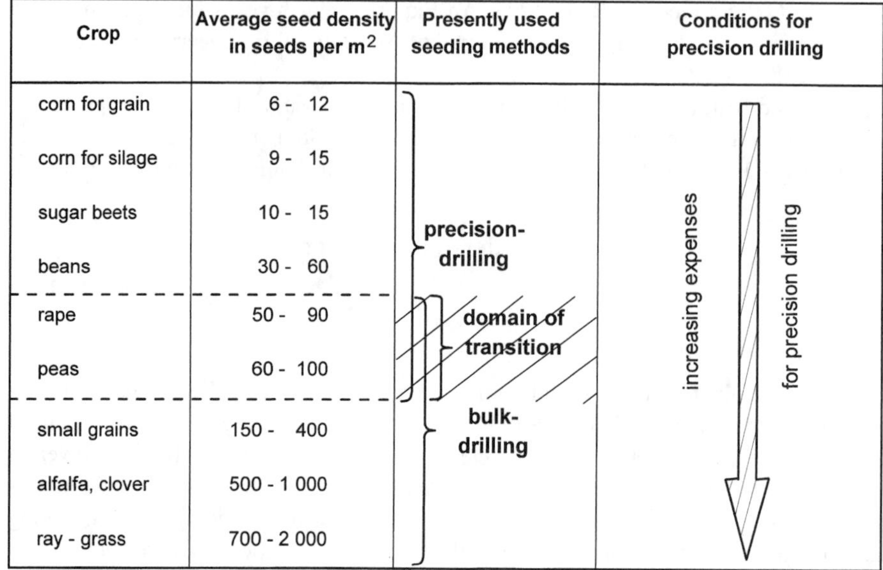

Crop	Average seed density in seeds per m²	Presently used seeding methods	Conditions for precision drilling
corn for grain	6 - 12	precision-drilling	increasing expenses for precision drilling
corn for silage	9 - 15		
sugar beets	10 - 15		
beans	30 - 60		
rape	50 - 90	domain of transition	
peas	60 - 100		
small grains	150 - 400	bulk-drilling	
alfalfa, clover	500 - 1 000		
ray - grass	700 - 2 000		

Figure 1.194. Conditions for seeding methods.

for row-crop harvesting. This is because the number of seeding and row-crop harvesting units increases for a given working width.

With bulk drilling, the row spacing varies between 8 cm and 25 cm. Narrow row spacings are common especially with grass, clover, alfalfa, and small grain seeds in humid areas. Here again average seed spacing within the row always is a fraction of row spacing.

Seeding depth needed depends very much on seed size and on water content of the soil. The larger the seeds and the dryer the soil, the deeper the seeds should be placed to ensure emergence, and vice versa. In most cases, grass, clover, and alfalfa seeds need a soil cover of 1–2 cm; sugarbeet seeds 2–3 cm; small grains, rape, peas, and sunflower beween 2–5 cm; and beans and corn between 4 cm and 10 cm.

Precision Drilling

Basically, the metering of the seeds with precision drilling consists of two functions: singling of the seeds and transporting them to the furrow. In many cases, singling of the seeds is accomplished purely in a mechanical way; however, sometimes the singling process is supported by airflow. Accordingly, mechanical as well as air-assisted precision drills must be dealt with.

Mechanical Precision Drilling

Singling methods are illustrated in Fig. 1.195. In most cases, a rotating disk with holes or cupped fingers at the circumference picks single kernels from the supply. Sometimes instead of a disk, a circulating belt with seed holes is used.

The shape as well as the dimensions of the holes and cups must be adapted to those of the seeds. Too large or too small dimensions promote doubles or skips, respectively. In most cases it is not possible to avoid doubles or skips completely. Therefore, the

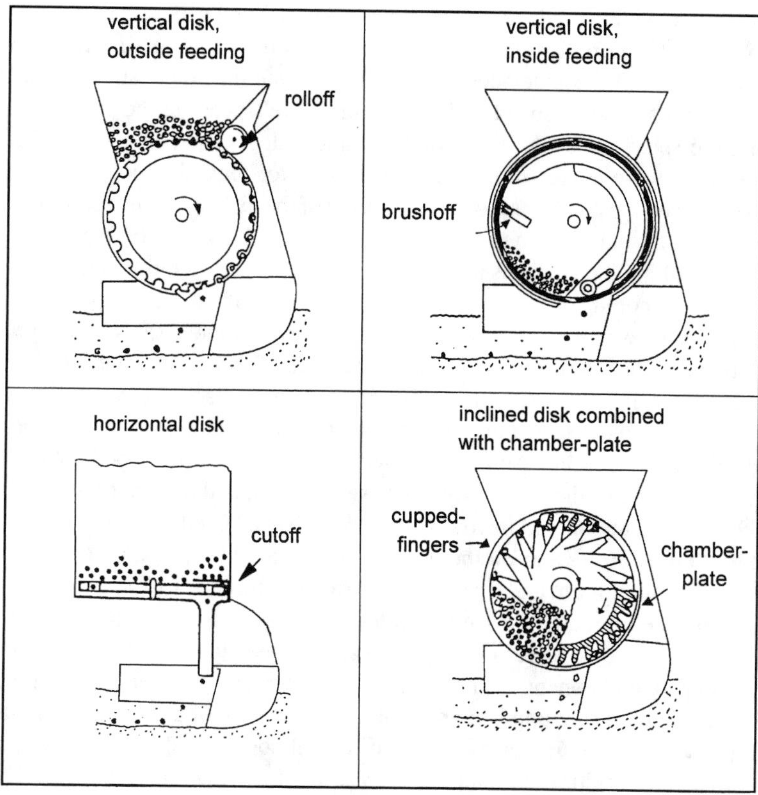

Figure 1.195. Mechanical precision-drilling.

adaptation should aim at an equal number of doubles as well as skips. This means that a balance in the total number of seeds is achieved.

It is important that the seeds get to the furrow in the original sequence. A long, free falling distance from the delivery point to the furrow is detrimental; it deteriorates an originally even seed sequence. This especially applies to small seeds.

In this respect, the orientation of the metering disk is relevant. With a vertically oriented metering disk, the falling height can be smaller than with an inclined disk and especially smaller than with a horizontally located disk. This explains the predominance of vertically oriented metering disks with small vegetable seeds and sugarbeet seeds. A disadvantage of vertical disks with larger seeds compared to horizontal disks can be the rather short peripheral sector available for filling. Therefore, in the United States for large seeds such as corn, horizontal disks still are in use despite their principal disadvantage in the falling height.

Inclined disks normally deliver the seeds at their highest point. Therefore, in order to get a low, free-falling height, they often operate with a parallel rotating chamber plate in the background. The seeds pass from the singling disk into the chambers of the parallel plate, which then delivers them closely above the furrow. In most cases, precision metering requires devices that remove doubles and triples from the respective cells. Counter-rotating rollers, stationary cuttoffs, or brushes are used for this purpose. Careful adjustment of these devices is essential. Despite this, there still is an influence of the peripheral seed disk speed and thus of the travel speed on the singling process. Increasing the speed raises the share of skips and reduces the percentage of doubles. Inclined disks often operate without special devices for removing doubles and triples. Excess seeds lie on top of other seeds. The rotation of the disk moves the gravity center of the excess seeds beyond the rim of the cup; thus, they fall back into the supply. However, this advantage of inclined disks is associated with less precise singling on sloped fields. Their use is not recommended on fields with more than 10% slope (1).

There can be uneven seed spacings in the furrow even if the kernels hit the ground in regular intervals. The main reason for this is bouncing of seeds along the furrow bottom, which occurs especially with spherical seeds if the trajectories are not vertical to the soil.

To some extent, these irregular spacings due to bouncing of the seed depend on the shape of the furrow-bottom. An acute-angled furrow-bottom often prevents excessive rolling and bouncing of the seeds, whereas a smooth rounded bottom promotes it. With new machines an acute-angled furrow-bottom, which is adapted to the seed dimensions, generally is formed. Yet wear on the lower pointed side of the opener can change the shape of the furrow-bottom and thus increase irregularities due to bouncing.

The travel speed imparts the seeds a velocity component in the direction of travel. Vertical seed trajectories, therefore, require that this velocity component is offset by an opposing velocity component, such as the peripheral velocity of vertical seed plates. If the traveling speed and the opposing peripheral velocity are of the same magnitude, the seeds drop vertically into the furrow, and detrimental bouncing along the furrow bottom decreases. However, with seed plates that are filled from the outside (Fig. 1.195), the peripheral velocities are much lower than the usual travel speeds. Peripheral velocities in the order of the travel speeds with these seed plates would be associated with centrifugal forces, which prevent most seeds from entering the cells.

This problem can be overcome by filling the vertical seed plates from the inside (Fig. 1.195). Now centrifugal forces support gravitational forces in the filling process. Therefore, a much higher peripheral velocity–downright in the order of the travel speed–can be used (Fig. 1.196). The seeds drop vertically into the furrow (1).

However, this method of *vertical seed dropping* requires peripheral cell distances at the seed disk, which correspond exactly to the desired seed distances in the furrow. Thus, the traditional method of varying the seed-spacing by adjusting the rotational velocity of the disk can be used only to a very limited extent; for major adjustments changing the seed disk is necessary.

Therefore, vertical seed dropping mainly is used with strictly spherical seeds such as sugarbeet pills, which tend to roll easily along the furrow-bottom.

Another approach to eliminate bouncing of the seeds along the furrow bottom is placing them into a hole instead of in a furrow. For this purpose so-called *punch-seeders* have been developed (Fig. 1.197). They are not yet used commercially except for punching through and seeding underneath a plastic mulch. Varying the seed spacing within the row with punch-seeders is even more difficult than with the vertical seed dropping method (2, 3).

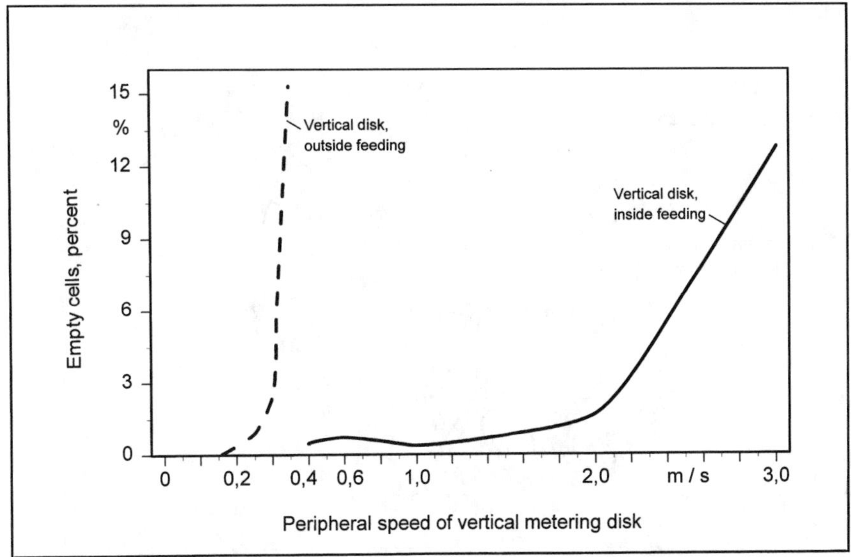

Figure 1.196. Outside- or inside feeding with vertical metering disks (Courtesy of W. Brinkmann, altered).

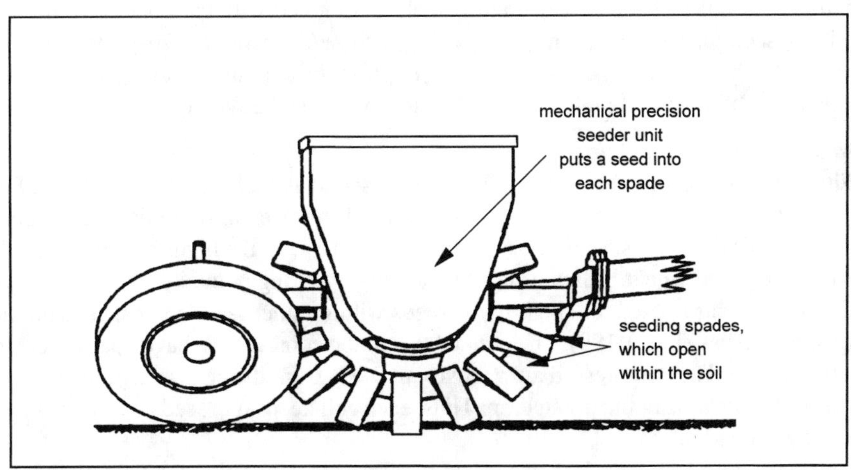

Figure 1.197. Operating principle of punch-seeder (Courtesy of L. N. Shaw, altered).

Generally, mechanical precision drilling requires a much closer adaptation in the dimensions of the singling elements to those of the seeds than air-assisted precision drilling. This applies especially to vertical or horizontal metering disks equipped with cells and slightly less with inclined disks equipped with cupped fingers. The cell diameters should be about 10% more than the largest diameters of the respective nearly spherical seeds (1).

There have been attempts to avoid the need of closely adopting the hole or cup dimensions to those of the seeds. A principle used with mechanical precision drilling is

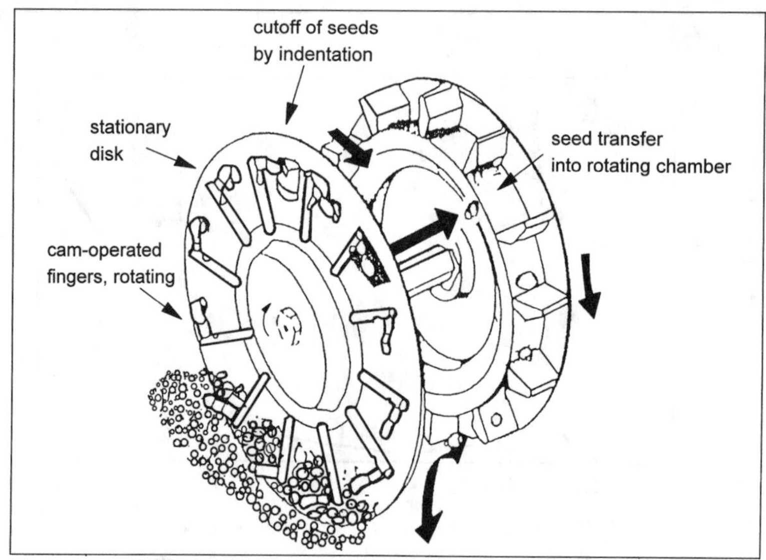

Figure 1.198. Precison-metering of corn by cam-operated fingers (Courtesy of John Deere Co., altered).

shown in Fig. 1.198. The vertical singling disk is equipped with cam-operated fingers. Within the seed supply these pointed fingers move close to the disk and respectively seize a seed. At the highest point, the seeds again pass from the singling disk respectively into the chambers of a parallel plate for delivery closely above the furrow.

Air-assisted Precision-drilling

Either the suction or the compression by air is used to assist in singling the seeds. For singling by suction, a vertical disk with suction holes passes through the seed supply (Fig. 1.199). Excess seeds per hole usually are brushed off. At the highest point the seeds pass to a sectioned wheel, which brings them to the dropping point.

Compressed air is used for singling the seeds within rather large conical cells on the periphery of a disk (Fig. 1.199). The pointed tips of the conical cells have open access to the atmosphere. The compressed air removes all seeds from the cells except the lowest, which covers the hole at the pointed tip. Thus, each cell keeps one seed and transports it to the dropping point.

Air-assisted precision drilling needs less precise seed calibration than all purely mechanical methods and therefore is used extensively for irregularly and non-spherically shaped seeds such as corn, sunflower, and some bean varieties. Custom operators especially appreciate that the seed disks seldom must be exchanged when moving from farm to farm with different seed varieties during the season. The investment for air-assisted precision drilling, however, is higher because of the PTO-driven blower.

The general concept of precision drilling is, that a metering unit is placed on each opener. This method allows for rather small distances from the delivery point to the furrow and thus favors the exact seed placement. However, this concept requires rather

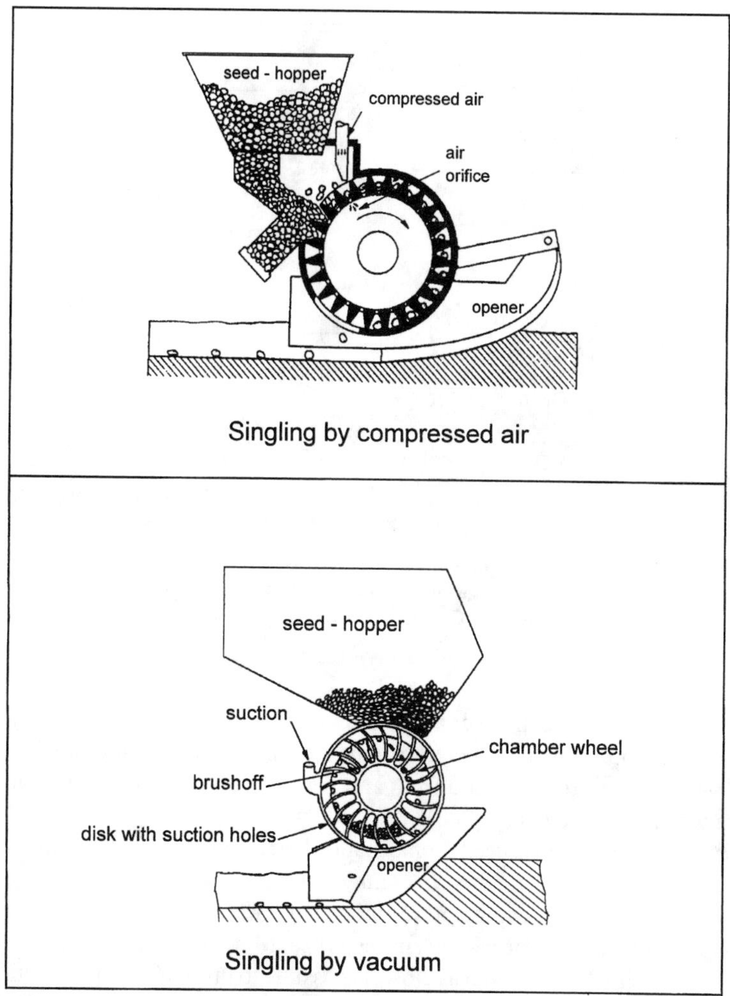

Figure 1.199. Air-assisted precision-drilling.

troublesome filling and monitoring of many seed hoppers. In case all openers of a machine are supplied with seeds from a central seed hopper, the supply can be refilled in bulk and monitoring it is facilitated.

But this concept of replacing the row of small hoppers and metering units with one central big hopper and one central metering system requires long seed tubes for gravitational and pneumatic seed transport to the openers. An originally even seed sequence deteriorates substantially during this seed transport (1, 4). Therefore, sometimes a central hopper provides for seed supply to the standard system of a small hoppers and a metering unit for each row. The investment for this method of twofold seed storage is high, yet it allows bulk filling as well as metering in place.

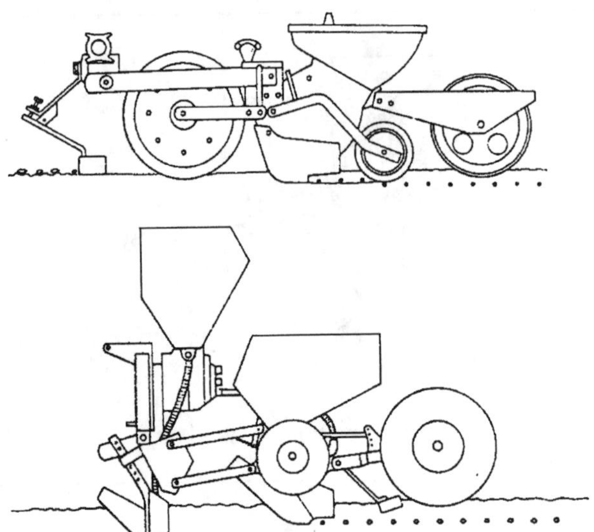

Figure 1.200. Depth control for precision drilling by either a front-plus rear gauge wheel (top) or only a rear-gauge wheel (below) (Courtesy Fähse and Tröster).

Seed Incorporation into the Soil

Maintaining a constant soil cover above the seeds requires depth control of precision drills. The vertical position of the metering units inclusive openers, therefore, is regulated by gauge-wheels. In Europe either a front and rear gauge wheel or alternatively only a rear gauge wheel are used (Fig. 1.200).

The former concept often is used for mechanical precision drilling of small seeds such as sugarbeets, and the latter method is common with pneumatic systems for corn and beans. In the United States, depth control by a gauge wheel located sideways from the opener traditionally is used for widely spaced crops. Morrison and Gerik (5) stated that the gauge wheels should be located as close as possible to the openers. They obtained the best depth control with either side wheels or linked front and rear wheels. Side wheels were the least sensitive to crop residues, but required higher down pressure levels.

Depth control can be deteriorated by clods, which lift the gauge wheels. Clod deflectors often precede front gauge wheels. Soil deflectors or rear gauge wheels with an inverted conical periphery close the furrow. The drier the soil is, the more it should be compacted around the seeds in order to attract water from beneath. Thus, also pressing the seeds into the furrow before closing it by a narrow press wheel can improve the emergence under dry conditions (6).

Sowing into non-ploughed land with crop residues on the surface requires special equipment in order to avoid clogging and uneven seed coverage. Where some soil cultivation has taken place, rotating finger-wheels, which run slanted to the direction of travel in front of the seeder along the row, can move the crop residues into the interrow area (7, 8, 9).

When soil cultivation is omitted completely, in most cases disk openers are needed instead of the traditional shoe openers. If operating under adequate pressure and in hard soil, these disk openers cut through plant residues. Side gauge wheels in many cases are useful for adequate depth control and for holding the soil around the opener in place. Closing the furrow often is attained by twin rear wheels, which are vertically slanted against each other and thus press the soil from both sides against the row.

Punch seeders hardly have any problems with crop residues (2, 3).

Drive and Monitoring

Traditionally, metering units are driven via ground contact wheels. This provides for constant seed distances independent of the travel speed. In the past, the drive often emanated from gauge wheels. Adjusting the transmission for every individual metering unit was necessary. Today the drive very often emanates from the wheels of the central machine frame. The seed spacing for all units can be adjusted by changing the central gear ratio.

An interesting new approach is to power the metering units individually by small electric dc motors (Fig. 1.201). The current is supplied by the electrical system of the tractor or with wide machines by a separate generator. The seed spacing can be adjusted easily from within the tractor cabin by an electronic control system, which varies the motor speed. Changing the seed spacing for site-specific farming as well as switching off of units for tramlines or when approaching the field borders are on the go possible.

Monitoring the seed metering especially is important with wide machines, since its operator cannot easily supervise all units. An optical detector located underneath the metering device indicates proper function and counts the number of seeds delivered. Precise detection of the seed number requires that the shares of doubles and empty cells are of the same order.

Figure 1.201. Electrically driven metering disk of a precision drill (Courtesy of Kleine Maschinenfabrik, Salzkotten).

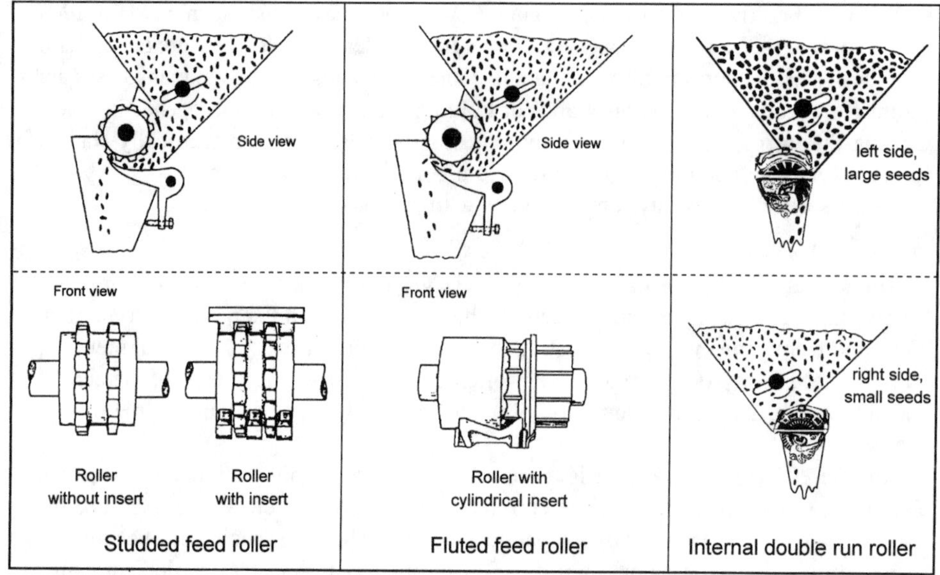

Figure 1.202. Bulk seed metering rollers.

Bulk Seeding

Metering of the Seeds

No singling of the seeds occurs; instead, a stream of seeds is bulk-fed into the conveying tubes. The metering of the seeds is done by studded feed rollers, fluted feed rollers, or by internal run rollers (Fig. 1.202). Long seed tubes are not detrimental to the seed sequence, since no initial even spacing exists. Therefore, instead of individual metering units–as with precision drilling–metering devices on a common drive shaft and a common seed hopper for all rows can be used. This facilitates loading the hopper in bulk. Conveying within the tubes to the openers is either by gravitational forces or in case of *air seeders*, pneumatically.

The air seeders allow for rather independent location of the seed hopper plus metering devices on the one hand and the openers on the other hand (Fig. 1.203). Thus, with combinations for soil cultivation and seeding the mass of the filled hopper can be used to load the cultivator tines. And for large machines, the seeds can be metered from a tender wagon, thus saving time for refilling.

Seeds can be fed into the airstream separately for each opener tube by the respective metering devices or centrally for all opener tubes by just one metering roller. With central feeding, all seeds initially are injected into a principal blower tube. The seed-air mixture is led upwards and hits a circular deflector manifold. The opener tubes emanate from this deflector manifold (Fig. 1.203).

Injecting the seeds into the air stream with separate feeding as well as with central feeding in most cases is realized with an air pressure below normal in order to achieve suction. For this purpose, the tube cross-section at the injection point is decreased.

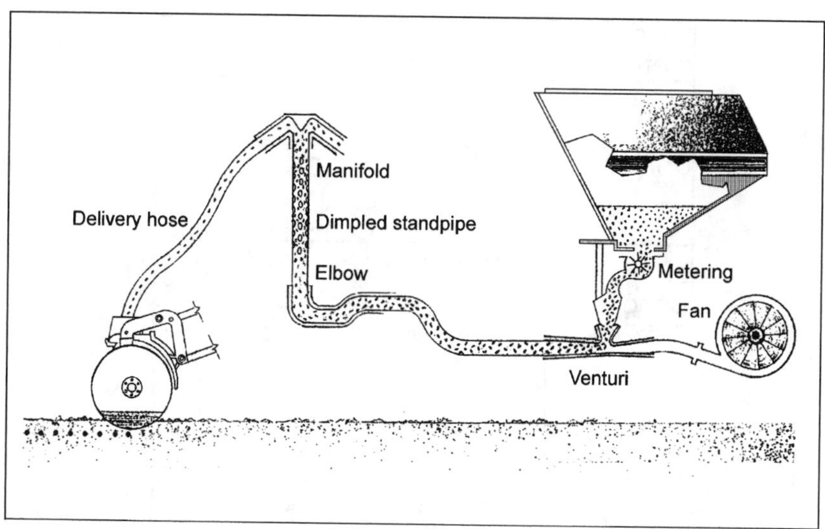

Figure 1.203. Operating principle of an air-seeder system Weiste (Courtesy: Case Corporation, altered).

Adjusting the seed rate is done either by changing the rotational speed or the volume of the feed rollers. In some cases, the volume of the feed rollers is adjusted to the size of the seeds (Fig. 1.202).

Normally, the feed rollers are driven via ground contact wheels. In the future, powering by an electric dc motor—as with precision drilling—may be important, since this facilitates the control for site-specific farming. However, varying the travel speed with this method still requires the input of a speed sensor into the control system.

Seed Incorporation into the Soil

Drilling is the dominant technique for seed placement into the soil. The seed may be deposited by shoe openers, disk openers, or sometimes in the furrows of packer rings (Fig. 1.204a).

Seeding in a band having a width of 3–4 cm can be accomplished in several ways (Fig. 1.204b). In some cases, plane disk openers slightly angled to the direction of travel are used to create the necessary furrow width. When soil cultivation and sowing are combined, seed tubes located in the soil flow of the tines can place the seed in a band of this width. For bandwidths ranging from 10–25 cm, special coulters such as wing openers or broad-shoe openers have been developed (Fig. 1.204c). Wing openers lift the soil, whereas broad-shoe openers move it sideways within the band width. At the bottom of these openers, the seed hits a baffle, which causes broadcasting within the coulter space. When the opener spacing across the width of the machine is decreased to fit the band width, broadcasting is achieved over the whole area.

Placing the seed in a broadcast pattern also can be achieved by sowing underneath the soil trajectories of a rotary cultivator (Fig. 1.204c). The soil is thrown over a full-width broadcast bar, which is positioned behind the rotor and slides over the uncultivated soil. Seed is conveyed pneumatically to the full-width broadcast bar. Occasionally, broadcast

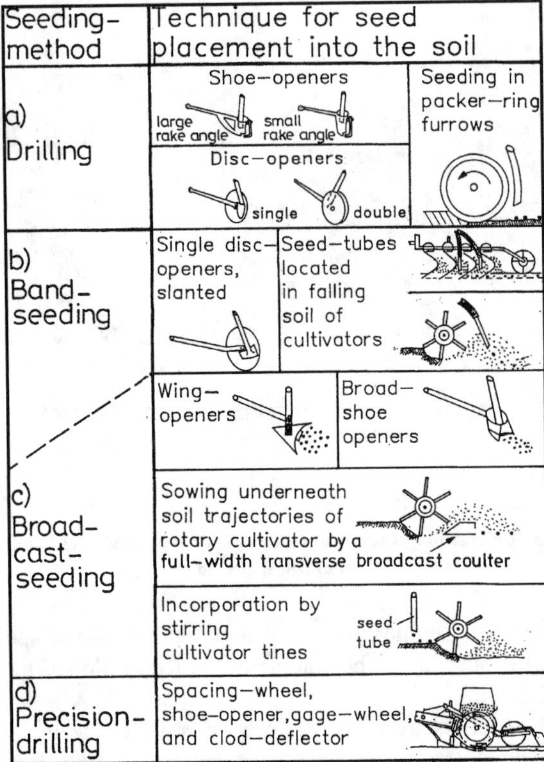

Figure 1.204. Seeding methods and seed placement into soil.

sowing is carried out by simply dropping seed freely onto the soil surface in front of a cultivator. Incorporation into the soil is done by the stirring action of the cultivator tines (Fig. 1.204c).

Sowing Depth Uniformity with Bulk Drilling

It has been shown several times, that gauge wheels attached to light openers for bulk seeding often do not improve the depth control, but instead change it for the worse (10, 11, 12). The gauge wheels tend to climb over coarse soil clods, lifting the light openers. With precision drills the situation is quite different. Their openers are usually weighted by the mass of the metering unit. The result is that most clods either are swept aside or pushed into the soil.

However, in soils with low water supply, wheels attached to bulk drill openers may have a positive effect because of soil compaction, since this attracts water from below.

With shoe-type openers for grain drills, the influence of the rake angle (angle between the soil surface in front of the opener and the opener front below the soil) on the precision of depth placement has been of interest. Gebresenhet and Jönsson (1992) have shown that a large rake angle is advantageous for the depth stability of the opener in light sandy soils (13). This seems to be in line with the experience of farmers in the Netherlands, where short-shoe openers with a small rake angle and long-shoe openers with a large rake angle (Fig. 1.204) are named *clay openers* and *sand openers*, respectively.

Sometimes shoe openers similar to cultivator tines, which point in the direction of travel, are used. These openers with an acute rake angle penetrate the soil more easily and are especially used in dry areas, where the seeds must be placed deeper.

For the future, the question arises whether depth control via automatic adjustment of the vertical opener pressure on the go should be used. A concept in this direction is the automatic opener depth control developed by Dyck, Wu and Lesko (14). The pressure on the opener is varied according to the height of the coulter lever above the soil, which is recorded by the ultrasonic sensor. Herbst (15) used a completely different approach. He assumed that differences in the opener penetration mainly result from changes in the soil resistance. The soil resistance is measured by means of a horizontally positioned penetrometer, is transduced, and used as the initial signal for controlling the depth by varying the opener pressure. Up to now, automatic pressure control is not used commercially.

Figure 1.205 shows the variation of the seeding depth obtained with commercial seeding methods from 42 field experiments (16). The results refer to a mean seeding depth ranging from 25–45 mm, as is commonly used when sowing small grains.

All drill openers, slanted disk openers, and broad shoe openers were used without gauge wheels. Wing openers were employed with, as well as without, gauge wheels. Since the use of gauge wheels with these openers did not alter the standard deviation of the seeding depth significantly, the data were pooled. The precision drills were equipped with gauge wheels in front of, as well as behind, the openers.

In seedbeds prepared by plowing and secondary cultivation, the use of shoe openers, disk openers, wing openers and broad-shoe openers gave a standard deviation of the seeding depth from 6–11 mm. With seeding in packer ring furrows, as well with the

Seeding–method		Technique for seed placement into the soil	Stand. deviation of the seeding–depth	
			soil ploughed	soil not ploughed
Drilling		Shoe–openers, Disc–openers	6–11 mm	12–17 mm
		Packer–ring furrows (only for sandy soils)	6–11 mm	
Band–seeding	Bandwidth 3 cm	Seed tubes located in falling soil of cultivating implements	12–17 mm	14–19 mm
		Disc–openers, slanted		
	Bandw.8cm	Wing–openers, Broad–shoe openers	6–11 mm	
Broadcast–seeding		Wing–openers, Broad–shoe openers		
		Seeding underneath soil trajectories of rotary cultivator (broadcast.–bar)	5–13 mm	5–11 mm
		Incorporation by stirring cultivator tines	15–20 mm	
Precision–drilling		Shoe–openers equipped with gage–wheels and clod–deflectors	4–9 mm	

Figure 1.205. Variation of the seeding-depth.

broadcasting bar underneath the soil thrown by a rotary cultivator, a rather even soil surface is required; it is not possible to follow the soil contour as closely with these techniques as with conventional openers. If the preceding cultivation provides this surface, the accuracy of seed placement can be as with conventional openers (Fig. 1.205).

In most cases, precision drilling gave the best results, which mainly is due to the openers pushing soil clods to one side rather than lifting over them. Therefore, the superiority of precision drilling to openers for bulk drilling depends on the cloddiness of the soil. Band sowing by seed tubes located in the falling soil of cultivating implements or incorporating the broadcast seed by stirring cultivator tines resulted in a very uneven sowing depth (Fig. 1.205).

It is generally known that following non-plowing techniques, the seeding depth is more uneven due to plant residues near the soil surface. Some seeding techniques do not function well under these circumstances because of clogging. However, when broadcast seeding underneath the soil trajectories of a rotary cultivator, the best results (Fig. 1.205) were obtained in unplowed soil, where only some leveling and straw incorporating had taken place by stubble cultivation. Thus, this technique lends itself when plowing is not advisable because of erosion problems.

Figure 1.206 shows the influence of the seeding-depth uniformity on the field emergence of small grains. The regression was obtained by using the results with different techniques under varying weather conditions. The negative effect of a high variation in the seeding depth is obvious, although it should be mentioned that it is less distinct with wet weather than under dry conditions (16).

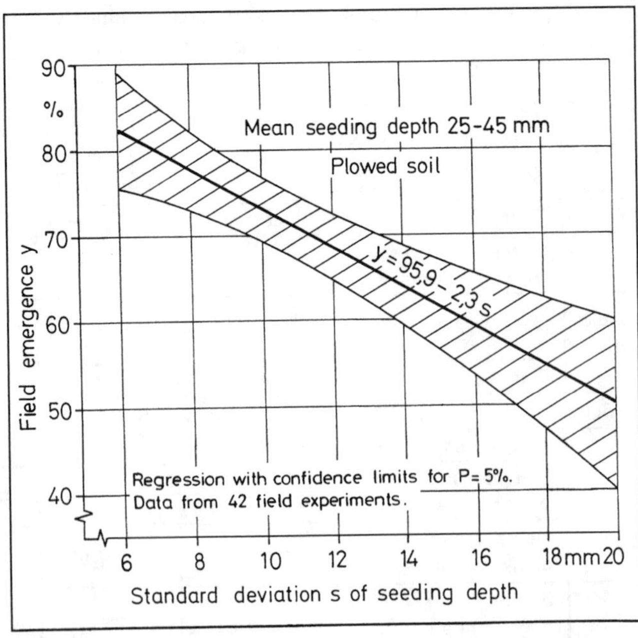

Figure 1.206. Variation of seeding depth and emergence of small grains.

Using a higher seed rate can only partly offset the negative effect of a low field emergence. It can, despite a low field emergence, provide for the plant density desired. Yet it results in staggered plant growth, which makes it impossible to treat all plants at the same stage with chemicals. Thus, despite the plant density obtained, the yield still can be affected. Furthermore, it is desirable that seeding techniques make it possible to use a low seed rate.

Lateral and Longitudinal Seed Distribution

The lateral seed distribution is recorded perpendicular to the direction of travel; thus, the basis of it is the seed mass metered to each opener of a machine. Contrary to it the basis of the longitudinal seed distribution are the seed distances in the row. The recording therefore occurs in the direction of travel.

In most cases the lateral distribution is no problem; generally, its coefficient of variation is below 4%. This applies to bulk drilling as well as precision drilling. Exceptions can occur with air bulk seeders on a slope, if central feeding is used. Since the vertical main blower tube and its circular deflector plate in this case are slanted, the seed rate for openers downwards is higher than for those upwards. How distinct this effect is depends not only on the slope, but also on the air velocity and on the seed sizes. The higher the air velocity is and the smaller the seeds are, the smaller is this slope effect (17).

The longitudinal distribution with bulk drilling based on the seed distances in the row has been found to follow a known statistical function (16, 18). The frequency of the seed distances Z at the end of the seeding tube, when entering the opener, or in the row, corresponds to the exponential density function:

$$p(Z) = \frac{1}{\bar{Z}} e^{-Z/\bar{Z}}$$

where $\bar{Z}$ is the mean distance.

This applies to volume seed metering by all known types of rollers and for gravitational as well as pneumatic transport in the seed tubes. It should be noted that computing the frequency of defined seed distances requires integration of the function $p(Z)$ within the limits of the respective distances. Thus, the frequency for distances between a and b is:

$$P(b) - P(a) = e^{-a/\bar{Z}} - e^{-b/\bar{Z}}$$

With drilling, this distribution does not result in any maximum in seed spacing within the range of the corresponding mean seed distance in the row (Fig. 1.207). Thus there is no controlled seed placement. Small mean seed distances in the row result in distinct seed clusters.

Since the coulters for band and broadcast sowing are fed with the same seed sequence, the frequency of the seed distances in the direction of travel within each bandwidth also follows an exponential distribution (Fig. 1.208). However, in this case, deviations in seed placement are created in the direction of travel as well as perpendicular to the direction of travel. The seed clusters of the exponential distribution are separated by kernel distances perpendicular to the direction of travel. With precise broadcast sowing, the seed distribution perpendicular to and in the direction of travel is the same. The

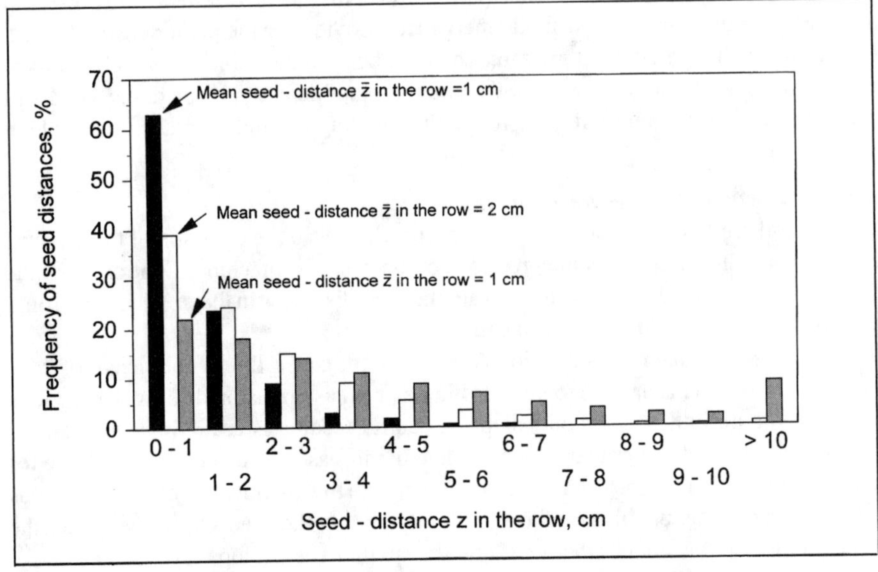

Figure 1.207. Seed distances in the row with bulk-drilling.

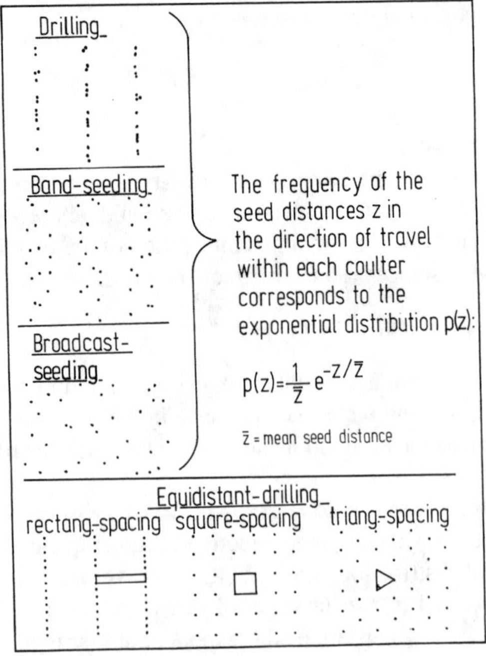

Figure 1.208. Seed spacing on the area.

seeds are distributed over the area in the same pattern as raindrops. It has been shown experimentally, that this pattern is defined by the Poisson distribution (18).

Precision drilling aims at equidistant spacing of the seeds, which could have a rectangular, a square, or a triangular arrangement. With present seed rates, the placement in a square or triangular arrangement would require very narrow row widths. Therefore, the square or triangular spacing is not used. If, on the other hand, larger row widths are utilized, the seeds must be placed a short distance apart within the row. This explains why precision drilling of small grains still is in an experimental stage. It should also be born in mind that precision drilling is merely an approach to equidistant spacing because of the technical deficiencies involved. The frequency of the seed distances in the direction of travel therefore follows a normal distribution, which has the adjusted spacing as its maximum. However, with proper use, the uniformity of seed spacing is much better than with bulk drilling. The coefficient of variation in the direction of travel with proper use of precision drilling is below 20%, whereas with the exponential distribution for bulk drilling it is 100% (18).

Seed Distribution over the Area and Yield

A visual observation indicates that the sowing methods differ in seed distribution over an area (Fig. 1.208). The worst seed distribution over the area generally results from bulk drilling with wide row widths and, therefore, small seed distances within the row. Narrow row widths improve the situation. In can be shown that bulk drilling is converted to broadcasting when its row width theoretically approaches zero (16, 18).

Since in practice also with precision drilling the seed distances in the row are much lower than the row widths, the seed distribution over the area with this method too can be improved by decreasing the row width. Figure 1.209 shows that, with small grains,

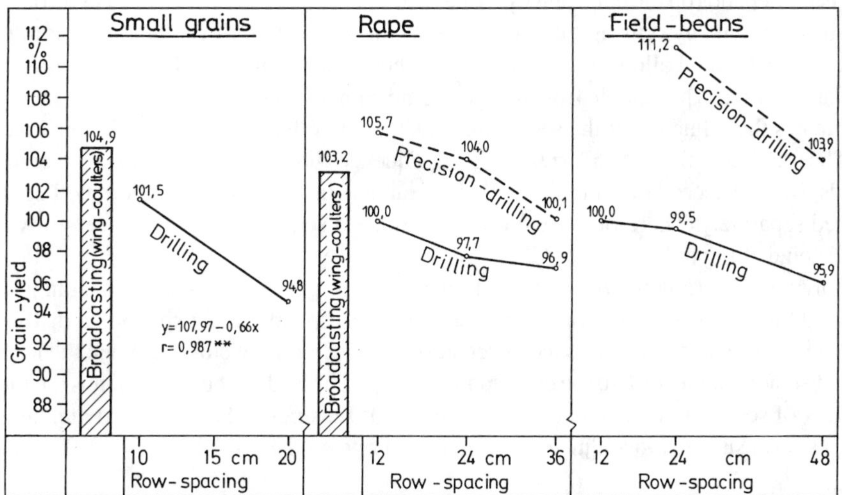

Figure 1.209. Seed distribution over the area and yield (yield when drilling with 12 cm row-spacing = 100).

rape, and field beans the yield increases, when the row width decreases. As expected, precision drilling always has a higher yield than bulk drilling. The yield for broadcasting with wing coulters surpasses narrow bulk drilling, but is below narrow precision drilling.

It should be noted, that the broadcasting method used had a sowing depth uniformity which is similar to that of conventional bulk drilling methods (Fig. 1.205). Grain yields were not higher than with drilling, when following broadcasting the seeds were incorporated into the soil by the stirring action of cultivator tines (19). The negative effect of the less precise seeding depth on the yield seems to be at least as important as the positive influence of the improved spatial seed distribution. Therefore, good yields with precision drilling may result from the improved seed distribution over the area, as well as from the more precise seeding depth. Increased yields with narrow row precision seeding also have been reported for corn and cotton (20, 21). There are of course limits for narrow row seeding. With widely spaced crops these can arise from mechanical weed control and harvesting methods. Clogging of crop residues or clods between adjacent openers can increase with close spacings.

Increasing the number of transverse opener rows by staggering the openers can help to overcome this problem. Some manufacturers offer grain drills designed for a row width down to 8 cm. The openers are arranged in three or four transverse rows.

However, it should be noted that, apart from the clogging problem, narrow row widths can impair the precision of the seeding depth. Because of the close spacing, the depth of trailing openers can be influenced by ridges from preceding openers. Therefore, broadcasting methods with an adequate uniformity in the sowing depth should be regarded as an alternative to bulk drilling for closely spaced crops such as small grains.

Sensing of the Seed Rate with Bulk-metering

Present day bulk metering systems make it difficult to adjust the seed rate precisely, since this depends on the bulk density of the seed as well as on the wheelslip. This problem can be overcome by sensing the seed sequence in the seed tubes and adjusting it to the speed. This method allows to control the number of seeds instead of the seed mass per unit area, which is preferable from an agronomic point of view. Furthermore, this method eliminates the influence of the seed bulk density and of the wheel slip on the seed rate.

The problem is the fast and irregular seed sequence with bulk metering of small grains. Seeds, which succeed one another without clearance in the falling direction, will not be sensed separately. Therefore, up to now optical seed sensing has been successful with precision drilling only.

Solving this problem with bulk metering of small grains is possible by taking into account that the seed sequence at the tube outlet corresponds to an exponential distribution. Therefore, it is not necessary to detect seeds which follow closely. When the larger seed distances are recorded, the corresponding exponential distribution and thus the total number of seeds can be computed. This method of sensing the bulk-metered number of seeds allows site-specific adjustment of the seed rate on the go (22).

Tramlines

It is useless to deposit seeds where no plants can grow. Therefore, it makes sense to provide for unsown tramlines there, where the wheeltracks for top dressing of fertilizer and pesticides will be. This applies to closely spaced crops such as small cereals and

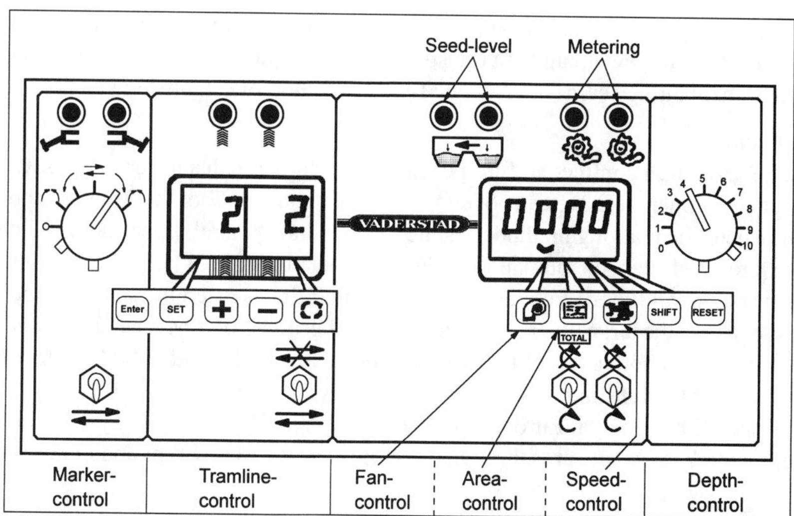

Figure 1.210. Computer-assisted monitoring and-control (Courtesy of Väderstad, Sweden [redrawn and altered]).

rape, which do not yield in wheel tracks. With widely spaced crops the row spacing can be arranged such that these wheeltracks just fit into the interrow area.

Tramlines can improve the precision of fertilizer and pesticide distribution, since driving deviations are minimized. The yield depression caused by the tramlines proportionally can be decreased by wide working widths of the top dressing machines. Therefore, the top dressing operations should be carried out with multiple working widths of sowing.

Arranging of tramlines with bulk metering is possible by closing the respective seed gates of the hopper, by locking the respective seed rollers, or, for air seeders with central feeding, by closing the respective tube inlets at the circular deflector plate. Shifting of tramlines is done either by hand within the tractor cabin or automatically while turning at the headlands in an adjusted sequence.

Computer-Assisted Monitoring and Control

Figure 1.210 shows the switchboard and monitor for electronic tramline shifting of an air-seeder. The shifting sequence is manually put in; accordingly each marker change causes a switch forward. Activating the markers is triggered by the hydraulic lift at the headlands. When the adjusted point in the shifting sequence is attained, the respective tube inlets at the circular deflector plate are closed or opened by magnets.

Further possibilities for computer-assisted monitoring result from sensors, which register the seed level in the hoppers or record the travel distance. Thus the need for refilling and the acreage done can be indicated.

1.3.3. Planters

Planters are used for vegetative reproduction of crops. Either parts of plants or whole small plants are put into the soil. Potato tubers can be regarded as parts of plants. On the other hand, with paddy rice, cabbage, and trees it is customary to put into the soil whole small plants, which grew up under a transparent cover in garden beds or in greenhouses. The term transplanting seems appropriate for this method of establishing a crop.

Potato Planters

Potatoes grown in continental Europe and in the United States differ in sizes. On the average, the continental European potato weighs only about 35% of the American potato. This influences the planting methods despite the fact that, generally, for planting smaller fractions are used than for human or industrial consumption. Whereas in continental Europe whole small tubers are planted, in the United States often the seed potatoes are cut in order to use the large tubers more effectively. The mass of tubers needed per ha still is high; it varies on both sides of the Atlantic between 1200 kg and 2000 kg. Therefore, the delivery and the refilling of seed tubers deserves attention.

The standard continental European potato planter singles the tubers by vertical belts with cups, which move upwards through the supply (Fig. 1.211). Skips and multiples can

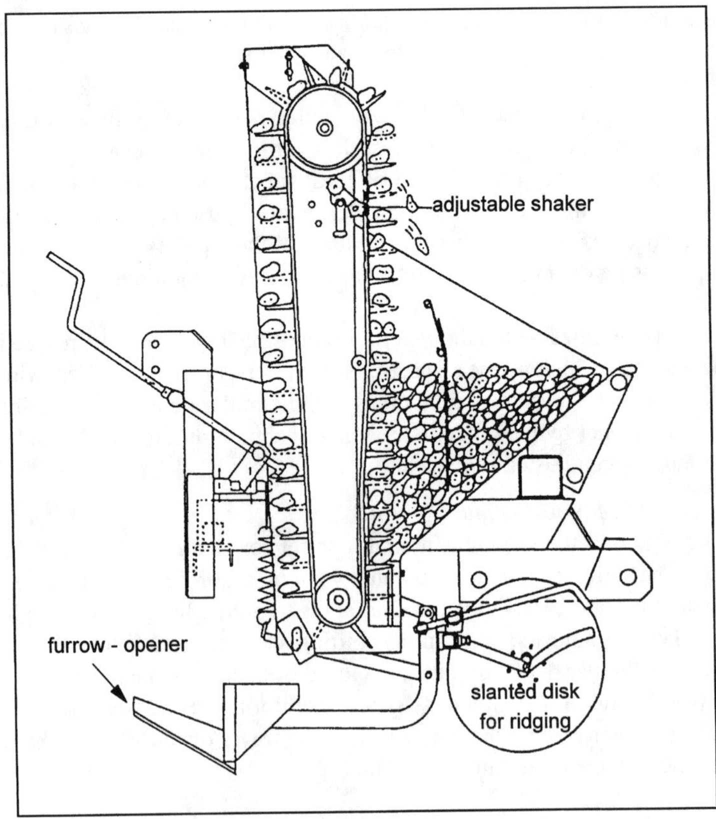

Figure 1.211. Continental European potato planter (Courtesy Tröster Maschinenfabrik, Butzbach).

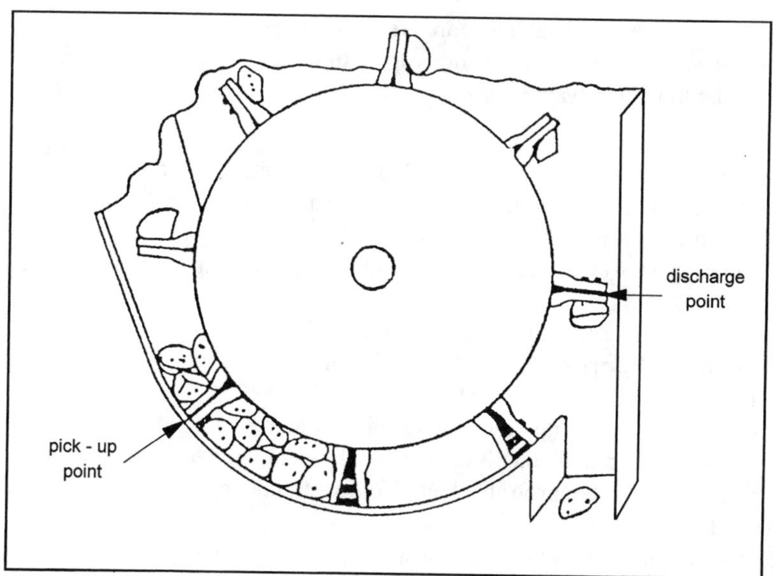

Figure 1.212. Singling of potatoes in American planter (Courtesy H. P. Smith, altered).

occur. The larger the potatoes in comparison to the cups and the higher the belt speed, the higher the share of skips and the lower the share of multiples is. Therefore it is possible to use the machines in such a way that either skips or multiples dominate. In the past, the machines were operated with skips and, for practical purposes, without multiples. The cups with skips were sensed and provided with a tuber from a compensating magazine. The present-day method is opposite the planters are operated in such a way that mainly multiples instead of skips are attained. These multiples are removed by shaking the belt (Fig. 1.211).

The shape of the American cut tubers makes the singling by cups rather difficult. On the other hand, these tuber pieces with flat surfaces can effectively be picked up by sharp pins that pierce the tubers (Fig. 1.212). For this purpose a picker wheel with arms at the periphery moves though the tuber supply. The split arms are cam-operated in such a way that the sharp pins pierce into the supply, pick up a piece of potato, and retract at the delivery point.

The tuber distances within the row with the American as well as with the European method are adjusted by the relation between the pickup speed and travel speed. The tuber distance within the row varies from 25–35 cm, and the row-width from 75–100 cm. The row width must be adapted to interrow cultivating and harvesting methods. Most American and European planters use two pickup units per row. This makes it possible to attain a high travel speed with a low speed of the pickup unit. Apart form this, the American method generally allows higher travel speeds without sacrificing accuracy than the European system. However, the American method may promote the transfer of virus- and bacterial infections between tubers.

In most parts of the world potatoes are grown in ridges, which facilitate the separation from the soil when harvesting. Exceptions apply to some dry regions, where a flat surface is left in order to reduce evaporation.

Transplanters

When the natural growing season is short, transplanting of crops results in higher yields. The seedling development occurs under artificial climate before the outside growing season starts. In some parts of the world, transplanting is used to make possible double-cropping. However, crop establishment by transplanting is much more expensive than by sowing.

Semiautomatic transplanting machines have mechanical transfer devices that are hand fed, yet mechanically place the plants in the furrow. These machines are more and more replaced by fully mechanized transplanters (23, 24, 25).

The seedlings with or without retaining soil on the roots are put in a tray. Sometimes the seedlings are grown individually in hexagonal paper pots, which provide for a uniform geometry. The paper pots can be attached to folded strands and are put in trays. The paper is biodegradable.

Figure 1.213 shows an automatic transplanting system. A wheel rotating behind the seedlings on a sloped platform is equipped with cam-operated planting fingers. These fingers grasp a seedling and move it into the furrow, where it is released. Transplanting machines are used extensively for paddy rice in Asia (Fig. 1.214). Special floating devices are used to provide for a constant planting depth independent of the sinkage of the propelling wheels. The row width is often about 30 cm. The plant distance within the row can be adjusted by the relation between the planting frequency and the travel speed and often varies between 10 cm and 20 cm.

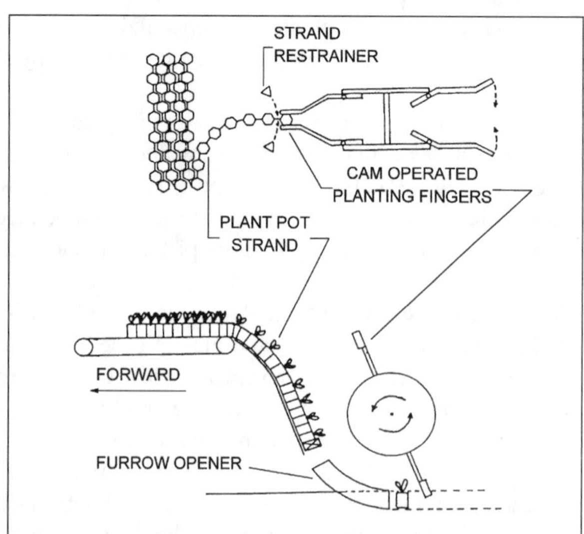

Figure 1.213. Automatic transplanting for seedlings grown in paper pots (Courtesy C. W. Suggs).

Figure 1.214. Transplanting of paddy rice (Courtesy Daedong Industrial Co., Korea).

References

1. Brinkmann, W. 1985. Precision drilling of sugar beets, fodder beets and corn (in German). In: Landtechnik, Heraugegeben von Horst Eichhorn: Stuttgart, pp. 279–297.
2. Kromer, K. H., Shaw, L. H., and Brinkmann, W. 1987. Planters for Hole Imbedding of Seed. ASAE Paper 87-1018, St. Joseph: Michigan.
3. Shaw, L. N., and Kromer, K. H. 1987. Revolving spade planter soil opener. ASAE Paper 87-019, St. Joseph: Michigan.
4. Fouad, H. A., and Brinkmann, W. 1975. Research on pneumatic conveying of singled seeds in a plastic tube (in German). Grundlagen der Landtechnik: Düsseldorf, 25(6): 177–186.
5. Morrison, J. E., and Gerik, T. J. 1985. Planters Depth Control. Part I and Part II. Transactions of the ASAE 28(5): 1415–1418, 28(6): 1745–1748.
6. Stout, B. A., Buchele, W. F., and Snyder, F. W. 1961. Effect of soil compaction on seedling emergence under simulated field conditions. Agricultural Engineering, St. Joseph: Michigan 42(2): 68–71.
7. Erbach, D. C., and Kaspar, T. C. 1994. Row Cleaners for No-Till Planting of Corn. ASAE Paper No. 94-1004, St. Joseph: Michigan.
8. Erbach, D. C., and Kaspar, T. C. 1993. Reducing Stand Establishment Risk in Conservation Tillage. ASAE Paper No. 93-1559, St. Joseph: Michigan.
9. Nelson, W. S., and Shinners, K. J. 1988. Performance of Rake Mechanismus for Creating Residue Free Seed Bands. ASAE Paper No. 88-1037, St. Joseph: Michigan.
10. Metzner, C. 1982. Effects of crop residues in the seedbed on seeding technique for grain (in German). Ph.D. thesis, University of Bonn: Germany.
11. Solie, J. B. et al. 1991. Design and field testing two ultranarrow row grain drills. ASAE Paper No. 91-1007. St. Joseph: Michigan.
12. Solie, J. B. et al. 1991. Reduced row spacing for improved wheat yields in weed-free and weed-infested fields. Transactions of the ASAE 34(4): 1654–1660.

13. Gebresenhet, G., and Jönsson, H. 1992. Performances of seed drill coulters in relation to speed, depth and rake angle. J. of Agric. Engineering Research 52(2): 121–145.
14. Dyck, F. B., Wu, W. K., and Lesko, R. 1985. Automatic depth control for cultivators and air seeders. In Proc. of the Agri-Mation I Conference and Exposition, 25–28 Feb. St. Joseph: Michigan.
15. Herbst, A. 1988. Experiments about automatic control of opener-depth with drilling (in German). Ph.D. thesis. Technical University of Dresden: Germany.
16. Heege, H. J. 1993. Seeding Methods Performance for Cereals, Rape, and Beans. Transactions of the ASAE, St. Joseph: Michigan 36(3): 653–661.
17. Heege, H. J., and Zaehres, W. 1975. Pneumatic seed distribution with bulk drilling (in German). Grundlagen der Landtechnik: Düsseldorf 25(4): 111–115.
18. Heege, H. J. 1985. Seed distribution over the soil surface with drilled and broadcast cereals. Tranlation No. 529 of the National Institute of Agricultural Engineering, Wrest Park, Silsoe: England.
19. Ball, B. C. 1986. Cereal production with broadcast seed and reduced tillage. J. of Agric. Engineering Research 35(2): 71–95.
20. Hoff, D. J., and Mederski, H. J. 1960. Effect of Equidistant Corn Plant Spacing on Yield. Agronomy Journal, Madison, pp. 295–299.
21. Williford, J. R. 1992. Production of Cotton in Narrow Row Spacing Transactions of the ASAE, St. Joseph: Michigan 35(4): 1109–1112.
22. Heege, H. J., and Feldhaus, B. 1997. The Go Control of Seeds per Unit Area for Grain Drills. ASAE Paper No. 97-3061, St. Joseph: Michigan.
23. Okamoto, T., Kitani, O., and Toni, T. 1993. Robotic Transplanting of Orchid Protocorn in Mericlone Culture. Japanese Society of Agric. Machinery 55(5): 103 ff.
24. Nambu, T., Masuda, A., and Tonimura, M. 1993. Studies of Automatic Transplanter Using Chein Pot. Japanese Society of Agric. Machinery, No. 6, p. 113 ff.
25. Suggs, C. W. et al. 1987. Self feeding transplanter for tobacco and vegetable crops. Applied Engineering in Agriculture, St. Joseph: Michigan 3(2): 148–151.

1.4. Fertilizer Distributors

J. W. Hofstee, L. Speelman, B. Scheufler

1.4.1. Introduction

The most common application technique for distributing dry mineral fertilizer is spreading with a fertilizer spreader. Other application techniques for fertilizer are injection of anhydrous ammonia and spraying liquid fertilizer. Spreading of manure will not be discussed in this context.

The first fertilizer spreaders were introduced at the end of the 19th century. Most of these spreaders were plate-and-flicker type spreaders. Spinning disc spreaders are described from this time on, but their share of the market remained small until about 1950. The main reasons for this small share where the unpredictable and disappointing results caused by the unknown influences of physical properties, wind and driving speed. Improvement of techniques and the availability of adjusting data made possible the introduction of the spinning disk type spreaders on a larger scale. The first spinning disc spreaders had only one disc. Spinning disc spreaders with two discs were introduced many years later in the 1960s.

A fertilizer spreader has to meet the following qualifications:

- Equal distribution of the fertilizer in both transversal and longitudinal direction, with full and partly empty hoppers, with different travel speeds and different application rates;
- Sufficient capacity of the hopper;
- Possibilities for adaptation to relative high crops;
- Easy and accurate adjustment of mass flow;
- Possibilities for border spreading;
- High resistance against corrosion;
- Easy cleaning;
- High capacity.

1.4.2. Fertilizer Distributor Types

There exist many types of fertilizer spreaders. Fertilizer spreaders can be subdivided in spreaders with a variable working width (spinning disk and oscillating spout) and spreaders with a fixed working width (pneumatic, plate-and-flicker, auger). Liquid fertilizer is usually spread with a field sprayer and anhydrous ammonia is injected in the soil with a chisel tine and an injection tube behind it.

Spinning Disc Spreaders

Description

A spinning disc spreader (Fig. 1.215) consists of one or two rotating discs with two or more vanes mounted on the disc. The spreader has a hopper, most times cone shaped, above the disc(s) and the fertilizer is fed in a stream, either by gravity or a conveyor

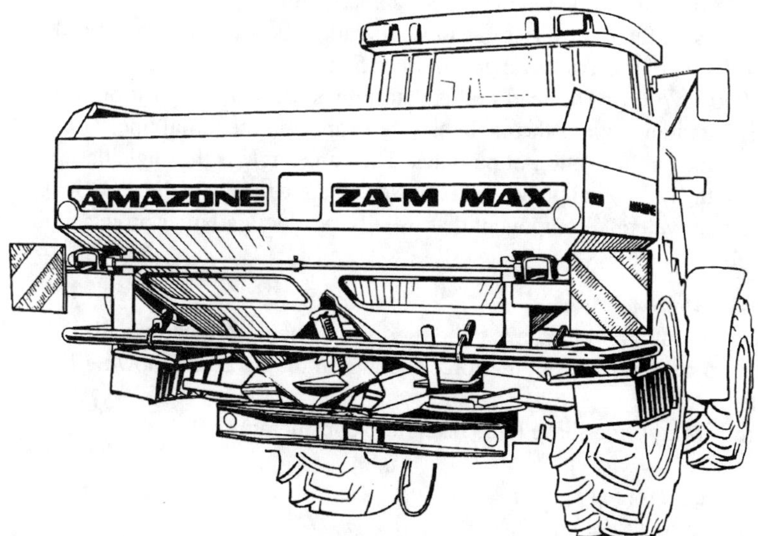

Figure 1.215. Example of a spinning disc fertilizer spreader.

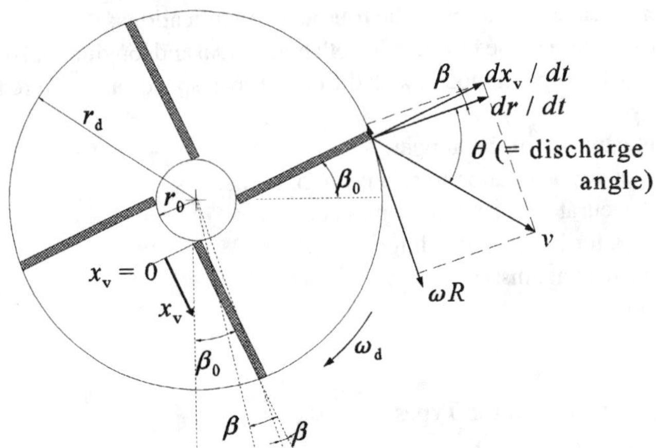

Figure 1.216. Important design factors of a spinning disc fertilizer spreader.

mechanism, to the disc(s). The disc distributes the fertilizer by throwing it away with velocities normally ranging between 15 m/s and 50 m/s; for some spreaders the velocities can reach values up to 70 m/s. These high velocities are required to obtain large effective working widths.

Disc Design

The important design factors of a spinning disc (Fig. 1.216) are the disc radius (r_d), the feed radius (r_0), the pitch angle (β_0) of the vane, the cone angle of the disc (α), and the shape of the vane (straight, curved, or combinations).

The radius of the disc varies for most spreaders between 0.30 m and 0.45 m. The angular velocity of most discs ranges between 55 rad/s (about 540 rev/min) and 100 rad/s (about 1000 rev/min). The feed radius determines the starting position of particles on the disc. The pitch angle is the angle between vane and the radial line. This angle has a large influence on the motion of particles. The cone angle is the angle between the disk surface and the horizontal plane. A positive cone angle results in the particles discharged with a vertical velocity component that usually results in a larger projected distance of the particles.

Equation of Motion

The motion of a particle on a spinning disc can be described by a differential equation that can be derived from the forces acting on the particle following D'Alembert's Principle.

The following forces have to be taken into consideration:

- Gravity ($F_g = m_p \cdot g$)
- Centrifugal force ($F_{ce} = m_p \cdot \omega_d^2 \cdot r$)
- Mass inertia ($F_I = m_p \cdot \frac{d^2x_v}{dt^2}$)
- Coriolis force ($F_{Co} = 2 \cdot m_p \cdot \omega_d \cdot \frac{dx_v}{dt}$)

- Friction (F_f) friction forces occur when one of the forces mentioned before has a component which is perpendicular to the direction of motion.

When the vane shape is circular there is also an additional centrifugal force, depending on the velocity along the vane and the virtual radius of the vane.

When analyzing the motion of the particles it is necessary to make a distinction between the radial position of the particle r (i.e., the distance to the center of the disc) and the path along the vane (x_v). Most fertilizer spreaders have straight vanes which are placed on the disc with an angle with respect to the radial line: the pitch angle β ($\beta > 0°$: backward pitched; $\beta < 0°$: forward pitched). The pitch angle for a straight radial vane depends on the radial position. The pitch angle at the beginning of the vane ($r = r_0$) is the initial pitch angle β_0. The instantaneous pitch angle is equal to:

$$\beta = \arctan\left(\frac{r_0 \sin\beta_0}{x_v \cdot \cos\alpha + r_0 \cos\beta_0}\right) \tag{1.86}$$

There exist vane curvatures for which the pitch angle remains constant for varying radial positions. Such a vane curvature is a logarithmic spiral.

Centrifugal Force. The centrifugal force (Fig. 1.217) always acts in the horizontal plane along a line through the center of rotation and the radial position of the particle. The centrifugal force can be resolved in:

- Direction of motion : $+F_{ce} \cdot \cos\alpha \cdot \cos\beta$
- Perpendicular to the disc : $+F_{ce} \cdot \sin\alpha \cdot \cos\beta$
- Perpendicular to the vane : $-F_{ce} \cdot \sin\beta$

Coriolis Force. The Coriolis force (Fig. 1.218) always acts in the horizontal plane and is perpendicular to the direction of motion. Only the horizontal component of the velocity along the vane is of relevance.

- perpendicular to the vane: $F_{Co} = 2 \cdot m_p \cdot \cos\alpha \cdot \omega_d \cdot \frac{dx_v}{dt}$

Mass Inertia. The mass inertia (Fig. 1.218) always acts in a line in the direction of motion.

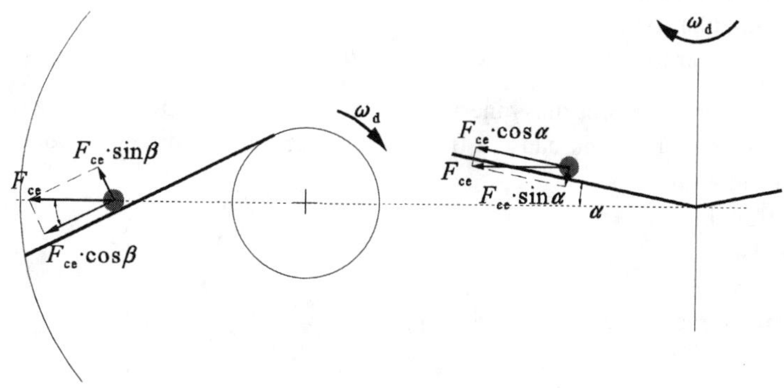

Figure 1.217. Centrifugal force.

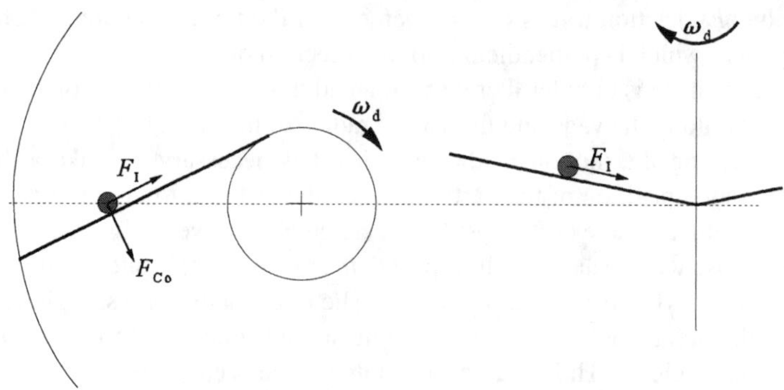

Figure 1.218. Coriolis force and mass inertia.

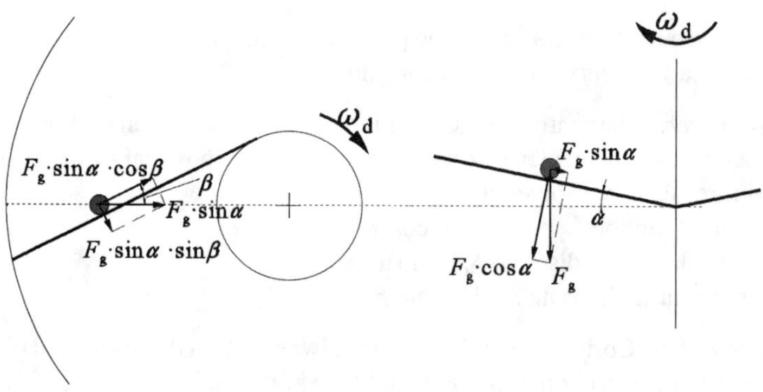

Figure 1.219. Gravity.

Gravity. The gravity (Fig. 1.219) always acts in the vertical plane along a line in the direction of the center of the earth. The gravity can be resolved in:

- Direction of motion : $F_g \cdot \sin\alpha \cdot \cos\beta$
- Perpendicular to the disc : $F_g \cdot \cos\alpha$
- Perpendicular to the vane : $F_g \cdot \sin\alpha \cdot \sin\beta$

Friction. All forces except mass inertia have at least one component perpendicular to the vane or disc surface and cause a friction force acting in a direction opposite to the direction of motion.

- centrifugal force
  with disc : $-\omega_d^2 \cdot r \cdot \sin\alpha \cdot \cos\beta \cdot \mu_d$
  with vane : $\omega_d^2 \cdot r \cdot \sin\beta \cdot \mu_v$
- Coriolis force : $2 \cdot \cos\alpha \cdot \mu_v \cdot \omega_d \cdot \frac{dx_v}{dt}$
- gravity
  with disc : $g \cdot \cos\alpha \cdot \mu_d$
  with vane : $g \cdot \sin\alpha \cdot \sin\beta \cdot \mu_v$

Equation of Motion. This results in the following general equation of motion for a particle on a cone-shaped disc along a pitched straight vane:

$$\frac{d^2x_v}{dt^2} + 2\cos\alpha\,\mu_v\omega_d\frac{dx_v}{dt} - \cdots$$
$$(\cos\alpha\cdot\cos\beta - \sin\alpha\cdot\cos\beta\cdot\mu_d + \sin\beta\cdot\mu_v)\omega_d^2 r + \cdots$$
$$(\sin\alpha\cdot\cos\beta + \sin\alpha\cdot\sin\beta\cdot\mu_v + \cos\alpha\cdot\mu_d)g = 0 \qquad (1.87)$$

It has to be realized that β is a function of the radial position as described by equation 1.86 and that the relation between r and x_v is as follows:

$$r = \sqrt{r_0^2 + (x_v\cdot\cos\alpha)^2 + 2\cdot r_0\cdot x_v\cdot\cos\alpha\cdot\cos\beta_0} \qquad (1.88)$$

Similar equations for circular vanes, logarithmic spiral vanes and parabolic vanes are given by Hofstee [9].

Particle Motion

The final result of the acceleration of particles along the vane of a spinning disc is that the particle leaves the disc at a certain position with a certain velocity in a certain direction. Relevant discharge variables are:

- Residence time: This is the time passed between the moment the particle starts the motion on the disc and the moment the particle leaves the disc;
- Rotation angle of the disc: This is the angular rotation of the disc during the residence time;
- Tangential velocity: This is the circumferential velocity of the tip of the vane;
- Velocity along the vane: ($\frac{dx_v}{dt}$);
- Radial velocity: This is the velocity of the particle in radial direction ($\frac{dr}{dt}$);
- Discharge velocity: This is the absolute velocity by which the particle is discharged and is equal to the vectorial addition of the velocity along the vane and the tangential velocity;
- Discharge angle: This is the angle between the radial direction and the discharge direction.

Feed Radius. Analysis of the equation of motion (Eq. 1.87) will show that there is a minimum distance at which the particles have to be fed to the disc. This distance is however small (less than 1 mm for rotational velocities greater than 340 rev/min for $\mu = 0.3$).

Pitch Angle. The influence of the pitch angle (combined with the influence of the coefficient of friction) is for a flat spinning disc ($\alpha = 0°$) shown in Figs. 1.220–1.223. Figure 1.220 shows that a change of the pitch angle β_0 from $-60°$ (forward pitched) to $+60°$ (backward pitched) first results in a relatively strong decrease of the residence time to a minimum value just beyond 0°, followed by a small increase towards +60°. This figure further shows that backward-pitched vanes are less sensitive for variation of the coefficient of friction than forward pitched vanes are.

Figure 1.221 shows that the velocity along the vane increases when the pitch angle increases from $-60°$ to $+60°$. Figure 1.222 shows that this does not result in an increase

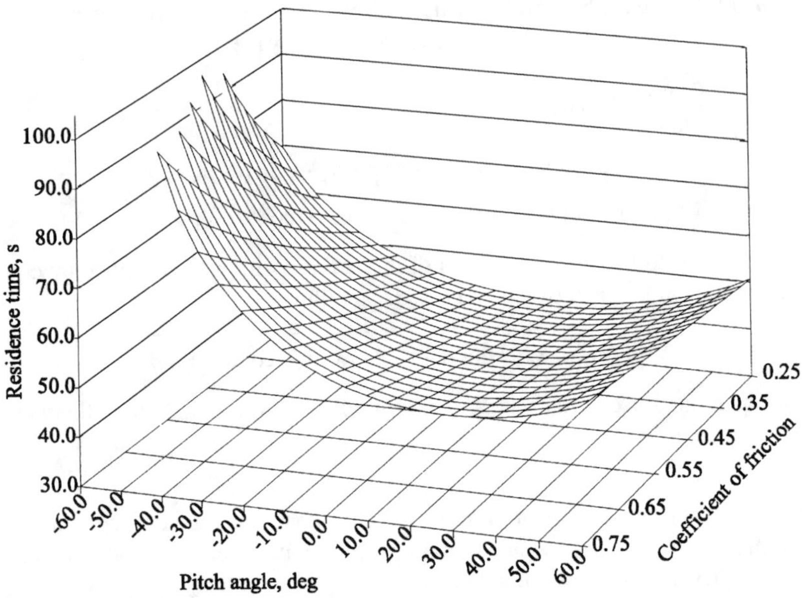

Figure 1.220. Influence of pitch angle and coefficient of friction on the residence time of fertilizer particles on the disc. ($n = 540$ rev/min, $r_d = 0.40$ m, $r_0 = 0.10$ m, $\alpha = 0°$)

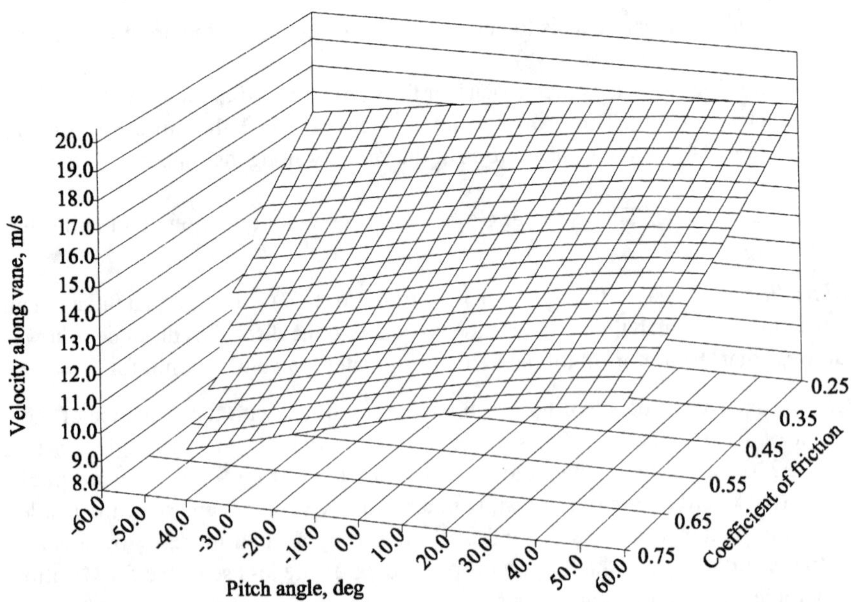

Figure 1.221. Influence of pitch angle and coefficient of friction on the velocity along the vane of fertilizer particles. ($n = 540$ rev/min, $r_d = 0.40$ m, $r_0 = 0.10$ m, $\alpha = 0°$)

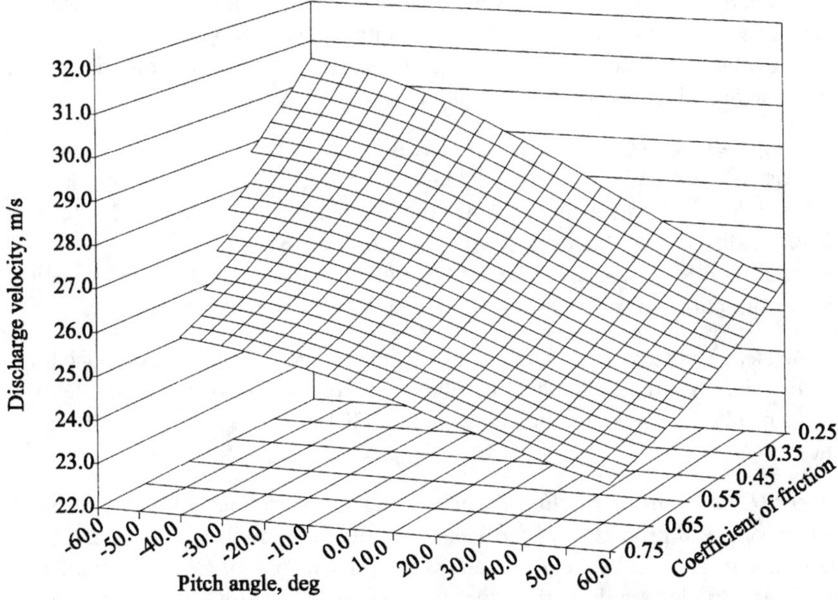

Figure 1.222. Influence of pitch angle and coefficient of friction on the discharge velocity of fertilizer particles. ($n = 540$ rev/min, $r_d = 0.40$ m, $r_0 = 0.10$ m, $\alpha = 0°$)

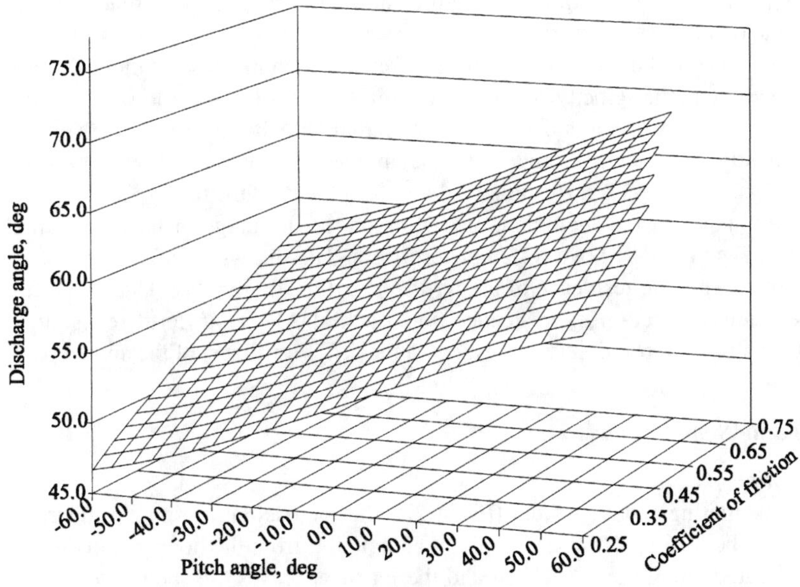

Figure 1.223. Influence of pitch angle and coefficient of friction on the discharge angle fertilizer particles. ($n = 540$ rev/min, $r_d = 0.40$ m, $r_0 = 0.10$ m, $\alpha = 0°$)

of the discharge velocity but in a decrease of the discharge velocity. The velocity along the vane of a backward pitched vane has a velocity component opposite to the tangential velocity of the tip of the vane. Figure 1.223 finally shows that the discharge angle θ increases with an increase of the pitch angle from $-60°$ to $+60°$.

Vane Length and/or Disc Radius. An increase of vane length or disc radius results in an increase of both radial velocity and tangential velocity and consequently in a larger discharge velocity. Further, the residence time increases and consequently the rotation angle of the disc during the residence on the disc of the particle.

A variation in vane length can be used to obtain a variation in discharge velocity and discharge directions in order to obtain a specific distribution of the fertilizer.

Cone Angle. A positive cone angle α gives the particles a velocity component in upward direction. The particles then are able to travel a larger distance through the air and, as a result, the spread width increases. Particle trajectories are further discussed in Section 1.4.3.

Fertilizer (Coefficient of Friction). The coefficient of friction (μ) is one of the most important physical properties of fertilizer that affect the motion of particles on a disc. Values vary in general between 0.2 and 0.5 (Hofstee [6, 7]). Coefficients of friction of fertilizer particles are difficult to measure because it is difficult to create measuring conditions similar to those when the particle moves along the vane.

The coefficient of friction is sometimes also used to incorporate side effects as mutual interactions of particles, suction by the vanes, rolling of particles, and so on (Dobler and Flatow [3]). In these situations the coefficient of friction is an apparent coefficient of friction and is indicated by μ^*.

An increase of the coefficient of friction results in an increase of the residence time and a decrease of the velocity along the vane (Figs. 1.220–1.223). A lower velocity along the vane further results in a decrease of the radial velocity, the discharge velocity and consequently the discharge angle. The magnitude depends on such variables as pitch angle (Figs. 1.220–1.223), rotational velocity and feed radius.

The influence of the coefficient of friction on the discharge of fertilizer particles is shown in Fig. 1.224 for two different fertilizers. The throw-off angle is the angle the disc turns during the residence of the fertilizer particle on the disc. Due to a difference in coefficient of friction the residence time of the DAP 18–46 fertilizer is larger and results in a relative rotation (with respect to the CAN fertilizer) of the spread pattern of about 10°.

Oscillating Spout Spreaders

Description

The oscillating spout spreader (Fig. 1.225) has an oscillating spout to distribute the fertilizer. The fertilizer flows, with aid of an agitator, from the hopper into the spout. It is accelerated along the spout wall and the fertilizer leaves the spout after one or two impacts on the wall.

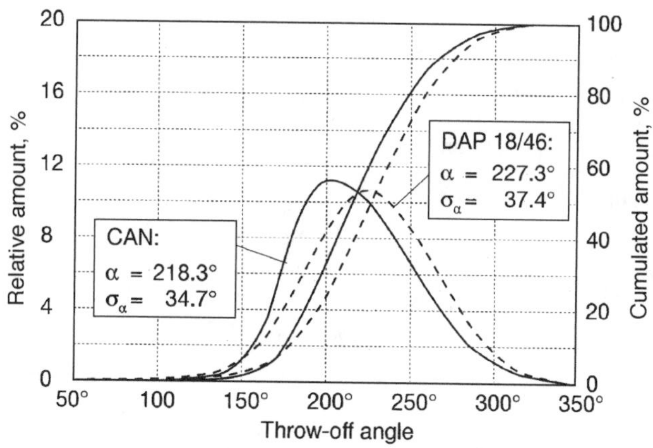

Figure 1.224. Influence of coefficient of friction on the discharge of the fertilizer for two different fertilizers.

Figure 1.225. Example of an oscillating spout spreader.

Kinematics and Spout Design

A schematic view of the working principle is shown in Fig. 1.226. The PTO of the tractor drives the flywheel and the crank with length l_c is an integral part of this flywheel. A rod with length l_r is connected at one end to this crank and at the other to an elastic coupling that drives the spout. The spout makes an oscillatory motion in the horizontal

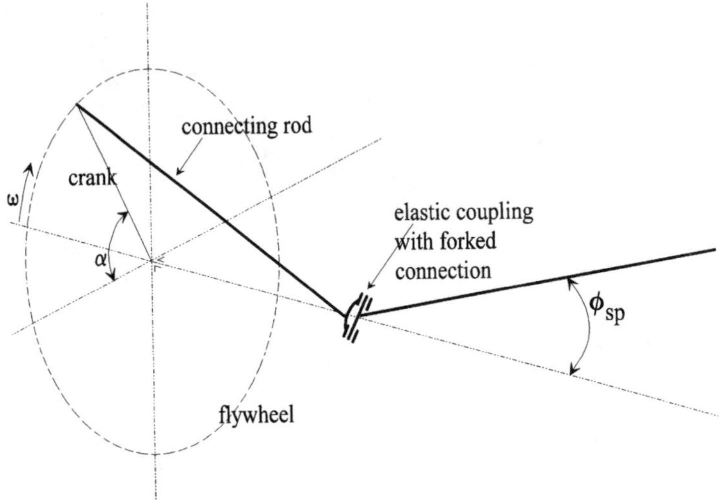

Figure 1.226. Schematic view of the propulsion of a spout. The flywheel rotates in the vertical plane and the spout oscillates in the horizontal plane.

plane. The instantaneous oscillation angle of the spout, $\phi_{sp}(t)$, is given by:

$$\phi_{sp}(t) = \arctan\left(\frac{\cos\alpha(t)}{\sqrt{C^{-2}-1}}\right) \tag{1.89}$$

$\alpha(t)$ is the instantaneous angle of the driving shaft and $C = l_c/l_r$. The angular velocity and angular acceleration of the spout can be obtained by successive differentiation of Eq. 1.89.

The spout of the spreader is a converging and slightly oval tube made of nylon strengthened with glass fibre. Inside the spout are two ribs to prevent the particles starting a circular motion along the spout wall during transport. At the end the spout is equipped with a bow that has the task of transforming the default two peak spread pattern into a one-peak spread pattern. The most important design parameters of the spout (Fig. 1.227) are the spout angle γ and the spout length, l_{sp}. Also of importance are the rotary frequency and the crank-connecting rod ratio of the driving mechanism.

Particle Motion

The motion of particles in an oscillating spout shows a large similarity with the motion along a vane of a spinning disc. The oscillating motion of the spout (comparable with the vane) makes it discontinuous. The fertilizer flows out of the hopper through the metering slice into a bowl and then flows into the spout: the sorting-in process. This sorting-in process determines the initial conditions for the motion along the spout (the position x_{sp} at $t = 0$ and the velocity v_{xsp} at $t = 0$). The fertilizer particles are then accelerated by the centrifugal forces due to the oscillatory motion of the spout. Owing to this oscillatory motion, the spout will decelerate and change its direction of motion. The fertilizer particles continue their motion, cross the spout, impact on the opposite spout

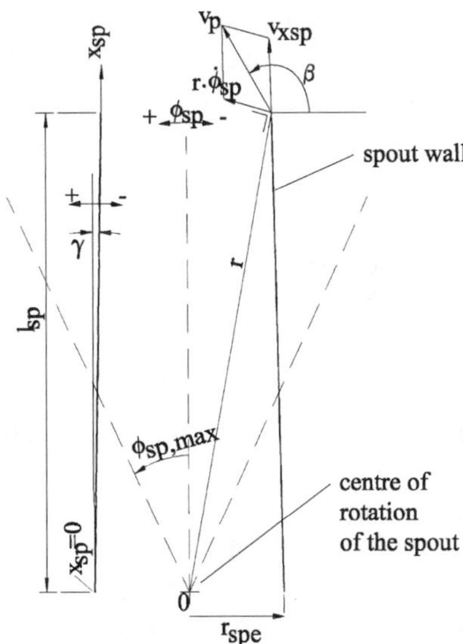

Figure 1.227. Important variables of the design of the spout. The path of a particle along the wall is identified by x_{sp}. The particle leaves the spout when $x_{sp} \geq l_{sp}/\cos\gamma$. The velocity of the particle along the wall is v_{xsp} and the resulting discharge velocity is the resultant of v_{xsp} and the tangential velocity of the spout end ($r d\phi_{sp}/dt$).

wall and continue their motion along this spout wall. This continues until they leave the spout or, when the spout decelerates again, they will recross the spout. During this second crossing they will either leave the spout or impact again on the opposite spout wall, slide along this wall and then leave the spout.

Equation of Motion. The general equation of motion of a particle sliding along the wall (Speelman [15]) is equal to:

$$\frac{d^2x_{sp}}{dt^2} - 2\mu\frac{d\phi_{sp}}{dt}\frac{dx_{sp}}{dt}p - \mu\frac{d^2\phi_{sp}}{dt^2}x_{sp}p - \cdots$$
$$\left(\frac{d\phi_{sp}}{dt}\right)^2 x_{sp} - \frac{d^2\phi_{sp}}{dt^2}r_{spe}p + \mu\left(\frac{d\phi_{sp}}{dt}\right)^2 r_{sp} = 0 \tag{1.90}$$

x_{sp} is the position along the spout wall and r_{spe} the radius of the spout at the entrance. $d\phi_{sp}/dt$ and $d^2\phi_{sp}/dt^2$ respectively are the angular velocity and angular acceleration of the spout. p identifies the spout wall and is -1 to describe the motion along the left wall and $+1$ to describe the motion along the right wall. μ is the coefficient of friction between spout wall and fertilizer. Simulation studies [8, 15] showed that beyond a certain oscillation angle of the spout the resulting force acting on the particles is directed inwards along the spout. This means, in practice, that the particles will stay at the entrance of the spout and will not start the motion along the spout wall until the spout reverses its motion.

Initial Velocity and Initial Position. The effects of both the initial velocity along the spout (v_{xsp}) and the initial position along the spout (x_{sp}) on the discharge velocity depend strongly on the oscillation angle of the spout at the moment particles start their motion. These three variables determine the moment of spout crossing and therefore the moment of discharge. The resulting discharge velocities depend on the velocities of the particles along the spout wall and the angular velocity of the spout at discharge time.

Propulsion and Spout Design. An increase of the crank-connecting rod ratio C generally results in shorter residence times and larger discharge velocities. There is an interaction with the oscillation angle of the spout at the moment the particles start the motion. Higher values of C result in higher discharge velocities that can be reached but not *a priori* in higher discharge velocities.

In general a higher angular velocity of the driving shaft results in higher levels for the discharge velocity of the fertilizer particles. The actual discharge velocity also depends on the oscillation angle of the spout at the start of the motion. Further it has to be mentioned that the forces acting on the spout increase with the second power of the increase of the angular velocity. This gives very limited possibilities for an increase of angular speed of the driving shaft to achieve larger spread widths. An increase of spout length does not always result in larger discharge velocities. There is a certain trajectory, corresponding with particles that cross the spout a second time and leave the spout during this crossing, in which spout length does not have any effect.

A decrease of the spout angle from $+5^\circ$ (divergent) to -5° (convergent) results for most situations in an increase of discharge velocity. When the particles cross the spout twice and leave the spout when sliding along the wall, the effect is opposite.

Physical Properties of the Fertilizer. The most important physical property for the motion in the spout is the coefficient of friction. The effect of the coefficient of friction on the discharge velocity is rather diverse. There is an interaction with the oscillation angle of the spout at the start of the motion. A higher coefficient of friction results initially in a lower discharge velocity but after a certain oscillation angle a higher coefficient of friction results in higher discharge velocities. There is hardly any effect of coefficient of friction for those particles that cross the spout twice and leave the spout when sliding along the spout wall.

Boom Spreaders

The particular identification of a boom spreader is that contrary to centrifugal and oscillating spout fertilizer spreaders the defined working width is determined by the size of the distributor boom. One has to distinguish implements with pneumatic and with mechanical lateral delivery of the material to be spread. Just like centrifugal spreaders, pneumatic spreaders require an overlap with the neighboring spread pattern. The overlapping area is considerably smaller. For worm-auger boom spreaders with mechanical

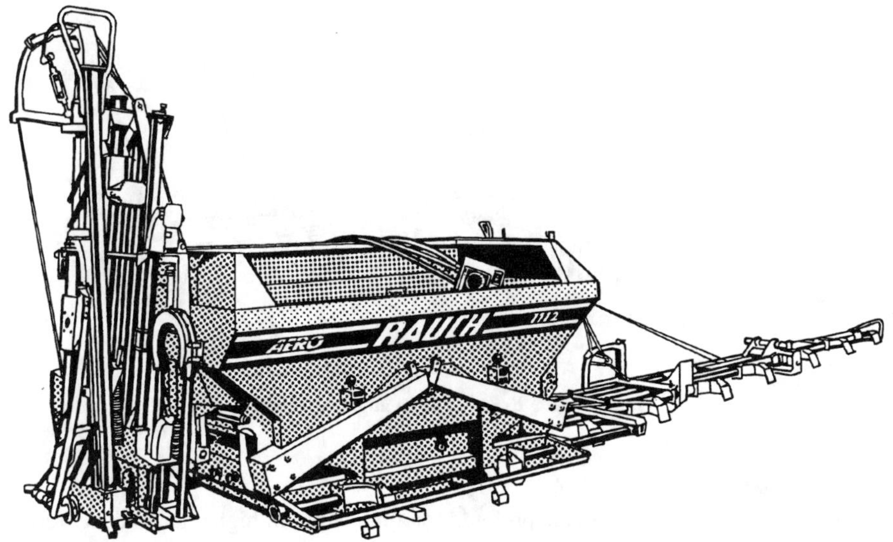

Figure 1.228. Example of a pneumatic fertilizer spreader.

lateral distribution, the spreading width corresponds to the effective working width, so that there is no overlapping area. A uniform fertilizer distribution with boom spreaders therefore requires accurate driving.

Pneumatic Spreaders

In a pneumatic fertilizer spreader (Fig. 1.228) the fertilizer is fed from the metering rollers via injectors into streams of air. These metering rollers are, for example, driven by a hydrostatic motor. With the aid of an electric remote control this way a simple, stepless rate adjustment during the operation is possible.

The metering rollers are divided into segments which can be switched on or off in any order. The working width is in this way variably changeable in part widths (part width control). It is possible to accurately spread plots, spreading losses along field sides and odd-shaped field side endings can be reduced to a minimum. The complete metering rollers can be exchanged for special metering units for the application of microgranules (pesticides) and fine seeds.

The fertilizer is transported by the stream of air through tubes which are fixed to the distributing booms. The bends with deflector plates mounted to the end of the tubes produce small overlapping spread fans and distribute the fertilizer evenly across the field surface. Depending on the working width and mechanical execution, the spacings of the jet outlets are between 50 cm and 100 cm. The conversion of the machine from normal fertilizing to late top dressing into taller crops can be done by turning around the deflector plates so that the fertilizer is deflected upwards and will therefore drop with reduced energy from above into the crop. For road transport the booms are folded into vertical position to the sides of the machine's hopper.

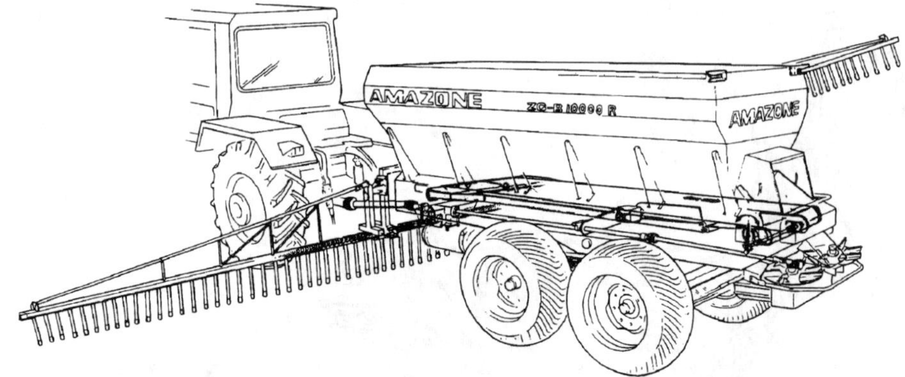

Figure 1.229. Example of an auger spreader.

Auger Spreaders

Auger spreaders (Fig. 1.229) are prefered for application of high rates of dusty fertilizers, e.g., for the basic fertilizers such as basic slag, potash, ground lime and lime marl. For the uniform distribution of granular nitrogen containing fertilizers they are not suited. Very often, therefore the auger unit of a spreader is designed to be exchangeable with a centrifugal disc spreading unit so that grain-like granular fertilizers can be distributed with a high work rate. There are also spreaders available with a reversible (redirectable) floor belt. The running direction of the auger belt can be switched from forward running to reverse running. This way it is possible without exchanging spreading units on the machine to spread powdery (dusty), soil moist, or granular fertilizer with a large working width. The distributor auger boom is mounted to the front of the machine, the twin disc spreader unit to the rear.

Auger spreaders are preferably available as trailed implements with payloads from 2–16 tons and working widths from 6–12 m. The fertilizer is delivered from the hopper by scraper floor chains, rubber auger belts or honeycomb floor belt chains to the spreading auger booms. The use of rubber floor auger belts is getting more popular as they react less sensitively to foreign particles in the spreading material.

Due to uneven loading or when traveling alongside slopes the rubber auger floor belts tend to shift sideways when they become prone to damage. By an automatic belt steering consisting of a steering frame with steering rollers and return roller with swivel pivot, it is possible to keep the floor belt aligned in central running position. The metering of the spread rate is controlled by adjustable shutter slides in front of the outlet openings and by the adjustable floor belt speed. If the floor belt and spreading auger boom is driven by a universal joint shaft, a synchronous run results whereby when changing engine speed the ratio of the spread rate remains constant in the worm auger spreading booms and on the floor belt. Alternatively the auger and spreading mechanisms can also be driven by hydrostatic motors. The effort in design is simpler than for the mechanical drive. At a ground wheel drive the floor belt is driven ground related. At varying forward speeds due to changing tractor gears, the spread rate remains constant.

The driven worm auger booms deliver the fertilizer into tubes with adjustable outlet openings. To achieve a uniform distribution of the fertilizer within the working width the opening diameters of the outlets must be adapted. In the outer range of the auger booms the pressure within the stream of fertilizer is lower than within the central area of the auger boom. That means, that the opening diameters of the outlets will have to be set progressively larger.

The dust development can be reduced considerably by such options as dust-protective covers, especially with drag hoses.

Aerial Spreaders

In some crops and under certain conditions, for example, rice, dry fertilizers and other chemicals are spread with aircraft (Fig. 1.230). The most widely used type of distributor is the ram-air or venturi spreader. These spreaders are mounted under the aircraft.

Venturi spreaders consist of an air entry section, a throat and material entry gate, and a vane expansion and redirection section. The throat-and-vane expansion area usually are referred to as venturi. Internally the spreader is broken down in to 5 to 13 ducts. These are straight in the front and curved toward the rear to direct the discharged material toward a lateral path. The discharge directions with respect to the flying path increase from about 10°–20° for the center vanes to 45° or more (up to 70° for wider spreaders) for the outer vanes. For most spreaders these rear vanes are adjustable.

The inlet vanes are adjustable for most venturi spreaders. This is necessary to fine-tune the spreader to a particular aircraft to achieve a uniform distribution pattern. Shifting the

Figure 1.230. Example of an aerial spreader.

front vane position influences the amount of material deposited in the area served by the particular opening and the volume of air to be forced into this vane, and hence influences the velocity of air and exiting material.

Liquid Fertilizer Spreaders

Liquid fertilizers can be subdivided in three groups: anhydrous ammonia, solution fertilizers, and suspension fertilizers. The use of liquid fertilizers is a relatively new (second half of the 20th Century) practice. In comparison with fertilizing granular material, high investments are required for the technical solution (storing and applying). Furthermore comprehensive special knowledge is required for the application technique.

Liquid fertilizers have several advantages:

- Liquid nitrogen fertilizer is partly lower in price than granular material;
- The fertilizing effect is achieved via the soil and leaves and therefore is independent of rain and humidity in the upper layer of the soil;
- Combination with some pesticides is possible;
- Simple filling procedure with the aid of pumps (working requirement approx. 10% lower than with the bulk fertilizer handling chain);
- With certain plant protective agents the spray rate can be reduced (up to 30%) by adding liquid fertilizer.

Despite the advantages, there are also some disadvantages:

- High storing costs for farm owned storing facilities;
- Liquid fertilizer has considerable corrosive properties;
- Official restrictions, advice etc. have to be adhered to;
- Handling of field sprayers with large working widths (24 m and more) is considerably costlier than if centrifugal broadcasters would be used with comparable working widths;
- Possibility of etching and resulting yield losses;
- Weather factors have to be considered.

Anhydrous ammonia has to be handled with care to prevent hazardous situations. It has a boiling point of $-33.3°C$. Non-refrigerated storage requires high pressure vessels. The gauge pressure is at 37.8°C equal to $13.6 \cdot 10^5$ Pa. The nitrogen content of anhydrous ammonia is 82%. This high nitrogen content and the low costs are important reasons for its high popularity, especially in the United States. Good transport and storage facilities also contribute to this popularity. Anhydrous ammonia is stored at atmospheric pressure in refrigerated tanks to maintain the liquid status. In Europe, anhydrous ammonia is applied only on a small scale.

Anhydrous ammonia is injected into the soil by knives. A tube is mounted on the back of the knife and at this point the ammonia turns from liquid into gas. To prevent loss of gas into the atmosphere it is important to properly close the injection channel. The moisture content of the soil is very important. The injection depth varies between about 15 cm and 23 cm but on sandy soils a deeper placement is required because of the lower ammonia retention capacity. The spacing between the knives is usually 30 cm to 45 cm. The high power requirement for applying the ammonia at the right depth may be a disadvantage of applying ammonia.

The difference between solution fertilizers and suspension fertilizers is that in solution fertilizers all nutrients are in solution. Aqua ammonia is a solution of anhydrous ammonia and water and contains normally about 20% N. It is applied in the same way as anhydrous ammonia. The required injection depth is less.

A popular nitrogen solution is a mixture of ammonium nitrate, urea and water. The nitrogen content varies between 28% and 32%. This type contains a corrosion inhibitor (a small quantity of anhydrous ammonia (5 kg of NH_3/ton) or ammonium phosphate (0.05% to 0.2% P_2O_5). Nitrogen solutions can be used as a leaf and as a soil fertilizer. Leaf fertilization shows the quickest effect as the nutrient solution penetrates directly into the leaf. Afternoon application therefore is especially effective as the nitrogen absorption via the leaf takes place most quickly during the night hours with high humidity of the air. This way even in dry weather conditions the fertilizer can be effective.

A major problem with solution mixtures was the relatively low nutrient concentrations. In the 1970s it became possible to produce a solution with a grade of 10-34-0 or 11-37-0. A major problem is that it is not possible to produce solution mixtures with a high potash (K) content.

Suspension fertilizers contain a suspending agent as gelling-type clay. The gel has to keep the usually finely divided solids into suspension. On one hand the gel has to be strong enough to support the solids without their settling, and on the other hand is has to be weak enough to allow agitating, pouring and pumping. The interest in suspensions has grown because of the need of including potassium in liquid fertilizers.

Solution and suspension fertilizers can be applied with field sprayers. Liquid fertilizer is applied in form of coarse droplets by special flat fan and multiple hole jets (Fig. 1.231), depending on whether the fertilizer has to become effective more in the soil or in the leaves. The combination of agents influences also the choice of the jets.

Danger of etching which could result from the wrong application technique should be considered. Etching could be caused by too-small droplets, too-high liquid pressure (above 2 bar), booms being set too low above the ground, or by a wrong time of application. In grain crops, therefore, drag-hose attachments to the booms should be used from the time of tillage. With advanced drag-hose units, a sprayer boom can be converted in an easy way. For example, the drag-hose attachments are equipped with stainless steel weights on the bottom end of the hoses to optimize the crop penetration by the drag hoses and stabilize the position of the drag hoses during the operation in taller crops or drag hoses with distributing tubes are used respectively.

The field sprayer has to be designed from the beginning for the application of liquid fertilizer. Suitable corrosion-proof, liquid fertilizer resistant materials such as polyester, polyethylene, stainless steel and aluminum as well as liquid fertilizer proof pumps and control units have to be used.

Also the high densities of liquid fertilizers have to be considered. Matching the lifting power by the tractor hydraulics resp. axle loads with the carrying ability of the nursing tractor tires has to be considered. When using drag-hose attachments the booms are exposed to considerably higher strain.

When combining the operation of fertilizing and crop protection, the blendability and the time of the treatment play a decisive role. The optimum effect of the crop protective

Figure 1.231. Nozzle for applying liquid fertilizer.

measure has priority. Separated spray applications are necessary if the time for fertilizing and plant protective measures do not coincide or if a high fine droplet proportion resp. coverage degree is required. For a combined treatment only flat-fan jets can be used. Comprehensive information about the production, handling, storage and application of liquid fertilizers is given by TVA [16].

1.4.3. Particle Trajectories

After being discharged by the disc, the spout, or the distribution plate of a pneumatic spreader, fertilizer particles have to travel a large distance through the air, especially for the spinning disc and the oscillating spout spreaders. This distance can be more than 30 m, depending on the desired effective working width.

The forces acting on a particle are the buoyancy forces, gravity forces, inertial forces, and frictional forces (air resistance). The buoyancy force can be neglected since the density of air (about 1.2 kg/m^3) is much smaller than the density of the fertilizer particle (between 900 kg/m^3 and 2200 kg/m^3). The aerodynamic resistance coefficient K is a widely used parameter in equations to describe the motion of particles through the air. The coefficient is equal to:

$$K = \frac{3}{8} C_{\mathrm{D}} \rho_{\mathrm{a}} \frac{1}{\rho_{\mathrm{p}} r_{\mathrm{p}}} \tag{1.91}$$

in which C_{D} the drag coefficient, ρ_{a} the density of air, ρ_{p} the density of the particle and

r_p the radius of the particle, or

$$K = \frac{g}{v_{pt}^2} \tag{1.92}$$

in which v_{pt} the terminal velocity. The value of the drag coefficient C_D depends on the Reynold number (Re):

$$\mathrm{Re} = \frac{v \cdot d \cdot \rho_f}{\eta_f} \tag{1.93}$$

in which v the velocity and d the diameter of the object and ρ_f the density and η_f the dynamic viscosity of the fluid. There is no exact analytical relation between Re and C_D. The corresponding values can be found with the aid of tables, diagrams (Fig. 1.232) or approximating equations.

Fertilizer particles are irregularly shaped bodies. To use the equations that apply for spheres a correction factor that reflects the deviation from a sphere has to be introduced. When the characteristic dimensions of a particle can be measured, the diameter can be adjusted by the factor $6Z/\pi$. Z, the volume shape factor, is equal to:

$$Z = \frac{\pi}{6} \cdot \left(\frac{d_g}{d_e}\right)^3 \cdot \text{(sphericity)} \tag{1.94}$$

where d_e, the diameter of the equivalent sphere:

$$d_e = \sqrt[3]{\frac{6}{\pi} \cdot \frac{m_p}{\rho_p}} \tag{1.95}$$

with m_p the mass of the particle, d_g the geometric mean diameter:

$$d_g = \sqrt[3]{d_{maj} \cdot d_{int} \cdot d_{min}} \tag{1.96}$$

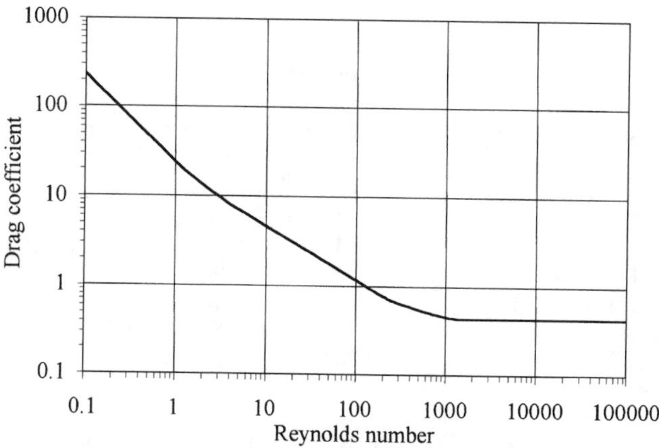

Figure 1.232. Relation between Reynold number (Re) and drag coefficient (C_D).

in which d_{maj}, d_{int} and d_{min} respectively the major, intermediate and minor diameter and the *sphericity* equal to:

$$sphericity = \frac{d_g}{d_{maj}} \tag{1.97}$$

The corrected aerodynamic resistance coefficient K then becomes:

$$K = \frac{3}{8} 0.44 \rho_a \frac{\pi}{\rho_p r_p 6Z} \tag{1.98}$$

Using this method requires substituting the value of 0.44 for C_D because the total adjustment is made by Z. However, determination of characteristic diameters of fertilizer particles is rather difficult. Hofstee [7] introduced the diameter coefficient q, equal to the ratio between the diameter of the particle according to which it behaves (d_{pm}) and the diameter of the equivalent sphere (d_e). Values for d_{pm} are determined in fall tests with different fertilizer types and dimensions. The resulting values for q for different fertilizer types are listed in Table 1.50. Using q, the aerodynamic resistance coefficient K becomes:

$$K = \frac{3}{8} C_D \rho_a \frac{1}{\rho_p r_p q} \tag{1.99}$$

The equation of motion of a particle traveling through the air can be obtained from an equilibrium of forces. Trajectories of particles through the air are three dimensional.

Table 1.50. Particle true densities and diameter coefficients for different fertilizers. A, B, and D identify the origin of the fertilizer

Fertilizer	Source	True Density, kg/m^3	Diameter Coefficient q
Ammonium nitrate (B)	7	1890	0.8
CAN 27N (A)	17	1820	0.88
CAN 27N (B)	17	1850	0.87
CAN 27N (D)	17	1830	0.83
CAN 27N	24	1860	0.88
CAN 22N+7MgO prills (B)	17	2000	0.97
Diammoniumphosphate (D)	17	1930	0.72
NP 23-23 (B)	17	1710	0.82
NP 26-14 (A)	17	1740	0.85
NP 26-7 (A)	17	1890	0.8
NPK 12-10-18	24	2090	0.76
NPK 12-10-18 (B)	17	1960	0.81
NPK 17-17-17 (B)	17	1780	0.82
Potassium 60	24	2030	0.63
Potassium Sulphate (D)	17	2230	0.74
Superphosphate (45%) (D)	17	2010	0.71
Triple Superphosphate (B)	17	2120	0.8
Urea granules (2.9 mm) (B)	17	1260	0.95
Urea granules (5–8 mm) (B)	17	1260	0.82
Urea prills (B)	17	1310	0.89

Consider therefore a three dimensional Cartesian space, with the positive Y axis in the direction opposite to the gravitational field. The XZ plane then corresponds with the horizontal plane. The equations of motions for respectively the X, Y, and Z direction are:

$$\frac{d^2x}{dt^2} = -K\,v_x\sqrt{v_x^2 + v_y^2 + v_z^2} \tag{1.100}$$

$$\frac{d^2y}{dt^2} = -K\,v_y\sqrt{v_x^2 + v_y^2 + v_z^2} - g \tag{1.101}$$

$$\frac{d^2z}{dt^2} = -K\,v_z\sqrt{v_x^2 + v_y^2 + v_z^2} \tag{1.102}$$

Either the X or Z direction can be left out when it is assumed that the particles move in a vertical plane perpendicular to the horizontal plane. It has to be realized that the aerodynamic resistance coefficient has to be based on the absolute particle velocity and not on the velocity of the particle in either X, Y, or Z direction.

A fertilizer particle with a lower value of K will travel farther than a particle with a higher value of K. The aerodynamic resistance coefficients of fertilizer particles vary in general between 0.5 and 0.03. Values for K can be calculated with Eq. 1.99 and the data in Table 1.50. Calculated particle trajectories for different discharge velocities, terminal velocities, and vertical discharge angles are shown in Figs. 1.233 to 1.236.

1.4.4. Spread Pattern Analysis

Irrespective of system/design of the fertilizer broadcaster, the lateral distribution is of decisive importance regarding an orderly way of operation. Fertilizer spreaders are

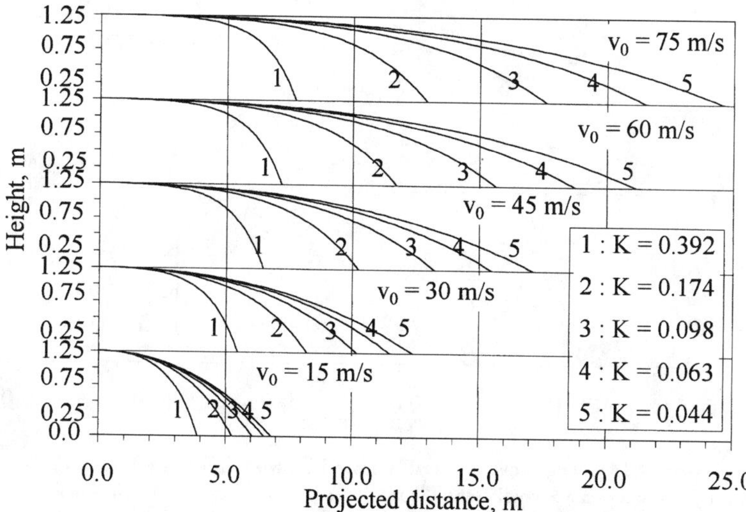

Figure 1.233. Trajectories of fertilizer particles with different values of the aerodynamic resistance coefficient K and discharge velocities for a vertical discharge angle $\alpha = 0°$.

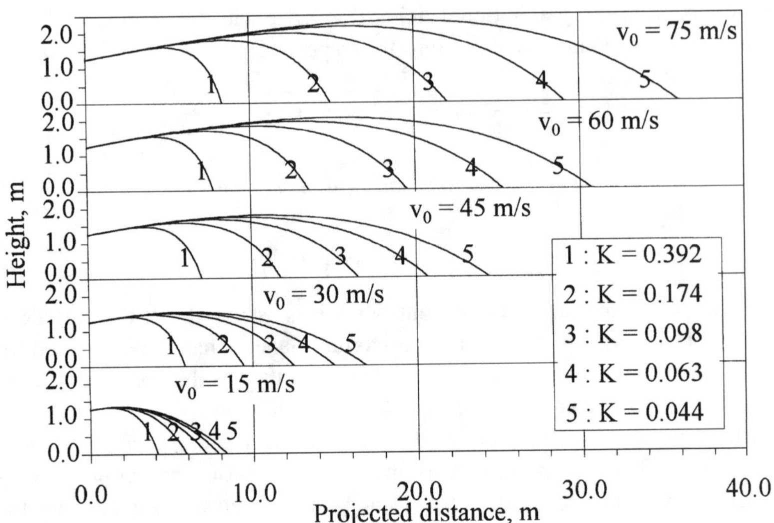

Figure 1.234. Trajectories of fertilizer particles with different values of the aerodynamic resistance coefficient *K* and discharge velocities for a vertical discharge angle $\alpha = 5°$.

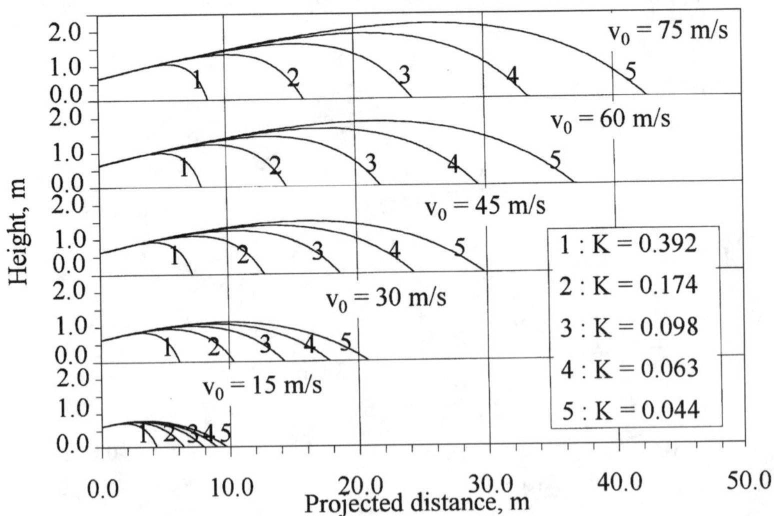

Figure 1.235. Trajectories of fertilizer particles with different values of the aerodynamic resistance coefficient *K* and discharge velocities for a vertical discharge angle $\alpha = 10°$.

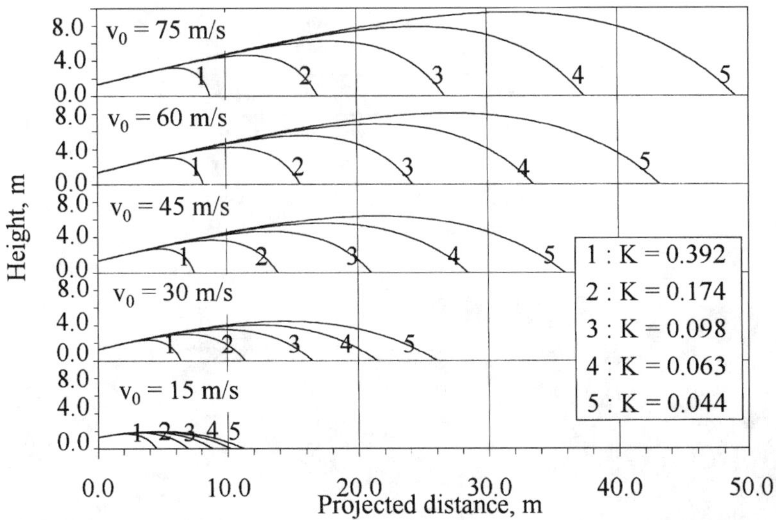

Figure 1.236. Trajectories of fertilizer particles with different values of the aerodynamic resistance coefficient *K* and discharge velocities for a vertical discharge angle $\alpha = 20°$.

tested by collecting the fertilizer in collecting trays (edge length of 50 cm × 50 cm) during a spreading run. This test can be done outdoors but, to exclude weather effects, it is prefered to conduct the test inside (Fig. 1.237). The individually collected amounts of fertilizer are separately weighed, and the results are computed into a spread pattern. The coefficient of variation (CV) (Eq. 1.103) is used as measure for the evaluation of the transversal distribution. The CV is calculated for the lapped spread pattern (Fig. 1.238).

$$\mathrm{CV} = \frac{100}{\bar{x}} \sqrt{\frac{\sum_{i=1}^{N} (x_\mathrm{i} - \bar{x})^2}{N-1}}\,\% \tag{1.103}$$

with:

$$\bar{x} = \frac{\sum_{i=1}^{N} x_\mathrm{i}}{N} \tag{1.104}$$

In these equations:

N = number of collecting trays corresponding with the effective working width
x_i = the amount of fertilizer collected in tray i

The procedure for determining a spread pattern is described in some standards (ISO 5690/2[10], ASAE 343.1[1]). A European standard is in preparation [2].

For evaluation of the spread patterns the following two criteria are decisive:

- Even lateral distribution within the working width
- High stability, insensitivity against outside influences

A general rule of thumb for the classification of a spread pattern is given in Table 1.51.

Table 1.51. Rule of thumb for the classification of spread patterns

CV	Classification
<5%	very good
5%–10%	good
10%–15%	satisfactory
>15%	not satisfactory

Figure 1.237. Testing facility for indoor testing of fertilizer spreaders.

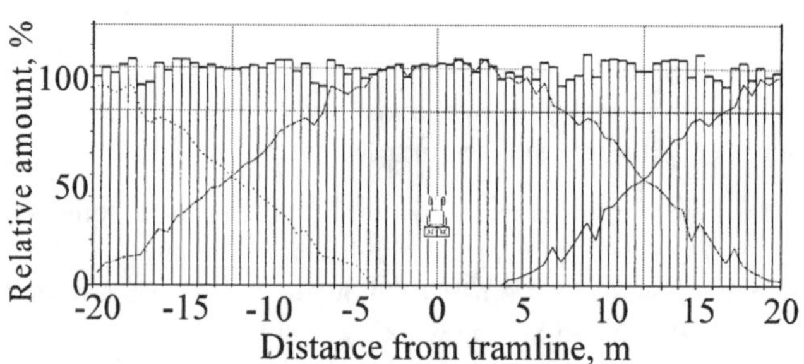

Figure 1.238. Example of a spread pattern.

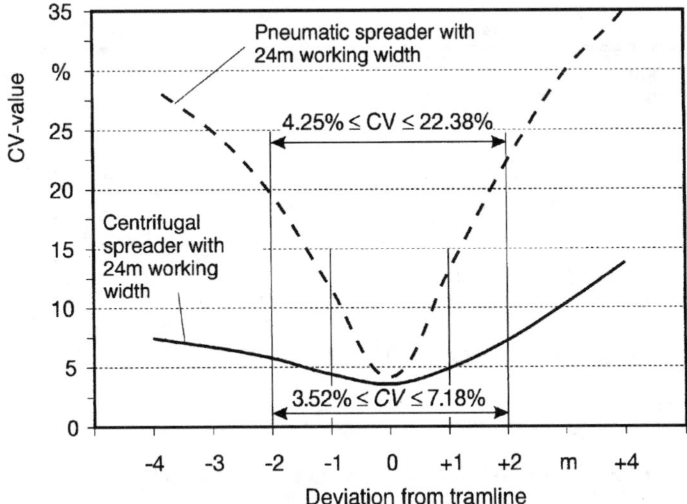

Figure 1.239. Influence of a transversal driving error on the CV for a spinning disc and a pneumatic fertilizer spreader.

A fertilizer broadcaster which produces a spread pattern of high stability reacts insensitively to outside influences such as inaccurate bout following, side wind, swinging of the tractor, inaccurate mounting height and changing material properties of the fertilizers. In general, fertilizer spreaders that produce a spread pattern with a small overlapping area (pneumatic fertilizer spreaders, auger spreaders and field sprayers) react sensitively to inaccurate following bout travel and swinging movements of the tractor. At spread patterns with large overlapping zones (centrifugal and oscillating spout spreaders) these influences are hardly noticeable (Fig. 1.239). Transversal driving errors up to 3 m result in only a minor worsening of the CV for a centrifugal or oscillating spout spreader, while for a pneumatic fertilizer spreader a transversal driving error of 1 m already results in a considerable increase of the CV.

With centrifugal broadcasters the lateral distribution is highly influenced by the material properties of the kind of fertilizer spread. With such broadcasters corresponding settings will have to be conducted to ensure the desired working width. [5]

Regarding outside influences the following statements can be made:

- For spreaders with distribution booms the distribution accuracy depends less on the type of fertilizer but in a considerable way on the way of driving and reaction at terrain undulances by the tractor operator.
- For centrifugal and spout type spreaders the material properties of the fertilizer will have an effect on the lateral distribution. At large working widths also the influence of wind is noticeable. Spread patterns with a shallow CV curve in the area of the working width react rather insensitive.

When fertilizing alongside the boundaries of a field, the interests of both environmental protection and economy have to be considered.

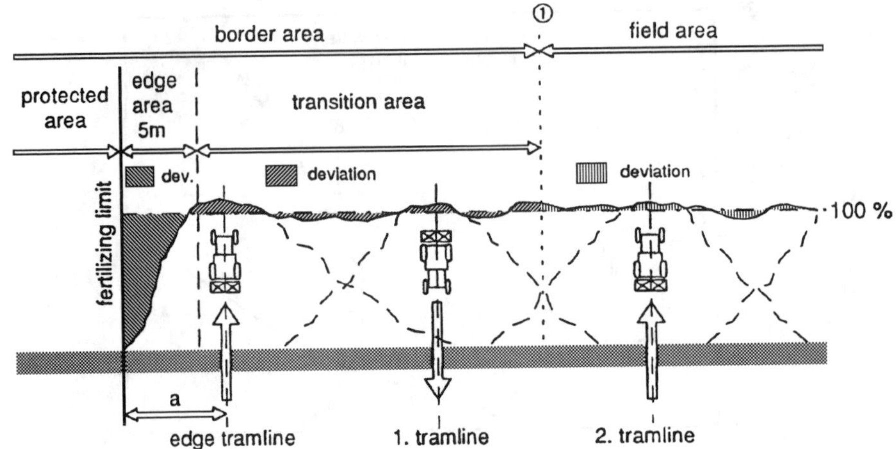

Figure 1.240. Identified areas in relation to boundary spreading of fertilizers.

In order to put up a generally valid assessment scheme for all fertilizer broadcasters and boundary spread systems, the operated area has to be split up into the main field area and a boundary area. For the main field area the previously mentioned guidelines for spread pattern classification are applicable. The boundary spread area must be split up into three sections (protected area, edge area and transient area), each of which with different requirements (Fig. 1.240).

No fertilizer must drop in the protected area, e.g., open waters. However, a threshold should be determined, e.g., 1% of the application amount. Such values occur in testing when some grains of fertilizer are deflected from the collecting trays. In the field's side area, about a 5 m-wide strip, the farmers prefer to achieve a full application rate to ensure the same yield level as in the field's inside. On the other hand, ecologists want to ensure that no fertilizer is being washed out to the protected area by rainfall or floods. How the fertilizer distribution in this area shall be finalized, e.g., a linear or a parabolic rise of the distribution, has still to be determined. In the transient area an overlaying occurs between the boundary spreading and the normal spreading for the field's inside. The transient area reaches from the boundary area to the first overlapping point of two spread patterns where the boundary spreading has no influence any more. In the transient area the spread rate of the fertilizer should be on the same level as in the field's inside and the lateral distribution should be as even as possible.

1.4.5. Developments

Multiple Hoppers

In order to improve the specific nutrient requirements there also are fertilizer spreaders with more than one hopper. Each hopper contains a specific fertilizer and, by mixing these fertilizers during spreading, the required amounts of the nutrients can be applied almost independently of each other. Different fertilizers have different physical properties and this may cause segregation traveling through the air. This may lead to an uneven distribution of one or more of the nutrients. To obtain good results a proper selection of the different fertilizers is required.

Site Specific Spreading

Most arable areas show a heterogenous structure regarding kind of soil, nutrient content, water supply, etc. Still, usually the applied amount of fertilizer during the application is kept constant. The simple position determination with the aid of a satellite detection (DPGS) now offers the possibility to react on specific local requirements. [11]

Location-specific fertilizer spreading requires the preparation of a fertilizing map by the management information system. This map has to be exchanged with the process computer on the fertilizer spreader. [12, 14] During spreading a "parallel thinking" of the fertilizer spreader is required. At possible working widths up to 36 m, being equivalent to 60 m throwing width, frequently situations may occur in which it is not sufficient to regulate the metered quantity according to a local determination. Simultaneously it has to be considered,

- Which sectors of the application map are influenced by the spread fan;
- In which travel direction and speed it is operated;
- Which system caused time constants are applicable for the variation of the application rate.

By intelligent tying up of the operational data an optimum fertilizer application is possible. [13]

Electronics

Process computers in connection with sensors and actuators are being used on fertilizer applicators to fully automate varying operational steps such as the regulating of the metering device; simultaneously it is possible to check various functions to provide information to the farmer/operator as well as to record and store management data. These multiple possibilities result in an improvement of the operation quality and in added comfort.

Controlling and regulating the application amount requires control of the flow out of the hopper. The flow depends on the required application rate, the driving speed and the effective working width. For regulating the application amount the following relation holds:

$$Q = 10000 \cdot Q/v \cdot b \tag{1.105}$$

where Q = flow rate in kg/s, v = driving speed, and b = working width. In general the flow rate Q is a function of the material and the shutter design. Presently there are no sensors available to record instantaneous flow. One therefore still has to conduct a calibration procedure whereby the relation between the setting figure (shutter slide position or speed of the metering roller) and the flow is determined. With the aid of frame-integrated weighing cells or the weight power recording in the tractor's three-point hydraulics, it is possible to weigh the amount of fertilizer inside the hopper. The change of weight during a determined covered distance during the spreading operation then allows a consequential determination on the set spread rate. More simple are the relations at field sprayers. The metered amount of liquid can be accurately determined and the measured value be provided to the regulating circuit by cellular wheels or wing radanemometers.

References

1. ASAE Procedure for measuring distribution uniformity and calibrating granular broadcast spreaders. ASAE Standard S341.2.
2. CEN Agricultural and forestry machinery–Full width distributors and broadcasters for solid fertilizer, Part 1: Requirements; Part 2: Test methods. European Standard (*in preparation*).
3. Dobler, K., and Flatow J. (1968) Berechnung der Wurfvorgänge beim Schleuderdüngerstreuer. Grundlagen Landtechnik 18(4): 129–134.
4. Grift, T. E., Walker J. T., and Hofstee J. W. (1997) Aerodynamic properties of individual fertilizer particles. Transactions of the ASAE, 40(1): 13–20.
5. Hebbler, K. (1995). Berechnung der Querverteilung von Zweischeibenstreuern. Landtechnik 50(2): 62–63
6. Hofstee, J. W., and Huisman W. (1990) Handling and spreading of fertilizers: Part 1, Physical properties of fertilizer in relation to particle motion. Journal of Agricultural Engineering Research, 62: 9–24.
7. Hofstee, J. W. (1992) Handling and spreading of fertilizers: Part 2, Physical properties of fertilizer, measuring methods and data. Journal of Agricultural Engineering Research, 53: 141–162.
8. Hofstee, J. W. (1995) Handling and spreading of fertilizers: Part 4, The reciprocating spout type fertilizer spreader. Journal of Agricultural Engineering Research, 62: 9–24.
9. Hofstee, J. W. (1995) Handling and spreading of fertilizers: Part 5, The spinning disc type fertilizer spreader. Journal of Agricultural Engineering Research, 62: 142–162.
10. ISO Equipment for distributing fertilizer–Test methods Part 1–Full width fertilizer distributors. ISO 5690/1.
11. Jürschik, P., and Schmerler J. (1995) 200 Hektar teilflächen-spezifisch gedüngt. Landtechnik 50(3): 148–149.
12. Marquering, J. and Scheufler B. (1997) Intelligente Verknüpfung von Betriebsdaten und gerätespezifischen Technologien. Landtechnik, 52(2): 78–79.
13. Marquering, J. and Scheufler B. (1997) Technische Probleme bei der teilflächenspezifischen Ausbringung von Produktionsmitteln. Zeitschrift für Agrarinformatik 5(1): 9.
14. Scheufler, B. (1997) Mineraldüngung. Jahrbuch Agrartechnik 9: 117–125.
15. Speelman, L. (1979) Features of a reciprocating spout broadcaster in the process of granular fertilizer application. Mededelingen Landbouwhogeschool Wageningen, pp. 217.
16. TVA (1984) Fluid fertilizers. Ed. J.M. Potts, TVA Bulletin Y-185, pp. 130.

1.5. Pest Control Equipment

R. Wilkinson, P. Balsari and R. Oberti

1.5.1. Pest Control Methods

In agricultural production *a pest* is anything that impedes or competes with the desired crop. The pest may be other plants (weeds), insects, fungi or diseases. Control of these pests is accomplished by chemical means, nonchemical treatments, or a combination of measures sometimes known as *integrated pest management* (IPM). IPM strives for pest control using biological, chemical and physical means that are effective, economical and environmentally friendly.

Non-Chemical Methods

These methods usually have low environmental impact but may be expensive, time-consuming or laborious. The high cost of chemicals, concerns about public health, environmental impact of chemical controls, and the growing popularity of *clean agriculture* and *organic farming* are all factors that have promoted the use of traditional and new nonchemical pest control methods.

In developed countries particularly for horticultural and fresh produce crops, there is growing interest in these methods, even though they may be less profitable than chemical controls. Traditional nonchemical pest controls have primarily focused on physical treatments.

Chemical Control Methods

Chemicals that control pests, i.e., pesticides, can be classed as either *contact* or *systemic*. Chemicals that control by contact must be in direct contact with the pest they are to control. Systemic chemicals are absorbed into the plant by roots and leaves, and cause interference with the ongoing growth process of the pest. Chemicals may be applied to the soil or to the foliage as liquids (sprays) or as solids (dust and granular).

1.5.2. Non-Chemical Techniques

Mechanical Weeding

Weeding equipment roots out and removes weeds from the soil. Its use is becoming increasingly widespread, with improved effectiveness, on row-cultivated crops, including winter cereals, vineyard and orchard crops.

The *hoeing machine* may be tractor-mounted or pull type. Tines or rotary hoes are fixed to the frame and penetrate the upper crust of the soil (≈ 5 cm deep) uprooting weeds from the soil portion between the rows. The treatment is effective on dry, compact soil, and a stable working depth is maintained by small wheels. Hoeing is usually combined with banding applications of herbicides in the pre- and post-emergence stages. Rows treated may vary from 2 to 8 and speeds from 2 km/h to 8 km/h.

Blade cultivators are equipped with spikes that terminate with a horizontal blade, often tine or ducksfoot shaped. These operate between rows, breaking the upper soil crust, and at a depth ≤ 5 cm beneath the surface, cutting weed roots and rhizomes. A

Figure 1.241. Brushing machine.

similar cutting effect is obtained with a vertical *disk weeder*. Care must be taken in setting the operating distance from the crop to avoid damaging its root system. These types of implements also are used for operating along the tree rows in orchards. Special sensing devices detect the presence of a tree trunk and cause the implement to be shifted to the side, so that it can circumvent the plant.

The *brushing machine* is effective during the early stages of weed development. This implement is mounted on a tractor whose PTO drives a set of robust brushes (15–50 cm wide or more), which rotate on a vertical or horizontal axis (Fig. 1.241). The gap between brushes is adjustable; they operate between rows on the upper soil layer (<5 cm deep), uprooting weeds and carefully removing their roots from the soil, thereby excluding all possibility of resprouting. Optimal results are achieved with low speeds of rotation, a tractor forward speed of 2–6 km/h, and soil that is not too dry or compacted. A special guard protects the crops from the soil that is removed. The operator must wear adequate protection against the dust that is raised.

The light spiked *chain harrow* also can be used on non-row-cultivated crops. It consists of a set of narrow, flexible, vertical steel bars. The spikes scratch the soil with a pressure that can be adjusted in accordance with the crop's stage of development. This makes it possible, with careful adjustment, to operate during the pre-emergence stage without damaging the crop. At post-emergence, when the crop is deeper-rooted than the weeds, the best results are obtained in the earliest stages of weed development. Harrowing is less effective on wet or stony soil and is totally ineffective against deep-rooted perennial weeds. An optimal tractor speed 7 km/h and a large working width (6–24 m) make it possible to achieve high work rates.

There also are a number of smaller and simpler traditional tools based on the same principles as the machines described above (Fig. 1.242) that can be manually driven or animal drawn: the single or two-row rotary weeder, the wheel hoe, the three-tine wheel hoe. These are all low-cost types of equipment used on small farms in developing countries; the resulting quality of work is good but they are highly labor-intensive (0.01 ha/h to 0.10 ha/h).

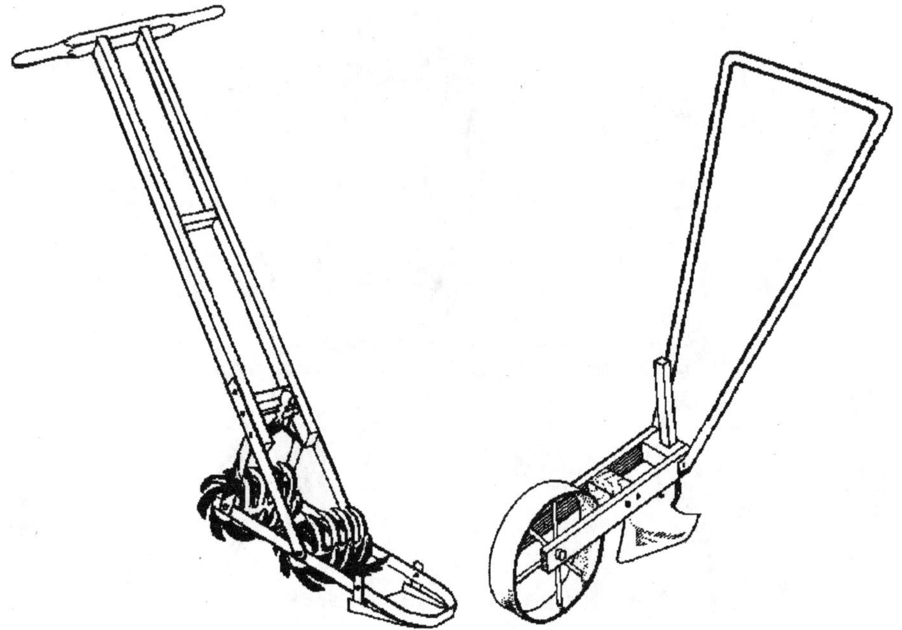

Figure 1.242. Manual Weeders.

Thermal Treatments

Heat can also be effectively used for weed and pest control. In the *flame weeding* technique, the area between rows and close to the crop is flamed using a gas burner that advances at a low-medium speed (2–7 km/h). The heat of the flame causes the breakup and subsequent withering of weed tissues. On the ground, the increase in temperature only extends to a depth of a few millimeters. The flame weeding equipment (Fig. 1.243) consists of a boom supporting the burners, a gas (liquid propane) tank and a delivery system.

The best results are obtained on weeds in the early stages of development. If the weeds are advanced, the treatment will require increased fuel consumption. Timeliness is an important constraint on this technique. If disregarded, costs will increase and it may not be successful.

The flame weeding operation has minimal environmental impact, but is too expensive to perform on the entire crop surface (gas consumption 150–200 kg/ha). It becomes economically feasible only when used in conjunction with other operations; flaming being restricted to the area near the rows that cannot be mechanically weeded (gas consumption 20–70 kg/ha).

Soil solarization consists of placing clear plastic mulch over moist soil for a period of 1–3 months during the summer. Solar radiation raises the soil temperature, which inactivates some fungi and kills most weed seeds and nematoda. The treatment is effective if the upper layers of the soil are held at high temperature for a total time τ that is

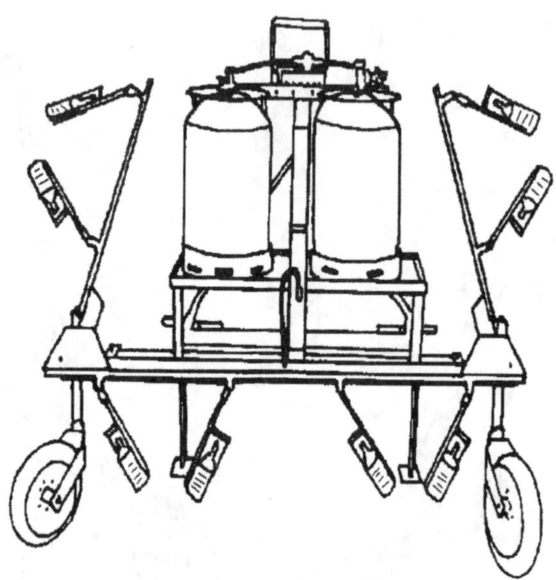

Figure 1.243. Flame weeding equipment.

sufficiently long. Typical values are: $T = 40°C$ for $\tau > 50\,h$; $T = 45°C$ for $\tau > 24\,h$; $T = 50°C$ for $\tau > 12\,h$.

Solarization fails if the soil does not acquire enough heat (in fact its use is recommended only in warm or tropical climates) or the soil moisture content is too low. Preliminary irrigation sometimes may be necessary.

The plastic film is laid using tractor-mounted equipment consisting of a frame supporting a spool and two lateral furrowing discs. The tractor moves at 1.5–3 km/h, unrolling the film from the spool onto the soil, while two cylinders press its edges into the furrow that has just been opened, and then cover it with soil.

Vacuum

Collection by vacuum has proved effective on insects and larvae of sufficiently large size, and in particular on potato beetles and grasshoppers. Tractor-mounted suction equipment has been developed: a hydraulically powered impeller blasts a violent air jet towards the upper part of the plants, while simultaneously generating a suction airflow at ground level. Insects and larvae are blown off leaves and fall onto the ground, where they are collected by vacuum suction. At a forward speed of 5 km/h, collection efficiency is $\approx$90% for insects and $\approx$80% for larvae. Suction is not effective against eggs.

Biological Pest-control Equipment

A more frequent pest-control practice is inundation with predators and parasites. Biological agents such as predators' eggs, larvae, mites, ladybugs etc. are commonly supplied mixed with inactive material or organic grit (bran, sawdust, hulls etc.). The mix is often distributed manually by walking workers, who gently shake a discrete amount onto the foliage at regular intervals. Application rates are low, ranging in general from a few dm^3/ha to 20–30 dm^3/ha of mix.

Equipment for ground and aerial application has also been developed, and performs the following functions: it maintains the homogeneity of the contained mixture, releases the mix at an adjustable rate, protects the biological agents from any consequential damage, and improves on the work rates achievable with manual labor. In general, such equipment consists of a reservoir without an agitation system that is chilled to reduce organism activity. The mixture is metered by a variable-speed electric motor.

1.5.3. Chemical Control Methods

Chemical application is, at present, the primary method used for the protection of crops (Fig. 1.244). However, the use of chemicals has a significant impact on the environment. It raises concerns about public health and the quality of food and drinkable water, and is an increasing expense on farm production costs in both developed and developing countries.

Thus, it is important to understand the principles of chemical application, and in particular how spraying equipment works, and the adjustments that it needs.

Formulations

Plant-protection products can have two different mechanisms of action:

By Contact: the effect is restricted to those areas on which a sufficient amount of active ingredient has been deposited; the overall efficacy of treatment depends on the degree of coverage of the target, and hence on the density and homogeneity of distribution.

Systemic: the active ingredient is absorbed by the leaves or roots and conveyed via the lymphatic system to the points of action; the treatment is effective if a sufficient amount of active ingredient has been absorbed overall.

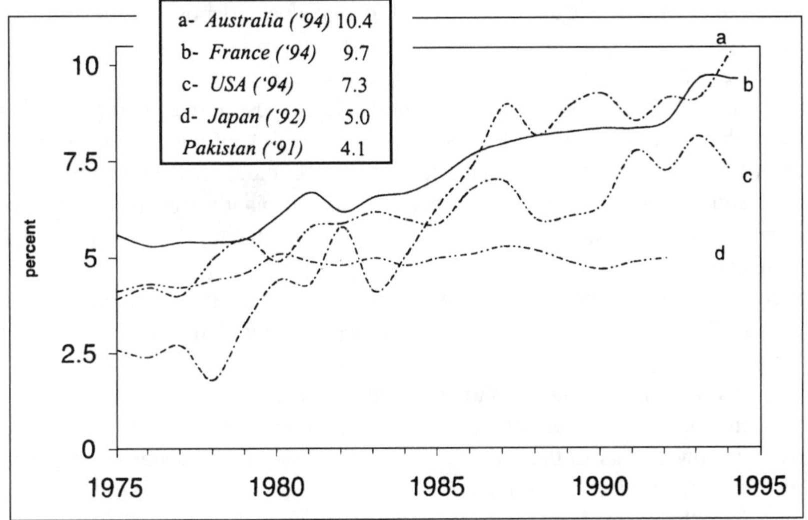

Figure 1.244. Percentage of final crop output spent by farms on chemicals.

Table 1.52. Formulation of active ingredients

	Dry Application
DP	Dust: fine particles of solid AI (30 μm) with inert powders
GR	Granules: solid AI dust mixed with powders in large particles (250 μm)
RB	Baits: chemicals mixed with edible products
	Spray Application
CS	Capsule suspension: microscopic capsules (30 μm) containing the AI
EC	Emulsifiable concentrates: liquid or solid AI not soluble in water, but in a special solvent or gel
SC	Suspension concentrate: suspension of an insoluble, solid or liquid AI in a water-dilutable fluid
SL	Soluble liquid: water-soluble liquid AI, or solid AI dissolved in a small quantity of water
SP	Soluble powder: micronized, water-soluble solid AI (30 μm), or effervescent tablet
UL	Ultra-low volume formulation: ready-to-use non-volatile, low-viscosity liquid
WG	Water-dispersible granules: compacted granules (1 mm) of insoluble solid AI
WP	Wettable powder: micronized, insoluble solid AI; is applied homogeneously dispersed in water

The active ingredients (AI) used for crop protection are classified according to their function as:

Herbicides: Substances that kill or inhibit weed development. Weeds damage crops by: competing for environmental resources (light, nutrients, water, space, CO_2, O_2), hosting parasites and diseases, and by directly releasing toxic substances. *Contact herbicides* destroy the shoot systems of the weeds on which they are deposited and retained. *Systemic herbicides* are absorbed by leaves or roots and transported inside the weed, where they interfere with its vital processes. *Soil-applied herbicides* act on weed seeds in the early stages of development; they can be used before the crop has emerged (pre-emergence) or in subsequent stages (post-emergence).

Fungicides: Substances that kill or inhibit fungi development. Fungi damage plants by causing: stunted growth, malformations, rotting or drying. *Contact fungicides* are active only on those surface areas on which they have been deposited; they are used as preventive treatments. *Systemic fungicides* are absorbed by leaves and distributed inside the plant; they can also cure internal suffering layers.

Insecticides: Substances that kill insect pests. These pests damage crops by: directly destroying the epigeal or hypogeal parts; restricting the flow of nutrients to certain parts of the plant; inoculating viral or fungal infections. Insecticides are active both directly on the insect, and also after they have been absorbed into the plant's organism. In both cases, the pesticide's action on the parasite is by passage through the scarf-skin, by swallowing, and by inhalation.

For commercial use, the AI is prepared in a variety of formulations designed to be applied "as is" or to be diluted in water. The most commonly used formulations are listed in Table 1.52.

The new formulations–water-soluble granules, capsules, gels, tablets–and the new types of containers–i.e., water-soluble bags that dissolve directly in the tank–are characterized by: lower risks to the operator, ease of dosage, ease of storage, virtually no problems with disposal of the containers.

To facilitate the water-dilution operation, a number of adjuvants are available. These include diluents to facilitate mixing with water, droplet spreaders, drift retarders, adhesion agents, foam inhibitors and more.

Dry formulations are used widely in areas where water is scarce and for some particular treatments (i.e., sulfur on vines). Granule formulations allow very precise application, avoiding the dangerous drift phenomena typical of powder applications. The characteristics of "dry" materials are discussed in Section 9, Granular Applicators.

Fogs and aerosols are generally used for the treatments in greenhouses. The small droplets of these formulations are characterized by a high level of drift, which is advantageous in enclosed environments where it facilitates penetration into all parts of the crop canopy. Soil fumigation is a treatment normally used for disinfecting the soil prior to planting a new crop. It is carried out by introducing the active ingredient in gaseous form beneath plastic sheeting spread over the ground, or by injecting into the soil, liquids or granules that subsequently emit vapors.

Characteristics of Droplets

In practice, a spray is made up of droplets whose size varies within a certain range. The range of droplet diameters depends on the physical characteristics of the atomized fluid and on the operating parameters of the nozzle (pressure in hydraulic nozzles; speed of rotation in rotary nozzles; air speed and flow rate in pneumatic nozzles).

Droplet size is usually expressed as the *volume median diameter* of a spray, $D_{V0,5}$ or VMD—i.e., the value which divides the diameter distribution into two parts containing equal volumes of liquid—or as *number median diameter*, $D_{N0,5}$ or NMD, i.e., the value which divides the diameter distribution into two parts containing equal numbers of droplets. The ratio VMD/NMD is a spray's dimensional homogeneity coefficient: the closer its value approaches 1, the more uniform the size of the droplets (Fig. 1.245).

The size of a droplet strongly influences its trajectory: after being emitted from the nozzle (at a speed of approx. 20–30 m/s, in the case of hydraulic nozzles) the droplet is rapidly decelerated by friction until it attains a velocity that is solely a function of its diameter. The motion of the air in which the droplet descends also influences its trajectory. Droplets with $\oslash \geq 200\,\mu$m are essentially unaffected by external currents. Droplets with smaller diameters, on the other hand, are transported by the air and can travel considerable distances before they are eventually deposited, as shown in Fig. 1.246. This is known as *drift*, and is responsible for product waste, phytotoxicity in adjacent crops, environmental pollution and potential damage to the operator's health.

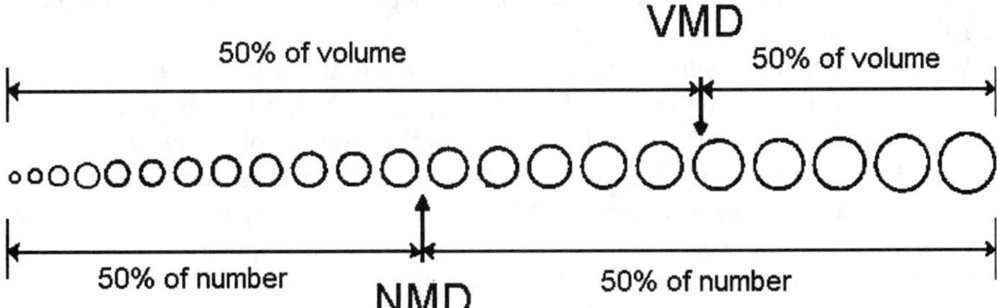

Figure 1.245. Definition of VMD and NMD.

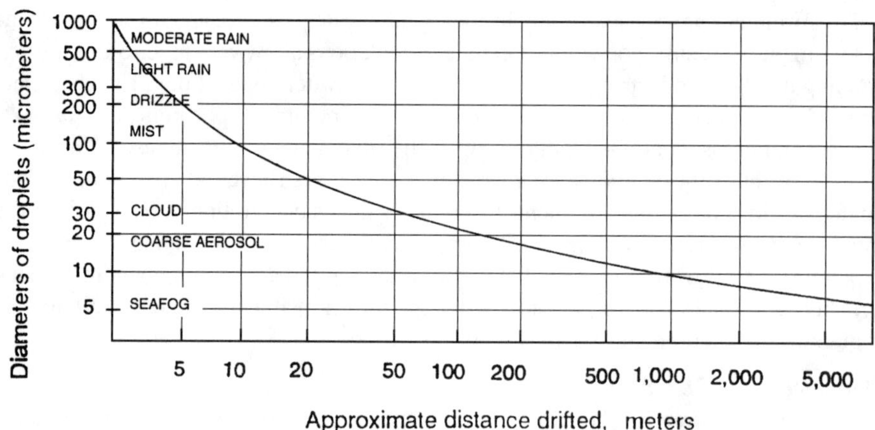

Figure 1.246. Drift vs. droplet size. (Drops falling 3 m in 5 km/h wind.)

In treatments of orchard crops with thick foliage, the gravitational descent of pressure-propelled droplets provides insufficient coverage of the innermost parts of the crop canopy. In such cases, a turbulent auxiliary airflow is used to: move the foliage about, exposing the shielded portions; transport the smaller-sized droplets, causing them to penetrate the canopy; and redistribute the droplets within the foliage. The optimal range and velocity of the airflow should have sufficient force to shift and replace the existing air inside the foliage and sufficient speed to move through the mass of foliage, emerging at a speed of 0.5–1 m/s. Airflows with lower speeds tend to over-deposit on the external portions of the canopy while higher air speeds cause a substantial portion of the spray to pass through the vegetation without being deposited.

For a given formulation, the smaller the size of the droplets the greater will be their adherence to the surface on which they have been deposited; with droplets greater than 400 μm diameter, there is risk of incurring runoff from the target.

As can be seen in Fig. 1.247, the droplet diameter also affects its evaporation rate. This phenomenon has three direct effects on the treatment: it reduces the wet surface, increases the spray's susceptibility to drift, and varies the concentration of active ingredient.

Methods of Application

The optimal method of application on the crop depends on the product's mechanism of action and on the nature of the target.

When making *soil applications*, in the pre-planting and pre-emergence stages, it is essential that the soil be sufficiently broken up to avoid the formation of shielded areas, and that soil moisture be sufficient to allow penetration of the plant-protection product. In the case of herbicide treatments, the formulation's mechanism of action is *by direct contact* with the weed seeds, inhibiting their germination or their early stages of embryonic development. This type of mechanism requires good homogeneity of distribution and a density of at least 25–50 droplets/cm^2. In treatments on flooded soil, the diffusion in water permits the use of very coarse drops at 3–5 droplets/cm^2.

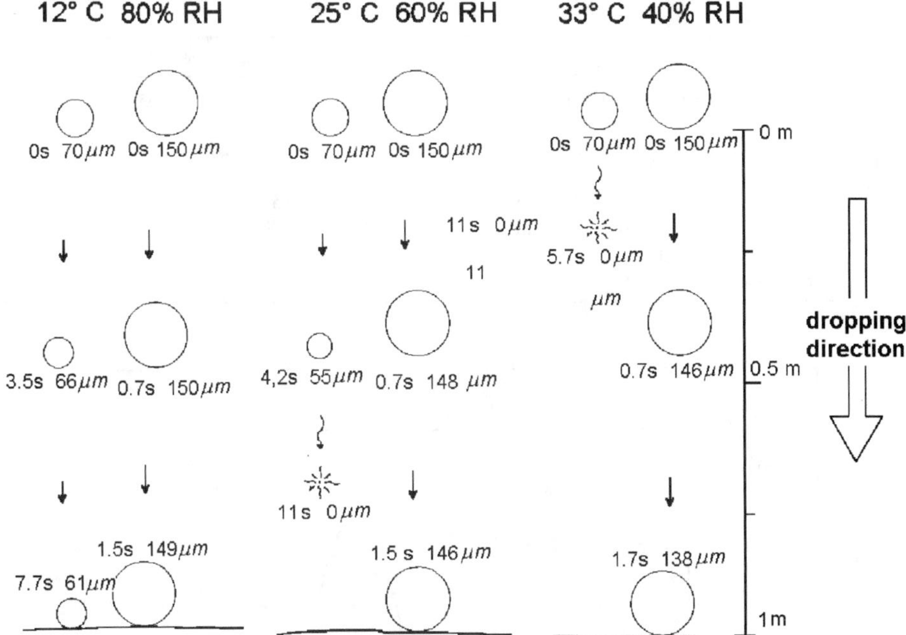

Figure 1.247. Evaporation and size reduction of free falling droplets—3 atmospheric conditions.

In *foliar applications*, to achieve complete coverage of the target (weed or crop) the droplets must be able to penetrate the crop canopy. When using products which act *by contact*, it is essential to cover the greatest possible surface area. Thus, fine and medium-fine droplets are used with high impact density (50–100 droplets/cm^2). Due to their small size, the droplets are easily moved by the lightest air currents, thus improving deposition on the underparts of the plant. However, this ease of movement also means that great care must be taken to avoid drift. The use of plant-protection products with *systemic action* does not require complete coverage of the target, but only that the plant receives and absorbs a sufficient amount of active ingredient. In this case, medium-large droplets are used with low impact density (20–30 droplets/cm^2).

Using this data, it is possible to calculate the volume to be distributed (Table 1.53), taking into account that, in the case of soil treatments the effective target surface is the area of the sprayed field, whereas for plant treatments it is the total surface of the epigeal portion of the plant. This is usually expressed as a *LAI (Leaf Area Index)*: the numeric ratio between the leaf surface of the target and the surface of the field in which it is growing.

Its value, for a given species, is closely related to the type of crop, its age, and the stage of growth, as shown in Fig. 1.248 and Table 1.54. Taking into account the droplet diameters $\oslash$ (μm) and impact densities n_i (drops/cm^2) given above, we can calculate the required volume of distribution:

$$V = \pi/6\, n_i\, \oslash^3\, 10^{-7} \qquad [\text{dm}^3/\text{ha}]$$

Table 1.53. Droplet dimensions and use

Droplets	VMD	USE	Indicative Volume
Aerosol and Mists	$<100\,\mu$m	Fogging; Greenhouse appl.	$10 \leftrightarrow 100$ dm^3/ha
Fine	$100 \leftrightarrow 200\,\mu$m	Orchard appl.; Contact products.	$100 \leftrightarrow 250$ dm^3/ha · LAI
Medium	$200 \leftrightarrow 300\,\mu$m	Contact products.	$150 \leftrightarrow 300$ dm^3/ha · LAI
Coarse	$300 \leftrightarrow 425\,\mu$m	Systemic products; Soil appl.	$100 \leftrightarrow 250$ dm^3/ha · LAI $150 \leftrightarrow 400$ dm^3/ha
Very coarse	$>425\,\mu$m	Soil flooded appl.	$100 \leftrightarrow 200$ dm^3/ha

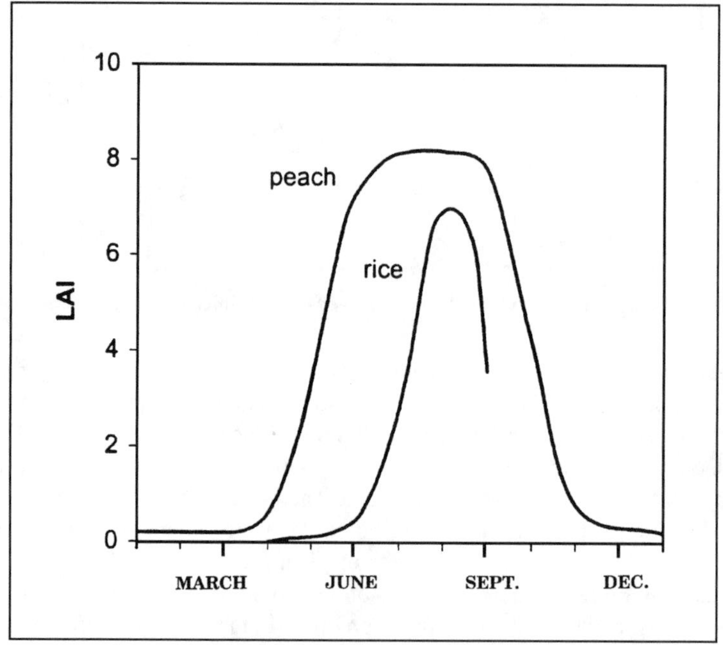

Figure 1.248. Seasonal progress and LAI.

Each droplet deposited on the target surface will cover a contact area that depends on the wettability of the target itself, and which is measured by an *expansion coefficient* c_e, defined as the ratio of the droplet's contact diameter to its real diameter $\oslash$ (μm) (Fig. 1.250). Its value is, to a good approximation, independent of droplet diameter. Therefore, the total surface S_{tot} covered by the spray proves to be:

$$S_{\text{tot}} = 1500\, c_e^2 V/\oslash \qquad [\text{m}^2/\text{ha}]$$

The wetting agents contained in the formulations increase the value of c_e. The same active ingredient AI, distributed on grasses, can have $c_e < 0.3$ when diluted only in water,

Table 1.54. Seasonal progress & LAI

Crop	LAI MAX
Wheat	9–12
Rice	5–8
Soy-bean	7–9
Maize	3–5
Potato	3–4
Sugar-beet	3–5
Cotton	0,5–1,5
Peach	6–9
Pear modern shape (spindel or pyramid)	2–4
Pear old shape (globe, sphere)	≥15
Apple modern shape	2–4
Apple old shape	≥12

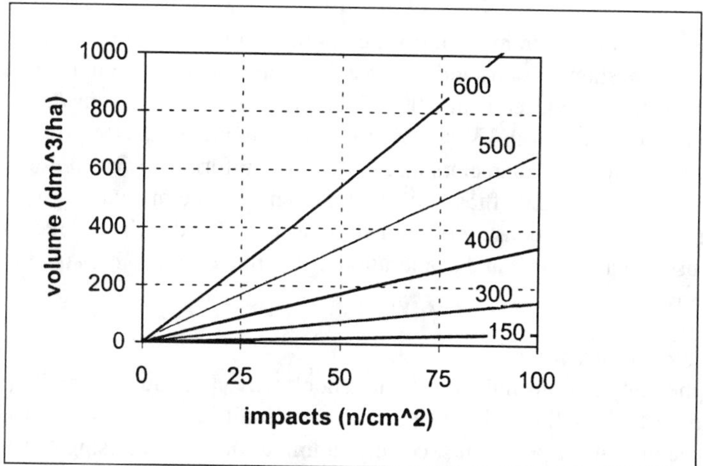

Figure 1.249. Volume to be distributed vs. impact density and droplet diameter.

but a value of $c_e \approx 2$ is possible if a suitable wetting agent is added. In dicotyledons, wettability is generally better and values of c_e between 2 and 4 are typical.

As the above equation shows, it is possible to cover a greater surface with the same distributed volume by reducing the size of the droplets. This point is also confirmed by the data shown in Fig. 1.249.

The current trend is to use lower amounts of water (carrier) while keeping the chemical dosage constant, through the use of treatments defined as *low volume* and *ultra low volume* (LV and ULV). LV techniques cut down on the time needed for tank filling. They reduce

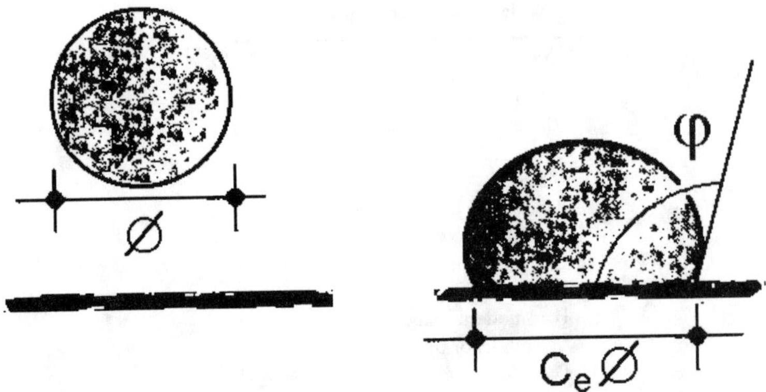

Figure 1.250. True and wetted (contact) diameter of a droplet. The contact angle ⊘ gives a measure of the wettability of the target surface high wet ⊘ ≤ 60° low wet ⊘ ≥ 120°.

the cost and difficulty of transporting water to the field, and permit better timeliness. Experiments have shown a substantial increase in spray deposition on inner parts of the canopy with volumes reduced from 2000–5000 dm^3/ha to 300–600 dm^3/ha.

However, it is essential in LV spraying that the atomizing device produce droplets with a restricted range of diameters and extremely uniform distribution on the target. Under these conditions, the efficacy of the treatment can be maintained with volumes of ≤30 dm^3/ha on herbaceous crops, and volumes of ≤200 dm^3/ha on tree crops. Care should be used to avoid drift and evaporation of the fine droplets that are characteristic of ULV spraying.

Meteorological Constraints

Atmospheric conditions influence treatment effectiveness. Strong winds ($v > 3$ m/s, corresponding to a continous leaves motion) can result in spray drift. Conversely, the total absence of wind ($v < 0.5$ m/s, corresponding to the vertical rising of smoke) also can be dangerous, especially in the temperature-inversion conditions that typically occur in the early hours of the morning or on very hot days: under such conditions the droplets are subject to floating transport phenomena, and subsequent evaporation.

Slight amounts of rain or dew can enhance the efficacy of the treatment, especially the LV type, by helping to redistribute the product on the treated surface. However, heavy or protracted rainfall can cause the product to wash off, especially from the more exposed parts of the plant: if it occurs too soon after application (a few hours) the efficacy of treatment may be totally compromised. In *ground treatments*, in-depth penetration of the product is greatly facilitated by even moderate precipitation (4–5 mm).

Temperatures less than 10°C reduce leaf absorption, impairing the efficacy of treatments based on systemic products. Excessively high temperatures combined with low air humidity levels cause a rapid decrease in droplet size, due to evaporation.

1.5.4. Spraying Equipment

Chemicals to control insects and diseases that affect plants can be applied in a great variety of ways. These include broadcast sprays, directed or band sprays, in-furrow application, soil injection, spot treatment, dust and granular application and more. Some methods such as air blast spraying, fumigation, aerial application and chemigation are highly specialized and require special training to use the equipment associated with these application methods. Only methods that are in general use are presented here.

Hand Sprayers

Hand sprayers are ideally suited to apply small quantities of pesticides. They can be used to spot spray; to apply chemicals inside structures such as greenhouses or to treat areas that are hard to reach.

The *compressed air sprayer* is usually a hand-carried sprayer with a capacity of 4–12 dm^3. Pesticides and water are proportioned according to label instructions and placed in the airtight tank. Air pressure in the tank is increased with the built-in pump until spray is freely delivered when the control valve is opened (Fig. 1.251). This type of sprayer typically has a shoulder strap or hand grip for the operator. The nozzle and control valve are manipulated by the free hand and directed toward the target pest. As nozzle output begins to diminish, the air pressure must be recharged by operating the air pump. Most hand sprayers do not have a pressure regulator to maintain constant application pressure.

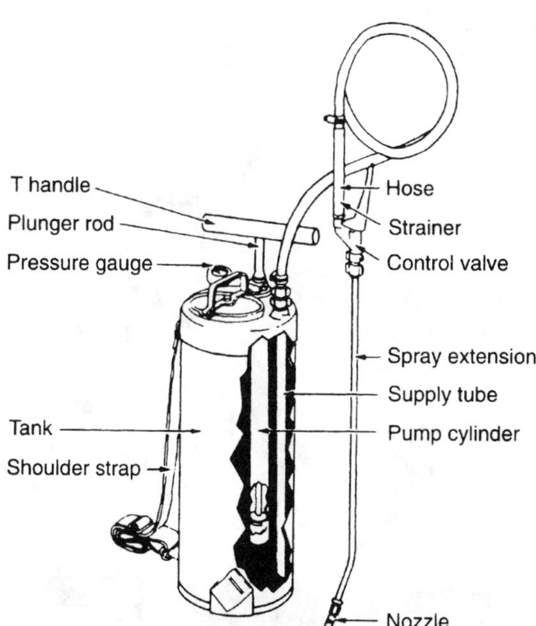

Figure 1.251. Compressed air sprayer.

However, models that do have this feature can apply a lower, more uniform rate that is more environmentally friendly.

Knapsack Sprayers

This type of sprayer also is limited to about 15–20 dm^3 capacity. A larger size becomes uncomfortable for the operator. The spray tank is fashioned so that it can be carried on the operator's back being secured by straps as with a backpack–hence the name. The pump is mounted inside the tank and operated by one hand that moves a pumping handle up and down to produce nozzle discharge as noted. The nozzle and control valve are handled by the alternate hand and the spray directed toward the target. Most operators find these units quite acceptable to use, but as with any hand sprayer, the discharge rate and walking speed are critical components for a predictable application rate.

Battery-operated Hand Sprayer

Battery-powered *controlled droplet applicators* (CDA) use a spinning disk type of atomizer to produce a uniform droplet size. These units are suitable to apply concentrates and ultra low volume (ULV). Thus the amount of material that must be carried by the operator is greatly reduced. The droplet size is inversely proportional to the disk RPM. Higher RPM reduces the droplet diameter. The units are lightweight, easy to use and usually powered with common flashlight dry cells; the batteries being positioned in the handle (Fig. 1.252). These sprayers are limited to treating small areas and their cost, battery expense and low work capacity may restrict their use.

Figure 1.252. Controlled droplet applicator.

Power Equipment

Small Motorized Sprayers

Some small sprayers have all the components of large sprayers but are mounted on a small cart or wheelbarrow. Tanks are usually 60–120 dm^3 capacity. Typically they are propelled manually or pulled by a small tractor. A small engine, 3–4 kW, provides power for the pump and agitation system. This greatly increases their capacity over hand-operated sprayers, but their small size still makes them unsuitable for general field use. Most models have an adjustable nozzle on a hand gun; others may include a small boom with multiple nozzles.

Boom Sprayers

These sprayers usually are low pressure units and are the principal type used for field spraying. Fig. 1.253. Tanks and booms may be mounted on tractors or trucks, designed as trailered units, powered and pulled by a tractor, or even self-propelled. The pumps are usually roller, centrifugal or diaphragm pumps. Tank capacity may range from 200–4000 dm^3. Booms may range in length from 8–36 m. Boom height must be easily adjustable from 30–180 cm above the target to insure good nozzle performance and spray pattern overlap. Some booms are self-leveling to reduce travel undulation and provide more uniform application.

Tandem axle wheel arrangements or large diameter wheels also are used to smooth out travel and improve application uniformity over rough ground. Booms are sometimes referred to as *wet*, where spray fluid is carried to the nozzle inside the boom structural members, or *dry*, with the spray material carried to the nozzles by connecting hoses. The

Figure 1.253. Typical boom sprayer—3 point mounted.

spacing of the nozzle on the boom may range from 20–150 cm depending upon the type of nozzle and its application.

Boomless Sprayers

Boomless sprayers as a class cover a wide range of tank sizes, pressures and delivery rates. One type of boomless sprayer has a center cluster of nozzles, operates at low pressure and projects a wide swath pattern without being encumbered by a wide boom. Uniformity usually is not as good as with a boom sprayer, but is very acceptable for many field and roadside applications. Where spray must penetrate thick foliage or animal hair or reach the top of tall trees, high pressure is required. Piston pumps usually are used to deliver pressure from 1–3 MPa.

Application rates may range up to 240 dm^3/min and are typically applied manually with a hand gun. Sprayers with tank capacities of 2000–4000 dm^3 may be truck mounted, trailered or self-propelled units.

Air-assisted Spraying

Air Blast Sprayers

Air blast sprayers have been used for many years to spray fruit trees. These units consist of a fan that is capable of producing a high volume of air flow into which the spray is dispensed. As the sprayer moves through the orchard, it blows the spray-laden air into the tree canopy, displacing the original air and depositing the spray material on the leaf surfaces Fig. 1.254. Most, but not all, horticultural sprayers use high pressure pumps (1.5–3 MPa) to produce fine droplets that are carried by the air. When low pressure

Figure 1.254. Air blast sprayer.

pumps are used, good atomization depends on high-velocity air shear to produce fine droplets. Typically, air blast sprayers direct the air up into the tree from a ground level machine position. Tanks vary in size from 400–4000 dm^3. They can apply either high or low volumes of material. Coverage will depend upon the machine and tree size. They may direct output to only one side of a row, both sides or cover double or more rows. Nozzles are typically selected to deliver a higher proportion of the discharge to the top half of the tree. This is discussed in Section 1.5.6, Orchard Spraying.

The latest air-carrier equipment uses diffusers of various shapes (turret, vertical slits, two fans–one in the standard position and a second, less powerful one, placed above) to shift the point of spray emission and the auxiliary air flow closer to the vegetation (Fig. 1.270, Section 1.5.10). On this type of equipment, it is possible to use nozzles with the same atomization characteristics across the entire boom.

Good coverage and penetration along with high spraying capacity are general characteristics of the air-blast sprayer. The high annual use rate of these sprayers requires a high level of manufacturing quality. The limitation of these sprayers is their high power requirement, and the potential for increased drift.

Tower and Air Curtain Sprayers

There is an increasing trend today to use tower sprayers when spraying orchard trees. These sprayers are constructed to deliver air and spray material along a narrow vertical band that forms an air curtain extending from near the base of the tree to the top. The nozzles discharge closer to the tree foliage than with the conventional sprayers. This improves spray deposition and coverage over that of the conventional air blast sprayers, which attempt to blow the spray up into the treetops from the ground.

The air curtain is produced in some sprayers by use of multiple, long narrow tangential fans mounted vertically (Fig. 1.255). Other sprayers use an air discharge manifold modified to form a long narrow slot. The objective is the same: to have a more even air spray discharge in the tree, and to reduce the distance between the nozzle and target.

The advantage of the tower sprayer is more uniform spray deposition and reduced drift. This is offset by more expensive construction and reduced maneuverability.

Air-assisted Boom Sprayers

Several models of boom sprayers now are equipped with a fan and air manifold with the specific objective to improve spray deposition and reduce drift. Some models are configured with a horizontal air manifold across the boom just ahead of the nozzles. The air curtain discharged from the manifold blow downward, impinges upon the spray and helps to carry it to the target, providing turbulence for good deposition and reducing drift caused by typical atmospheric wind.

Another approach uses flexible air ducts to conduct air from a central fan to the plants being sprayed. A spray nozzle at the end of the air duct mixes spray with the directed air, which carries it to the target plant, improving coverage and reducing drift.

Aerial Application

Aircraft are used to apply pesticides as sprays or dust, and to seed certain types of crops. A primary advantage of aircraft spraying is the speed with which fields may be covered. Under certain soil or crop conditions, applying chemicals by aircraft may be the

Figure 1.255. Tower-air curtain orchard sprayer.

only possible or reasonable alternative to ground applications. A disadvantage of aerial application is an increased opportunity for pesticide drift.

This method requires highly trained professional pilots and applicators and reliable aircraft equipment. Aircraft may be either fixed-wing craft, i.e., mono-wings or biplanes, or rotary wing types (helicopters). Aircraft sprayers typically have effective spray swath widths of 12–20 m. They fly at speeds of approximately 150–240 km/h at an altitude of 2–6 m while spraying.

1.5.5. Sprayer Components

Pumps

Pumps are classed broadly into two groups: positive displacement pumps, where the output is directly proportional to the pump speed, and nonpositive displacement pumps, where output is inversely proportional to the pump pressure. A piston pump is an example of a positive displacement pump. A centrifugal pump is a nonpositive displacement pump.

The pump in itself does not produce pressure; rather, it is the pump's ability to produce flow against the resistance of the spraying system that creates the pressure in the spray material. The higher the resistance that the pump must overcome, the greater the pressure.

The pump selected for a sprayer must have sufficient capacity to supply the volume needed by the nozzles, the hydraulic agitators, where they are used, maintain desired

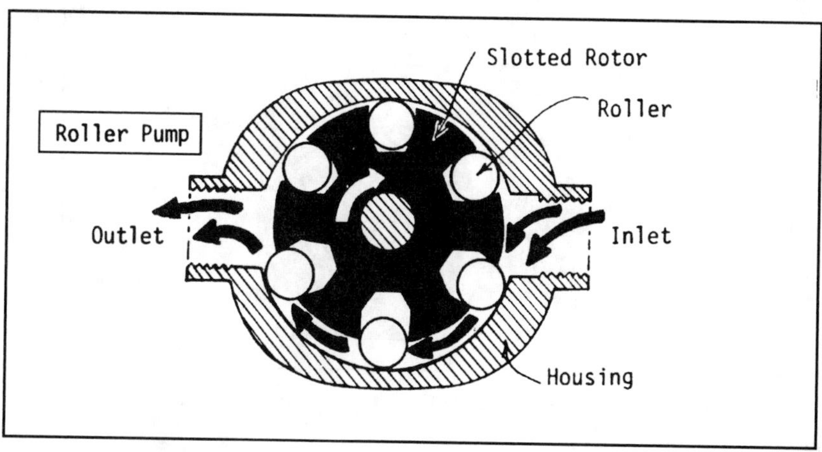

Figure 1.256. Roller pump.

pressure and compensate for wear. Other factors that influence pump selection are: cost, maximum operating pressure, priming ease, resistance to corrosion and wear, and type of power source.

Although there are many types of pumps available, those most used today to apply agricultural pesticides are of four general types: (1) roller, (2) centrifugal, (3) piston and (4) diaphragm.

Roller Pumps

The roller pump consists of a rotor with resilient rollers that rotates within an eccentric housing (Fig. 1.256). These pumps are popular because of their low initial cost, compact size, ease of repair and efficient operation at tractor PTO speeds. They are easy to mount and have good priming characteristics. Moreover, their flow rate (30–120 dm^3/min) and pressure ranges (0–2.0 MPa) are adequate for most spraying jobs. Pump output is determined by the length and diameter of the case, its eccentricity and the speed of rotor rotation. Accelerated wear when pumping abrasive materials is a limitation of this type pump.

Centrifugal Pumps

Centrifugal pumps have become increasingly popular in recent years. They are long-lasting and handle wettable powders and abrasive materials very well. Their high flow rate (250–500 dm^3/min) at low pressure provides plenty of flow for operation of hydraulic agitators in the tank, but they must be driven at higher speeds than is normal for tractor PTO shafts. The impeller is the only moving part and it must operate between 2000 r/min and 4200 r/min to give rated performance (Fig. 1.257). This requires a speed-up drive of belts, gears or a hydraulic motor. Because centrifugal pumps are non-positive displacement pumps, their volume falls off rapidly above 200–300 kPa, and permits the pump output to be controlled without a relief valve. Typical maximum working pressure is about 400 kPa for single stage pumps.

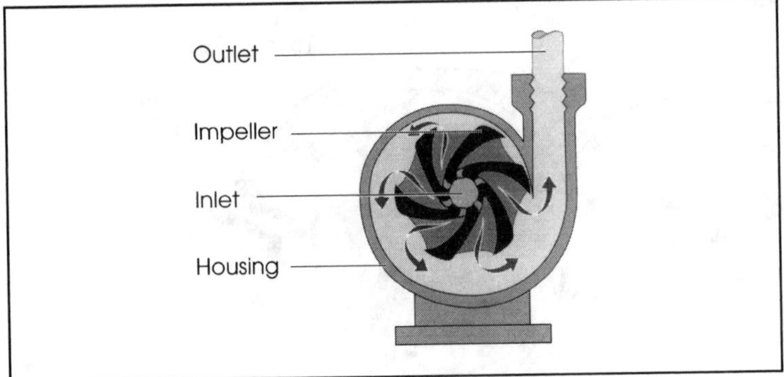

Figure 1.257. Centrifugal pump.

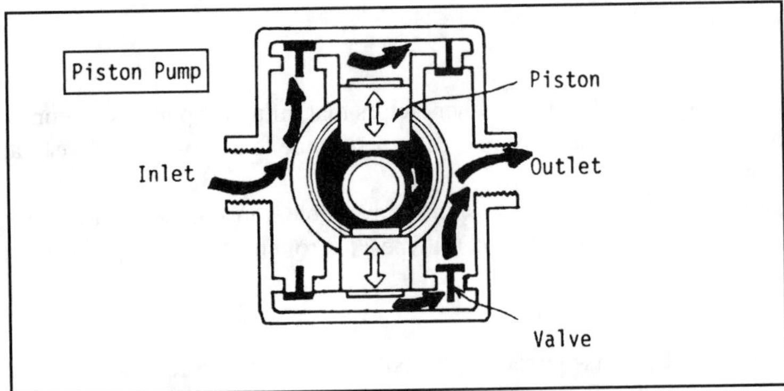

Figure 1.258. Piston pump.

These pumps are not self-priming and must be mounted below the tank (for gravity priming) or equipped with a priming system. Their primary use is for low-pressure applications.

Piston Pumps

These pumps are popular because of their priming ease, high pressure capability and efficient operation at tractor PTO speeds. They are dependable, long-lived and highly adaptable. Most operate efficiently at tractor PTO speeds and are capable of developing high pressures. However, their initial cost is higher than that of roller pumps of comparable capacity.

A piston pump is a positive displacement pump. Its output is proportional to speed and virtually independent of pressure.

With a piston pump, liquid is propelled by a piston moving in a cylinder. The intake stroke draws the liquid in through one valve, and the output stroke forces the liquid out through another valve (Fig. 1.258). Piston pumps should have either an internal or external air chamber (surge tank) to dampen pulsations in the liquid flow associated with each pump stroke.

The small to medium sizes, with capacities of 20–80 dm^3/min, are adequate for most crop spraying.

Larger sizes with capacities to 100–140 dm^3/min are well adapted to separate gas-engine drives or may be driven by a tractor PTO shaft. Horticultural sprayers and professional applicators commonly use these larger sizes.

The large pumps achieve higher flow rates with two to eight cylinders, and the multi-cylinder design produces a smoother flow than pumps with fewer cylinders.

Figure 1.262 shows a typical connection diagram for a piston pump. It is similar to that for a roller pump except that a surge tank is often added at the pump outlet and a glycerine-filled gauge is recommended.

Diaphragm Pumps

Diaphragm pumps are popular in the horticulture and agriculture market because they can handle abrasive and corrosive chemicals at high pressures. They operate efficiently at tractor PTO speeds of 540 r/min and permit a wide selection of flow rates.

Diaphragm pumps have modest initial cost and are lightweight and compact. They are also self-priming and require less horsepower than other pump types with similar flow and pressure ratings.

Like piston pumps, diaphragm pumps are positive displacement pumps–output remains constant with a fixed pump speed, regardless of pressure changes.

The diaphragm pump has its pumping chamber sealed at one end by a flexible membrane or diaphragm. The other end has an inlet and an outlet valve. The diaphragm is connected to a piston. As the piston moves to increase the chamber volume, liquid is drawn by suction through the inlet valve. Return of the piston forces the diaphragm inward, shrinking the chamber and propelling the liquid out (Fig. 1.259). A pulsation

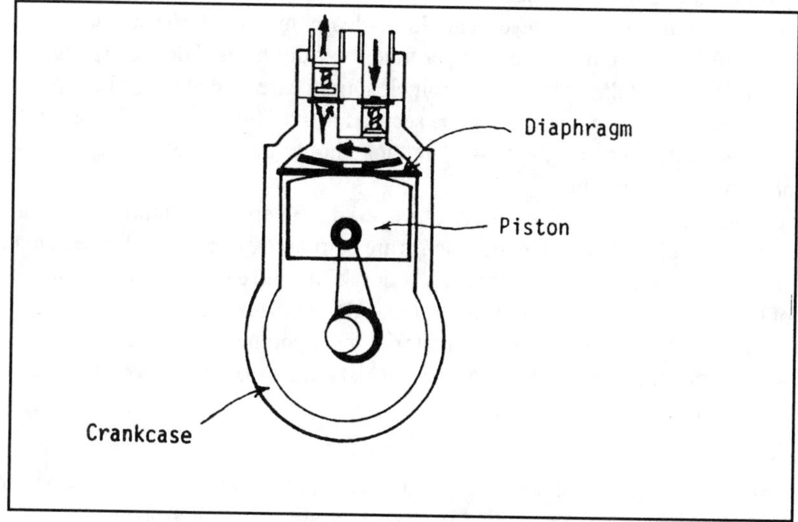

Figure 1.259. Diaphragm pump.

chamber is required to smooth out line pulses. These pumps can operate dry and service is easy when diaphragms need to be replaced.

Pump Capacity and Relative Characteristics

The pump should have sufficient capacity to supply all the nozzles, provide jet agitation when it is used, and compensate for pump wear.

Original pump capacity should be approximately 20% higher than that required for nozzles and agitation flow to allow for pumping loss and wear.

Table 1.55 summarizes characteristics of various pumps. Consider the relative strengths and weaknesses when selecting a suitable pump.

Sprayer Tanks

The function of the spray tank is simply to hold a suitable quantity of spray material for delivery to the spray pump. There are several choices in size, shape and tank materials to be considered.

The size of the spray tank is a compromise between the weight to be transported (resulting in soil compaction) and the time and frequency required for refilling. The sprayer's tank capacity must guarantee a sufficient operating range so that any plot can be treated, distributing the maximum volume ($V\text{max}$), without the need for intermediate refilling. Therefore, taking into consideration the plot of greatest length $l\text{max}$, we must choose a *capacity* that is a multiple of:

$$C\text{min} = 2\,l\text{max}\; b\; V\text{max}\; 10^{-4}$$

where: $l\text{max}$ is plot length (m), b is boom width (m), $V\text{max}$ is max application rate in dm^3/ha.

Tanks also should be corrosion-resistant (choice of material), easy to fill and clean, shaped suitably for mounting and effective agitation, and have adequate openings for fitting pumps and agitation connections.

The tank capacity at various levels should be clearly marked. If the tank is not translucent, it should have a sight gauge to determine the fluid level. The sight gauge should have a shutoff valve at the bottom to permit closing in case of damage. The top opening of the tank needs a fitted cover that can be secured to avoid spills and splashes and large enough to facilitate cleaning the tank. The tank should have a bottom drain to permit easy and complete emptying.

Spray tanks are made from a variety of materials. Fiberglass, aluminum, galvanized steel, stainless steel and polyethylene plastics are commonly used. The choice of material is based primarily on its corrosion resistance, durability and ease of repair. Stainless steel is one of the best materials for spray tanks, but its high cost is usually not justified except for sprayers with high annual use. The material most commonly used today for sprayer tanks is polyethylene plastic. This material resists corrosion, is reasonably tough and durable, and its cost is modest.

Agitation

Agitation is the continuous mixing of materials in the spray tank to ensure a uniform chemical mix. Most materials applied by a sprayer are mixed or suspended in water or

Table 1.55. Pumps for sprayers-characteristics compared

Pump Types	Roller	Centrifugal	Turbine	Piston	Diaphragm	Gear	Flex-Impeller
Materials Handled	Emulsions and non-abrasive materials	Liquids or suspensions.	Most; some turbines may be damaged by abrasives.	Liquids or suspensions.	Most; some chemicals may damage diaphragm.	Oil emulsion, non-abrasive material–not wettable powder.	Most materials and wettable powders.
Relative Purchase Price	Low.	Medium.	Medium.	High.	Variable.	Low.	Low.
Durability	Pressure decreases with wear.	Long life.	Long life.	Long life.	Long life.	Limited life under adverse conditions.	Cannot be operated dry.
Pressure Ranges MPa	0–2.0	0–0.5	0–0.4	0–7.0	0–5.8	0.275–7	0–0.275
Operating Speeds (RPM)	300–1,000	2,000–4,200	600–1,200	600–1,800	200–1,200	300–1,800	300–1,500
Typical Flow Rates (I/m)	135	460	300	230	230	250	115
Advantages	Low cost. Easy to service. Operates at PTO speeds. Medium volume. Easy to prime.	Handles all materials. High volume. Long life.	Can run directly from 1,000 rpm PTO. High volume.	High pressures. Wear resistant. Handles all materials. Self-priming.	Wear resistant. Resistant to most chemicals. Self-priming.	Self-priming. Semi-positive displacement. PTO speeds. Easy to replace. Operates in either direction.	Low cost. PTO speeds. Built-in pressure relief. Operates in either direction.
Disadvantages	Short life if material is abrasive.	Low pressure. Not self-priming. Requires speed-up drive or high-speed hydraulic motor.	Low pressure. Not self-priming. Requires speed-up drive for 540 rpm PTO shafts.	High cost. Needs surge tank.	Needs surge tank.	Cannot handle abrasive materials. Pressure lost with wear.	Low pressure. Impeller subject to wear.

liquid fertilizer, and uniform application depends on a uniform tank mix. The agitation, or mixing, can be produced by a hydraulic jet, mechanical apparatus or air sparging.

Hydraulic jet agitation uses part of the pump flow to create a mixing action in the tank. The pump must be large enough to provide the extra flow volume required by the jet agitator. The increased pump capacity is approximately 5% of tank capacity when the flow is discharged through a standard sparger line. When a venturi-siphon nozzle is used, the discharge from the agitator is increased 2.5 times and the mixing action can be obtained with increased pump capacity of only 2%–3% of tank capacity.

Mechanical agitation is produced by paddles or propellers in the bottom of the sprayer tank. These are driven mechanically by the power source for the pump, or by a 12-volt electric motor connected to the tractor's electrical system. Mechanical agitators are often used on sprayers with piston pumps because jet agitation requires a larger pump and cost increases rapidly with the size of the piston pump.

An air sparger is sometimes used to agitate spray mixes in systems that use compressed air. It is similar in design to the hydraulic jet agitator, but the air sparger uses air from a compressor instead of fluid from the pump.

Strainers and Screens

A plugged nozzle is one of the most frustrating problems that farmers experience with sprayers. Strainers and screens usually are of three types on an agricultural sprayer: basket or tank-opening strainers, line strainers and nozzle strainers.

A *basket strainer* in the tank opening is the first screen that material will pass through. This is usually a coarse (16–20 mesh) screen. This coarse screen prevents leaves, debris and large objects from entering the tank.

A *line strainer* is the most critical strainer of the sprayer. It usually has a screen size of 16–80 mesh, and is positioned between the tank and ahead of the pump to remove material that would damage the pump.

Nozzles are the third place screens are located. Small-capacity nozzles must have screens to prevent plugging. Typically 50–100-mesh screens are used (Fig. 1.260). An

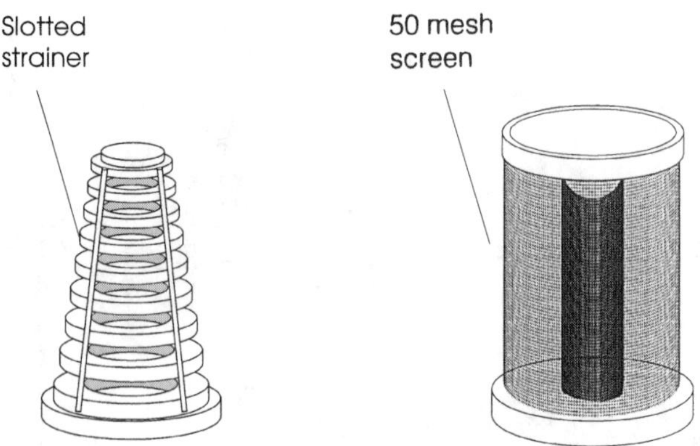

Figure 1.260. Nozzle strainers.

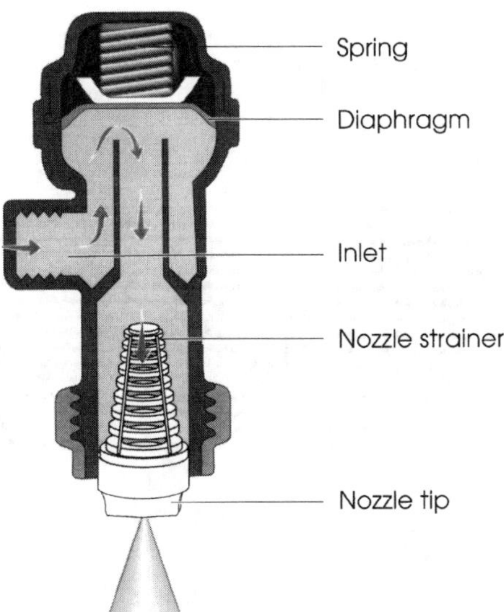

Figure 1.261. Diahragm check nozzle.

alternative to the nozzle screens is to put a second suitable line strainer in each section of the boom.

Diaphragm check valves are another nozzle feature that facilitates spraying. The diaphragm check (Fig. 1.261) is less restrictive than ball check strainer, prevents nozzle drip when the boom is shut off and allows nozzles to be serviced in the field without boom leakage.

Plumbing and Controls

Pump Connections

The typical nonpositive displacement pump (i.e., centrifugal, et al.) does not produce pressures that are high enough to require a pressure relief valve. Pressure is controlled by throttling the pump output; zero pressure at full flow to maximum pressure of no flow. Lack of suitable valves to completely control the discharge (shut off all output) is usually the problem when difficulties are experienced in controlling these pumps. Typical connections are as shown in Fig. 1.262, but with the relief valve, bypass line and surge tank omitted.

In contrast, positive displacement pumps continue to deliver a constant output for each revolution of the pump. As pressures can become very high, a pressure relief valve and bypass line is required to avoid damage to sprayer components. Pumps with distinct pulsations (i.e., piston and diaphragm) may also require a surge tank and dampening on gauges, as shown in Fig. 1.262.

Pressure Gauge

The pressure gauge should be a type intended to measure liquid pressure and have a range of $1\frac{1}{2}$–2 times the expected maximum pressure. It is impossible to calibrate a

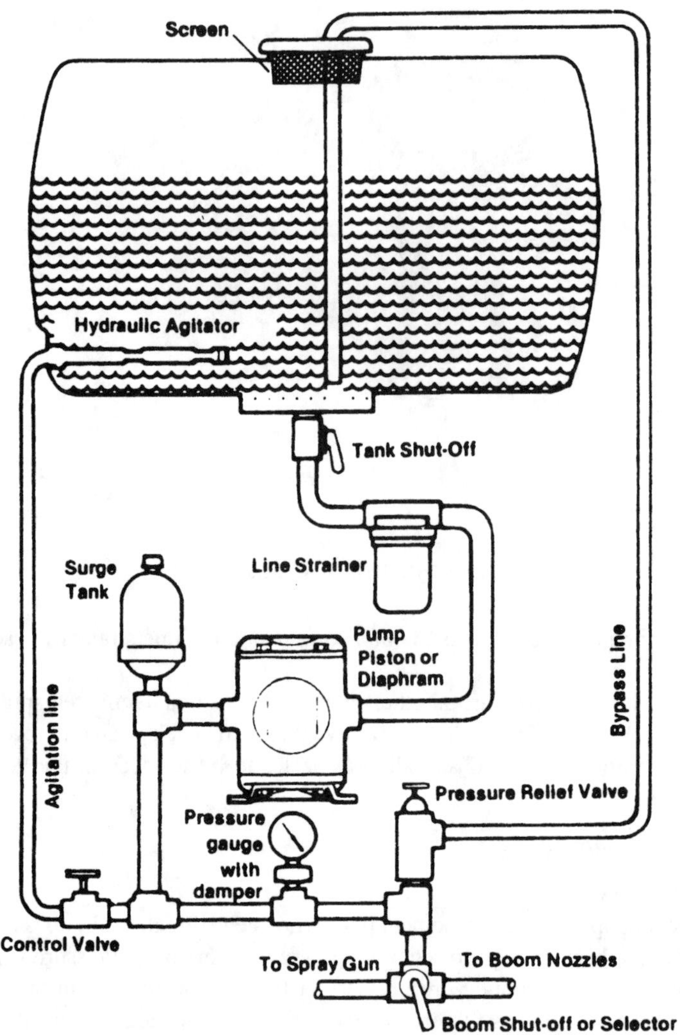

Figure 1.262. Typical sprayer plumbing for positive displacement pumps.

sprayer properly if the pressure gauge does not read correctly. A liquid-filled gauge (oil or glycerin) is recommended as it is more reliable than the dry type. The pressure should be measured in the boom near the nozzles.

Control Valves

Valves to control and limit pressure and to control flow volume and direction are vital to a proper functioning sprayer.

Relief valves limit maximum pressure of the sprayer and prevent damage. In addition to limiting pressure, the unloader valve *unloads* the pump (full flow at low pressure) to save energy and wear when the sprayer is idling and not calling for sprayer discharge. Throttling valves are used to control the amount of flow volume and boom selector valves control the active section of the sprayer boom. Flow control valves are available as manual or as electric valves.

There is a great variety of control valves available, and the catalogs of spray equipment manufacturers are recommended for details.

Control Monitors

Control monitors may be either nozzle monitors or system monitors. The nozzle monitor will alert the operator to a malfunctioning or plugged nozzle so corrections can be made promptly and skips in the field avoided.

System monitors detect the operating conditions of the total sprayer. They are sensitive to variations in travel speed, pressure and flow rate. Some monitors also control the flow rate and pressure automatically to compensate for changes in speed or flow. Flow compensation is usually done by changing the pressure setting. Monitors are helpful in precise chemical application work and may result in better pest control, more efficient distribution and reduced chemical costs.

Nozzles

The nozzle performs three functions on the sprayer: regulates flow; atomizes the mixture into droplets; and disperses the spray in a specific pattern.

Nozzle flow rate is a function of nozzle type, orifice size and pressure. The working pressure p (kPa) is related to the nozzle emission capacity q according to the following equation:

$$q = md^2\sqrt{p} \qquad [\mathrm{dm^3/min}]$$

where m (0.15–0.65) is an efflux coefficient related to the construction characteristics of the nozzle and d (mm) is the orifice diameter. Hence the distributed volume may be varied by changing the nozzle (i.e., by varying m and d) or by adjusting the working pressure p.

Manufacturers' catalogs list nozzle flow rates and various pressures and discharge rates per hectare at various ground speeds. Although the discharge rate increases with pressure, it is not linear. The discharge rate for any nozzle is proportional to the square root of the pressure.

$$\frac{q_1}{q_2} = \frac{\sqrt{p_1}}{\sqrt{p_2}}$$

This characteristic is used in many spray controllers to achieve a constant spray rate as sprayer speed varies. However, large changes in nozzle flow are achieved most effectively by changing the orifice diameter. The nozzle orifice diameter, its operating pressure and other design features influence the size of the droplets that are dispersed. These characteristics are discussed in more detail in Section 1.5.3 under the heading "Characteristics of Droplets."

The shape of the spray pattern is the third function that nozzles perform. Each pattern has two characteristics, the spray angle and the shape of the pattern. The typical spray

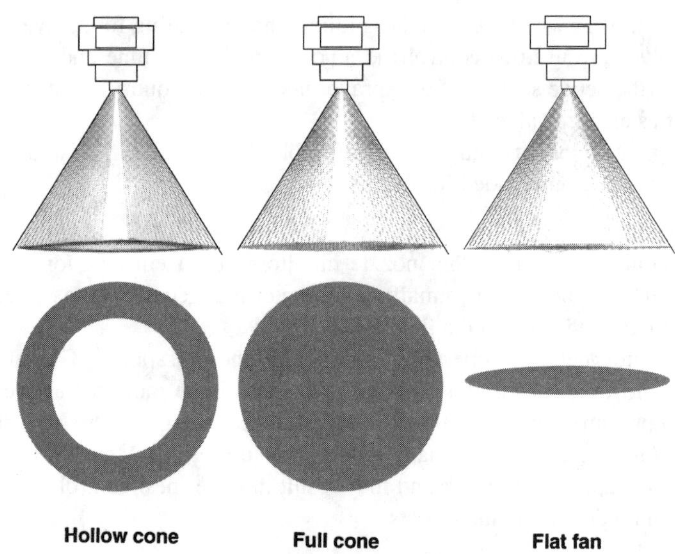

Figure 1.263. Basic nozzle spray patterns.

angle is 65°, 80°, 110° or 120° to 130°. The shape of the pattern can be reduced to a basic three: the flat fan, the hollow cone and the solid cone (Fig. 1.263).

Flat Fan Nozzles

Flat fan nozzles are used widely for broadcast spraying of herbicides and some insecticides. Spray droplets form a flat fan shape as they leave the nozzle. Because the fan shape has a tapered edge with less spray material at the outer edges, adjacent fan patterns must be overlapped to give uniform coverage (Fig. 1.264). For maximum uniformity, fan overlap must be approximately 30%–50% at the target point. Nozzles are usually spaced 50 cm apart on the boom. The usual pressure is 0.1 to 0.4 MPa. Lower pressures produce large droplets and reduce drift. Higher pressures produce the opposite.

Several manufacturers have recently introduced modified flat fan nozzles that have *low drift* features. Most of these nozzles have a primary orifice, an internal mixing chamber to allow fine droplets to recombine into larger drops and a final secondary (discharge) orifice.

Even Flat Fan Nozzle

The even flat fan spray nozzle is similar to a regular flat fan nozzle but applies spray materials uniformly across the entire pattern rather than tapering off at the edges. These nozzles are used mainly for banding herbicides over row crops and have spray angles of 40°, 80° or 95°. This nozzle is not suitable for broadcast applications.

Flooding Fan (Deflector) Nozzle

Flooding fan nozzles produce a wide-angle, flat spray pattern and are used for applying herbicides and chemical fertilizer mixtures (Fig. 1.265). Their wide angle–110°–130°—

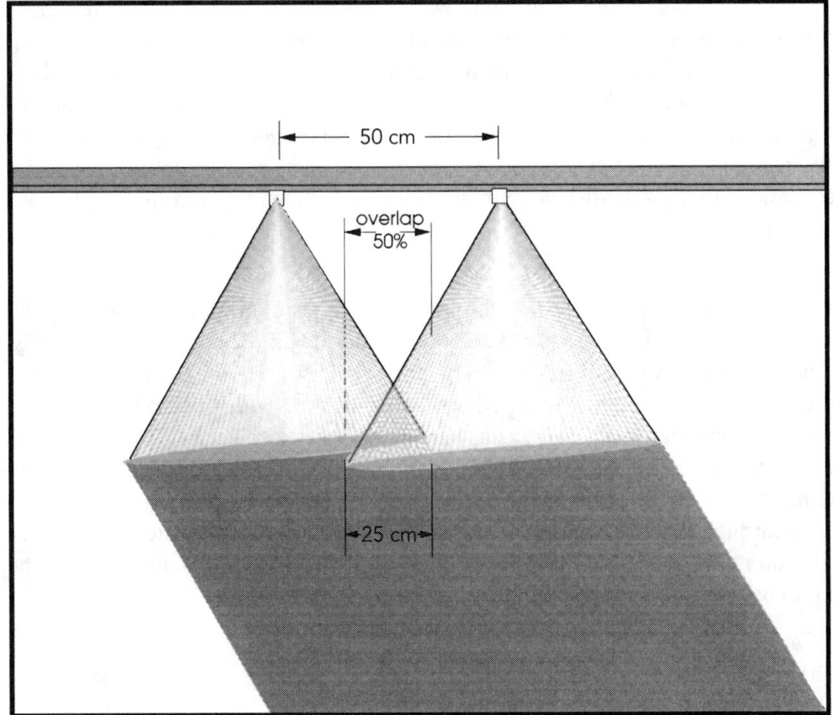

Figure 1.264. Overlap and offset of flat fan nozzles.

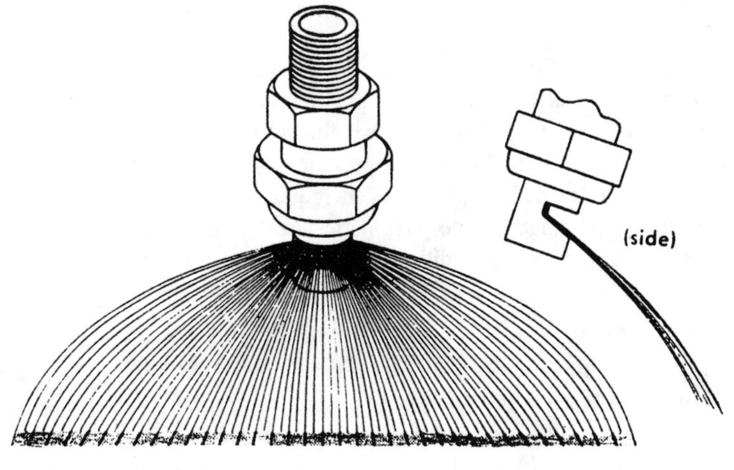

Figure 1.265. Flooding (deflector) flat fan nozzle.

allows wide spacing on the boom and lower boom height. Their large orifices produce large droplets and so reduce drift and clogging. They give best results in a pressure range of 0.07–0.25 MPa. The pattern is not as uniform as that of regular flat fan nozzles, and the width of the pattern changes more with pressure than it does with the regular flat fan. Coverage uniformity is improved when the nozzles are mounted to give 100% overlap (100% overlap is double coverage, i.e., each nozzle spray reaches to the center of each adjacent nozzle). Nozzle discharge may be directed at any angle from straight down to straight back.

Hollow Cone Nozzle (Disc and Core Type)

The hollow cone nozzle produces the second of the basic spray patterns, with the liquid concentrated on the outside edge of a conical pattern. Hollow cone nozzles usually are used to apply insecticides or fungicides to crops when penetration or complete coverage of the leaf surface is important. Drift potential is high with this nozzle because the droplets are very small. These nozzles usually operate from 0.25–1.0 MPa, depending on the nozzle type and the material being applied. The nozzle consists of a disc and a core that can be easily changed. The core gives the fluid a swirling motion before it is metered through the disk orifice, which results in the hollow cone pattern. Disc and core nozzles do not produce a uniform distribution for broadcast application and are better suited to banding or directed spraying.

Full Cone Nozzles

The full or solid cone is the third of the basic spray patterns. The nozzles produce a swirl and a counter swirl inside the nozzle that result in a full cone pattern. Full cone nozzles produce large, evenly distributed droplets and high flow rates. A wide full cone tip maintains its spray pattern through a range of pressures and flows. It is a low-drift nozzle, often used to apply soil-incorporated herbicides.

Pneumatic Nozzles

Pneumatic nozzles differ from hydraulic nozzles in that they combine two fluids, air and spray solution. Each fluid has a separate circuit that controls pressure and flow rate independent of the other fluid. The two fluids are combined inside a special tip (Fig. 1.266). By proper choice of control pressures, one pneumatic nozzle can emit a wide range of flow rates and droplet sizes. It requires a number of hydraulic nozzles to equal the discharge range of one pneumatic nozzle. To retrofit an existing sprayer with pneumatic nozzle requires additional air lines, an air compressor, controller and expense.

Rotary Nozzles

The rotary nozzle or *controlled droplet applicator* (CDA) has been used for years in aerial application but is a relatively new device for ground applications. The rotary nozzle atomizes the spray by using a rotating spinner and centrifugal force rather than forcing material through a hydraulic orifice. Rotary nozzles are of two basic types: the spinning disc, or cone, suitable for low flow (Fig. 1.267), and the spinning cage or screen basket, used for flows of 0.5 dm^3/min or greater. An attractive feature of the rotary nozzle

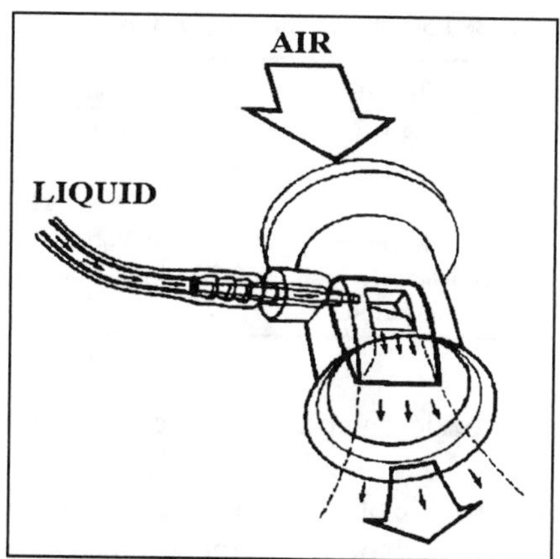

Figure 1.266. Pneumatic nozzle.

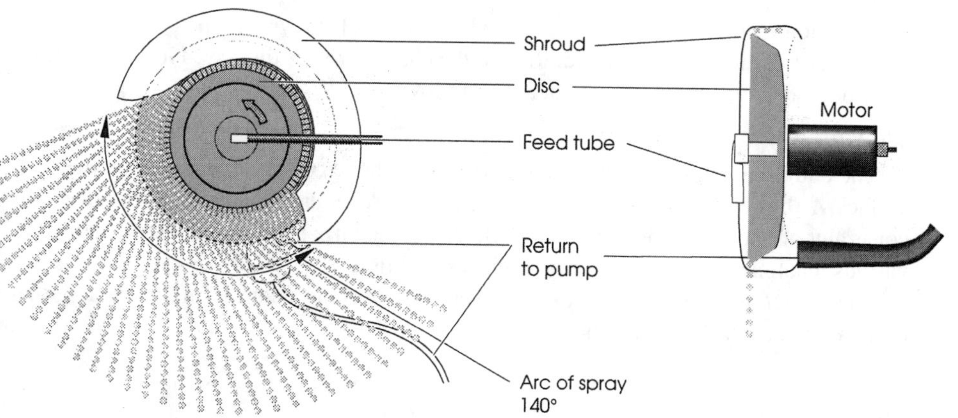

Figure 1.267. Vertical disc CDA.

is that it forms relatively uniform droplets. This controlled droplet size minimizes drift and evaporation of the small droplets, and because there are few large droplets, reduces chemical loss from runoff and ineffective placement.

The droplet size depends on the rotational speed of the spinning cone or disc. 2000 n/min produces droplets of approximately 200 μm. Higher speed produces smaller droplets. By controlling the droplet size, the nozzle can be used to apply insecticides, herbicides, fungicides or growth regulators.

Nozzle Materials and Wear Life

Nozzle life depends upon many factors: the hardness of the nozzle material, the type and size of the nozzle, the operating pressure, and the abrasiveness of the spray material.

Table 1.56. Nozzle life compared to brass

Material	Life
Aluminum	Same
Plastic or Nylon	2 times
Stainless Steel	3.5 times
Hardened Stainless	10 to 20 times
Ceramic	75 to 120 times
Tungsten Carbide	150 to 200 times

In general, the harder the nozzle material, the longer its expected life. Also, the higher its cost.

Table 1.56 shows the relative life of various material used for nozzles as compared to brass, a soft inexpensive material. Nozzle life for soft material may be as short as 15–20 hours under poor conditions–check flow rates often.

1.5.6. Choice of Equipment

The choice of equipment must be based on a careful analysis of the type of farm, its characteristics and its needs. In addition to the characteristics discussed in Section 1.5.5 the work capacity should be considered as follows:

Work Rate

The equipment must be capable of performing the necessary treatments on the crop surface A (ha) within the useful time available UTA (hours): this is the most stringent constraint on the choice of equipment. Therefore, the farm must choose from among the various machines that meet this preliminary condition.

The *actual work rate*, AWR, of a sprayer with working width b (m) operating at an optimal mean speed v (km/h) is:

$$\text{AWR} = 0.1\, b\, v\, \eta \qquad [\text{ha/h}]$$

The *field efficiency* η (0.3–0.95) takes into account: the partial overlap of machine passes during treatment, slowdown at turns and time needed for filling, mixing and transfer. Its value can range from $\eta < 0.4$ (difficult conditions) to $\eta \geq 0.9$ (LV and ULV treatments) on plots close to the farm center. For normal-volume treatments on plots that are not too distant from the farm center, we can assume $\eta \approx 0.6$–0.7.

Setting AWR equal to the actual work rate needed by the farm AWN $= A$ /UTA we can determine the required working width b of *machines for the treatment of herbaceous crops*. Having established, *a priori*, approximate values for η and v (4–12 km/h) based on the type of farm and the equipment in question, b turns out to be:

$$b = (10A)/(v\, \eta\, \text{UTA}) \qquad [\text{m}]$$

In the case of *machines for the treatment of tree crops*, because the working width is fixed by the distance between trees b_{ir} , we can calculate the necessary equipment characteristics, determining v (2–12 km/h) and η (and hence the tank size and performance characteristics):

$$\eta v = (10A)/(b_{ir}\text{UTA}) \qquad \text{[km/h]}$$

Fans and Capacity

Fans provide the air jet in air-assisted equipment. The air flow produced must displace and replace the existing air within the foliage. Hence, the required flow rate Q_a depends on crop size and can be calculated by:

$$Q_a = 1000\, h_c\, w_c\, v \qquad [\text{m}^3/\text{h}]$$

where h_c (m) is the maximum height of the canopy, w_c (m) is its maximum width, v (km/h) is the forward speed of the sprayer.

Axial fans, which deliver an airflow characterized by a high volume/speed ratio, are suitable for penetration of orchard foliage. Centrifugal fans, which deliver a lower volume of air at high speed, are more suitable for twin-fluid nozzles and knapsack air-assisted equipment.

Crop Spraying

Among hydraulic nozzles, those of the fan type are adequate to meet most *crop spraying* requirements. Therefore, if only a single type of nozzle can be used, this is the family of choice. Cone nozzles are better suited for treatments whose action is by contact, and for operations on particularly thick vegetation. Flooding (deflector) nozzles are used for treatments on bare soil, especially for the distribution of products in suspension. Table 1.57 indicates some typical applications.

Table 1.57. Nozzle applications

	Nozzle Type					
Application/Use	Full Cone	Flat Fan	Hollow Cone	Flooding Fan	Pneumatic	Rotary
Bare soil	A (0.15–0.2 MPa)			A (0.1–0.2 MPa)	B	A (low speed)
Contact herbicide Post-emergence	B (0.2–0.4 MPa)		A (0.3–0.4 MPa)		A	B (high speed)
Systemic herbicide Post-emergence	A (0.15–0.2 MPa)	B (0.15–0.3 MPa)			B	A (low speed)
Herbicide on row	B (0.15–0.3 MPa)	A (0.15–0.25 MPa)	B (0.3–0.4 MPa)			B
Contact fungicide and Pesticide	B (0.2–0.4 MPa)		A (0.3–0.4 MPa)		A	B (high speed)
Systemic fungicide and Pesticide	A (0.2–0.3 MPa)		B (0.3–0.4 MPa)		B	A (medium-high speed)
Orchard spraying (air-assisted)			A (0.5–2 MPa)		A	

Note: A = Excellent; B = Acceptable.

Rotary nozzles, which permit a wide range of variations in droplet diameter, can adequately meet a variety of crop spraying requirements, even though they are generally characterized by poor penetration into vegetation. In the case of LV and ULV treatments, it is necessary to use rotary or pneumatic nozzles. The spray produced by the pneumatic nozzle penetrates even thick vegetation very effectively, due to the action of the airflow whose speed must not be excessive.

To achieve a sufficiently uniform spray pattern, it is necessary to keep the boom at an optimal height above the target, taking into account the spacing between nozzles, the degree of overlap between the jets produced and the spraying angle. Boom height is calculated as:

$$h = 0.5 \cdot [1/tg(\gamma/2)] \cdot [d/(1-(s/100))] \qquad [\mathrm{m}]$$

where d (m) is the distance between two adjacent nozzles; γ is the spraying angle; $s(\%)$ is the degree of overlap between jets ($s = 40\%$ for fan nozzles; $s = 25\%$ for cone nozzles; $s = 100\%$ for flooding (deflector) nozzles).

Orchard Spraying

In *orchard spraying*, when operating with hydraulic spraying equipment, cone nozzles generally are used. On traditional sprayers with a semicircular boom placed close to the fan, the nozzles must have different atomization characteristics depending on their position on the boom, and varying distance from the target. More specifically, low-capacity nozzles are chosen for the lower part; nozzles that produce fine droplets which are easily transportable by the air jet are selected for the central part, and high-capacity nozzles with a small spraying angle that produce medium-coarse droplets are placed on the upper part. The capacity should be proportioned to deliver about 2/3 of the discharge to the top half of the tree canopy and 1/3 to the lower half when using a traditional orchard sprayer.

1.5.7. Spraying Techniques

Drift reduction

The ideal objective of spraying is to put all of the chemical on the target. In practice, this is almost impossible to achieve.

Drift can occur in two ways. *Vapor drift* occurs when highly volatile chemicals vaporize after being applied to a target area. Vapors then move to another area, where they may cause damage. The amount of vaporization depends on the temperature, the relative humidity and the characteristics of the pesticide formulation used. Choosing the correct formulation can be very important in reducing vapor drift.

Droplet drift is the physical movement of spray particles away from the target area. Many factors, including wind velocity and direction, height and direction of spraying, temperature and relative humidity, and temperature inversions, will affect drift, but the most important is droplet size. Small droplets fall through the air more slowly than large droplets. Figure 1.246 shows how far droplets of various size can be carried by a wind as they fall from a 50-cm height. Drift can be minimized by controlling the following factors:

1. Use the lowest spraying pressures recommended for the nozzles in use.
2. Keep the boom height as low as possible while still maintaining uniform coverage.
3. Use ample carrier, and larger droplets.
4. Use nozzles resistant to drift, such as flooding, low-pressure flat fan, Raindrop, or the newer low-drift nozzles.
5. Service nozzles that are plugged to avoid fogging.
6. Use chemical formulations that are low in volatility and avoid spraying on hot days.
7. Use an antidrift additive to increase spray viscosity and droplet size.
8. Spray when winds are below 8 km/h. Never spray in winds that exceed 16 km/h.
9. Avoid spraying when there is a temperature inversion (i.e., air temperatures increase with elevation).
10. Use special sprayers designed to minimize drift, i.e., shielded nozzles, air curtain spray assist, electrostatic-charged sprays.

Other techniques to control drift are discussed in Future Trends, Part 10.2.

Chemical Mixing and Disposal of Excess Pesticide

All agricultural chemicals should be handled carefully to avoid accidental spills and contaminations. Because some minor spillage and runoff of sprayer wash water is almost inevitable when working with pesticides, it is wise to load and clean the sprayer on a mixing-loading pad. The pad will contain any spills and rinse water and allow it to be pumped into a holding tank for later application or disposal. The pad can be constructed of sealed concrete or made from appropriate fabric when portability is desired.

To minimize disposal problems, buy and mix only the quantities of chemicals that you need.

Cleaning Equipment

Rather than return the empty sprayer to the loading pad for washing and cleaning, a practice that is gaining acceptance is to carry an auxiliary tank of clean water on the sprayer that can be used to wash down and rinse the sprayer in the field. This leaves the diluted spray material in the sprayed field and permits the sprayer to return to the pad clean, thus eliminating an accumulation of chemical wash water that would need to be disposed of later.

When doing in-field washing, be careful to vary the washout location in the field to prevent a chemical concentration buildup in any particular place. Also, never drain surplus pesticide or rinsate where it can run off into streams, lakes or other surface water, or where it can contaminate wells and groundwater.

Injection metering of spray chemicals is another approach to solving many of the handling and disposal problems of surplus spray tank mix and rinsate (Fig. 1.269). This is discussed in Section 1.5.10 under, "Future Trends."

Closed-Transfer Systems

In the closed-transfer system, chemicals are transferred from the concentrate containers to the sprayer by a pump and suction line. Because these closed-transfer systems are safe and convenient, relatively easy to build and adaptable to existing sprayers, we can expect to see their use increase in the future.

Ergonomics and Operator Safety

All pesticides should be handled and treated with care and respect. Minimize your exposure to pesticides when mixing and spraying by using protective clothing and equipment. Wear a long-sleeved shirt, long pants, rubber gloves, rubber boots, a hat that protects the neck, goggles and a respirator. Research has shown that ordinary denim cloth in good condition gives reasonably good protection and can be washed and reused. A closed cab provides the best operator protection.

Never smoke or eat while working with pesticides, and wash hands well before eating or using the toilet. Carrying a container of clean water on the spray rig is a good safety measure. If you spill a pesticide on yourself, wash it off IMMEDIATELY and change clothes. Know the signs of pesticide poisoning and where, when and how to get help from a poison center. Keep emergency phone numbers clearly posted near the phone where they are easily found if needed.

1.5.8. Sprayer Calibration

Many methods are available for calibrating sprayers, but some are easier to use than others. Two methods that work well are given here. All calibration techniques rely on three variables that affect the amount of spray applied per hectare: the nozzle flow rate, the ground speed, and the effective width of each nozzle. Any change in one of these variables will have a direct effect upon the others.

Sprayer Ground Speed

Knowing the travel speed and maintaining a constant rate are crucial to good calibration and application. To reduce errors, it is advisable to check tachometer and throttle setting speeds by timing the travel over a measured distance. Do this speed check in the field similar to the one to be sprayed, with the sprayer at least half full of water and using the throttle setting and gear that will be used for spraying.

$$\text{Speed (km/h)} = \frac{\text{distance (m)} \times 3.6}{\text{travel time (s)}}$$

Timed-flow Method for Calibrating Boom Sprayers

This is an excellent method that gets you close to the needed nozzle size and flow rate quickly and easily.

This calibration procedure allows you, the grower, to make the management decisions of: the type of nozzle to use (based upon experience and preference); the rate per hectare to apply; the travel speed of the sprayer; and the nozzle spacing on the boom.

Using these variables in the equation below, the size of the nozzle is calculated and then identified in the appropriate nozzle catalog.

$$N\ (\text{l/min}) = \frac{A\ (\text{l/ha}) \times S\ (\text{km/h}) \times W\ (\text{cm})}{60{,}000}$$

N = nozzle flow (l/min)
A = application rate (l/ha)
S = speed (km/h)
W = nozzle spacing on boom or spray width (cm)

With the correct nozzles installed on the boom, conduct a flow check to see if the output is correct. Set the throttle at the setting for spraying. Adjust the boom pressure to conform to the specification given in the catalog for the selected nozzles. Use a measuring cup to determine the flow rate (l/min) and make any necessary adjustment in pressure to get the flow rate exactly correct. Check the flow rate for each nozzle and correct any that are more than ±5% off the average flow of all boom nozzles.

If the viscosity of the spray mix is different from that for water, it will affect the calibration flow. Use a correction factor given in nozzle catalogs. This will allow the viscous flow rate to match the water based flow as determined in the catalog.

Calibration Flow Check

Milliliter flow = Liters/hectare applied.

This method of checking calibration is very easy and can be used to fine tune a sprayer quickly, but it does require driving a distance in the field.

This is not a particularly good method for determining new nozzle sizes when first setting up a sprayer. With this method, as with any, check that the nozzle flow rate is uniform and correct any nozzles that are more than ±5% from the average. Then proceed as follows:

1. For broadcast applications, determine the distance, in cm, between nozzles. For banded applications, determine the bandwidth in centimeters.
2. Locate this width in Table 1.58 and read off the corresponding course distance.
3. In the field to be sprayed, mark off a course of the proper distance. You might set permanent markers in the field or paint fence posts to make this step easier next time.
4. Fasten a container to one nozzle on the sprayer so that it will catch all of the discharge from that nozzle. This assumes that all the nozzles are uniform in their discharge, as described above.
5. Start a distance back from the beginning of the course to get up to operating speed, then turn the sprayer ON at the beginning of the course and OFF at the end of the course.
6. Measure the volume collected in the container in milliliters. Do this several times to be sure the results are reliable. You may average the output from several nozzles to get a more reliable reading.
7. MILLILITERS COLLECTED = LITERS PER HECTARE BEING APPLIED.

REMEMBER: When testing, use the same speed and pressure that you will use in the actual spraying operation. If you change either the ground speed or the spray pressure, the l/ha will change and you will need to recalibrate the sprayer. This is true for changes in nozzle types, as well.

Table 1.58.

Milliliter flow = l/ha	
Nozzle Spacing or Row-Band Width W (cm)	Dist. (D) Meters (for 1/1000 ha)
10	100
20	50
30	33.3
40	25
50	20
60	16.7
70	14.3
80	12.5
90	11.1
100	10
110	9.1
120	8.3

Note: W = nozzle spacing on boom cm; D = distance in meters; ($10\,m^2 = 1/1000$ ha) $10{,}000\,m^2$ = ha; Drive the course distance (meters); Collect from one (representative) nozzle; No. of ml = l/ha (application).

1.5.9. Granular Applicators

The bulk of agricultural chemicals applied in granular form are fertilizers. The granular pesticides are a small proportion of total pesticides. Nevertheless, granular pesticides offer some distinct advantages over sprayed liquids. Some of the advantages are: granules are dry and premixed so hauling water to the field is not required; application equipment is simpler and less expensive; there are fewer drift problems with granules than with spraying; granular materials are generally safer to use than sprays; and accidental spills are far easier to clean up than liquid spills.

The disadvantages are: materials must be kept dry in storage; they are more bulky to store and handle; and they are generally more expensive than other formulations.

Granular agricultural chemicals cover a wide spectrum in size from very coarse (size of small grains) to dust particles (less than 100 microns). Granular pesticide applicators fall into two major types: gravity or drop type, and pneumatic or air driven applicators. Fertilizers are usually applied by a spinner spreader and are not discussed here.

Metering Devices

Gravity flow devices are the most common and cheapest (Fig. 1.268). They consist of a variable size opening in the bottom of the hopper. Material flows through the gate by gravity into a drop tube or diffuser for uniform ground application. An agitation in the hopper keeps material from bridging and blocking flow. Gravity flow metering devices are more sensitive to changes in applicator travel speed, decreasing application rate as speed increases.

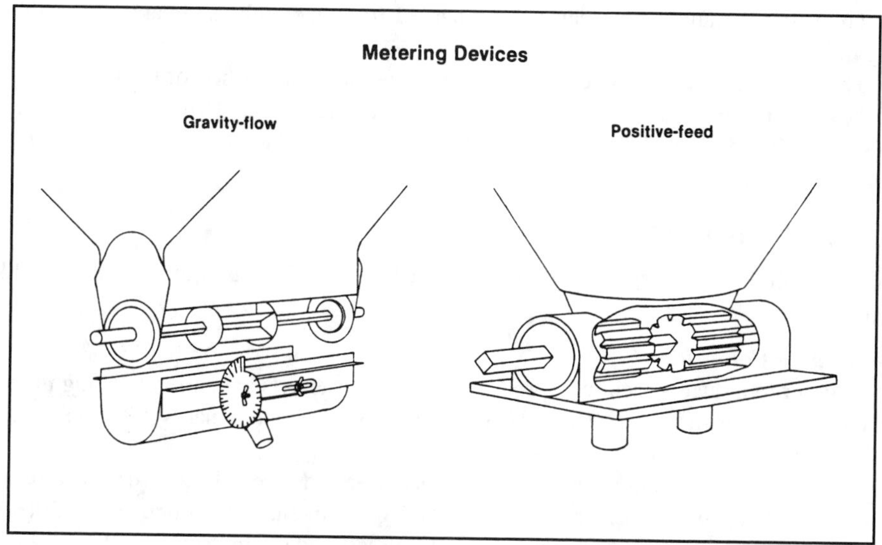

Figure 1.268. Granular metering devices.

Positive feed devices have an auger or a notched metering wheel that rotates in the bottom of the hopper (Fig. 1.268). The amount of metered material depends upon the number, width and depth of the exposed notches, and/or the speed of rotation of the metering rotor. Positive feed metering is more expensive but gives a more consistent flow of material with minimum effect due to variations in travel speed.

Calibrating and Using Granular Applicators

Manufacturers' instructions should be carefully followed in caring for, calibrating and using any granular applicator. Details will vary depending upon the make and model of the equipment. However, the flowability of the granular material affects all applicators. Granulates flow at different rates depending upon size, density, type of material, temperature and humidity. The same material made by different manufacturers, or used on different days, etc., may have widely different flow rates. Also, individual units on row applicators need to be calibrated independently, even when set the same.

Application Rate

Granular rates usually are indicated as *kilograms of active ingredient per hectare* (kg AI/ha). The kilograms of active ingredient per kilogram of product as given on the label is used to determine the kilogram of actual product to apply per hectare, using the formula:

$$\text{kg of product per ha} = \text{kg AI/ha} \times \frac{100\%}{\%\ \text{AI of product}}$$

With the rate of kilograms per hectare established (for either broadcast hectare or band row hectare) the metering device is set to deliver this amount. Calibration is usually done by collecting the amount of material discharged from each tube and carefully weighing it.

The weight of the product collected must be the same proportion of the rate (kg/ha) as the test run area (width × length) is of a treated hectare. Any metering devices that are more than 5% off from the desired rate per hectare will need to be adjusted accordingly and rechecked.

1.5.10. Future Trends

The evolution of the spraying sector is shaped by two distinct requirements in world agriculture:

Developing Countries

In order to reduce agricultural product losses, great efforts are currently being made to promote the diffusion of chemicals, and to develop equipment that is suitable for and economically accessible to small farms.

Difficulties with water transport and supply have led researchers to develop manual sprayers capable of performing LV and ULV treatments. In particular, hand-held battery-powered spinning disc sprayers, as discussed under hand sprayers (Fig. 1.252, Section 1.5.4), permit accurate control of distributed volume and droplet size, which can be adapted to varying needs. The liquid flows onto a disc driven by a low-current electric motor. These systems, designed to distribute $V \leq 10$ dm^3/ha, are extremely lightweight and easy to handle, with a tank capacity of $C \leq 3$ dm^3.

Electrostatic charging of spray has been applied to equipment of the above type, and to conventional manual sprayers with hydraulic atomization. Flashlight (stick) batteries provide the energy supply for a high-voltage generator of small size and very low cost, suitable for use on manual equipment. The advantages of electrostatic charging, on these systems, are particularly evident in ULV treatments on small-sized, well-spaced crops.

The use of lightweight two-stroke engines has enabled the development of knapsack air-assisted sprayers. The air-stream is used both to transport the droplets, generated by a rotary or hydraulic nozzle, and for the pneumatic atomization of the liquid. These systems, despite the problems of high costs and lack of skills and spare parts, make it possible to exploit the advantages of air-conveyance for treating thick vegetation even when using manual equipment.

Developed Countries

One of the goals of advanced agricultural engineering development is to reduce pesticide use and dependence and to reduce the emissions of pesticides into the environment.

The concept of *precision spraying* makes it possible to instantaneously change and adapt the chemical application to the needs of the target that is actually being sprayed.

For tree crops, profile spraying systems have been developed based on signals received from a laser or ultrasonic scan of the target location. From the information collected, profiles are constructed and solenoid valves control nozzle flow, emitting spray only where a target is sensed.

Global positioning system (GPS) technology is now available to automatically determine field position. Maps indicating problem areas to be treated are compiled

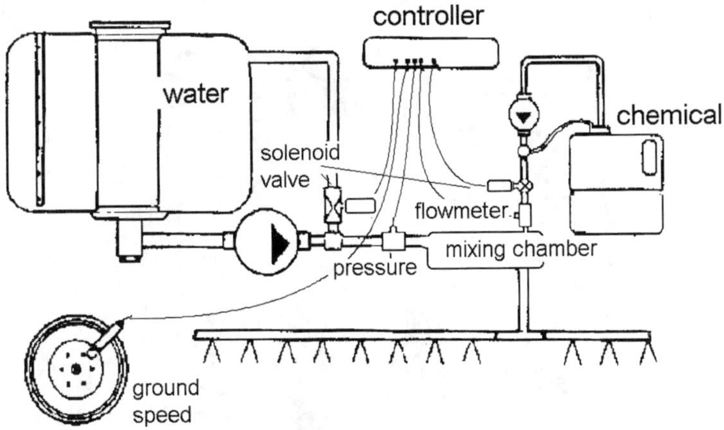

Figure 1.269. Direct injection system.

beforehand. Machine position from GPS data can then be used to make the required treatment.

The GPS technology is already a great asset for locating points for aerial application work, and recording the track record of the actual chemical application machine. And its use in ground operated machines is growing rapidly.

To avoid the need for beforehand mapping of infected areas, systems are now being developed that will identify weeds "on-the-go" by image analysis or reflectance and use on-line control of spraying. Future application for these intelligent systems is especially promising for non-chemical treatments.

Both researchers and manufacturers are showing a great deal of interest in the *direct injection* system. In this sprayer the formulation and the water are kept separated, each in its own tank and hydraulic circuit. They are only mixed prior to emission, in a chamber positioned near the boom, where the formulation is injected into the water in the correct proportions for making up the plant-treatment mixture (Fig. 1.269).

This system offers various advantages over traditional machines: it provides substantial time savings by eliminating the mixture-preparation step; it prevents the operator from coming into contact with the concentrated plant-protection product; it is possible, during a single treatment, to use different types of treatment products in different dosages, as required. After the treatment, the tank contains no residual diluted treatment product, thus eliminating disposal problems.

The great concern to control off-target drift has led to some simple but effective mechanical devices being developed to combat drift phenomena: rotating fans, vertical diffusers and modular turrets that move the spray-carrying air current closer to the foliage canopy (Fig. 1.270). By directing the spray and air stream and optimally adapting it to the shape and volume of the plants, which varies with the season, the age of the plant, and the different species, drift can be reduced.

The *tunnel sprayer* has proved extremely effective in reducing drift losses. It is a tunnel a few meters in length (<3m), generally made of plastic. Spray nozzles are placed

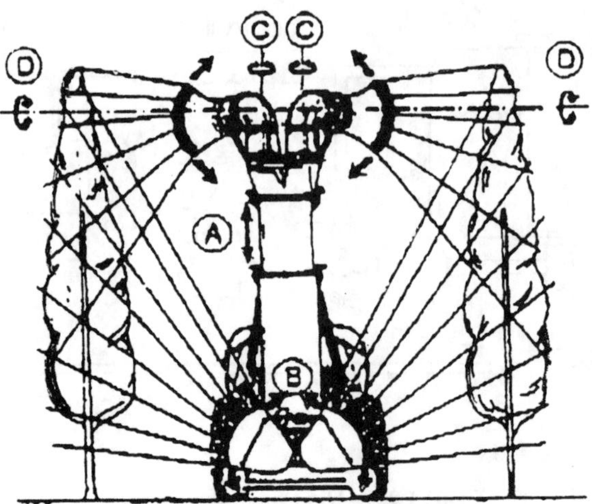

Figure 1.270. Adjustable air diffuser.

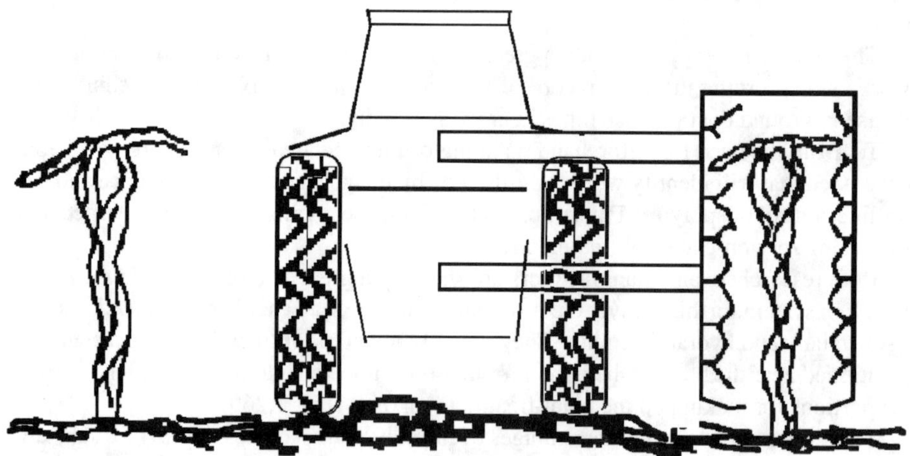

Figure 1.271. Tunnel sprayer.

inside (Fig. 1.271). As the machine advances along the row, the tunnel encloses the vegetation to be treated. Spraying takes place inside this "chamber," and is in some cases assisted by a fan-generated air current. The droplets not deposited on the foliage are intercepted by the tunnel walls, recovered and fed back into the circuit. Operating with low volumes (200–300 dm^3/ha), 90% of the emitted spray is deposited on the target.

The use of shielding screens placed along the boom is also being tested for treatments on field crops. The aim is to shield the nozzle from wind, isolate the field portion being treated, reduce drift and, simultaneously, open the vegetation by means of a *combing* action, thus promoting better penetration of the spray.

1.6. Harvesters and Threshers

Grain

H. D. Kutzbach and G. R. Quick

Combine harvesters equipped with the right attachments can harvest grains and seeds of a wide range of types and sizes, from mustard seeds to broad beans, from ground-hugging clovers to corn over 2 m tall. They recover the grains from the field and separate them from the rest of the crop material, the *material other than grain* (MOG), which is dumped back on the field. The history of grain harvesting is described impressively by Quick and Buchele [1]. Recent combine harvester development is represented by Kutzbach [2] in the LAV Yearbook Agricultural Engineering. Since the transition from hand harvest to the use of combine harvesters, productivity of harvesting has increased from 10 kg/man-hour in 1800 to about 60,000 kg/man-hour with stripper header today, [3, 4] and losses are reduced to as low as 1%–3%. Different crop properties and different harvest conditions make great demands on combines. The combine has to match these demands by design, and the driver has to continually adjust the combine to the optimum for each crop and harvest condition.

1.6.1. Functional Components of Combine Harvesters

The main processes in a modern combine harvester are gathering and cutting, threshing, separating, cleaning, and materials handling. Figure 1.272 shows the main elements of a conventional combine with straw walkers.

- The header (cutting platform) divides, gathers and cuts the crop with a reel and cutterbar then augers the crop into the feederhouse for presentation to the threshing unit.
- The threshing unit consisting of threshing drum and concave (or other systems described later), which detach the grain from the ears.
- The separator unit or straw walker separates the remaining grain from the straw.
- The cleaning unit does the final separation of grain from the chaff and broken straw pieces delivered by the concave and straw walker. The clean grain is conveyed

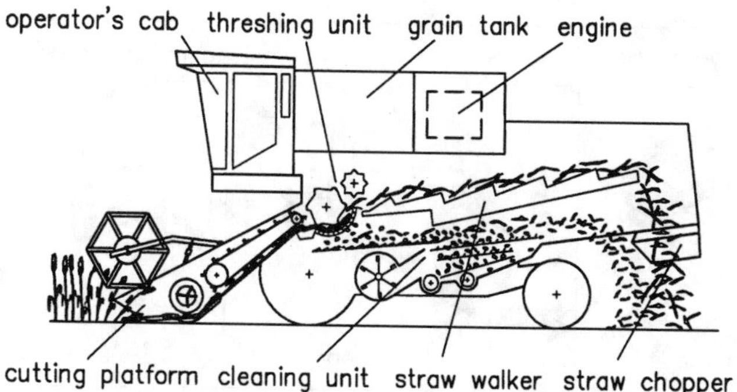

Figure 1.272. Main units of a conventional combine.

into the grain tank. Unthreshed grain heads or pieces of cob (tailings) are returned for rethreshing.

Apart from self-propelled combines, there is a small market for pull-type combines and a few tractor-mounted combines. These are powered by PTO drive, and are either tractor-drawn or affixed onto the tractor. Since they need no separate engine, drive train or operator compartment, these combines are cheaper, but have some major disadvantages, such as: restricted view of the header, less maneuverability, and lower speed, depending on the tractor available. More preparation time is required than for self-propelled combines and the pulling tractor is unavailable for other work. In hilly areas, there are specially built hillside combines on the market, which keep the processor body level on slopes of 10%–12% by adjusting the drive axle as well as the rear steering axle.

1.6.2. Threshing and Separation

Threshing—detaching the kernels from the ears or pods—is accomplished by a combination of impact and rubbing action. While the conventional tangential threshing unit threshes mostly by impact, other threshing devices like rotary threshing units act more by rubbing. A survey of a number of threshing devices is given by Caspers [5]. Rotary threshing units in which the crop is fed axially or tangentially into the rotor are becoming more popular (Fig. 1.273). Additional tasks of the threshing units are the separation of the grain through the concave and transferring the straw to the straw walker or separating section. In rotary combines, generally, the front part of the one or two rotors threshes and the rear part separates the grain from the straw, making use of higher g-forces without the need for gravity-dependent walkers.

Tangential Threshing Unit

Tangential threshing units with raspbars (Fig. 1.274) are versatile and can be used for most crops. A spike-tooth threshing cylinder on the other hand (Fig. 1.275) offers some advantages for rice; wire-loop cylinders are used for hold-on threshers and Japanese rice combines with head feeder systems. The feed elevator, usually a drag chain, supplies the crop to the threshing cylinder at about 2.5–3 m/s. The ideal feed is with the ears first. The threshing cylinder running at about 30 m/s peripheral speed can work intensively on

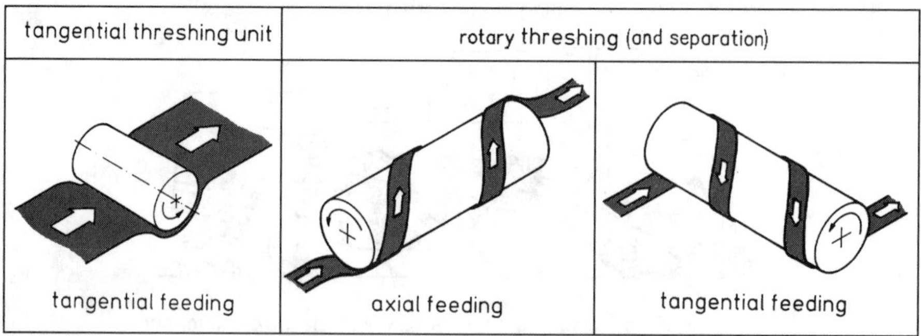

Figure 1.273. Feeding of threshing cylinders.

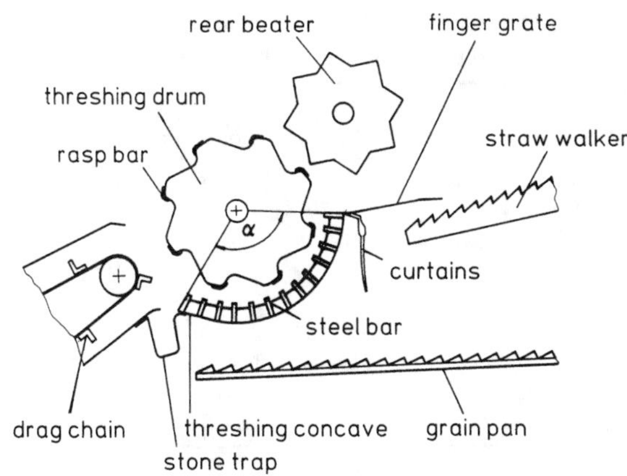

Figure 1.274. Main parts of tangential threshing device.

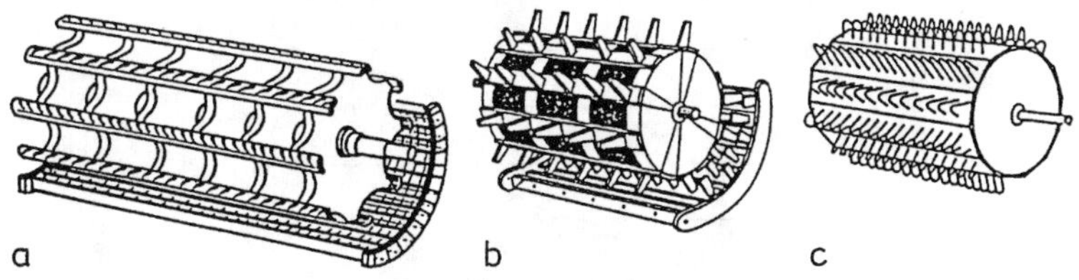

Figure 1.275. Threshing cylinders. a) raspbar; b) spike-tooth; c) wire-loop.

the crop held tight on the stem. The rear beater transfers the straw and any grain that is not separated by the threshing cylinder, delivering it onto the front of the straw walkers, and preventing backfeeding at the cylinder. Adjustment for varying crop conditions is made by changing peripheral speed and concave-to-cylinder clearances. With higher cylinder peripheral speeds threshing losses decrease, but grain damage increases (Fig. 1.276). Some typical peripheral speeds are shown in Table 1.59. To adjust speed, the threshing cylinder is driven by a variable-speed power belt drive or hydrostatic transmission, often in association with a reduction gearbox for speed-sensitive crops like peas or soybeans.

Grain separation increases along the length of the concave but at a diminishing rate (Fig. 1.282). Under good conditions, grain separation at the concave can be as high as 90%. It drops down to 50%–60% under worse conditions, with the consequence of a high burden on separator or straw walkers. Large cylinder diameters allow long concaves, but the centrifugal forces needed for good separation decrease with increasing threshing cylinder diameter. Typical diameter is 600 mm. To improve threshing and grain separation with the conventional design for high capacity combines, most manufacturers have turned to multiple cylinder threshing units (Fig. 1.277). Additional separation cylinders may be placed behind the beater (New Holland, MF, Fiatagri, Deutz-Fahr) or in front of the threshing cylinder (Claas). Between the drums the material is loosened up for improved grain penetration.

Table 1.59. Typical conventional raspbar cylinder settings for a range of crops

Crop	Peripheral Speed [m/s]	Clearance Front [mm]	Rear
Barley	27–34	10–18	3–10
Beans	7–20	20–35	10–18
Maize (Corn)	10–20	25–30	15–20
Oats	27–35	12–20	3–10
Peas	7–18	20–30	10–18
Rapeseed	15–24	20–30	10–20
Rye	25–35	12–20	3–10
Rice	20–30	14–18	3–6
Wheat	24–35	12–20	4–10

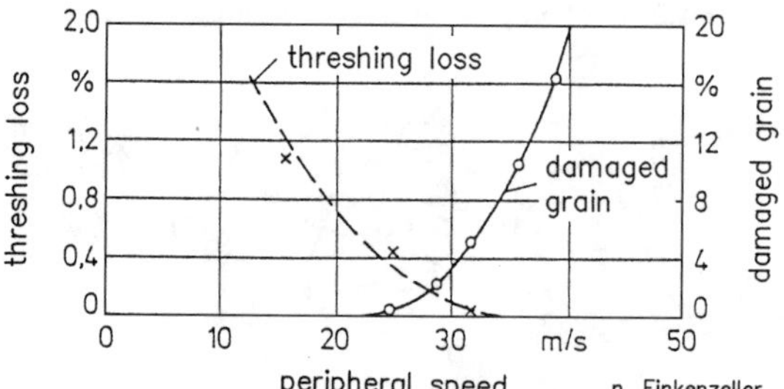

Figure 1.276. Threshing loss and grain damage versus peripheral speed.

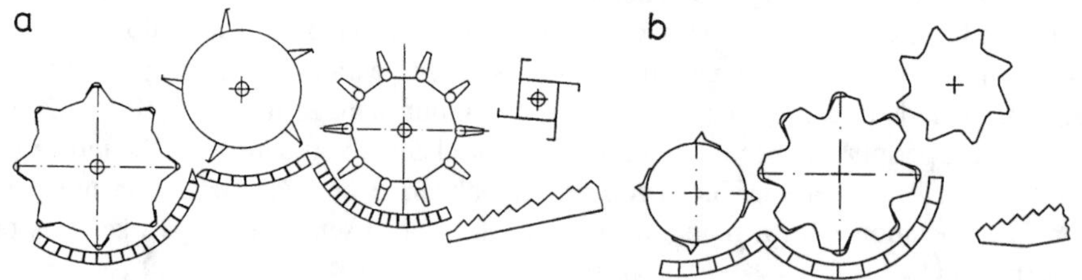

Figure 1.277. Additional cylinders and concaves. a) New Holland; b) Claas.

The increased concave length of the multiple cylinder threshing unit causes higher stress on the crop and leads to higher levels of straw breakup. For very dry conditions, some manufacturers provide the option to expand the clearance under the additional cylinder, or to swivel the concave completely out of the way.

Straw Walker

Straw walkers take up a lot of space but have a low power requirement for grain-straw separation. Four to eight long, permeable channel sections with a width of 0.25–0.35 m and a length up to 6 m are mounted on two crankshafts that rotate at approximately

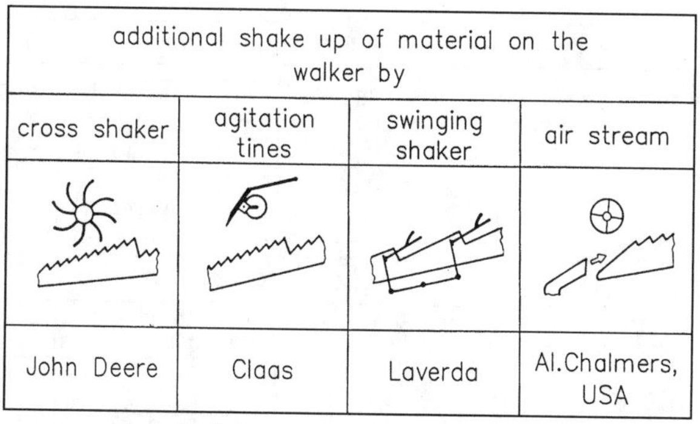

additional shake up of material on the walker by			
cross shaker	agitation tines	swinging shaker	air stream
John Deere	Claas	Laverda	Al.Chalmers, USA

Figure 1.278. Walker auxiliaries.

200 rpm. The crank throw is about 0.05–0.08 m. Amplitude and frequency of channel movement are coordinated [6], but they are not adjustable. Straw bounces on top of the channel sections; grain and some chaff are sifted down and separated from the straw. Four to five steps assist this operation by further loosening the straw layer. The separated grain is conveyed to the grain pan of the cleaning unit by a set of augers or by a return pan below the channel sections. The motion of the walkers moves the MOG rearward.

Separation efficiency of straw walkers decreases rapidly with increasing MOG throughput because the straw layer cannot be loosened enough and grain gets caught in the straw. To improve separation efficiency of straw walkers, a range of walker auxiliaries have been tried and some are in commercial use (Fig. 1.278).

Rotary Separators

With rotary separators, the principle separating force is the centrifugal action, compared to gravity only with straw walkers. The centrifugal force field caused by the rotation of the straw mat together with the rotor can be 50 to 100 times that of gravity. Different designs and arrangements of rotating separation elements in comparison with gravity separation are shown in Fig. 1.279. Several manufacturers use rotary separators in conjunction with stationary cylindrical screens (Fig. 1.280). Crop path can be tangential. In this case several drums are necessary to get sufficient separation length (CS, Claas). With an axial separator, the crop moves in axial and tangential direction along a helical path. Thereby the separation length will be long enough for nearly complete separation of the remaining grain, using one or two rotors (TF, New Holland; CTS, John Deere; Lexion, Claas). Crop motion in rotary separators is forced by paddles or beaters on the rotor and stationary helical guide vanes, resulting in higher capacity per unit volume of separator, and independent of ground slope, but at a higher power requirement.

Rotary Combines

Rotary combines with axial threshing units perform the threshing and separation with one or two axial rotors (Fig. 1.280). This leads to a simple design and quite different performance characteristics. The rotors are equipped with raspbars and a concave in

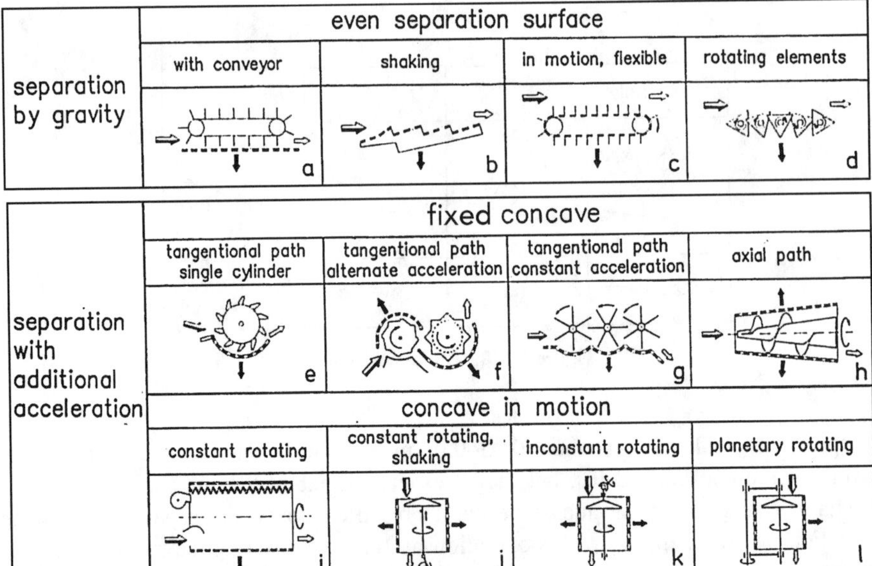

Figure 1.279. Separation systems.

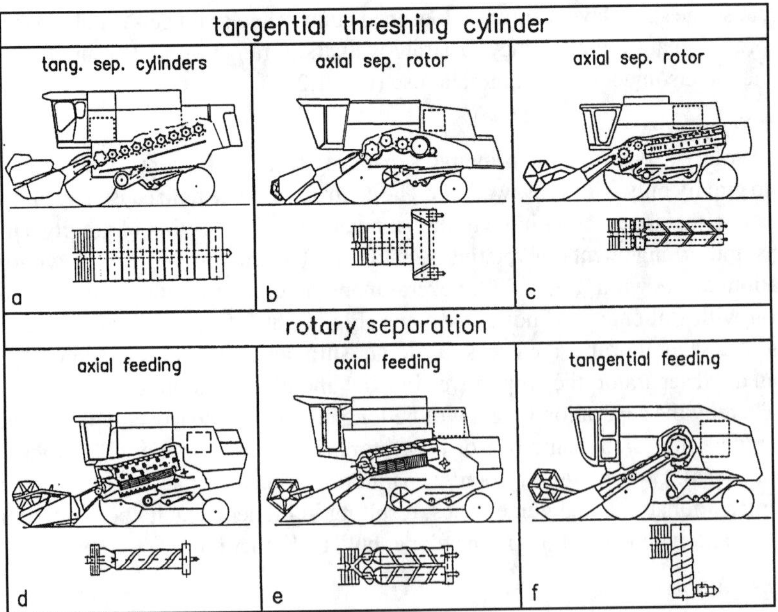

Figure 1.280. Principles of nonconventional combines. a) Claas; b,e) New Holland; c) John Deere; d) Case; f) AGCO.

the front section, tines or beater-bars and separation grates in the rear section. The first rotary combine since World War II was introduced by New Holland in 1975, utilizing twin rotors (TR, New Holland). The TR rotors are installed longitudinally in the combine. This requires axial feeding and dividing the crop equally to the two rotors. Using only one rotor avoids problems with crop distribution (AF, Case; White/MF and others). Transverse-mounted rotors lead to a good and uniform tangential feeding of the crop (Crop Tiger, Claas; Gleaner, AGCO). Fiatagri had the rotor integrated into the header (MX, Fiatagri).

Comparison of Tangential and Axial Threshing Units

The performance criteria of threshing devices are primarily:

- Throughput
- Threshing losses
- Grain separation
- Grain damage
- MOG-separation and power requirements

Threshing loss is defined as the percentage by weight of whole grains not detached from the ears; grain separation as percentage by weight of separated grain at the concave; grain damage as percentage by weight of damaged grain in the sample compared with the total grain entering the threshing unit [7].

Figure 1.281 shows the influence of MOG-throughput, peripheral speed, MOG-moisture content, and amount of green content on working quality of a rotary threshing unit. These studies compared this threshing unit with a tangential threshing system with walkers [8]. This type of axial threshing unit had higher power requirements, higher MOG-separation, a lower percentage of grain damage, and higher throughput in a given machine envelope.

Generally, rotary threshing units have advantages in corn, soy, and rice threshing, but they cannot thresh as many kinds of crop as tangential units can. Rotary threshing units are more sensitive to crop moisture content than the tangential threshers. Axial threshing units show no effect of ground surface irregularity (slopes). Grain loss characteristics are also different (see Fig. 1.291).

Separation Theory

The characteristic grain separation versus length of the concave can best be described by an exponential function [9, 10]. Detachment of grain from ears can also be described by an exponential function [11, 12]:

$$f_1(x) = \lambda e^{-\lambda x} \tag{1.106}$$

The proportion of unthreshed grain s_n is given by:

$$s_n = 1 - \int_0^x \lambda e^{-\lambda x} ds = e^{-\lambda x} \tag{1.107}$$

For a constant throughput, at every cross-section of the threshing unit, the sum of proportion of unthreshed grain s_n, free grain s_f, and separated grain s_S is:

$$s_n + s_f + s_S = 1 \tag{1.108}$$

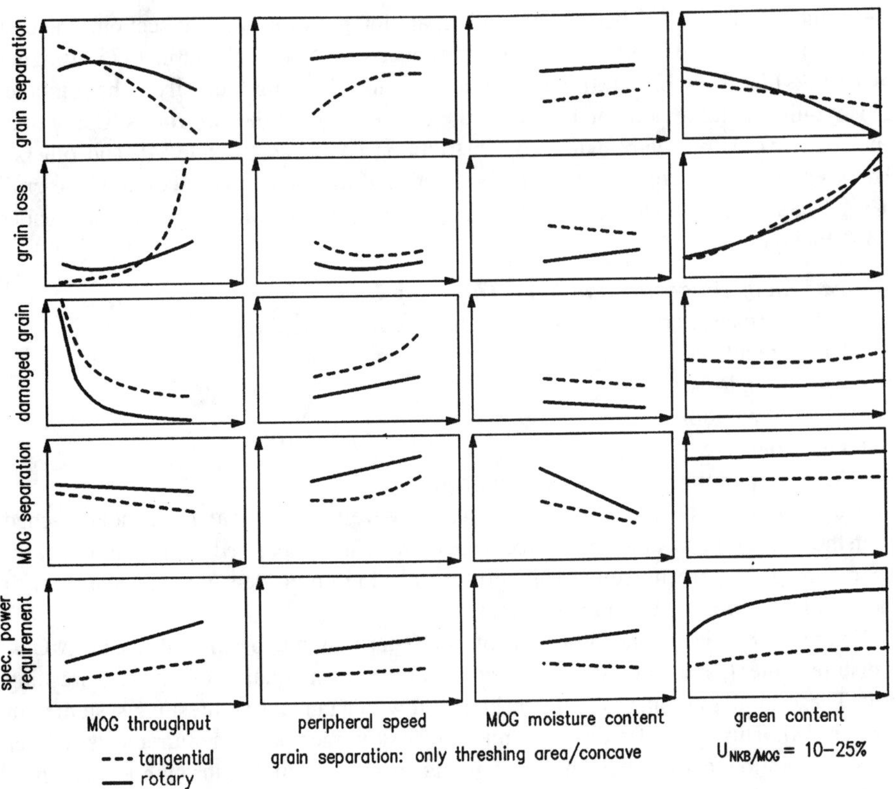

Figure 1.281. Performance characteristic of rotary and tangential threshing devices.

With the assumption that the frequency of grain separation s_d is proportional to the amount of free grain:

$$s_d = \frac{ds_s}{dx} = \beta s_f \tag{1.109}$$

The cumulative proportion of separated grain s_S is:

$$s_S = \frac{1}{\lambda - \beta}[\lambda(1 - e^{-\beta x}) - \beta(1 - e^{-\lambda x})] \tag{1.110}$$

and the frequency of grain separation s_d is:

$$s_d = \frac{\lambda\beta}{\lambda - \beta}(e^{-\beta x} - e^{-\lambda x}) \tag{1.111}$$

Figure 1.282 shows unthreshed grain s_n, free grain s_f, cumulative separated grain s_S, and frequency of grain separation s_d plotted against rotor length. For tangential and axial threshing units, there are different values of L, λ and β.

According to [11], the linear rate of threshing λ is given by:

$$\lambda = k_t(\varrho v^2 LD)/(q_p \delta_m v_{\text{ax}}) \tag{1.112}$$

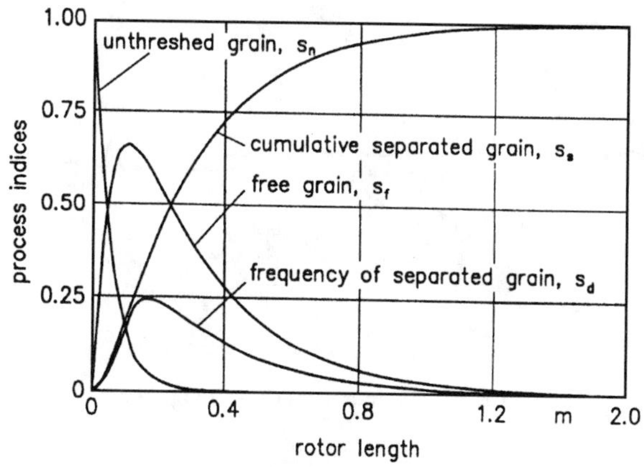

Figure 1.282. Grain separation versus separation length (winter barley, total throughput 5 kg/s).

where:

ϱ = bulk density of MOG [kg/m^3]
v = peripheral speed of the rotor, [m/s]
v_{ax} = crop speed [m/s]
L = cylinder (rotor) length [m]
D = cylinder (rotor) diameter, [m]
q_p = throughput of MOG, [kg/s]
δ_m = concave clearance, [m]
k_t = threshing factor

Threshing factor k_t relates to: machine type, crop variety, moisture content, etc.

The rate of separation β is proportional to the probability of a kernel passage through an opening in the concave and depends on the ratio of kernel diameter to opening size.

Cleaning

The main cleaning device (cleaning shoe) takes care of the final separation of grain from other crop material such as chaff, broken straw pieces, dirt, or weed seeds. The grain and MOG separated at the concave and separation unit are fed to the grain pan or a set of augers which deliver the material to the chaffer or top sieve (Fig. 1.283). The cleaning shoe consists of the grain pan, one or two winnowing steps, two or three oscillating sieves, and an aerodynamic means of blowing air through the adjustable sieve openings. At the end of the top sieve (chaffer) the chaffer extension separates grain and broken ears containing grain out of the material. Together with the material not separated from the lower sieve, these tailings are rethreshed and cleaned once more. The clean grain is conveyed to the grain tank.

Separation of clean grain on the cleaning shoe occurs due to differences in the terminal velocities and dimensions respectively of grain, broken straw pieces, and chaff material under the action of both mechanical forces (oscillation of the sieves) and pneumatic

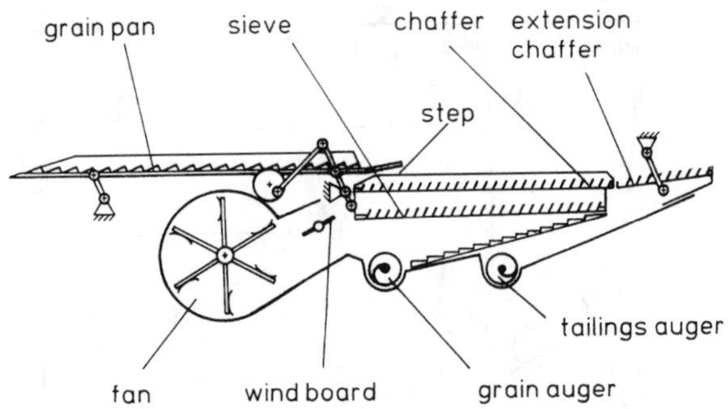

Figure 1.283. Main parts of cleaning shoe.

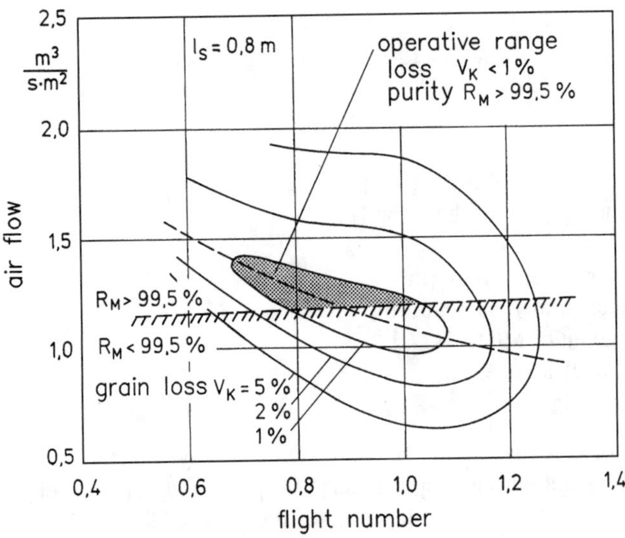

Figure 1.284. Interaction of mechanical and pneumatic parameters.

forces (direction and air velocity). To some extent mechanical and pneumatic forces are exchangeable [13]. For sufficient purity, however, a minimal air velocity is essential (Fig. 1.284). One parameter for mechanical performance is the flight number $Fr_v = a\omega^2 \sin(ß - \alpha)/(g \cos\alpha)$, where a, $\omega = 2\pi f$, and ß are oscillating amplitude, frequency, and direction; α is sieve inclination, and g the gravity constant. Flight number is the relationship between components perpendicular to the sieve, oscillating acceleration and gravity. For modern cleaning shoes the flight number is about $Fr_v = 1$. Typical values for the mechanical parameters are: $a = 17$–38 mm, $f = 4.3$–6 Hz, $\alpha = 0$–$5°$ and $ß = 23$–$33°$.

Pneumatic parameters play an important role in cleaning shoe performance. Airflow should be even across the width of the sieve and decrease strongly from front to rear of the sieve. Two winnowing steps (Fig. 1.285) result in a significant capacity increase of a

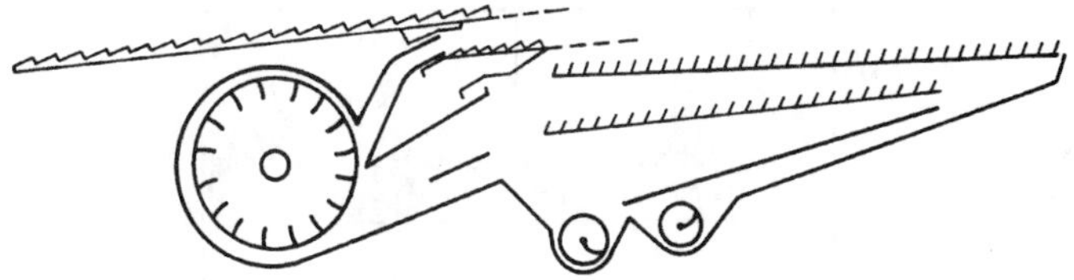

Figure 1.285. Cleaning shoe with two winnowing steps.

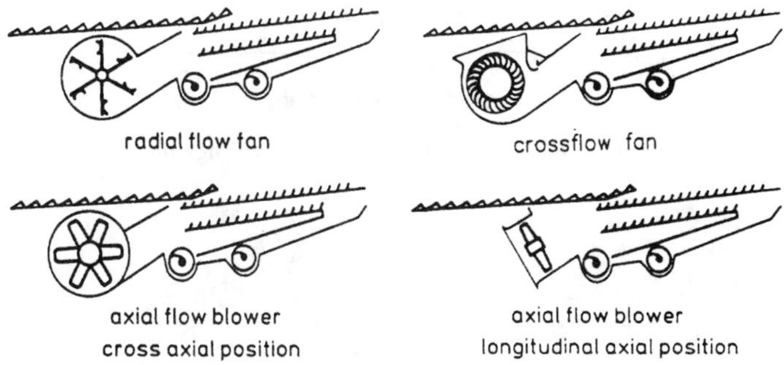

Figure 1.286. Fan types of cleaning units.

cleaning shoe. The airflow should ideally be angled as steeply as possible, i.e., 30° in the winnowing steps and 20–30° on the sieve; in practice, however, it is smaller. Values for the air velocity in the winnowing steps are 6–8 m/s, 5 m/s at the beginning of the loaded sieve, and about 3 m/s at the end. A crossflow fan delivers the best distribution across the sieve width. However an appropriate design has to have a stable and the necessarily steep pressure-volume characteristic. In wide cleaning shoes (1.7–2.0 m), the radial fan, drawing air from each side inlet, delivers too little air in the middle. Therefore 3–4 blowers are used side by side with small gaps for air inlets. Axial fans are also used under certain conditions (Fig. 1.286).

Air velocity (controlled by fan speed) should be set in relation to the grain throughput of the cleaning shoe, to maintain fluidization of the material and allow good penetration of the grain through the MOG layer (Fig. 1.287). With a constant fan speed, losses increase sharply at low throughputs because some grain is blown out (*flight phase*), but losses increase also at a high throughput, because the material layer cannot be loosened enough and grain penetration is hampered (*bulk phase*) [14]. Fan speed should increase with grain throughput for lowest losses (Fig. 1.288).

Cleaning shoes are particularly sensitive to the angle of tilt of the combine when harvesting on side slopes. The material floats to the downhill side, the air escapes on the upper side with high losses in both the flight and bulk phases. Different designs are marketed that greatly improve the behavior of the cleaning section on slopes (Fig. 1.289). Deutz-Fahr and New Holland offer partial-width leveling of the sieves. Additional lateral oscillation conveys the material to the upper side of the inclined sieve on the Claas design.

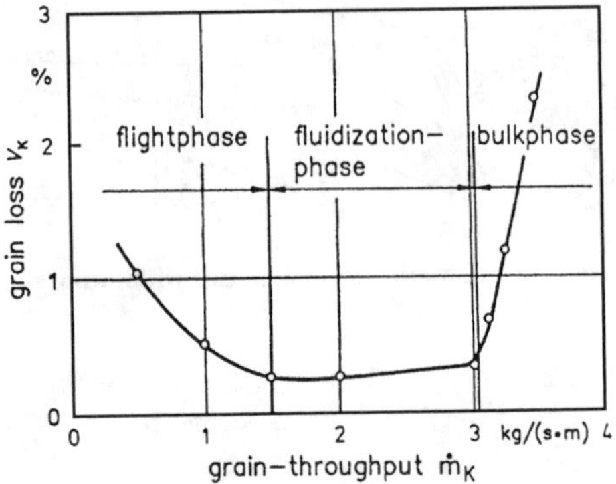

Figure 1.287. Typical loss curve of a cleaning unit.

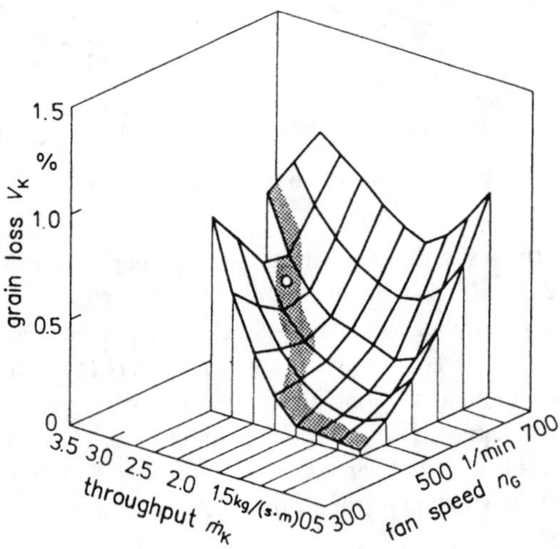

Figure 1.288. Performance characteristic of cleaning unit (wheat, $MC_{grain} = 12\%$, $MC_{MOG} = 18\%$).

With hillside or sidehill combines, the whole combine processor body is leveled for side slopes and, on some models, for driving up or downhill by adjustment of the bull gear (portal) drive axle and by hydraulic cylinders on the steering axle. This way the straw walker and cleaning unit operate independent from slope and the danger of overturning is reduced. The extra expense of hillside combines is only justifiable on steep hilly terrain.

Cleaning Theory

A mathematical model for the grain/chaff separation on grain pan and cleaning shoe is based on physical laws [15]. Diffusion leads to a well mixed grain/chaff layer. With the force of gravity acting on the kernels, they penetrate the mat towards the sieve surface in a process which may be described by the physical law of convection.

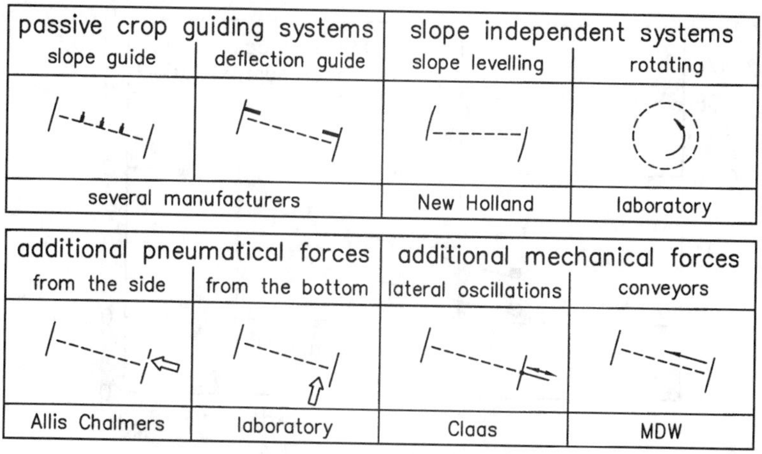

passive crop guiding systems		slope independent systems	
slope guide	deflection guide	slope levelling	rotating
several manufacturers		New Holland	laboratory

additional pneumatical forces		additional mechanical forces	
from the side	from the bottom	lateral oscillations	conveyors
Allis Chalmers	laboratory	Claas	MDW

Figure 1.289. Possibilities to reduce slope sensitivity of cleaning units.

In the basic equation of segregation, the distribution function of the grain mass $u(y, t)$ is determined by the diffusion component with the diffusion constant D_y and the convection component with the average sinking velocity v_y.

$$\frac{\partial}{\partial t}u(y,t) = D_y \frac{\partial^2 u}{\partial y^2} - v_y \frac{\partial u}{\partial y} \tag{1.113}$$

Since the sieve is effectively an obstacle to grain separation, a stochastic separation model is combined with the convection-diffusion model. All model parameters depend on process parameters of the grain pan or cleaning shoe. For a standard set of cleaning parameters, Fig. 1.290 shows typical values of the average sinking velocity and the diffusion constant.

1.6.3. Combine Harvester Performance

The performance of the combine (area or throughput capacity) depends on the following parameters: desired work quality, acceptable loss level, crop properties and harvest conditions, combine type and design, engine power, MOG-throughput, field shape and size, harvest management, and especially straw length.

Work Quality

Essential criteria for the work quality of the combine are: losses; grain damage; and grain dockage, which relates to trash level.

Overall combine losses must include: header losses (shatter and cutterbar losses); and processing losses.

For cereals the processing losses at the prescribed throughput should not exceed:

For threshing losses (grain in ears), 0.1%–0.2%

For separating losses (grain in straw), 0.3%–3%

For cleaning losses (grain in chaff), 0.1%–0.5%

While the shatter losses depend on crop variety, ripeness, forward speed, and reel action, and the cutterbar losses on the distance of the ears from the ground and the

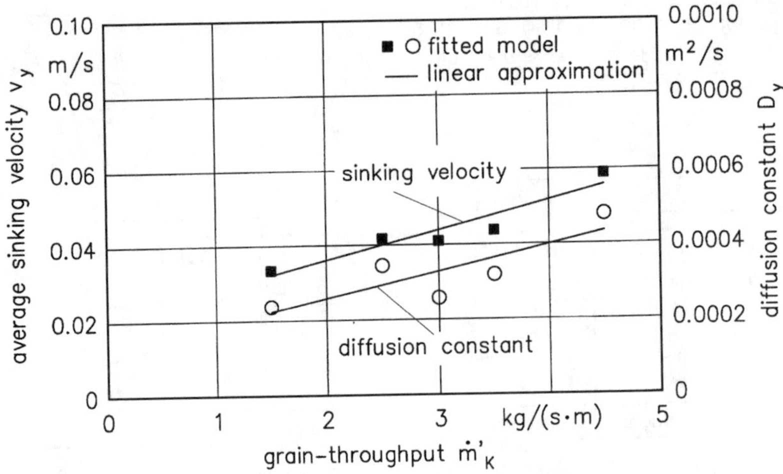

Figure 1.290. Effect of grain throughput (wheat) on average sinking velocity and diffusion constant for typical cleaning parameters.

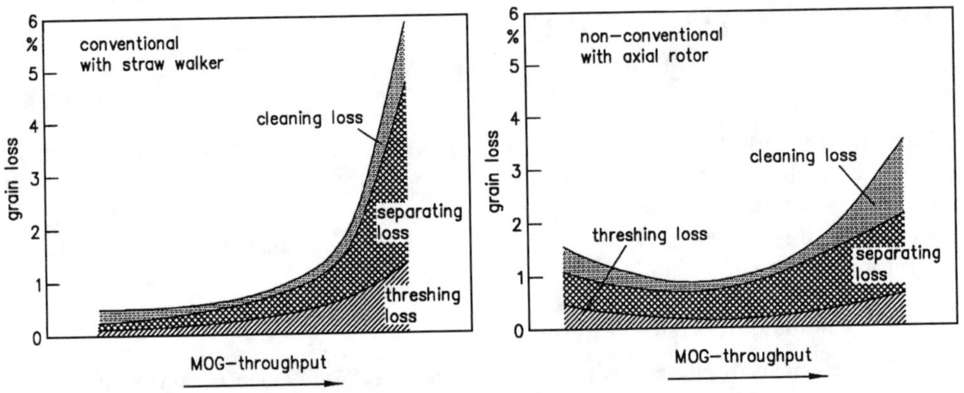

Figure 1.291. Grain loss behavior of conventional and nonconventional combines.

position of the cutterbar, the losses of the separation units rise rapidly with higher MOG-throughput. Figure 1.291 shows the typical loss behavior of conventional and rotary combines, showing somewhat higher losses at low throughput and lower losses at high throughput for the rotary designs.

The typical steep loss characteristic at high throughputs is due to reduced separation efficiency or overload/performance breakdown of the threshing unit, as well as of the straw walker or rotary separator and the cleaning unit, when a thicker crop mat provides higher resistance to grain penetration.

The accepted amount of losses serves for specification of the optimal throughput. International standards for wheat set the acceptable loss level at 1% at the measured MOG throughput [7]. In practice, however, monetary losses relate to harvest cost. The combine operator has to know the level of losses to set the optimal speed v [km/h] at harvest. This depends on optimal throughput $\dot{m}_{\mathrm{MOG}}$ [t/h] at accepted losses, grain/straw-ratio: $R = m_G/m_{\mathrm{MOG}}$ [−] effective cutting width b [m] and yield m'_G [t/ha]. The following relationship applies:

$$v = 10 \cdot R \cdot \dot{m}_{\mathrm{MOG}}/(b \cdot m'_G) \tag{1.114}$$

The effective area capacity $\dot{A}_{\mathrm{eff}}$ [ha/h] (spot rate of work) can be calculated from speed v and effective cutting width b (2%–5% less than cutterbar width because of overlap).

$$\dot{A}_{\mathrm{eff}} = 0.1 \cdot b \cdot v \tag{1.115}$$

The overall workrate is less than the spot rate because of turning time at the end of each row, unloading time when not on the move, time spent on machine adjustment and unclogging, as well as the operator's rest breaks. This is considered in general by the field efficiency η_F [−] obtained from the relation of productive time to operational time on the field. Then the overall area capacity is

$$\dot{A}_{\mathrm{ov}} = \eta_F \cdot 0.1 \cdot b \cdot v \tag{1.116}$$

Values of field efficiency for combine harvesting are between $\eta_F = 0.65$ to 0.8.

Grain damage includes broken kernels, seed coat damage and internal invisible damage, which occur mostly during threshing but also by auger conveying. High mechanical damage results in poor germination, poor storability of grain because of molding, and poor processing characteristics. Very small pieces are blown out by the cleaning shoe and thereby lower the harvested mass in the grain tank.

Grain dockage refers to trash in the bin sample: parts of chaff and small straw pieces, foreign matter and weed seeds, etc. Caused by the low density of these MOG-parts, 0.5%–2% of dockage (related to the mass of grain) is clearly visible. Purity is defined as proportion of clean grain to the total mass (grain and MOG). Values are between 98% and 99.5%.

Crop Properties and Harvest Conditions

Crop properties of the harvested crop, such as moisture content, external friction behavior, angle of repose, structure of straw, grain dimensions, and attachment strength in the ears, have essential and complex effects on combine performance. Crop properties depend on crop type, crop maturity, crop variety, process of crop growth, nutrition and water status during growth, etc. These different crop properties cause essential differences in the performance of the same combine in different crops, but also within different years in the same crop (Fig. 1.292). One way to standardize combine testing is by comparison of the performance of the test machine against a reference combine.

Harvest conditions such as grain and straw moisture, lodging, weed density, yield, slope, and evenness of the plot all influence combine performance in a given variety. Climatic factors, such as relative humidity, wind, and temperature also affect combine performance especially as they influence grain surface moisture content and, thereby,

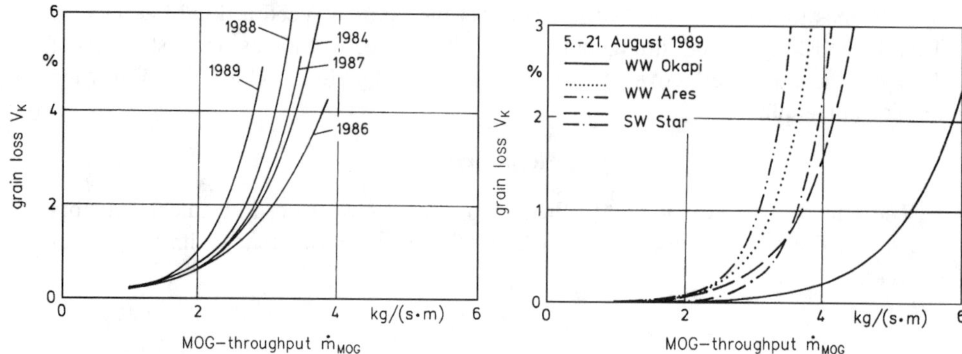

Figure 1.292. Influence of year and variety on loss curves.

friction behavior. Generally, combine performance has a tendency to decline with higher friction values.

The operator can deal with different crop properties and harvest conditions by different combine adjustments. These are: forward speed; reel position; reel speed; cutterbar height; cylinder peripheral speed and front/rear clearance of the concave; sieve openings of chaffer; chaffer extension and sieve; fan speed; and position of wind deflectors. A combine operator has to have considerable skill to stay in the crop, avoid obstacles, monitor sample quality, unload on the move, and keep the machine running at optimal field efficiency.

Combine Type and Design

Capacity of a conventional combine increases approximately linearly with increasing threshing cylinder width, if walker and cleaning shoe areas are in the right proportions. Maximum width limitation for the threshing cylinder of about 1.7 m is set by the given total combine body width of 3–3.5 m governed by road and rail transport regulations.

Higher capacities can be achieved by additional separation cylinders before or behind the threshing drum especially for harvest under more moist conditions. Rotary combines with stripperfronts are capable of much higher spot rates even in excess of 60 metric tons of grain per hour. If the capacity of the conventional combine with 1.7 m threshing width is set to 100% in cereals, capacity increase in cereals may be estimated at an additional 10%–25% each in this order: additional cylinder, axial flow, rotary separators.

Performance is influenced by the interplay of the different conveying and separating elements. Combine performance is vitally affected right from the front. Coordinated capacity is important. Comparing just the sieve and walker areas in relation to the threshing cylinder width is not adequate. Additional crop layer loosening, good design of the transitions, particularly the transition of threshing cylinder, beater, and straw walker can determine overall combine performance.

Engine Power

Figure 1.293 shows a powerflow diagram for a smaller conventional combine. Besides power for cutting, threshing and cleaning, power is necessary for the drive train, straw

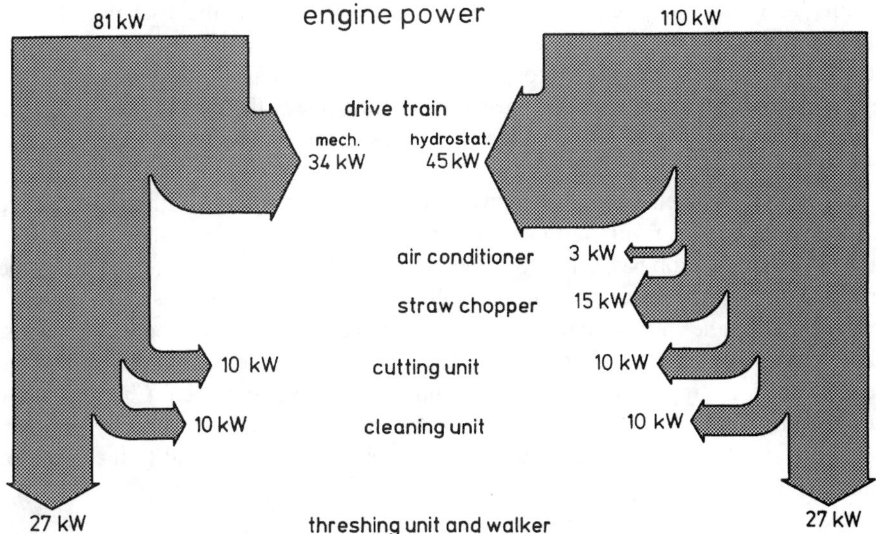

Figure 1.293. Power flow diagram of combines.

chopper, unloader and air conditioner. Hydrostatic drives need more power for the drive train, especially on slopes with a full grain tank. The drive train on 10% slope with 8 t grain and 12 t combine weight requires 75 kW, for example. In general, the power requirement can be calculated roughly in proportion to the MOG throughput at around 9 kW·h/t (specific power index).

Nonconventional combines, especially rotary combines, have a higher engine power requirement (up to 275 kW in 1997). Conventional combines had approximately 100 kW per m width of the cylinder in 1997, compared with 30 kW per m in 1956, when the first all-steel bodied combines were sold.

Harvest Management

If the harvest is well organized, the overall combine work rate can be increased by providing grain transport capacities in time, unloading on the go, minimizing repair time by good daily maintenance and preventive service, having relief operators during the main operator's rest time, avoiding changing fields during harvest times, providing sufficient storage and, if necessary, sufficient grain drying capacity.

Combine Testing

Combine testing is performed in the laboratory as well as in the field. The objectives of combine field tests are to determine functional behavior and performance characteristics, for example the loss/MOG or grain throughput curves. According to [17, 18], field tests are carried out on even fields, and in comparison with a reference combine. Five to seven different speeds and, therefore, different throughputs generate into a loss curve. A field length of 9–25 m with steady-state conditions is necessary to determine the grain and MOG throughput as well as the grain losses. Material from the separator and the cleaning unit might be discharged separately into canvas bags at the discharge chute at rear of the combine, or the material might be collected on a cloth [18]. The material is weighed to evaluate the MOG throughput and the grain cleaned out to determine the separating and cleaning losses. To measure threshing losses the material is rethreshed. Prairie Agriculture Machinery Institute, Canada, (PAMI) and a few other test stations offer results of combine testing.

Design improvements can also be evaluated in laboratory tests, with the advantage of off-season work, uniform crop and better control of test conditions. But the disadvantage is that the crop has to be stored and crop properties may change and affect the performance characteristics. The crop (usually as unthreshed sheaves) is delivered via a conveyer belt to the cutting platform of the combine over a period of about 20 seconds. Also, in this case, data should be collected only at steady-state conditions. To test the cleaning section, crop is supplied to the grain pan by a conveyer belt. Chopped straw (20–30 mm) can be reliably used as MOG. A distributor adds the necessary grain on top of the supplied MOG-layer. The supply speed should correspond to the speed of conveying of the cleaning unit (about 0.5 m/s).

1.6.4. Information and Control Systems

Information and control systems are used on combines to: increase productivity, decrease harvest costs, improve harvest quality, reduce operator stress acquire crop yield and moisture data, and improve harvest management.

Cabs and Controls (Operator Compartment)

Dust proof and sometimes air-conditioned cabs, along with well-arranged, easily reached operating controls make work easier for the operator. If these are provided, the operator can concentrate more on the harvest process and optimize combine capacity. By integrating essential electrically operated functions in a multifunction lever or joy stick control it is possible to adjust the reel and the cutterbar by touch (Fig. 1.294). The work of the operator in changing fields or crops is simplified by systems to adjust the threshing and separating elements. Adjustment settings of several elements according to different crop and harvest conditions are saved in a microprocessor. When required, the threshing and separating elements are set to the stored values in memory by electro-hydraulic controllers. The stored values can be upgraded by operator experience.

Information Systems

Information systems contain supervision and warning devices, decision help for adjustment of the combine, and the registration of management data. Supervision and warning devices inform the operator about such factors as: the technical condition of the engine (temperature, oil pressure etc.); the rotational speed of the most important shafts; level of grain in the grain tank; tendency for straw walker blockage and grain loss indication.

Grain loss sensors are an essential guide to combine forward speed and therefore the throughput set according to the loss throughput curve. The loss sensors register grain that has bypassed cleaning unit or straw walkers and falls onto the sensors. Grain impacts are converted into impulse signals by a microphone, and the frequency of the impulses

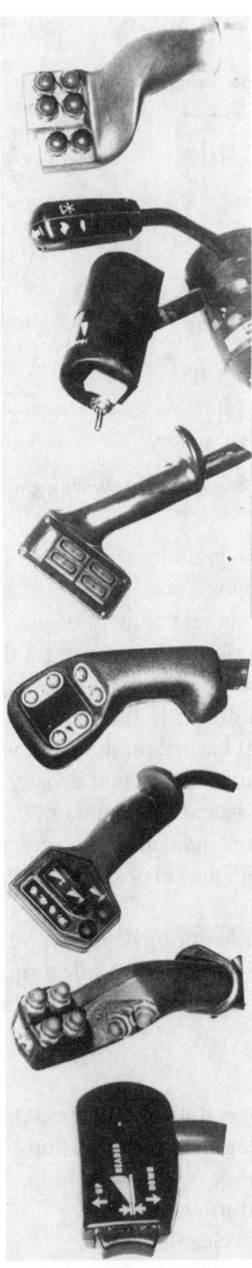

Figure 1.294. Speed control levers with integrated switches.

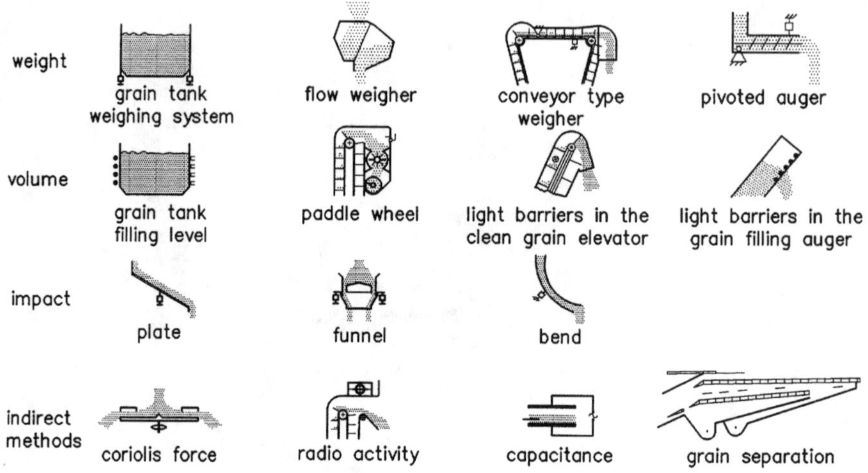

Figure 1.295. Sensors for measuring grain throughput.

is indicated to the operator on the grain loss monitors, which accordingly are providing relative loss values, not absolute loss level.

Area capacity, grain throughput and amount of grain harvested per day or per field are important management data. Recording the grain throughput is also necessary for yield mapping. Measuring MOG throughput is important for the forward speed control. Measuring grain flow is carried out with different sensors (Fig. 1.295). Grain-volume flow devices (paddle wheel, light barriers in the clean grain elevator) are commercially available which require the input of the actual density of the grain for calibration; in addition grain throughput is measured by gamma radiation or capacity changes of the electrical field. The information is indicated on a screen (AgroCom, Claas; Field Star, MF; Green Star, John Deere). With the aid of satellite GPS, field-yield can also be plotted from stored data.

Although volume and torque measuring devices have been examined for determining MOG throughput, the signals are influenced by idle torque and actual crop density. There was only one MOG measuring system in production in 1997 (belt slip of threshing cylinder drive, MF).

Control Systems

Control systems relieve the operator from the need to precisely track the adjustment parameters according to the actual harvest conditions. Important control systems can manage:

- Cutter bar position and platform height
- Leveling of hill combines or cleaning shoes
- Automatic steering down the row for corn harvesters (Fig. 1.296)

One type of sensor for position control of the cutterbar uses two end-mounted height sensors, although there may be problems driving backwards. Contactless working ultrasonic sensors for height control are also available.

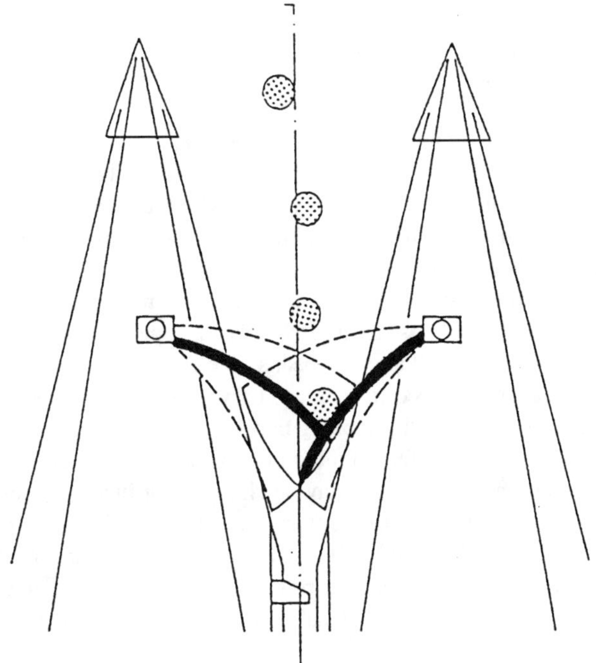

Figure 1.296. Sensor bars for automatic steering of corn harvesters.

Automatic steering needs stringent safety requirements. Automatic steering in corn rows reduces strain on the operator. Working in the evening in the twilight is possible. Operating speed with automatic steering is about 10% higher than without.

Automatic adjustment of forward speed dependent on the throughput of MOG and on the losses according to the loss/throughput curve had not been completely solved by 1997. The investigations by Eimer in 1973 [19] have shown fundamental advantages. Huisman and McGechan [20, 21] judge the economic advantages critically. Measuring the MOG throughput inside the combine, however, is subject to the fundamental fault that the measurement is carried out too late, so keeping the throughput at a constant value is difficult to achieve. Recording crop density in front of the combine is important in order to adjust a constant throughput by changing forward speed. Yield mapping, which analyzes previously harvested rows, allows estimates of crop densities in front of the combine [22].

1.6.5. Combine Attachments and Variants

The basic combine can harvest a great variety of different crops. Special equipment or attachments may be necessary for some crops, for example, specialty header fronts, cylinder speed controllers, special sieves, closed threshing cylinder, and chopper for stalks of corn under the header have been marketed.

Header

Windrow pickup headers are used in some regions to pick up crops which have been cut and windrowed for additional crop drying such as for rape seed, field peas, wheat and barley.

Grain headers are used for small grains. Bat or slat type, pickup, or even air reels gather the crop. Shatter and cutterbar losses are affected by cutterbar height, reel position with respect to the cutterbar, and reel peripheral speed, which is recommended to be about 25%–50% faster than the forward speed of the combine. Some examples for add-on attachments at the header are: side cutterbars and extended platform on the header for rape seed; trays to catch shatter losses for sunflower; and crop lifters or flexible floating cutterbar for crops with grains close to ground (soybeans, peas etc.). Different knife and guard combinations are also available (soybeans, rice).

Row crop corn headers are used for corn, to harvest three to eighteen rows at a time (Fig. 1.297). Gathering chains with snapping bars or snapping rolls grab the corn stalks, pull them down, and snap the ear free of the stalk. The ears are conveyed to the threshing unit, the stalks remain on the field and may be chopped for incorporation. Special row crop headers are also available for sorghum, sunflower and soybeans.

Stripper headers (Figs. 1.298 and 1.301) are increasingly popular for small grains and rice in certain regions. A combing rotor strips the ears from the stalks and supplies them to the threshing unit [23]. Because of the small volume of MOG, this leads to a substantial

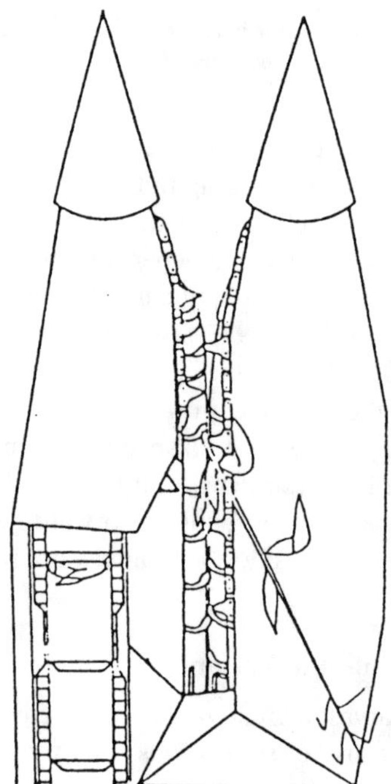

Figure 1.297. Corn header.

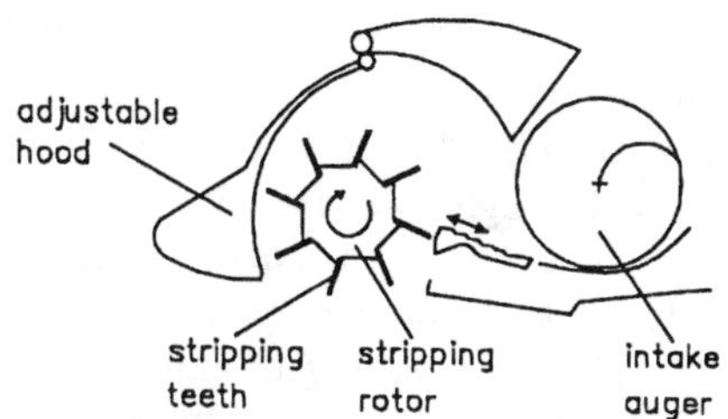

Figure 1.298. Stripper header.

increase of the combine throughput (up to double grain throughput). However, there are harvest conditions in which the stripper performs poorly and the standard header must be available as well.

Chopper

If the straw is not to be baled but incorporated into the soil, a straw chopper attachment is used. The chopper catches the straw falling from the straw walker and cuts this with high-speed rotating knives. For a good mulching of straw in the soil, the chopped straw is intentionally frayed and split by the design of the chopper hood. The straw and the chaff should be redistributed over the whole cutting width by rotating distributors. Sheet metal shields are able to compensate for crosswinds.

1.6.6. Rice Harvesting

Rice is the staple food of more than half of humanity. All of the rice grown in the industrialised nations is combine-harvested. Worldwide, however, only 31% of rice paddy is harvested by combines, the majority by hand-sickle and threshers (Quick, 1997, based on an assessment of IRRI worldwide rice statistics).

Field rice is known as *paddy* or *rough rice* when harvested, whereas milled grain is simply known as *rice*. From the broadacres of California or the Australian Murrumbidgee to the distant hills of Bhutan, the timing of the rice harvest is critical everywhere. Rice farmers are anxious to reap the rewards of their season's effort, but the harvest is a bottleneck, especially in the developing world. Mature paddy is highly susceptible to losses and serious downgrading if the harvest is drawn out too long. The higher the grain moisture level at harvest—up to a limit—the higher the whole grain yield when the rice is milled. Ideal grain moisture at harvest is around 20%. Harvest delays risk having the grain reabsorb moisture. Rewetted grain rapidly loses mill yield, i.e., there is less *whole grain* after it has been milled. The aim is to recover as much whole grain because rice has its highest value as whole grain, unlike other grains, which are ground to flour or processed before sale.

Rice Combines: Why a Rice Combine?

There are several justifications for rice-special combines:

Paddy Field Conditions. Traction and flotation conditions for machinery operation are usually worse for rice fields than for other grains.

There is a compensation—a rice combine does not need to be designed for hillside operation. But a rice combine does need high underframe clearance and traction aids such as *rice tires*, halftracks, or even full crawler ground drive.

Since rice must be harvested at high Moisture Content (MC) for best milling quality, the soil should not be allowed to dry out so much that the rice kernels on this perennial species will dry too rapidly before being harvested. Whether irrigated rice fields are drained or not, the straw is still at a much higher moisture content than the grain, for example 50% to 70% wet basis, compared with a paddy moisture level of say 14% to 27%.

MOG/GRAIN Ratio. The volume of straw or MOG that is taken in by a cutterbar-equipped combine is higher for paddy than for other cereals. The weight ratio for paddy, for example, is 1.5 metric tons of MOG per metric ton of paddy, compared with 0.8–1.0 t of MOG per metric ton of wheat. If the paddy is lodged—and it often is—then the MOG/GRAIN ratio taken into the combine goes up to 3 or 4 to 1. Combines equipped with stripperfronts take in far less straw—under good conditions.

Rice Straw and Paddy are Highly Abrasive. The paddy (unhulled grain) is tightly encased in layers of husk that are protective, but high in siliceous material—compounds similar to the key ingredient of sandpaper. Rice *straw* is also high in silica. Rice combine models need to have selected components made from stainless steel, alloys or abrasion-resistant coatings.

Rice Crops are More Susceptible to Lodging. Prior to the adoption of high-yielding varieties and laser-levelled rice fields, rice was traditionally a long-strawed crop, bred to grow in standing water of variable depth. It was easily knocked down by storms. But even modern varieties (MV) with their faster maturations and shorter straw lengths still tend to be green at harvest time. Patches or even whole fields of lodged crop are not uncommon. The machine needs to be designed to pick up and process this lodged material if excessive losses are to be avoided.

Threshability. There are over 60,000 varieties of rice worldwide. Some types are tough to thresh, all are vulnerable to mechanical damage. The harvesting process needs to beat, comb, rub or otherwise ease the grains off the panicle and out of the straw, yet not exacerbate grain breakage. Hidden cracks or skinning from *overthreshing* show up as brokens later at the rice mill, with subsequent price penalties for the rice grower.

On a global scale, less than one-third of the world's rice is combine-harvested, but over ninety percent of the rice that is *traded* across international borders is harvested by combines. The rest is manually harvested. On the other hand, only 4% of the world's rice crop crosses national boundaries. Most rice is used in the country where it is grown [24].

Custom-Made Rice Combines

The mechanization of the rice harvest in the industrialised world has been complete, rapid and comparatively recent. In the U.S. and Australia for example, where the world's highest-yielding rice crops are grown—8 to 14 t/ha—there never was a stage where commercial rice crops were manually transplanted or cut by hand-sickle.

Combine-harvesters were contemporaneously adopted in the U.S., Australia, and similarly in Europe in the 1930's, initially as tractor-drawn machines. There was a brief phase of mechanical reaping and binding on rice fields. The early combine harvesters that were built for other cereals would not long withstand the adverse field conditions and the wiry and abrasive nature of the rice plant. As a result, indigenous manufacturers of combines, some made by farmers themselves, were the mainspring for emergence of the first self-propelled combines on US rice fields in the 1950's. These were made by local entrepreneurs to meet the peculiar demands of rice. As a market emerged, the full-line manufacturers gave attention to *rice special* combines.

Combine Categories

Rice combines are categorised into two general classes according to the threshing mechanism: conventional *tangential* feed with walkers, and axial flow or *rotary* combines without walkers.

A study of the power and weight characteristics of rice combines of all kinds across the rice world, from Russia through Asia to South America indicates a trend for combines to require 36 kW of engine power per meter of gathering width; a power to weight ratio of around 14 kW per metric ton at typically 1.7 metric tons per meter of gathering width, and specific power around 12 kW/t/h grain throughput (12 kWh/t). This is 33% higher than for other small grains and is reflected in the smaller gathering fronts on rice combines; the Case 2188 in rice, for example, is typically fitted with a cutterbar rice head of 20 ft, compared with 30 ft or wider cutting width for wheat on the same combine.

Red and Green Dominate

The American and Australian rice harvest is dominated nowadays by two combine brands: Case-IH and Deere & Co. The new price of late model rice combines was between US$210,000–250,000 on half-tracks in 1997.

Other brands such as Claas, Massey Ferguson, New Holland, Agco, Laverda, Lova, etc., are marketed for rice along with Deere and Case.

John Deere's CTS Rice Combine (Fig. 1.299) was designed primarily for rice, although it is successful in other cereals. Its launch in 1991 represented a significant departure for Deere & Co., which had adhered to straw walkers until *cylinder tine separation* (CTS). CTS refers to the cast iron tines on twin contra-rotating cylinders, which operate at 700 rpm. The cylinders are mounted in a modular separating assembly that can be rolled out to the rear of the combine for service. In 1993, a Deere CTS combine equipped with an 18-foot Shelbourne Reynolds stripper front set a new world record harvest rate of 45.66 metric tons per hour, sustained over eight hours on a California rice field.

The launch of International Harvester's Axial-Flow combine in 1977 made that company a strong contender on US and Australian rice fields. The 1400-series single rotor principle was exhaustively tested in rice before launch and was well accepted. Successive models of the International (later Case-IH) Axial-Flow machines and the White Rotary design established beyond any question the capability of rotary combines in rice fields. Advantages gained were high capacity yet gentle threshing action for higher quality rice, due to the larger clearances and the single rotor's repetitive action (Fig. 1.300). Straw

Figure 1.299. John Deere's 'CTS' combine, a walkerless machine designed primarily for rice, can be used in other crops. It represents a significant departure for Deere & Co., which had previously eschewed rotary separation in favor of straw walkers. Deere's CTS combine for rice is equipped with 260 hP (201 kW) engine, 7200 L grain bin. GR Quick photo shows a CTS with Shelbourne Reynolds stripper front on an Australian rice field in 1997; this machine's owner was successfully using a raspbar cylinder.

breakage is much higher than walkers but is of little concern. Transport or guide vanes in the rotor cover control the rearward motion of the crop around the rotor as it progresses over the concave and separator grates. Transport vane angle settings are adjustable. With the axial-flow concept, processor body width can be narrower for the same throughput, grain damage is lower, and grain is less sensitive to rotor speed and concave settings than for a conventional, tangential-feed threshing cylinder.

A high-inertia rice specialty rotor with multiple wear-resistant elements is needed for processing tough-stemmed and abrasive crops.

Stripper Fronts Gaining on Rice Fields

Some 10%–25% of US, Australian and South American rice is harvested with combines equipped with stripper fronts. The stripper front suits the rice varieties grown in those countries, with their dense crop canopy and ease of threshability (Fig. 1.301). Spot-harvest rates, exceeding one metric ton a minute, have been recorded in rice from combines with stripper fronts. Grain loss at the front is much higher than for a cutterbar but is still at a generally acceptable level in rice. Contractors appreciate the increased throughput, reduced fuel consumption and lower processor component wear with stripper fronts on combines (Fig. 1.302). Harvesting lodged crops is easier and much faster

Figure 1.300. Case 2188 combine with cutterbar rice front; in Australia a Vibramat attachment to the cutterbar enhances crop flow across the platform. GR Quick photo on an Australian rice field in 1997.

with a stripper front (at least later in the rice season) than for a cutterbar. In some conditions the amount of trash in the bin sample can be higher than with a cutterbar-equipped combine. Recent research on rice trash showed that trash could be reduced with higher fan speed and faster travel but is aggravated at slow forward speeds, and in lodged crop conditions [27].

Traction and Flotation Assistance for Combines in Rice Fields

The heaviest Class 7 combine may weigh up to 24 metric tons with full grain bin. A fully loaded and bogged combine harvester in a paddy field is not an elegant sight—and very difficult to extricate. Some operators have damaged their harvesters trying to pull them out of a bog. Expensive combines have been pulled in half, or at least the rear axle torn away from the frame. Then there is the task of tidying up the deep furrows left by the harvest machines and grain transporters. A level ground surface is essential to minimize water loss from flooded paddy fields.

Forestry logger tires are an option for rice fields to reduce rutting—these are tires with generous width and low inflation pressures. Steel full-track undercarriages and half-tracks with rear-steering axle have been options for combines for years, but more recently, prospective combine buyers have available the choice of *rubber* tracks, such as Caterpillar's MobilTrac full rubber track/friction driven system. Yet another option appeared in 1994 when Deere & Co. offered a rubber-belted half track cog-driven system which they developed for combines in conjunction with Waltanna of Australia. Deere &

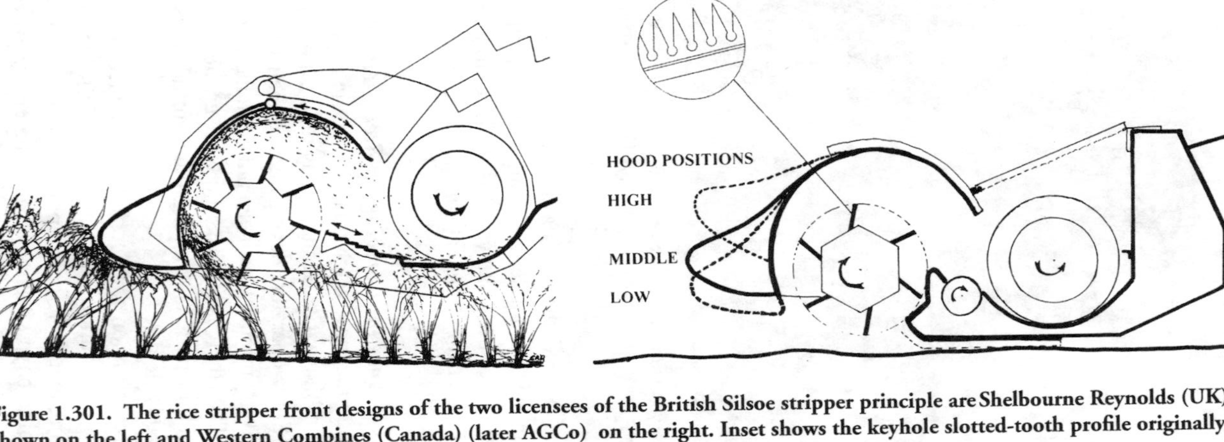

Figure 1.301. The rice stripper front designs of the two licensees of the British Silsoe stripper principle are Shelbourne Reynolds (UK) shown on the left and Western Combines (Canada) (later AGCo) on the right. Inset shows the keyhole slotted-tooth profile originally developed by Silsoe's Wilf Klinner.

Figure 1.302. Showing the teeth and wear plates on the Shelbourne Reynolds stripper front. As the teeth wear, their harvesting characteristics change; new teeth bring in more trash, worn teeth cause losses. A set of teeth can harvest 10,000 metric tons of rice in good conditions, but much less in lodged crop.

Co. was the first full-line manufacturer to offer auxiliary rear wheel assist on combines, with a hydrostatic system patented in 1973. Deere's auxiliary combine drive was plumbed in such a way that the torque applied to the individual hydraulic wheel motors in the rear wheels was proportionate to the torque applied to the main drivewheels, yet if spinout occurred at the rear, the mainwheels would not lose power. Today most contract harvesters purchase their combines with rear wheel assist as standard, and several systems are available from shortline makers.

Each of these and other traction alternates are options for reducing the soil damage caused by heavy combines in wet fields—and reducing the high costs of land preparation and leveling after fields have been compacted by combine tires.

A Question of Teeth

Most rice combines with tangential-feed cylinders use spike-tooth threshing cylinders with peg type concaves. Some use several cylinders in sequence. The spikes or peg teeth pull the straw bulk through the machine, to comb out the grain as gently as possible and to maximize the opportunities to thresh all the grain out of the heads as it passes through the combine body. Even the axial-flow rotor on the Case combine has a specialty rotor for rice. All the combines have a returns system that brings back from the cleaner any unthreshed heads for a second threshing operation. The aim is to keep processor grain losses below 3% (ASAE Standard S 343.3).

Peg teeth have a long ancestry and, despite over 200 years' development, they may look surprisingly similar to the earliest patented designs which were originally intended for wheat. Today, however, pegs or spike teeth are seldom used in wheat fields. This is because modern wheats mature faster and, at harvest, have stiff and brittle straw easily broken or chopped up by peg teeth—this causes very high grain losses, deterring adoption of stripper fronts on Australian wheat fields. On the other hand, the more aggressive and greater number of contacts with pegs is needed for the fine, tough and wiry pedicels and fibrous stems of rice, *Oryza sativa spp.*

Straw pieces put an excessive load on the combine cleaning system in particular. This is reduced if the smoother raspbar thresher system is used. *Raspbars* can be used for a rice crop—if the crop is dry enough, but very few seasons are *that* dry. On the other hand, the adoption of stripper fronts makes it possible to use the raspbar drum successfully, which facilitates changeover to harvest other crops if conditions are favorable to a stripper front.

Rice Combines in Asia

In Japan, compact combines have proliferated; there is almost one on each farm, even though average farm size is under two ha. The trend has been followed by Taiwan and South Korea—each of these being countries where industrial growth has taken labor away from agriculture and rice prices are kept high by government policy. Most of the compact combines are head-feeding self-propelled units developed specifically for rice. They are unsuited to other crops. Several Japanese makers are now producing a few rotary combines with multicrop capability.

Chinese Combines

The world's cheapest combine (in $ per metric ton) is probably the Beijing Combine Harvester General Works combine. In 1995, the 90 hp diesel-powered Model 4LZ-3 with 3 m front could be procured from the factory for *only US$8700.* China has had as many as 20 state-operated combine factories, and their total production peaked at 11,913 units in 1992. The most popular Chinese combines are tractor wraparounds, rice-wheat machines mounted onto locally available tractors. Not included in the total are *hand-tractor-mounted combines* (which have no equivalent in the Western world), these are usually of the head-feed type. Some European combine models are produced in China under license, as well as other modified Western-style machines.

Combines in Southeast Asia

Asian countries which experienced rapid economic growth in the decade of the 1990s, such as Malaysia and Thailand, saw rapid adoption of combines for much of the harvesting on the larger rice farms. In Malaysia the trend was for contractors and repair shops to acquire secondhand European-built combines, then repair them for a "second life" on local rice fields. A roadside combine restorer in Malaysia could fully recondition the machine and sell it to local contractors for one-third of new price.

In Thailand, on the other hand, a score of manufacturers have been selling 3 m combines with Western-style headers and an axial flow thresher on top of a crawler-tracked undercarriage. The first Thai combine was fabricated at Pathum Thani in 1987. By the end of 1995 around 3000 had been produced. The largest maker, Kaset Pattana, made 250 per year, and equipped them with 120–175 hp engines driving through mechanical or hydrostatic transmission option. Some contractors build their own and harvest at a price between 8% and 14% of paddy value, depending on the proximity of the field and crop conditions. Some 80% of Central Thailand and 40% of the entire country's crop is combine-harvested. Flotation is aided by the 4 ft wide wooden grousers fitted to the track plates of the D2 Caterpillar tracks that are utilized in the undercarriage. There are problems with the first crop with its very long straw, and fields that are small or distant from roads become inaccessible to this combine. The machines are heavy—up to 5 t—slow and cumbersome, but the labor shortage is so acute that the Thai rice industry in central Thailand has become dependent on the combines. A crew of four accompanies the machine and the paddy is unloaded by sack.

Unless small-area farmers can reduce their harvest costs, they will be unable to compete and may be forced out of farming, which increases out-migration of rural workers. Contract harvesting is the solution. The small-scale 0.8 m wide stripper-gatherer developed at the International Rice Research Institute was designed as a machine that can be hand-carried into small fields (Fig. 1.303). This design is being manufactured in several Asian countries [25] but it needs a separate power thresher.

Summary

In the industrialised nations, rice is combine-harvested. Most of the rice that is *traded* internationally is harvested by combine harvesters, by combines modified or designed specifically for paddy field conditions. Those crop and field conditions for rice are demanding enough to justify rice-special versions.

The rice combine market is shared between walker type and rotary designs, the tangential-feed machines usually being equipped with spike-tooth cylinders. Conventional raspbars that are used in other cereals *can* be used for a rice crop—if the crop is dry or if a stripper front is fitted but, for best quality, rice should not be allowed to dry too much in the field, or quality suffers.

An overview of the specifications of combines used across the rice world indicates that, on average, they require a specific power of 12 kwh/t grain throughput, depending on conditions. Custom charge rate for combining rice is typically around 10% of paddy price.

In the Western world, the market for combines has diminished to one-fifth the number sold twenty years ago. Combines have become larger, with far higher capacity, and the number of farms has steadily diminished. Deere and Case have the major slice of the market for rice combines, with both now offering walkerless designs especially developed for the rice crop [26].

1.6.7. Power Threshers as Precursors of Mechanization

Power threshers played an important role as the forerunners of farm mechanisation in the Western world. The same was true of Japan and is proving to be the case yet again in nations of the developing world.

Western-style or even modern, small Japanese combine harvesters have been shunned by low-income farmers with fields that are often too small or inaccessible for self-driven equipment. Manual threshing is probably the most tedious and least-attractive field activity. Rising labor costs and manpower scarcities for harvesting accordingly have hastened the development of a wide range of power thresher designs across the rice world in the last 40 years. Threshing machines—power threshers—have enjoyed an illustrious history. In the industrialized world they are completely outdated, but in their heyday they

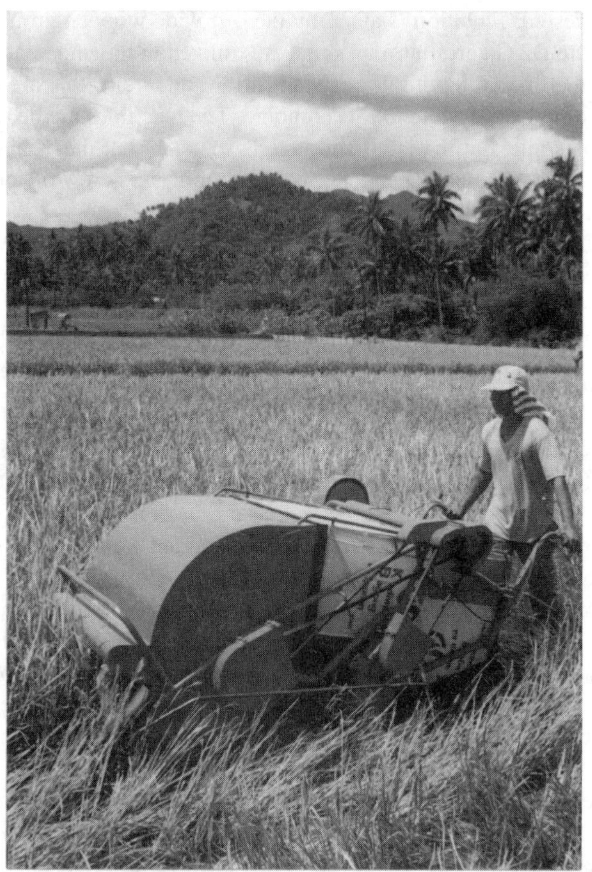

Figure 1.303. The SG 800 stripper-gatherer developed by IRRI in the Philippines for small-area rice farms. The walk-behind rice stripper weighs just 240 kg and is propelled with 11 hp air cooled engine, can harvest one ha/day with a team of seven and a thresher (25).

were as important as they were large. Antique threshers can still be seen at show days. They are an impressive sight, especially when connected by an enormous belt drive to a steam traction engine. But they are no more than nostalgic museum or show pieces today. By contrast, in developing countries, there are places where threshing is still done using muscle power. Hand or foot treading, either human or animal, or manual bundle-beating still goes on. This type of work takes weeks, even months to get through the harvest. The big Western-style threshers were tried in some developing countries, but they were just too costly, too heavy, and ill-suited to small or inaccessible fields. It was only after the 1950s that Japan, South Korea, Taiwan, China, and other Asian countries began to adopt small power threshers. This came about with rising labor costs and harvest labor shortfalls that intensified with rapid industrial growth during reconstruction after World War II.

In Japan threshers progressed from hand-feed to mechanical head-feed, and from stationary to self-mobile machines.

Other Countries in Asia

In some low-income parts of Asian nations, labor costs are cheap enough that the use of the sickle and manual threshing still predominates. But in these places the use of the power thresher is the necessary first step to enhance labor productivity and reduce drudgery—this might be called *labor-intensive mechanization.* In the Philippines, for example, 57% of the people are involved in farming, and rice is the major commodity. Ninety percent of the rice crop is mechanically threshed in that country. There are only a handful of combine harvesters. The recent development at IRRI of small pedestrian-controlled stripper harvesters has raised the level of interest in this approach to the rice harvest, with six manufacturers each turning out a small number of these stripper-gatherers to test the market. IRRI's SG800 stripper, however, requires a stationary thresher [25].

Chinese Threshers

China, with a 66% rural population, is the world's largest rice producer, making up a third of the world's total. Foot-treadle-operated threshers are still to be found in China, but electric motor-driven models of the same type, i.e., hold-on thresher with wire loop cylinder, are widely used. The Chinese have introduced rural electrification to an extent unprecedented in history. It is not uncommon to see in remote fields, far from any buildings, a thresher brigade manhandling their sled-mounted thresher with its electric motor over to a power pole socket connected to the overhead power line.

Many thresher designs have been manufactured in China, such as conical throw-in threshers, twin-drum through-flow types, vertical-shaft as well as horizontal axial-flow and fan-type power threshers. The main factor holding back axial-flow designs has probably been Chinese farmers' diligence in making use of the whole crop, in which case they did not want to break up the rice straw. The more recently developed axial flow and rotary thresher designs do not leave the straw intact.

IRRI Axial-Flow Thresher Developments

A comprehensive USAID research project funded between 1965 and 1973 precipitated the successful redevelopment and commercialization of the axial-flow threshers at IRRI with designs that were compact, elegantly simple, and intended for small-area farming in Asia. The project placed emphasis on the need for equipment designs that could be manufactured in-country. Outreach engineers on the project sought to popularize the design and had some impact in Indonesia and India. But in the Philippines and Thailand, particularly, the adoption rate of the axial-flow thresher was rapid and extensive. In 1975 Thai manufacturers took the concept from trailer-mounted models through to power-hungry, truck-mounted models, with engines up to 100 kw, complete with power feeders and sacking elevators. In the Philippines, numerous manufacturers fabricated smaller models that could be manhandled on bamboo poles into remote fields, and they also made compact trailer-mounted models with full cleaning systems.

IRRI Engineers continued research and promotional efforts on axial-flow thresher developments. A German GTZ project has assisted with a study of crop movement

Figure 1.304. Axial-flow and rotary threshers have been around for over two centuries, but what the IRRI team was able to achieve in some developing countries was to popularize an axial design that was well-matched to small-area farming conditions and that could be locally fabricated at modest cost. This photo shows a TC 800 axial flow thresher, the 0.5 t/h unit which is used in conjunction with the SG800 Stripper-Gatherer (B Douthwaite, IRRI photo).

around the thresher rotor by using magnets and induction sensors. The straw may travel from 4–11 times around the rotor (Fig. 1.304). Considerations of straw flow lead to a Vietnamese design which utilizes less teeth and can thresh rice bundles that are dripping wet from flooded fields. By 1990, the government estimated that there were 1000 thresher manufacturers in the Mekong Delta of Vietnam alone, and that they had built over 50,000 units (Fig. 1.305). Following the popularization of power threshers in the Mekong, Vietnam joined the top three rice-exporting nations, with most of Vietnam's export grain coming from the Mekong Delta.

Throw-in Threshers That Chop the Straw for Stockfeed

Many farmers in Asia want cereal straw—wheat straw in particular—to be chopped up fine for use as animal feed. In some seasons chopped wheat straw has almost as much

Figure 1.305. Trailer-mounted threshers are the most popular style in Vietnam, some are mounted on a self-driven chassis; this unit from Cantho is powered by a locally made, 15 hp water-cooled diesel and has a capacity of up to 2 t/h of cleaned paddy. The thresher is driven down the road to each pile of rice bundles stacked there ready for threshing, near a canal on the Mekong Delta. The road is also used as a paddy-drying surface when the sun shines. (GR Quick photo).

value as the grain. Indian threshers achieve a simultaneous threshing and straw-chopping action: all of the crop material is forced through the concave and the rotor is equipped with chopping blades similar to those on a hammer mill. Often a fan is mounted integrally on the rotor axle to provide the air blast for the cleaning system underneath the concave, along with heavy flywheels to maintain the momentum required for chopping.

Where wheat and paddy are grown in rotation, such as in Egypt, parts of China and in India, farmers need threshers capable of use in either crop with a minimum of modification. An adapted thresher with a set of adjustable louvers inside the cover of the thresher drum can be fixed at 90° to the rotor axis and the straw outlet blanked out for threshing *wheat* in the beater mode. The louvers are set at 75° to the rotor axis and the straw outlet door opened for separated straw discharge for threshing *paddy*.

A Quantitative Assessment of Power Threshers

Information from commercial sources from eleven countries across the rice world was accumulated to calculate power versus throughput and, in the case of axial-flow threshers, throughput versus rotor width. Least-squares regression yields the result that for this group of machines as a whole, specific power (power requirement per unit of throughput, also measured by slope of the regression line for the dataset) was 6.125 kW·h/t of rough rice grain throughput.

For the *axial-flow* thresher data subset only, the slope coefficient of the trendline for throughput versus *rotor size* was 4.79 t/h per meter of rotor length. Head-feed threshers were found to require only half the power of the other types as a whole. This type of compact power thresher was not intended for capacities much above 1 t/h, because the feed tray is usually loaded by hand, which restricts feed rate.

Capacities above 2 t/h of grain throughput for throw-in or whole crop power threshers are possible in rice with a power feed mechanism. Power feeders add safety. Unless the feed apron is of sufficient length, there is always a risk of the operator getting a hand or arm pulled in to the drum or rotor of manually fed threshers. The Indian milling type threshers are the most hazardous in that regard and Indian standards have been established to try to minimize injuries to operators feeding the thresher.

There are still huge numbers of power threshers made in Asia. In China for example, a 1992 estimate was that there were 5.9 million threshers on farms and 200,000 were produced (over sixteen times the annual combine production); in India the estimate was 2.2 million on farms with an annual production of around 60,000; while for Japan the peak number of power threshers was 3.3 million in 1967 and production peaked the next year at 372, 263 units. In 1996, annual production of Japanese power threshers was 12,400—less than one-fifth of the shipment of combines that year.

The average contract charge rate for use of a power thresher has been about 4%–5% of paddy value on farm for the Asian countries surveyed.

Summary

Local demands for threshing equipment have encouraged unprecedented production of power threshers in many developing countries. The thresher has proven to be a precursor to mechanization and a needed item that can launch indigenous manufacturing capabilities in low income countries.

References

1. Quick, G. R. and Buchele, W. 1978. The grain harvesters, pp. 268 St. Joseph: ASAE.
2. Kutzbach, H. D. 1988–1998. Combine harvesters: Yearbook Agricultural Engineering No. 1–10, Münster: Landwirtschaftsverlag.
3. Busse, W. 1981. Mechanisierung der Getreideernte-Revolution für die Kornkammern der Welt. VDI-Bericht Nr. 407: 19–26. Düsseldorf: VDI-Verlag.
4. Segler, G. 1956. Maschinen in der Landwirtschaft. Hamburg: Paul Parey Verlag.
5. Caspers, L. 1969. Systematik der Dreschorgane. Grundlagen der Landtechnik 19(1): 9–17.
6. Baader, W., Sonnenberg, H., and Peters, H., 1969. Die Entmischung eines Korngut-Fasergut-Haufwerks auf einer vertikal schwingenden, horizontalen Unterlage. Grundlagen der Landtechnik 19(5): 149–157.
7. ANSI/ASAE/S 343.3. Terminology for Combines and Grain harvesting. ASAE St. Joseph, MI, 1997.
8. Wacker, P. 1989. Quality of work of axial and tangential threshing unit. Proc. CIGR Congress, Dublin 3: 1863–1868.

9. Arnold, R. E. 1964. Experiments with raspbar threshing drums. J. Agric. Engng. Res. 9: 99–134.
10. Caspers, L. 1973. Die Abscheidefunktion als Beitrag zur Theorie des Schlagleistendreschwerks. Landbauforschung 19: pp. 154. Braunschweig: FAL.
11. Huynh, M. V., Powell, T., and Siddal, N. J. 1982. Threshing and separating process—A mathematical model, Transactions of the ASAE 25(1): 65–73.
12. Miu, P. I., Beck, F., and Kutzbach, H. D. 1997. Mathematical modeling of threshing and separating process in axial threshing units: ASAE Paper No 971063.
13. Freye, Th. 1980. Untersuchungen zur Trennung von Korn-Spreu-Gemischen durch die Reinigungsanlage des Mähdreschers. Forschungsbericht Agrartechnik. (47): 114, Hohenheim: Agrartechnik.
14. Segler, G. and Freye, Th. 1977. Vibro-pneumatische Trennung von Stroh, Korn und Spreu im Mähdrescher. Grundlagen der Landtechnik 27(4): 101–108.
15. Beck, F. and Kutzbach, H. D. 1996. Computer simulation of grain/chaff separation on grain pan and cleaning shoe of the combine harvester: ASAE Paper No. 961051.
16. ANSI/ASAE/S 392.2 Combine Capacity and Performance Test Procedure. ASAE St. Joseph, MI, 1997.
17. ISO 8210 Equipment for harvesting–Combine harvesters–Test procedure. ISO Genève, 1989.
18. Beck, Th. and Kutzbach, H. D. 1990. Messung und Beurteilung der Leistung von Mähdreschern. Landtechnik 45(6): 218–222.
19. Eimer, M. 1973. Untersuchungen zur Durchsatzregelung von Mähdreschern. Habilitationsschrift. Göttingen: Agrartechnik.
20. Huisman, W. 1983. Optimum cereal combine harvester operation by means of automatic machine and threshing speed control. Doctoral thesis pp. 295, Wageningen: Agricultural Engineering.
21. McGechan, M. B. and Glasbey, C. A. 1982. Benefits of different forward speed control systems for combine harvesters. J. Agric. Engng. Res. 27: 537–552.
22. Reitz, P. 1997. Untersuchungen zur Ertragskartierung während der Getreideernte mit dem Mähdrescher. Forschungsbericht Agrartechnik (305): 147, Hohenheim: Agrartechnik.
23. Klinner et. al. 1987. A new concept in combine harvester headers. J. Agr. Engng. Res. 38(1): 37–45.
24. IRRI. 1993. *World Rice Statistics, 1993–1994*. International Rice Research Institute, Manila, Philippines.
25. Quick, G. R. and Douthwaite, B. 1994. A bright spot on the farm equipment scene. Development of stripper harvesting on small farms in Asia reveals unforeseen advantages. *Resource* 1(2): 14–17. June 1994.
26. Quick, G. R. et al. 1996. Rice Harvesters Research Report. Tests show harvest management is crucial. *Farming Ahead with the Kondinin Group* 57: 32–45, September 1996.
27. Quick, G. R. and Hamilton, G. R. 1997. Recent evaluations of grain harvester combinations in Australia. Paper 97-1066. *ASAE meeting*, Minneapolis, August 1997.

Forage Crops

A. G. Cavalchini

1.6.8. Foreword

Forage crops can basically be divided into two main categories: meadow-type forages and forage cereals.

Although meadow-type forages—grasses and legumes—continue to play a fundamental role in livestock farming on a global scale, in the high intensive agricultural systems they have been largely replaced by whole-plant harvesting of cereal crops (maize, barley, wheat) (Fig. 1.306) which permit:

- A substantial simplification of the harvest mechanization chain.
- Generally a much lower cost per *forage unit* (FU) and higher unit productivity (UF/ha).
- A more consistent level of forage quality, with a higher energy content, which is therefore better suited to modern methods of feeding beef and dairy cattle.

In connection with this last point, it should be noted that feed-transformation efficiency has improved considerably over the last decades in all those husbandry systems that have introduced rations with increasingly high energy content, both in unit terms (kJ/kg of feed) and absolute terms (kg of feed/day) (Table 1.60). In less developed countries, where the cost of cereals remains high and there is competition between humans and animals for food, grazing is still the primary resource and feed supplementation is often limited to byproducts.

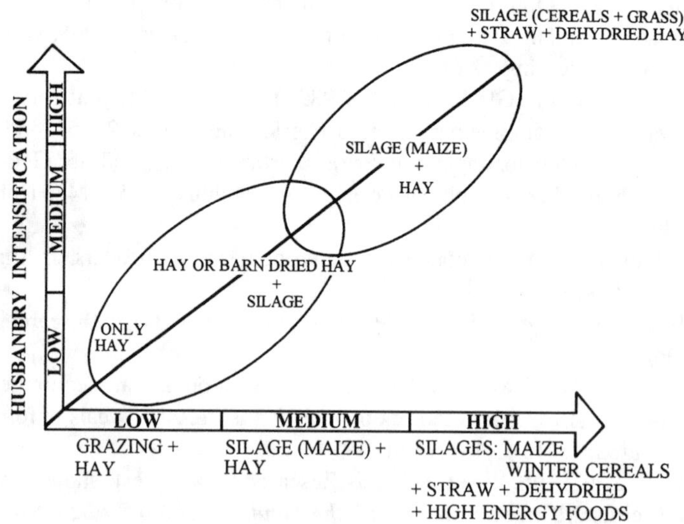

Figure 1.306. Different feeding systems depending on land and husbandry intensification.

Table 1.60. Feed efficiency of dairy cattle (U.S. & Europe)

Year	Milk/Year (kg/cow, year)	Feed/Day (kg/day)	Gross Energy Efficiency (%)
1945	2200	10	10
1970	4500	13	14
1990	7000	18	18
Superior animals	13000	28	22

Source: Modified from Van Soest 1995.

Table 1.61. Energetic efficiency of different techniques for alfalfa and maize harvesting

Crop	Harvesting Technique	Input (GJ/ha)	Output (GJ/ha)	Efficiency: Output/Input
Alfalfa	Hay making	9	25–44	2.8–4.9
	Barn hay drying:			
	With conventional energy	28	36–55	1.3–2.0
	Solar energy	10	36–55	3.6–5.5
	Haylage (harvested at 50% m.c.)	10	31–51	3.1–5.1
	Partial dehydration (from 65 to 15% m.c.)	100	46–58	0.46–0.58
	Total dehydration (from 80 to 15% m.c.)	180	55–65	0.30–0.36
Maize	Whole plant silage	30	70–80	2.3–2.65

Moreover, we must consider that meadow forage species have not matched the yield increases attained by other crops over the past 30 years. Also, for this reason, meadow forages are not so attractive in intensive systems, and many farmers consider hay to be a second-class crop.

It should be kept in mind that the *energy efficiency* of the process for producing and harvesting meadow forages is generally higher than for forage cereals, which require greater inputs, particularly of agrochemicals (Table 1.61).

1.6.9. Meadow-Type Forages

The influence of the vegetative stage on the quantity and quality of forage must be emphasized (Fig. 1.307). This problem has repercussions on the planning of harvest activities and on the sizing of harvest mechanization chains, which must be capable to guarantee the timeliness of operations. In addition during the various stages of harvesting, treatment and storage, *physiological*, *mechanical* and *fermentation* losses incur: these are often very substantial, both in terms of absolute value FU/ha, and unit value (FU/kg of dry matter). Physiological losses occur during field drying, mainly before the moisture content (MC) has fallen below 35%–40%, although they continue till 15% mechanical losses occur during the various harvesting operations and can be of two types: losses due to product that is not collected; and losses resulting from mechanical manipulation, particularly of the delicate leaf portion which tends to break off. Fermentation losses

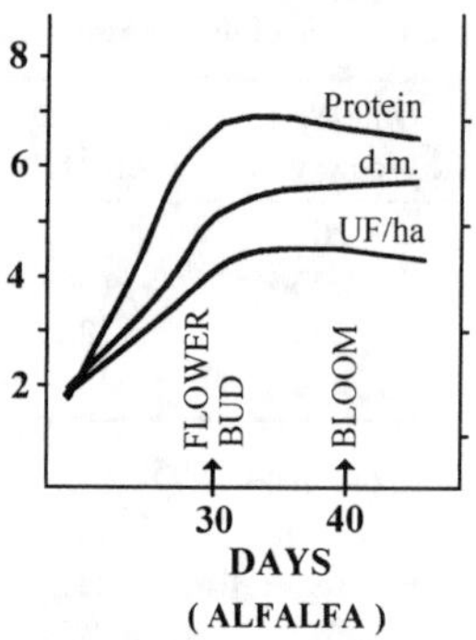

Figure 1.307. Dry matter (DM), forage unit (FU) and protein yields depend on the stage of plant growth (based on INRA, 1996).

occur during storage in the barn or silo; they are caused by the transformation of proteins, sugars, etc., which is accompanied by a release of heat and, consequently, by a loss of energy content.

Harvest, Treatment and Storage Methods

Meadow forage species are mainly harvested and stored in the form of *hay*, but new methods have been developed as alternatives to *traditional haymaking*, which requires natural drying in the field. The *wilted silage*, or haylage, technique has found widespread application in countries with wet, rainy climates (northern Europe), and is now also spreading to countries with drier climates. The *barn drying* method was introduced in mountain areas where frequent rains during harvest time occur. This technique involves partial drying in special drying barns, using natural or heated air ($\Delta t = +10°–20°C$). Also in the 1950s, the *dehydration* method was proposed, which is based on the complete artificial evaporation of the moisture using air at high temperature (150°–900°C). The various alternatives are shown in Fig. 1.308.

Harvest and Use of Green Forage

This is the least advantageous solution for intensive animal-husbandry systems, because it requires a daily organization of the harvest mechanization chain, it involves the distribution of high volumes of product to the animals, and the availability of the feed is only seasonal. However, it offers the advantage of very low or null losses. The implementation of this method requires the sequence of operations shown in Fig. 1.309. Certain operations can be grouped together. For example, mowing and windrowing can be easily performed by a single machine, the cutter-windrower, while a more advanced level of mechanization would require the use of cutter-loader or a flail harvester. The use of a forage harvester, as an alternative to a cutter-loader, is justified only in cases where such a machine is already present in the farm for other purposes. The work times and energy requirements are given in Table 1.62.

Traditional Haymaking

This is based on natural drying in the field of the forage from 80% down to 20%–30% of MC on harvesting (Fig. 1.308). Certain manipulations (spreading, tedding, raking) must be performed after mowing to facilitate drying of the forage. The frequency of such manipulations depends on the product mass and the climatic conditions. Windrowing the forage for the night and then tedding are justified only when the relative humidity of the night air is very high. However, this procedure is best avoided as it increases mechanical losses, especially with delicate forages such as legumes. To improve drying, by uniforming the moisture in the leaves and stems, and reducing field-drying time, it is possible to perform conditioning in conjunction with mowing. There are also a great

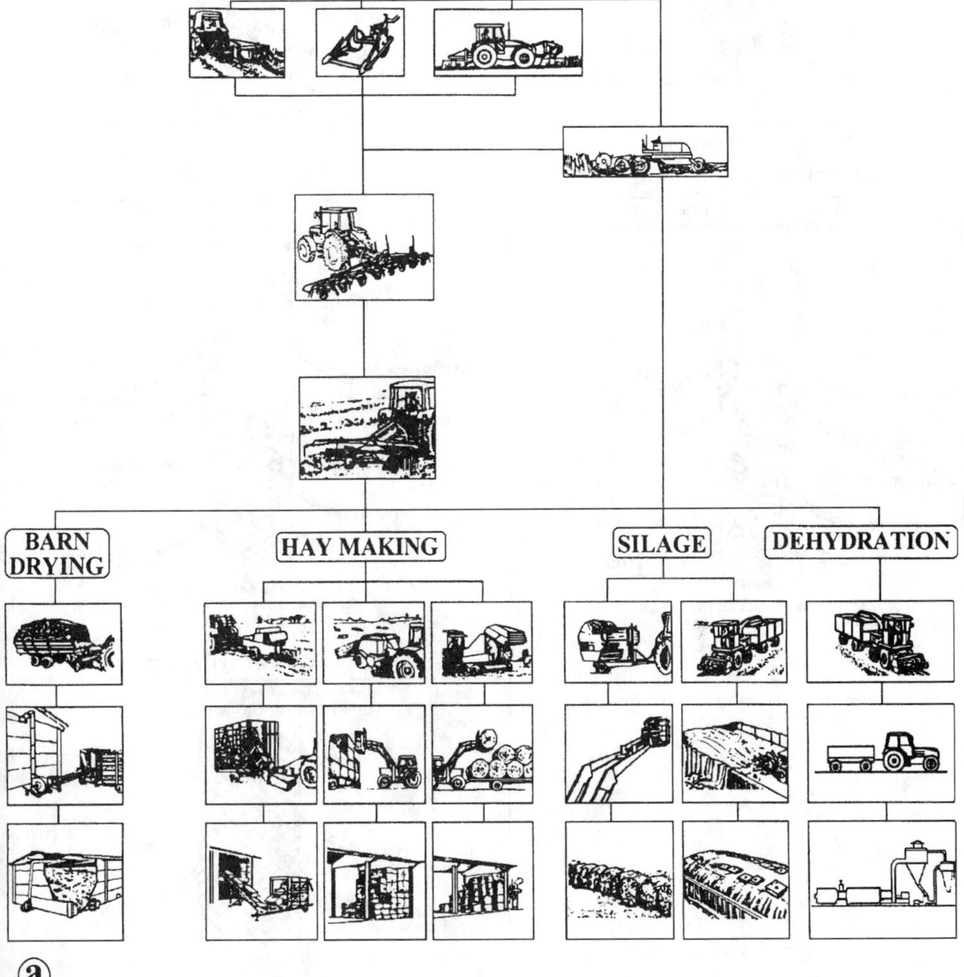

Figure 1.308. **(a) The different meadow forage possibilities and (b) the net obtainable yield.**

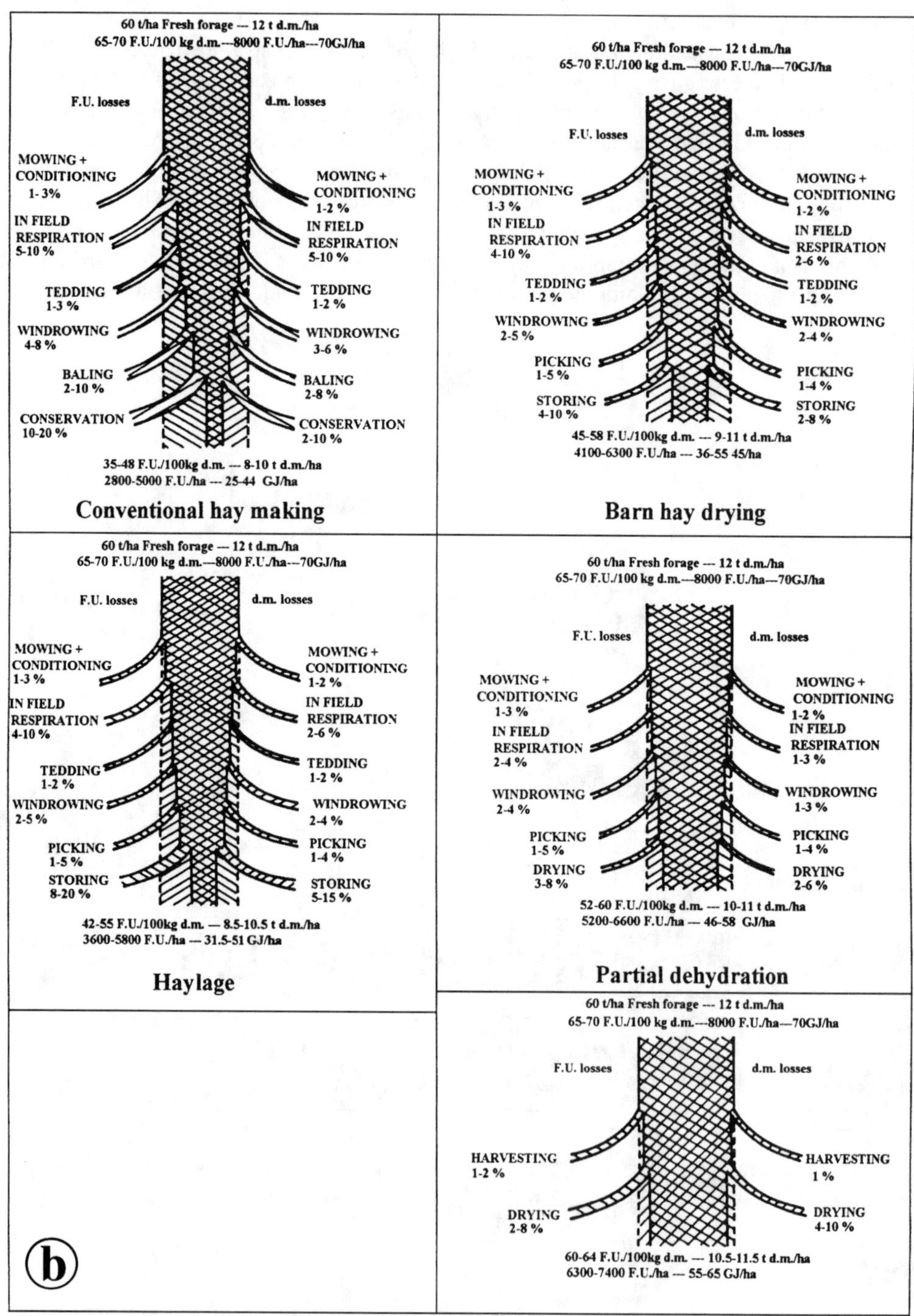

Figure 1.308. (Continued)

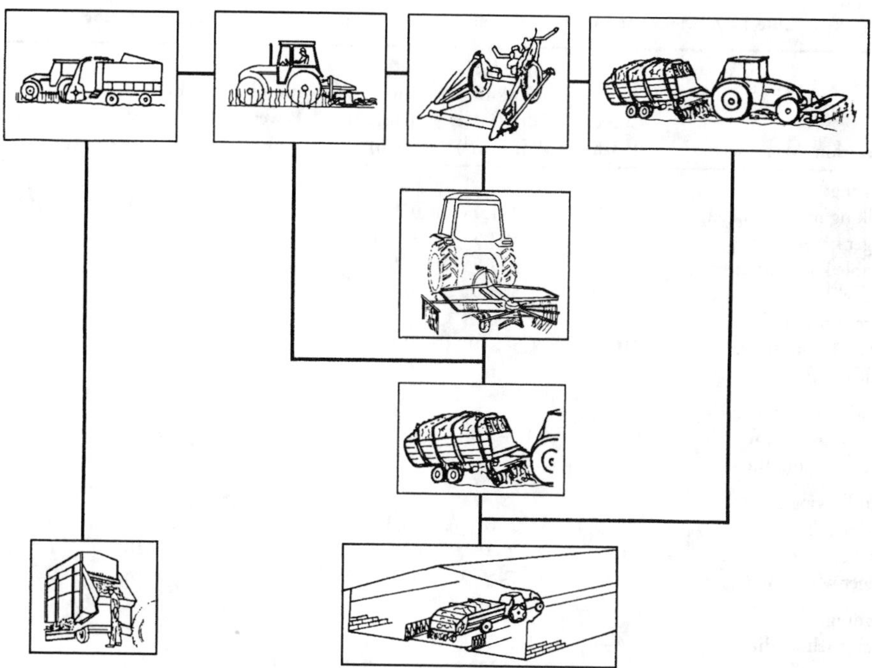

Figure 1.309. Different work site layout in green forage harvesting.

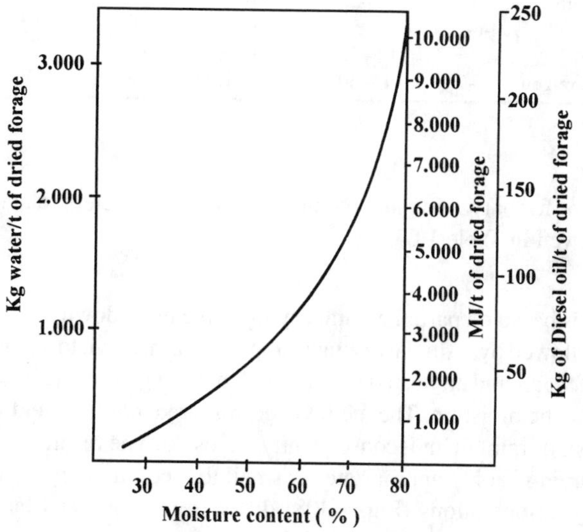

Figure 1.310. Water quantity to be evaporated and energy consumption referred to 1 t of dried forage (15% dry matter) for different moisture content of the forage when conveyed to the drier.

Table 1.62. Main parameters of the machines involved in the green forage harvesting

Operation, Machine	Speed (km/h)	Average Working Width (m) or Volume (m^3)*	Capacity (ha/h,m) (t/h)*	Minimum Power (kW)	Energy Consumption (kW·h/ha)	Time (man-h/ha)
Mowing:						
Walking mower (finger bar)	2–3	1.2–1.4	0.2–0.25	8	18–20	4–5
Finger bar mower	5–7	1.5–2.5	0.4–0.5	15	18–20	0.8–1.6
Double knife cutter bar	6–9	1.5–2.5	0.5–0.7	15	18–20	0.6–1.2
Rotary disc mower driven from below	9–10	1.5–3.0	0.7–0.8	25	20–25	0.5–1.0
Rotary drum mower driven from the top	10–12	1.5–2–0	0.8–0.9	30	20–25	0.6–0.9
Mowing + Windrowing:						
Rotary drum mower driven from the top	10–12	1.5–2.0	0.8–0.9	30	20–25	0.6–0.9
Windrowing:						
Rotary rake	7–9	3–6	0.3–0.4	25–30	18–20	0.6–1.5
Parallel bar rake	6–7	2–3	0.25–0.35	15–20	15–18	1.0–2.0
Finger wheel rake	6–8	2–5	0.25–0.35	15–20	15–18	0.7–2.0
Loading:						
Forage self-loading wagon	4–6	1.2–2.0 15–30*	0.6–1.2 15–20*	30	15–20	0.8–1.5
Mowing + loading:						
Self propelled forage wagon equipped with mowing system	4–7	2–2.5 15–30*	0.6–1.0 10–15*	30	1 5–20	1–1.5
Flail forage harvester + forage wagon	4–7	1.5–2.0 15–30	0.6–0.8 10–12*	40	30–50	1.2–1.6

number of options for harvesting and baling. The main parameters related to the various operations are given in Table 1.63.

Barn Drying

This method is based on partial natural drying in the field, down to a moisture content of 40%–45%, followed by artificial drying in a special barn, down to a moisture content of 15%. Both ambient air and air heated with a small thermal gradient ($\Delta t = 10°–20°C$) are used to evaporate the moisture. The method permits a reduction in field-drying time and its associated risk of rainfall and, consequently, in losses. The result is an higher-quality product. Barn drying is only appropriate for small to medium dairy farms.

The barn drying operations (Fig. 1.308) differ from traditional haymaking in that they do not require so much tedding, and the harvest operations are carried out using hay wagons. The productivity of the entire harvest mechanization chain is limited by the capacity of these machines, and by that of the systems for filling and unloading the barns. Productivity and energy requirements are given in Table 1.64.

Table 1.63. Main parameters of the operations and machines involved in the hay making

Operation, Machine	Speed (km/h)	Average Working Width (m)	Capacity (ha/h,m)	Minimum Power (kW)	Energy Consumption (kWh/ha) (kWh/t)*	Time (man h/ha) (man h/t)*
Mowing: See Table 1.62						
Mowing + Windrowing: See Table 1.62						
Teddding:						
Rotary tedder	10–14	2–6	1.5–6	20	10–15	0.2–0.7
Windrowing: See Table 1.62						
Baling:						
Conventional baler	4–6	1.0–1.8	2–8	25–45	1.8–2*	0.13–0.5*
Round baler	5–7	1.5–2.0	2–8	40–60	2.0–2.2*	0.13–0.5*
Big baler	6–8	1.8–2.2	10–15	70–100	1.9–2.1*	0.07–0.1*
Stack-wagon	4–6	1.6–2.0	10–15	40–70	1.5–1.8*	0.007–0.1*

Table 1.64. Main parameters related to the operations and machines involved in the barn hay drying. The main difference with the hay making technique are: tedding is normally not practiced; mowing is performed jointly with windrowing; hay (at high moisture content −40%–45%) is harvested by self-loading forage wagons

Operation, Machine	Speed (km/h)	Average Working Width (m)	Capacity (t dm/h)	Minimum Power (kW)	Energy Requirement (kWh/t d.m.)	Time (man-h/t d.m.)
Mowing Mowing; + Windrowing; Tedding; Windrowing: See Table 1.62						
Harvesting:						
Self-loading wagon	4–7	1.5–2.0	4–12	40	1.5–2.5	0.08–0.25
Dischargement + Loading the barn:						
Pneumatic loader	=====	=========	5–8	15	2–2.5	0.12–0.5

Dehydration

The primary objective of this method is to reduce, in order to obtain both high outputs (UF/ha, or kJ/ha) and high nutritive value of the feed (FU/kg d.m. PD/kg d.m.), retaining—in particular—the high vitamin A and D content. These characteristics make dehydrated alfalfa hay an important component in the diets of poultry and swine. The dehydration technique involves: direct cutting of the forage using a forage harvester, high-temperature (150°–900°C) evaporation of the moisture content from

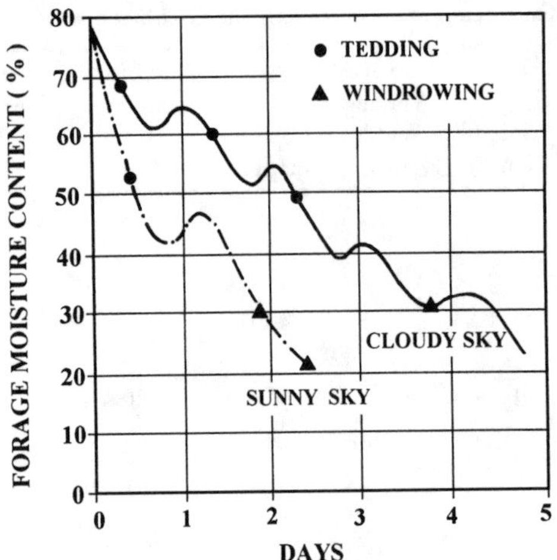

Figure 1.311. Typical natural drying in the field in different weather conditions.

80% to 15%, mainly in drum dryers. Consequently, the energy consumption is high and only the highest-quality forages, such as alfalfa, can justify the expense of artificial drying.

Today the preferred method is *partial dehydration*, which permits substantial savings in energy, increases the work capacity of the driers, and consequently reduces costs drastically. With this method, the forage is mowed using a windrower and left to pre-cure in the field down to a moisture content of 50%–65%. It is then collected by a forage harvester equipped with windrow-pick-up and finally it is dehydrated in a drier plant. Energy consumption is substantially reduced, though it remains high in absolute terms (Fig. 1.310). The quality of the dried forage remains excellent and is only slightly inferior to that obtained with total dehydration, thanks to the limited field time: the natural drying curve is in fact very steep in the first stages (Fig. 1.311). Consequently, the risk of rainfall is also fairly low. The main parameters are given in Table 1.65.

Silage

The forage is cut at high moisture content and stored in an anaerobic environment (silos or other containers) reducing field losses and permitting some freedom from climatic constraints. Wilted silage does not involve too many manipulations, nor does it require complex or costly mechanization chains, nor direct energy utilization. Abnormal fermentation phenomena can occur during storage, leading to the development of molds and listeria and degradation of the product, which can become toxic for the animals. During storage, fermentation is more easily controlled for graminaceous species, whereas legumes, due to high protein content, can easily give rise to fermentation of a butyric nature, i.e., to rotting of the forage.

Because of the advantages mentioned above, silage is becoming increasingly widespread, with harvest moisture levels that are continuously rising, thanks to the development

Table 1.65. Main parameters related to the operations and machines involved in the dehydration

Operation, Machine	Speed (km/h)	Average Working Width (m)	Capacity (t/h)	Installed Power (kW)	Energy Required (kWh/t)	Time (man-h/t)
a Total Dehydration (Direct Harvesting)						
Harvesting*:						
Trailed, or front mounted models	5–6	1.5–2.5	10–20	70–120	2.5–3.0	0.05–0.1
Self-propelled models	6–8	3–5	20–50	150–350	2.5–3.0	0.02–0.05
b Partial Dehydration (Forage Harvested After a Curing Time (4–10 h) in the Field)						
Mowing + windrowing:						
Self-propelled windrowing	8–10	3–5	2–4	50–100	15–20	0.3–0.6
Front mounted rotary mower	10–12	2.5–4.0	2–3.5	50–100	25–30	0.4–0.6
Loading:						
Forage harvester + pickup: trailed or front mounted models	5–7	1.5–2.0	10–15	70–120	2.5–3.0**	0.08–0.014
Self-propelled models	6–8	2.5–4.5	20–40	150–350	2.5–3.0**	0.03–0.07
Special selfloading forage wagon	5–6	1.5–2.5 2–4**	10–30	70–120	2–2.5**	0.03–0.01

* Forage harvester equipped with cutting bar.
** Loading capacity (t).

of new machines and techniques. Whereas in the past, forage was harvested at moisture levels of 40% and 50%, for legumes and grasses, respectively, harvest today often takes place at values of 60%–70%. The mechanization operations are shown in Fig. 1.308. The ideal machine for harvesting is the forage harvester which chops the forage into 40–60 mm length, in order to facilitate a compact mass and to promote an anaerobic environment. Alternatively, it is possible to use haywagons equipped with knives, which handle both the picking up and cutting, as well as the transport. In recent years, the ensilage of cylindrical bales wrapped in plastic film has also become popular. This is in fact a versatile solution which, however, entails additional operations, requires a great deal of skill on the part of the workers, and consumes high quantities of non-biodegradable plastic materials. The productivity and energy consumption are given in Table 1.66.

Machines and Equipment

Mowers

Mowers can be divided into *rotary mowers* and *cutterbars*. The former cut the forage by means of impact forces, while the latter employ a shearing action. Both types can be either *self-propelled* or *tractor-powered.* The former are generally machines of small

Table 1.66. Main parameters related to the operations and machines involved in the haylage. Mowing normally is performed by a mower-windrower; tedding normally not performed

Operation, Machine	Speed (km/h)	Average Working Width (m)	Capacity (t/h)	Installed Power (kW)	Energy Requirement (kWh/t)	Time (man-h/t)
Mowing: Mowing + Windrowing Tedding; Windrowing: See Table 1.62						
Harvesting:						
Forage harvester						
Equipped with pickup						
Front or trailed models	5–7	1.5–2.0	10–20	70–120	2.5–3.0	0.05–0.1
Selfpropelled models	6–8	2.5–4.5	20–50	150–350	2.5–3.0	0.02–0.05
Special selfloading						
Forage wagon	5–6	1.5–2.5	10–30	70–120	2–2.5	0.03–0.1

size, whereas the larger machines also perform other functions, such as conditioning Small walking mowers, which represent the first step of agricultural mechanization are self-propelled; they adopt single action cutterbars no wider than 1–1.5 m. Mower can be mounted to the tractor: rear, the simplest, most economical and therefore mos frequently-used solution; mid, no longer commonly used, because it requires specia mounting kits; front, which is gaining popularity because front three-point linkages ar now available on many modern tractors. This last solution has the advantage of bette control of the machine and superior maneuverability. However, it also requires comple hitching systems capable of disengaging or raising the implement in case of impact wit an obstacle. Rear- or mid-mounted mowers have the bar connected only on one side hence they can easily incorporate a breakaway feature, which allows the bar to pivo rearward to clear an obstruction without damage.

The principles of operation cutterbar mowers can be divided into: those with a oscillating knife and fixed finger bar; and those with dual-oscillating elements. In th former type, a simple, consolidated and reliable solution, the cutting element consist of a fixed part (bar with guards, fingers, or teeth) and a moving part (the cutter blade composed of many knives) (Fig. 1.312). The limit is speed, no greater than 5–7 km/h The need to obtain higher working speeds (8–9 km/h), cleaner cutting, and reduce vibrations led to the development of mowers with dual oscillating knives, or with a oscillating knife and fingerbar. The latter is more robust and better suited for cutting crops close to the ground, whereas dual oscillating knives—more vulnerable becaus unprotected by guards—should not come into contact with the ground.

There are also different solutions for rotary mowers, whose cutting element can rotat on a horizontal or vertical axis. Those with a horizontal axis of rotation are now falling into disuse due to their poor cleanness of cutting. Mowers with a vertical axis of rotatio are very popular due to their high working speed (10–12 km/h), robust construction and low maintenance requirements, but require more energy than cutterbars (Table 1.62).

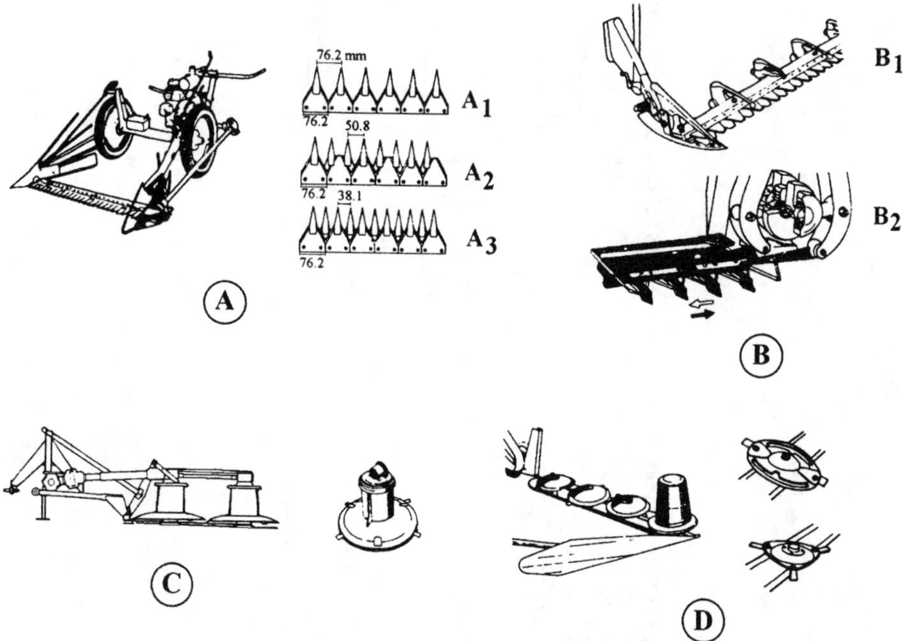

Figure 1.312. Different mower solutions: (A, B) cutterbars; (C, D) rotary mowers. (A) single cutterbars present different finger intervals (A_1, A_2, A_3); (B) double movement cutterbars can have double knife bars without fingers (B_1), or one cutterbar and one fingerbar moving in opposition (B_2); (C) two drum mower with top drive operates also windrowing action; (D) disc mower can have several discs with 2 or 3 knives.

Conditioners

Conditioners perform a mechanical treatment of the forage in order to reduce losses. The treatment does accelerate the evaporation of water from the coarser fraction of the forage (stems), thus aligning the drying time of the stems with that of the leaves. This permits a slight reduction in the field drying time of the forage and, most important, reduces losses during windrowing and baling because the leaves are not so dry, and consequently less delicate (Fig. 1.313). Nutritive value can increase by 10%–15%. It is clear that species with high leaf-to-stem ratios (legumes: alfalfa = 1.2–1.3) benefit more than graminaceous species, which are characterized by low leaf-to-stem ratios.

There are two fundamental principles of operation for conditioners, based on either *rollers* or *flails*. The former solution is more suitable for legumes, whereas the second, more vigorous, is appropriate for grasses. The roller solution consists of two rollers on a horizontal axis, placed parallel to the cutterbar, through which the freshly cut forage passes and receives the conditioning treatment. The treatment can be regulated by adjusting the contrast of the upper roller. The intensity of conditioning also depends on the construction of the two rollers, which can be metal, rubber-coated, smooth, corrugated or grooved (Fig. 1.314). To increase the effectiveness of the treatment, rollers with slightly different peripheral speeds (0.5%–10%) have been used in order to exert an

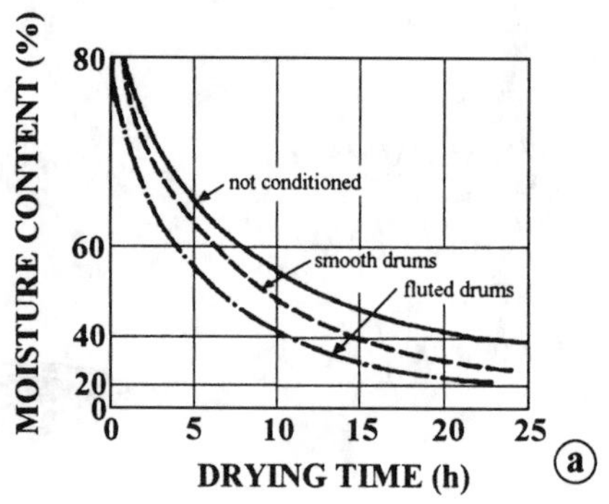

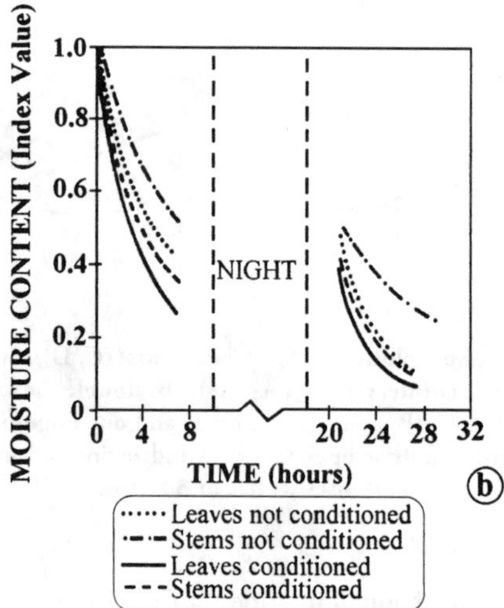

Figure 1.313. Field drying process (alfalfa): a) roller conditioner; b) conditioning reduces differences between leaves and stems drying time.

additional action on the forage. Inflatable rollers (pressure 4–10 atm) in conjunction with a grooved roller, can perform different intensities of treatment by adjusting the inflation pressure. This last solution, together with those that use fluted interlocking rollers, has proven particularly effective on alfalfa. To obtain uniform and effective conditioning, it is also important that the width of the conditioning implement be comparable to that of the cutting implement. The diameter of conditioning rolls generally ranges from 170–220 mm, while the speed of rotation is in the order of 700–1200 rpm: this corresponds to a peripheral speed of 7–12 m/sec.

Flail conditioners employ a different principle of operation. Conditioning action is produced by a rotor turning on a horizontal axis, on which special flails are mounted that strike the forage as it comes in from the cutting element. The flail rotor is partially

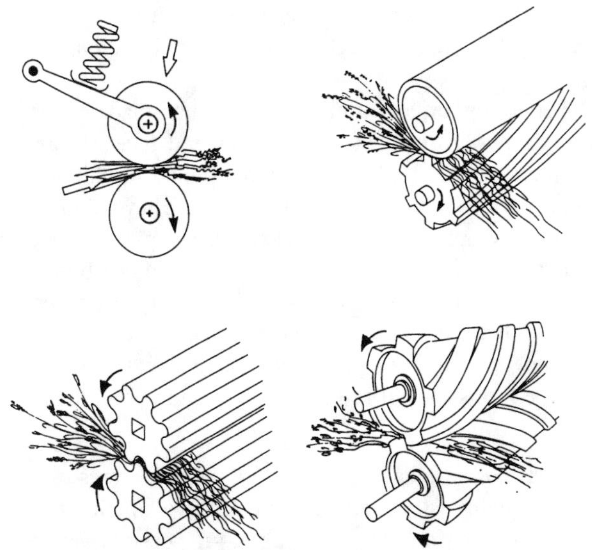

Figure 1.314. Different solutions of conditioning rolls.

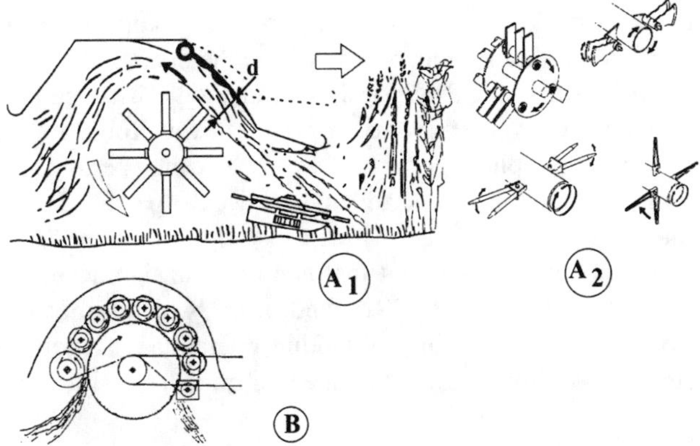

Figure 1.315. A): flail conditioner can perform differently varying the distance d between flails and the peripheral housing and, or adopting different flails. B): in the mat conditioning system the forage is even more vigorously treated due to the peripheral speed difference between the central big drum and the peripheral small drums.

enclosed (about 120°–150°) in a pheripherial housing whose distance from the rotor is adjustable. This distance determines the intensity of the conditioning action (Fig. 1.315).

The conditioning operation today is associated with both mowing and windrowing. This latter operation is more common in large-size machines, generally self-propelled, simply called windrowers. To help convey the forage to the conditioning element, there is generally a reel. The reel becomes unnecessary with rotary mowers whose high working speeds allow the forage to reach the conditioning elements without any assistance.

Concerning the coupling, broad differentiation can be made between self-propelled and tractor-powered types. Among the tractor-powered types, in addition to the rear and/or front three-point linkage and the traditional pulled machines, there are also sophisticated hydraulically driven pull-type solutions which have a hitch over the platform. This solution is normally employed on large machines with platform widths of 4–6 m. The advantages of the hydraulically driven pull-type, in terms of flexibility and performance, can also be obtained with the front three-point linkage types, or better with tractors equipped with reverse drive. In this case, maneuverability becomes comparable to that of extremely sophisticated, large self-propelled machines with high working capacities, justified only for large farms specialized in the production of alfalfa hay.

Tedders

This operation does not pose any particular problems in terms of quality or productivity. As far as work quality is concerned, there are two simple precautions that must be observed: avoid operating on too dry forages—if possible, use moisture contents higher than 50%–55%—to minimize mechanical losses; and avoid contact between the implement (flexible, metal or synthetic forks) and the ground, as this will contaminate the forage with soil particles. Large tedders must have the width subdivided into several reciprocally articulated working sections (rotors) in order to follow irregularities in the terrain. Traditional animal-drawn tedders reproduced the operation exactly as it is performed manually with a pitchfork. Today a simpler method is used, consisting of rotors on a vertical axis inclined slightly forward, on which flexible forks are mounted to lift the forage off the ground and pitch it over backwards. The high peripheral speed, on the order of 9–12 m/sec, to which should be added the advance speed (2.5–3 m/sec), results in vigorous manipulation of the forage, which causes losses that remain negligible so long as the moisture content is high, but that can become significant if the forage is drier (Fig. 1.317). With large self-propelled windrowers, which form a relatively fluffy windrow after mowing and conditioning, the tedding operation is not performed. Tedders are exclusively tractor-powered and rear mounted, or trailed.

Rakes Windrowers

This operation, though apparently simple, poses problems in terms of both quality and labor productivity. Often, it proves to be the bottleneck of the entire harvest mechanization chain, and puts a limit on the surface that can be covered by other machines. Today, rakes are expected to gently lift the forage from the swath and place it in a loose, fluffy windrow, without knocking off the leaves.

Modern side delivery rakes can be classified as *parallel-bar rakes*, *wheel rakes*, and *rotary rakes* (Fig. 1.316). A further classification can be made according to the method by which they are attached to the tractor: *trailed; rear-mounted; front-mounted*. The rear-mounted solution is the one most commonly used today. However, on slopes, frontal versions are preferable. If the requirement is for a high workrate—obtainable only by

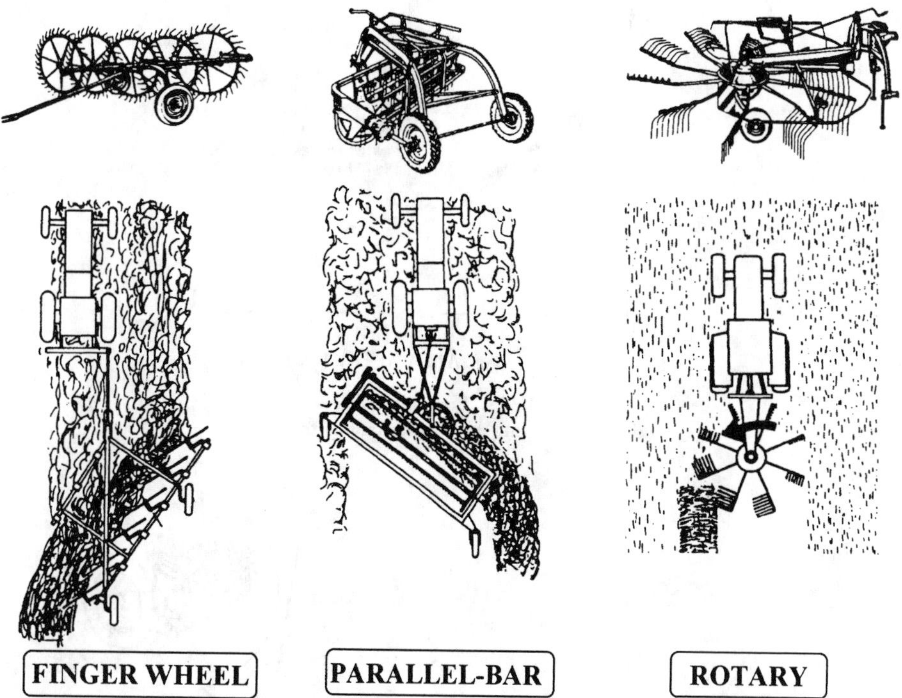

Figure 1.316. The most popular types of rake-windrowers and their operating principles.

substantially increasing the working width, it is necessary to use a trailed version with two or more implements mounted independently on a special frame.

Work quality is difficult to achieve, particularly as concerns: the ability to rake the product, even on irregular terrain, without contaminating it with soil; ensuring delicate contact between the implement (forks) and the product, in order to limit manipulation losses; and creating loose and fluffy windrows. The first point can only be achieved by reducing the working width, or by subdividing it into several independent elements capable of following the irregularities (profile) of the terrain. The best solutions are those based on rotors with tine arms whose diameter does not exceed 3–4 m. The solution in which idle wheels are individually connected to a frame follows irregularities in the terrain very well, but the dragging of the wheels increases soil contamination and forms too compact and tightly rolled windrows (Fig. 1.317). Loose and fluffy windrows make it possible to reduce losses by anticipating the windrowing when the forage is still quite moist and less vulnerable. At this respect the preferred system is again the one based on rotors.

Self-loading Wagons

This machine was widely used for gathering loose forage, but today its use is restricted to poorly mechanized agricultural systems or to particular applications. Self-loading

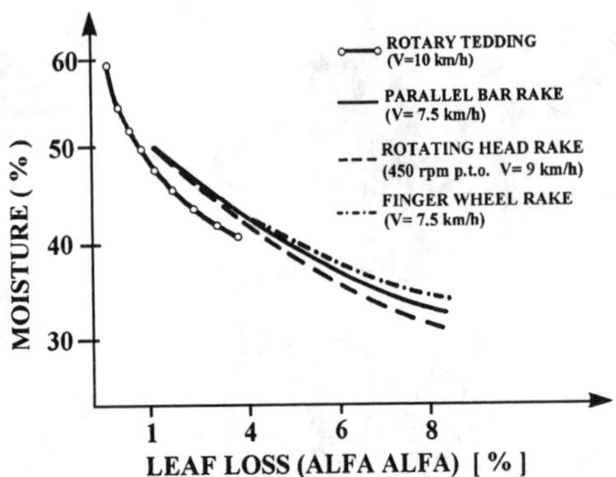

Figure 1.317. Loss in tedding and raking-windrowing due to mechanical treatment.

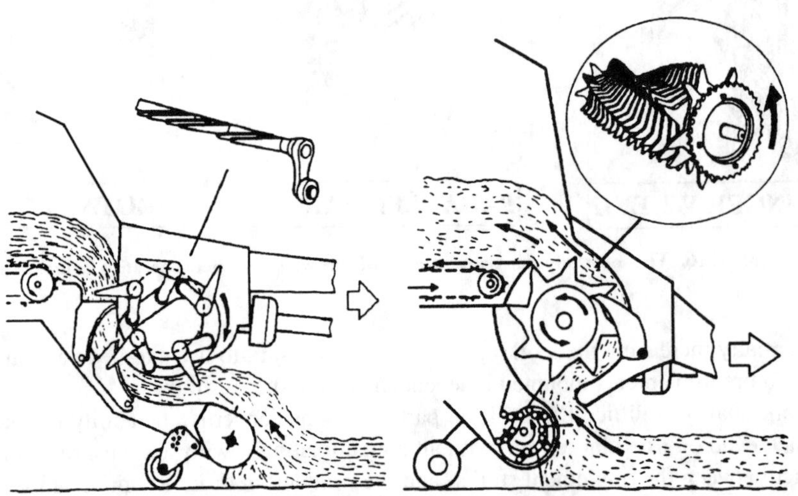

Figure 1.318. Different pickup and cutting apparatus employed in modern self-loading trailers for wilted silaging, or barn drying.

wagons consist of a trailer with high sideboards, capable of carrying large volumes of product (15–25 m^3), a chain-slat floor conveyor and a traditional cylinder type pickup. Modern self-loading wagons, equipped with a forage chopping system that guarantees lengths of 20–40 cm for barn drying, and 8–10 cm for silage, are used for barn drying and silaging (Fig. 1.318). These techniques, if performed correctly, guarantee high-quality forage with lower total losses; if opportunely organized, the work rates and productivity for the harvest, transport and storage phases are also entirely satisfactory. Special high-capacity hay wagons (30 m^3) capable of chopping the forage to 8–12 cm lengths are used in highly organized harvest mechanization chains for forage (50%–60% MC) intended for dehydration.

Baling and Packaging

The problem of baling agricultural products, primarily hay and straw, was solved by the traditional alternative balers which produce hay bales of 20–30 kg, and became widespread during 1950s and 60s, revolutionizing forage harvesting. In fact, the manpower demand for harvest operations was reduced by half, dropping from an average of 40 man-hours to 20 man-hours per ha. However, by the late 1960s the level of mechanization achieved with alternative balers was becoming insufficient, particularly due to the bottleneck of bale handling not solved by the proposed systems for manipulating bales: from simple bale accumulators to sophisticated bale wagons, which were too complex, costly and relatively unreliable. At the time, the solution to the bale manipulation problem seemed to lie in one of two possible directions: aggregating the forage into dense, bite-size packages of hay, with characteristics approaching those of a fluid, so that it could be transported using augers or conveyor belts; or, on the contrary, making the bales large enough to justify their individual manipulation with dedicated lifting equipment. The former concept led to the development of field hay cubers, soon abandoned due to their high energy costs and their limited field of application. The latter concept led to the development of various types of big balers—stack-wagons, round balers, rectangular big balers—which rapidly established themselves in the agricultural world (Fig. 1.319).

Conventional Balers. These are intrinsically complex, but generally reliable, are trailed and powered through the PTO of medium-power tractors, and operate laterally. The forage is lifted from the windrow by a pickup, conveyed to an auger or feed fork mechanism and then forced into the compression chamber where a plunger, driven by a crankarm and pitman, moves at about 80–100 strokes per minute. The section of the compression chamber is generally 36 × 46 cm, but there are also sections of 36 × 48 cm and 40 × 52 cm. The bale length is adjustable between 0.60 m and 1.2 m, while the weight is 20–30 kg, corresponding to a specific mass of 120–170 kg/m^3. The complexity of the baler is already evident: its numerous mechanisms must work together in a synchronized manner, the forage must be rotated through 90° twice, and must be uniformly fed into the entire section of the chamber (Fig. 1.320). The greatest complexity, however, lies in the tie mechanism—today of the twine type only, the wire system having been abandoned—which is the true heart of the machine. Despite the complexity, balers have achieved consolidated reliability, as well as high quality levels and work rates that can exceed 10 t/h of hay. As far as quality is concerned, the dimensions and density of the bales allow the hay to be picked at 25%–28% moisture, thereby reducing leaf losses without generating excessive high temperatures during storage, as is instead the case for round and big rectangular bales (Fig. 1.321).

Round Balers. The first patent on round balers dates back to 1885 and is credited to T. Workman (USA). In the 1970s, to solve the handling problems, large round balers capable of packaging bales weighing up to 0.7 t were manufactured. In addition the system offer the main advantage of producing weather-resistant bales, allowing hay storage to be postponed until completion of the harvest or, in certain situations, the creation of open-air feed stores for livestock raised *au plein aire.* Ground rolling baler

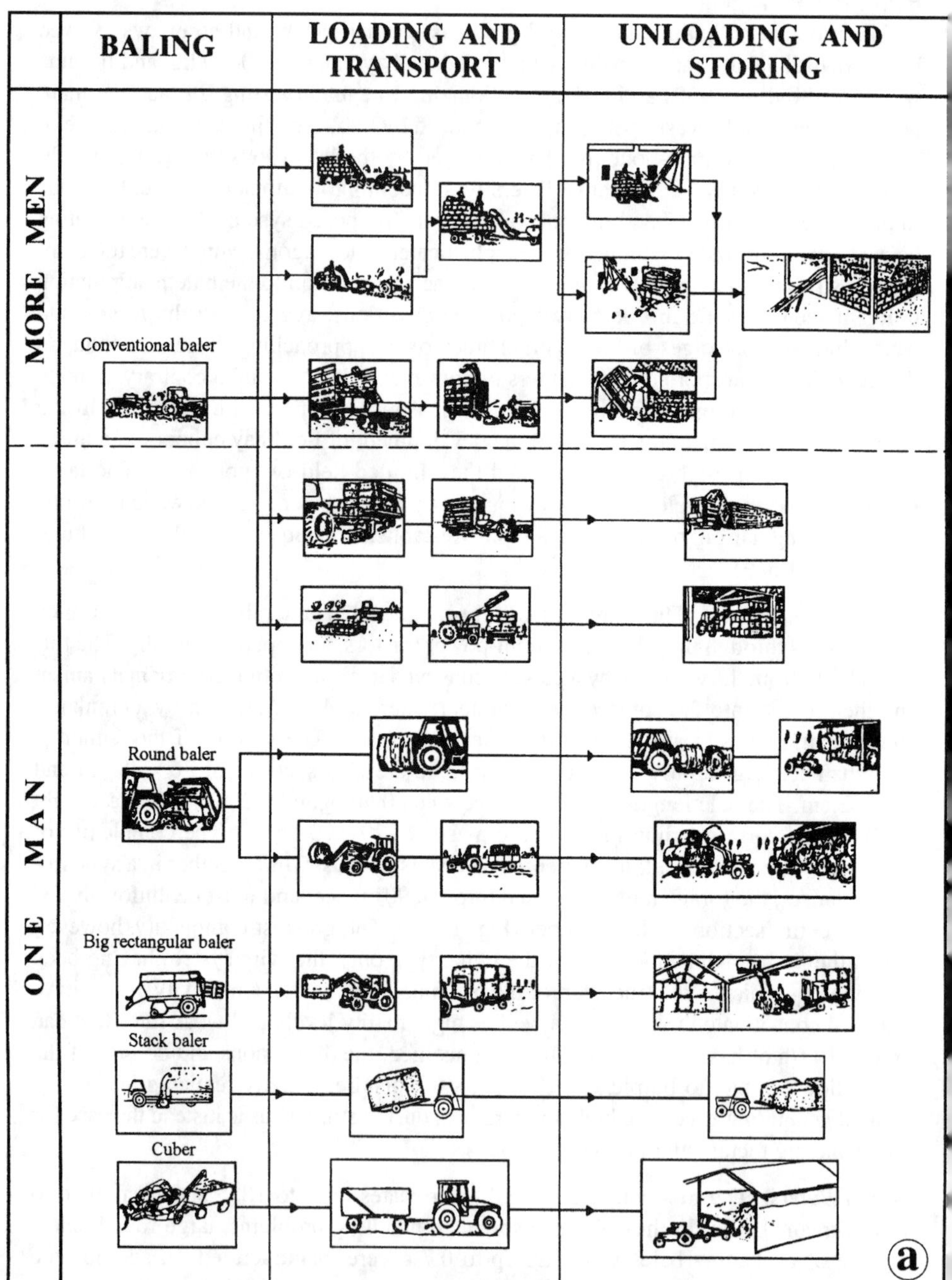

Figure 1.319. (a) The alternatives in hay baling and (b) the main parameters.

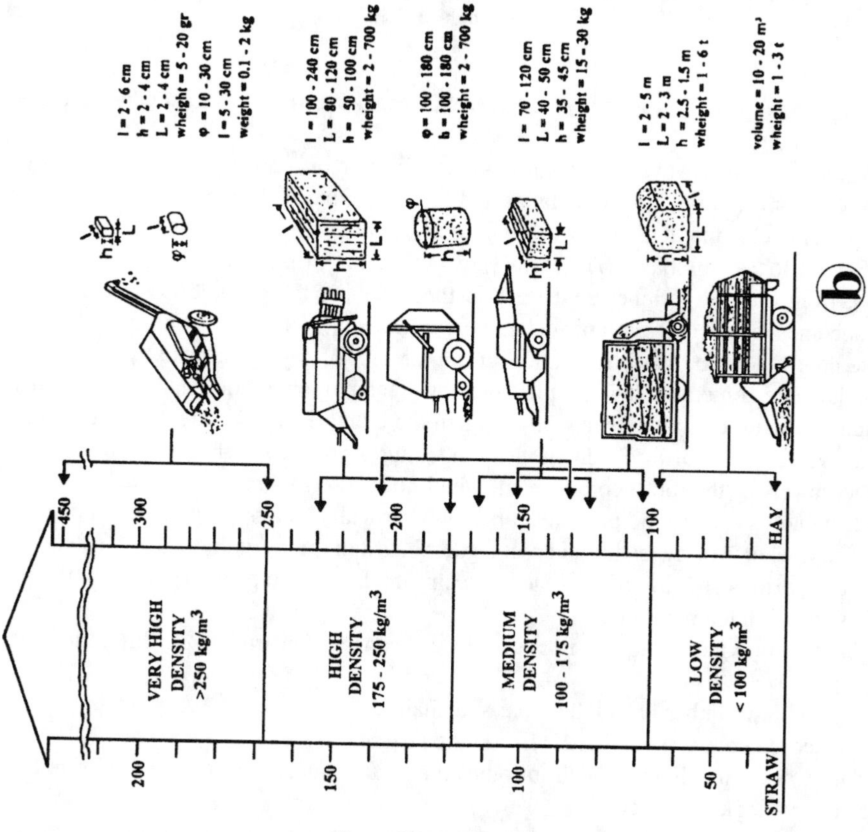

VERY HIGH DENSITY >250 kg/m^3

HIGH DENSITY 175 - 250 kg/m^3

MEDIUM DENSITY 100 - 175 kg/m^3

LOW DENSITY < 100 kg/m^3

STRAW 50 100 150 200

HAY 100 150 200 250 300 450

l = 2 - 6 cm
h = 2 - 4 cm
L = 2 - 4 cm
wheight = 5 - 20 gr
φ = 10 - 30 cm
l = 5 - 30 cm
weight = 0.1 - 2 kg

l = 100 - 240 cm
L = 80 - 120 cm
h = 50 - 100 cm
wheight = 2 - 700 kg

φ = 100 - 180 cm
h = 100 - 180 cm
wheight = 2 - 700 kg

l = 70 - 120 cm
L = 40 - 50 cm
h = 35 - 45 cm
wheight = 15 - 30 kg

l = 2 - 5 m
L = 2 - 3 m
h = 2.5 - 1.5 m
wheight = 1 - 6 t

volume = 10 - 20 m^3
wheight = 1 - 3 t

Energy requirement (hay) [kWh/t]	Capacity (hay) [t/h]	Minimun power (of the tractor) [kW]
20 - 28	4 - 6	150
5 - 8	5 - 10	80
2 - 2.5	10 - 20	100
2 - 2.3 * 2.2 - 2.5 **	4 - 12	50
1.5 - 2.5	4 - 8	30
1.5 - 1.8	10 -15	60
1 - 1.2	4 - 6 (transport excluded)	30

*compact core models **loose core models

b

Figure 1.319. (Continued)

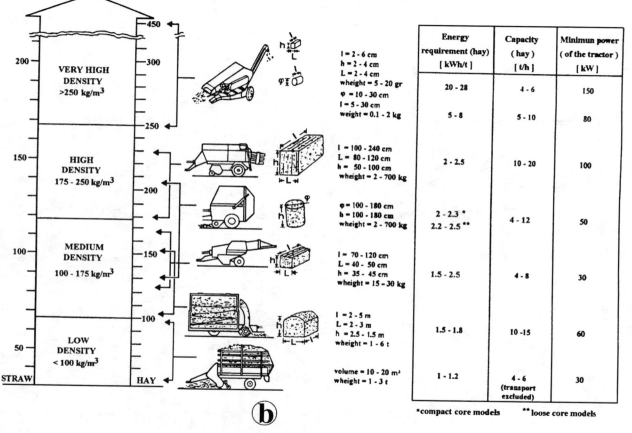

Figure 1.319. (Continued)

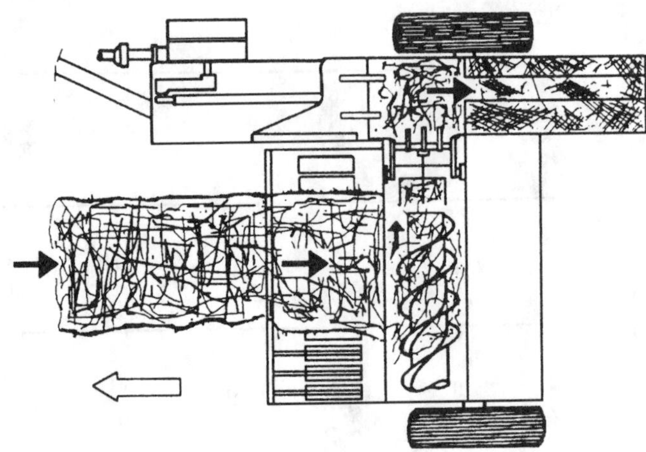

Figure 1.320. Traditional rectangular pick-up baler. The figure shows the way done by hay.

models, which produce round stacks rather than round bales, were intended for this last application.

Round balers—always trailed in axis with the tractor—can be classified into two main categories: core compacted bales and loose core bales. The difference lies in the compression chamber which has a variable section for core compacted bales and a fixed section for loose core bales. The former guarantees uniform compression of the whole mass, from core to periphery, whereas in the latter the forage is loose in the center but becomes increasingly compacted moving towards the periphery. Consequently, with loose core balers forage fermentation or completion of drying, especially if baled at high moisture content (>25%), is promoted by greater air circulation in the central area, which produces a chimney-stack effect when the cylindrical bales are piled vertically: for a given forage moisture content, the temperatures produced during fermentation are in fact lower for the loose core system, which therefore generally guarantees superior product quality. The round bales must be piled vertically to achieve good fermentation. Similarly, when storing traditional rectangular bales, the sides with the tie twine should not be superimposed, in order to facilitate air circulation and prevent the formation of molds and high temperatures.

The compression chamber in round balers can be constructed in different ways (Fig. 1.322):

- Variable-chamber balers: belt, bar and chains.
- Fixed-chamber balers: belt, roll, bar and chains.

Recently, new models have been introduced that are capable of operating with both a fixed and variable chamber.

To guarantee the forage harvest in the presence of adverse meteorological conditions, in the late 1980s the technique of wrapping the cylindrical bales in plastic film was introduced, thereby transforming them into microsilos. This technique has met with

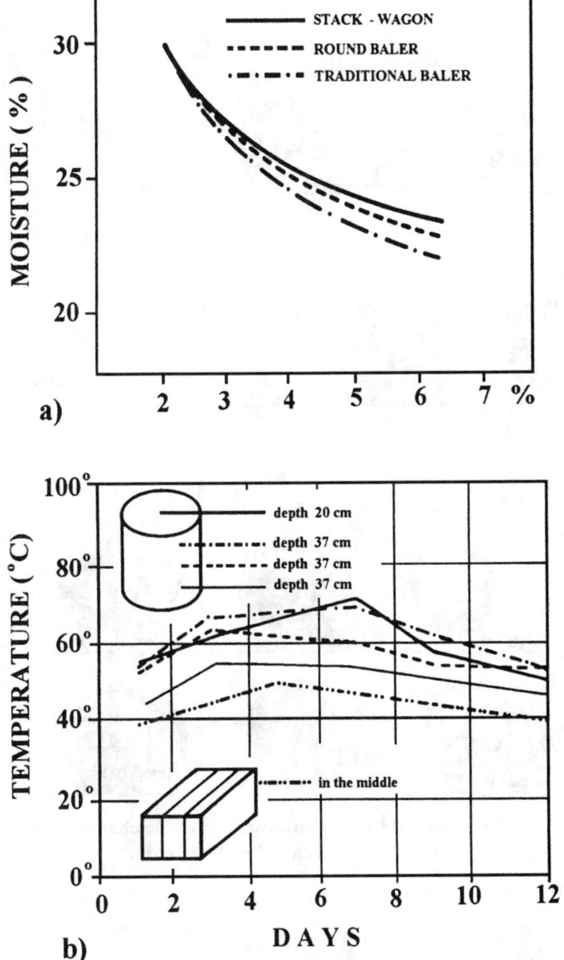

Figure 1.321. **a) Mechanical loss in baling; b) temperature development after baling in a round bale (core compacted) and in a small rectangular bale. Moisture at baling: 28%.**

substantial success, so much so that special round balers have been developed specifically for harvesting moist forage intended for wrapping. The main peculiarities of these machines are: a fixed-section compression chamber; the presence of a chopper (based on rotor and knives, or flails); and the feeding of the chamber from the top (Fig. 1.323).

Typical dimensions, work rates and absorbed power values are given in Fig. 1.319b. It can be seen that the variable-chamber system requires less power, and that the size most commonly used today is 1.2 × 1.5 m, because it matches the dimensions of road transport vehicles whose 2.4 m width is a multiple of 1.2 m.

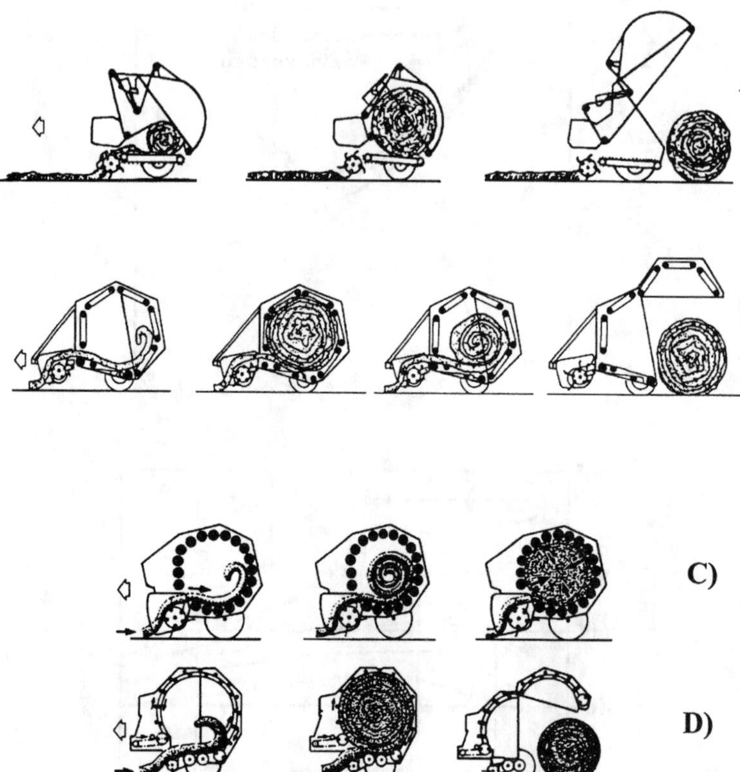

Figure 1.322. Different round baler solutions: A) variable chamber, core compacted; B, C, D) fixed chamber, loose core.

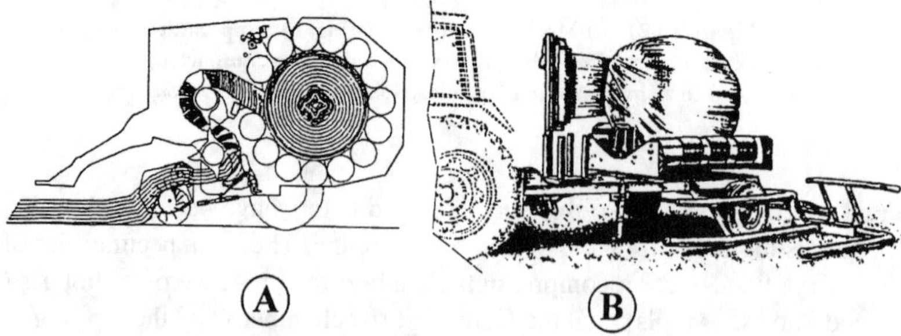

Figure 1.323. A) round baler specifically designed for bales to be wrapped; B) a wrapping machine in operation. Figure (A) shows the presence of a dual chopper and the feding of the chamber from the top (modified from Wolvo document).

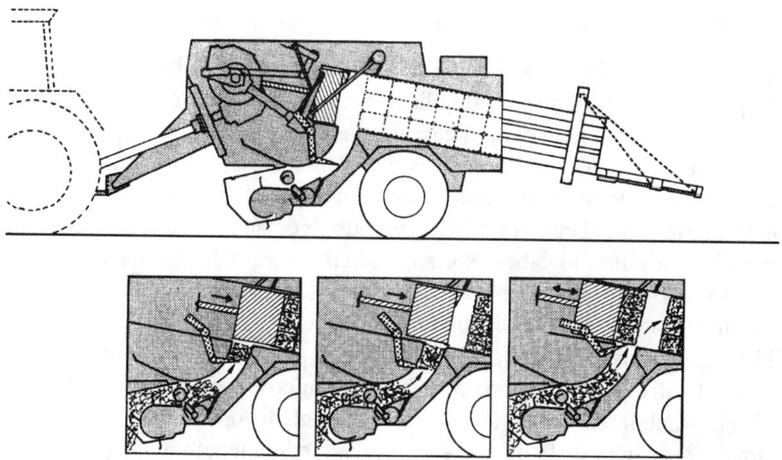

Figure 1.324. Rectangular big baler with compression prechamber (modified from New Holland document).

Big Balers. These offer another alternative, geared to improving performance both during the harvest and in the manipulation and transport phases. The bales are rectangular in shape; their size and density (Fig. 1.319b) are optimized for transport, with a bale length equal to that of the trucks (2.4 m) and a density (180–230 kg/m^3) higher than that of traditional rectangular bales, and comparable to that of round bales. Big balers are trailed in axis with the tractor. The compression is usually effected in two phases: in a pre-chamber the forage is accumulated and partially pressed before entering the upper compression chamber. In the main chamber a plunger completes the compression, while binding is performed by 5–6 binders (Fig. 1.324). The losses and product quality are satisfactory, and comparable to those of traditional balers. Offsetting the high performance (work rates in the order of 15 t/h) are the high power requirements (60–90 kW at the PTO), which make this type of machine suitable only for large-scale farms or contractists.

To meet the needs of medium-sized farms, new balers have recently been introduced which operate on the same principle described above, but on a smaller scale: the bales (volume: 0.7 m^3; weight: 150–200 kg) are easier to manipulate in small spaces.

Stack Wagons. These form fairly compressed, weather-resistant stacks characterized by a sloping top to assist in shedding water. Stack wagons, trailed in axis with the tractor, are simple and consist of a rectangular compression chamber, with vertical sheetmetal side walls, and a mobile canopy on top which acts as a compression element. These are generally equipped with a flail type pickup which allows the product to be conveyed into the chamber by pneumatic means. The flail type pickup also can harvest stover but causes excessive leaf losses if used with too dry (<25%–28%) alfalfa. Therefore, it is essential to avoid the hottest hours of the day. With these precautions, product quality is satisfactory even when the stacks remain out of doors for long periods. For these reasons, stack wagons are the ideal solution to create winter feed stores left on the field, to be used by animals raised permanently out of doors.

The stacks are very large (1–6 t), and the work rates area also very high (10–15 t/h).

Hay Cubers. These were introduced with the aim of facilitating handling, transport and storage operations. Field cubers today have been all but abandoned due to their high energy costs and limited field of application, whereas stationary cubing is still practiced in concentrated feed plants and drying plants. Nonetheless, it is worthwhile to discuss this method briefly because its future reintroduction is not to be excluded. The most suitable forage to be field cubed is alfalfa, because it contains a natural soluble adhesive that helps bind the cubes together. Normally alfalfa is ready for cubing after it has been field cured to 12%–15% moisture, but to achieve sufficient adhesion is necessary to spray water into the forage to activate the natural adhesive.

Field cubers consist of (Fig. 1.325): a cylinder type pickup; sprayers; feeding rolls; a chopper, and finally the cubing apparatus. This is composed of a rotary auger and a heavy press wheel, which forces the forage into and through die openings in a ring. The cubes have a square section of 2–4 cm × 2–4 cm and a random length in the order of 4–8 cm, whereas their specific mass is 350–400 kg/m^3. The formation of the cubes is an extrusion process based on sliding friction, and this is the reason for the high energy costs: 25–30 kW·h/t. In order to retain the advantages of field cubing while reducing its two principal drawbacks—high energy costs and limited field of application—alternatives have been proposed which compress the forage into small cylinders (*roll wafers*) having a diameter of 10–20 cm, random length (5–20 cm) and high specific mass (300–350 kg/m^3). In this case, the cylinders are formed by rolling the forage inside a circular section compression chamber shaped like a truncated cone, and delimited by rotating cylinders. In this way, rolling friction is employed instead of sliding friction, making it

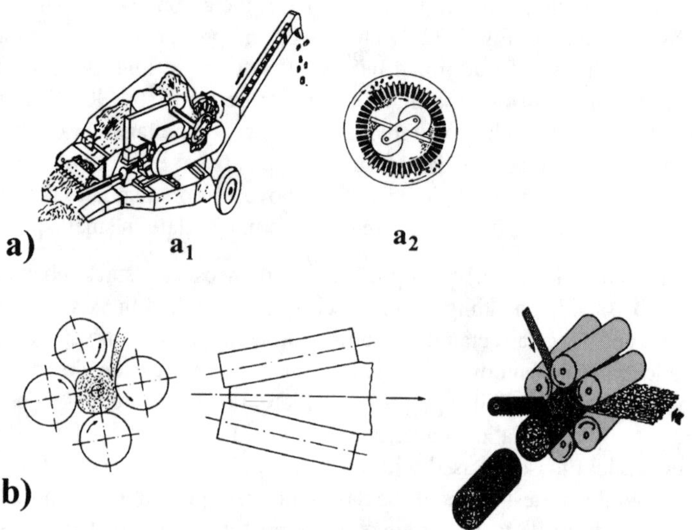

Figure 1.325. A trailed field cuber (a_1) and its cube forming wheel (a_2) based on the extrusion principle.

possible to reduce energy costs (8–10 kW·h/t) and use most types of forage grasses at moisture contents in the usual range (20%–28%). These systems are once again being taken into consideration by some major manufacturers.

Wrapping and Ensilage

During the past decade the practice of ensiling forage wrapped in plastic film has become popular. The moisture content of the harvested forage is in the order of 40%, but special round balers equipped with a chopper can operate even at moisture contents of 60%–70%. There are two alternatives: wrapping bales individually (Fig. 1.323b), or wrapping several bales together in a single plastic tunnel. Both methods guarantee results of excellent quality, provided all operations are correctly carried out by skilled workers. The method has obvious advantages, particularly in rainy areas and at extreme latitudes, in terms of overcoming adverse climate conditions and permitting the recovery, late in the season, of products which would otherwise be unable to dry sufficiently. The high consumption of non-biodegradable plastic material, however, makes this technique unfavorable from the environmental standpoint.

Handling and Storage

This has a significant influence on the timeliness of haymaking and on the overall productivity of the harvest mechanization chain. In fact, the incidence of handling and storage operations has an order of magnitude that ranges from 70%–80% (over a total time—baling + handling—of 1–1.2 man-h/t) for traditional balers, to 40%–50% for round balers (over total time of 0.2–0.35 man-h/t), and big balers (over 0.15–0.20 man-h/t). Moreover, the timeliness of operations significantly affects the final result of a harvest which is largely dependent on climate conditions.

The round and big or medium squared bales are handled by mechanical lifting implements. These are generally front-mounted tractor loaders, although today large-scale farms and contractors are attempting to reduce work times by using dedicated self-propelled industrial loading vehicles. In the case of round bales, however, the pickup timeliness requirement is lower because their shape and compact surface layers make them able to withstand precipitation. This reasoning also applies to stack wagons.

A more thorough description will be devoted to traditional small two-twine bales, for which many methods and types of equipment have been proposed. The *bale accumulator* consists fundamentally of a frame attached to the baler, into which the bales are dropped immediately after being formed; they are then arranged by various methods inside the frame, generally in groups of 8 bales. The group is then deposited in the field, ready to be handled (loading, unloading and storage operations) by a special tractor front-end loader with bale fork (Fig. 1.326a). A front-end loader loads the bales onto a trailer, but unloading and stacking must be performed manually. *Fully mechanized systems* (automatic bale wagons) consist of a trailer equipped with a mechanical device that picks the bales up from the field, after which they are arranged on the loading bed. Fig. 1.326d shows the most well-known and widely used system, which is also capable of mechanically unloading the bales in a stack. Finally, *bale ejectors*, toss the newly formed bales into a trailer drawn by the baler. They consist of two rubber belts moving at high speed, and capable of throwing the bale a distance of 4–5 m. Small bales, no longer

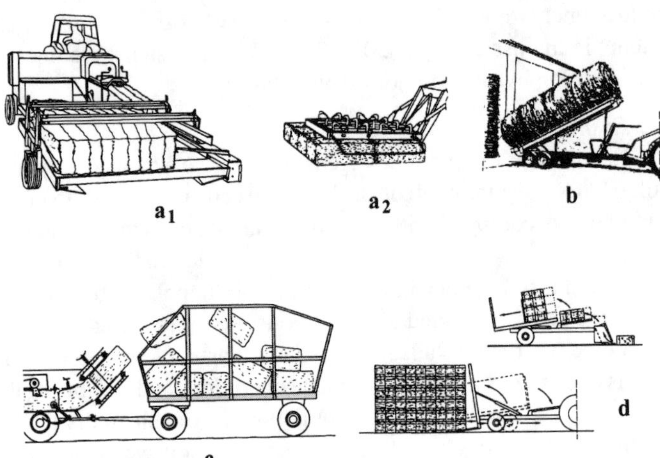

Figure 1.326. Some special means for bale handling: a) bale accumulator (a_1) and its specific fork lift (a_2); b) round bale trailer capable to perform both bale loading and unloading; c) bale ejector; d) automatic bale wagons load orderly the bales and store a group of 60–100 bales in a stack.

than 70 cm, are normally recommended. This method eliminates the loading operations, but not the unloading and storage operations (Fig. 1.326c).

1.6.10. Forage Cereals

The Methods

The harvesting of forage cereals offers no alternatives to silage. This technique presupposes storage of the forage in an anaerobic environment, and therefore requires that the product be chopped into small pieces (5–10 mm) to guarantee adequate compaction of the product. Cereals, thanks to their high content of soluble sugars which produce good lactic acid fermentation, are better suited for storage in silos than meadow forage crops. Therefore, this method is also viable for less developed farming systems.

The most commonly used crop is maize, but winter cereals (barley, wheat, triticale) are also gaining popularity. These, in fact, make it possible to:

- Obtain two crops in more fertile, well-irrigated areas: winter cereal + maize.
- Obtain a satisfactory mass of product for livestock feed in less favorable areas.

In the former case it is possible to achieve 25,000 UF/ha (10,000–12,000 UF/ha from the first crop + 12,000–15,000 UF/ha from the second crop), albeit with substantial use of agrochemicals and other technical aids. The second case generally yields no less than 8,000–10,000 UF/ha, a value that is entirely satisfactory when compared with the yields obtainable from the traditional meadow. The values shown, minus the storage losses which we can assume to be 10%, correspond to the feed available to the animals. Moreover, the product is harvested standing and therefore is not subject to soil contamination. The chopped product is stored in silos which consist of simple side walls (2–4 m in height) for containing the product, but in less-developed farm systems even

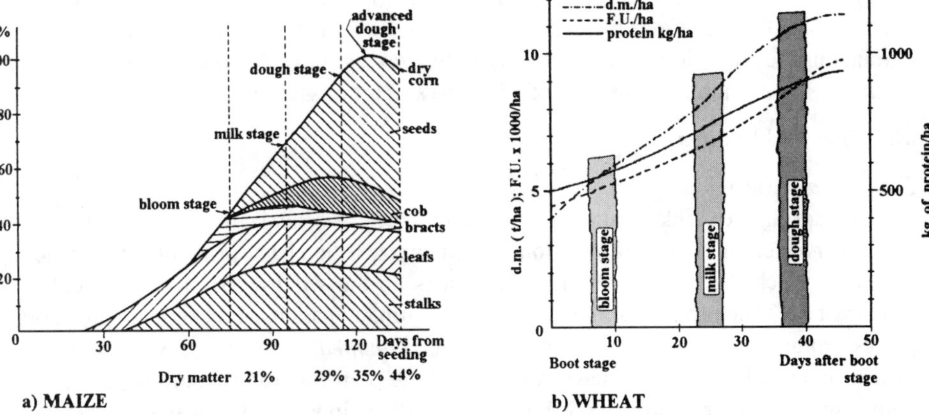

Figure 1.327. Maize composition related to the stage of plant growth (A); wheat plant yield of dry matter, feed unit and protein related to the stage of plant growth.

more rudimentary methods can be used, such as pits dug in the ground. On the contrary, vertical tower silos are no longer feasible due to their high investment costs and slow filling times, which are incompatible with the high workrates of forage harvesters. To obtain high-quality silage, it is sufficient to follow a few simple rules:

1. Select the optimal stage of maturity of the crop.
2. Perform timely harvesting and filling of the silo.
3. Thoroughly compact the entire product mass, taking particular care along the silo walls, using wheel tractors.
4. Carefully cover the top of the forage mass with a plastic sheet or, in certain climates and for large product quantities, plant certain grasses (lolium) which produce a waterproof turf layer.

For maize, the optimal stage of maturity is the waxy stage, with a dry matter content of 35%–38% (Fig. 1.327a). For winter cereals it is preferable to anticipate harvesting to the milky or early-waxy stage of maturity (Fig. 1.327b), corresponding to 32%–35% dry matter, due to the lower compressibility of these crops. In some specialized animal-husbandry systems, the trend is towards cutting the maize plants at 40–50 cm above the ground in order to obtain a product that: has a higher energy content; is more digestible because the more lignified lower portion of the plant is removed; and is less contaminated with the soil and/or molds which are always found on the lower parts of the crop.

Another consolidated practice is the corn silage (maize mash) consisting of either just the grains or the entire ear (grains + corn − cob + bracts), suitably ground. Whereas whole-plant cereal silage is only used for feeding ruminants, the maize-mash is also successfully employed in swine-breeding. To make the mash, the maize is harvested at an advanced stage of maturity, with a dry matter content of 65%–70%. It is possible to use either specially equipped complete forage harvesters, or normal combine harvesters with subsequent milling of the grains. For ensilage, horizontal silos are still valid, but it is also possible to justify vertical tower structures, given the lower mass of product. Compared with traditional harvesting of grains, which require moisture contents below 15% (dry matter > 85%) to be stored by traditional methods, this practice has the advantage of eliminating the grain-drying operation, permitting savings of 15%–20% on *gross saleable production* in economic terms, and of 150–150 kW·h/t of grain in energy terms.

Machines and Equipment

Forage Harvesters

They are the key link in any forage cereal harvest mechanization chain. Although forage harvesters are available in various types and sizes, their basic principle of operation is always much the same. A forage head picks up the windrowed product or cuts a standing crop. The product is then carried to a cutting apparatus and chopped into short pieces. Forage harvesters may be classified as: *mounted*, *pull-type* or *self-propelled.* Tractor-mounted machines attach to the tractor three-point hitch and are usually equipped with a single row, or pickup head. Front-mounted versions are also available: they are particularly interesting when coupled to the rear axis on reversible tractors, a solution that guarantees high maneuverability and efficiency. There are also high-capacity models available which attach to high-power tractors (100–150 kW) equipped with 3–4 row heads, 3 m or longer cutterheads and 2–2.5 m pickup heads. Pull type models are available in sizes to match most 70–100 kW tractors. Self-propelled forage harvesters generally offer excellent maneuverability, many operator conveniences and high capacity, with installed power ranging from 200–350 kW. Like most harvesting machines, it has rear-wheel steering and front-wheel drive. However, certain models also offer the option of four-wheel drive.

The trend is towards using large, self-propelled machines hired from contractors in order to reduce farm costs and to complete harvest operations in a very timely manner, thereby limiting exposure to air, and hence ensuring good product quality.

Concerning the cutterhead (Fig. 1.328), instead of the flywheel type rotor, manufacturers today prefer the cylinder solution which guarantees higher performance, improved reliability, lower energy costs and greater simplicity of construction. More up-to-date models incorporate metal-detection systems, positioned in the feeding apparatus.

Schematically, forage harvesters consist of a *base machine* and a *harvest head.* The base machine is made up of the following main elements (Fig. 1.329): *chopping apparatus* (cut-and-throw cutterhead); *feeding-dosing apparatus*, consisting of a set of 4–5 superimposed feed rolls. There are also optional accessory elements such as: *knife sharpener*, to sharpen the knives without removing the knives or cutterhead from the machine; *auxiliary blower* to improve throwing of the chopped material when the cutterhead action is insufficient; *corn cracker* to break the corn, which is useful for improving the storage and digestibility of crops harvested at high d.m. content (>40%); *metal detector*, to prevent the entry of dangerous metallic objects. As far as the base machine is concerned, attention should be directed to the feed rolls and cutterhead. The *feed rolls* consist of 4–5 rolls mounted on top of each other (2–3 below, 2 above), through which the forage passes in a uniform mat that advances at a velocity equal to the peripheral speed of the rolls. The speed of rotation is varied by changing sprockets or shifting gears on simpler machines, or by means of full-fledged power-shift transmissions actuated from the driver's seat.

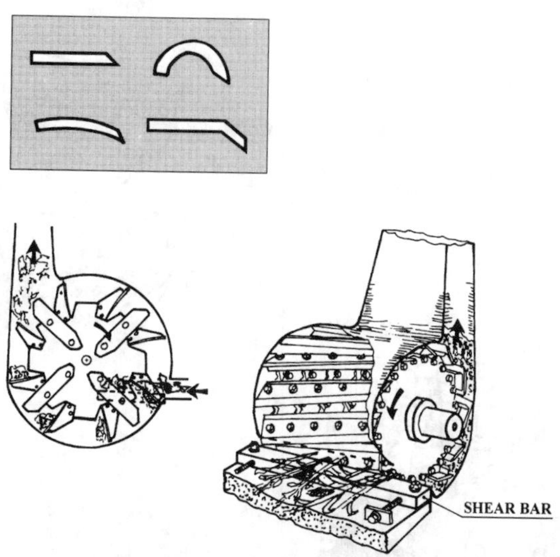

Figure 1.328. Flywheel (a) and cylinder type cutterhead. In the figure are shown the main knife shapes employed in the cylinder cutterheads.

The lower rolls are generally mounted on fixed bearings, whereas the upper rolls are spring-loaded in order to feed different quantity of forage. The lower front and upper feed rolls generally have deep flutes, as they must firmly grip the crop mat at all times. The lower rear roll, on the other hand, is generally smooth: if it were fluted, parts of the crop might be caught and carried through past the cutterhead, and dropped on the ground. There should always be the possibility of inverting the direction of rotation of the feed rolls, in order to empty them in the event of the machine jamming.

However, the most significant element of any forage harvester is the *cutterhead*, which determines the capacity, efficiency, and quality of cutting. Cutting capacity, efficiency, and quality are interlinked, and fundamentally depend on the shape and condition of the knives and the stationary knife or shearbar. The cutterhead consists of a rotor, along whose periphery a set of knives is mounted, varying in number from 8 to 12. The length and diameter of the rotor depend on the size of the machine. The diameter ranges from a minimum of 500 mm to a maximum of 800 mm on some self-propelled machines, although cutterhead diameters in the order of 600–650 mm are most usual. The width, on the other hand, varies from a minimum of 250 mm for one-row side-mounted machines, to a maximum of 800 mm for higher-capacity self-propelled machines. The speed of rotation is high, ranging from 1000 to 1200 rpm, which corresponds to peripheral speeds in the order of 15–20 m/s. The most common shape of the knives are the flat and "J" shapes, stiffer and more efficient than spiral and cupped knives, especially when a recutter is used (Fig. 1.328). Spiral knives are used only on higher-capacity machines, whereas small mounted or pulled machines generally use straight knives, which are more economical. It is also possible to vary the cutting length by changing the number

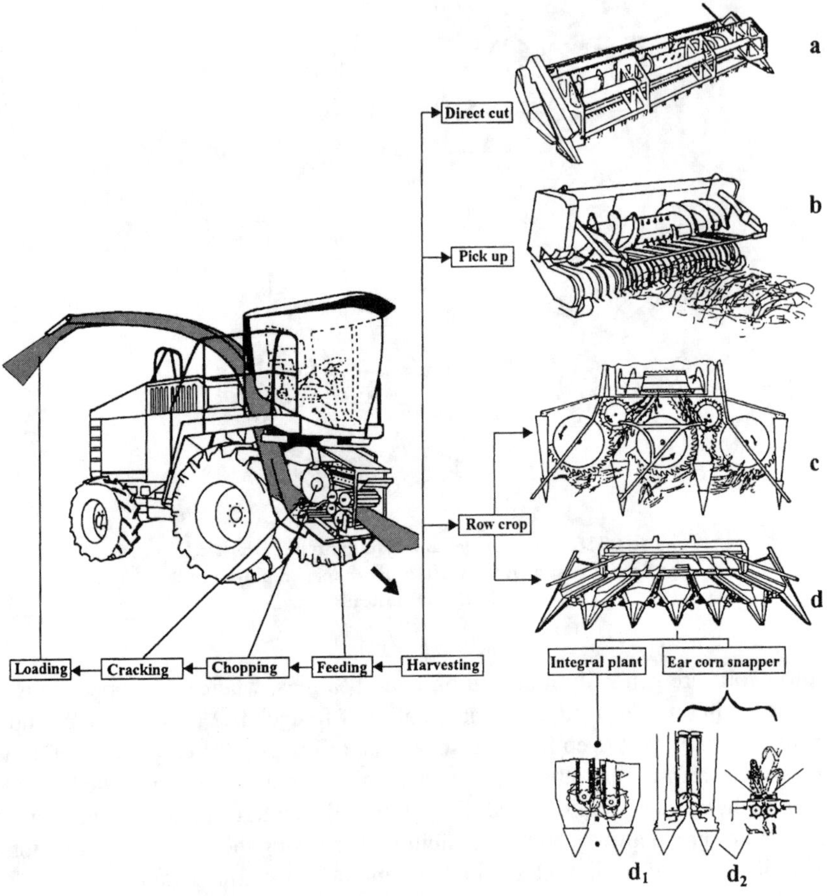

Figure 1.329. Self-propelled harvester can be equipped with different heads: mower bar (a); pickup (b); row crop (c,d); and ear-corn snapper (d_2). (modified from New Holland and Claas documents).

of knives on the rotor. However, it is preferable to vary only the speed of rotation of the feed rolls: this in the interests of safety, as well as to ensure optimal balancing of the rotor. Also for these reasons, the cutterhead has a fixed speed of rotation.

As far as forage heads are concerned, the most common types include (Fig. 1.329):

- Direct-cut or mower bar for winter cereals or standing grasses
- Windrow-pickup for haylage and dehydration
- Row crop, especially for corn silage
- Ear-corn-snapper for silage

Mower bars for direct-cutting of grass and winter cereals, present cutting width from 1.5 m to more than 4 m. Direct-cutter heads use single or double cutterbars and are equipped with a reel to gather crop material into an auger which feeds material into the feed rolls. Normally the reels used on forage heads are not adjustable in length, height and rotary speed. For this reason, in the case of self-propelled models and with laid crops, it is preferable to use the more sophisticated combine heads. The smaller sized models (rotor width <300 mm) often are unsuitable to harvest winter cereals, because the long plant stalks (>1m) positioned crosswise are unable to enter the feed rolls.

Windrow-pickups are used for silage, haylage and dry hay. Various pickup widths are available to match harvester capacity and windrow size, ranging from >1.5 m for the smallest mounted models to 3–4 m for big self-propelled models. Retractable fingers and auger-flight extensions feed forage into feed rolls.

Row-crops are usually used to harvest maize or sorghum and sorghum-sudan cross silage, are available in one-to six (eight) row sizes and with different row widths. After the plants have been cut, gathering chains or belts grab the stalks and feed them into feed rolls. For laid crops, belts are usually more efficient. Moreover, recently the main manufacturers have also adopted a special head for row crops which permits harvesting independently of row width and is particularly useful at the beginning of the harvest because it is possible to operate perpendicularly to the rows. The solution, effective also for laid crops, is based on a series of rotors ($\phi = 1$–1.5 m) which cut the plants and then grab the stalks up to the feeding elements (Fig. 1.329c).

Snapper heads operate on same principle as corn heads for combines. Basically, these row heads are equipped with two contrasting rolls which pull stalks through snapping bars under the gathering chains to snap off the ears. Gathering chains carry the ears back to a cross auger which carries corn to the cutterhead for chopping. Nevertheless, to obtain good fermentation and digestibility of the silage, the chopped corn exiting the rotor is fed into two small corn crakers useful also for whole-plant maize silage. To improve the cracking effect, the lower part of the cutting rotor housing can be fitted with recutter screens, which act as additional stationary knives. Corn crackers and recutter screens require additional energy per ton of forage.

The energy demand is dependent on the type of product (crop) and its stage of maturity (% of dry matter), the length of cutting, the sharpness of the knives and the distance between the knives and the fixed shear bar. Typical cutting lengths are: 6–9 mm for maize and winter cereals, 20–30 mm for lolium and other grasses for silage or haylage, 40–50 mm for alfalfa for dehydration. If the cutting elements are in good condition, the energy requirements are in the order of 2–3.0 kW·h/t depending on the crop and the cutting length (Fig. 1.330). However, energy requirements can be much greater when the knives are not sharp, or when there is a large gap between the knives and shearbar (Fig. 1.331). It follows that the *working capacity* is primarily a function of the available power. Taking the case of maize, typical values would be 100 t/h for simple mounted one-row models, 300 t/h for trailed front-mounted two-row models and 6–100 t/h for the large self-propelled models. These last-mentioned machines offer ever-increasing powers which can reach values as high as 350–400 kW. In all cases, the working capacity depends on the available power, and may be roughly calculated by multiplying the power by a utilization coefficient of 0.6–0.7 and dividing the result by the energy consumption: working capacity (t/h) = available power [kW]×0.6–0.7/energy requirement [kW·h/t]. This approximate evaluation is valid provided the limiting value

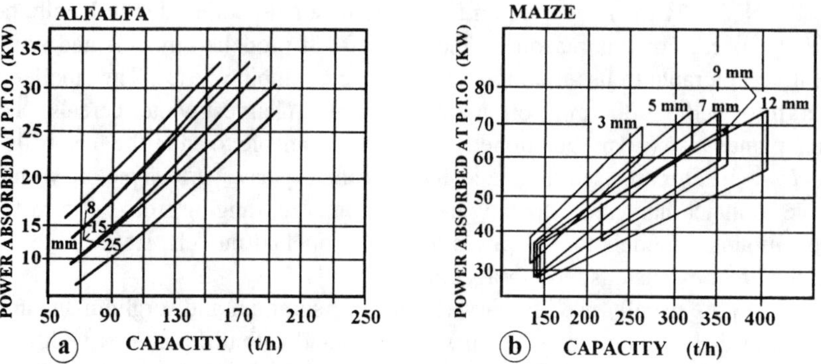

Figure 1.330. Power required at PTO and capacity for different crops and cutting length.

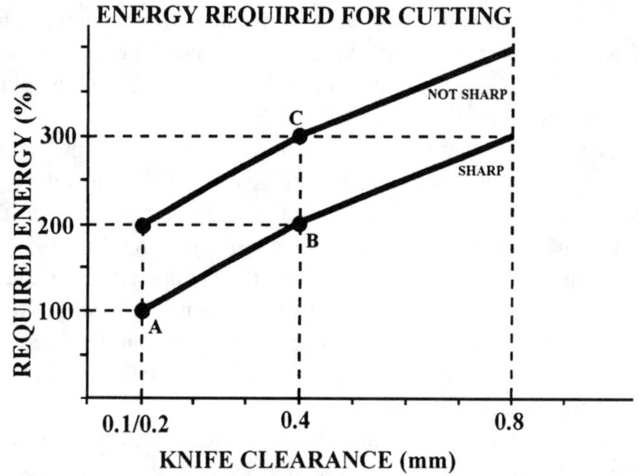

Figure 1.331. Energy requirement if the cutting head is operating not correctly (from New Holland document).

for forward velocity is taken into account: this should preferably not exceed 7–8 km/h with direct cutting heads, 6–7 km/h with pickup heads, and 6–7 km/h with row-crop heads.

Flail harvesters also fall into the category of forage harvesters. Their basic element is the rotor, a horizontal shaft with flails or knives attached to it. The flails exert a shearing action on the crop, carrying it round the outer circumference of the flails, under the rotor housing, and cutting and shattering it into small pieces. The same rotor, or an auxiliary blower, is then used to blow the chopped material onto a wagon through a small discharge spout. However, flail harvesters are now considered to be out-of-date, and in any case are suitable only for harvesting green or pre-cured forage crops intended for ensilage. They are not suitable for harvesting forage cereals, for which modern cutterhead forage harvesters are more appropriate.

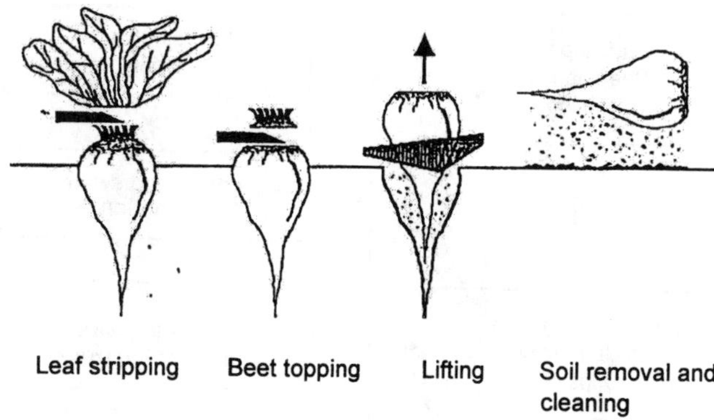

Figure 1.332. The main operations of sugar beet harvesting.

Root Crops

E. Manfredi and R. Peters

1.6.11. Sugar Beet Harvesting

Sugar beet harvesting requires a high degree of specialization, with several stages of coordinated mechanization culminating in delivery of the product to the sugar beet factory.

Sugar beet harvesting currently involves four main operations (Fig. 1.332)

1) Leaf stripping and chopping.
2) Beet topping with collet cutting.
3) Removal of the root crop from the soil.
4) Cleaning the product to remove impurities (soil stuck to the product, loose soil, dry leafstalks, collets, pieces of root, stones, etc.).

These basic operations are followed by other, optional ones that can be used in various combinations, e.g.:

- Depositing the root crop in swaths.
- Direct loading onto a vehicle beside the loader (multistage harvesters) or into the harvester (complete harvesters).

Main Stages of Mechanical Harvesting

Types of Machines

Multistage Harvesters. The operations are carried out in two or three harvesting stages. For three-stage operations, each operation is performed by a towed machine. Stage one involves leaf stripping and beet topping; stage two, lifting and swath formation; stage three, loading, which may be directly onto a truck running parallel to the loader or into a self-loading truck with its own body (Fig. 1.333).

For two-stage operations, the first stage is carried out with the leaf stripper-beet topper and the lifter-swather, respectively fit at the front or rear of the tractor, or self-propelled. The second stage involves loading of the root crop from the swath, with a trailed loader

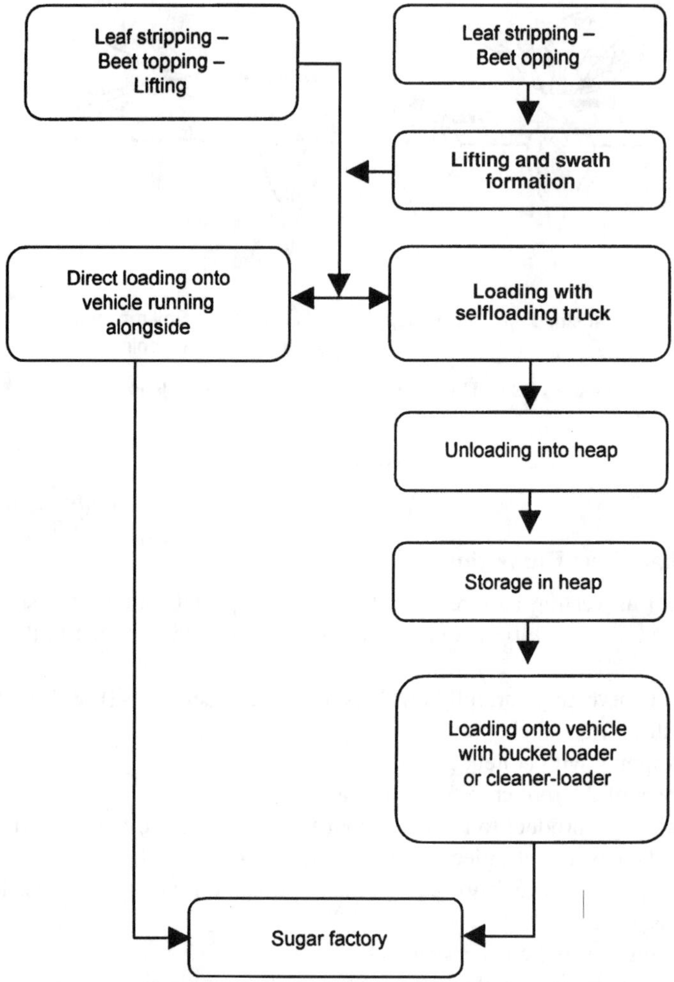

Figure 1.333. Sequence of operations in multistage harvesters.

or a self-loading truck. In the first case, the root crop is loaded directly onto a vehicle; in the second the product is loaded then unloaded in heaps at the head of the fields.

On machines used for the above-mentioned stages, the leaf stripper has a single wheel (with 60–90 flails and speed of 800–1100 rpm), and a swath auger (with diameter 0.30–0.45 m and speed of 290–360 rpm); the beet topper has a driven drum feeler with fixed trapezoidal blade, slide and scalper. The lifter-swath auger consists of a series of self-centering ploughshare supports, which may mount fixed or oscillating ploughshares. Before being deposited in the swath, the root crop is cleaned by passing it over two soil removing wheels, or over soil removing augers also designed for root crop feed, fit on a frame which is divided into two parts and jointed at the center.

The *loader* operates by means of chain conveyors with the aid of star wheels, for improved soil removal.

The *self-loading trucks* have a twin-axle frame and a hopper capacity of from 20–23 m^3, the height for unloading to the ground being from 1.3–1.6 m, and the root crop being cleaned by conveyor belts (with surfaces varying from 1.4–2.4 m^2) and by star wheels.

The approximate operating capacity of these machines is:

- 1.8 ha/h for six-row machines in the leaf stripping/beet topping and lifting/swath formation stages.
- 2.0 ha/h for six-row machines in the leaf stripping/beet topping/lifting/swath formation stages.
- 2.0 ha/h for the loader.
- 3.0 ha/h for the self-loading truck.

Complete Machines. These machines carry out leaf stripping/beet topping, lifting, and cleaning in a single pass and provide various possibilities for storage and loading onto vehicles to transport the product to the sugar beet factory (Fig. 1.334).

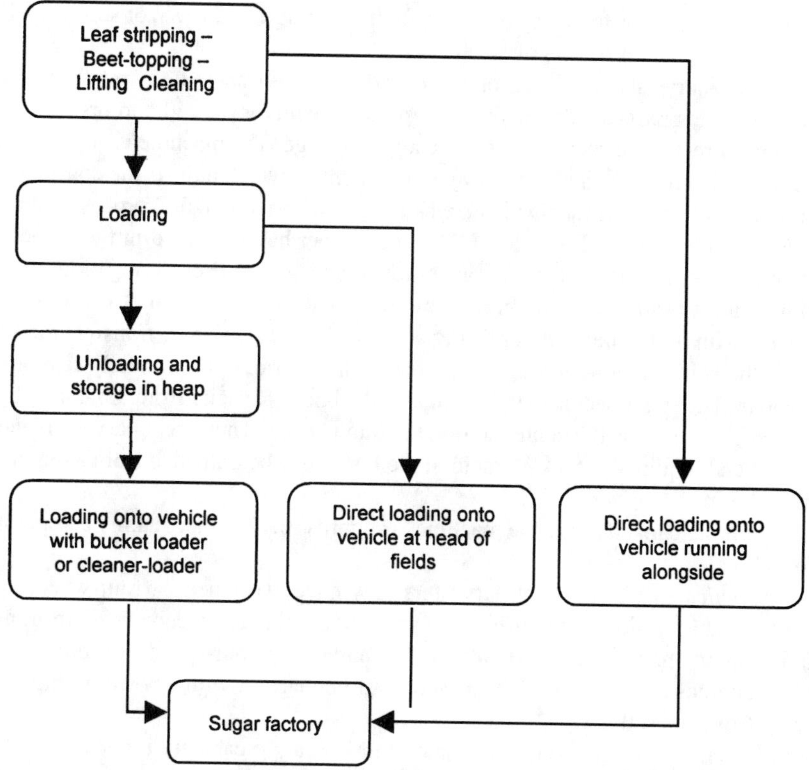

Figure 1.334. Sequence of operations with complete machines.

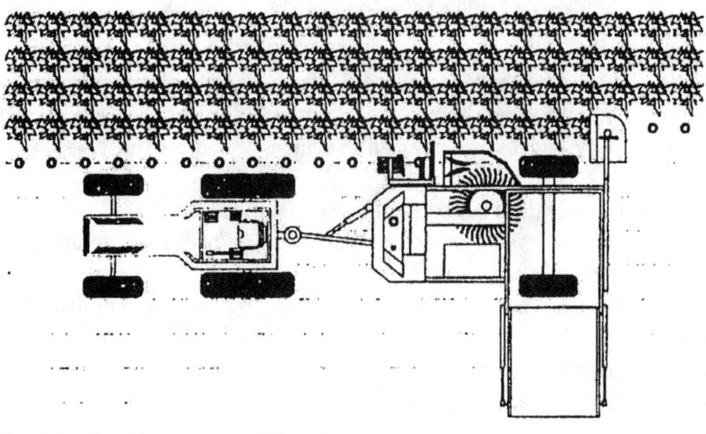

Figure 1.335. Single-row trailed machine, with leaf stripping on the row adjacent to that harvested.

The machines may be towed by a tractor, for harvesting 1 or 2 rows, or self-propelled, for harvesting 1, 2, 3, or 6 rows of sugar beet.

Trailed beet harvesters have only one axle, and motion is transmitted from the tractor's PTO. The working parts are driven by hydrostatic transmissions, with motors which can reverse their direction of rotation to relieve any blockage. The machine is automatically aligned along the row by an electrical-hydraulic control which adjusts the coupling bar. The optional leaf stripping mechanism, consisting of a horizontal disc, processes the row adjacent to that harvested (Fig. 1.335). The beet topper has a driven drum with inclined trapezoidal blade. The lifter is of the conventional type, with self-centering ploughshares which may have oscillating motion upon request. The sugar beet collet is removed by two wheels with rubber flails. After lifting, a vertical wheel with rods transfers the sugar beet onto the soil removing wheel, which has a diameter of approximately 1 m and an oblique axis. The sugar beet is then loaded into the body (capacity from 3.5–5 m^3) by an elevator, angled towards the center so that loading is even. These beet harvesters weigh approx. 3 t and require 40–50 kW tractors; the body may be unloaded from a maximum height of 3.5 m.

The operating capacity of these machines is approx. 0.12 ha/h, equaling an operating time of 8 h/ha.

Self-propelled single-row beet harvesters may have two-wheel or four-wheel drive, with mixed transmission to the driving wheels (hydraulic and mechanical) from a hydrostatic engine, mechanical differential axles, and two to four-speed gear box.

Diesel engines are used, either aspirated or turbocharged, with a power rating which may vary from 90–120 kW.

These machines weigh from 5–6 t, and their operating capacity ranges from 0.15–0.2 ha/h, equaling an operating time of 5–7 h/ha.

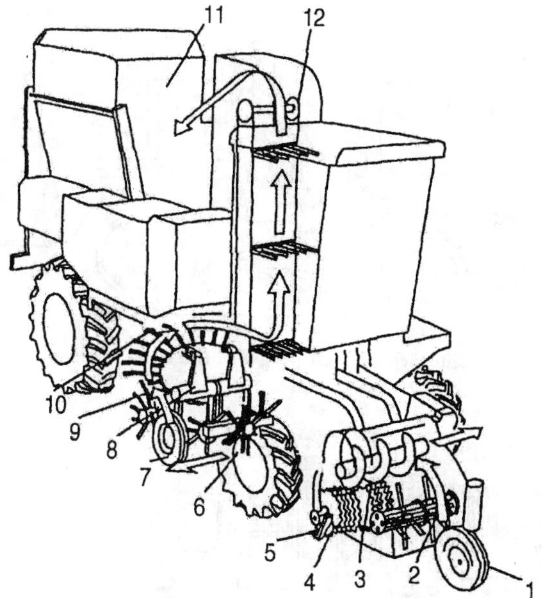

Figure 1.336. Diagram of a self-propelled two-row harvester.

Self-propelled two-row beet harvesters have an operating capacity of 0.25–0.35 ha/h (3–4 hours/hectare), a power rating of approx. 150 kW, weight 8–10 t and body capacity 7–9 m^3.

This category of machine maintains the distribution of the working parts of the single-row machine, with one or two main soil-removing wheels, and the elevator fit at the front of the body, alongside the driver position (Fig. 1.336).

1) Wheel for adjustment of leaf stripper height, 2) single wheel leaf stripper with steel flails, 3) swath auger for chopped leaves, 4) driven drum feelers, 5) beet topping blade, 6) collet-cutter blade with rubber flails, 7) wheel for adjusting ploughshare depth, 8) lifting mechanism with self-centering, oscillating ploughshares at negative angle, 9) feed wheel with horizontal axis, 10) soil removing wheel, 11) loading body, 12) elevator.

Self-propelled three-row beet harvesters (Fig. 1.337) have a power rating of 170–190 kW, weigh from 9–13 t, have 10–13 m^3 body capacity, and may be fit with a steering rear axle. With the change in harvesting from two to three rows, the arrangement of machine parts has partially changed as well. The longitudinally positioned engine has been fit transversely (although still behind the driver's cabin), while the elevator has been mounted on the rear of the body. This has permitted insertion of three soil removing wheels, which not only clean the beet, but also transfer it from the lifters to the elevator. These machines have an operating capacity of 0.35–0.55 ha/h, equaling an operating time of 2–3 h/ha.

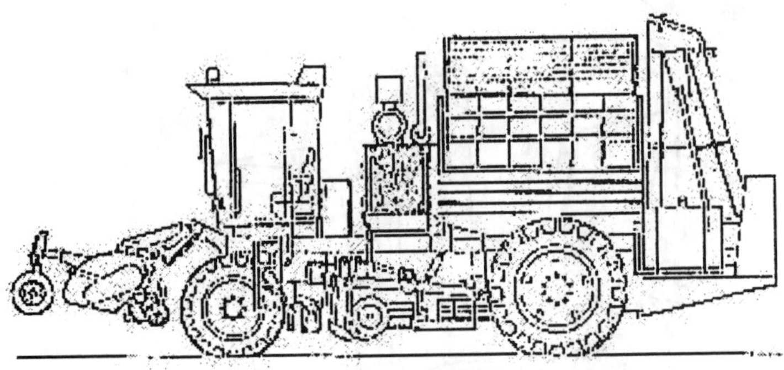

Figure 1.337. Side view of a self-propelled three-row one pass machine.

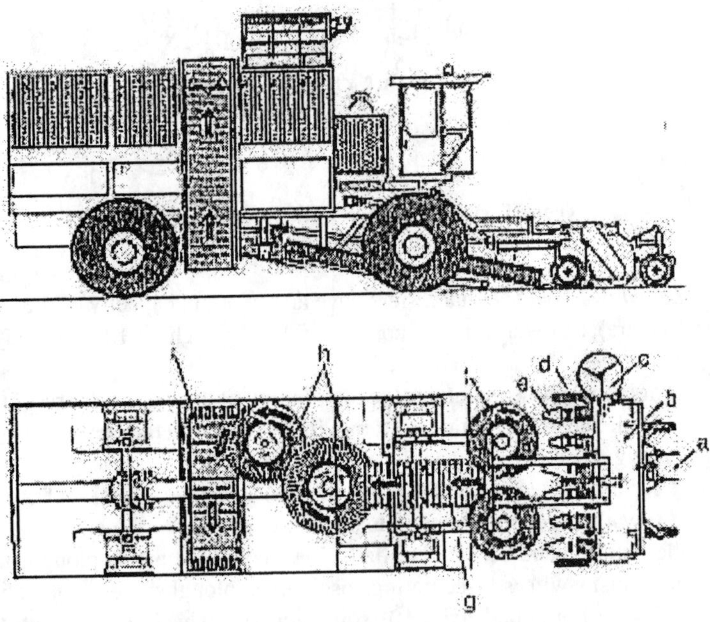

Figure 1.338. Self-propelled six-row complete machine (side and top views).

The basic innovation in *self-propelled six-row* beet harvesters (Fig. 1.338) is that all of the working parts for the topping, lifting and initial cleaning of the product are positioned at the front of the machine. As a result, the sugar beet is lifted before the wheels pass over it. This important advantage permits work on soil that has not been compacted, and thus easily removed by the machine's cleaning parts.

The harvester may also be fit with a self-drive system (operated by a row finder with electrical-hydraulic control) positioned in front of the leaf stripper, which operates on the two central strips between rows. The engines on these machines have a power rating ranging from 210 kW–270 kW. Motion is transmitted to the wheels by a mixed

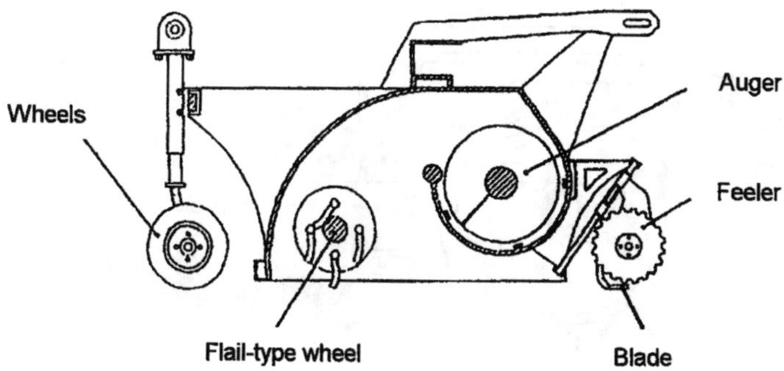

Figure 1.339. Leaf stripping-beet topping mechanism on a self-propelled six-row machine.

hydraulic-mechanical transmission, consisting of a hydrostatic engine coupled to a power divider and mechanical gears, which transfers motion to the two differential axles (front and rear).

The various operating parameters (lifting depth, alignment of working parts on rows, product flow on cleaning mechanism, etc.) are in part controlled and adjusted automatically by an onboard computer, which also stores such operating data as area processed, average forward speed, etc.

The leaf stripper may have one (Fig. 1.339) or two wheels (the first with metal flails and the second with rubber flails) and an upper swath auger, to which a leaf-spreader disc or mechanism to recover chopped leaves may be fit. The beet topper may be either of two types: with driven drum feeler or with fixed slide feeler. The collet is cut by an inclined trapezoidal blade. During operation, the distance between the blade and feeler may be fixed or proportional to the root crop outgrowth from the soil. The maximum distance between feeler and blade can be adjusted from the driver position to adapt the degree of topping to the average conditions of the cultivated land.

On these machines, the lifting mechanism (Fig. 1.340) again consists of self-centering, oscillating ploughshares, which are often leveling. Oscillation applies either to the entire ploughshare support (concordant oscillation) or to the individual ploughshares (discordant oscillation). The sugar beet extracted by the ploughshares is transferred onto the initial cleaning mechanism by rubber blades mounted on a rotating transverse axis. On some models, there are feeler wheels between the ploughshare support units on the same axis, coupled to an electrical-hydraulic actuator, which keep ploughshare excavation depth constant (Fig. 1.341).

After lifting, the product is conveyed to a first cleaning mechanism, which may consist of four or five augers or two soil-removing wheels. Both systems are designed to reduce the section of product flow toward the main cleaning parts in the lower zone of the tractor frame. The main cleaning parts normally consist of a series of soil-removing wheels (there may be two or three, with a diameter of 1.5–2.0 m). In some cases they are connected by conveyors which also have a cleaning function. Alternatively, the last wheel may be replaced with a rubber tined roller conveyor. The speed of these parts is

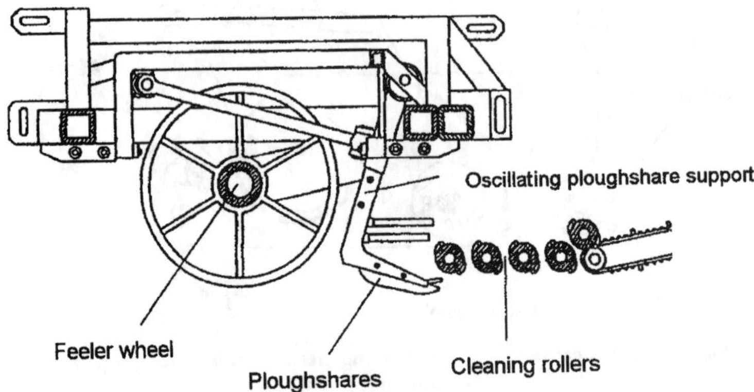

Figure 1.340. Lifting mechanism on a self-propelled six-row machine fit with three feeler wheels coupled to an electrical-hydraulic actuator, to keep ploughshare excavation depth constant even on soft soil.

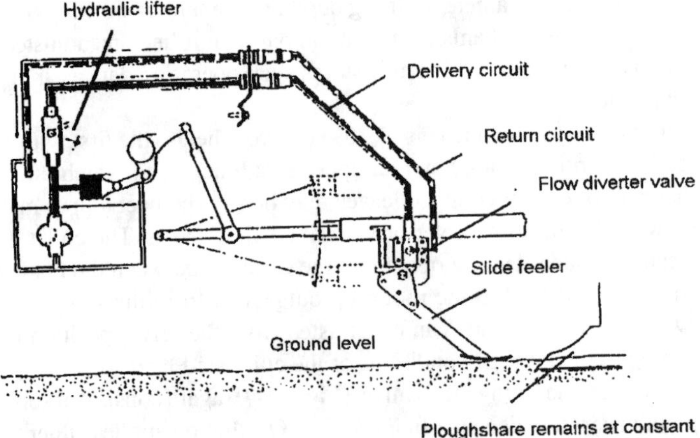

Figure 1.341. Diagram of an electrical-hydraulic actuator. When the feeler moves downwards, the valve diverts oil flow to the lifter jack, transferring part of the load onto the machine's front axle; this resets the depth of the ploughshares in the soil.

adjustable, while the direction of rotation can be inverted to relieve any blockage. Finger rollers may be fit to clear away any soil which may block the mesh of the soil-removing wheels.

The product may be loaded into the body or directly onto a vehicle running parallel to the machine. In the latter case, the body may be used for temporary (no-stop) storage of the product between one trailer-truck and the next. The elevator, which may also operate in reverse, can be used as an unloading conveyor. In this case, the sugar beet passes through a series of cleaning augers before being lifted. The body, with capacity of 13–25 m^3, has a mobile base consisting of one transverse conveyor and one or two longitudinal conveyors, designed to transport the product to the unloading conveyor. The product can be unloaded at the head of the plot of land, either into heaps or directly onto the means of transport, at a height varying from 2–5 m. The operating capacity of these machines varies from 0.65–1.05 ha/h, equal to an operating time of 1–1.5 h/ha.

On some models, the leaf stripping-beet topping and lifting unit may be removed and the tractor used as a self-loading truck for the swaths.

Loaders and Cleaners. These machines can be divided into two categories:

- *Cleaners-loaders*, fit at a fixed point near the heaps of root crop (between the bucket loader and the vehicle) which then go to the factory (Fig. 1.342).

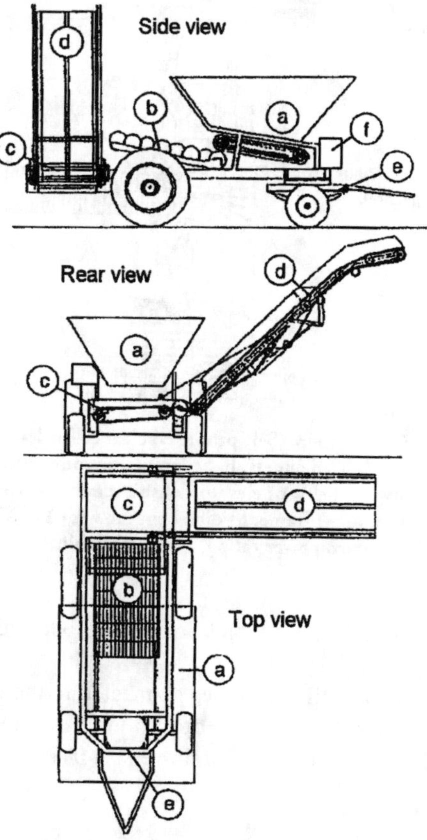

Figure 1.342. Trailed cleaner-loader:
a) hopper, b) cleaning mechanism with rubber star wheels, c) conveyor belt, d) elevator, e) frame, f) independent diesel engine with 40 kW max. power rating.

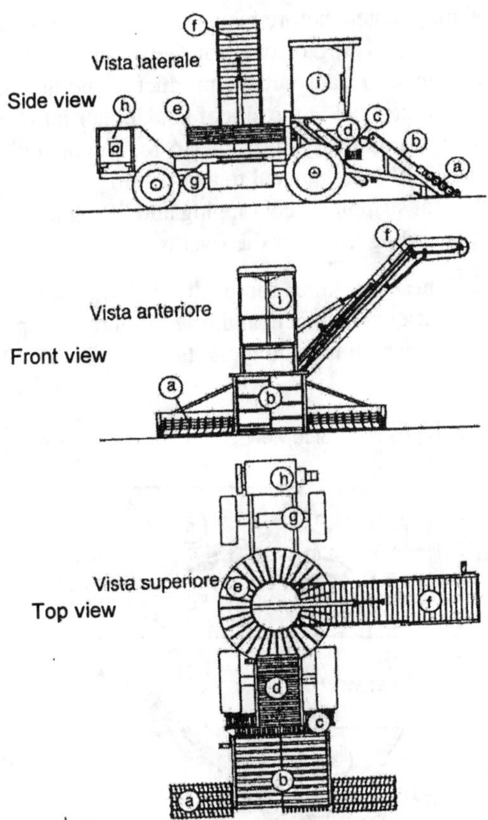

Figure 1.343. Self-propelled cleaner-loader: a) side-feed augers, b) accumulator conveyor, c) conveyor belts, d) elevator, e) soil removing wheel, f) elevator, g) frame, h) diesel engine with 114 kW max. power rating, i) driver's cabin.

- Self-propelled *cleaners-loaders* which load the product directly from the heaps (Fig. 1.343).

The *bucket loaders* are versatile machines connected to the tractor or self-propelled machine, and may be used in various farm operations, replacing the bucket with other types of equipment, such as forks, grippers, shovels, etc. (Fig. 1.344).

Mechanisms for Basic Operations

Topping. The leaf stripping mechanism can be divided into two main categories:

1) Single-wheel
2) Twin-wheel

In the first case, the wheel, with horizontal axis and metal flails, turns in the direction opposite that of forward movement (800–1200 rpm) propelling the chopped product,

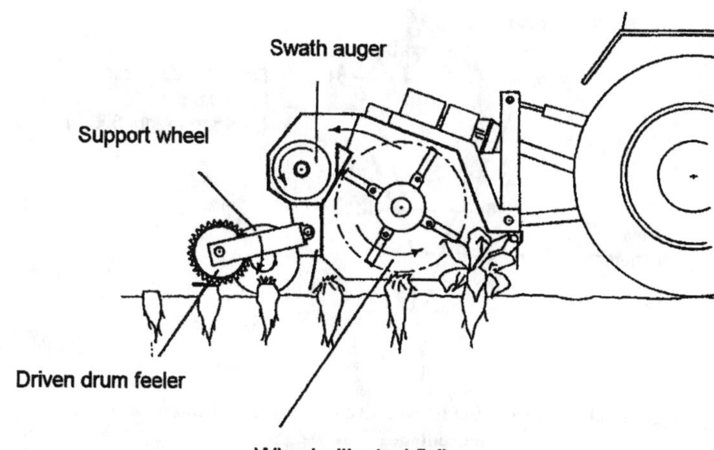

Figure 1.344. Single-wheel leaf stripper with driven drum beet topper.

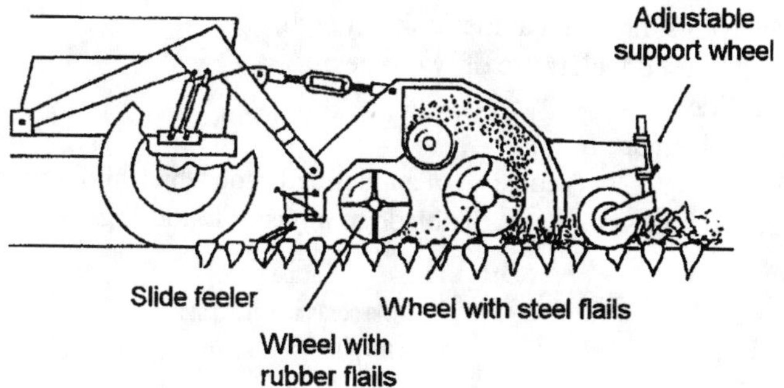

Figure 1.345. Twin-wheel leaf stripper with fixed-slide beet topper.

by centrifugal force, onto an auger designed to unload it sideways onto the ground in a swath, or to transfer it onto conveyor belts or augers so that it can be recovered (Fig. 1.344). A leaf spreader disc is sometimes fit at auger outfeed to evenly spread the chopped material over the field, providing organic substances that can be worked into the soil during primary tilling.

In contrast, the twin wheel leaf stripper (Fig. 1.345) consists of three main parts, the first wheel and auger being similar to those of the single wheel leaf stripper, while the second wheel, with speed of 400–500 rpm, has flexible rubber flails along its entire working width. These are designed to clean residue leafstalks, especially dry stalks, from the collet. The two wheels may turn in the same direction or in opposite directions.

The collet must be cut horizontally at the correct height, as defined by agreements between sugar beet factories and beet growers associations. The cut must be one centimeter above the leaf insertions closest to the ground, these being the first to develop during the root crop's growth (Fig. 1.346).

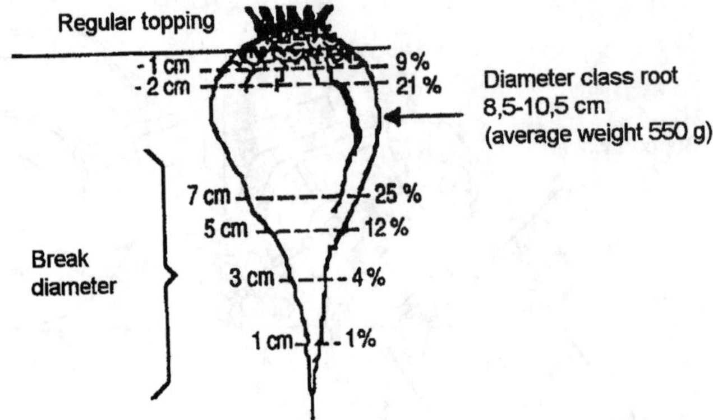

Figure 1.346. Losses due to excessive topping and distal breaks, as a percentage of the weight.

The beet-topper has the following main parts:

- Feeler, which identifies the position of the head of the root crop
- Blade, which cuts the collet at the desired height

The most widely used feelers are of two types:

- fixed, with slide
- rotary, with driven drum.

The slide feeler (Fig. 1.347) consists of an inclined shoe which slides over the head of each sugar beet and raises or lowers the blade to adjust it to the sugar beet's outgrowth.

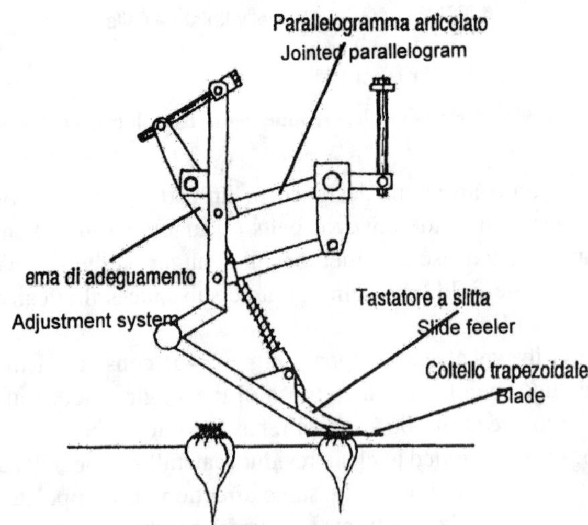

Figure 1.347. Beet topper, with distance between feeler and blade proportional to root crop outgrowth.

The driven drum feeler (Figs. 1.336, 1.339, and 1.344) consists of a drum with notches cut into its edge to increase adherence to the head of the root crop.

The motion required to drive the feeler may be provided by a hydraulic motor, with speed proportional to the speed of forward movement, or by mechanical transmissions, directly from one of the machine's front wheels.

The feeler touches the collet before the blade to keep topping height constant during cutting.

During operation, the distance between the feeler and blade may be either fixed (this being the usual method) or proportional to the size of the root crop, by using a system that changes the distance between the feeler and blade depending on the height of the feeler from the ground (Fig. 1.347).

Lifting. On some machines, before the root crop is lifted, the top is cleaned of any dry leafstalks which could jam the lifting parts. This operation is performed by wheels with rubber flails fit at the front of the lifting mechanism. These wheels have a horizontal axis that is longitudinal to the direction of forward movement (Fig. 1.348).

The sugar beet is lifted out of the soil by parts fit on self-centering frames (that may move a few centimeters transversely to the direction of forward movement). In some cases, these frames are leveling, i.e., they oscillate to adjust to small lateral slopes (Fig. 1.348).

The lifting mechanism often has oscillating harmonic motion transmitted by a camshaft driven by mechanical transmissions or directly by a hydraulic motor; oscillation may be concordant, in which case the ploughshare support frame oscillates, or discordant, with each ploughshare mounted on its own oscillating support (Figs. 1.349–1.350).

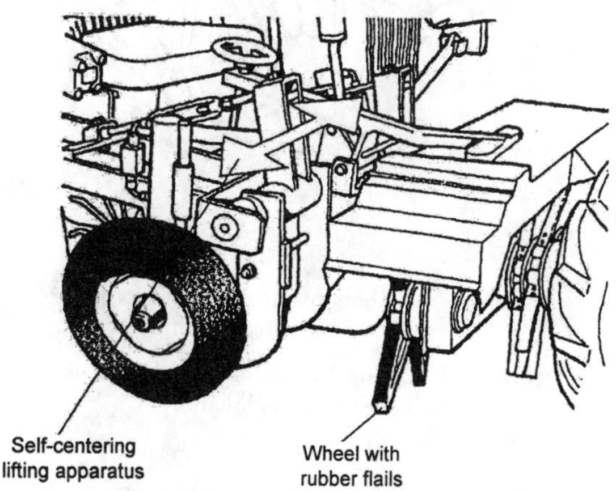

Figure 1.348. Self-centering lifting mechanism (aligns itself with the rows of sugar beet) with wheel and rubber flails for collet removal.

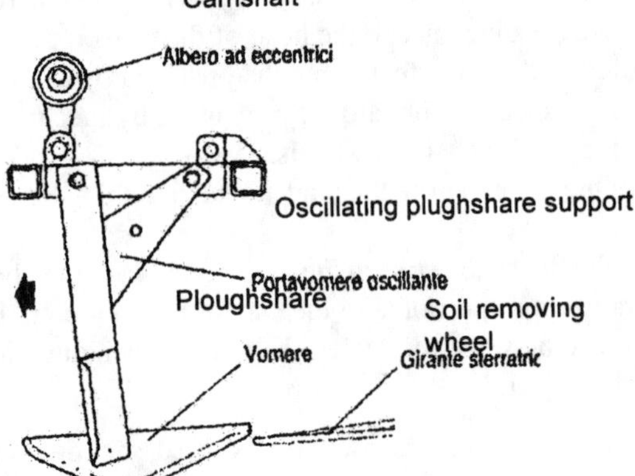

Figure 1.349. Lifting mechanism with concordant oscillation of ploughshares.

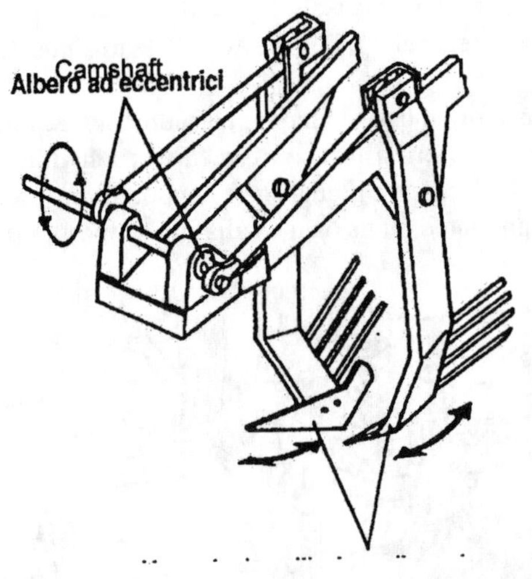

Figure 1.350. Lifting mechanism with discordant oscillation of ploughshares.

Cleaning. A large amount of soil mixed with the root crop is collected during lifting, and must be eliminated from the machine.

The root crop is cleaned not only by conveyor belts and elevators, but also by special soil removing wheels driven by hydraulic motors. In this way, surface speed can be varied continuously from 17–25 m/s to adjust to product conditions. These wheels consist of steel rods which radiate from the axis of rotation, and are angled several degrees on the

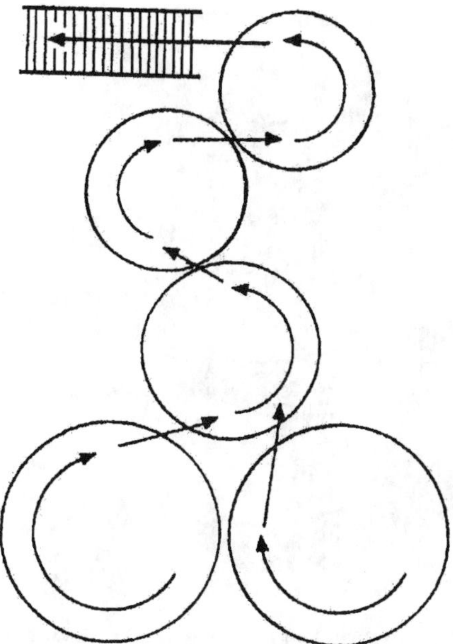

Figure 1.351. Path of product in a system with five soil removing wheels (sugar beet transfer and cleaning).

longitudinal plane so that they are closer to the ground near the lifting parts and further from the ground near the conveyor parts. Therefore, depending on type of machine and operating capacity, there are from one to five soil-removing wheels (Fig. 1.351) in a variety of configurations which guide the lifted product along a path of varying length and complexity. The soil-removing wheels have a vertical peripheral grille, also made of rods, which retains the product (Fig. 1.352).

On dry ground, the wheels must skim the ground in front, so as to harvest all of the root crop, while on damp ground, the wheel must be raised to provide greater cleaning action, even if this means risking some loss. In this case, you also have to increase the speed of rotation or reduce the machine's forward speed.

Loading. The root crop is loaded into the harvesters by chain conveyors which also have a cleaning function. The conveyors most suitable for this purpose are those with rods mounted on steel belts coated with rubber to avoid damage to the root crop being transported.

The sugar beet is lifted and loaded into the body by an elevator. This may have spring-loaded tines with rubber tips, or tines curved into a U-shape to reduce damage to the root crop (Figs. 1.336, 1.342, and 1.343).

Controls and Adjustments for Quality Harvesting. Manufacturers of sugar beet harvesters have concentrated their efforts on obtaining an acceptable compromise between

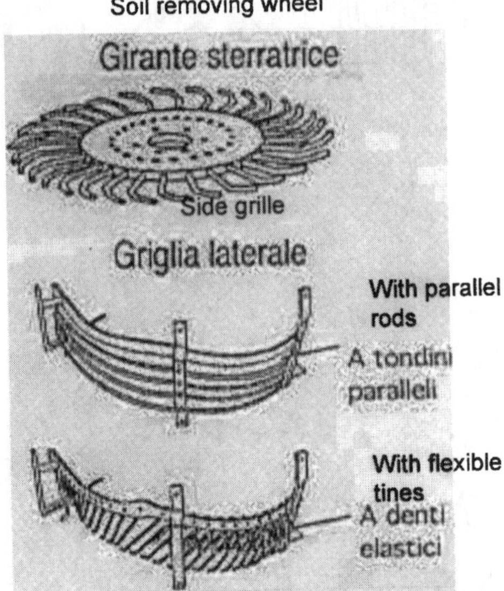

Figure 1.352. Soil-removing wheel with various product retaining grilles.

the following requirements:

- Reduction of harvesting time.
- Highest possible production.
- Elimination of soil lifted with root crop during lifting.

For a quality harvest, the equipment must be quickly adaptable to a variety of operating conditions. For this reason, new machine models have centralized adjustment of main work parameters: ploughshare operating depth, leaf stripper height, wheel speed, etc., are easily adjusted from the driver position.

Manufacturers currently supply ploughshares, wheels and grilles with different specifications to be used under different operating conditions, so that the operator may replace some parts to maintain high work quality when moving from dry to damp ground. In order to limit tares, preparation of the machine based on the type of ground to be worked is therefore essential.

Topping. The *leaf stripping* operation has significantly increased the precision of topping, and thus the quality and quantity of the harvest.

The combined leaf stripping/topping mechanism has acquired much more importance.

It should be remembered that correct topping produces a good harvest and reduces direct and indirect losses of saccharose, including those caused by long storage in the field or in silos at the factory.

A larger percentage of correctly topped sugar beet corresponds to a smaller percentage of beet that is broken and/or cut obliquely.

Lifting. *Soil tare*, especially in clayey areas, is a major problem during the harvest and for the sugar industry alike. The harvest is made more difficult due to wasted time; the sugar industry has to invest time and money in washing and removing soil and residual water.

With respect to lifting elements, the introduction of ploughshares with combined connection profiles having a convex front followed by a straight blade angled forwards (positive angle), has led to fewer base breaks, reduced soil tare, and reduced energy absorption.

With regard to soil tare and energy absorbed during lifting, results are further improved by using ploughshares with a straight blade and positive angle profile (as opposed to the horizontal profile which is common on the negative angle convex blade), and by inducing vibrations in them. In fact, the vibrating lifting mechanism, driven by special transmissions (mechanical with cams, or combined with hydraulic driving of the camshaft), reduces the soil tare.

Transmissions. The transmissions used are mainly hydraulic (from the internal-combustion engine, drive pump hydraulic unit), from those for driving power to those for the various moving parts (rotary or linear and oscillating).

Some versions of self-propelled complete machines have hydraulic units, *load sensing* and *automotive devices* for:

- Adjustment of power supply.
- Energy saving.
- Simplicity of construction, with fewer parts in the hydraulic system.
- Automatic control of the hydraulic transmission system in forward movement and work depending on diesel engine specifications (torque and rpm).
- Reduced control requirements for the operator.

Moreover, the latest self-propelled six-row machines have four steering driving wheels, a self-drive system with row finder device and electrical-hydraulic control, automatic excavation depth control and maintenance of the same excavation depth on the various rows (feeler wheels coupled to electrical-hydraulic actuator), forward speed control, calculation of area processed, etc.

1.6.12. Potatoes

In the main potato cultivation areas, two-row complete harvesters are the most common kind of implement, whereas on smaller farms, single-row complete harvesters or elevator diggers are used (Table 1.67). The essential differentiation criteria between the complete harvesters are the number of rows, the size of the sieving area, the efficiency of the separator and picking equipment, and the bunker-hopper capacity or the discharge performance.

Lifting

The offset intake has prevailed in the single-row complete harvesters, and machines with offset intake are becoming increasingly available also in the case of the two-row machines, especially the bunker-hopper harvesters (Fig. 1.353). This allows the tractor

Table 1.67. How to choose the right potato harvesting conditions and equipment

Criteria	Circumstances	Solutions
Capacity	Sufficient working people – Small fields – Large fields	Elevator digger – 2-row, mounted or drawn – 2-row or 4-row, drawn
	Insufficient working people ~15 ha potatoes/year 15–30 ha potatoes/year 30–50 ha potatoes/year 70–100 ha potatoes/year 70–125 ha potatoes/year Up to 50% more capacity >125 ha potatoes/year	Complete harvester – 1-row, <2 t hopper capacity – 1-row, 2–3 t hopper capacity – 1-row, >3 t hopper capacity – 2-row, >4 t hopper capacity – 2-row, unmanned, drawn or self-propelled Two-stage harvesting, 2- or 4-row – 4-row, bunker-hopper or discharge elevator, drawn or self-propelled
Variety	Big tuber size	– Increase gap of the web bars – Slow rotation speed of the webs – Soil cushion until the end of the web – Rubber-hedgehog belt and deflector – Low falling distances
	Small tuber size	– Decrease gap of the web bars
	Round tubers	– Slow rotation speed of the webs – Soil cushion until the end of the web
Soil	With clods	– Shallow digging depth – Increase gap of the web bars – Web agitator system (damage!) – Separation rollers – Rubber-hedgehog belt and deflector – Irrigation if to dry
	With stones	– Shallow digging depth – Increase gap of the web bars – Soil cushion until the end of the web – Rotating brushes or turning brush belts – Rotating rollers (rebound) – Air vacuum unit – No two-stage harvesting (damage!)
	Dry conditions	– Decrease gap of the web bars – Increase driving speed – Soil cushion until the end of the web
	Wet conditions	– Increase gap of the web bars – Web agitator system (damage!) – Separation rollers – Rubber-hedgehog belt and deflector – Two-stage harvesting
Haulm	Natural, green	– Wide-meshed haulm web – Slow rotation speed of the webs
	Green, haulm cut	– Wide-meshed haulm web – Counter rotating rubber hedgehog belt

(Cont.)

Table 1.67. (Continued)

Criteria	Circumstances	Solutions
	Haulm desiccation	– Wide-meshed haulm web – Counter rotating rubber hedgehog belt – Haulm rollers
	Fine haulms	– Counter rotating rubber hedgehog belt – Haulm rollers – Separation rollers – Blower
Storage	Clamp	– Dry potatoes – No soil – Elevator digger – Two-stage harvesting – Complete harvester + dry soil
	Indoor storage without forced ventilation	– Dry potatoes – No soil – Elevator digger – Two-stage harvesting – Complete harvester
	Indoor storage with forced ventilation	– No soil – Elevator digger – Two-stage harvesting – Complete harvester

Figure 1.353. Two-row bunker-hopper harvester with offset intake.

to drive beside the ridges on the area already lifted and enables the use of more powerful tractors, whose track width need not correspond to the potato row width. A hydraulically adjustable axle allows the harvesters to be both side- and mid-mounted. An automatic off-set steering facilitates the use of the machines and can be combined with further functions.

Sieving

The amount of soil separation is determined both by the gap of the web bars as well as the sieving area and the drop steps that are in most cases present in the sieving channel. Web soil separation can be improved by the use of agitator systems. To minimize the additional risk of potato damage, however, powered rotary tools are increasingly used. These aim to counter potato jumping on the web via high-frequency, low-amplitude oscillations. An appreciable reduction of the risk of potato damage can be achieved basically by means of the web-agitator, the intensity of which can be set from the tractor, and the adjustment of the speed of the web via a multistep transmission or an infinitely variable hydrodynamic drive.

The principle of horizontally oscillating webs for low tuber damage soil separation has yet to be developed to the series production stage. An undulating web is aimed at optimizing tuber protection during the soil-separation process. Lateral pairs of guiding rollers divide the web into ranges whose gradients are infinitely variable and can thus be adapted to the required separation capacity. This design also prevents the tubers from rolling down the web over a longer distance. This reduces the number of collisions between potatoes and clods or stones, which often cause damage to the tubers.

In some harvesters, the web is supplemented by conveyor systems surrounding the web or revolving above it. The conveyor pickups separate the web surface into individual sections. Even if the web is mounted at a steep angle, this prevents potatoes from rolling back, dispensing with the need for special elevators. The somewhat greater rotational speed of the pickups, which can set to varying speeds in some systems, enhances soil separation in conjunction with a steeper web angle. Haulm separation is the third function of the conveyor systems surrounding the web. The pickups are cross-linked by rubber belts so that the haulms remain on the web whilst the potatoes can fall through. For soil separation in stone pickers and some complete harvesters, new separation devices such as different types of plastic rollers are being tested or used as alternative to webs.

Haulm, Clod, and Stone Separation

Haulm-separation

The haulm separation method most frequently used in nearly all complete harvesters today is the wide-meshed haulm web, which operates in a highly crop-sparing fashion. It is often used in combination with haulm rollers or a counter rotating rubber-hedgehog belt for the fine haulms that get past the wide-meshed stage. Another solution for the debris removal is a blower at the end of the sieving channel which blows the fine haulms out off the harvester. In the case of good a haulm desiccation haulm rollers and a counter rotating rubber-hedgehog belt could be enough for the haulm separation.

Separation Rollers

For the separation of clods, remaining soil and other admixtures, differing forms of separating rollers are offered, in particular in connection with the two-row and four-row machines. In this regard the modular construction system of the complete harvesters facilitates the combination of different aggregates.

Following the sieving webs or at the end of the side elevator spiral, smooth, and rubber finger rollers, familiar from conditioning technology, are used at right angles to the direction of travel. For wet lifting conditions on heavier soils, axial-roller separation attachments are supplied as an universally employable solution. The separation unit, which is attached transverse to the direction of travel, consists of several counter rotating pairs of rollers, in which a smooth roller is always combined with a profiled rubber roller. In the event of an interruption, the hydraulic drive alters the direction of the roller rotation for a brief time and thus expels, for example, trapped stones.

Differences between manufacturers result from the design and material of the rollers and of the bearing system, from the form of the spirals, and from the systems which control the dwell time on the separating unit. However, there are fundamental restrictions regarding the material standing times and the potato quality, when the axial rollers are used in dry lifting conditions, when the harvested potatoes are relatively small, and when the proportion of stones is relatively high.

Separation Units

The combination of the transverse rubber-hedgehog belt and deflector today constitutes the standard attachment for the separation of tuber-similar admixtures for many single-row machines. It is suited for both clods as well as for a small to medium stone stock. A further development of the oscillation-operation tools are the horizontally or vertically rotating deflectors, which combine one to three finger rows. These implements, of various designs and sometimes height-adjustable, can be quickly and selectively adjusted by hydraulic drives to meet specific harvesting conditions and achieve a high working quality.

A higher proportion of stones in the crop can be separated by rotating brushes or turning brush belts. Larger-dimensioned versions of the deflector and stone-separation fittings that have proven their value with the single-row machines are being used more and more in connection with two-row complete bunker-hopper harvesters. Another separation unit for two-row harvesters uses the differing rebound characteristics of the potatoes, and stones and clods at rotating rollers. However, separation performance is impaired when product characteristics overlap, e.g., small stones, large potatoes. In addition the risk of tuber damage has still to be studied adequately.

In some regions of Nothern America the harvesters are fitted with an air vacuum unit for stone separation. These separation equipments have their own powerful drive motor on the harvesters to run the heavy duty radiator. The heavier stones remain on the web while the potatoes are transported on another conveyor through the vacuum.

Electronic Separation Units

Electronic separation units have yet to become established features in potato harvesters. With the frequent and rapid changes in lifted material and the properties of the crop, such as skin color, clinging soil or moisture, their separation performance is not always satisfactory. However, when harvesting potatoes for industrial use it is advantageous to work with mechanical, electronic, or pneumatic-hydraulic separation units at the field edge. These separation units can also serve as a loading station for truck haulage.

Hoppers/Discharge Elevators

Among single-row complete harvesters, the varyingly sized moving-floor hopper (2–4 metric tons) dominates. In all machines, a better exploitation of the hopper capacity can be achieved by the use of active side panels running all the way to the end of the discharge section. The large loading heights of single-row machines also permit a safe loading of higher-walled trailers. In two-row complete harvesters, stronger demand for hopper harvesters is evident. The hopper capacity of 4 to 6 t facilitates organization and, in most cases, also allows field and road transport to be separated. Therefore pedogenic as well as logistic demands can be satisfied. On larger fields, the hopper is used only as an intermediate store, substantially reducing the time for actual transfer.

Most two-row and the four-row complete harvesters are supplied with a discharge elevator. To bridge waiting periods, the discharge elevator of some two-row unmanned harvesters can be set to run in the opposite direction so that the potatoes can be deposited on the other side between the two neighbouring ridges. The potatoes are then collected on the next pass together with the two ridges. This requires the harvester to be fitted with a full-length share, semi-offset ridge rollers and a lateral extension discharge elevator in the lower, horizontal section of the machine. For this enhanced harvesting method a windrower with a cross-elevator can also be used (Fig. 1.354), feeding all ridge pairs to be lifted by the harvester with potatoes from two to eight rows.

Figure 1.354. Four-row windrower with cross elevator.

Figure 1.355. Single-row complete harvester with offset intake, separation unit and bunker-hopper (2 t).

Drawn four-row harvesters with a discharge elevator can be partially equipped with a driving axle and a drawbar suitable for half-offset loading so that they can deliver the necessary tractive force even at a row width of 75 cm.

Single-row, Two-row and Four-row Harvesters

In the case of the single-row complete harvesters, the machines available can be classified into three performance categories differentiated primarily according to their respective sieving area and hopper capacity (<2 t, 2–3 t, and >3 t) (Fig. 1.355). The modular system permits a multitude of variants to be constructed from the individual basic types of complete harvesters that cover the requirements of the widest conceivable range of farms.

The concentration of potato cultivation on fewer, yet in terms of their size expanding farms means that the large single-row complete harvesters are reaching the limits of their potential. The transition to two-row complete harvesters is made easier by the growing offer of two-row bunker-hopper harvesters with active clod and stone separation equipment. The use of a two-row bunker-hopper harvester enables the potatoes to be driven away as hitherto organized by means of trailers parked at the end of the field.

A precondition for the use of the two-row and four-row unmanned harvesters are conditions with a low content of clods and stones (Fig. 1.356). Otherwise these admixtures

Figure 1.356. Two-row unmanned harvester with discharge elevator.

constitutes an early source of damage potential for the tubers during harvest, transportation and store filling. In many cases, therefore, a unmanned harvester is fitted with a picking platform as a measure to ensure that, whenever necessary, too high proportions of clods and stones and/or rotten or mother tubers can be picked out before discharged (Fig. 1.357).

Self-propelled Harvesters

In the case of the two-row, three-row, and four-row self-propelled machines, both bunker-hopper as well as discharge elevator equipped versions are supplied (Fig. 1.358). A self-propelled harvester which lifts crops from unspoilt tilth offers advantages under unfavourable harvesting conditions. However, their high performance requires permanent availability of transport vehicles and a store filling technique with sufficient capacity if the potatoes must be stored afterwards. The daily and annual use is limited by the potato quality required for processing. This also affects the cost of the individual mechanization strategies. In contrast to the sugar beet harvest, for example, the high purchase price of self-propelled machines cannot be offset by a clearly higher campaign output.

Two-Stage Harvesting

An alternative method to direct ridge-lifting is two-stage harvesting—depositing the potatoes in longitudinal swathes and collecting them at a later time with a complete harvester. The combination in particular of four-row windrowing and loading using two-row complete harvesters or special windrow loaders enables very high lifting performances to be attained. When the work is organized well, the use of a two-row windrower and a

Figure 1.357. Self-propelled four-row harvester with an air vacuum unit for stone separation (factory photo).

Figure 1.358. Four-row complete harvester with picking platform.

Figure 1.359. Two-row elevator digger used for windrowing.

single-row harvester for loading the windrows also enables performances to be attained that are better than when the direct harvesting technique using a single-row complete harvester is employed.

Due to the widespread use of two-stage harvesting, special two-row and four-row windrowers are offered. In the case of two-row machines, both mounted as well as trailed windrowers are supplied (Fig. 1.359). In addition to four-row windrowers, two-row machines with adjustable depositing belts also are used for the four-row windrowing. Special window-intakes are offered for almost all single-row and the larger part of the two-row complete harvesters. In these cases a depth wheel adjustable from the tractor is generally used for the depth control. The powered roller share has proven itself as an feed tool. In addition to regular single-row and two-row harvesters a special windrow loader with an elevator is available for this purpose (Fig. 1.360). This machine allows manual separation by distributing the potatoes over two sorting tables. Special hopper harvesters for windrow loading have not been available so far.

As a result of the potatoes being allowed to dry a brief time in the swathe, the original light colour of the skins of the potatoes at the time of harvest is maintained. At the same time, the dry and clean potatoes also exhibit a better storage stability. This is all the more important in potato varieties that are difficult to store and in storages with relatively simple ventilation systems.

Figure 1.360. Special windrow loader with two picking tables.

Damage to Potato Tubers

Damage of potato tubers due to mechanical forces is among the most important causes of loss of quality throughout the world. The causes of such potato damage are mechanical injuries of the potatoes, which may occur at several of the production stages, and especially during harvesting and grading operations (Table 1.68). The susceptibility of tubers to at harvest is mostly determined by the cultivar, growth

Table 1.68. How to avoid potato tuber damage at harvest time

Driving speed	Increasing while maintaining the motor speed reduces the risk of damage.
Digging depth	Set as shallow as possible without the potatoes being damaged by the share.
Rotation speeds of the webs and transport belts	These should be kept as low as possible to prevent the potatoes from jumping or rolling about.
Web agitator	As a rule, this should be used with caution and set only when actually required.
Filling of the hopper	Make full use of the height-adjustment features of the picking belt and the cushioning of the roller floor.
Falling distance	Keep as low as possible at all transfer points, also when loading the transport vehicles.
Tuber temperature	Potatoes should not be harvested at tuber temperatures below 8–12°C. If the potatoes are lifted at too low a temperature, a raised incidence of tuber damage can be reckoned with.

conditions of the plants, operating conditions at the time of harvest, and the technology employed.

A new harvesting technique can, however, be employed successfully and with a low degree of damage to the tubers only when the lifting conditions are conducive to this end. The creation of favourable conditions starts early on with, for instance, properly timed cultivation of the soil. The soil and weather conditions at the time of harvest also play a decisive role regarding the quality of the potatoes.

Bruise-free harvesting operations also mean a fine adjustment of the complete-harvester equipment to the conditions of use on the one hand, and a quality-conscious attitude of the operating personnel on the other. It is readily apparent that the damage to the tubers can occur at any point in the harvester, and that the tractor driver plays a key role when it comes to the avoidance of tuber damage. Not only does damage lead to directly visible lesions, it also increases the risk of infection and the loss of water during storage and also the tendency towards black spots in later handling.

Bibliography

Bentini, M., E. Biavati: La raccolta meccanica della barbabietola da zucchero. Macchine e motori agricoli. No. 7–8, 1995.

Culpin, Claude: *Farm machinery. Twelfth edition.* Blackwell scientific publications.

Herrenschwand, Willy: *Arrachage des betteraves: un travail soigné est payant.* Technique des champs, 12, 1991.

Manfredi, E., : Macchine per la qualità. Macchine e motori agricoli. No. 7–8, 1996.

Walter, Jean-François: *La récolte des betteraves.* Cemagref, Btmea.

Fruits and Vegetables

M. Ruiz-Altisent and J. Ortiz-Canavate

1.6.13. Introduction

Fruits and vegetables have a high importance in world food production and human nutrition and health. Mechanical harvest of fruits and vegetables shows special problems like:

- Products to be harvested are enormously variable regarding agronomic, physiological, structural characteristics, size and shape, detachment, etc.
- Harvesting machines have to be very specialized and they are used a low number of hours in a year.
- Fruits and vegetables have been, and still are, harvested manually even in highly developed countries, so that labor problems usually appear when trying to introduce mechanization with the aims of improving economy and quality.
- Factors regarding adequate varieties, planting systems and scheduling, soil and irrigation management, materials handling, grading and sorting, processing, and others, which in themselves need considerable know-how and technification, impose strict conditions on the viability of mechanical harvest of any fruit or vegetable species.

Most operations which are coincident with the ones used in other crop productions like: soil tillage, fertilizing, seeding or planting, spraying, etc., are generally solved using mechanical equipment, in most fruit and vegetable productions. Operations involving: cleaning, handling and transportation, which can be performed in fixed installations, are also generalized with the application of mechanization and in some cases of automation equipment [1].

Mechanical harvesting of only those fruit and vegetable products destined to processing can be considered generalized and economical in developed countries. Products for fresh market can in many cases be harvested using mechanized aids, which have attained very diverse level of sophistication for different species and locations. And in later years, robotic harvesting is being developed aimed to solve fresh fruits harvesting with the same quality as manual harvesting.

Manual harvesting of fruits and vegetables accounts for 30%–60% of the total production costs, with a high net share in the final price of the product. Therefore, mechanization of harvest operations has a high potential for input reduction.

This section deals mainly with the existing principles and functions that make up the mechanical harvesting equipment for temperate fruits and open-air grown vegetables.

1.6.14. Harvesting Functions

While detachment and removal, control, cleaning and selection, conveying and loading are the required functional operations for a harvester, the order in which these functions are achieved is determined by the requirements of the specific product or commodity (as an example, hand harvesting always begins with selection) [2]. In mechanical harvesting systems, detachment is seldom as selective as desirable, therefore the selection function is achieved after detachment, in the form of sorting devices (or even manual sorting in the machine), or else at a later processing (sorting, cleaning, grading, and packaging) inside fixed premises. *Detachment* is the actual separation of the harvested portion of the plant: fruits, buds, tubers, roots, leaves, etc. Catching and control: padding of catching surfaces is required to gain or maintain product control during harvesting operations [3]. Good padding materials can absorb the impact energy of the product, thus preventing its absorption and damage to the fruit; these materials have to be easy to keep and to clean, and durable.

Selection is the process in which only the ripe, correctly sized or desirable product is obtained from the entire recovered crop material, while the remainders are rejected. Size is often associated with product quality. Harvesting machines are sometimes equipped for size and color grading in the field. Product maturity requires special attention as the main property of the product that decides harvest date, and defines the product susceptibility to damage.

Transportation: whenever possible, bulk handling systems are preferred for transportation of the product from the field to the grading/marketing station. Bulk handling (trucks or tractor trailers) are used for industry products such as tomatoes, green beans, onions, potatoes, peaches, wine grapes, olives. Standard pallet containers can be handled with standard forklift equipment and are used also for fresh market products like apples, melons, cucumbers, etc. Delicate fruits like strawberries need small, market-ready containers; these are also used in appropriate sizes, for example for ripe peaches and apricots. Vegetables like lettuce, cauliflower, broccoli, etc., are sometimes wrapped and packed in the field. Today, infield grading and packaging of produce is contemplated as a good solution for reducing costs and quality losses in many fresh-market products (vegetables, fruits, especially tropical fruits).

Damage is an important consideration [3]. Product bruising, cutting, scuffing and direct damage to the remaining plant can be a consequence of mechanical harvest. Damage reduces the value of the commodity in the market; damage to the remaining plant can affect future crops or life of the plant itself [4].

Harvesting functions often interact with each other. For example, if inertial detachment is used by interacting with the plant, the separated commodity often has an associated kinetic energy, which aggravates the problem of product control (damages) when compared to other types of detachment and removal procedures [3] (Table 1.69).

1.6.15. Principles and Devices for the Detachment

The application of well-directed energy is necessary to effect detachment; the procedure in which this energy is applied is the first and basic consideration in the aim of mechanical detachment and removal and depends on the commodity in question. Severing the attachment forces requires that the ultimate fatigue, tensile, or shear strength threshold must be exceeded [2] (Table 1.70).

Low-height Herbaceous Structures (Vegetables, Strawberries, etc.)

We differentiate between root and surface crops.

Table 1.69. Comparative damage thresholds for selected fruits

	Drop Height onto Hard Surface Causing Visible Bruise[a] (mm)	Average Mass (g)	Observations
Apple	10	220	High turgidity, increased damage susceptibility
Pear	20	175	
Peach	25	150	More susceptible in compression
Apricot	150	60	
Tomato	200	75	Skin most important
Orange, lemon	400	200	Rough surface, increased damage susceptibility due to friction
Olive (table)	200	30	
Small fruits			
Rubus	50	10	
Ribes	120	10	
Strawberries	50	30	

Sources: Various authors and ref. [6].

[a] 10 mm in diameter; wide variations for different fruit conditions; time of harvest was selected when data were known.

Table 1.70. Detachment forces for selected fruits

	Detachment Force (N)	Observations
Apple, pear	40[a]	
Almond	1.5[a]	
Cherries	12–16	8–10 N[a]
Grape	3–5	Single berry
	25–50	clusters
Peach, apricot	1–5	9–15 N[a]
Tomato	20–30	
Orange	20–30	
Lemon	30–40	
Strawberries	4–9	
Small fruits	0.5–2.5	
Olive	2–8	

Sources: Various authors and refs. [6–8, 12].
[a] Limb-peduncle detachment, shearing action.

Root Crops

Two different principles are used for the harvesting of these crops: digging (potatoes, carrots, and so forth) and pulling (carrots, leeks, etc.) [1, 5].

a) Digging consists of the uprooting and lifting of the crop together with considerable amounts of soil, from which it must be separated, by the thrust of a digging or plowing share. Working parameters of this system are length and inclination angle of the share and working speed and cutting depth, determined by the location of the recoverable product in soil, together with the above-mentioned parameters; these constitute the regulations of the system. Following the coulters, a vibrating rod-chain conveyor is needed to clean and to feed the product to the complementary operations of handling, cleaning, and loading units. The aboveground plant parts, very fragile in comparison with the subsurface parts, may be recovered or dispersed, before or at the same time as the harvesting of the roots. This harvesting system is appropriate for any roots or tubers (e.g., carrots or potatoes), below-ground fruits (e.g., peanuts) or bulbs (e.g., onions and garlic).

b) Pulling the aerial portion of the plants often is used for the harvest of some root crops (e.g., carrots) and also some surface crops (e.g., leeks and salad greens); the structure and the strength of the plants must permit in this case the engagement of the above ground leaves and the uprooting of the entire plant, aided by a subsurface coulter; the big advantage is that very little soil is extracted with the product (Fig. 1.361).

The equation for the pulling force is as follows:

$$F_p = 2\mu N. \tag{1.117}$$

For pulling speed:

$$\bar{v}_p = \bar{v}_a + \bar{v}_r \tag{1.118}$$

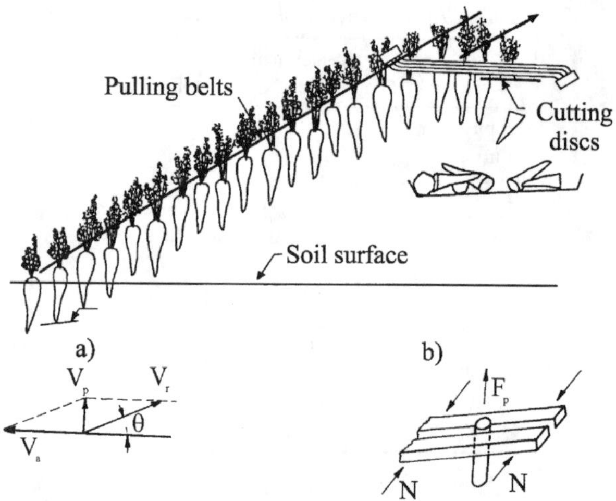

Figure 1.361. Pulling principle and units. Velocities (a) and forces (b) in the pulling belts [1].

where

$\mu =$ friction coefficient
$N =$ normal force exerted by the belts on the plant
$\bar{v}_a =$ working speed
$\bar{v}_r =$ speed of belts

The aerial parts have to be well aligned in the row and uniformly spaced, and they are engaged and grasped by a pair of elevating belts, which, due to the combination of the effects of advance speed and velocity and inclination of the belts, exert a vertical pulling force to the plant. Functional parameters of this system are then belt speed, working speed, belt angle of inclination, and the pulling force exerted by the belts due to friction and compression between the belts and the plants. At the upper end of the belts, a pair of topping implements (counterrotating bars or rotating disks) are placed to remove the tops.

Surface Crops

Cutting, combing, stripping, vibration, and threshing are different functions used for the detachment of the very wide range of aerial parts found in the diverse commodities to be harvested.

a) Cutting action is applied mainly to the so-called leafy products: cabbage, lettuce, spinach, endive–escarole, Brussels sprouts, celery, broccoli, cauliflower, artichokes, chard, chicory, mustard, parsley, watercress, and any other so-called greens or similar plant parts. Each of these products needs multiple harvests for maximum yield for fresh market, except perhaps spinach, which is mostly grown and harvested for processing. Their common properties are that they have to be cleanly removed and handled

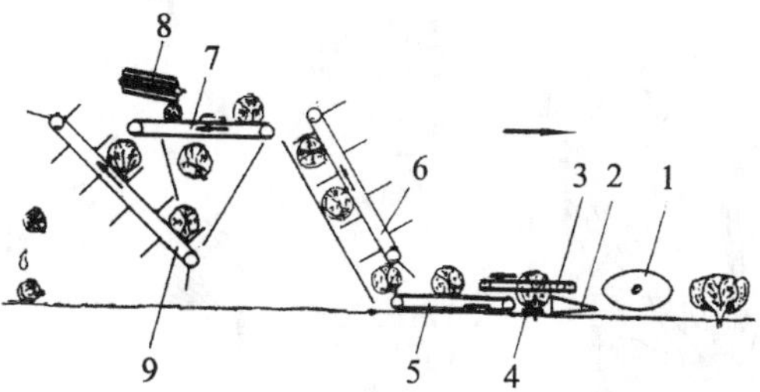

Figure 1.362. Cabbage harvester. (1) guiding discs, (2) Guiding surfaces, (3) driving belt, (4) coulter bar, (5), (6), (7) band conveyors, (8), (9) separation devices [1].

softly for minimum leaf loss. Cutting is the most effective method of removal, and it is also used in combination with other harvesting systems (such as the cutting of whole plants of tomato; cutting undesired aerial parts of root crops; cutting of Brussels sprouts stalks for subsequent stripping of the sprouts, etc.). The conventional cutting devices are applied for this function: rotating discs with dented edge, flat or with some concavity, and cutting bars, simple or double. In some cases the objective is to remove entire plants, even below the surface (tomatoes, pickling cucumbers) to be subjected to further detachment of the fruits; in others, to recover the tender edible green parts (spinach, cabbage) (Fig. 1.362). Frequently, a powered reel or rotating side cylinders are added to help feeding and further conveying of the product up the band or bar-chain conveyor. Functional parameters of cutting devices are

- Cutting speed
- Advance speed
- Sharpness of the cutting edge
- Speed and position of the feeding reel or cylinders, with respect to the cutting elements

b) Combing is based on particular properties of the plants:

- The plant is firmly attached to the soil by its root system and grows erect.
- The portions to be detached and removed have different size, shape, or rigidity from the leaves and stalks (e.g., green beans or pea pods).
- The portions to be detached possess a zone of abscission, susceptible to being severed by traction or by flexural forces.
- The parts to be recovered (e.g., strawberry fruits, green-bean pods) are capable of resisting the action of the combing fingers.

Functional parameters of combing are

- Speed of the extreme of the combing fingers, be around 4 m/s for a forward speed of approx. 0.5 m/s.
- Speed and position of the conveying device situated behind the combing fingers, to assure the removal of the product from further impacts.
- Speed of the feeding reel, conveying band, or brushes, which are installed in front of the combing fingers (Fig. 1.363).

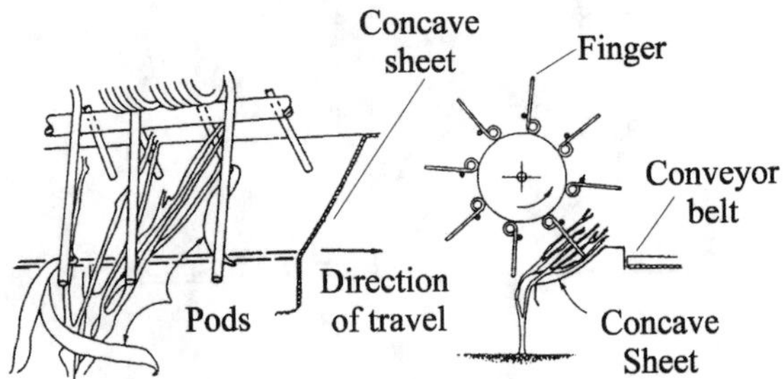

Figure 1.363. Cylindrical reel-type snap-bean harvester. (a) Pods are removed by contact of fingers with the pedicle (b) Detachment originated by the cylindrical reel and the concave sheet in the plant [1].

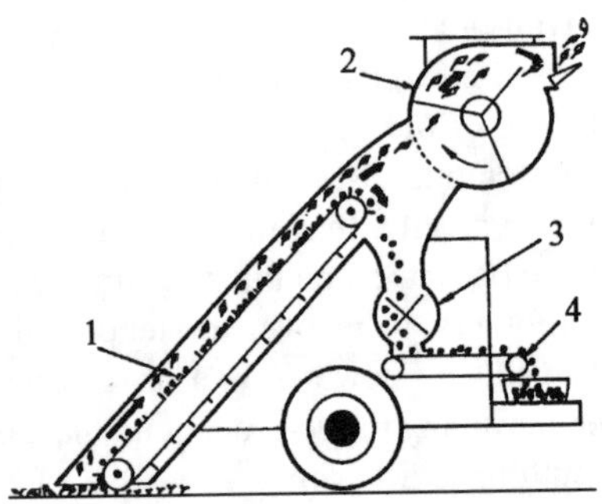

Figure 1.364. Strawberry harvester by the combing principle. (1) Finger band to comb the strawberry plant. (2) Fan. (3) Discharge valve. (4) Conveyor and filling device [1].

As mentioned above, harvesters for strawberries (fruits that must be classified as aerial herbaceous commodities) may be provided with combing fingers attached to slow-moving elevating bands, and the operation is aided by air currents that elevate fruits and leaves to the fingers for easier combing (Fig. 1.364). Instead of combing, some strawberry harvesters use a cutterbar–reel principle, which cuts the entire plant and separates the fruits in the machine by again air-lifting and cutting, or by stripping. The combing effect attained by two counterrotating brushes is used in harvesters for some varieties of paprika pepper for the detachment of the fruits (the concept derives from the combers/strippers for harvesting cotton capsules).

c) Stripping by the action of counterrotating rollers pulling the plant down, stripping is a procedure by which the commodity (fruit) is separated from the plant, based on the differential properties between fruit and plant: size, shape, and attachment strength. This principle is used to detach cucumbers (pickling) from their already cut supporting plants, in pepper harvesters, and in combination with other functions in selected harvesters:

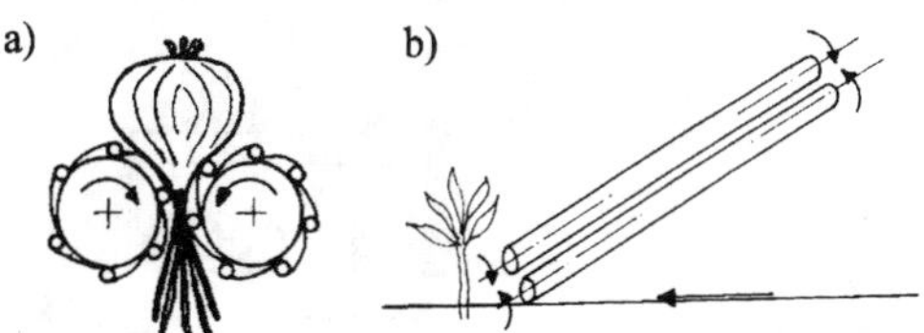

Figure 1.365. Stripping machine with counterrotating rollers. (a) Front view, (b) side view [1].

destemming of onions, strawberries, olives, and cherries; and detachment and deleafing of (sweet) corn cobs. The principle consists of the pulling action of two counterrotating rollers, provided with some roughness condition: It can be the surface itself (rough rubber rollers), or helix-shaped structures attached to the rollers. The rollers trap the long and thin plant parts (stalks, with leaves) and sever them from the commodity by a pulling action (Fig. 1.365, see also Fig. 1.364). The product is conveyed in different ways, such as inclination of the rollers or conveying by the helicoidal attachments on them, or by falling onto separate conveyors (as in some onion harvesters). Functional parameters of this system are turning speed of the rollers, separation between the rollers, and engagement speed (forward speed) to the plant. Helix-shaped open cylinders, counterrotating, have been used in paprika-pepper harvesting machines, having the effect of combing rather than stripping, as the helixes comb the fruits rather than stripping the plant. Stripping also is used for the detachment of Brussels sprouts from the previously cut stalks. In this case, the effect is achieved with rubber cords.

d) Shaking is done to plants previously cut, in tomato harvesters. Industrial mechanical tomato harvesting is the most advanced mechanical vegetable harvesting, in that world-wide there are harvesters used of numerous makes and designs, which harvest a very high percentage of the industry tomato surface. Tomato plants are cut at the soil surface and plants with fruits are conveyed to the top of a vibrating platform, on which, with the tomatoes in a hanging situation, a shaking energy is exerted on them (Fig. 1.366). Because the values of the detachment forces, Fd, of tomato fruits are in the range of 20 to 30 N, and the masses, m, of the fruits are 0.05 to 0.20 kg, the minimum applied accelerations for fruit detachment are on the order of 10 to 60 times the acceleration of gravity (see also Table 1.70).

$$n\cdot g\,(\mathrm{m/s^2}) = F_d\,(\mathrm{N})/m\,(\mathrm{kg}) \tag{1.119}$$

This relationship means that only those products with a relatively high mass and relatively low detachment force are easy to detach by vibration, as is the case with the present tomato varieties. In general, fruits are easy to detach by shaking if n is between 1 and 10 [6]. Inertial application of vibratory energy for detachment results from accelerating the plant commodity with a suitable machine to attain a pattern of vibration and a frequency that are suitable for that commodity. In this sense it has to be designated as "the only noncontact" principle for detachment of fruits [1]. In the case of tree fruits (see further ahead) the necessary acceleration has to be applied at the lowest-force abscission point

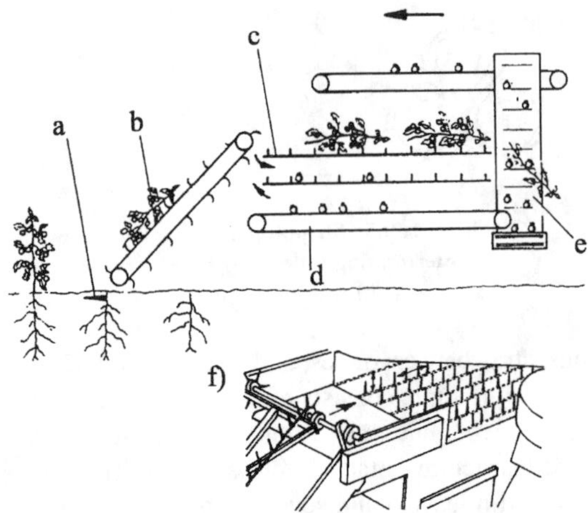

Figure 1.366. Tomato harvester. (a) Cutting unit, (b) elevating chain, (c) shaker, (d) selection band, (e) loading conveyor, (f) shaker unit [1].

of the fruit, and that is difficult for small fruits. These considerations can be applied to other vegetable crops for which shaking could be a good solution for detachment.

The rest of the functions and units of tomato harvesters combine complementary operations and manual or automatic optical VIS (visible, color) or NIR (nearinfrared) sensors for color grading and for soil-clod separation, respectively.

e) Threshing. This separation procedure uses a rotating drum to remove peas (also some beans) from their pods in pea harvesters. Similarly to one of the effects used for grain threshing, friction is used to separate the pea grain from their hulls. Friction is applied by slow-rotating rubber rollers or bands against a drum, which sometimes also is itself rotating and behaves as the concave as it sieves the peas through it (Fig. 1.367). From there, peas are cleaned by air, conveyed, and loaded in refrigerated bins.

Bushy Structures (Small Fruits, Wine Grapes)

As mentioned above, small fruits are at the same time difficult to detach by shaking (noncontact) due to their small mass, and damage-susceptible if detached by combing or stripping (contact), as described for some vegetable harvesters. Therefore, the detachment of small fruits has been accomplished by a combination of contact and noncontact actions applied by vibrating tools.

Small-Fruits Harvesters

A number of species are referred to as "small fruits:" raspberries, blackberries (*Rubus* species), and blackcurrant and redcurrant (*Ribes* species), along with gooseberries, blueberries, boysenberries, and so forth. They have in common their way of growing in bushy structures, their small size, and their use mainly for processing. In the past 10 years there has been considerable activity in the application of mechanical harvesters to collect these fruits, which are the only fruits produced in some cold (northern) areas. Contact

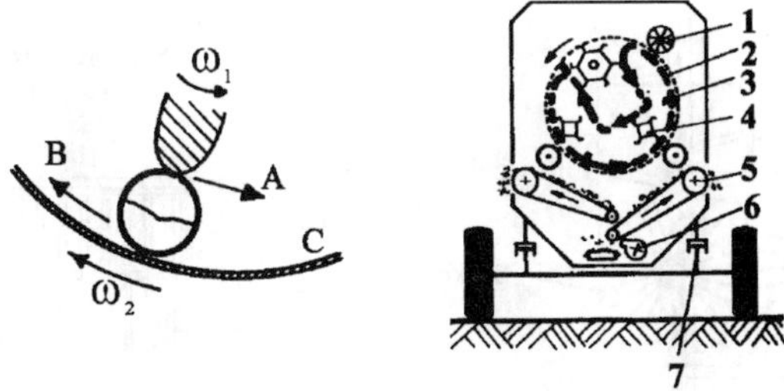

Figure 1.367. Pea-threshing principle (dehulling) and pea harvester. C, concave drum; ω_1, ω_2, rotating speeds of bars and concave drum; AB, shearing effect on pod; 1, cleaning brush; 2, concave sieving drum; 3, pulling bolt; 4, threshing cylinder; 5, inclined band; 6, air cleaner; 7, slope-adjustment mechanism [1, 6].

principles (combing and stripping) have been tested for the detachment of small fruits in bushy as well as tree structures, with little success. Present small fruit harvesters use a combination of shaking with soft combing, based on vertical, tilted, or horizontal drums provided with fingers or spikes and oscillatory motion, which applies a shaking effect on the fruiting structures. They work on plants trellised in different systems: "T," "V," or "Y" (Fig. 1.368) [7, 8]. The trellis system to be used and the shaking functional parameters depend on the fruit species and the conditions of fruiting. Drums are 50 to 100 cm in diameter, and the shaking is created by inertia. The frequencies depend on the fruit species: 5 to 10 Hz for raspberries (*Rubus* spp.) and 10 to 25 Hz for currant (*Ribes* spp.) with amplitudes 40 to 75 mm. Travel speed are 1 to 1.7 km/h (lower speeds for lower shaking frequencies). The quality that one may expect for the mechanically harvested small fruits is highly dependent on fruit species and variety, yield, maturity, trellis system, and the functional control capabilities of the harvester units.

Grape Harvesters

Mechanical grape harvesters were designed in the United States during the late 1950s. The fruit-detachment system consists of a series of vibrating horizontal rods that are free to move on their rear end and mechanically driven with an oscillatory motion on their front end (Fig. 1.369). The number of rods depends on the height of the plants, with 8 to 16 rods for vines that are 1.2 to 1.8 m high. The effect of the rods on the vines is a combination of vibration and impact on the branches and on the vinegrapes themselves. The frequency of the vibration is 9 to 10 Hz, with an amplitude of 88 to 140 mm [9].

In the United States, the latest grape harvesters have a detachment unit consisting of two sliding bars, provided with horizontal stroke. The bars engage the vine trunks at a height of 50 to 70 cm; the vines are forced to vibrate at a higher frequency (10 to 20 Hz), causing the detachment of clusters and also, more often, of individual grapes [10].

In the late 1980s, in France, a new design was introduced that is capable of a gentler handling of the vine shoots. In a similar arrangement to the former system, the rods are substituted by arched rods, which in their rear end are all fixed to the same member, which

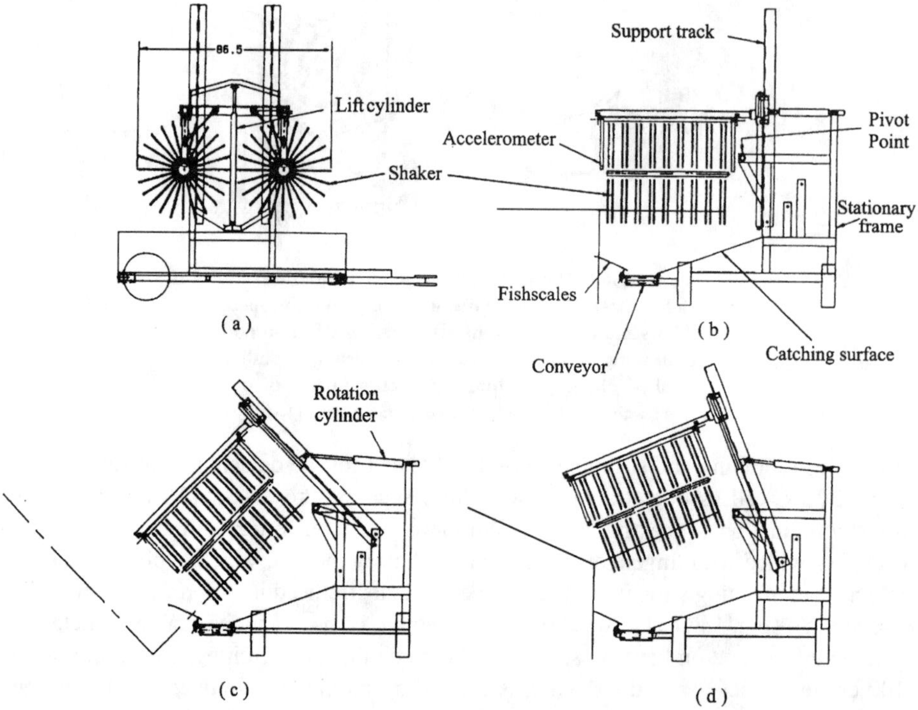

Figure 1.368. Blackberry harvester. Side (a) and front (b) view of "T" trellis, (c) "V" trellis, (d) "Y" trellis [8].

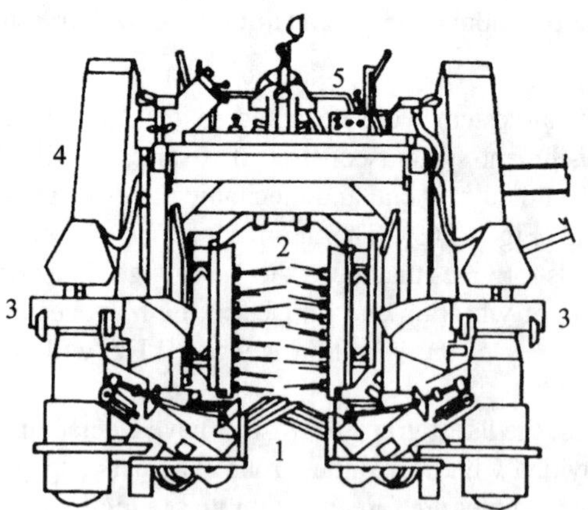

Figure 1.369. Over-the-row grape harvester. (1) Overlapping spring-loaded plates; (2) shaking rods; (3) leaf blowers; (4) grape conveyors; (5) diving port [17].

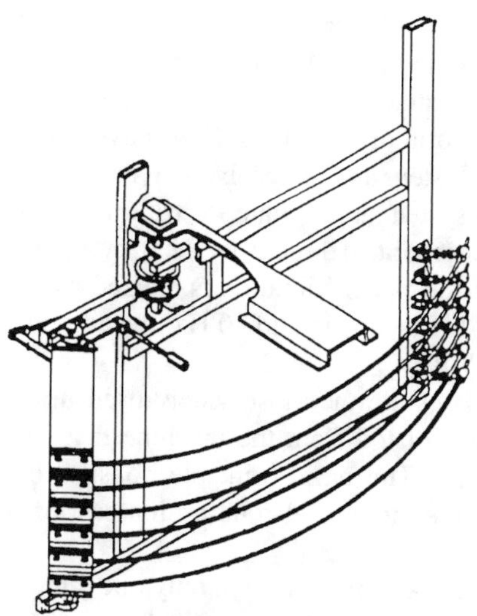

Figure 1.370. Oscillatory shaker with arched rods [2].

is pivotally mounted on an axis (Fig. 1.370). The motion of these rods corresponds to the connecting bar of a bell-crank four-bar linkage. In this type of shaker, the frequency is higher, 15 to 23 Hz, and the amplitude smaller, 25 to 70 mm [11]. This means that whereas the former system applies peak accelerations of 130 to 180 times the force of gravity in the three perpendicular directions, on the vine trunk, the new types apply lower accelerations, 50 to 100 times gravity. The result is a reduction in deleafing, broken woods, and torn wood, of 50% to 90%, as well as a better quality of grapes. Additionally, these harvesters can be driven at a higher speed along the row, therefore increasing their productivity. All these grape harvesters are straddle type (over-the-row) (see Fig. 1.369) machines.

Functional parameters of the detachment system are therefore frequency and amplitude of vibration, and travel speed. In these harvesters, the shaking (detachment) unit is mounted on a hanging frame, which engages the trellised vines loosely. In the lower part of this frame, a series of overlapping plates forms a catching platform. As the harvester moves forward, the retractable plates embrace the trunks, closing the catching surface tightly. The system has been developed to a very efficient level, where no more than a 5% of the crop is lost for most grape varieties. The machine incorporates one or two lateral conveying bands and a loading elevator to load the grapes into the hopper, after one or two steps of air cleaning, which eliminates leaves, dust, and shoot pieces. The harvesters work over-the-row, and some are provided with sensing elements in front, to align them to the vine rows. Present harvesters are able to work in vineyards with narrow row spacing (down to 2 m) and are able to engage vines as low as 15 to 20 cm from the ground [9].

Tree Structures (Fruits, Nuts)

We differentiate between open-trained trees and high-density orchards.

Tree Shakers for Open-Trained Trees

The most widely used method to harvest fruits mechanically is the use of inertial trunk or limb shakers that attach to the pertinent wood and are able to transfer large amounts of energy in the form of vibrations. Nowadays, the shake–catch method is the only mechanical harvest system used extensively in deciduous tree fruits.

The simplest shaking system appropriate for fruit trees is the tractor-mounted cable shaker. In it, the motion is generated directly by the tractor power-take-off (p.t.o) through an eccentric that powers the cable. Vibration is created by the returning movement of the tree branch or trunk. Frequency is 5 to 10 Hz; the amplitude is large, 20 to 60 mm; and power is 10 to 30 kW [6].

Eccentric rotating masses are the most widely used in tree-shaking machines. Inertial shakers have to be isolated from the machine that carries them, so that no vibration is transferred to it. The basic principle consists of transmitting to the tree the forces generated by one or several rotating masses, or by a slider-crank mechanism.

The slider-crank shaker transmits forces in only one direction. The magnitude of the force depends on the rotation speed and the mass of the housing of the shaker. Slider-crank mechanisms are applied exclusively in limb shakers. The frequency is 10 to 20 Hz; amplitude, 20 to 40 mm; and power, 20 to 40 kW. Reciprocating housing mass is 100 to 200 kg, and clamping force is approximately 5 kN. The diameter of the branches can be a maximum 30 cm, and the clamping surface is $2 \times 30\ cm^2$.

In shakers provided with eccentric rotating masses, centrifugal forces are generated. The distribution of these forces can be varied by changing the sizes of rotating masses, their eccentricity, and their rotating speeds. Normally, forces are multidirectional, although two equal-sized masses, rotating at the same speed (in opposite directions), generate a one-directional oscillating force. Multidirectional shakers generally are used as trunk shakers (Fig. 1.371). The frequency is 20 to 40 Hz; amplitude, 5 to 20 mm; power, 30 to 70 kW. Eccentric masses are 20 to 60 kg; total mass of the shaker, 600 to 1000 kg maximum diameter of the trunks, 40 to 50 cm; clamp force, 5 to 7 kN; and clamp contact surface, 2×40 cm. Trunk shakers are faster and easier to operate than limb shakers (Fig. 1.372). The structure of the trees must be adapted to shaking (3–4 main limbs at maximum). The use of trunk shakers is not well suited for large trees (>50 cm in diameter) or for trees with hanging branches, which lead to low fruit detachment; in these cases, limb shakers are preferred.

a) Vibration of Fruits. When a fruit vibrates, there simultaneously appear traction, twisting, bending, and shear forces, and also fatigue effects. As mentioned for tomato shaking, an acceleration has to be applied to the fruit (expressed in number of 'g's') to produce its detachment from the limb or the peduncle. Table 1.71 summarizes vibration parameters typically used for shaker design.

Damage to the fruits is a limiting factor for the application of the shaking method, and also in the case of processing fruit. The use of abscission chemicals has not had widespread application due to the high cost and to problems related to timeliness of application, as well as to concerns about chemical residues on the fruits.

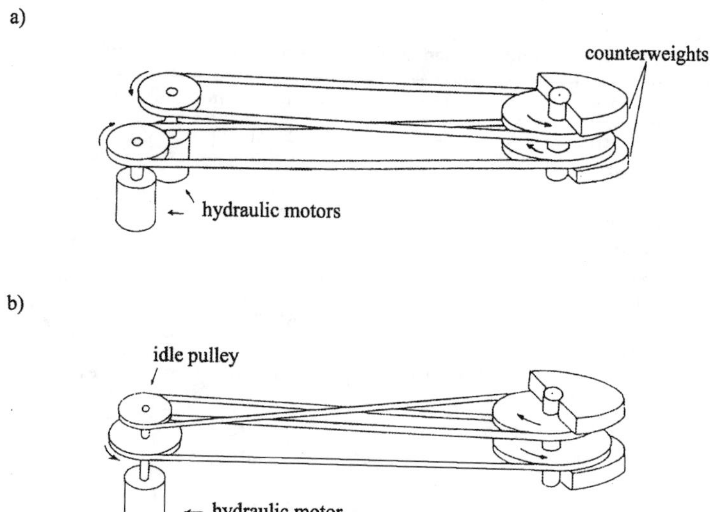

Figure 1.371. Drive of the rotating counterweights by two hydraulic motors with two bolts (a) and by one hydraulic motor and one idle pulley with one revolving belt (b).

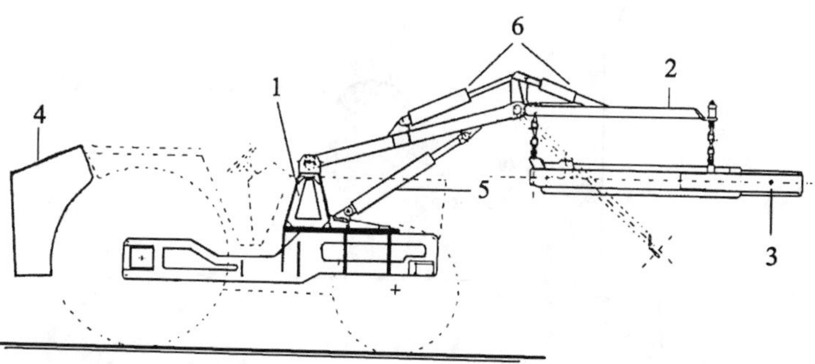

Figure 1.372. Tractor-mounted trunk shaker. (1) Supporting frame, (2) articulated arm, (3) shaker, (4) oil reservoir, (5), (6) hydraulic cylinders [1].

b) Kinematics of the Inertial Shaker. If only one rotating mass, m, is considered (Fig. 1.373a) [1], rotating at a constant angular speed ω, the centrifugal force has the following expression:

$$\bar{F} = mr\omega^2 \exp(i\omega t)$$

where r is the eccentricity of the rotating mass. This type of orbital shaker is appropriate for some fruits, such as almonds.

If two eccentric masses are considered, m_1 and m_2 (Fig. 1.373b), rotating at different angular speeds, ω_1 and ω_2, in the same or in opposite directions and mounted on one or

Table 1.71. Frequencies and amplitudes used in fruit and nut harvesters

Harvested Crop	Frequency (Hz)	Amplitude (mm)	Observations
Strawberries	5–15	20–40	Mass: 6–9 g
Cherries	10–20	15–60	
Apricots	15–30	8–12	
Almonds	15–30	8–12	
Apples	15–30	8–12	
Prunes	15–30	10–14	
Peaches	15–30	12–16	
Olives	20–35	50–75	
Oranges	10–15	12–16	
Grapes	9–10/10–20	80–140	See text
	15–23	25–70	
Tomatoes	5–10	30–50	
Small fruits			
Rubus	5–10	40–75	
Ribes	10–25	40–75	

Sources: [1, 6, 11, 12, 14].

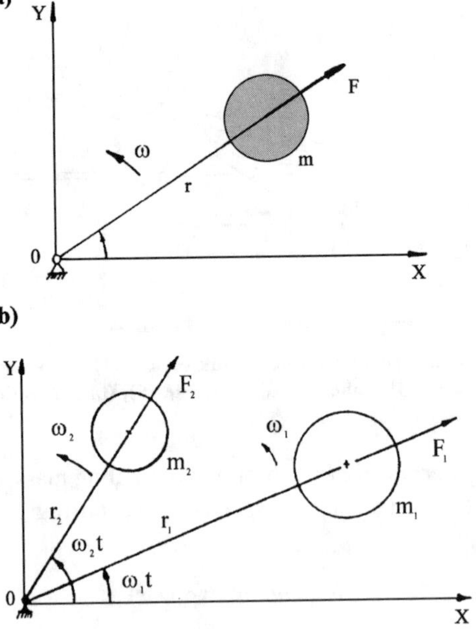

Figure 1.373. Multidirectional shaker with one (orbital shaker) (a) and two (b) independent rotating masses [6].

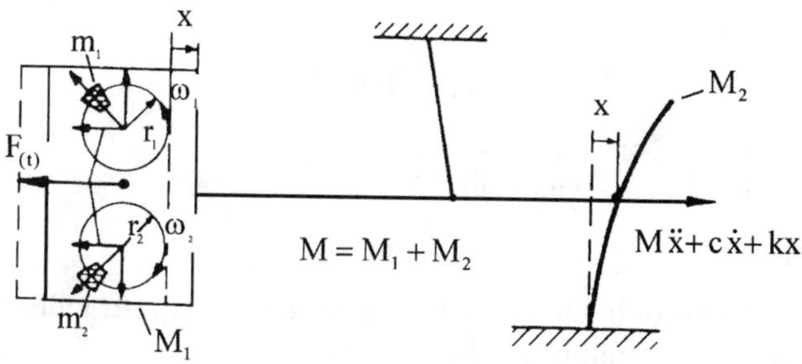

Figure 1.374. Forces created in a two rotating-mass inertia shaker [6].

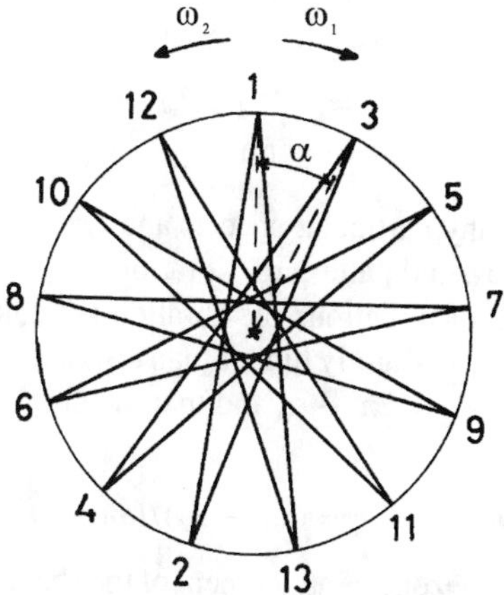

Figure 1.375. Directions of vibration, depending on ω_1 and ω_2 [1].

on two different axles, the forces generated by their rotations are:

$$\bar{F}_1 = m_1 r_1 \omega_1^2 \exp(i\omega_1 t) \tag{1.120}$$

$$\bar{F}_2 = m_2 r_2 \omega_2^2 \exp(i\omega_2 t) \tag{1.121}$$

The usual output of inertia shakers is a relatively high frequency of vibration (12–40 Hz), and a short stroke (5–20 mm) delivered to the tree. The resultant force $\bar{F}_1 + \bar{F}_2$ can be adjusted for magnitude and pattern in any desired way by changing the relative masses of the eccentrics and by modifying their respective rotating speeds (Fig. 1.374).

The force generated by the shaker (Fig. 1.374) is

$$F(t) = M\ddot{x} + c\dot{x} + kx \tag{1.122}$$

where

M = total mass (equivalent mass of the tree + mass of the shaker) (kg)
c = damping factor (N·s/m)
k = elastic constant (N/m)

The maximum force is

$$F_{\max} = 2mr\omega^2$$

where $m_1 \cong m_2, r_1 \cong r_2$, and $\omega_1 \cong \omega_2 = \omega$.

The resulting amplitude of vibration [12] is approximately

$$s = 2mr/M$$

The equivalent mass of limbs M can be estimated as 0.2 kg per millimeter diameter and for trunks as 2 kg per millimeter diameter.

Natural frequency, damping factor, and elastic coefficient of trees change with time of the year, between the following approximate values:

$$f = 0.5 - 3.5\,\text{Hz}$$

$$c = 0.1 - 0.7$$

$$k \leq 2000\ \text{N/cm}$$

Fruit detachment is, for any particular species and type of trees, the effect of a combination of stroke, frequency, and pattern of vibration.

To calculate the number of directions of vibration (shaking pattern) let us consider (Fig. 1.375) that the speed ω_1 is slightly higher than ω_2, that they are rotating in opposite directions: $\omega_1 t = 2\pi + \alpha; \omega_2 t = 2\pi - \alpha$, and that the number of directions of shaking [1] is

$$n = 2\pi/\alpha = (\omega_1 + \omega_2)/(\omega_1 - \omega_2)$$

c) Damage to the Trees. The zone of attachment of the shaker to the tree (trunk or limb) has been found to be the most dangerous for tree damage. During shaking applied by the clamp to the wood, there are not only radial stresses but also longitudinal and tangential [6]. The stresses that can be absorbed without damage under the shaker clamp are very much related to bark humidity. For fruit trees, the admissible values for tangential, radial, and longitudinal stresses are, respectively, $\sigma_t \leq 600$ kPa, $\sigma_r \leq 1700$ kPa, and $\sigma_l \leq 1800$ kPa.

Figure 1.376 shows the comparative variation of bark resistance (N/cm^2) for different times of tree growth (times of the year) and irrigation, both of which influence bark humidity.

Excessive radial stresses are likely to cause damage in selected fruit trees. The causes of excessive stress can be bad design of the clamp, as well as careless operation. For example, damage can be caused by inadequate adjustment of clamping pressure or inadequate contact area between the bark and the clamp. To avoid tangential stresses (the worst for the tree) some manufacturers place a flap on the clamp, keeping the clamp-flap contact surface well lubricated.

Harvesting Fruits in High-density Orchards with Over-the Row Machines

A new possibility for fruit production is the high-density dwarf-tree plantation. (Trees are planted at 2×1 or 2.5×1 m; this means 4000–5000 trees/ha, with a higher and earlier cropping). Complete mechanization of cultivating, pruning, spraying, and also harvesting is possible, implemented with over-the-row machines. Maximum height of

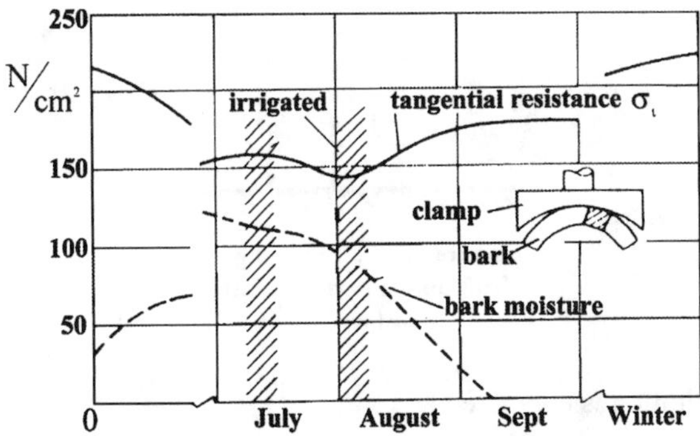

Figure 1.376. Bark-resistance (tangential-resistance) variation along time of the year and irrigation (plum tree) [6].

the trees must be maintained at 2.4 m for the clearance of the machine (<2.3 m). If the fruits are processing-oriented, adapted grape harvesters can be applied for the harvest of apples, peaches, apricots, plums, and so forth.

Pruning machines for hedging and topping the trees are used extensively in high-density and also conventional orchards.

1.6.16. Complementary Operations

Operations that have to be performed by harvesters after product detachment are very similar for all of them. A comprehensive view may include all systems designed to catch, convey, separate, clean, grade, and load the product.

These have been shown as parts of the different harvesters and include chain-and-band conveyors, aerodynamical separators, sieves and sizers, horizontal disk-roller cleaners to separate small fruits and foreign materials, manual grading platforms, electronic color and stone and soil-clod sorters, band conveyor loaders, and decelerating sleeves. Specially needed in fruit harvesters based on shakers are catching structures, frames, and umbrellas; the rest of complementary operations are similar in concept to those for vegetable harvesters.

A common characteristic of these elements is that they have to be designed to apply as low as possible energies to the products, so as to avoid damage. This objective is achieved by avoiding unnecessary accelerations, drops, and sharp turns; by adding decelerating elements such as sleeves and strip courtains; and by covering with energy-absorbing materials (damping materials). Sharp edges have to be avoided carefully. Solutions applied now to fruit-handling equipment should be studied for their possible use in the development of new harvesting equipment.

Most damage to detached fruits is caused during their fall inside the tree; to avoid this problem, fruiting branches should be located in different vertical planes whenever possible. Processing fruits should be harvested as early as possible and be processed in the shortest time after harvest. Padding should be included on all surfaces likely to be in contact with the fruits. Padding materials are designed to absorb the kinetic energy of the falling fruit. Figure 1.377 shows the effect of bouncing of a fruit on any surface. The

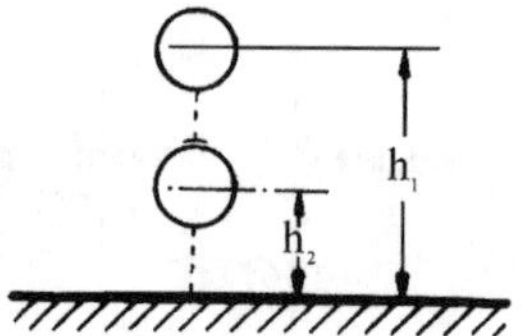

Figure 1.377. Bouncing of fruit onto a partially elastic surface [1].

coefficient of restitution is expressed as

$$e = v_2/v_1 = [h_2/h_1]^{0.5}$$

For $e = 1$, the impact is perfectly elastic, and for $e = 0$, the impact is perfectly plastic. Padding effect must be obtained by the deformation of the padding material, which must be soft and resilient (i.e., slow in deformation).

For the cleaning and separation of leaves, twigs, and other foreign materials from collected fruits and vegetables, blowers and vacuum blowers are mounted on the harvesters. For round fruits, flotation airspeed can be expressed by

$$v_{fl} = [4\delta_{fr} g d_{fr}/3\delta_a C_r]^{0.5} \qquad (1.123)$$

where

δ_a, δ_{fr} = density of air and of fruit (kg/m^3)
d_{fr} = diameter of fruit (m)
C_r = air resistance coefficient
g = gravity acceleration (m/s^2)

Flotation air speeds for selected fruits including peaches, apricots, and plums can be found in the literature; for example, for cherries, 16 to 19 m/s; grapes, 18 to 20 m/s; and leaves 3 to 7 m/s.

1.6.17. Mechanical Aids to Manual Harvest

Many products can be selectively (manually) detached, and the rest of operations subsequently can be performed by mechanical units. Such products include vegetables such as lettuce, all cole species (cabbage, cauliflower, broccoli, Brussels sprouts), zucchini, green and bell peppers, melons, small fruits, and tree fruits. Different types of carriers and conveyors have been designed and are used extensively [1].

Carts. Three-wheeled, manually driven carts are used for filling boxes in the field with celery, leeks, and many types of greens. The plants are cut manually by pickers.

Carriers. Tractor-pulled or self-propelled carriers are used. The handpicker sits and picks the fruits, filling the boxes, which are removed by another handworker. These are much used for strawberries, and also for radishes, leeks, and many other fresh-market vegetables.

Table 1.72. Working capacities of mechanical aids and harvesters for some vegetables

	Machine Type	Capacity
Mechanical Aids		
Lettuce	Platforms, simple	1.5–2 × hand-harvest capacity
	Platforms, self-propelled	25,000 heads/d
Cauliflower	Interrow platform	100–200 heads/man h
	Selective harvester	300–400 heads/man h
Artichoke	Platform, simple	300–350 heads/man h
Celery	Platform, simple	1.5–2 × hand-harvest capacity
Pepper	Platform, simple	120–210 kg/man h
Pickling cucumbers	Man-carrier 14 m-w	300 kg/h
Root crops	Platform tractor trailed	100 kg/h
Harvesters		
Cabbage	One-row	1000–1300 kg/h
Celery	One-row	3000–5000 plants/h
Green beans	Trailed, one-row	2000–2500 kg/h
	Front picker, 2–3 m wide	6000–10,000 kg/h
Green peas	Front picker, 3 m wide	1500 kg/h
Picking cucumbers	Self-propelled, 2 m wide	3500 kg/h
Green onions	Self-propelled, 1–2 m wide	2500–3500 kg/h
Root crops	Trailed, self-propelled per row	3500 kg/h
Tomato, industry	One-row, manual selection (8–10 workers)	
	• Trailed	3000–4000 kg/h
	• Self-propelled	5000–6000 kg/h
	• Self-propelled, optical selection	6000–8000 kg/h
Pepper, industry	One-row, trailed	500 kg/h

Sources: [1, 5, 6, 13].

Harvesting Platforms. These can be self-propelled or tractor-pulled. They carry 8 to 15 workers for the wrapping, size grading, and box filling and often are used for lettuce harvesting. Width is 5 to 14 m; speed, 0.1 to 1.5 km/h; and power, 20 to 45 kW. The same type of platforms often are used for harvesting products such as cabbages, cauliflower, broccoli, melons, and pineapples. Fruits can be either prepared, sized, and boxed in the platform, as described for lettuce, or just conveyed to bins or trailers for bulk transportation to the plant (simple platforms) (Table 1.72, Fig. 1.378). Work capacity is substantially increased with these aids (20%–100%). In general, harvesting aids result in only moderate increases in productivity, as workers' capacity is very interdependant [13]. Table 1.72 shows values of work capacity of harvest aids and harvesters for different vegetables.

Positioners for Harvesting Tree Fruits. These are one-man platforms, self-propelled, which can be used both for picking and for pruning fruit trees. Work capacity is not greatly improved [6, 13, 14] as compared with the investment needed. They can be appropriate for small orchards devoted to quality fruit where no high-efficiency equipment is justified.

Multilevel Picking Platforms. Picking of fresh-market fruit definitely is solved by the introduction of the latest multilevel picking platforms [15]. Orchard ideally must be formed in hedgerows. Picking is the only operation that is performed manually; the rest

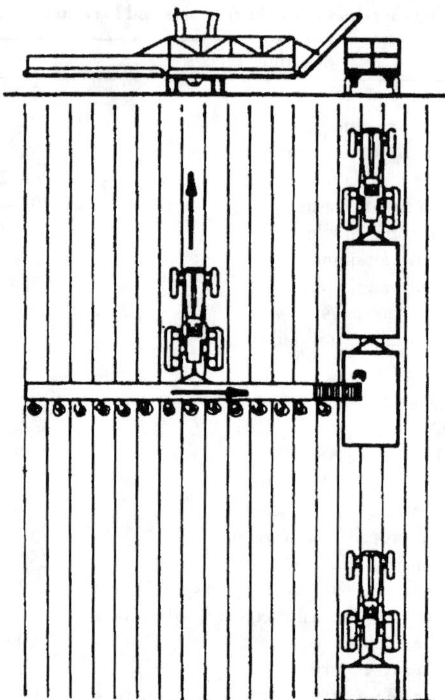

Figure 1.378. Tractor-trailed platform for bulk manual harvesting of produce [1].

are mechanized. Each fruit is placed individually on soft rubber–fingered conveyors, up to a soft central conveyor and to the bins. This system is able to preserve the quality of the fruits even better than totally manual ladder-and-bag procedures. Working capacities of these manual fruit-picking aids are summarized in Table 1.73. The fruit is placed in conveniently located conveyors, which load the fruit in bins, transported by the same unit. Travel speed is very low and harvesting capacity is only slightly higher than manual, up to 30% higher, but the operation is greatly improved in terms of work organization and handworker comfort.

1.6.18. Electronic Sensors and Robotic Harvesting

Mechanical (contact) and electronic (γ-ray) sensors have been developed for selective harvest, such as that of lettuce by ripeness. In this case head compactness is the criterion for quality.

Optical sensors have been applied for color recognition and soil-clod separation in tomato harvesters. The development of electronic sensing has been great, but the application is mainly in handling equipment.

Robotic harvesting has been oriented for soft, fresh fruit that cannot be harvested with conventional machines, because of excessive mechanical damage. Therefore, it represents an alternative to the current mechanical harvesting systems, superior from

Table 1.73. Working capacities of mechanical aids and harvesters for some fruits

	Working Capacity	Losses (%)
Mechanical aids		
Tree-fruits		
Single-man positioner	130 kg/man h or 0.95 ha/100 h	3
Multiple (hedgerow) platform	140–220 kg/man h or 2–10 ha/100 h[a]	3–6
Strawberries, man carriers	10–12 kg/man h	5
Small fruits, hand combs	35 kg/man h	8
Harvesters		
Grapes	2000–6500 kg/h	5–15
Nuts, ground sweepers	6000–7000 per meter width	2
Tree-fruit	40–60 trees/h	5–8
Trunk Shakers		
Limb shakers	8–10 trees/h	5–15
Strawberries	200–600 kg/man h	10–20
Small fruits	500–1000 kg/man h	10–30

Sources: [1, 6–9, 11, 12, 14, 15].
[a] 6–8 pickers.

the point of view of fruit quality and with a clear projection to the future consisting of automated fruit picking with a robotic system that emulates the human picker.

Although much research has been oriented towards this goal, no commercial robot is yet (as of 1998) capable of replacing manual labor for picking fresh fruits. A fruit-picking robot should possess the capability to locate fruits on the tree in three dimensions, and in any light conditions (including the dark); to approach and reach for the fruit; to detach the fruit and transfer it to a suitable container; and to move by itself in the orchard from tree to tree and row to row without human help [16]. Several up-to-date technologies are involved in achieving these goals, such as artificial vision, image processing, robot kinematics, sensors, control, and computerized signal analysis. Prototypes for the robotic harvesting of commodities different from tree fruits, such as melons and grapes, also are being developed.

Looking into the future, it seems possible that the robot, equipped with the appropriate sensors, would be able to select the most suitable fruits on the tree itself (in terms of quality), to be picked concurrently with the fruit-identification process. In the case in which the system would be able to obtain the best quality, and operating in controlled spaces (e.g., high-density orchards and greenhouses, rather than in the open), there could be commercial fruit-picking robots available in some years' time.

List of Symbols

c : damping factor (N·s/m)
C_r : air resistance coefficient
d : diameter (m)
e : coefficient of restitution

f : frequency (Hz)
F : force (N)
g : acceleration of gravity (m/s^2) ($\approx$9.8 m/s^2)
k : elastic coefficient (N/m)
m, M : mass, kg
NIR : nearinfrared
r : eccentricity (m)
VIS : visible
v : speed (m/s)
ω : angular speed (rad/s)
δ : density (kg/m^3)
σ : pressure, stress (kPa, N/cm^2)

References

1. Ortiz-Cañavate, J. and Hernanz, J. L. 1989. *Tecnica de la Mecanización Agraria* [Technique in Agricultural Mechanization]. Madrid: Ed. Mundi-Prensa.
2. Srivastava, A. K., Goering, C. E., and Rohrbach., R. P. 1993. *Engineering Principles of Agricultural Machines*. ASAE Textbook No. 6. St. Joseph, MI: ASAE.
3. Ryall, A. L. and Lipton, W. J. 1972. *Handling, Transportation and Storage of Fruits and Vegetables*. Westport, CT: AVI Publishing.
4. Ruiz-Altisent, M. 1991. Damage mechanisms in the handling of fruits. In *Progress in Agricultural Physics and Engineering*, J. Matthews (ed.), pp. 231–257. CAB International.
5. Gracia, C. and Palau, E. *Mecanizacion de los Cultivos Horticolas* [Mechanization of Horticultural Crops]. Madrid: Ed. Mundi Prensa.
6. Moser, E. 1984. Verfahrenstechnik intensivkulturen. In *Lehrbuch der Agrartechnik*, Band 4. Hamburg and Berlin: Verlag Paul Parey.
7. Salamon, Z. 1992. Influence of basic work parameters of the harvester on the precision of harvest and on damage of currant bushes. Proceedings of the 3rd International Symposium on Fruit, Nut and Vegetable Harvesting Mechanization. KVL. Denmark, Sweden, Norway.
8. Peterson, D. L., Takeda, F., and Kornecki, T. S. 1989. New shaking concepts for brambles. Trans. ASAE 32(4): 1165–1168.
9. Gil Sierra, J. 1990. *Maquinaria para el Cultivo y Recoleccion de la Vid* [Machinery for the Culture and Harvest of Grapevine]. Madrid: Ed. Mundi-Prensa.
10. Kepner, R. A., Bainer, R., and Barger, E. L. 1972. *Principles of Farm Machinery*. Westport, CT: AVI Publishing.
11. Barbe, P., Chaber, J., Sevila, F., Leppert, B., and Carbonneau, A. 1992. Characterization of various fruit detachment systems for grape mechanical harvesters. Proceedings of the 3rd International Symposium on Fruit, Nut and Vegetable Harvesting Mechanization: 2–12. Denmark.
12. Fridley, R. B. 1983. Vibration and vibratory mechanisms for the harvest of tree fruits. In Principles and practices for harvesting and handling fruits and nuts, O'Brien, M., Cargill, B. F., and Fridley, R. B. Westport (eds.), CT: AVI Publishing.

13. Ruiz-Altisent, M. 1989. Vegetable harvesting. In *Agriculture and Mechanization* [in Spanish]. Banco Bilbao Vizcaya: El Campo.
14. Ortiz-Cañavate, J., Gil Sierra, J., and Ruiz-Altisent, M. 1994. Mechanization of fruit harvest [in Spanish]. *Hortofruticultura*, HF 3: 51–56.
15. Tecnical-economical evaluation and optimization of the mechanized harvest of fruits [in Spanish]. 1997. Department of Agroforestry Engineering, University Lerida, Spain.
16. Sarig, Y. 1993. Robotics for fruit harvesting: A state-of-the-art review. *J. Agric. Eng. Research* 54: 256–280.
17. Ortiz-Cañavate, J., Hernanz, J. L., and Ruiz-Altisent, M. (eds.), 1995. *Las Máquinas Agricolas y su Aplicación* [Agricultural Machinery and its Application]. Madrid: Ed. Mundi-Prensa.

Tropical Crops

R. Pirot

1.6.19. Introduction

The use of harvesting machines results from practical considerations of the farming system involved: using power-driven harvesters depends on the level of mechanization of the other operations and on the factors required to carry out harvesting. It makes it necessary to develop convenient infrastructures such as roads and access works, and to use crop varieties suitable for mechanical harvesting, seed drilling and row planting with interrow spacings matching the technical features of the harvesters, and so forth. Most of the time, introducing harvesting machines has failed because such requirements were not taken into account.

1.6.20. Sugar Cane

Harvesting includes cutting, loading, and transport. These are the most important operations, with the highest costs and the biggest work input. Organization of the harvest, which must be matched to the throughput capacity of the factory, is fundamental. Furthermore, the time between cutting and crushing of the cane must not exceed 24 hours if the cane has been burned, or 48 hours if it is cut green; otherwise there will be a loss of sugar.

Nowadays, two main harvesting methods are used:

- As whole cane, manual or usually with power-driven equipment
- Cut into sections, usually with power-driven equipment

In many South American, Asian, and African countries, hand harvesting still is practiced extensively, while loading and transportation have been mechanized. This has the advantage of supplying relatively clean cane to the factory. Elsewhere, the cost and lack of availability of hand labor have led to an increase in mechanized harvesting, while the daily quantities passing through the factories have increased considerably.

With hand cutting, the cane is cut at the base and the top trimmed off, after which it is laid in swaths (windrows) or heaps; it is these successive operations that mechanization attempts to imitate. Of a large number of prototypes adapted to particular cropping systems, very few have emerged as operational, commercialized machines.

The attempt to adapt machines to local conditions was long ago abandoned, and instead the field and the crop are managed for mechanized harvesting. This involves removing stones; leveling ridges or filling in furrows, depending on the situation; and choosing the straightest, lodging-resistant cane varieties.

The cost of cutting machines and their requirements for use are such that in most cases they are only used on large estates and are out of reach for small planters unless they group together.

Whole-stick Harvesting

This is always made easier by burning the crop in the field or by prior manual or mechanized trashing. Burning also has the effect of clearing the field of undesirable animals (wasps, ants, snakes, etc.). It should not be done, however, without due care, for although it facilitates the work by removing the leaves and clearing the stems, it disrupts the harvest if the fire is badly controlled, increasing the risk of starting other fires. Also, it speeds up the inversion of sucrose. From that moment the cane must be quickly delivered to the factory to avoid serious degradation of the juice.

The interest in cutting the cane whole is to bring about a progressive evolution, quantitatively and technically, from hand cutting to mechanized harvesting without having to alter the procedures for loading, transport, and delivery at the factory. In this way, it will be possible to deal with the problems of lack of hand labor for cutting without major upheavals.

Although it is true that it is always important to organize the transport to match up with the cutting, whole-stick cutters have the advantage of allowing the operation to be divided up and of not being completely dependent on transport, as is the case with chopper-harvesters.

Semimechanized "Cutting"

Cutting consists of three main operations (done by the worker during hand cutting):

- Cutting the cane at the base
- Cutting the top off, and sometimes trashing
- Placing the cane in swaths or heaps

Semimechanized cutting always includes mechanization of the actual cut, and that of one or more of the other operations mentioned. Because of this, it necessitates a second operation, either by hand or another machine.

Some Examples of "Cutting" Machines. Studies were carried out in Asia (Japan, Thailand) on very small machines. One of them was based on the 8-hp Iseki maize harvester (Asian Institute of Technology, Bangkok, Thailand). A Japanese machine (Bunmei Noki NB 1 1) is powered by a 9-hp diesel engine and has a cutting output of 1 ha per 10 hours. The Howard Mantis Canequip "cutter" (Fig. 1.379) is a small disk cutter, mounted in front of a conventional 70 to 80-hp agricultural tractor. Its throughput is about 15 to 20 tons/h of cane cut and windrowed. The machine is controlled by a hydraulic system whose pump is driven from the tractor's power take-off. Other machines work on much the same principles.

Figure 1.379. Howard Mantis Canequip cutter.

The Mac Connel "cutter" has two parts: helical dividers are mounted in front of the tractor. At the rear is the cutting part (drum and cutting blades). The machine also has a topping device. The front components (divider and topper) are driven hydraulically, while the cutting drum is driven from the PTO.

Each of these machines has its place in particular situations, such as on small fields, steep slopes, or where access is difficult, for straight-stemmed, low-yielding cane. They reduce the hardship of hand labor and provide a partial solution for mechanizing small estates.

Mechanized "Cutting"

Mechanized cutting eliminates all the hand work between the actual cut and loading. It reproduces all the operations carried out by human cutters: cutting at the base, topping and possibly trashing, and dropping in windrows or heaps. At this level of mechanization, there is a large number of machines that are developed to various degrees. Those mentioned have reached the stage of operational use and are, or could be, commercialized.

Self-propelled Full Stick Harvesters

Cutter windrowers. These are cutters such as Cameco (Broussard, Thomson, etc.) of at least 150 hp (Fig. 1.380). Originating from Louisiana, they are available in one- or two-row form. The cane is straightened up, topped, cut at the base, carried up by a conveyor, and then windrowed at either three or six rows per windrow, by means of a directional conveyor.

This high-performance machine can reach average hourly throughputs of 60 tons (range 50–70 tons). To achieve this, operational conditions must be perfect. The machines were designed for Louisiana conditions in which the cane is planted on ridges 1.7 m apart. Hence, it speeds cutting if the plantation is ridged.

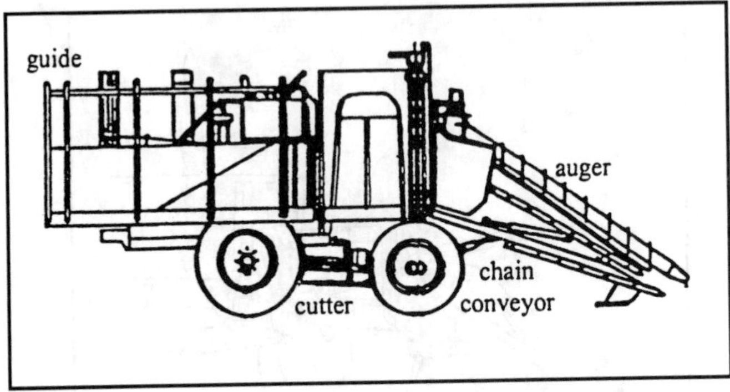

Figure 1.380. Cameco cutter–windrower.

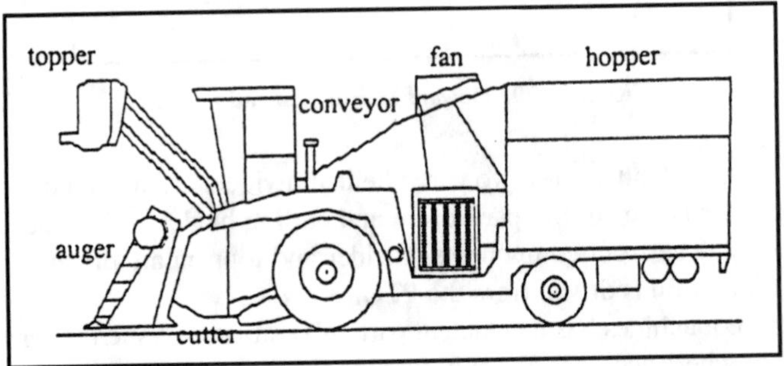

Figure 1.381. Legras cutter–stacker.

Its use is limited to crops of up to about 100 to 120 t/ha, with the cane fairly erect. There is no leaf-stripping system on the machine: the leaves are burned in the windrow. Windrowing means that it is possible to load the crop in the same way as when it is cut by hand.

Cutter-stackers. These are cutters like the Austoft "Tiger" (Artoli "EG 500 MD 103 Super," Carib "Centurion", Legras "RCL 4000", etc.) of at least 200 hp (Fig. 1.381). They work on a single row.

The cane is straightened up, topped and cut at the base. It is then conveyed, horizontally, to a hopper, where bundles of 500 to 1500 kg are formed. At the outlet of the conveyor, an air draft expels some of the leaf straw and non-cane material. Once the bundle is ready, the operator controls its drop from the cab. The distribution of bundles in the field has to take account of the method of loading in order to facilitate, but also to minimize, the movements of the loader. These machines can achieve hourly throughputs of 50 tons.

These machines are more versatile than the previous ones. They can be used where yields are high, 150 tons/ha and above, and in less erect crops. It should be noted, however, that the quality of the topping depends partly on the shape and height of the plants.

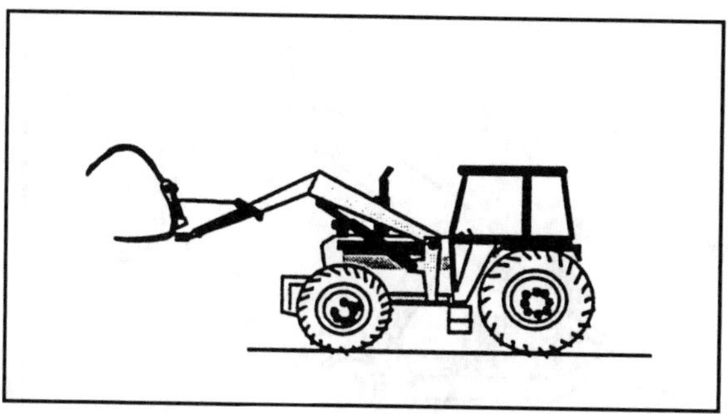

Figure 1.382. Front-mounted tractor loader.

Loading Whole Cane. This is an operation that was mechanized early on and in a particular way because of the onerous nature of hand loading and the problem of finding the necessary labor. There are two basic systems of loading: discontinuous and continuous. Discontinuous loading is preferably done starting with heaps but also by picking up the windrow. Continuous loading applies only to windrows. This is when a large amount of unwanted material (stones, soil, etc.) becomes mixed with the load.

Discontinuous Loaders

Front-mounted Tractor Loader. This is a loader mounted on an agricultural tractor by means of a frame that is specifically adapted for each hitch of tractor (Fig. 1.382).

The grab, or drag, whose working width is 1.4 m, is made up of a fixed part with three teeth, mounted on a tool frame, and a moving part, also with three teeth, controlled by two rams. Under good conditions, the hourly output can reach 15 tons.

This arrangement has certain disadvantages such as the damage caused to the cane stumps by the operations. It is nevertheless an economical solution that is worthwhile for small farms already possessing a tractor, which both loads and transports the cane.

Self-propelled Front-end Loader. Besides the adaptations to tractor-mounted loaders, there is a self-propelled machine specially designed for sugar cane (Bell) whose performance is excellent due to its exceptional maneuverability (Fig. 1.383).

The Bell loader is a three-wheeled fixed-arm self-propelled loader. The front wheels are driven; the rear wheel is free. It is fitted with an air-cooled diesel engine, driving hydraulic pumps. The oil sump is cleverly fitted into the tubular chassis.

The transmission is hydrostatic. The machine is steered by using the feet to operate the drive to the wheel motors. This leaves the driver's hands free to operate the arms and the grab. Its design makes it a very maneuverable machine, at ease on both flat land and slopes. Throughput can be 30 to 40 tons/h. It works equally well with cane in heaps or in windrows.

Like the previous machine, it suffers from having to move around while loading and thus risks damaging the cane stumps.

Swivel Loaders. This is a mechanical shovel fitted with a grab. The Atlas loader, for example, is an 83-hp self-propelled loader. It consists of a chassis mounted on

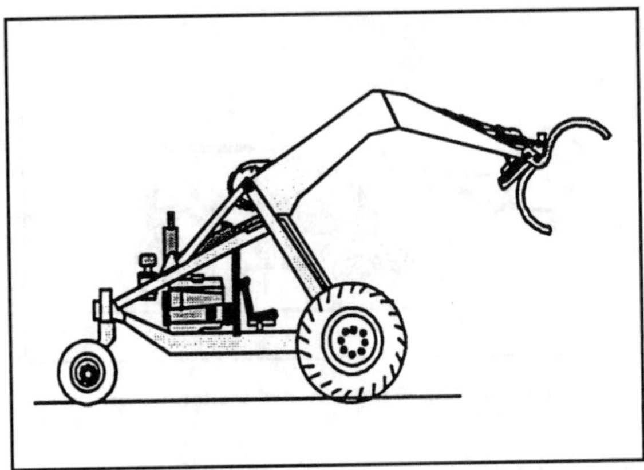

Figure 1.383. Bell self-propelled front-end loader.

rubber-tired wheels, with a cab fitted to it capable of turning through 3600 degrees. This is fitted with an arm with a maximum reach of 8.3 m with a lifting force at the end of the arm of more than 2 tons.

This type of loader is mainly used in humid zones. It is used for loading cane stacked beside roads, but more and more it is being used on the land itself. Capacity reaches 40 to 45 tons/h while loading in the field and 60 to 65 tons/h when loading from the headland.

Power-loading Trailers. These are both a loader and a means of transport. The bundles can be made up by hand or with a loader. The attraction of this arrangement is that the work can be split, allowing one of the machines to continue working while the other stops, thus offering extra flexibility.

The bundles are lifted over the side (side-loader) or, most often, the back of the trailer using a winch. The winches, mounted on the trailer, are driven from the PTO of a tractor, so that it needs numerous linkages and gears, universal joints, and pulleys driving chains. More modern trailers have winches driven by hydraulic motors, which makes for a simpler and more reliable system.

Continuous Loaders. These only work on windrowed cane.

Pushloader. The first machines designed to pick up cane were ordinary front-end loaders on which were mounted forks working close to the ground and pushing the windrows so as to form a heap in front of the tractor. This arrangement, which still involves a lot of maneuvering, has been improved by incorporating a pivoting arm. The claw gathers the heap formed by the fork (piler) while the tractor is moving forward, picks it up, and swings round to load it into the truck. The tractor thus remains in line with the windrow.

The capacity of these machines is high: 60 to 80 tons/h.

Continuous Loaders. These have an arrangement of very wide chain elevators that pick up the whole windrow. This elevator carries the stream of canes to a chopping device with circular saws that cuts the canes into 40 to 50-cm lengths. The pieces fall back onto a conveyor, which delivers them to a following trailer.

The models on the market are of very high capacity (200 tons/h) but they are handicapped by a significant percentage of crop loss, which often necessitates gleaning.

Transport and Delivery of Whole Cane. Transport is the third operation of prime importance for harvest, as it influences the regular supply to the factory. The method varies according to the distances to be covered. Thus, transport tends more and more towards a mixed system: by agricultural tractor and trailer for short distances (up to 10 km), by high-capacity (70 m^3) road trailers or articulated lorries (trailer trucks) for longer distances, and delivery to the factory via intermediate reception platforms.

This still leaves the private railway networks. The cane is taken from the collection points to be carried by rail to the factory.

Whatever the method of transport, it is necessary to provide a means of weighing and sampling the cane at the reception area.

Unloading is done in various ways according to the method of transport:

- As chained bundles, unloaded by a suitable crane on a traveling gantry
- Chain nets
- Tipping from the side or the rear
- Tipping after elevation for emptying into a trailer
- Slat conveyor

The cane then is picked up again by grabs or by a loader, the stacker, to be thrown onto the feed table, which leads to the crusher at the factory entrance.

Harvest of Chopped Cane

Chopper-harvesters

These are machines that carry out all the operations including loading (Fig. 1.384). The cane is cut at the base, then chopped into lengths of 20 to 40 cm, and finally loaded directly into a trailer that accompanies the machine. These cutter-choppers are very powerful machines that have either pneumatic tires or tracks, depending on the type of land. They are differentiated by the point at which the cane is chopped:

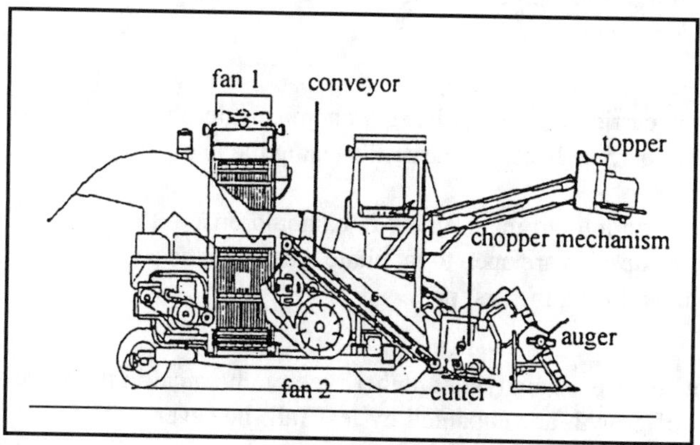

Figure 1.384. Claas cutter–chopper loader.

immediately after cutting ("bottom chopping") or after conveying through the machine ("top chopping").

When the cane is chopped straight after cutting it has the advantage of reducing the power requirement of the machine, but the chopping blades, in this position, are more exposed to damage by stones. Conversely, conveying the unchopped cane uses more power, but the stones are eliminated as the cane advances.

Transport

When trailers have to move around the field being loaded while traveling alongside the harvester, their capacity depends entirely on the terrain. Another solution is to have two methods of transport, one for the field and the other for the road.

The chopped cane is transported in trailers with crates with either plain or steel-mesh walls. The walls may have openings cut out to allow for sampling when the cane arrives at the factory. The capacity of these trailers is usually 6 to 14 tons of chopped cane. The tonnage increases in areas of flat terrain, giving rise to "cane trains," which may carry over 100 tons (e.g., in Australia).

Unloading at the Factory

For chopped cane, the trailers with crates empty their load by tipping it straight into the reception hoppers. These in turn feed an elevator, which empties the chopped cane onto the main table. At this level the chopped cane rejoins the route for whole cane at the entrance to the crusher.

1.6.21. Cotton

Cotton harvesting can begin when the first capsules open, but they do not all mature at the same time; under favorable conditions the plant continues to develop and the leaves remain green.

In areas with a dry season or a cold winter, vegetative growth slows down or even stops; the leaves wither and fall to the ground, facilitating the harvest.

Defoliation can be brought about artificially using chemicals, a technique that became widespread with the development of mechanized harvesting, but one that also makes it easier to harvest by hand.

Defoliation

Defoliation is carried out by applying a chemical (defoliant) either by a sprayer or from aircraft, 10 to 15 days before harvest, causing the leaves to fall without killing the plant.

The capsules continue to mature, and one should ensure that at the treatment date, 60%–70% of the capsules are open to facilitate the use of the cotton picker and to obtain the maximum pick during the first pass.

Desiccation

Desiccation consists of applying another type of chemical in the same way, causing total cessation of growth accompanied by leaf fall: however, at the treatment date the maximum number of capsules should have matured (70% to 80% of open capsules).

This technique is used when the crop is to be harvested with a cotton stripper, which is done in a single pass.

Actual Harvest

The operation, which is easy to do by hand, is very labor-intensive, needing 25 to 40 man-days' work per hectare.[1]

The increasing shortage of hand labor in certain countries has lead very quickly to almost complete mechanization (e.g., United States, Australia). It also has developed in many other countries alongside hand picking (CIS, ex-USSR, European countries, etc.). The most common machines are usually large, relatively expensive, and quite sophisticated, with a high output (more than 1 ha/h for a four-row harvester), but they require special growing conditions, with meticulous attention to cultural operations:

- Removal of all stumps, stones, crop residues, and so forth from the land
- Uniform sowing of suitable varieties
- Uniform weed-free crop
- Even ridging, not too high, and with no dips along the top of the ridge
- Well-cleared land with plenty of turning space on the headland
- Special high-capacity trailer to transport the seed cotton to the ginnery

However, the product is not as clean as with hand picking and requires extra cleaning systems during ginning. Two kinds of machine currently are used extensively:

- Rotating drum harvesters with rotating spindles that gather the seed cotton; these machines called "cotton pickers" produce a relatively clean sample
- Harvesters with inclined rollers or fingers which pull off the capsules, called "cotton strippers," these produce a sample with far more impurities

The Cotton Picker

This machine consists essentially of:

- A harvesting head made up of two rotating drums, situated on either side of the row being harvested, and fitted with rotating spindles that penetrate the vegetation and remove the cotton from the capsules; the cotton is then taken off the spindles by a special mechanism
- A system to convey the seed cotton to the hopper
- A large wire-mesh hopper with an emptying arrangement

The rotating spindles (cylindrical or conical) are the novel feature of this system of harvesting.[2]

A moistening device keeps the spindles clean so that the fibers wind around them easily; it uses either water (pure or with a detergent added) or a light oil, which has the advantage of not wetting the cotton so much.

The purpose of the extractor is to remove the fibers that are wound round the spindles. After extraction, the seed cotton is sucked away by a strong current of air that helps to eliminate foreign material. The product is then air-conveyed to the hopper.

[1] The daily yield of harvesters increases with the potential yield of the fields: 25–50 kg for yields of 1 ton/ha, 50–75 kg for yields of 2 ton/ha, 100–150 kg for higher yields.

[2] The first patent for a cotton picker was taken out in the United States in 1850.

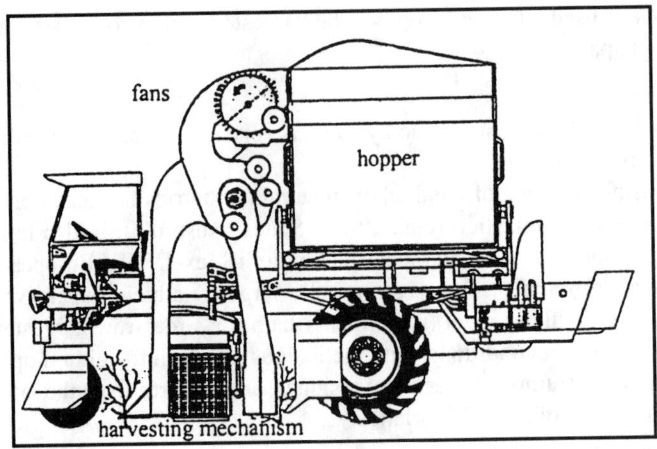

Figure 1.385. Cotton stripper.

Most of the harvest is picked at the first pass. The product obtained from the second pass, 1 to 2 months later, only represents 5% to 15% of the total harvest.

The most popular are self-propelled, two- or four-row machines, fitted with a powerful engine (100–150 kW), with a hydrostatic transmission system (working speed 0–6 km/h) and a large-capacity hopper (20–30 m^3). Three-wheeled models (with one steering wheel at the rear) are designed for harvesting small plots.

However, the harvesting cost remains high when the purchase and maintenance costs of these machines are taken into account, and losses in the field can be substantial (10%–25% according to the harvesting conditions).

The Cotton Stripper (Fig. 1.385)

This machine harvests the crop in a single pass, pulling off both mature and immature capsules and yielding a product with a much bigger proportion of impurities than is the case with the picker. The harvesting heads are of two types:

- Those with flexible metal fingers 60 to 90 cm long, inclined from the horizontal, spaced about 2 cm apart, forming a comb which pulls off all the capsules; the product is then conveyed towards the center where the heaviest material (green capsules, large impurities) is separated out
- Others with rollers, with a lifter-conveyor mechanism channeling the vegetation, fitted with a system to remove the capsules, made of two inclined ribbed rollers made of steel or rubber, or brushes made of synthetic material (nylon); the product is conveyed towards the central sieve where separation of waste takes place, after which the cotton is blown to the hopper in a rising air current.

The green, heavier capsules that are separated by the air current are collected in a 300-L tank, which is periodically emptied onto the headland.

The seed cotton is stored in a high-capacity hopper (15–23 m^3) with an emptying arrangement and sometimes a compacting device, which increases its capacity by 20%.

Roller machines are either tractor-mounted (one or two rows) or self-propelled (two, four, five or six rows). They are used for harvesting small-sized varieties (less than 1 m in height) that mature over a short period.

The finger machines, which are less common, are used for harvesting crops sown at high density.

The throughput is generally higher than with a picker because the forward speed is faster (5–8 km/h), and the product can be held in the field, lightly compressed, if the weather conditions are favorable. But the main attraction of harvesting by stripper is its relatively low cost compared with the cost of harvesting with the picker, due to lower purchase and maintenance costs. The amount of foreign material can be very high (25%–30%), calling for special cleaning equipment at the ginnery, and the ginning percentage is usually poor (about 25%).

Mechanized harvesting always results in bigger crop loss than with hand picking, sometimes with a large amount of cotton falling onto the ground as the machine passes through the crop. Gleaning machines have been introduced to pick up the cotton fallen on the ground. Various systems have been considered, of which some are directly adaptable for the picker (belts with slits or metal fingers, nylon brushes).

These machines have been developed mainly for use with the picker, and they succeed in recovering up to 50% of the cotton which falls during the harvest.

1.6.22. Groundnuts

Groundnut harvesting is subject to two important considerations:

- It is important to harvest at a very precise stage of maturity, to assure either good quality seed or else the maximum yield of oil and, for Spanish-type varieties, which have low dormancy, to avoid germination of the seeds in the soil if harvest is delayed.
- Premature soil hardening can cause much of the crop to remain in the ground and results in excessive wear on implements.

However, determination of maturity is difficult, as there is no characteristic indication allowing it to be reliably established. In Virginia-type varieties, it is the appearance of the leaves (which begin to turn yellow) and the brown color of the inside of the shell that are the best indicators.

The two main harvesting techniques used for groundnuts are lifting and uprooting. With lifting, the taproot is cut just below the layer formed by the pods: the earth around the plant is fairly well loosened, and the plant is separated from the soil. With uprooting, the whole plant is lifted from the soil and put aside, sometimes in windrows.

After a period of drying in the field, the plants are gathered into heaps or windrows followed directly by threshing (green-pod removal) or for prolonged drying followed by threshing elsewhere with a stationary thresher.

Lifting

Lifters

Mechanized harvesting developed while the area cultivated was increasing, but with varying degrees of success according to the country or the region, the farming system, and the intended use and purchase price of the product.

The lifting blades are designed to fit on a frame mounted on the tractor's three-point linkage. They are rather similar in shape and design to those used for animal draft, but more robust, and the width is adapted for lifting two rows.

The most popular model is an oblique blade with a cutting angle of about 3 degrees, bolted at the front end onto a plate attached to the standard. Straight blades with one or two standards are now rarely used.

A 25- to 30-hp tractor is sufficient to lift two rows 60 cm apart, or double rows, with an average harvesting output of 3 h/ha.

With a more powerful tractor (45–50 hp) it is possible to harvest four rows at a time by fitting two blades, with a working width of 1.5 to 2.0 m.

Metal finger-shaped bars fixed along the back of the blade act as shakers and throw off some of the soil. Blockages are reduced by mounting the standards to the side.

Shaking Lifters

A machine adapted from a vegetable harvester was used experimentally for harvesting groundnuts during the 1980s: it shook the plants and thus helped them to dry out on the soil. It was a Simon carrot lifter adapted to work behind low-power tractors (20 hp).

It consists of a straight blade, supported at each end by two standards, and extended backwards into a toothed shaking platform driven by the PTO (Fig. 1.386). Several adjustments can be made—depth, angle of penetration, working width (blades of 0.9, 1.2, or 1.4 m), vibration amplitude of the shaker, and so forth—in order to cope with different working and soil conditions.

With the 0.90 m blade a low-power tractor (at least 20 hp) is able to pull a two-row lifter at a rate of 4 h/ha. This simple machine is rarely used.

Uprooting

Direct uprooting, which consists of extracting the plant from the soil by pulling at the foliage, is the oldest method. Machines have been designed to do this, with a device in front to loosen the soil and thus reduce pod losses.

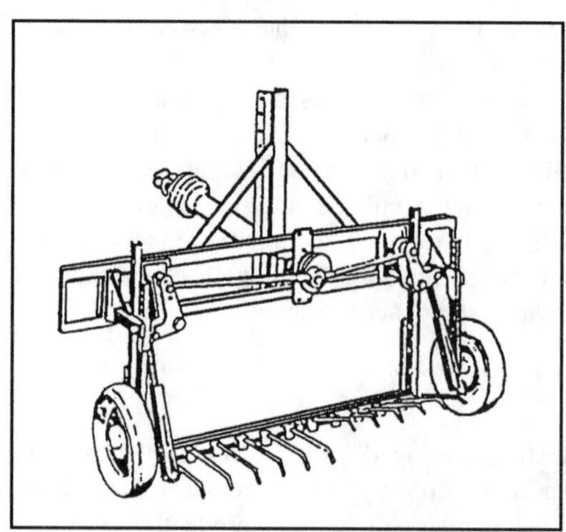

Figure 1.386. Simon shaking lifter.

Gripping Uprooters

On the first uprooters, the lifting blades were placed under the tractor and two rotating rear-mounted agricultural chains acted as uprooters. On current machines the chains are replaced by wheels with pneumatic tires.

The lifting blades are fitted with a safety device to cover them if they encounter an obstacle. Uprooting is done by gripping the tops between the rubber tires, which are turning (driven from the PTO) in a semihorizontal plane.

These machines, mainly used in Australia where they were first introduced, are designed to harvest two rows, and have seven rubber-tired wheels. These tires, inflated to a low pressure, remain supple and in contact. The groundnut plants, guided by metal deflectors, are gathered into a single windrow at the rear. These machines are used mainly on very sandy soils which flow easily at harvest.

Uprooter-Shaker-Windrowers

These machines were designed for large-scale, fully mechanized farming systems and are widely used in industrialized countries (United States, South Africa, Australia, etc.) (Fig. 1.387).

Several makes and models are available on the market, fitting onto the three-point linkage behind the tractor for harvesting two rows (1.2–1.7 m) or four rows (2.5–3.0 m).

The main difference in design lies in the shaking system, using rollers with several rows of notched discs or an inclined moving apron made of metal bars covered in spikes. The machines are fitted with front-mounted lifting equipment usually with two oblique blades with attached metal fingers, with an elevator that also serves to shake a lot of soil off the plants, and sometimes with a windrowing device composed of metal deflectors that form a single windrow from several crop rows and thus facilitate collection for threshing. These are, at present, the most popular machines.

The power needed depends on the working width and the soil hardness at the time of harvest. For example, for a tractor of average power (50 hp) on light soil, four rows (total width 1.5–2.0 m) can be harvested at a rate of 1.5–2.0 h/ha.

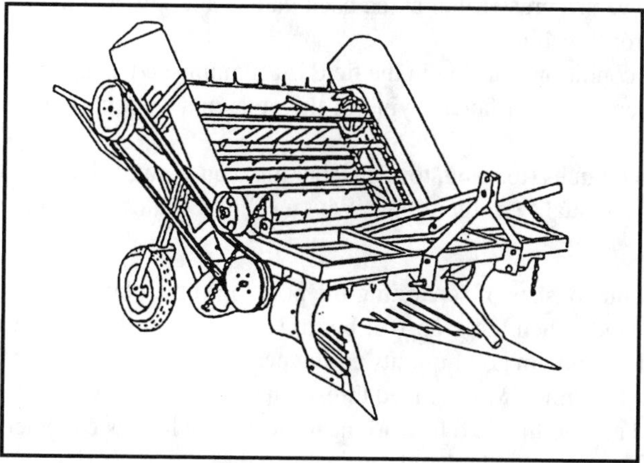

Figure 1.387. Lilliston groundnuts uprooter–shaker–windrower.

However, the quality of earlier operations (soil loosening, evenness of sowing) greatly affects the ease of harvesting. It is imperative that the land be completely cleared of roots in order to avoid risk of seriously damaging the equipment.

Certain models offer different additional features such as:

- A circular coulter at the front to cut creeping stems just before uprooting
- A heavy rear-mounted roller to flatten out the soil on which the windrow is laid
- Different devices to adjust the angle and orientation of the shaking apron
- Sometimes, the combination of two methods of shaking (apron and rollers) so as to provide more vigorous agitation

Drying on the Soil

When they are uprooted, groundnut plants contain 60% to 80% water, and the pods still contain 35%. In relatively dry climatic conditions, drying on the soil brings the water content of the pods down in 2 or 3 days to 20% to 25% which is favorable for mechanized threshing. Under wetter conditions, drying is slower and there is a risk of molds and deterioration of the shells; various techniques are recommended.

Topping

Prior topping of the plants by cutting off the aerial parts 1 or 2 days before harvest by lifting gets rid of the leaves, facilitating the drying process later; but for mechanized uprooting, it is necessary to retain plenty of stems.

This method, which is recommended for wet regions, has been applied in certain parts of the United States using brushwood clearers with horizontal knives. However, it is not recommended for certain eating varieties of groundnuts.

Tossing the Windrow

To accelerate drying, the Americans practice the method known as "turned windrow," which involves bringing the pods to the top. This way they do not take up moisture from the soil, and the collection losses from the collector–thresher are reduced.

Tossing the windrows is done with machines, the most common of which are adapted from the apron harvesters by taking off the lifting device at the front. On the apron shaker–windrowers, a press-roller at the back flattens the soil, making it easier to lift the new windrow for threshing.

In very wet conditions, drying in the field is accomplished using drying rakes, sometimes supplemented by artificial drying of the pods in a stationary drying facility after threshing.

All these techniques (topping, tossing, shaking) cause much defoliation of the stems, which reduces the quality of the forage, which is highly valued in some regions.

Threshing

This operation consists of separating the pods from the plants. It may be done after drying on the land, when the crop will be threshed green, which is sometimes recommended in order to obtain good-quality eating peanuts; or after prolonged drying, which is a case of dry threshing. Mechanized threshing is done either with stationary threshers that are moved around the field to work near the crop, which is collected into stacks; or continuously, by gathering the windrows using a thresher–picker.

Threshers

Groundnut threshers, as used currently as stationary machines, are of relatively simple design and low capacity (less than 1 ton/h). They are mobile and are used by individual farmers and cooperatives of small farmers.

These machines consist of:

- A threshing device (cylinder and concave) that separates the pods and breaks up the stems into short lengths
- A winnowing system, sometimes with sieves
- A device to gather the pods and deliver them to the sacking device

The lightest models, of Asiatic origin, are of low capacity (100–150 kg/hour). They are driven by small engine (1–2 hp) and are readily transportable.

Heavier models have a larger throughput (500–800 kg/h). They are semimobile, that is, mounted on an axle with rubber-tired wheels, and driven by an auxiliary engine of 8 to 10 hp. They can be pulled by a pair of oxen or a tractor. In the latter case they may be driven from the PTO.

Thresher-Pickers

The windrow can be processed continuously using a collector–thresher mounted on the back of the tractor and driven from the PTO (Fig. 1.388).

At work, the tractor straddles the windrow, which has already been partially dried. The crop is gathered up by a pick-up whose height above the ground is adjustable. The feed cylinder, flued with fine flexible metal teeth, channels the product towards the threshing system, where progressive separation of the pods takes place. The gynophores are cut after separation is complete.

The threshing device is made up of several cylinders, on which are fitted metal bars that have pairs of fine flexible teeth fitted to them. For each cylinder there is a concave made of perforated metal, the first having a fine mesh, with one of larger mesh behind, through which the pods pass.

The arrangement for separating the straw from the pods after threshing consists of separating blades fitted with flexible teeth, and counterblades composed of a wide-mesh

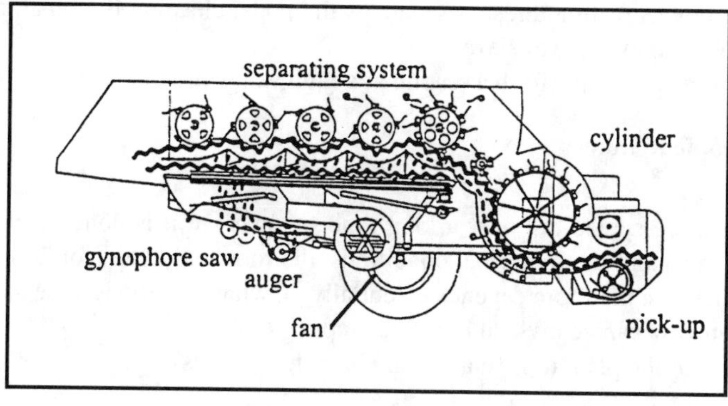

Figure 1.388. Lilliston picker–thresher.

grid. On other models, separation is achieved on articulated shakers, which also expel the straw at the rear.

The cleaning mechanism, situated under the separating blades (or under the shakers) is composed of a set of reciprocating sieves for grading the pods, a fan to expel light waste material, and a three-row table of gynophore saws. The system for storing the pods consists of a transverse auger at the end of the sieves, a pneumatic conveyor driven by a fan and a large-capacity hopper with an emptying device.

The straw is discharged onto the ground at the rear. The recommended working speed varies from 3 to 7 km/h. The throughput of these machines is high (more than 1 ton/h, with a tractor of at least 50 hp), but they require the work to be organized so as to synchronize the lifting and the threshing.

They can be used for both harvesting green (after drying on the land) and for dry threshing, given certain adjustments. In fact, the quality of the work depends largely on the precision of these adjustments, which depend on the state of the crop at the moment of threshing (e.g., moisture content of pods, amount of straw, etc.).

For the production of groundnuts for eating, it is essential to have a high proportion of intact pods that are not damaged by the thresher (1st quality exportable). However for the production of groundnut oil, threshing can be more vigorous, producing a higher yield.

The pods are then dried or simply stored depending on their water content.

Modified Combine Harvesters

Certain models of cereal combine harvester can be adapted for harvesting groundnuts by installing a pickup, threshing cylinder, and so forth. But these self-propelled high-throughput machines are a big investment and can only be justified in highly developed mixed farming, or as shared equipment that can be used on cereals and other oil crops.

Uprooter-Thresher

Machines have been introduced for harvesting the green crop directly,[3] uprooting the plants and threshing them immediately. The principle is as follows:

- A blade at the front lifts the plants (usually two paired rows)
- The plants are pulled out and conveyed by gripping them between two belts.
- Further downstream a thresher separates the pods; cleaning the separated pods is done by winnowing over sieves

However, this technique for harvesting at a high water content is as yet little used.

1.6.23. Tropical Tubers

Cassava

Harvest can be spread out over a long time: on average it is done between 10 and 20 months after planting. In normal conditions, the roots only keep for 24 to 48 hours. The time of harvest therefore depends essentially on what the crop is to be used for. The harvesting process can be divided into four steps:

- Removal of the plant tops (main stem, branches, leaves)
- Digging up the root

[3] Work of the SRI: Silsoe Research Institute (UK) and North Carolina State College (US).

- Removing the soil from the roots
- Loading and transport

Removal of the Plant Tops

This is usually done by hand using a machete. The stems can be used to provide cuttings for a new planting.

Brush cutters can be used for the purpose. They leave a carpet of fine particles on the soil, which does not, however, hinder the subsequent mechanical lifting. This is an important point to consider, as the use of rotary brush cutters leaves large branches behind, which hamper the subsequent lifting operation. Furthermore, these crop pieces behave as cuttings and take root, which causes problems for the following crop. Outputs are in the order of 3 to 4 h/ha.

The use of front-mounted brush cutters where the rotor is heavy cup–equipped and anticlockwise running has been tested with success and can be recommended for mechanizing this operation.

Lifting the Roots

The first attempts at mechanizing lifting were made using a heavy plough; the technique necessitated extra hand labor to shake the soil off the roots and lay them out in rows. Blade subsoilers have been used.

These techniques, although rudimentary, dispensed with the use of hand labor for a job that is slow and arduous, particularly on heavy soil.

Separating the Soil

Separation of the soil is done after lifting. It consists of removing the soil from the roots and lining them up while combining several rows to facilitate later collection. At present, manufacturers offer different implements for lifting and shaking.

There are three types of machine:

- Machines with reciprocating shaking bars (Fig. 1.389) consist of a blade extended by rods, which are made to move by the PTO. The roots are lifted and then shaken free of soil before being deposited on the ground behind the machine.

Figure 1.389. GMD cassava lifter-shaker.

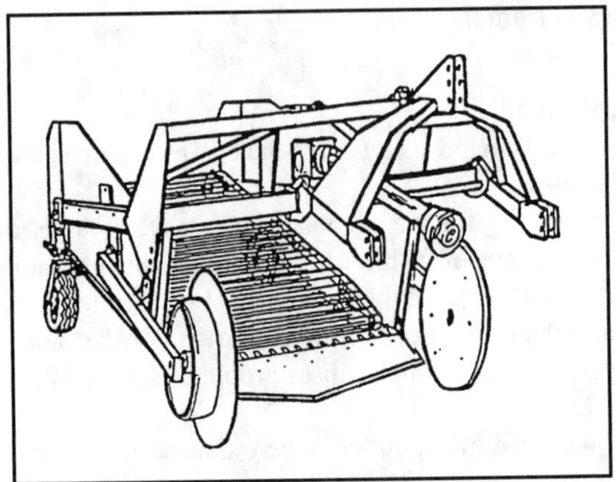

Figure 1.390. API cassava lifter.

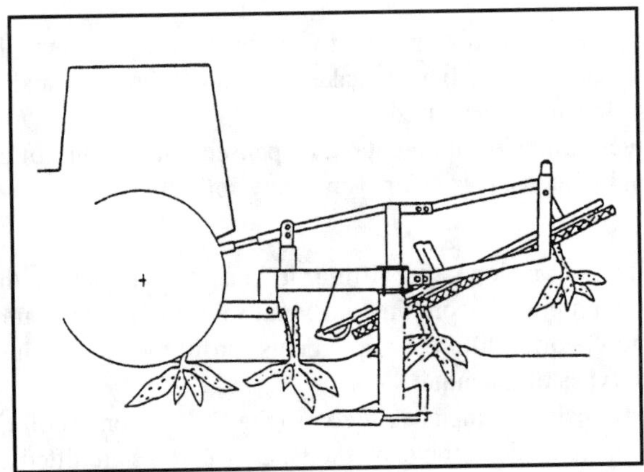

Figure 1.391. ITL cassava belt-lifter.

- The "shaking table lifter" (Fig. 1.390) is derived from potato harvesters. The principle is the same, but the machine itself more solid; after being lifted, the roots are dropped onto a riddle through which the soil falls. They then are deposited on the soil behind the machine ready for loading. Although under normal harvesting conditions these machines leave very little of the crop behind in the soil, they have the disadvantage of damaging the roots, which cannot therefore be used fresh and must be quickly processed. To minimize this defect the University of Leipzig (Germany) has designed a machine that lifts the roots without damaging them.
- Machines with belts (Fig. 1.391) make use of the principle found in vegetable harvesters: the stems are cut 20 to 25 cm above the soil, and the harvester gently lifts the cassava plant while shattering the soil. At the same moment the stem stump is gripped between two angled rubber belts. The roots are pulled from the soil and dropped behind the machine. The quality of the roots is as good as with hand

harvesting. The problems that remain at present are the impossibility of using the machines for shredding the plant tops, and the fact that the system requires the plant to have a single stem at the point where it is grasped by the belt in order to work effectively.

The output for a one-row machine is more or less identical for the three types, between 5 and 7 h/ha.

Loading and Transport

There are no specific machines for the loading operation. It is probable that for large ventures, sugar-beet equipment would be satisfactory.

A Brazilian manufacturer (Lorenz) is offering a "lifter–shaker–loader" based on a sugar-beet harvester and drawn behind a tractor in front of which is fixed a brush-cutter. In ideal working conditions it harvests and loads 1 ha in 4 hours. This machine is aimed exclusively at large agricultural ventures.

Yam

The decision to harvest yams mechanically requires not only that they be planted at a constant depth, but also that there be an appropriate choice of varieties: it is important not to damage deformed or badly positioned tubers or leave any in the ground. Varieties with rounded, grouped-together tubers are preferred to others.

In order to prevent the machine from clogging, it is necessary to remove the plants before the lifter passes through the soil. This is done by hand for creeping varieties and sometimes with a brush-cutter for semierect varieties.

The first trials were made with plows with deflectors attached to the moldboard. The tubers were brought to the surface; some handwork was needed to remove the soil.

As with cassava, more complex machines have been developed from potato harvesters. They are either simple lifters composed of a blade about 1 m wide supported by a standard on each side, with metal rods behind the blade to separate the soil from the tubers; or "shaking harvesters"—the one which has been used in the Côte d'Ivoire has two groups of rods, driven from the PTO, behind the lifting blade.

This machine has been improved by:

- Reinforcing the blade supports
- Making the blade in a single piece with a steep angle of penetration
- Reducing the working width (from 1.2 to 0.8 m).
- Increasing the working depth (to 50 cm)
- Reducing the number of rods

The quality of the work was very satisfactory when the soil structure was good. The quantity of undamaged tubers could reach 90%. This proportion fell rapidly on heavy or dry soils (to 40%).

The unharvested tubers reached a maximum of 5% and could be considered insignificant under ideal harvesting conditions.

Planting should not be deeper than 10 cm to avoid the risk of cutting a large number of tubers and should be in the middle of the ridge to avoid cutting them with the standards.

Although the tubers are then collected by hand, mechanized lifting reduces the labor requirements substantially (to 1 man-day compared with 40 for hand lifting).

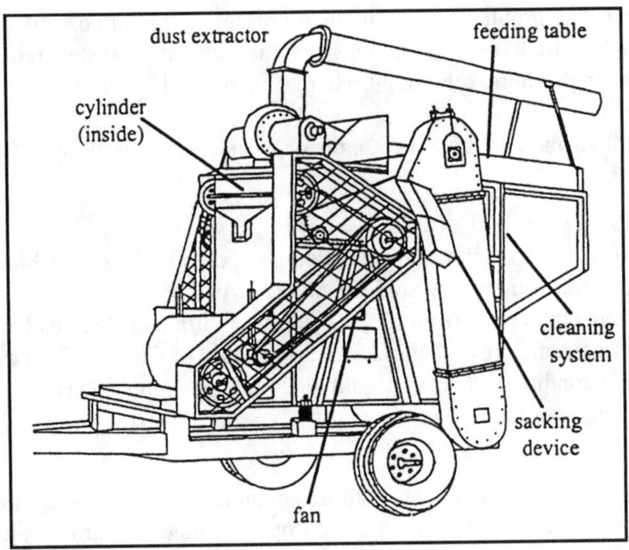

Figure 1.392. SISMAR millet thresher.

1.6.24. Millet

The introduction of purpose-made threshers has proved to be necessary, as the ordinary cereal thresher (with teeth or fingers) has insufficient grip on the very compact millet ears, which are composed of very densely packed spikelets with very small seeds. Two thresher models now being used in Africa are proving satisfactory. The BS 1000 (Fig. 1.392) was introduced and built in Senegal by SISMAR (The Sahelian Industrial Society for Mechanization, Agricultural Machinery and Representation). It has an output of 800 kg/h for millet and 1000 kg/h for sorghum, and requires a tractor of 35 to 40 hp to drive it and move it between the harvesting site and the stacks.

On this machine, threshing is achieved by forcing the ear to pass between two surfaces, which pull the grains from the rachis. The thresher consists of a feed table, a feeder, a cylinder, a cleaning drum, and a trough. The feed table is situated above the machine. The ears are fed in by hand. A feeder (an unperforated drum turning at 30 rpm) whose axle is parallel to that of the cylinder assures a steady supply of ears and forced feeding of the cylinder. Both the cylinder and the concave are covered in expanded metal (with the projections on the beater and concave being opposed to bring about a vigorous brushing of the ear) that has received surface treatment to increase its resistance.

The degree of enclosure of the cylinder within the concave (180 degrees) is greater than for ordinary threshing machines. The grains are collected in a rotating cleaning drum made of perforated sheet metal through which all the threshed material passes. It is subjected to a strong air current produced by a fan turning at 650 rpm; the grain, which is heaviest, falls through the holes, while the rachises, which are more voluminous, are blown out of the machine, and the dust and light particles are sucked out at the level of the cylinder and expelled through an air shaft. The grains themselves fall into a trough fitted with an auger leading to the sacking device.

Figure 1.393. Bamba millet thresher.

Six people are needed to work it. Losses due to damage are about 3%; the finished product contains 90% to 95% of good grain and 5% to 10% of waste.

The Bamba (Fig. 1.393) developed from the Bamby (a maize shelter), made in France by Bourgoin, is a model of smaller capacity than the BS 1000 (300 kg/h for millet). It can also be used for maize (1500 kg/h) and sorghum (600 kg/h) with specialized parts on the cylinder and appropriate cylinder speed and choice of cleaning sieves. Weighing about 300 kg and fitted with a 10-hp engine, it can be moved from one stack to another or from one working site to another by draft animals.

The threshing principle of the Bamba is derived from that for maize. The machine is composed of a feed table, a cylinder, and a system for cleaning and collecting the threshed grain.

The feed table, fitted with a safety hood, is fixed opposite one side of the beater. The ears are fed in by hand under the safety hood. A knife, located at the bottom of the feed table at the entrance to the thresher, cuts the ears into small pieces.

These are threshed by a special cylinder (three beater bars arranged spirally at the entrance). The concave is divided into two parts: an upper part made of woven metal and a lower part that is a sieve in the shape of a drawer. Beating takes place axially (the grains being thrown against the concave and also rubbing each other), and the waste (rachises) is thrown out opposite the feed side and passes over a vibrating sieve to recover the grains thrown out with the waste. The grains pass through the sieve of the concave and fall onto a cleaning sieve winnowed by a fan. The clean grain falls into a hopper and is elevated to the sacking device.

For millet, the Bamba needs three or four people to operate it. Losses in the waste are less than 3%, and the finished product contains between 95% and 99% of good grain and 1% to 5% of waste.

The machines are used by hire from companies, and the small quantities to be threshed on each farm means that some organization is needed among the farmers concerned.

1.6.25. Coffee

To obtain good-quality coffee, the berries must be ripe when picked: red to reddish-yellow. Maturity depends on the period over which flowering is triggered by the first rains. It often results in one main phase of ripening followed by several smaller ones.

Considering this spaced-out flowering and the influence of external factors, the berries will not all be ripe at the same time, and several pickings (two or three) will be needed.

This problem limits the scope for mechanization and results in poor quality of the product thus obtained. The advantage of being able to harvest all the berries in a single pass, either manually or mechanically, has led to research into a means of inducing maturity. Ethephon sprays have been tried, but they have not solved the problem because they induce maturity of the outer skin without having much effect on that of the bean. The harvested product is lighter and of inferior quality to that from a correct harvest at maturity.

Mechanical harvesting has given rise to several attempts to solve these problems. One of these machines, designed on the principle of causing the branches to shake by shaking either the trunk or the branches themselves did not prove satisfactory because of the damage done to the trunk and branches, and the risk of uprooting the tree. The Brazilians have introduced systems to assist manual harvesting, which increase the workrate of the pickers but often to the detriment of quality (Table 1.74).

- Use of a special bag hanging from the picker's belt and allowing quick emptying ("Jamaican" method).
- System of winnowing by suction ("Sugrao" method): the pickers let the berries fall onto canvas sheets spread on the soil. They are then sucked towards a winnower mounted on a tractor and moving between the crop rows. The winnower blows out the lightest material (leaves, etc.). The green coffee is then bagged at its outlet. In these two cases, picking is still done by hand; on the other hand, other methods make use of harvesters which have a vibrating picking device.

Table 1.74. Comparative results from different types of harvest

Type of Harvest	Efficiency after One Pass (a)	Impurities (%)	Hand-Labor Factor (%)	Hand-Labor Requirement (b)	Comparative Cost (%)
Manual	99	0.63	100	1.3	100
"Jamaica"	61	0.23	100	1.65	115
"Sugrao"	97	0.67	78	0.90	94
"Coco"	98	0.61	65	0.66	63
"Jacto"	96	0.60	26	0.40	81

(a) (%) of ripe berries picked
(b) 1 person = 600 h/harvest

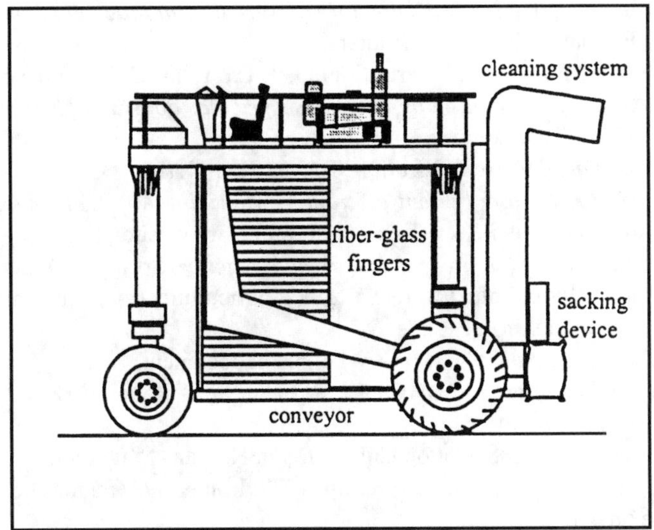

Figure 1.394. Jacto self-propelled coffee picker.

- The "Coco" method: canvas sheets are spread under the trees, and the material detached by the machine, which straddles the row, falls into them. The picking system consists of rotating drums, fitted with fiberglass fingers turning at a speed of about 1000 rpm. The harvested product is treated as before with a mobile winnower.
- Finally, a complete machine (Jacto) that straddles the row, with exactly the same picking system as the Coco and a series of mobile blowers underneath which gather the product, which is transported in two chains of lateral buckets and air cleaned (Fig. 1.394).

Certain imperfections in the latest machines (picking height and an inability to work on slopes) have been put right by the Australians, who have invested heavily to deal with this problem. The desire to avoid hand labor has been pushed to the extreme in some cases; the sides of the machine are huge water tanks that enable the beans to be pulped in water immediately after picking; the running water and freshly picked berries are delivered to four pulpers at the top of the machine.

It should be borne in mind that, considering that picking represents a very high proportion of the total cost of coffee growing, the use of these machines is certain to increase on large plantations.

Research programs aimed at foreshortening maturity are in progress all over the world. They make use of agricultural technology (appropriate pruning), growth regulators, and genetics (in the search for suitable varieties).

References

CEEMAT. 1974. *Manuel de Motorisation des Cultures Tropicales*, Vol. 1 and 2. Coll. Techniques rurales en Afrique. Paris, France: Ministère de la Coopération.

Sing, G. and Bhardwaj, K. C. 1985. *Directory of Agricultural Machinery and Manufacturers.* Bhopal, India: Jagran Printers.

Dias Gadanha, C., Molin, J. P., Duarte Coelho, J. L., Yahn, C. H., and Wada Tomm, S. M. A. 1991. *Maquinas e Implementos Agricolas do Brasil.* Campinas, Brasil, Emopi-Graphica e Editoria.

Kishida, Y. 1993. *Farming Mechanization.* Tokyo: Shin-Norin.

Rizzoli, B. 1986. *Costruttori Italiani e Stranieri Macchine Attrezzature Agricole: Di Possibile Impiego nei Paesi Tropicali.* Firenze, Italy: Litografla.

Hunsigi, G. *Production of Sugarcane: Theory and Practice.* Springer-Verlag.

La canne à sucre 1989. Coll. Le Technicien d'agriculture tropicale. Paris, France: Maisonneuve et Larose.

Sugar Cane Mechanisation. 1986. Massey-Ferguson document.

Mécanisation de la culture de la Canne à sucre à la Réunion. Hand report by Mr. Laude.

Bahl, Jain, and Sharma. 1988. Cotton Cultivation in Haryana State India.

Gall, Delatre, Braud, Cheze, Bals, and Kemp. 1977. Mat N'598. La mécanisation de la culture du coton.

Dissard. 1975. Le battage motorisé de 1'arachide de bouche. 01éagineux–30(2).

Havard. 1987. Les acquis en matière de semis et de récolte mécanique de 1'arachide.

Gillier, and Sylvestre. 1969. L'arachide. Collections Techniques Agricoles et Productions Tropicales. Editions Maisonneuve et Larose. Acct.

Selvestre, P. and Arraudeau, M. 1983. *Le manioc*. Paris: Maisonneuve et Larose.

Degras, L. 1986. *L'igname*. Paris: Maisonneuve et Larose.

Pouzet, D. 1988. *AméIioration de la culture mécanisée du manioc en Côte d'Ivoire.* Montpellier, France. Mémoire et travaux de I'IRAT.

Cheze. 1983. *Mécanisation intégrale de la récolte du manioc.* Antony, France: CEEMAT.

Bergeret, A. 1983. *Mécanisation de la culture du manioc en Guyane: Essai de la souleveuse API. Modification de la planteuse GMD-Mouzon.* Antony, France: CEEMAT.

Bergeret, A. and Militzer, A. 1990. Mécanisation de la culture du manioc en Guyane: Premiers essais de 1'arracheuse souleveuse de manioc MEM 02 de I'ITL. Montpellier, France: CEEMAT.

Guia Rural n. 5. 1988. Revolução na raiz. Colhedora mecânica pode ajudar mandioca a acabar com a fome no pais. Abril, pp. 68–69.

Vandevenne, R. and Tourtzevitch, Y. 1977. *Essais de mécanisation de la culture de 1'igname en Côte d'Ioire.* Antony, France: CEEMAT.

Improved farm implements and processing machineries. 1984. College of Agricultural Engineering, Tamil Nadu, Agricultural University, Coimbatore, India.

Plessard, F. 1975. *Contribution à la mécanisation de 1'agriculture en zones tropicales sèches: Le battage mécanique du mil pénicillaire.* Bambey, Sénégal: ISRA.

Costes, R. 1989. *Cafiers et café*. Paris: *Maisonneuve et Larose*.

Muller, R. A. *Caféiculture d'Arabica en Australie: Images d'un futur*? Paris: IRCC.

Guia Rural. 1991. Todos os systema de colheita de café. Brasil, Abril.

1.7. Transportation

T. Taniguchi

1.7.1. Introduction[1, 2, 3, 4]

It can be said that agriculture is an industry of conveyance, because various types of vehicles are required for transportation tasks ranging from carrying fertilizer and chemicals to getting the actual crops to market. It is not always evident that a high proportion of the work on farms consists of transporting products and supplies such as feed, manure, and other production aids. The task of moving this varied type and amount of material requires different and specialized vehicles. A look at the history of farm mechanization in Japan before the advent of the tractor and power tiller reveals that compact three- or four-wheel trucks were prevalent. This suggests that even farmers whose main products were grain crops, with a volume smaller than that of potatoes and other root crops, were looking for ways to make the task of transportation less labor-intensive and more efficient. The increase in female and aged labor shouldering the physically demanding task of farming, coupled with advances in science, technology, and environmental changes, means that methods of transportation also must change to keep step. The technology described in this chapter refers mainly to East Asia, Japan, and South Korea.

1.7.2. Powered Farm Vehicles for Use in the Field[3, 4]

Monowheelers

Monowheelers have an engine output of 1.5 kW and are capable of carrying 0.12 to 0.15-Mg (Mega-gram) loads. Standard tire size is 3.50 to 8.2 Ply Rating (PR) (Fig. 1.395). Because this vehicle type has only one wheel, loads that are not uniformly distributed can be unstable and difficult to control. Therefore, the transport distance is short.

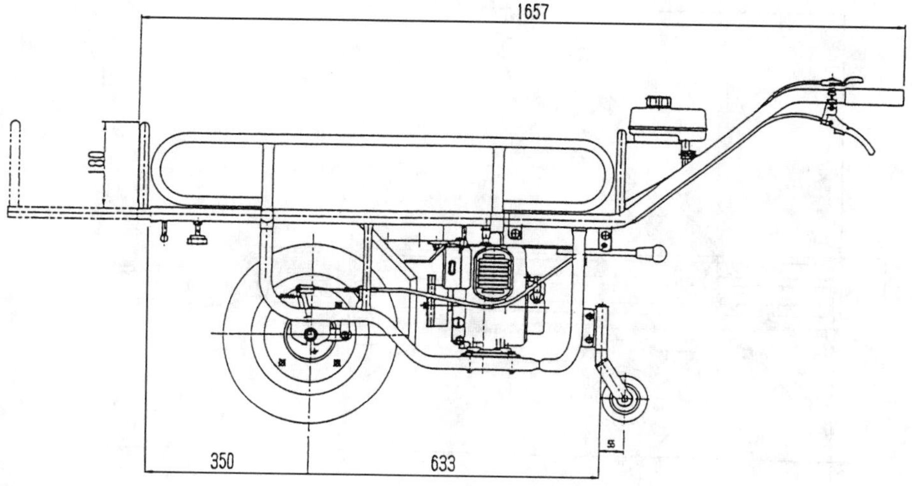

Figure 1.395. Monowheeler.[5)]

Tri-wheelers

Walking Triwheeler

Triwheelers are designed for stability, maneuverability, and sharp turns. These units are classified into two types: front-wheel steering and rear-wheel steering. Rear-wheel steering types are clutch-driven and are capable of carrying 0.15- to 0.35-Mg loads, depending on engine power that ranges from 2.6 to 3.7-kW. Models capable of carrying loads of 0.3-Mg have a manually operated hydraulic power cylinder used for dumping. Speed is set to approximate human walking speeds of 2.8 to 5.7 km/h.

Tri-wheelers That Serve as Both Walking and Riding Types

Almost all front-wheel steering–type triwheelers can be operated both while walking and riding. As shown in Fig. 1.396, the steering wheel is situated at the front of the vehicle and can be moved from either the driver's seat or the outside front of the vehicle, which allows the operator to walk in front of it. Loads are placed behind the driver to ensure that the driver's view is never obstructed. The dimensions of the front-wheel steering type are a little larger than the rear-wheel steering model. Consequently, the former is capable of carrying loads of up to 0.8-Mg. However, the 0.5-Mg capacity class is the most popular model. The speed for each model ranges from 3.0 to 10.0 km/h, in order to accommodate an operator who is either walking or riding. A manually operated hydraulic dump mechanism can handle loads of up to 0.45-Mg. On models capable of carrying loads over 0.45-Mg and with the 3.7-kW engine, a power-operated hydraulic dumping mechanism is used.

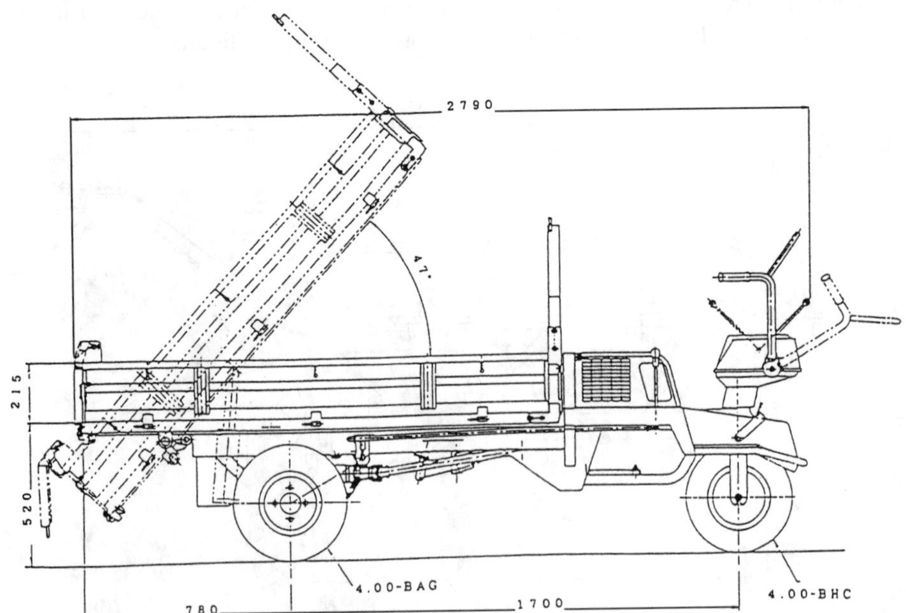

Figure 1.396. Triwheeler, walking and riding.[5)]

Four-wheel Carriers[3, 4]

Four-wheel Carriers That Serve Both as Walk-along and Ride-on Vehicles

The four-wheel carriers are an alternative to the triwheel, although the design is similar. One wheel is added to the front-steering triwheel model, and the rear wheels become the drive wheels. Four-wheel transport vehicles are more stable than triwheelers. Load capacity and dimensions of the two are nearly the same. Some models are equipped with a differential lock mechanism. The four-wheel models can be driven while walking or riding by adding a wood plate to use as a driver's seat.

Ride-only Four-wheel Carriers

Dimension and loading capacity are almost the same for both the four-wheel ride-on model and four-wheel walk-along model. Standard engine output is 3.7 to 5.2-kW, while some models are made larger and have a load capacity of 1.3-Mg. Top speed is set under 15 km/h but can vary depending on the regulations of each country.

Articulated-Steering Four-wheel Carriers

The most popular articulated-steering model has a body length, width, and height of 3.18, 1.22, and 1.33 m, respectively. The loading platform is 1.74 m long, 1.08 m wide, and 0.24 m high. These models are capable of carrying a 0.5-Mg load. The articulated-steering type is more maneuverable than the front-wheel steering model, but stability varies greatly depending on the steering angle (articulated angle) and is considered inferior to that with the Ackermann steering system. When driving on a hillside, downward forces generally act on the vehicle, but in case of an articulated steering, moment forces act on the articulated joint, and as a result, the vehicle experiences twisting, which diminishes the steering ability. However, the mass of these vehicles is large enough that there is little possibility of turning over.

Models mounted with a 6.3-kW engine are able to run at speeds of up to 15 km/h. The tires on these vehicles are wider, allowing them to move easily across soft or muddy ground.

Multiwheel Carriers[3, 4]

Walking Multiwheel Carriers

There are two models of multiwheel carriers on the market: ride-on types and models that are operated while walking. Walking types have six wheels, which are driven by a 2.6-kW engine. The operator controls the vehicle from the rear. A typical six-wheel model is 2.00 m in length, 0.84 m wide, and 0.94 m high and is capable of carrying a 0.25-Mg load. They are easy to run on narrow roads, due to their narrow width. Some models have a swing mechanism located in the front four wheels, which makes it possible to follow profiles on the field.

Riding Multiwheel Carriers

There are six-wheel and eight-wheel models. Most six-wheel models on the market are capable carrying 0.2-Mg or 0.3-Mg loads. Those with a 0.3-Mg capacity are 3.34 m long, 1.68 m wide, and 1.99-m high and are powered by a 21.3-kW engine.

A standard eight-wheel model is 3.66 m long, 1.98 m wide, and 1.95 m high. The eight-wheel model runs on a 30.9-kW engine. These vehicles are sometimes called *self-propelled, multitask farm vehicles.* They use a hydraulically operated three-point linkage and PTO that allows them to mount a forage harvester, corn harvester, manure spreader or high dump wagon. The riding type six-wheel and eight-wheel vehicles are equipped with hydrostatic transmission (HST), so that acceleration, braking, and turning are done by moving two levers. The levers work independently to control revolutions of the right and left wheels. A maximum speed of 15 km/h can be reached without changing gears if the HST and two-gear transmission are used in combination. These vehicles can climb inclines up to 30 degrees. The turning radius of the six-wheel model is 1.9 m, while the eight-wheel model has a turning radius of 2.0 m. Those models with the HST are able to turn much better, as they pivot in the center while the wheels on one side move opposite the others depending on the direction of the turn. This is known as a super pivot turn. The six- and eight-wheel vehicles handle soft or muddy terrain much better than the triwheeler or the four-wheel vehicle, but they are not able to move as well over uneven fields or ridges, and the steering ability is reduced when running on a slope.

Crawler-type Carriers[3, 4]

Walking Crawler

There are many models that are operated while walking. They differ mainly in the size of the load each can carry (from 0.2-Mg to 0.8-Mg). The track is made of flexible rubber and is usually 130 to 250 mm wide, with a lug pitch of 60 to 72 mm and 32 to 48 links. The engine ranges in size from 2.9 to 5.5-kW, with a top speed ranging from 2.9 to 6.3 km/h (Fig. 1.397). Some models are equipped with HST. Steering is done through a combination of engaging side clutches and brakes. For example, to make a right turn, the clutch for the left side is engaged while the brake for the right side also is pressed, at the same time that the clutch for the right is disengaged. This system works well on level ground, but performance diminishes on uneven fields, as is the case with multiwheeled vehicles. These vehicles have many advantages in that they are superior when climbing slopes and are very stable. Dumping is done using either a manually operated hydraulic cylinder or power-operated hydraulics. Some models are able to lift the loading platform up to 50 to 90 cm high using hydraulics and link mechanisms. This is a convenient feature when transferring loads to another vehicle or in place of a stepladder (Fig.1.398).

Walking and Riding Crawler Vehicles

The walk-along model can be converted to a riding type by simply attaching a removable seat and a step plate to the rear of the vehicle. Vehicle body dimension and loading capacity are almost the same as with a walking type (Fig. 1.399). The crawler contact length of these models is short (only 750 to 1,100 mm). That and their light weight, coupled with the fact that the driver is in the rear, cause them to be unstable and flip backwards easily. This is especially true when no load is being carried to counter the driver's weight.

Riding-type Crawler Carriers

Riding-type crawler vehicles usually are used in situations in which loads of 0.5 to 1.0-Mg must be carried. Models carrying 2.0-Mg loads are available with 5.7- to 7.4-kW engines. There are three or four forward gears, as well as reverse. Top speed is set at 10 km/h. Crawlers with a 1.0-Mg loading capacity have 37 links, a pitch of 90 mm, and

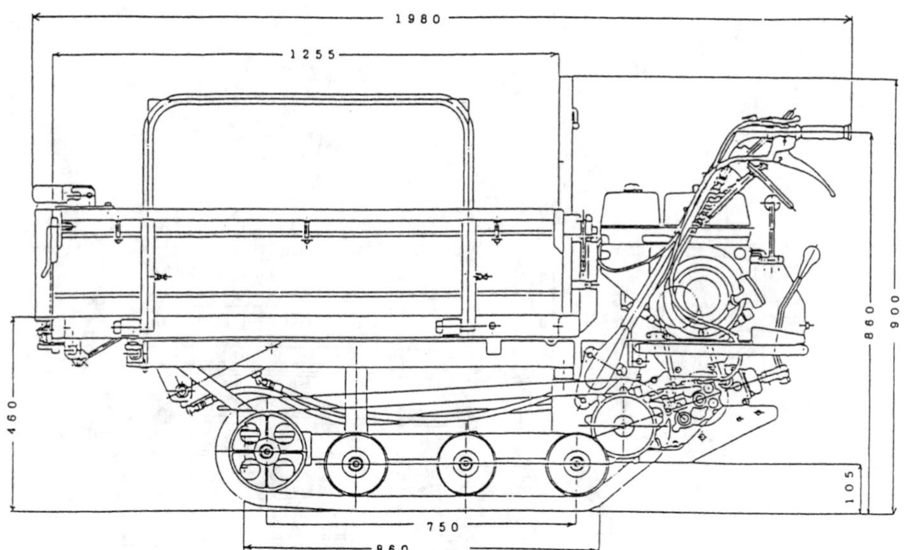

Figure 1.397. Walking crawler carrier.[6]

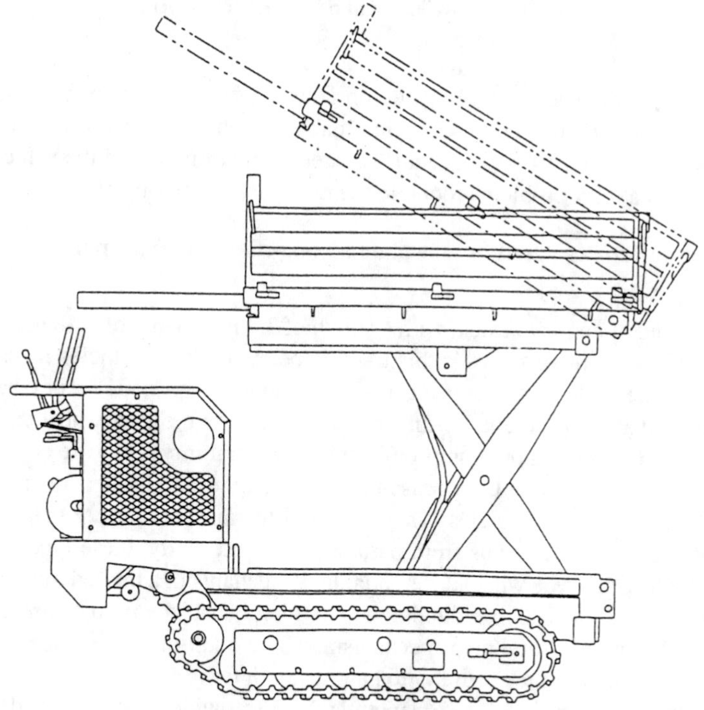

Figure 1.398. Walking crawler carrier; used as a carrier or a platform for high position tasks.[7]

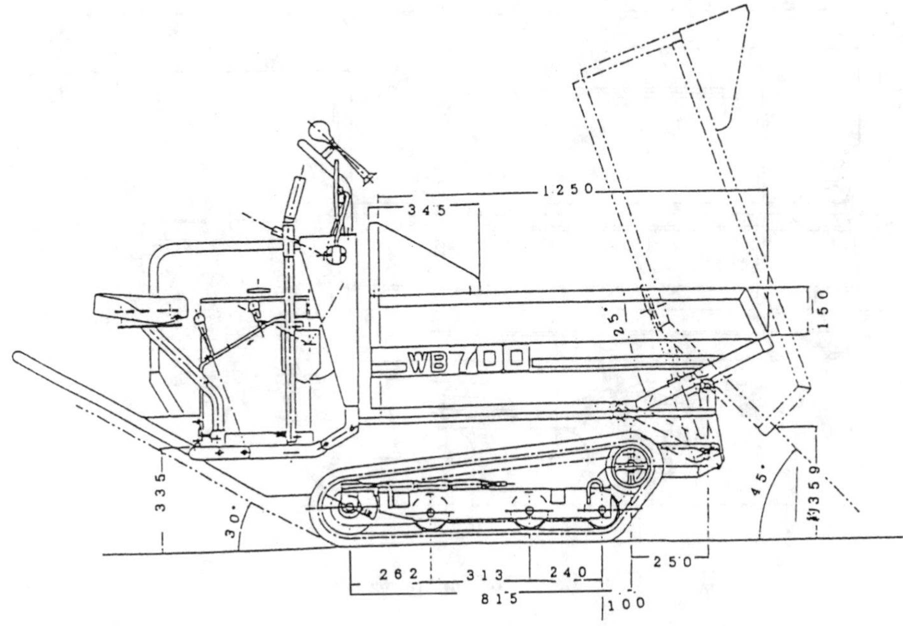

Figure 1.399. Walking and riding crawler carrier.[5]

a width of 300 mm. The track on a ride-on crawler is a little longer than that of the walking and riding combined type, which makes it better suited for soft or swampy fields. As shown in Fig. 1.400, the seat is situated in the front part of the vehicle, which eliminates the stability problems found in the ride-on/walking models.

1.7.3. Motor Trucks (Used for Local and Long-distance Transport)[3, 4]

Subcompact Trucks

This is a kind of compact truck, with a length of 3.4 m, a width of 1.48 m, and height of 2.0 m. Engine displacement for this truck is below 660 cc. Vehicles meeting these specifications are called "subcompact trucks" in Japan (Fig. 1.401). They have many advantages in that they are able to turn more sharply and the four-wheel drive allows them to handle terrain without compromising maneuverability. They are economical to maintain and very cheap. For these reasons, in Japan, almost all farmers managing small farms make use of these vehicles. They are used to transport farm tools, small farm instruments, and farm materials from barns to fields. Recently, these type of vehicles have been built with a rear-wheel differential lock mechanism, a limited-slip differential gear mechanism, a super high-reduction gear (which is half the ratio of first gear), and a hydraulically operated dumping mechanism. These additions have gone a long way towards, improving the truck's ability to handle rough terrain.

The loading platform is 2.0 m in length and 1.33 m wide. The truck bed is 0.66 m from the ground on the two-wheel drive models and can handle a 0.35-Mg payload. The

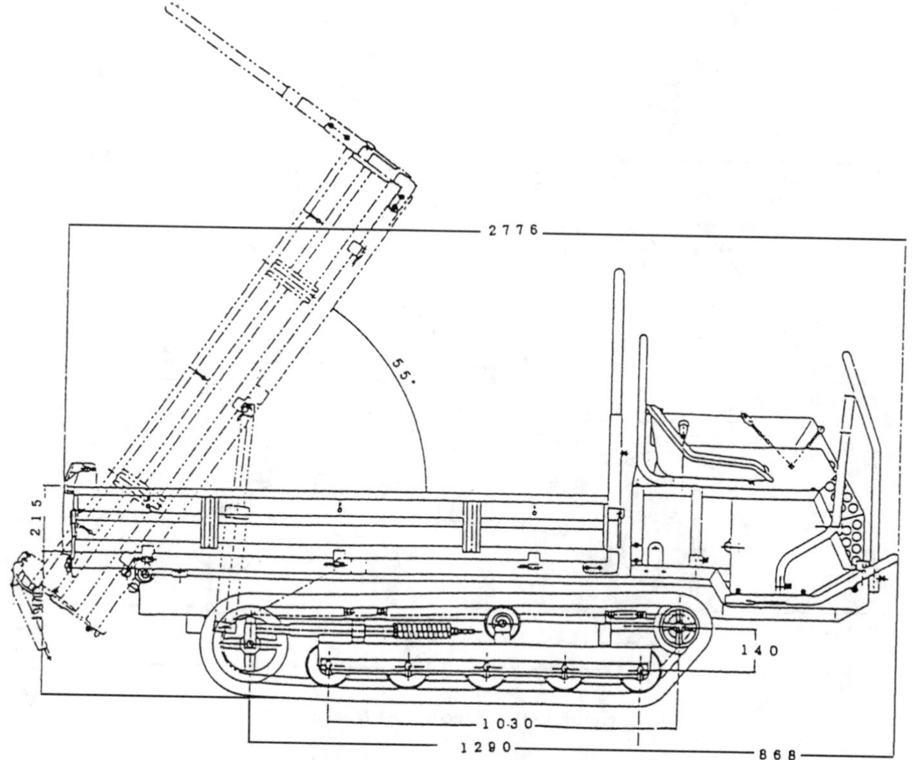

Figure 1.400. Riding crawler carrier.[5]

sides of the box usually fold down for ease in loading, while the front of the box is a guard frame.

Farm Trucks[3, 4, 18]

Most farms use a 3- to 5-Mg truck with a long wheel base, fitted with drop sides and a hydraulically operated dumping mechanism. For jobs such as transporting sugar beets or green crops, modifications (increases in height) to the sides and rear gates of the truck bed are permitted on vehicles used only to transport farm products, because farm products are often lighter than industrial materials (Fig. 1.402).

Recently, there has been renewed interest in organic farming. For upland farms, this means that manure must be brought from dairy farms, often over very long distances. A farm truck can easily serve this purpose. The farm truck is also a four-wheel drive vehicle, and recent advances in tire technology have meant that it performs very well in all field conditions. A manure spreader can be attached to the truck bed. The standard model is not equipped with a PTO but uses a hydraulic motor instead to run the beater component of the manure spreader. For this purpose, vehicles with a tank capacity of 6 to 9 m^3 have been developed. The size of these trucks means that three axles (10 wheels) are used to drive it. This drive train means better handling in rough terrain and in soft or muddy

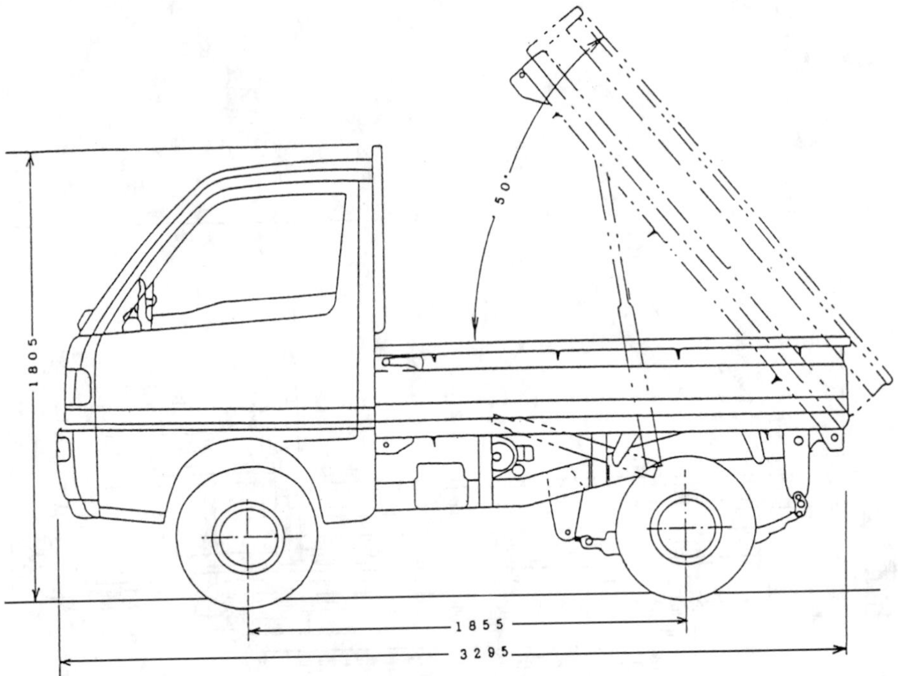

Figure 1.401. Subcompact truck.[8]

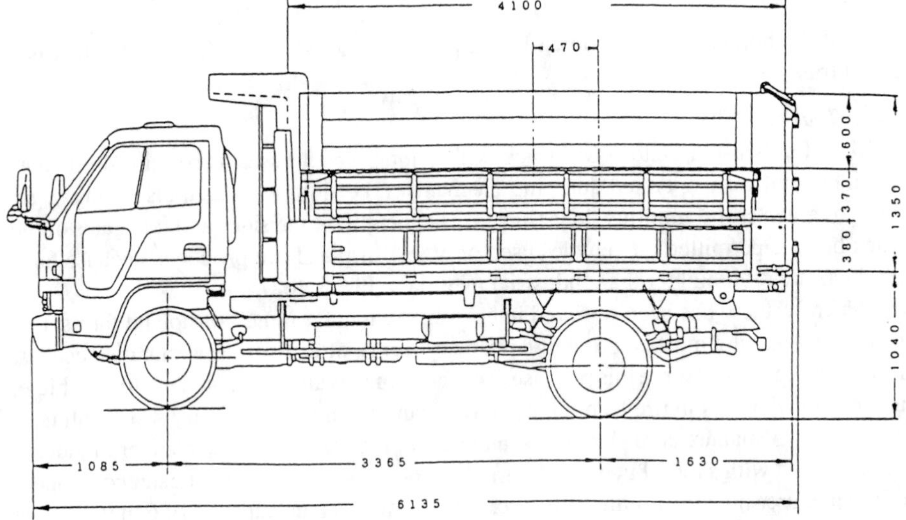

Figure 1.402. Farm dump truck (limited for farm-product transportation).[9]

fields. It also enhances the truck's ability to climb slopes and steep hills. Two types of tanks are available: a stainless steel tank for the sanitary transport of milk and a tank for moving stock feed to a silo.

1.7.4. Trailers[3, 4]

Pickup wagons, forage wagons, and crop carriers all are types of wagons, but for purposes herein, only wagons used for conveyance are covered. It is said that in western Europe, transportation tasks make up 30% to 40% of a tractor's operating time. In developing countries, a similar trend is emerging, and more tractors with trailers can be seen on public roads. Given the duties tractors are being called upon to perform, special attention must be paid to ensure they operate well in all conditions.

Trailers for Use with Walking Tractors[3, 4, 10, 11]

These kinds of trailers are used in conjunction with walking-type tractors (power tillers) and can carry a 0.5-Mg load. The construction, size, configuration, and performance of these trailers all must meet the industrial standards set by each government. The length and width dimensions range from 1.3 to 1.95 m and 0.85 to 0.95 m, respectively. These trailers are grouped into many classes, depending upon the dimensions. The industrial standard requires that these trailers be equipped with a service brake (foot brake) as well as a parking brake. However, if the service brake also acts as a parking brake, there is no need to have a separate parking brake.

New models are being built with several plate springs between the box bed and frame to reduce the shock to the load when moving over rough ground.

Trailers for Use with Four-wheel Tractors[3, 4, 12, 18]

The design and construction of a trailer for use with four-wheel tractors is almost the same as for trailers used with two-wheel tractors. Most two-wheeled tractor trailers have wooden platforms with lengths and widths that vary from 3 to 4.25 m and 1.8 to 2.0 m, respectively. The axle is located in the rear so that the trailer is heavier in the front. The hinged sides and tailboard can easily be detached. These types of trailers are normally fitted with hydraulic tipping mechanisms and a suitable ring for hitching purposes.

Trailers can be categorized into four groups: one axle with two wheels, two axles each with two wheels, one axle with four wheels (dual tires), and a tandem axle with four wheels (dual tires). There are two methods of hand-hitching the trailer to the tractor: the fixed hitch and the swing drawbar hitch.

Hydraulically powered auto-hitches (pick-up hitches) can be controlled from the operator's seat and are used to connect the trailer to the tractor. The hydraulic hitching method is certain to supersede the hand-hitching method due to its numerous obvious advantages.

Most tractor trailers are fitted with braking mechanisms. In fact, most countries have made it compulsory that trailers for use on public roads are to be fitted with efficient braking mechanisms. Two types of brakes may be used: the overrun, self-acting type or the driver-operated type. In the situation in which hand-operated brakes are present, it is necessary that the lever be within easy reach of the driver. It is also necessary that a ratchet device be fitted to keep the brakes on when required.

Trailers with Hydraulic Tippers[3, 4, 18, 19]

In various farms on which there is a need to haul sugar beets, sweet potatoes, and corn in bulk quantities, two-wheeled hydraulic tippers are suitable. These types of trailers are also preferable to four-wheelers for hauling manure and several other jobs.

The transportation of green crops to silos requires a steep angle of tip for ease of emptying. Some manufacturers provide a 50-degree tip together with a top-hinged rear door, which is automatically opened by a linkage during tipping.

The use of the tandem four-wheel arrangement with a pair of wheels on an auxiliary axle may be necessary in some situations. These situations include the transportation of heavy loads over soft and wet lands.

The factors governing the size of the wheel load of transport equipment mainly include the empty vehicle weight, the payload capacity, and the vertical load on the trailer hitch. Under normal agricultural purposes, at speeds of up to 30 km/h, 16/70-20 (418 mm wide and 1075 mm in diameter)–class tires at 80 kPa inflation pressure will accommodate a load of about 3.1-Mg. Table 1.75 gives a summary of the maximum possible field loads for single-, tandem-, and triple-axled trailers at a tire inflation pressure of 80 kPa. As an example from this table, a combination of a tractor and a slurry tank with three axles makes it possible to transport 22 m^3 (22-Mg) slurry over a field using a tire-inflation pressure of 80 kPa.

Grain Trailers[19]

Most of the grain trailers used in the American corn belt have an axle load of 25 Mg. There is also a larger model whose axle load is 40-Mg, which may be used from time to time depending on demand. Grain trailers that are equipped with rubber-belt tracks have a mass of 23-Mg when filled with grain.

It is necessary that heavy vehicles with more than one axle and wide tires with low ground contact pressures have steerable axles, as in the case of bulk trailers, with a load-carrying capacity of 10-Mg.

General guidelines for maximum subsoil stresses and axle or wheel loads of off-road vehicles should be established. For example, in Sweden a load limit of 6-Mg per axle

Table 1.75. Typical payload capacity and maximum load on wheel for trailers and slurry tanks with 16/70-20 tires (after Tijink, 1991)[19]

Payload Capacity (Mg)	Maximum Load on Trailer Wheel (Mg)		
	1 Axle	2 Axles	3 Axles
4	2.4	—	—
6	3.3	—	—
8	4.3	2.3	—
10	5.2	2.8	—
12	—	—	—
16	—	4.5	3.2
20	—	—	4.0
25	—	—	4.9
30	—	—	5.8

for farm vehicles was recommended by soil scientists sometime back. This load limit has since been adopted by many farmers and manufacturers.

Trailers Used for Transporting Combine Harvesters and Heavy Equipment[3, 4, 18]

Transportation of combines and other heavy equipment is usually a serious problem, demanding careful handling. Various types of pneumatic-tired trailers may be used in such situations.

There are three methods of loading a combine and implements onto the trailer. The first is by using outriggers and auxiliary bridges. The second involves the use of an inclining platform like a hydraulic see-saw but without an auxiliary bridge. The third is a combination of both. The smaller trailers (with a platform of 2.5 m in length and width of 1.28 m) are capable of carrying a load of 1.2-Mg. They are used for carrying the Japanese-type combine (axial type, Jidatsu combine). The bigger models have a low platform, whose length and width are 4.2 m and 1.9 m, respectively, and a loading capacity of 4.5-Mg. Some models have an authorized payload of 0.75-Mg.

1.7.5. Loaders[3, 4, 13, 18]

Hoist and Truck Loaders

Raising heavy farm products in sacks to high places may be accomplished by the use of friction hoists. These hoists may be driven by an engine or by an electric motor. The most commonly used hoist is the two-way unit–type, which is capable of carrying 0.13-Mg at a time.

Portable, self-contained, engine-driven hoists with a light and swinging jib are available. These may be used for stationary work. They may also be mounted on a trailer or truck platform for purposes of lifting sacks and bales or farm products in the field.

Other types of truck loaders consist of a lifting platform attached to the rear or side of the truck. The lifting platform usually is operated by a hydraulic cylinder.

Universal Elevators

These are general-purpose elevators. They comprise a simple conveyor of chains and slots running on a bed, which may be solid or in the form of a grille and usually have an adjustable angle. With some elevators, instead of having a bed with an adjustable angle, the end section of a long elevator may be hinged and adjusted, allowing for various loading heights. Two pneumatic-tired wheels are, in most cases, used to carry small-sized elevators. However, some machines with a high lift have four wheels. Most universal elevators can deliver up to 0.1-Mg of sacks at a height ranging from 2.7 to 3.7 m per minute.

Tractor-mounted Loaders[14]

Tractor-mounted loaders are classified as front-end loaders and back-end loaders. Front-end loaders often are used for loading manure onto a manure spreader, or the bucket can be changed to a manure fork for turning over manure. Two specifications that are important, with respect to the tractor-mounted loader, are the "dumping reach" and the "dumping clearance." The dumping reach is the distance between the end of the bucket and the front of the tractor when the bucket is in the dumping position. The dumping clearance is the distance from the bucket in the dumped position and the ground (Fig. 1.403). On tractors with 75 kW, a 0.6-m^3 bucket with a load of 1.9-Mg or

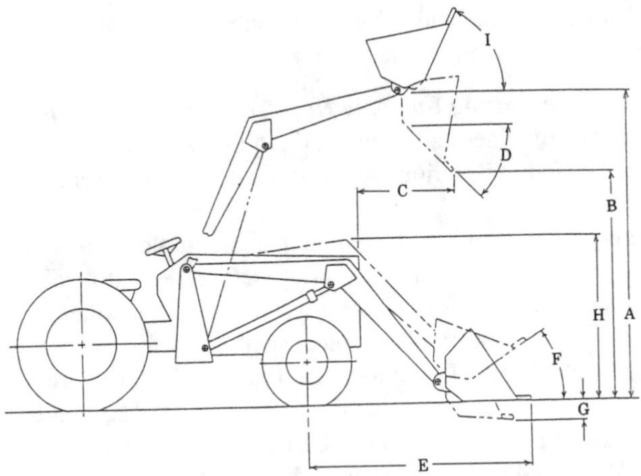

Figure 1.403. Front-end agricultural loader.[14] (A) maximum lift height; (B) clearance with bucket dumped; (C) reach at maximum height; (D) maximum dump angle; (E) reach with bucket on ground; (F) bucket rollback angle; (G) digging depth; (H) overall height in carry position; and (I) rollback angle at full height.

a 1.8 m–wide manure fork can be lifted to the top position. There are various additional attachments that can be used with the front-end loader.

Back-end loaders also have a variety of attachments including a loading hook, a rear-mounted manure loader, forks used to grab and carry big bales, and tractor-mounted fork lifts.

Slewing tractor-mounted loaders are also available for handling many kinds of work on farms, especially for moving manure, silage, and sugar beets.

Self-propelled Loaders

Self-propelled all wheel drive loaders with 9.6 ~ 40.8 kW engines, and a bucket capacity 0.14 ~ 0.38 m^3 are multi purpose vehicles in a farm setting. These types of loaders use a skid type steering that allows them to turn very tightly making them an excellent tool for cleaning barn yards and other feeding buildings. A standard loader has a 0.31 m^3 bucket capable of carrying a 0.6-Mg load. The engine is 22.1 kW and the vehicle can reach a top speed of 10.0 km/h. Special attachments can be attached to the loader, making them multipurpose vehicles that can be used to move, load or pile farm products (Fig.1.404). The loaders use articulated steering and have a mechanical transmission combined with a torque converter, and power shift transmission.

The front-loader usually provides an attachment for handling equipment. In some cases, however, equipment mounted on the tractor's three-point linkage system may be used as a carrier for loaded boxes.

***Forklifts*[15, 18]**

These vehicles are mainly used to move loads vertically. An example is the process of moving a load from the ground to the bed of a truck.

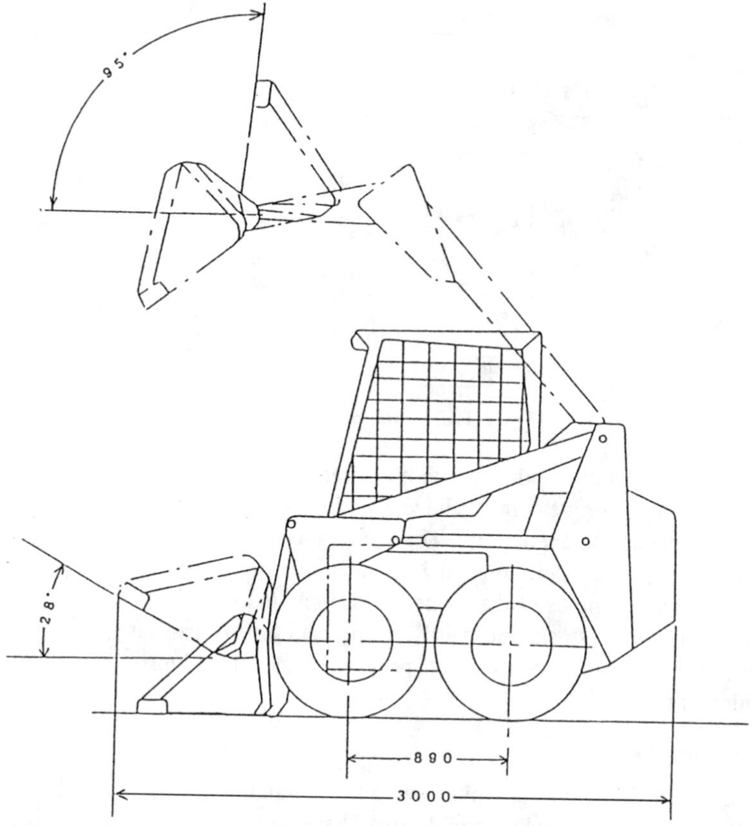

Figure 1.404. Skid steering loader.[16]

The use of containers that can easily be picked up and carried using hydraulic equipment on a forklift is considered to be a development of great importance. There are many models of forklifts ranging in sizes from 0.5 to 37-Mg. However, in the agriculture industry, 1.0- to 2.0-Mg forklifts are very common.

The maximum weight that can be handled by forklifts with front-loaders on medium-powered tractors is usually about 0.5-Mg. There is a need to have a parallel-lift action; a tilt-back mechanism is also recommended. The tractor rear-mounting types of forklifts can be designed to handle weights ranging from 0.5 to 1.0-Mg. Vertical lift equipment with a lift ranging from 3.5 to 4.85 m is available. Simple attachments that can be used with a three-point linkage system and are suitable for lifting and transporting either pallets or bulk boxes also are available.

On large-scale farms, self-propelled forklift trucks usually are used. These trucks have capacities of 1-Mg or more and lifting heights up to about 4.6 m. The self-propelled forklift trucks usually are designed for short distances duties on smooth, hard surfaces. Self-propelled forklift trucks have an advantage over the tractor forklifts in that they are

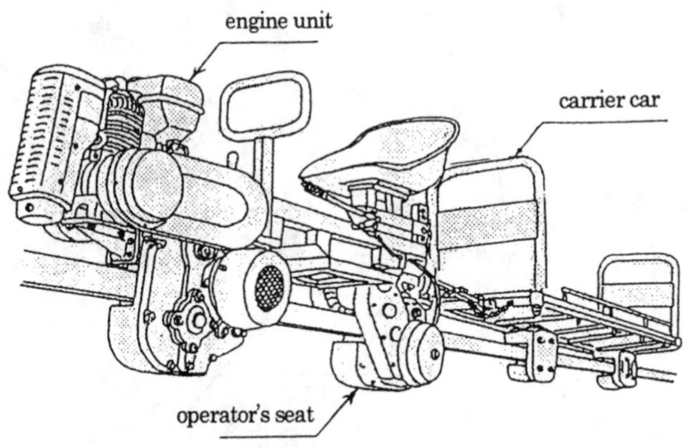

Figure 1.405. Monorail carrier.[17]

much more maneuverable and hence more suitable for use in confined spaces such as refrigerated storage. Torque converts have been incorporated in some forklift vehicles in an attempt to speed up maneuvering. The use of such vehicles on farms is increasingly becoming common for operations such as all kinds of lifting, loading, and carrying activities. Typical examples of these operations include loading and stacking big bales, transferring bales form stacks into yards, filling clamp silos, handling bagged fertilizer on pallets, loading pallets onto road transport vehicles, and loading and unloading heavy farm equipment.

1.7.6. Monorails[17]

More than 100,000 monorail units having a total rail length of around 15,000 km are in operation in the citrus orchards of Japan. These monorails are preferred because of their ease of installation, even in an already established orchard. They are capable of carrying 0.15- to 0.20-Mg loads and can handle inclines of up to 45 degrees. Carrying speed is usually 30 to 40 m/min when fully loaded.

The engine unit is hitched to the carrier wagon. Four rollers, two on the top and two on the bottom, hold the monorail on track, while driving wheels mesh with holes in the rail and pull the unit along. This system ensures that there is no danger of slippage or derailment (Fig. 1.405).

References

1. Ishihara, S. 1978. Transportation in upland farming (An extract from the proceedings of demonstration and investigation meeting for high performance farm machinery = The proceedings for agriculture and transportation). 1–5, Association of Tokachi Farm Mechanization, Obihiro Hokkaido: Japan.
2. Honda, T., et al. 1978. Change of farm management and history of truck utilization in agriculture (An extract from the proceedings of demonstration and investigation meeting for high performance farm machinery = The proceedings for agriculture and transportation). 19–26, Association of Tokachi Farm Mechanization, Obihiro Hokkaido: Japan.

3. Association of Agricultural Machinery (Japan). 1984. *Handbook of Farm Machinery*. Corona: Tokyo, pp. 365–403.
4. Association of Agricultural Machinery (Japan). 1996. *Handbook of Farm Machinery, Handbook of Bioproduction Machinery*. Corona: Tokyo, pp. 360–370.
5. Courtesy of Winbull Yamaguchi Co., Ltd., Yashiro Hyogo: Japan.
6. Courtesy of Fujii Corp., Ltd., Tsubame Niigata: Japan.
7. From a catalogue of Yunika Co., Ltd., Kiyosu Aichi: Japan.
8. From a catalogue of Suzuki Co., Ltd., Hamamatsu Shizuoka: Japan.
9. From a catalogue of Isuzu Motor Co., Ltd., Tokyo: Japan.
10. Japanese Industrial Standard JIS B9209, 1-7.
11. Japanese Industrial Standard JIS B9208, 1-5.
12. Shippen, J. M., et al. 1984 *Basic Farm Machinery*. Pergamon Press, Oxford: England, pp. 74–77.
13. Shiozawa, J. 1978. Material handling of farm products as a material (An extract from the proceedings of the demonstration and investigation meeting for high performance farm machinery = The proceedings for agriculture and transportation). 27–36, Association of Tokachi Farm Mechanization Conversation, Obihiro: Japan, pp. 27–36.
14. ASAE Standard. 301,2, 115–116.
15. Murai, N. 1980. Research bulletin of Hokkaido Agricultural Experimental Station, Nagayama Hokkaido: Japan, vol. 32, 185–222.
16. From a catalogue of Toyo Umpanki Co., Ltd., Osaka: Japan.
17. Association of Monorail Engineering (Japan). 1995. *Monorail Handbook (The Construction of the Monorail and its Utilization)*, no. 1358. Association of Farm Mechanization of Japan, Tokyo, p. 29.
18. Culpin, C. 1976. *Farm Machinery*, 9th ed. Granada Publishing, London, pp. 300–312.
19. Soane, B. D., and van Ouwerkerk, C. ed. 1994. *Soil Compaction in Crop Production*. Elsevier: Amsterdam, pp. 471–481.

1.8. Specific Equipment Used for Cultivation Inside Greenhouses

H. Tantau

1.8.1. Introduction

Protected cultivation is one of the most intensive methods of plant production. This means that both investment and operating costs are quite high. Labor costs still form the main part of production costs in protected cultivation; therefore, it is necessary to use specific equipment for seeding, transplanting, transportation, irrigation, and plant protection. In some areas of protected cultivation (e.g., seedling and pot-plant production), industrial-like production systems are now practiced with automatic handling and transportation of plants. These types of systems are useful for specialized nurseries.

1.8.2. Seeders and Seeding Lines

For the production of seedlings, specific machines have been developed and are combined with seeding lines using a belt conveyor. The main sections of the line are a filling station, a roller dibbler, the seeder, and a covering and watering station. The height of the conveyor is adjustable so as to match up with most proprietary filling and watering lines. Single-sow seeders are mostly used in greenhouses.

Requirements

Accuracy of up to 99.9% singulation is possible, depending upon seed and seed variety. Seeds ranging from begonia to melon can be sown with the seeder, and accuracy increases with more regularly shaped seeds. Once set, a high degree of reproducible accuracy is ensured.

The seeder should be capable of sowing into almost any plug tray, seed flat, or bedding plant container on the market today. Pots in carrying trays also can be directly sown, and changing from one tray to another takes only a few minutes.

With an optimum set up, seeders will single-sow seeds at the rate of about 100,000 seeds per hour. Depending on the application, seeders with up to 20 rows are available.

Principles of Operation

A nozzle bar, with a maximum of 20 optional rows, picks the seed up by vacuum from a vibrating seed tray. The vacuum is created either by a simple venturi, a vacuum pump, or a vacuum cleaner. The seeds are held until the seed flat is correctly positioned underneath the seeder. They then are discharged down tubes into the receiver cups, from which they drop a short distance into the seed flat. This helps eliminate seed bounce. The nozzle bar automatically returns to the seed tray and collects the next line of seeds. The cycle is repeated automatically until the flat is completely seeded, ready for the next flat. The majority of bedding plant seed can be sown with a 0.3-mm nozzle bar.

A range of nozzle bars with different-sized holes is available for single, double, or triple sowing. Using a self-clearing nozzle bar, a wire is automatically ejected through each nozzle hole, ensuring a clear nozzle on each operation of the seeder. This feature helps to eliminate visual inspection during long seeding runs. Nozzle bars are available from 0.3 to 1.2 mm single-hole nozzle sizes.

Special nozzle bars are available for begonia and lobelia. These nozzle bars are based on the self-clearing type but are different in that the wire retracts into the nozzle hole just far enough to form a pocket, which will accept two begonia or five lobelia seeds. When the nozzle bar goes down into the seed tray, it picks up the seeds inside the pocket, and a clump of seeds around the tip. As the nozzles return to the discharge position, they pass through a brush, wiping off all excess seed around the nozzle tip. Special rubber-tipped nozzles can be used for pelleted seed.

Machines are usually air-operated and only need a connection to a suitable air compressor in order to run. A compressor output of about 100 l/min at 5.5 bar is recommended.

A tray-indexing mechanism is used in conjunction with the seeder to index the seed trays under the seeder for precise sowing of seeds. The mechanism also is powered by compressed air and operates from the same supply as the seeder, so both units are synchronized. During operation, the growing container is loaded under the front of the seeder. An air trip valve starts the cycle, and the mechanism pushes the tray forward a row at a time, seeding each time until the tray is fully seeded. Then the mechanism stops, ready for the next tray.

Seeders also can be fitted to the majority of peat block–making machines to seed the blocks as they come along the conveyor after being pressed. The seeder is synchronized with a simple lever trip valve, which is mounted on the blocking machine mechanism. The seeder is supplied with a set of row-spacing plates, punched to suit most standard block sizes.

The roller dibbler is a simple device for dibbling filled plug flats prior to seeding. Mounted on a base board or conveyor, the plug flats pass under a roller, which has dibbler pegs protruding from it that produce neat, uniform depressions. Roller dibblers are available for most popular plug trays.

The vermiculite dispenser provides a quick, easy, and reliable method of covering seeds sown by the seeder with vermiculite. According to the vermiculite grade and depth required, the amount dispensed can be easily varied with only two adjustments. The coverer is air-operated and synchronized with the seeder and belt conveyor.

Liquid chemical formulations can be injected into the water supply inside the watering station and applied with a minimum of wastage. When using the watering bar there is virtually no run-off.

One of the most important aspects of the entire production line (i.e., from the seeder to the transplanter) is the trays. Several systems of cell trays are available. Seedling trays should be easy to fill both manually and mechanically, easy to wash and disinfect after use, easy to stack for transport and storage, and easy to separate.

Seedling trays are made of recyclable polyethylene or polypropylene and should last through repeated handlings over many years. The seedlings should grow homogeneously with a robust root system. A hole in the cell bottom allows easy drainage of excess irrigation water. Plants are easy to extract from the trays, making them suitable for both manual and mechanical planting. Growing and transport racks made from metal make it possible to handle several trays at a time.

Common tray dimensions are 40×40 cm with a height of 4 to 10 cm and cell dimensions of 2×2 to 8×8 cm. This results in a cell volume of 12 to 380 cm^3. The number of cells per tray lies between 25 and 325; that means 156 to 2025 cells/m^2.

1.8.3. Transplanters

Various systems are available for transplanting into a wide variety of flat and pot combinations. Different possible methods are available:

- The transplanter can be moved from one greenhouse to another and down the aisle when transplanting. This method saves the labor of transporting planted flats or pots to different houses for finishing off.
- The transplanter can be situated at a fixed place and the planted flats and pots transported to the greenhouses. This setup is useful if transportation has been automated (e.g., using container for pot plants).

Transplanters include ejector pins that push the plug up out of the plug tray through the drain hole, one row at a time. The plugs are received by the stainless-steel fingers waiting overhead. The fingers then close clutching the root ball.

Accurate spacing is then performed as the trolley finger assemblies move across the rail, carrying the plugs to the correct position above the flat. The fingers then plant the plugs into the flat at a consistent and uniform depth. As the fingers leave the soil, the plant shedding rod comes down to hold the plant in position. Then the fingers return and collapse over the plug tray to repeat the cycle.

Production rates of up to 20,000 plugs per hour are possible. Changeover from one flat combination to another also should be possible.

1.8.4. Transportation

As transportation can decrease productivity, it is very important to reduce the need for transportation as much as possible or to use automatic transport systems.

The transport system depends very much on the crop being grown in the greenhouse and the production system in use. For example, for production of cucumber and tomatoes, double-rail carts are used. For transportation of rose and gerbera, container monorails are useful. The monorails can be driven automatically. For pot-plant production, different transport systems are available. Danish trolleys are most commonly used for transportation inside the nursery and also to the market. For more advanced transportation within a greenhouse, conveyor belts can be installed if static or rolling benches are used. The conveyor belts are modules of about 6 m length, which can be put together quite easily. Special transporting systems are possible for container benches, allowing automatic transportation.

1.8.5. Benches

Several different bench systems are available. The main groups are static benches, movable or rolling benches, and container benches.

The cover of the benches depends on the irrigation system. For ebb-and-flood irrigation, normally plastic material (e.g., Danish top) or aluminum is used. In this case the system should be completely adjustable. A special system includes benches with gutters for irrigation. These benches may need a slope of 0.5% to 1%.

For overhead or manual irrigation it is also possible to use styrofoam and plastisol segments. The benches are located on two roll pipes, which can be connected to the heating system. In growing areas, mobile benching primarily is used, allowing access to several benches from only one path, thus making greater use of the greenhouse itself. A space savings of up to 25% is possible. That results in a better usage of the expensive greenhouse area, a higher production rate, and lower energy costs.

By using a container system, it is possible to separate the growing and labor areas. This improves labor conditions significantly. An optimal usage of greenhouse space is possible. Automation of the container system can significantly reduce the number of labor hours. In practice there are many possible set-ups for a container system. In Fig. 1.406, one example is shown. Potting, spacing, bench-washing, grading, and packing machines are all placed within the labor area. Containers are then transported by different systems from the labor space to the growing area. There the containers are arranged closely together in order to save space (Fig. 1.407).

1.8.6. Irrigation Systems

Key components of any irrigation system are the well, pump, and proper-sized main and lateral lines. During the initial engineering and design stages of a watering system, it

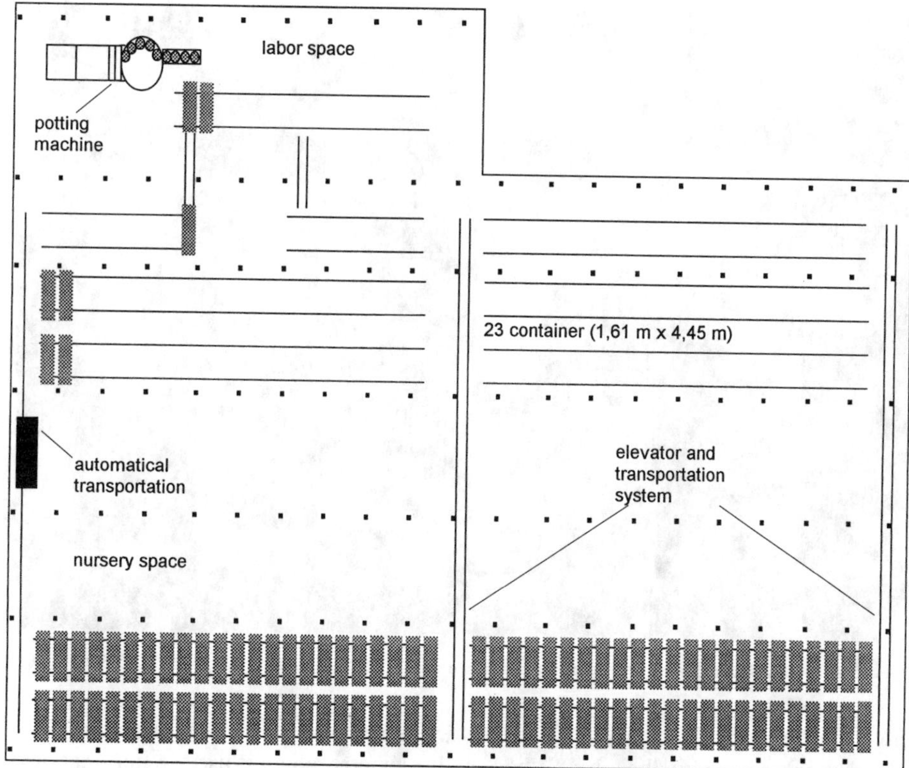

Figure 1.406. Container system (principle set-up).

is necessary to carefully determine the size of the area to be irrigated with consideration towards increasing the capacity.

Overhead sprinkling is one of the more commonly used methods for irrigating greenhouse crops. Other irrigation methods include drip or trickle and subirrigation.

Overhead Irrigation

Overhead irrigation has relatively low initial and maintenance costs. A major drawback, however, is an uneven distribution of water. This becomes a much more serious problem if the system is also being used for the application of fertilizers, herbicides, and pesticides. Another negative aspect of overhead irrigation is that it promotes the development and spread of foliar diseases. Efficiency of overhead sprinklers is influenced by the type of spray head and spacing used. The plant tops may deflect some of the water, causing it to fall outside the container rather than entering the growing medium.

Drip-and-Trickle Irrigation

Drip-and-trickle irrigation is more expensive to install but provides irrigation to individual containers or plants without wetting the foliage. Several types of irrigation are available that deliver various volumes of water. One main problem is clogging due to the

Figure 1.407. Greenhouse with container system and artificial light for production of pot roses.

high calcium and magnesium levels in most irrigation waters, as well as from growth of algae or bacteria within the tubes. When small-diameter tubing is used, water treatment (e.g., filtering) is desirable. These systems also can be used for injecting fertilizer and in the application of systemic insecticides and fungicides.

Drip emitters are the most precise option, delivering water directly to the soil at the root zone at very slow application rates.

There are three basic types of drip irrigation:

- *Emitter:* The water must pass through an emitter consisting of thin channels. These emitters or "drippers" reduce the pressure and flow rate of water applied from plastic tubing. Emitters are easy to install but more expensive and require maintenance when adjusting and unplugging.
- *Drip tape:* This consists of a black plastic collapsible tube with emitter holes spaced 20 to 30 cm apart. Pressure flow-rate reduction is achieved by sealed chambers that tend to plug even when the irrigation water is filtered. The tubing can be relatively inexpensive but generally has a longevity of only one growing season.
- *Porous hose:* Made from scrap tires, this product emits water through small holes in the hose wall and suffers from nonuniformity of water distribution.

Subirrigation

This method is frequently used for growing potted plants. One disadvantage of subirrigation is that a continuous sheet of water may be present, which can act as a carrier for root-rot organisms. Continuous capillary watering also may lead to salt build-up in the media.

Subirrigation is used in ebb-and-flow benches. These benches are specifically designed to supply water and nutrients through subirrigation. Via a pump and piping, the fertilizer solution is pumped into the ebb-and-flow bench. The solution then is automatically drained after a given time frame. If a closed cultivation system is used, the water is recycled. This makes the system economical and environmentally friendly. Problems may occur if there is a transmission of disease; therefore, the width of the system should not be too large.

Stagnant areas must be avoided. Coarse media aerate well but hold little liquid reserve. Fine media can hold up to 40% liquid but may not aerate well. The medium and frequency of fill cycles must be matched to the specific crop.

Soil-less Culture

Different systems have been developed to grow plants without soil. One of the reasons for this development was disease problems occurring in soil. Additionally, a higher production rate is possible with soilless culture. One of the goals of these systems must be to supply a regularly exchanged fully oxygenated solution to the roots and to simultaneously provide physical support for the plants. A closed system is possible if the drainage water is collected. A problem may occur by transmission of diseases as the water is passing the roots. Several systems have been developed for sterilization of recycled water using heat, ultraviolet radiation, ozone, special sand, or Rockwool filters.

Nutrient-Film Technique

In this system the roots form a mat at the bottom of a trough and a shallow solution meanders through the mat. The roots themselves create localized dams that reduce the fluid-exchange rate in some spots. The top roots are well oxygenated but the bottom ones may not be. Nutrient-film technique (NFT) can work very well but requires matching the trough shape, fluid depth, and flow rate to the specific crop. The small amount of liquid in the trough at one time is usually sufficient to carry over a few hours of power failure. Most NFT systems start with bare-root plants or plants in Rockwool cubes. Support comes from a collar around the base of the plant or from above. The root mass can support only little weight without crushing the free space out of the bottom roots.

Aeroponics

In Aeroponics, a mist of solution is sprayed onto free-hanging plants. Because so little solution is left on the roots, after a power failure they can die quickly.

Aerohydroponics

Aerohydroponics uses the same kind of trough as NFT, but the fluid is sprayed into a layer of liquid several inches deep at one end. The fluid overflows at the other end of the trough. Many so-called NFT systems are shallow aerohydroponics. The roots tolerate the deep solution because the sprayer oxygenates it so well. Because the roots are allowed to float in a deeper solution, mats can be avoided. The plants actually are located in cups filled with gravel or growrock. The cups and media support the plants as well as if they were grown in soil. Large seeds can be started in the cups, while smaller ones can be started in soil or in Rockwool cubes. Cuttings can be placed directly in the cups and will root. There is enough liquid and it is well-enough oxygenated to tolerate hours of power failure. The eventual limit depends on whether the roots run out of solution or oxygen first.

Raft System

In a raft system, a slab of foam is floated on a tub of solution, and plants grow through holes in the raft. An aquarium air pump and air stones pass a stream of bubbles around the roots. Even succulents like lithops, whose roots suffocate easily, grow well. After a power failure, the roots can grow for many hours before consuming the dissolved oxygen. This is a cheap way to grow small specimen plants and lightweights such as lettuce, but it fails to support large plants such as peppers and tomatoes.

Rockwool

Rockwool and special plastic foams are quite commonly used for many different crops (e.g., tomatoes, cucumber, roses, gerbera). It is important that the solution flows fast enough through the wool to avoid pockets of unbalanced and nonoxygenated solution. The Rockwool mat normally is wrapped up in a white plastic film with holes on top for the plants. The fertigation is done by triple irrigation. A closed system is possible through putting the Rockwool mats in gutters in order to collect the drainage water. A problem may occur with the transmission of diseases.

1.8.7. Artificial Lighting

Northern Europe endures poor light conditions during the winter. Light is a limiting factor for plant production. Artificial light can significantly improve climate conditions, resulting in much better quality and a reduction of growing time during the winter months by more than 50%. However, investment and running costs are considerably high. Electrical power consumption from a common installation is about 50 W/m^2. Because more than 90% of electrical energy is converted into heat this can contribute to the heating of the greenhouse, if there is also a demand for heat.

If a combined power- and heat-generation unit is used, about 150 W/m^2 is available for heating the greenhouse, when the artificial light is switched on. In this case the heat demand of the greenhouse normally would be lower. The result is a surplus of heat, which must be either ventilated or stored in a storage tank.

High-pressure sodium lamps usually are used to artificially light greenhouses. Use of incandescent lamps is discouraged because the red light emitted from these lamps causes the plants to stretch. Fluorescent lamps are used in growth chambers. These lamps are rich in blue light, which produce short and squat seedlings.

The lamps should be configured for a uniform distribution of light over the entire growing area. Light intensity mainly depends on the crop's requirements and economy.

Light intensity should be maintained at about 150 to 250μ/m^2·s of photosynthetically active radiation. The photoperiod (or day length) is important for several crops. Shorter photoperiods are acceptable if the light intensity is increased to provide the same total daily accumulated light (approximately 22 mol/m^2·d). Light output of Cool White Fluorescent (CWF) lamps decays over time. Thus, it is important to measure the light output of the lamps regularly. If the light intensity drops below an acceptable level (e.g., 200μ mol/m^2·s), new lamps should be installed. A quantum sensor can be used to measure the amount of photosynthetically active radiation.

Replacing only a portion of the total lamps at a time means some lamps are working at their maximum light output and some are declining in efficiency, and the fluctuation in total light output is minimized. Such an exchange system makes for more uniform lighting, which is critical for uniform production.

1.8.8. Greenhouse Climate Control

One of the goals of horticultural production in greenhouses is to increase the sustainable income of the grower. The investment costs for greenhouses as well as labor and energy costs are much higher compared with conventional plant production. This can only be balanced out with a better utilization of the yielding potential of plants, higher labor productivity, and higher energy efficiency.

Higher plant productivity and quality, in combination with a reduction of energy consumption, require a better control of the environment. In addition to temperature the air humidity, CO_2 concentration, and light intensity are controlled in commercial greenhouses. For reducing the pollution of groundwater, the control of water and nutrition supply is becoming more and more important. Closed irrigation systems, including soilless culture such as NFT and Rockwool, can solve the pollution problem but require an improved control of nutrition (e.g., using ion-selective sensors).

As more and more microcomputers are used in commercial greenhouses, it is possible to increase the accuracy of environmental control by highly sophisticated control algorithms.

Modern control strategies lead to dynamic control. Optimal conditions must be identified based on the knowledge of plant physiology and ecology.

The optimizing of horticultural production requires plant-growth models for the description of plant reaction to the environmental changes. These models consist of a series of mathematical equations and open a large field for application of mathematics in horticulture.

Research and development have been done to improve the control accuracy by more sophisticated control algorithms and new control strategies. As new tasks have been added, the complexity of the system is increasing. For the use in commercial greenhouses, the system must be as flexible as possible in order to accomplish individual demands. This can be realized by a modular system. Furthermore, dividing the system into different optimizing and control levels creates a hierarchical system. Figure 1.408 shows a system with three main levels. Each level may be divided into several sublevels.

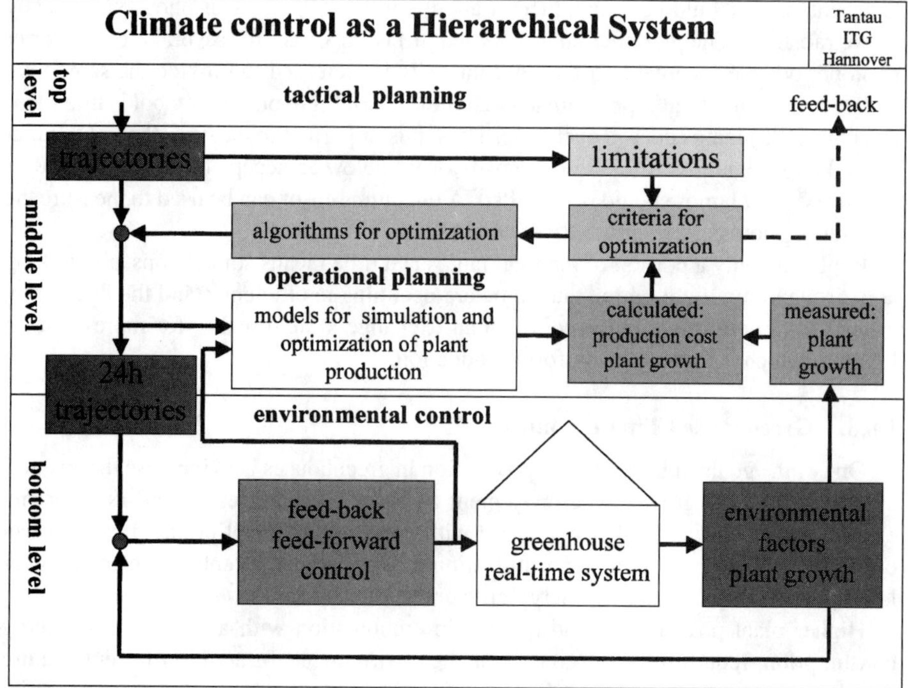

Figure 1.408. Structures of environmental control.

The top level is the level of production planning and management. The middle level represents the control of plant production. Decision rules and limitations for the on-line optimization are the input of this level.

The output gives the set points for the bottom level. There, the environmental control is realized using feedback–feedforward control.

In order to increase flexibility, the system should be modular; that means that each level is divided into several modules or components that are principally independent of each other. In such a control system the modules must be able to communicate to form a necessary a network.

In the past, most research and development has been carried out to improve control on the bottom level. For environmental control, commercial climate computers for climate control and substrate computers for control of water and nutrition supply are available.

From the engineering point of view, exact control of the climate in greenhouses is difficult because of time delay and the long time constant of the system. The very thin covering material is necessary to get a high light transitivity, but disturbances such as solar radiation, outside air temperature, and wind velocity can influence the climate inside a greenhouse very rapidly. Besides this, the heat load of a greenhouse can be reduced to 50% or even less within a few minutes by using thermal screens.

Several investigations have been carried out to improve the control accuracy. One possibility is the use of adaptive control algorithms.

For this aim, two approaches are of interest: the self-tuning regulator and the model reference adaptive system. The self-tuning controller consists of a combination of a recursive parameter-estimation algorithm and a control algorithm. Normally a parameter-optimized controller is used. However, several problems restrict the application of parameter-adaptive controllers to real processes, including the necessity to provide a number of specification parameters in advance and unsatisfactory control behavior during the adaptation. One example for the use of such models is the optimized control of the thermal screen. By such a strategy the screen is closed whenever the amount of energy exceeds the value of growth which is lost due to light reduction. Temperature control by temperature integration is another example. For control and optimization of plant production, feedback from the plants is favorable. This feedback can be used for feedback control and for an on-line parameter adaptation of the growth models.

Several methods of measurements of plant growth and plant development have been investigated:

- Net CO_2 uptake (net photosynthesis) using the greenhouse as a cuvette
- Remote sensing of plant temperature as an indicator of stomatal opening
- Transpiration rate using lysimeter or an electronic balance
- Growth rate of leaves with image processing
- Fresh weight with an electronic balance
- Height of the plants with mechanical sensor
- Stem diameter to detect the water status
- Plant development by image processing

These methods are applicable in experimental greenhouses, especially for development of new control strategies. The frequency for measurements of plant growth will be in the range from 1 per hour to 1 per day. This approach has been called the "speaking plant approach."

For the on-line optimization of plant production, numerical optimization techniques offer opportunities to derive optimal control from complex simulation models. The performance of different methods depends largely on the nature of the problem. If little is known about the surface or the solution space, random strategies mostly perform better than directed search methods. One problem is to find the global optimum and not local optima, especially when an on-line optimization is gained.

The other approach is the use of expert systems or artificial intelligence in order to make heuristic knowledge available for the control system. In this case, the expert system works as an interface between the user and control system. For this the explanation unit is very important but unfortunately not included in most of the commercial expert systems.

Several researchers are working in the field of artificial intelligence and expert systems for the control of plant production.

The combination of heuristic and procedural knowledge is of high importance for optimal control strategies on the middle level.

1.9. Forest Engineering

P. Abeels

Forest engineering involves the timber-extraction process and afforestation or reforestation practices. It also involves attention to techniques and safety aspects for sound operating methods. The steps of the extraction of logs from the forest are:

- Felling
- Log processing (delimbing debarking, and slashing)
- Skidding (bunching, sorting, and piling)
- Transportation (forest roads, trucks, etc.)

The extraction or removal conditions of some or of all the trees from a forest area resemble harvesting of agricultural products or mining of other raw materials. However, the biological and environmental impacts of forestry operations may have detrimental consequences for air, soil, and water, so that particular ways of thinking and practices must be added to engineering science for the engineer managing forests. Included in this section are some forest-engineering recommendations that best promote biodiversity of vegetation, environmental protection, and sustainability of nature and human-activities conservation.

1.9.1. Harvesting in General

Tree Harvesting

Trees in a forest must be harvested during their lifetime to meet fundamental biological requirements such as space to grow in height and thickness in accordance with crown and root development. These cuttings are called *thinnings* [16]. They are executed selectively, stem by stem, or systematically by groups of stems following lines or in places. The forest biological management rules, or *silviculture*, define the way to operate. At the end of trees' lifetimes (final cut) or in cases determined by the forester, large areas may be felled at once. This is called *clearcutting*.

Tree harvesting occurs after the forester determines soil and environmental conditions involved in the biology of the forest. A consultation must take place between the forester and the "harvesting operator" about concerns for the techniques used, access to the forest, and transportation aspects of the operations. A straight and direct relation exists between the working methods and the type and dimensions of the timber being extracted. Main modes of harvesting are full-tree (with top and branches), tree-length (cross-cutting at commercial diameter), log-length (boles), tree-section (trimmed), short-wood (cut to length), chunk, and chip.

Processing Techniques

Tree harvesting includes the main processing techniques such as cutting and handling. The cutting process must consider the main characteristics of wood [1, 15]. Cutting implies an impact of a tool plus the removal and the transportation of a chip or of sawdust. The fact that wood is a heterogeneous and anisotropic material means that the cutting tools must have a given geometry following specific angles and must be used at a given speed or force. The cutting edge of a tool can be positioned either parallel or

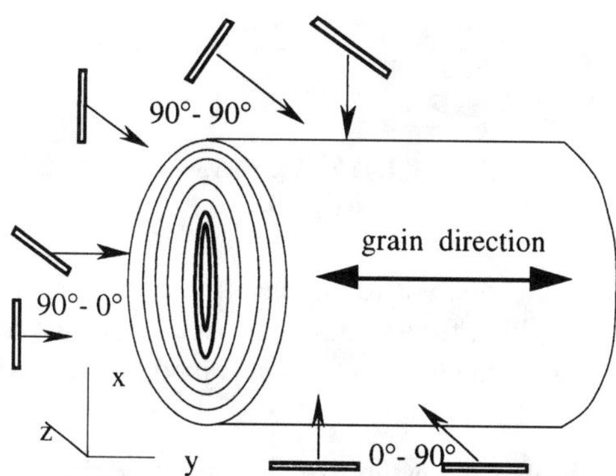

Figure 1.409. Wood-cutting directions. Bars indicate position of the tool cutting edge, arrows the tool working direction.

perpendicular to the wood being cut. Displacement of the tool may happen following the same two main directions with respect to the orientation of the vessels. Cutting and related angles are defined after the processing used. Cross-cutting and slicing happen as a 0°–90° cut while splitting a log for fuelwood is a 90°–0° cut. Felling with an axe, sawing a bole into planks, and tree shearing are 90°–90° processes (Fig. 1.409).

For each case, typical rake, sharpness, and clearance angles must be given to the tools. Working speed and force have specific values, depending on the wood species and the state of the timber (e.g., wet, dry, frozen).

Handlings and Maneuvers

The following are the steps involved in handling and maneuvering: locating, identifying, taking, moving, positioning, and putting down or dropping. Each step introduces requirements about visibility, room, clearance, weight, balance, grasp, and safety. The special case of manual handling must be considered from an ergonomic and safety point of view. The use of tools such as small grapples while wearing gloves reduces strain on the back and prevents the hands from touching the soil under the laying logs. Grapple claws must fit the proper dimensions and angles for grappling, while the crane arms must have the necessary lifting capacity and action range. Ergonomic and operational conditions must be appropriate for safety and quality of the products.

1.9.2. Felling

Felling Process

Felling happens at a given stump height following local uses. Such a height should be reconsidered when the stumps left may create hindrances or introduce a biologic or physical danger (e.g., fungi, fire) [1]. Manual felling occurs after a visual inspection all around the tree to be harvested and after removing all obstacles so fellers can get out of the way of the falling tree. The potential risk of hanging the tree crown must be evaluated. The steps for felling include location, recognizing and approaching the tree, determining the

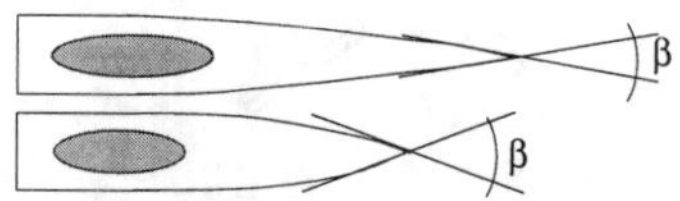

Figure 1.410. Axe-sharpness angle (β).

felling direction, clearing the close surroundings of the stump for escaping, making the notch, preparing the cut, felling, forcing down the tree if it is hanging, and cleaning the stump. After felling, the log can also be cleaned, sniped, or nosed (beveled, dressed).

Felling Tools

Manual felling uses tools such as an axe, a straight or circular saw, or a chainsaw (chisel, chipper saw). The shape and weight of the axes are related to the fiber type and the hardness of the wood species worked.

Heavy axes are used for hardwoods, while light ones with very sharp cutting edges are fitted for softwoods. The way the edges are sharpened is of the highest importance, to alleviate the working forces and to improve the quality of the cut. The axe handle is shaped for and its length should be appropriate to the ergonomic requirements of the human body and in accordance with the motion possibilities of the arms, legs, and bust.

The saws are tools with shaped teeth having an appropriate geometry for the work to be done following the already-mentioned cutting rules. Such geometry implies the definition of the rake, sharpness, clearance, and top-bevel and front-bevel angles, together with other data such as gullet shape and depth and swage width. The main types of straight saws have triangular teeth regularly or irregularly distributed and a series of cutting teeth mixed with rake (plane) teeth over the blade length.

The saws are characterized by the pitch or the distance between the cutting edges or teeth tips and by tooth height. For the cross-cutting saws the spacing between the raking teeth must be specified. Some general characteristics include sharpness angles (between 35° and 40°), setting (soft woods, 2 to 3 mm; hard woods, 1.2 to 2 mm); and depth gauge between cutting and rake tooth (softwoods, 0.5 to 0.7 mm; hardwoods, 0.1 to 0.2 mm). The gullet in front and at the rear of the rake tooth must be adequate in volume for removing chips and sawdust.

Circular saws may have triangular teeth or other main shapes such as: lying, hook, or parrot. Side clearance is also necessary for facilitating the passing of the body of the saw. It is obtained by twisting (spring-setting) the top of the teeth alternatively to the left and to the right or by swage-setting or widening the tooth top.

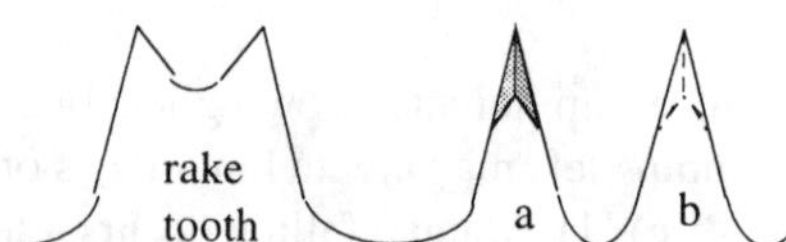

Figure 1.411. Rake tooth and top beveled cutting teeth. Alternatively left is a and right is b.

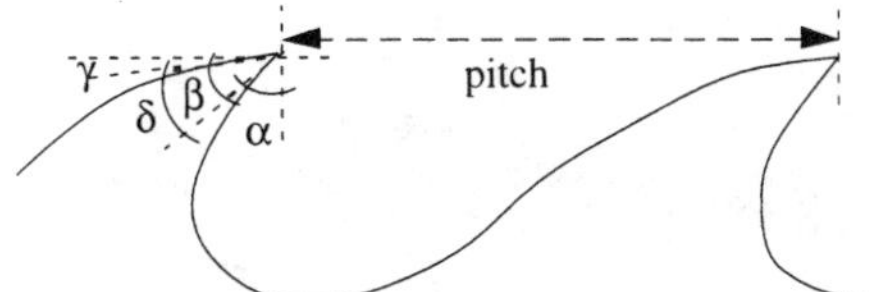

Figure 1.412. The saw-tooth angles: α, rake angle; β, sharpness angle; γ, clearance angle; δ, chip-deflection angle.

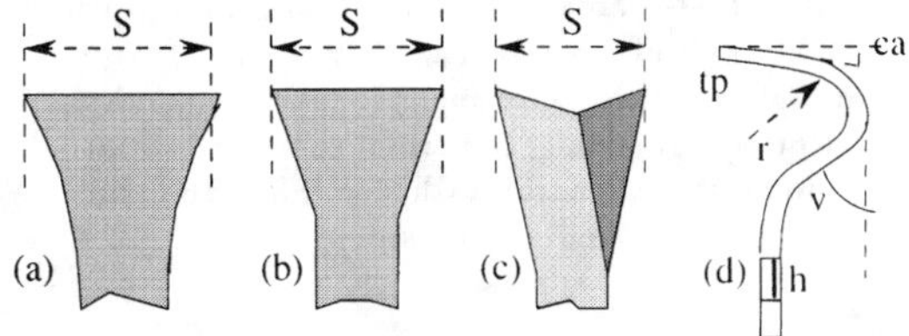

Figure 1.413. Cross-sections in the saw teeth. (a) and (b), swage set; (c), spring set; (d), chain-saw cutting link. ca: compensation angle at corner curve; h: rivet hole with chamfer and countersink; r: radius under top plate; tp: top plate; s: set width; v: vertical side plate relief angle.

Special iron parts may also be welded on the top face of the teeth to get a wider and more resisting cutting edge.

Particular care must be brought to the regularity of the swaging or setting. The tip of the tooth can be beveled laterally following the side of the setting. For the swage-set tooth, the top front face must be flat (no holes or curvature) and must be smooth to get the best possible sawn surface quality and low energy needs. The portable hand-held chainsaw is a main tool for felling, delimbing, bucking, slashing, and in some cases for splitting [10, 11]. It is a power-driven tool consisting of an integrated compact unit of power source and a cutting attachment with handles designed to be supported with two hands and an engine monitored by one of the two hands. The cutting equipment includes a chain with drive and cutting and connecting (side) links running on the side of a guide bar fitted with groove for the guidance. Chain running is obtained from the drive sprocket fitted directly on the engine axle or mounted with a reduction system. All details about chainsaw functioning, use, and safety are standardized, and it is essential to refer to them [10, 11].

The cutting link of a chainsaw cutter typically has a formed head, designed to perform cross-cutting and raking (or planing) functions. In cross-section, they offer a variety of shapes from a full round "chipper" shape to a square "chisel" shape, each with specific characteristics with regard to working conditions and maintenance. The different parts of those links are the top plate, side plate, depth gauge, gullet, and body with heel, toe, and rivet holes.

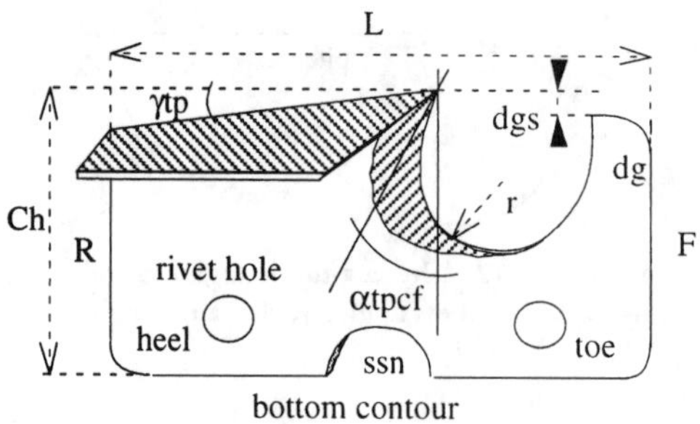

Figure 1.414. Side view of the cutting link. αtpcf: hook angle of top-plate face (cutting angle); γtp: top-plate top clearance angle; r: gullet radius; dgs: depth-gauge setting; B: bottom contour; F: front contour; R: rear contour; Ch: cutter height; L: link overall length; ssn: spur sprocket notch.

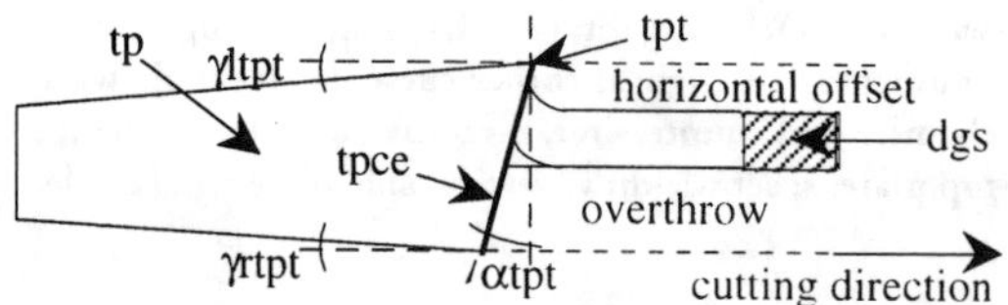

Figure 1.415. Top view of a cutting link. tp: top plate; tpt: top-plate tip; dg: depth gauge; tpce: top-plate cutting edge; αtpt: top-plate cutting angle; e: overthrow edge; γltpt: lateral left clearance angle of top and side plates; γrtpt: top plate overthrow edge (angle on right side).

The cutting edge must be sharp, and the cutting face must offer a very smooth surface at a correct rake angle (α_{tpcf} in the vertical plane) for an easy sliding of the removed chip all along the gullet. Each portion of the link must be seen in three dimensions (i.e., following three planes of reference at a given location) to define its geometry.

The clearance angles must reduce or avoid friction between tool and wood while cutting. Tool speed must be appropriate, and the use of the chain saw must be learned from experienced operators. Safety must be a main concern because heavy physical wounds may occur when working with the chain saw. Special protection equipment must be worn while operating the saw, even if some climate conditions would make it uncomfortable to do so.

Mechanical Tree Fellers

Self-propelled machines for felling trees have tipping devices, circular saws or discs, shearing claws, or chainsaws mounted on working heads at the end of crane booms.

Felling and bucking are operated by either chain or circular saws. Few shears are still in use for the harvesting of quality products.

Ergonomics

Work science must be considered as the most important concern for the human physiology in these operations. The main physics principle applied to the use of hand held forestry tools is $e = mv^2$. The mass of the tool is given by manufacturing, while speed is given by the limbs and body movements for manual operations.

Most tools are hand-carried in front of the operator's body, and they introduce dangerous constraints for the backbone, pelvis, and related muscles. Therefore physical training is essential and will contribute to more safety. When tools are motorized, the speed is monitored by an engine and its mechanical transmission. Strict operating conditions must be applied for safety and, again, professional training must be imposed for operating the forest machines.

1.9.3. Delimbing

The way the branches are implanted on a tree makes their removal often very dangerous, especially when they are bent by the felled tree lying on the soil. Manual limbing may use spades or portable motorized drilling heads. Very often, the chainsaw is used for manual limbing. The operator must be extremely focused because kickbacks of the chain saw bar may cause very great damage to the human body. Therefore, delimbing chain saws especially must be equipped with devices for reducing the risks of kickbacks (special chain links or an adapted profile for the depth gauge). They should be required always. Basic rules for manual operations include creating the best conditions for the work, piece by piece; good work-safety position of the body; and minimization of the risk of kickbacks. The limbing of thick branches is close to the cross-cutting process. More attention should be paid to pieces under stress or tension. Often it is necessary to saw the branch underneath first and then to finish with a top cut to avoid splitting. The whipping of a thick, sawn branch must be anticipated. When the branches are in verticils, two main methods of cutting are applicable. The first one is the lever method, where the chainsaw body rests on the log and the cutting bar is moved as a lever. The movement of the chain saw follows a crankshaft-like path over the length of the log. When the verticils are dense and close together, the pendulum method can be applied. The cutting bar in flat position on the tree cuts the branches on the upper portion of the tree for 60 to 80 cm (length of swing). Then the right side and part of the lower right side is processed to finish by the left side and its lower zone. The operator's feet must always stay behind (minimum 15 cm) and away from the saw's working place.

Delimbing should mainly use the saw at full speed and on the opposite side of the log to the operator. Walking is allowed when the log is between the operator and the chain saw.

Mechanized limbing tools include gate grids (static tool), curved cutting knives (static tool), rotary drills (dynamic tool as in the sawmill) and flails (dynamic tool). Most of the delimbing devices mounted on the felling heads are curved cutting knives. The curvature of these limbing knives fits the tree thickness. Cutting angles (rake, sharpness) are chosen depending on the tree species, dimensions, and climatic conditions.

The logs are driven through the delimbing devices by various types of feed rolls provided with spikes, dies, sharp edges, steel running tracks, or rubber fittings and even rubber tires. A choice is made following the log-user requirements. The nature and the state of the bark must be taken into consideration for such choice because very often the real wood of the log must not be altered by marks from the working tools.

1.9.4. Additional Interventions

Buttresses or basal swellings often must be cut away before felling, also taking into account the possible hollowness of the trunk. Liana must be removed due to the links they establish between trees in a stand, endangering any felling operation. Lopping or removal of some branches may be necessary up to a given height. By using driving wedges, lever (claw-bar), or log jack (peavey, cant dog), it is possible to lift the butt end during the cut and to guide the fall of the tree. After felling, lodged (hanging) logs must be carefully rolled off to avoid splitting or breaking. The use of a log swinger (lever with a side claw) may be helpful to force and lay the tree on the ground.

1.9.5. Topping, Bucking

After felling, the top of the tree is removed (because of too-small wood dimensions or too many branches in the crown). The topping of conifers happens mostly at a given minimum diameter of the stem (e.g., 5 to 8 cm). Then the log is bucked, following lengths required by the timber market. The most frequent tool for cross-cutting is the chainsaw. Some circular saws are used, also. The same rules as for the felling must be applied.

1.9.6. Debarking[1]

The removal of the bark happens at the cambium layer (between bark and real differentiated cells or cells in process of differentiation). The processing modes include peeling, drilling, and flailing. Spades of adequate width and form are handled manually. The sharp edge of the spade can remove bark and small branches all together when knots are cut or broken by the tool's impact. Therefore the handle of the spade ends with a bulb, the thickness of which must fit into the palm of the worker's hand in order to allow the application of significant impulses when debarking. Rotary cylinders with knives, manually or mechanically guided along and around the log, remove the bark. The pressure of the tool on the log must be adapted to the bark type and the state of the wood. The tool must avoid scratching the real wood underneath. The same process happens with the rotating diaphragm, which has cutting and raking teeth and through with the log is pushed. Flail drums are able to brake the branches and to remove bark at once. Here the state of the log is of the most importance for the process. More and more, debarking is operated at processing stations close to the forest (a problem for removing residues) or at the mill.

[1] Drum debarking is an industrial process, mostly used at the mill, not in the forest.

1.9.7. Wood Comminution

Short-rotation plantations and trees for energy are felled and bunched or directly comminuted into chunks or wood chips. Chunking and chipping are executed by whole-log chippers and cutters of various types and sizes. Whole-log chippers are made with either an inclined shoot for gravity feed or a horizontal spout for mechanized self-feeding. The cutters are heavy rotating steel discs, drums with knives, or cutting screws. Chunking devices are numerous. They all have cutting knives or cutting screws. The sizes of the chips and chunks are chosen following the requirements of the mills or plants.

1.9.8. Harvesters

More and more forest machines [9] are mounted with multifunction heads on crane booms. Most machines have several processes integrated on the same head, including felling, limbing, bucking, and cleaning the butt ends. Measuring and working devices are integrated, while automatic functions are systematically monitored by on-board computers that record all the data. Processing heads are fitted on booms mounted on various frame bases such as farm and forest tractors or, more often, excavators on rubber tires or steel/rubber tracks. The main progress provided by these machines relies on the lightness and the robustness of the working heads, which improves productivity and timber quality [1, 9–11].

1.9.9. Operation

The organization of the felling [16] is directed depending on the project and mainly by anticipating the skidding phase using horses or another given equipment. The felling patterns may be at an angle to the road system (30°, 45°, or 90°); top or butt to the road system; parallel or alternate (position of the tree tops with regard to the strip road); or combined top–butt patterns. This last pattern can increase the distance between skidtrails for a given crane reach. A dispersed felling is only acceptable if the trees are left on the ground.

Felling at 30° to 45° to the roadside frequently is done in skidding trees either one-by-one or after bunching using a horse, winch, or grapple. Such a pattern avoids too much swing or flip of the long logs. The 90° pattern is more suited for the forwarding (full carrying) by assortments (tree sections) or for chunking or chipping in the stand.

1.9.10. Security and Safety

Forest work is dangerous, and strict security and safety rules must be applied [11]. The area of operations must be clearly marked and access must be restricted to experienced personal. Wearing hard hats, caulked boots, and professional gloves should be required. No one should work out of hearing distance of a partner.

Power saws and tools or equipment must be kept in safe conditions at all times. It is always necessary to evaluate the way the felled log will swing or drop. No overhead cut can be allowed. The log must lay on the ground in a way in which it can be bucked safely. Operations should stop if winds are making working hazardous.

All workers should wear highly visible and recognizable overalls so that they can be quickly located. This is opposite to the fashion of forest inspectors and hunters [1, 13].

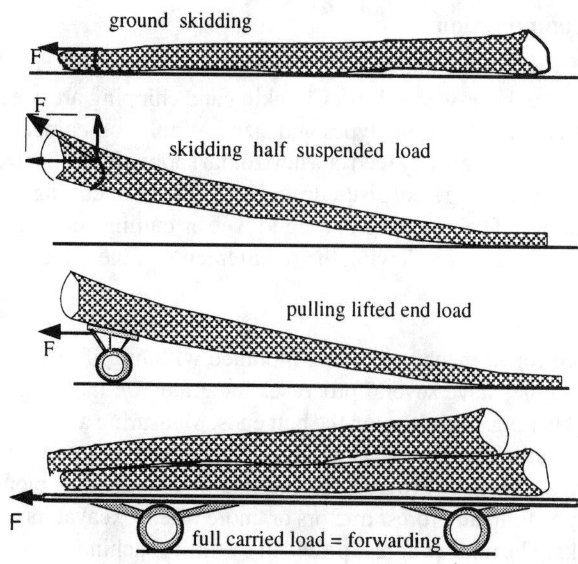

Figure 1.416. Types of skidding.

These are only few of the safety and security rules. No doubt complete sets of rules are available from insurance companies.

1.9.11. Timber Transportation

Log Extraction

Transportation starts at the stump or at a temporary pile prepared in the stand after a limited manual bunching as in the first thinnings. In a few countries, pit sawing converts logs into planks on the felling site to make the transportation easier.

In other cases manual extraction can use dry timber chutes or more sophisticated plastic chutes for sliding the logs downhill, especially for fuel wood. A hand sulky can help workers skid small to midsized logs. Motorized sulkies have been proposed, and small equipment on tracks or adapted horticultural machines are available, too.

Ground transportation of timber implies hauling, skidding, or forwarding followed by forest and road transportation. Midsized and large hauling, skidding, and forwarding machines are mounted on wheels equipped with tires or tracks, and on legs. All these machines and specialized trucks have to penetrate into the forest stands. Therefore, locomotion on soft soils is a main preoccupation due to the weight of the machines, and their overall dimensions can damage trees left in the stand, especially in thinnings.

Skidding

Transportation may bunch or move the felled logs over a distance on soft soils. Consideration must be given to soil conservation and to the protection of the remaining trees, more especially after the first and second thinnings. The sliding of logs behind a horse, a winch, or a grapple on a tractor requires traction. When the log lies on the

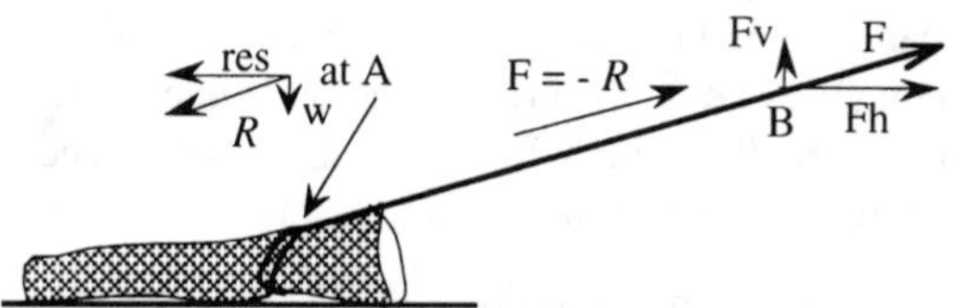

Figure 1.417. Traction line in timber hauling. A: choker bell; B: traction device; res: skidding resistance; F: traction; F_h: horizontal force; Fv: vertical force; w: weight; R: resultant.

ground, its diameter, length in contact, total weight, and type of bark and the terrain slope along with the nature and state of the soil underneath define the required traction (horizontal, vertical and frictional forces). A skidding coefficient has been proposed [4]:

$$C_s = P_t / W_l$$

where C_s is the coefficient of skidding, W_l is the total tree load, and P_t is the force component parallel to the skidding surface. The C_s value is used for determining the necessary drawbar pull (choice of a winching capacity). The tractive coefficient (drawbar pull/gross vehicle weight) guides the choice of a tractor. When horses are used, the way they are conducted may largely influence their tractive capacity. Here not only the weight but also the character of the animal, the food given, the ability of the driver, and some other environmental or psychological factors act on the work done. The main fact to take into consideration in the hauling is the traction line.

Any location on the traction line offers a decomposition of the vector F, given the forces in the different directions at that point (recall that a vector has direction, orientation, and size). So when a harness is used, A (log) and B (collar) must be on a same straight line. When w increases, the vertical component F_v increases as well, and the reaction to it is applied on the nape of the animal's neck. When only res increases due to the bulldozing in front of the butt end for example, the angle of the resultant action line is modified and must be combined with the traction line AB. Further developments are applications of that principle.

Likewise, harnessing is very important in animal skidding [1]. For example, a webbing on the crupper acting on the traction rein may shorten the length AB and lift the butt end. It is recommended to alleviate the animal's traction mainly for the start (tug). Different devices are suggested for hauling by animals such as pans, sledges, sulkies, and small arches. They reduce the value of the coefficient of friction on the soil and must be chosen according to conditions such as rockiness, gravel, wet mud, snow, ice, and so forth. In turn, the loads are greater so the productivity is increased. Taking into account the rather constant travel speed, empty and loaded, of the animal, it is essential to optimize the loads and to reduce handlings and maneuvers.

When mechanical skidding devices are used, either the rollers of the winch, the arch of the skidder, or the grapple (fixation or center) is the B location. When an adjustable height is available for B, it can lift the end of the towed log. The log is in semisuspended position [1, 6]. Statically, the log (length $= L_T$ and weight $= W_L$) is bent and part of its length (L_G) and weight (W_G) lie on the ground and will slide on it. The other part of the log weight (W_B) hangs at B. P_T is the horizontal pulling force. The following ratios are the most important for the loading of the skidder:

$$\text{Load weight on ground ratio: } \xi = W_G/W_L$$

$$\text{Butt weight ratio: } \zeta = W_B/W_L$$

$$\text{Ground length ratio: } \gamma = L_G/L_T$$

$$\text{Slenderness ratio: } \lambda = L_T/d_B(\text{butt diameter})$$

It may be useful to underline that if W_G defines the resistance to the sliding, it is also involved in the soil disturbances. Dynamically, the coefficient of resistance to skidding, $C_{RS} = P_T/W_G$, and W_B are acting on the traction device, tractor, or skidder and winch. They define rolling resistance, tire deflection, and effects on the soil (disturbance, sinkage, compaction).

The hauling of whole trees and tree lengths show similar deflections when they are suspended at the same heights. Shorter lengths show less deflection. The pull of the line increases linearly with the load size. However, when the load size comes from skidding multiple trees, the line pull rate of increase is lower than for single trees. Then W_B decreases with the increase in the number of logs in the load. This could result from the fact that part of the bunched trees are supporting other ones that are not in direct contact with the soil. The drag on the soil can introduce dirt or stones in the logs, which is not good for the tools and may require some more inspections in the industry.

Winch Skidding

The pull is operated by flexible steel ropes, which are assembled by twisting wires helically together around a core. This core is made of fiber, plastic rope, or other wires. Steel wires are twisted following a given technique (Seale, Warrington) to make the rope. For forest operations the cables must keep flexibility and resistance to shocks. Their use will avoid kinking and tight curls. Cables are listed by figuring the strands and the wire composition together with the mounting in lays. An example is given here: 6×7 (6/1) + fiber core means six strands each of seven wires assembled by six on one around a fiber core. The diameter of the wire used to build the cable determines the minimum diameter (thickness) of the drum of the winch (drum diameter = minimum $500 \times$ wire diameter, mm). Such a drum allows the cable to carefully wind in successive lays (guiding device) [14].

The allowed number of layers comes from the difference of size between flask and drum diameter $(a-c)$ divided by 2 times the cable diameter (d)—that is, $n = (a-c)/2d$. The length of the drum (b) mainly defines the admissible length of the cable and, in turn, the range of action of the winch. So:

$$L = \pi 1/d(an + dn^2)$$

and

$$b = Ld/\pi(an + dn^2)$$

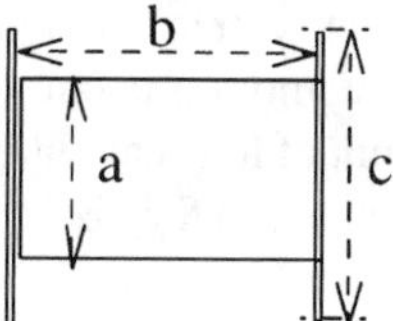

Figure 1.418. The drum of the winch. a: winch diameter; b: winch length; c: flange diameter.

where a is the drum diameter (mm), d is the cable diameter (mm), n is the number of layers, b is the drum length (mm), and L is the cable length (mm).

The traction resistance of a cable comes from:

$$\text{cable traction resistance} = \text{cable section} \times \text{wire specific resistance} \times \text{safety factor}$$

The specific resistance of the wire is given by the cable manufacturer and is related to the type of steel used for a given section. The safety factor (μ) is a correction depending on the mounting of the cable (twisting) and on the use foreseen for the cable (shocks, friction, kinking, etc.). In forestry μ can be one sixth to one tenth of the nominal value. The winching speed is in the range of 0.5 to 2 ms^{-1} ($\pm$2–7 km/h) and is compatible with the worker's walking speed. Cable unwinding must be very easy for manual operation. The life of a cable depends on its use and maintenance. For hauling, 500 working hours may be assumed. Appropriate chokers and chains must be used to attach the logs to the hauling cable. Other cables will be reviewed for cable yarding systems. The operation with cable occurs in the following sequence: pulling out winchline, fastening the end choker to the farthest log; sliding other chokers along the slack winchline; fastening the second-farthest log, then the third and the fourth one; reeling in the winchline; gathering the end choker; and pulling the other chokers. All the chokers are pulled to the rollers of the arch on the skidder and skidding starts. The winch operators and those people handling the ropes such as chokersetters (chokermen) must wear appropriate gloves (because of danger from protruding broken wires) with all the other body protections as for the lumberman. At breakage under traction, a cable may swing around. Workers must stay away from the cable line or cable way when in operation. A particular use of the cable system on the ground is found when using static running lines. A net is installed between the trees in the stand to be harvested. After felling and manual processing into logs, the logs are hanged on the running cable and conveyed to a landing at a distance.

Grapple Skidding

Grapple skidding is the normal evolution of the timber extraction practice by winch. The driver can remain in the cab and pick up the logs one by one (hydraulic remote commands) to gather the load. The skidder hauls a long distance to some place such as a landing or to the processing unit (chunking, sorting, chipping). The grappler handles large bunches of trees. Such a grapple skidder [11] will travel and maneuver all over the forest stand. Such equipment is often seen as a dangerous tool for forest sustainability and soil conservation. Here, too, locomotion is one of the most important factors. The size of the grapple defines the amount of logs caught by turn. The opening of the grapple and the angle of the claw tips are related to the size of the logs to handle (angle of entry in wood piles).

Forwarding

The full carrying of the logs is a practice introduced essentially for the short-wood harvesting systems. Over the years, the combination of a farm tractor with a trailer has evolved to the all-terrain six- and eight-wheel vehicles (three and four axles) with various motorizations and mechanical or hydrostatic transmission. These can be fitted for longer loads. The typical forwarder 4 $((1^\circ - 1^\circ) + (1^\circ - 1^\circ))$ has four axles (first number) underneath two bodies (between brackets) assembled by an articulation (sign +) to make a whole (double brackets). All the wheels are driven (sign $^\circ$ coupled with number). The forwarder [10] is equipped with a long-reaching and full-swinging crane with grapple. Loading and unloading times are crucial for the rentability of the forwarding. Training is required for the operator's ability and safety. These vehicles also are driven everywhere in the forest stands and must adhere to soil- and tree-protection requirements.

Locomotion

Optimum locomotion is an essential principle when operating with mobile equipment and machines on soft soils [2] because the forest conditions are not renewed yearly as with farm practices. This should be kept in mind. Forest machines are mounted on tracks or tires. Both mountings offer advantages and disadvantages in trafficability, stability, safety, productivity, and rentability. The effects of the rolling devices on the soil are related to the pressure and shear forces exerted following different directions [1, 2].

The graph in Fig. 1.419 shows typical distributions of the loads by tires (thick lines in black). These graphs vary, depending on the tire manufacturer. The axle load is transferred over the rolling device to the support, while the engine torque is transmitted over the wheels or tracks to the soft soil. Anchoring pieces like tire cleats or track tiles act tangentially to the soil surface. The "vertical" load squeezes the soil, which gives way under the strain. The soil sinks, packs, and even consolidates, preparing a firm "road." The all-terrain forest vehicle must avoid soil packing, especially consolidation. The wheel load distributed all over the surface of contact of the rolling device must be even and as low as possible.

Tire manufacturing and inflation pressure must be taken into consideration to achieve this. More effective information about tire deformation under load and about load distribution on typical soft soils should be available for making a correct choice. Rim mounting on the disk, tire sidewalls, and tire tread monitor the behavior of the tire. Rim seats must rest on tire flanks and correctly load the tread in order to evenly distribute the load supported by the wheel across the tread width.

In a very simplified approach the evaluation of the tire ground pressure should consider the distribution following a cone under the wheel axle that defines the real active length within the real contact area.

The loading cone opening is greater for small than for big wheels and is influenced by tire pressure (30° to 24°). The estimated active length of the contact area can be

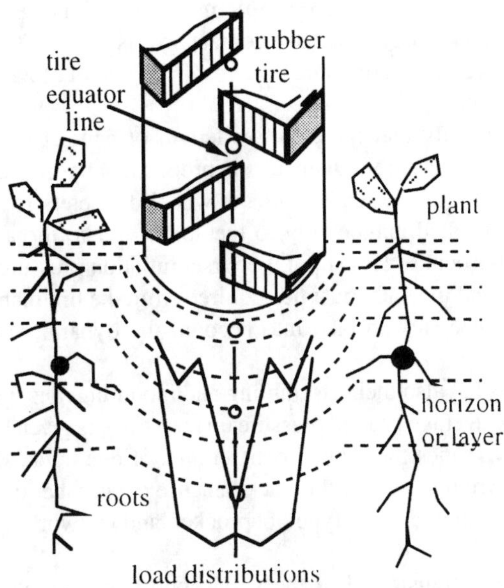

Figure 1.419. Axle and wheel load transfer to the soil.

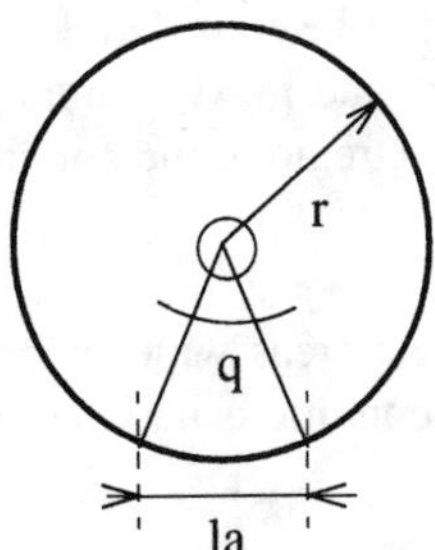

Figure 1.420. Real active length as part of the total length of contact of the tread on the soil. r: wheel radius; q: loading cone opening; la: active length of the contact area (part of the total length).

calculated as follows: $l_a/2 = r\,\theta \cdot \text{tg}\,\theta/2$ (for the given angles, the values of tg and sin are similar and turn around 0.21 to 0.25). For a wheel diameter of 1600 mm (18,4R34), the length to consider is $l_a = 2(800 \cdot 0.23) = 368$ mm. The width of the considered tire is 470 mm (inflated at 100 kPa), so the active contact area is: $368 \cdot 470 = 172{,}960$ mm^2 or 1730 cm^2. The rough estimated pressure on the soil for a load of 2200 kg is 2200/1730 = 1.28 kg/cm^2 or 128 kPa.

Knowing experimentally that the effective length of contact for the same sinkage into the soil for wheels of different diameters is proportionally greater for small wheels explains the interest in rather small or average wheel diameters for forest equipment. The wheel torque is applied tangentially to the soil closer to the active length of the contact area, which introduces less slip. Of course, inflation pressure plays a large role in tire deformability, and the data must be acquired from the tire manufacturer. In these considerations, it is necessary to take into account the dynamic constraints when in operation.

Not only the tracks but also their assembling and mounting must be considered to reduce the impacts. The average ground pressure under a track is given by total operational weight divided by two times the ground contact area of one track. However, this does not reflect the real distribution under the track because the number of shoes of the track, the number of rollers and idlers, the type of sprocket, and the working loads transferred affect the data.

The wheel torque is transferred to the soft support by friction and by anchoring prominent parts. Such forces applied tangentially to the soil or to the vegetation cover can scalp the upper lay or horizon. Impact on and disturbance of forest soils is a serious worldwide problem, and avoiding soil damage is the best option. This must receive the highest priority: Soil restoration must be considered as a bad "end-of-pipe" method [2]. The wheel slip must be controlled, and maximum tangential force application on soft soil should be fixed according to the regions and conditions (best forest practices).

1.9.12. Special Logging

The exploitation of remote forest areas with major hindrances to accessibility and also special soil or environmental conditions may require the use of special techniques.

Semiaerial Systems

Cable Systems

Numerous typical cable logging installations [18] in the forest may skid or lift the logs from the stump and haul semisuspended or full hanging trees. Heavy-cable logging techniques have been used in clearcuttings. Significant improvements have been introduced over the years in order to offer lighter cable systems for the thinnings, not only on mountainous or hilly terrain but also on rather more flat areas. Forests inaccessible due to topographic restrictions or watershed protection and areas where road building is too expensive may require cable logging.

The components of a cable system are the yarder (motorized winch or winches, also called a *donkey*), the spar pole, the tail support, the carriage, and the cable lines with their blocks. The spar and tail poles are anchored by four to six guy lines to the ground. Guy

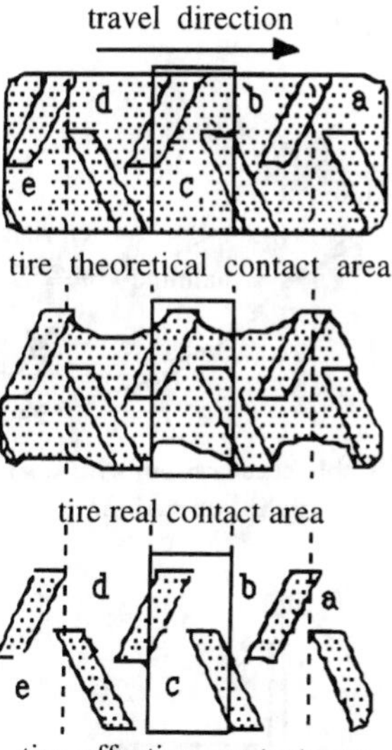

Figure 1.421. The distinctions on the contact area. a: initial pressure; b: increasing pressure; c: maximum pressure; d: pressure releasing; e: end pressure and possible partial recovery of softness (the portion marked wears the maximum of the load).

lines and blocks are fixed by short straps to anchoring trees, stumps, or other fixation possibilities [1, 18]. One or more suspended cables are used between spar and tail point to support and drive the ride of the carriage. A skyline is a supporting cable; the mainline is a carriage-driving line. In some cases a haulback line may manage the position of the carriage together with the mainline. Then the carriage movement is controlled in both directions. In some cases a movable stopper is located on the skyline to block the carriage at a given location for rigging up and pulling. In most cases, the carriage can be stopped and blocked on the skyline by traction applied to the mainline as the blocking mechanism is released by the lifting hook. Some sophisticated cable systems are also monitored by radio signals. When necessary, the skyline goes over intermediate supports or shoes. The carriage can be seen as a displaceable crane. Logs are attached to the yarding cables by chokers.

The systems may operate up- or down-hill. The line (skyline) has a range in length, and the carriage will offer another range in width that defines main operation characteristics. Road spacing is related to the length of the line, while the spacing between landings is defined by the carriage's lateral reaching possibilities. The location of the effective timber landing must be correctly chosen to facilitate the transfers to road-transportation

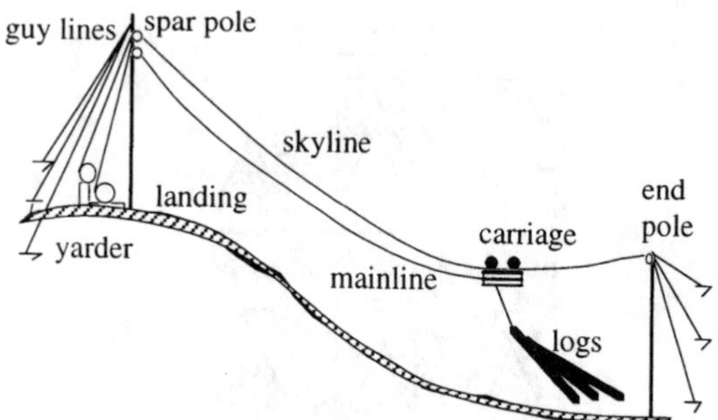

Figure 1.422. Cable crane system with suspended load in uphill operation.

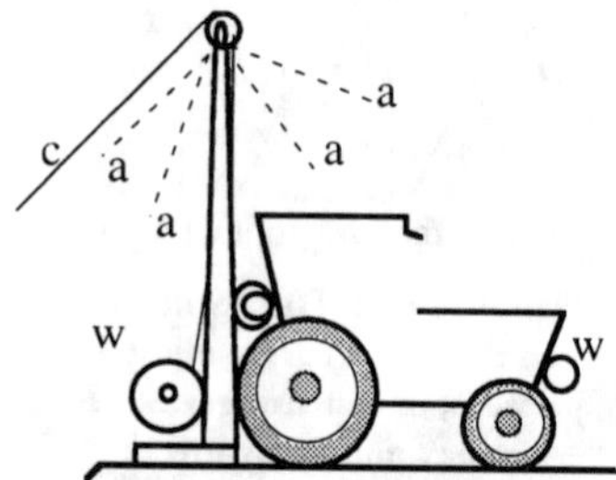

Figure 1.423. Light yarder for cable logging. a: guy lines; c: sky line; p: spar pole and block; w: winch.

equipment. Operational safety rules are crucial, and signals given or received by the chokerman and by the yarder must be clear and strictly observed. Personal safety equipment is another obligation. Special attention must be given to protecting against atmospheric thunderstorm and electrostatic phenomenon. All the installations must be effectively connected to earth. At the end of the cable line (spar or tail pole/tree), handlings must be foreseen to keep a clean temporary landing.

Air Cushion Vehicles

Air-cushion vehicles (ACVs) have been used in marsh lands for harvesting, logging, fertilizing, and seeding or planting. Platforms with peripheral skirts are supported by air inflated underneath. Motorized ventilators blow air in sufficient quantities to lift the platform and compensate for air leakage under the peripheral skirts. Air is inflated in a lateral compartment, and the device has a peripheral jet. When inflated fully under the whole platform it is called a *plenum chamber*. Propulsion by air propellers requires very high-powered engines. Some nice ACV developments have been used on trailers where the ratio of weight to inflated air pressure is more acceptable.

The operations with ACV do not introduce much erosion into the soil, especially if the soil is more or less wet. The loading, fertilizing, and seeding or planting devices mounted on the ACV platforms are classical.

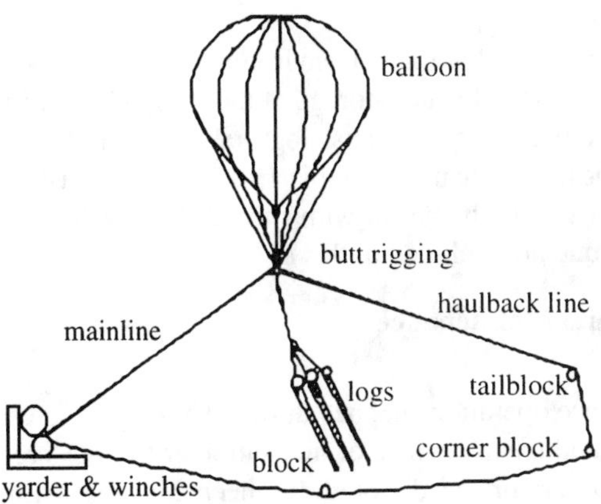

Figure 1.424. Balloon logging operation.

Aerial or Off-the-Ground Systems

Balloon Logging

Balloon logging is a practice whereby the felled trees or the logs can be lifted up and transferred from the stump to the landings at the road, rail, or river for further transportation.

The use of the balloon in logging can be seen as a substitute for the poles in cable logging. The shape of the balloon is of importance, especially on account of climatic local conditions.

Meteorology is a main constraint in balloon operations. The volume of the helium-filled balloon defines the load capacity. The volumes of the balloons are in the range of 2000 to 6000 or 7000 m^3 (or even more) to lift and yard loads from 2000 to 6000 kg.

Cables are adapted to the load capacity. The obtainable turnaround cycle is the key to rentability, and so the experience of the operating team is of most importance. The same safety rules must be applied in operation as for cable logging.

Helicopter Logging

The use of the helicopter as a flying crane for timber extraction started in the 1950s. Since then, technological improvements have introduced more opportunities for using such a harvesting technique. The team that handles the operation with a helicopter varies from 12 to 28 people. The steps of the operation are felling and grouping timber by load, tagline settings and chokering, helicopter positioning, hooking at safety/emergency release hook, lifting, flying travel, lowering and positioning the load, unhooking or automatic release, leaving the landing, and flight back. The helicopter has its place usually in inaccessible areas in mountainous zones that have overmature trees and possibilities of land and snow slides. The cost of the use of a helicopter in such circumstances must be compared with a conventional method, including special road building.

Water Transportation

For different technical, technological, environmental, and economic reasons, floating often is restricted to the actual water storage of the logs in large rafts close to the plants or transfer decks. Boom and tug boats push logs, take long rafts down the rivers, and sort timber at the mill. Log loaders handle and splash the logs at the dry docks. Inland shipping or transportation on rivers by barges or with special ships (self-dumping, self-propelled log barges) is technique not included in this overview.

1.9.13. Forest-Stand Maintenance

Special Comments

Several maintenance operations happen during the life of a forest stand, and the biological productivity and the avoidance of infestation or fire risks require some repeated interventions where operators need thoughtful mechanical help. Environmental protection and conservation are other constraints, because the public pressure on the forest often adds more restoration necessities. All operations are foreseen either over the entire area (100%), in lines (straight or sinuous openings, 30%–70%), in trenches with lateral mounds (20%–50%), or in spots (10%–30%).

Cleaning and Thinning of Young Stands

Thinning of very young stands must happen to give more space and to slow down the growth of wild concurrent vegetation. The circular brush cutter mechanizes the manual work. Its use must look at the handling of the portable machine (Fig. 1.425). When more than the cutting of small nondesirable trees is required, powered brush cutters (cross-axle types) are driven through the stand and clear the strips. The equipment works superficially, must be very resistant, and should be light to avoid soil and root compaction.

Pruning

Tree pruning is another operation that is mainly done manually. Pruning equipment includes sharp cutters, shears, (pruning scissors), and saws. They are mechanical, pneumatic, or hydraulic when assisted. The use of ladders or small lifting platforms has been introduced for the most valuable species to be pruned in the forest. The main safety rules for felling must be applied also in the pruning of trees.

Fertilization

Fertilization is an operation that first appeared in forest installations on very poor lands. More recently, foresters have asked to apply some correcting or compensating fertilizers because of bad atmospheric conditions or pollution. Powder or fine granular fertilizers are mostly blown (pneumatic distributors) or centrifugally broadcasted. The fertilizer-blowing device must be driven between the trees; therefore locomotion conditions and restrictions need to be considered. Liquid or viscous mixtures from animal husbandry are very seldom applied in forest stands. Foresters request adapted spraying machines, mainly because of size and maneuverability.

Fire Protection

Fire protection is obtained mainly by reducing the amount of slashes. Frequent soil scarification or even the ploughing of strips along roads or openings (e.g., electric lines,

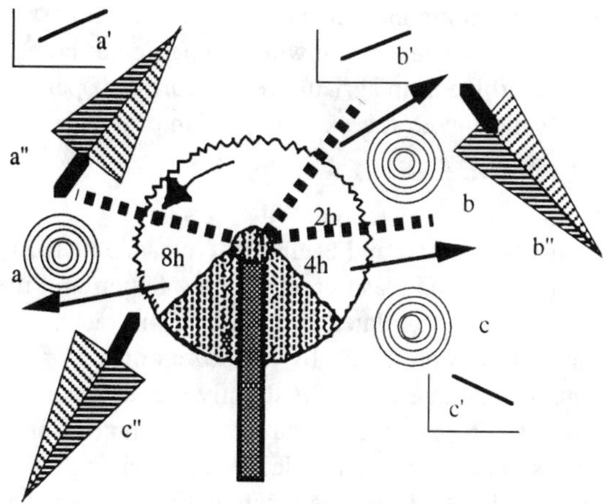

Figure 1.425. The use of circular brush cutters. a, b, c : stumps left and direction of moving the circular saw; a′, b′, c′: position of the saw when felling; a″, b″, c″: direction of felling following used portion of the circular saw (2- 4- and 8-hour sectors).

pipelines) may prevent fire extension. The opening of the forest by road for harvesting, fire control, or management, however, allows also the penetration of inexperienced people. A kind of compromise between interests must be the aim. The installation of ponds or waterholes for forest game can be part of the fire-protection program.

Soil Restoration

Mechanization of the processes of forestry is an absolute necessity. However, forest-stand damage and soil disturbance due to badly managed operations are a serious world-wide problem, which may escalate with increased use of heavy machinery. Therefore strategies for avoiding soil damage are the utmost priority. Assessment about damages relies on exact measurements (aeration, water conductivity, bulk density, soil strength, and pore volume, size, distribution, and continuity) and practical evaluations (visual evidence of rut, loss of soil structure, removal of forest floor). The techniques available [2] are mechanical (subsoiling/loosening; disking/mixing; ploughing/inverting, roto-tilling/mixing), chemical (fertilization, ash spreading, use of slashes), and biological (grass seeding, introduction of mycorhiza, multiplication of fauna). The results must include restoration of soil structure, hydrologic drainage, nutrient status, and biological activity. The goals are erosion prevention and water-infiltration capacity. Techniques are similar to soil preparation.

1.9.14. Forest-Stand Establishment

General Comments

Soil is the unconsolidated mineral material on the immediate surface of the earth that serves as natural medium for the growth of plants. A vertical section of the soil shows a succession of horizons extending into the parental material. Forest soils are especially developed under continuous vegetation and when no particular human intervention has imposed a given artificial profile as in agriculture. All forestry operations should protect and help in soil conservation, with the exception of improvements that are necessary such as drainage.

Site Preparation

Harvested areas must be cleared and even partly cleaned at least to allow replanting [2]. Large and heavy tree and slash crushers may flatten and reduce nonvaluable stands and brushes. This must be related to land-clearing techniques. Otherwise, after classical harvesting, slashes are raked by bulldozers or heavy skidders. The rakes or blades simultaneously penetrate the earth slightly, cut the roots, push other fibrous growth, and stack materials in piles for spontaneous decomposition or for destruction. The burning of the stacks is no longer the rule. Slashes may be reduced on the site by rotary chain or hammer flails, leaving chips for mulching. In some particular economic situations uprooting and root recovery can be applied. Stumps and roots are crushed, pinched, or kinked by grapples mounted on knuckle-booms for loading in vans for conveying to the mill. Very heavy wing plows are used for soils with a thick raw-material surface layer. Shallow or deep furrows simultaneously are opened where drainage is necessary. Disk plows or free-wheel revolving disks with aggressive rippers are dragged at the rear of heavy skidders or crawlers. The working pressure of the disks on the soil is hydraulically controlled. Powered rotary tillers are mounted on hydraulically monitored arms for adjusting the spacing between the prepared rows. Spot scarifiers use buckets with teeth in the front or the back and adapted to a track excavator crane arm. Balancing the bucket forward and backward cleans the spot and scratches the soil.

Planting

Artificial replantation of forest areas is in opposition to natural regeneration techniques by regulated silvicultural thinnings [1]. Following local general conditions and wild-game constraints, the planting uses different types of plants. Seedlings have bare roots or are in containers for hole or surface planting. The preparation of the planting site may clear individual small plots from slashes, operate a continuous scarification in strips, or scarify patches. Planting also may be done on mounds along the scarified strips or ditches. Soil scarification, or heavy scratching, breaks up the mineral soil close to the future plant. The depth of the opening for planting must correspond to 140% of the maximum length of the root system of the plant, because placement with a small raising up contributes to a good placement of the root system. No curl of the roots can be accepted. The planting hole must not have compacted bottom and walls. Manual planting tools can help operators. These include hoes, planting spades, and pipe sticks with articulated shovels at the end. Tree-planting machines open a planting trench with one or two slant discs, v-profiled ploughs, or v-shaped teeth. Seedlings between 20 and 60 or 70 cm tall are manually introduced in the trench at a rhythm corresponding to the plant spacing. A compacting device afterwards presses the soil around the plant and improves the root–soil contacts. Plants in containers are used with fully automated machines. These are often much more complex because they not only plant but also add some fertilizers or chemical treatment to the plantation. Surface planting using seedlings produced in a "pillow" lays these out on the soil without digging. Mechanization of the surface planting uses adapted forwarders, and it may even be done by helicopters in some cases in remote, inaccessible areas.

1.9.15. Forest Roads

Access to the forest includes primary skid trails (natural rough terrain), tracks (compacted soil), secondary forest roads (unpaved roads), and primary roads (paved structures) for connections to the highway system [19]. If skid trails must keep the soft-soil characteristics, one must consider off-road conditions, ensuring that the tracks and roads are permanent, consolidated, and stabilized. The tracks can be spaced so that they contribute to an economic use of horses (100–200 m), skidders (200–400 m), forwarders (350–1000 m), light cable systems (400–700 m), and so forth. Drainage is a main factor for consolidated roads, including the surface (run-off) and the subgrade (swelling and bringing down). For forest track and road building, wet areas, stream beds, rocky areas and bogs should be avoided. Stream crossings, dead ends, and switchbacks should be used at a minimum. The width of clearing should be of 9 to 10 m minimum, with 2 m beyond the edges of cut or fills. All debris and slashes or roots must be removed or burned before earthmoving. The balance between cuts and fills should be optimum in the length of the laying out. Road grades less than 7% to 8% are advisable, and adverse grades should be kept below 10%. Slopes should not exceed 12% to 15%. Too much drainage is always preferable and includes outslopes, culverts, and dips. Drainage needs are slightly reduced when the road profile foresees short rises and falls. A compacted ditch must always collect the water run-off when a road slopes in. A crowned road section may divert water runoff. Relief ditches must divert the water for the road ditches into the forest stand for infiltration into the native litter and soil. Sharp curves must be avoided. The outside part of a turn must be 2 m wider than the inside lane. Passing lanes or turnouts must be installed if circuit traffic is not imposed. Stream crossings must happen with as little disturbance as possible. Specific works are advisable for culvert and stream crossings. Specialized literature on the subject must be consulted. Periodic and frequent maintenance of forest roads are absolutely necessary. The sediment and debris catchments of the culverts must be regularly cleaned; drainage facilities must stay clear and working. Traffic should be stopped during wet periods, and temporary special mats should be installed where the road may wash out. Winterizing is recommended when a road is not maintained over several rainy seasons. Culverts especially should be removed, original stream cross-profiles should be restored, and road surfaces should be ripped and seeded to restore infiltration. On forest roads it is necessary to reduce rutting, washboarding, and many other mechanical damages due to the truck tires. The central tire inflation system allows deflation of the tires when the local conditions require adaptation of inner tire pressure. Such systems installed on the log trucks can reduce the risks of damages to the road, improve traction, extend the haul season, and lessen driver fatigue and back problems. Truck wear and tear and damage are reduced. On paved highways, the tire pressure is increased as needed. On soft soils in the forest stands (skid trails, tracks), central tire inflation systems require data from the tire manufacturers, if applicable.

1.9.16. Forest Regulations

Ordering Forest Machines

The ordering of forest machines should always be written following standards for denominations, characteristics, ergonomics, and economic data. For instance the United Nations (UN), International Standard Organization (ISO), Food and Agriculture Organization (FAO), and Economic Commission for Europe in collaboration with the International Labour Organization (ILO) have edited several protocols for the testing of devices and other technical instructions where most of the data are available for the correct definition of machines and tools for forestry.

Machines and Environment

Facing the detrimental conditions in several forest areas in the world, the FAO has edited the model code of forest-harvesting practice [2, 3], especially if mechanized operations are applied.

Best Management Practices

Several countries have produced and distributed "best management practices" handbooks to which it is imperative to refer before any operation in the forest zones. Thinning is becoming a primary forest-harvesting and maintenance rule everywhere. Best management should be considered especially for riparian zones and wetlands and does not exclude other areas. Slope stability of the watersheds must be a main objective, too. Renewable water resources must be protected everywhere for quality and quantity.

Techniques and Future

New objectives for forest production emphasize sustainability and biodiversity. The techniques for using available tools and the design of new processing units must now take into account several biotechnical constraints, especially in matters of soil conservation and hydrologic protection. Problems of air quality also require reviewing the available techniques and more particularly the engines and other power sources used in the mechanized processes. Testing procedures must be implemented and standardized for the benefits of people.

References

1. Abeels, P. F. J. 1995. *Génie Forestier* [in French]. Paris: Ed. Masson.
2. Abeels, P. F. J., et al. 1994. Soil, tree, machine interactions. In Proc. FORSITRISK, ECE/ FAO/ ILO/ IUFRO Div. P3.08.0, European Union Concerted action CEET-AIR2-CT93-1538, pp. 514. Bonn, Germany: Ed. Bundesministerium für Ernährung, Landwirtschaft und Forsten.
3. Dykstra, D. P. and Heinrich, R. 1995. *Model Code of Forest Harvesting Practice*. Rome: FAO.
4. European Standard EN 608. 1994. Agricultural and forestry machinery: Portable chain saws—safety, pp. 20.

5. Hafner, F. 1971. *Forstlicher Strassen und Wegebau* (in German). Wien, Germany: Ed. Östereichischer Agrarverlag.
6. Hassan, A. E. and Sirois, D. L. 1983. Weight distribution characteristics of semi-suspended trees. *Trans. ASAE* 26: 1291–1297.
7. Iff, R. H., Koger, J. L., Burt, E. C., and Cilver, E. W. 1984. C-A-R-T-S: Capacity analysis of rubber-tired skidders. *Trans. ASAE* 660–664.
8. ISO Standard 3339. Tractors and machinery for agriculture and forestry: Classification and terminology.
9. ISO/CD 6814. 1994. Machinery for forestry: Mobile and self-propelled machinery—ident. vocabulary, p. 4.
10. ISO/CD 13860. 1994. Machinery for forestry: Forwarders—specification definitions, p. 6.
11. ISO/CD 13861. 1994. Machinery for forestry: Skidders—specification definitions, p. 6.
12. ISO/DIS 6531. 1996. Machinery for forestry: Portable hand held chain saws—vocabulary, p. 10.
13. ISO/DIS 11850. Machinery for forestry: Self propelled machinery—safety, p. 12.
14. ISO 6687. 1994. Machinery for forestry: Winches—performance requirements, p. 4.
15. Koch, P. 1964. *Wood Machining Processes*. New York: The Ronald Press Company.
16. Kofman, P. D., et al. 1985. A manual on felling and bunching small trees from thinnings with small scale equipment on gentle terrain. Proc. IEA-FE- CPC7, pp. 116, Copenhagen, DFLRI.
17. Nagy, M. M., Trebett, G. V., and Welburn, G. V. 1980. *Log Bridge Construction Handbook.* Vancouver and Pointe Claire, Canada: FERIC.
18. Samset, I. 1979. *Forces and Powers in Winch- and Cable Systems* [in Norwegian, summaries in English]. As, Norway: Meddelelser fra NISK.
19. Samset, I. 1981. *Winch- and Cable Systems in Norwegian Forestry* [in Norwegian, summaries in English]. As, Norway: Meddelelser fra NISK.

1.10. Standardization

R. Hahn and B. Cheze

Standards have played an important role since the 1920s in the design, testing, marketing, and use of agricultural machinery. For purposes of this section, *standards* are defined as specifications or engineering practices developed through a consensus process to address a common problem or specific need identified by machinery manufacturers or farmers. Further, conformance is voluntary rather than mandated by a regulatory or enforcement agency of government.

The advent of the power take-off (PTO) on agricultural tractors to transfer engine power for the operation of towed implements and stationery machines serves to illustrate the key elements in the definition of a standard. In the absence of standardization, tractor manufacturers designed proprietary PTO drives for early-century tractors that operated at different speeds, rotated in different directions, and had different nominal shaft diameters and spline sizes. Soon there was a "need" in the marketplace, as farmers could only use tractor and implement combinations with the same PTO configuration and design. It was realized that a standard would provide an economic benefit both to manufacturers of tractors and implements and to their customers, the farmers. Engineers representing tractor manufacturers and others with technical expertise relating to the use of mechanical power for agriculture met in 1925 and 1926 to draft a common set of specifications—a standard—for the agricultural tractor PTO. This was a consensus process requiring give and take during the discussions and some degree of compromise. As a result, the American Society of Agricultural Engineers adopted a standard in April 1927 to describe a 540 r/min, 6 spline, 35 mm ($1\frac{3}{8}$ in.) nominal diameter, clockwise-rotating (when facing the projecting end of the shaft) PTO for agricultural tractors. The current version of this standard is ASAE S203.13 DEC94, Front and Rear Power Take-Off for Agricultural Tractors. Tractor manufacturers and implement driveline manufacturers voluntarily conform to this standard without a regulatory or legislative requirement to do so.

Standards are developed for the following reasons, and development is usually driven by an identified, common need:

- To provide interchangeability between similarly functional products and systems manufactured by two or more organizations, thus improving compatibility, safety, and performance
- To reduce the variety of components required to serve an industry, thus improving availability and economy
- To improve personal safety during the operation of equipment
- To establish performance criteria for products, materials, or systems
- To provide a common basis for testing, analyzing, describing, or informing regarding the performance and characteristics of products, methods, materials, or systems
- To provide design data in readily available form
- To develop a sound basis for codes, education, and legislation, and to promote uniformity of practice
- To provide a technical basis for international standardization
- To increase efficiency of engineering effort in design, development, and production

The focus today is clearly on the development of and conformance to one worldwide standard. This is especially important to multinational manufacturers and others developing global markets. Therefore, standards developed through the International Organization for Standardization (ISO) and the International Electrotechnical Commission (IEC) are increasingly important. Both organizations came into existence following World War II and are headquartered in Geneva, Switzerland. ISO has a series of technical committees (TCs) on various subjects. Of particular interest to agricultural engineers is ISO TC23, Tractors and Machinery for Agriculture and Forestry, formed in 1947. The administrative secretariat for TC23 is held by the Association Francaise de Normalisation (AFNOR) in Paris, France. A series of subcommittees (SCs) and working groups (WGs) have been formed to conduct international standards development. These are listed in Table 1.76 along with the country holding the secretariat.

ISO standards on agricultural tractors and machinery developed through ISO TC23 are listed in Table 1.77.

Table 1.76. Structure of ISO TC23: tractors and machinery for agriculture and forestry

Subcommittee	Working Group	Title	Secretariat
SC2		Common tests	United States
	WG1	Tests of couplings	Germany
	WG2	Electromagnetic compatibility (EMC)	Germany
SC3		Safety and comfort of the operator	United Kingdom
	WG8	Operator control	United States
	WG9	ROPS for narrow track wheeled tractors	Germany
SC4		Tractors	Germany
	WG4	Hydraulics	United States
	WG5	Electric power transmission	Germany
	WG6	Hook-type connections on towing vehicles	United Kingdom
	WG7	Braking systems for tractors and towed vehicles	Germany
SC6		Equipment for crop protection	France
	WG1	Terminology	France
	WG2	Test methods for air-assisted sprayers	France
	WG3	Test methods for nozzle bar suspension	France
	WG4	Drift measurement	United Kingdom
SC7		Equipment for harvesting and Conservation	Italy
	WG7	Grain dryers—test method	United Kingdom
SC13		Powered lawn and garden equipment	United States
	WG5	Lawn trimmers	United Kingdom
	WG12	Operator's controls	United States
	WG13	Lawn mowers	United States
SC14		Operator controls, operator symbols and other displays, operator manuals	United States
SC15		Machinery for forestry	Finland
SC17		Manually portable forest machinery	Sweden
SC18		Irrigation and drainage equipment and systems	Israel
	WG1	Pressurized irrigation equipment—Definitions	Israel
	WG3	Irrigation sprinklers	United States
SC19		Agricultural electronics	Germany
	WG1	Mobile equipment	Canada
	WG2	Stationary equipment	Netherlands
	WG3	Animal identification	Netherlands

Development of one worldwide standard through ISO can be a difficult task. Frequently, the development or adoption of an ISO standard follows the introduction of new designs or practices to the marketplace, often by many years. Manufacturers introduce design differences (or users demand them) such that worldwide standardization, after the fact, becomes exceedingly difficult without major compromises and concessions. For example, the common mast height for three-point hitch-mounted implements in Europe is greater than the mast height for similar implements in North America. Reconciling the difference is a complex matter. Additionally, differences in cultural practices around the world and differences in required or expected levels of human safety in the workplace can complicate ISO standard development and adoption. When these factors exist, the need to compromise in areas of major disagreement sometimes results in exclusion of important specifications on which agreement cannot be reached or in optional specifications, which in essence is no standard. For these reasons, national standards organizations often

Table 1.77. International standards pertaining to agriculture

ISO/TC 23, TRACTORS AND MACHINERY FOR AGRICULTURE AND FORESTRY TERMINOLOGY	
ISO 3339-0:1986	Tractors and machinery for agriculture and forestry—Classification and terminology—Part O: Classification system and classification
	Bilingual edition
COMMON TESTS (SC 2)	
ISO 2288:1997	Agricultural tractors and machines—Engine test code (bench test)—Net power
ISO 3462:1980	Tractors and machinery for agriculture and forestry—Seat reference point—Method of determination
	Amendment 1:1992
ISO 3737:1976	Agricultural tractors and self-propelled machines—Test method for enclosure pressurization systems
ISO 3795:1989	Road vehicles, and tractors and machinery for agriculture and forestry—Determination of burning behaviour of interior materials
ISO/TR 8953:1987*	Tractors and self-propelled machines for agriculture and forestry—Test method for performance of air-conditioning system
SAFETY (SC 3)	
ISO 3463:1989	Wheeled tractors for agriculture and forestry—Protective structures—Dynamic test method and acceptance conditions
ISO 3776:1989	Tractors for agriculture—Seat belt anchorages
ISO 4252:1992	Agricultural tractors—Operator's workplace, access and exit—Dimensions
ISO 4254-1:1989	Tractors and machinery for agriculture and forestry—Technical means for ensuring safety—Part 1: General
ISO 4254-2:1986	Tractors and machinery for agriculture and forestry—Technical means for providing safety—Part 2: Anhydrous ammonia applicators
ISO 4254-3:1992	Tractors and machinery for agriculture and forestry—Technical means for ensuring safety—Part 3: Tractors
ISO 4254-4:1990	Tractors and machinery for agriculture and forestry—Technical means for ensuring safety—Part 4: Forestry winches
ISO 4254-5:1992	Tractors and machinery for agriculture and forestry—Technical means for ensuring safety—Part 7: Combine harvesters, forage and cotton harvesters
ISO 4254-7:1994	Tractors and machinery for agriculture and forestry—Technical means for ensuring safety—Part 5: Power-driven soil-working equipment
ISO 4254-9:1992	Tractors and machinery for agriculture and forestry—Technical means for ensuring safety—Part 9: Equipment for sowing, planting and distributing fertilizers
ISO 5007:1990	Agricultural wheeled tractors—Operator's seat—Laboratory measurement of transmitted vibration
ISO 5008:1979	Agricultural wheeled tractors and field machinery—Measurement of whole-body vibration of the operator
	Amendment 2:1984
ISO 5395:1990	Power lawn-mowers, lawn tractors, lawn and garden tractors, professional mowers, and lawn and garden tractors with mowing attachments—Definitions, safety requirements and test procedures
	Amendment 1:1992
ISO 5395:1994	Power lawn mowers, lawn tractors, lawn and garden tractors, professional mowers, and lawn and garden tractors with mowing attachments—Definitions, safety requirements and test procedures

(Cont.)

Table 1.77. (Continued)

ISO 5674:1982	Tractors and machinery for agriculture and forestry—Guards for power take-off drive shafts—Test methods
ISO 5674-1:1992	Tractors and machinery for agriculture and forestry—Guards for power take-off (PTO) drive shafts—Part 1: Strength test
ISO 5674-2:1992	Tractors and machinery for agriculture and forestry—Guards for power take-off (PTO) drive shafts—Part 2: Wear test
ISO 5697:1982	Agricultural and forestry vehicles—Determination of braking performance
ISO 5700:1989	Wheeled tractors for agriculture and forestry—Protective structures—Static test method and acceptance conditions
ISO 8082:1994	Self-propelled machinery for forestry—Roll-over protective structures—Laboratory tests and performance requirements
ISO 8083:1989	Machinery for forestry—Falling-object protective structures—Laboratory tests and performance requirements
ISO 8437:1989	Snowthrowers—Safety requirements and test procedures
ISO 8437:1989/	
AGRICULTURAL TRACTORS (SC 4)	
ISO 500:1991	Agricultural tractors—Rear-mounted power take-off—Types 1, 2, and 3
ISO 730-1:1994	Agricultural wheeled tractors—Rear-mounted three-point linkage—Part 1: Categories 1, 2, 3 and 4
ISO 730-2:1979	Agricultural wheeled tractors—Three-point linkage—Part 2: Category 1 N (Narrow hitch)
ISO 730-3:1982	Agricultural wheeled tractors—Three-point linkage—Part 3: Category 4
ISO 789-1:1990	Agricultural tractors—Test procedures—Part 1: Power tests for power take-off
ISO 789-2:1993	Agricultural tractors—Test procedures—Part 2: Rear three-point linkage lifting capacity
ISO 789-3:1993	Agricultural tractors—Test procedures—Part 3: Turning and clearance diameters
ISO 789-4:1986	Agricultural tractors—Test procedures—Part 4: Measurement of exhaust smoke
ISO 789-5:1983	Agricultural tractors—Test procedures—Part 5: Partial power PTO—Non-mechanically transmitted power
ISO 789-6:1982/ DAM 1	Agricultural tractors—Test procedures—Part 6: Centre of gravity—Amendment 1:1995
ISO 789-7:1991	Agricultural tractors—Test procedures—Part 7: Axle power determination
ISO 789-8:1991	Agricultural tractors—Test procedures—Part 8: Engine air cleaner
ISO 789-9:1990	Agricultural tractors—Test procedures—Part 9: Power tests for drawbar Amendment 1:1993
ISO 789-10:1996	Agricultural tractors—Test procedures—Part 10: Hydraulic power at tractor/implement interface
ISO 789-11:1996	Agricultural tractors—Test procedures—Part 11: Steering capability of wheeled tractors
ISO 2057:1981	Agricultural tractors—Remote control hydraulic cylinders for trailed implements
ISO 2332:1983	Agricultural tractors and machinery—Connections—Clearance zone for the three-point linkage of implements
ISO 3965:1990	Agricultural wheeled tractors—Maximum speeds—Method of determination
ISO 4004:1983	Agricultural tractors and machinery—Track widths
ISO 4253:1993	Agricultural tractors—Operator's seating accommodation—Dimensions
ISO 5673:1980	Agricultural tractors—Power take-off drive shafts for machines and implements
ISO 5675:1992	Agricultural tractors and machinery—General purpose quick-action hydraulic couplers

(Cont.)

Table 1.77. (Continued)

ISO 5676:1983	Tractors and machinery for agriculture and forestry—Hydraulic coupling—Braking circuit
ISO 5692:1979	Agricultural vehicles—Mechanical connections on towed vehicles—Hitch-rings—Specifications
ISO 5711:1995	Tractors and machinery for agriculture and forestry—Wheel-to-hub fixing dimensions
ISO 5721:1989	Tractors for agriculture—Operator's field of vision
ISO 5721:1994	Tractors for agriculture—Operator's field of vision
ISO 6097:1989	Tractors and self-propelled machines for agriculture—Performance of heating and ventilation systems in closed cabs—Test method
ISO 6489-3:1992	Agricultural vehicles—Mechanical connections on towing vehicles—Part 3: Tractor drawbar
ISO 7072:1993	Tractors and machinery for agriculture and forestry—Linch pins and spring pins—Dimensions and requirements
ISO 8759-1:1985	Agricultural wheeled tractors—Front—mounted linkage and power take-off—Part 1: Power take-off
ISO 8759-2:1985	Agricultural wheeled tractors—Front-mounted linkage and power take-off—Part 2: Front linkage
ISO 8935:1990	Tractors for agriculture and forestry—Mountings and apertures for external equipment controls
ISO 10448:1994	Agricultural tractors—Hydraulic pressure for implements
ISO 11001-1:1993	Agricultural wheeled tractors and implements—Three-point hitch couplers—Part 1:U-frame coupler
ISO 11001-2:1993	Agricultural wheeled tractors and implements—Three-point hitch couplers—Part 2:A-frame coupler
ISO 11001-3:1993	Agricultural wheeled tractors and implements—Three-point hitch couplers—Part 3:Link coupler
ISO 11001-4:1994	Agricultural wheeled tractors and implements—Three-point hitch couplers—Part 4: Bar coupler
ISO 11374:1993	Agricultural tractors and implements—Four-point rigid hitch-Specifications
ISO/TR 12369:1994*	Agricultural tractors and machinery—Electrical power transmission connectors
TILLAGE EQUIPMENT (SC 5)	
ISO/TR 4122:1977*	Equipment for working the soil—Dimensions of flat disks—Type A
ISO 4197:1989	Equipment for working the soil—Hoe blades—Fixing dimensions
ISO 5713:1990	Equipment for working the soil—Fixing bolts for soil working elements
ISO 6880:1983	Machinery for agriculture—Trailed units of shallow tillage equipment—Main dimensions and attachment points
ISO 8909-1:1994	Forage harvesters—Part 1: Vocabulary
ISO 8909-2:1994	Forage harvesters—Part 2: Specification of characteristics and performance
ISO 8909-3:1994	Forage harvesters—Part 3: Test methods
ISO 8910:1993	Machinery and equipment for working the soil—Mouldboard plough working elements—Vocabulary
ISO 8912:1986	Equipment for working the soil—Roller sections—Coupling and section width
ISO 8945:1989	Equipment for working the soil—Rotary cultivator blades—Fixing dimensions
ISO 8947:1993	Agricultural machinery—equipment for working the soil—S-tines: test method
CROP PROTECTION EQUIPMENT (SC 6)	
ISO 4102:1984	Equipment for crop protection—Sprayers—Connection threading
ISO 5681:1992	Equipment for crop protection—Vocabulary Bilingual edition

(Cont.)

Table 1.77. (Continued)

ISO 5682-1:1981	Equipment for crop protection—Spraying equipment—Part 1: Test methods of sprayer nozzles
ISO 5682-2:1997	Equipment for crop protection—Spraying equipment—Part 2: Test methods for hydraulic sprayers
ISO 6686:1995	Equipment for crop protection—Anti-drip devices—Determination of performance
ISO 8169:1984	Equipment for crop protection—Sprayers—Connecting dimensions for nozzles and manometers
ISO 9357:1990	Equipment for crop protection—Agricultural sprayers—Tank nominal volume and filling hole diameter
ISO 10625:1996	Equipment for crop protection—Sprayer nozzles—Colour coding for identification
ISO 10626:1991	Equipment for crop protection—Sprayers—Connecting dimensions for nozzles with bayonet fixing
ISO 13440:1996	Equipment for crop protection—Agricultural sprayers—Determination of the volume of total residual
ISO 13441-1:1997	Air-assisted agricultural sprayers—Data sheets—Part 1: Typical layout
ISO 13341-2:1997	Air-assisted agricultural sprayers—Data sheets—Part 2: Technical specifications related to components
HARVESTING AND DRYING EQUIPMENT (SC 7)	
ISO 4254-7:1995	Tractors and machinery for agriculture and forestry—Technical means for ensuring safety—Part 7: Combine harvesters, forage and cotton harvesters
ISO 5687:1981	Equipment for harvesting—Combine harvester—Determination and designation of grain tank capacity and unloading device performance
ISO 5702:1983	Equipment for harvesting—Combine harvester component parts—Equivalent terms Trilingual edition
ISO 5715:1983	Equipment for harvesting—Dimensional compatibility of forage harvesting machinery
ISO 5718-1:1989	Harvesting equipment—Flat blades for rotary mowers—Specifications—Part 1: Type A flat blades
ISO 5718-2:1991	Harvesting equipment—Flat blades for rotary mowers—Part 2: Specifications for type B flat blades
ISO 6689-1:1997	Equipment for harvesting—Combines and functional components—Part 1: Vocabulary
ISO 6689-2:1997	Equipment for harvesting—Combines and functional components—Part 2: Assessment of characteristics and performance defined in vocabulary
ISO 8210:1989	Equipment for harvesting—Combine harvesters—Test procedure
ISO/TR 10391:1992*	Forage harvesters—Method of determining by screening and expressing particle size of chopped forage materials
VITICULTURAL EQUIPMENT	
ISO 3835-1:1976	Equipment for vine cultivation and wine making—Vocabulary—Part 1 Bilingual edition
ISO 3835-2:1977	Equipment for vine cultivation and wine making—Vocabulary—Part 2 Bilingual edition
ISO 3835-3:1980	Equipment for vine cultivation and wine making—Vocabulary—Part 3 Bilingual edition

(Cont.)

Table 1.77. (Continued)

ISO 3835-4:1981	Equipment for vine cultivation and wine making—Vocabulary—Part 4 Bilingual edition
ISO 5704:1980	Equipment for vine cultivation and wine making—Grape-harvesting machinery—Test methods
ISO 7224:1983	Equipment for vine cultivation and wine making—Mash pumps—Methods of test
PLANTING EQUIPMENT (SC 9)	
ISO 4002-1:1979	Equipment for sowing and planting—Part 1: Concave disks type D1—Dimensions
ISO 4002-2:1977	Equipment for sowing and planting—Disks—Part 2: Flat disks type D2 with single bevel—Dimensions
ISO 5690-1:1985	Equipment for distributing fertilizers—Test methods—Part 1: Full width fertilizer distributors
ISO 5690-2:1984	Equipment for distributing fertilizers—Test methods—Part 2: Fertilizer distributors in lines
ISO 5691:1981	Equipment for planting—Potato planters—Method of testing
ISO 6720:1989	Agricultural machinery—Equipment for sowing, planting, distributing fertilizers and spraying—Recommended working widths
ISO 7256-1:1984	Sowing equipment—Test methods—Part 1: Single seed drills (precision drills)
ISO 7256-2:1984	Sowing equipment—Test methods—Part 2: Seed drills for sowing in lines
ISO 8524:1986	Equipment for distributing granulated pesticides or herbicides—Test method
ISO 10627-1:1992	Agricultural sprayers—Data sheet—Part 1: Typical layout
ISO 10627-2:1992	Hydraulic agricultural sprayers—Data sheets—Part 2: Technical specifications related to components
TRANSPORTATION AND HANDLING EQUIPMENT	
ISO 5669:1982	Agricultural trailers and trailed equipment—Braking cylinders—Specifications
ISO 5670:1984	Agricultural trailers—Single-acting telescopic tipping cylinders 25 MPa (250 bar) series—Types 1, 2, and 3—Interchangeability dimensions
ISO 5696:1984	Trailed agricultural vehicles—Brakes and braking devices—Laboratory test method
FARMSTEAD EQUIPMENT	
ISO 3918:1996	Milking machine installations—Vocabulary Bilingual edition
ISO 5707:1996	Milking machine installations—Construction and performance
ISO 5708:1983	Refrigerated bulk milk tanks
ISO 5709:1981	Equipment for internal farm work and husbandry—Metal grids for cattle stalls
ISO 5710:1980	Equipment for internal farm work and husbandry—Continuous manure scraper conveyors for stalls
ISO 6690:1996	Milking machine installations—Mechanical tests
WHEELS	
ISO 8016: 1985	Machinery for agriculture—Wheels with integral hub
POWERED LAWN AND GARDEN EQUIPMENT (SC 13)	
ISO 9190:1990	Lawn and garden ride-on (riding) tractors—Drawbar
ISO 9191:1991	Lawn and garden ride-on (riding) tractors—Three-point hitch
ISO 9192:1991	Lawn and garden ride-on (riding) tractors—One-point tubular sleeve hitch
ISO 9193:1990	Lawn and garden ride-on (riding) tractors—Power take-off
ISO 10517:1993	Portable powered hedge trimmers—Definitions, mechanical safety requirements and tests

(Cont.)

Table 1.77. (Continued)

ISO 10518:1991	Powered walk-behind and hand-held lawn trimmers and lawn edge trimmers—Mechanical safety requirements and test methods Amendment 1:1992
ISO 10884:1995	Manually portable brush-cutters and grass-trimmers with internal combustion engine—Determination of sound power levels—Engineering method
ISO 11449:1994	Walk-behind powered rotary tillers—Definitions, safety requirements and test procedures
CONTROLS, SYMBOLS, MANUALS (SC 14)	
ISO 3600:1996	Tractors, machinery for agriculture and forestry, powered lawn and gerden equipment—Operator manuals and technical publications—Content and Presentation
ISO 3767-2:1991	Tractors, machinery for agriculture and forestry, powered lawn and garden equipment—Symbols for operator controls and other displays—Part 2: Symbols for agricultural tractors and machinery Amendment 1
ISO 3767-3:1995	Tractors, machinery for agriculture and forestry, powered lawn and garden equipment—Symbols for operator controls and other displays—Part 3: Symbols for powered lawn and garden equipment
ISO 3767-5:1992	Tractors, machinery for agriculture and forestry, powered lawn and garden equipment—Symbols for operator controls and other displays—Part 5: Symbols for manual portable forestry machinery
ISO/TR 3778:1987*	Agricultural tractors—Maximum actuating forces required to operate controls
ISO 3789-1:1982	Tractors, machinery for agriculture and forestry, powered lawn and garden equipment—Location and method of operation of operator controls—Part 1: Common controls
ISO 3789-2:1982	Tractors, machinery for agriculture and forestry, powered lawn and garden equipment—Location and method of operation of operator controls—Part 2: Controls for agricultural tractors and machinery
ISO 3789-3:1989	Tractors, machinery for agriculture and forestry, powered lawn and garden equipment—Location and method of operation of operator controls—Part 3: Controls for powered lawn and garden equipment
ISO 3789-4:1989	Tractors, machinery for agriculture and forestry, powered lawn and garden equipment—Location and method of operation of operator controls—Part 4: Controls for forestry log loaders
ISO 3971:1977	Rice milling—Symbols and equivalent terms Bilingual edition
ISO 11471:1995	Agricultural tractors and machinery—Coding of remote hydraulic power services and controls
FORESTRY MACHINERY (SC 15)	
ISO 6687:1994	Machinery for forestry—Winches—Performance requirements
ISO 6688:1982	Machinery for forestry—Disc trenches—Disc-to-hub flange fixing dimensions
ISO 6814:1983	Machinery for forestry—Mobile and self-propelled machinery—Identification vocabulary
ISO 6815:1983	Machinery for forestry—Hitches—Dimensions
ISO 6816:1984	Machinery for forestry—Winches—Classification and nomenclature Technical corrigendum 1:1990
ISO 7918:1995	Forestry machinery—Portable brush-cutters and grass-trimmers—Cutting attachment guard dimensions

(Cont.)

Table 1.77. (Continued)

ISO 11169:1993	Machinery for forestry—Wheeled special machines—Vocabulary, performance test methods and criteria for brake systems
ISO 11512:1995	Machinery for forestry—Tracked special machines—Performance criteria for brake systems
ISO 11684:1994	Tractors, machinery for agriculture and forestry, powered lawn and garden equipment—Safety signs and hazard pictorials—General principles
PORTABLE FORESTRY EQUIPMENT (SC 17)	
ISO 6531:1982	Machinery for forestry—Portable chain saws—Vocabulary Trilingual edition
ISO 6532:1982	Machinery for forestry—Portable chain saws—Technical data
ISO 6533:1993	Forestry machinery—Portable chain saw front hand-guard—Dimensions
ISO 6534:1992	Portable chain-saw—Hand-guards—Mechanical strength
ISO 6535:1991	Portable chain saws—Chain brake performance
ISO 6687:1994	Machinery for forestry—Winches—Performance requirements
ISO 7112:1982	Machinery for forestry—Portable brush saws—Vocabulary Trilingual edition
ISO 7113:1991	Forestry machinery—Portable brush saws—Saw blades
ISO 7293:1983	Forestry machinery—Portable chain saws—Engine performance and fuel consumption
ISO 7505:1986	Forestry machinery—Chain saws—Measurement of hand-transmitted vibration
ISO 7914:1994	Forestry machinery—Portable chain saws—Minimum handle clearance and sizes
ISO 7915:1991	Forestry machinery—Portable chain saws—Determination of handle strength
ISO 8334:1985	Forestry machinery—Portable chain saws—Determination of balance
ISO 8380:1985	Forestry machinery—Portable brush saws—Circular saw-blade guard—Strength
ISO 8893:1989	Forestry machinery—Portable brush saws—Engine performance and fuel consumption
ISO/TR 9412:1991*	Portable chain saws—Automatic chain brake and cutting equipment—Operator's safety test
ISO 9467:1993	Forestry machinery—Portable chain-saws and brush-cutters—Exhaust system-caused fire risk
ISO 9518:1992	Forestry machinery—Portable chain-saws—Kickback test
ISO 10726:1992	Portable chain saws—Chain catcher—Dimensions and mechanical strength
ISO 13772:1997	Forestry machinery—Portable chain-saws—Non-manually actuated chain brake performance
IRRIGATION EQUIPMENT (SC 18)	
ISO 7714:1985	Irrigation equipment—Volumetric valves—General requirements and test methods
ISO 7749-1:1986	Irrigation equipment—Rotating sprinklers—Part 1: Design and operational requirements
ISO 7749-2:1990	Irrigation equipment—Rotating sprinklers—Part 2: Uniformity of distribution and test methods
ISO 8026:1985	Irrigation equipment Irrigation sprayers—General requirements and test methods
ISO/TR 8059:1986*	Irrigation equipment—Automatic irrigation systems—Hydraulic control
ISO 8224-1:1985	Traveller irrigation machines—Part 1: Laboratory and field test methods
ISO 8224-2:1991	Traveller irrigation machines—Part 2: Softwall hose and coupling—Test methods
ISO 9260:1991	Agricultural irrigation equipment—Emitters—Specification and test methods
ISO 9261:1991	Agricultural irrigation equipment—Emitting-pipe systems—Specification and test methods

(Cont.)

Table 1.77. (Continued)

ISO 9635:1990	Irrigation equipment—Hydraulically operated irrigation valves
ISO 9644:1993	Agricultural irrigation equipment—Pressure losses in irrigation valves—Test method
ISO 9911:1993	Agricultural irrigation equipment—Manually operated small plastic valves
ISO 9912-1:1993	Agricultural irrigation equipment—Filters—Part 1: Strainer-type filters
ISO 9912-2:1992	Agricultural irrigation equipment—Filters—Part 2: Strainer-type filters
ISO 9912-3:1992	Agricultural irrigation equipment—Filters—Part 3: Automatic self-cleaning strainer-type filters
ISO 9952:1993	Agricultural irrigation equipment—Check valves
ISO 11545:1995	Agricultural irrigation equipment—Centre-pivot and moving lateral irrigation machines with sprayer or sprinkler nozzles—Determination of uniformity of water distribution
ISO 11678:1996	Agricultural irrigation equipment—Aluminium irrigation tubes
AGRICULTURAL ELECTRONICS (SC 19)	
ISO 11784:1996	Radio-frequency identification of animals—Code structure
ISO 11785:1996	Radio-frequency identification of animals—Technical concept
ISO 11786:1995	Agricultural tractors and machinery—Tractor-mounted sensor interface—Specifications
ISO 11787:1995	Machinery for agriculture and forestry—Data interchange between management computer and process computers—Data interchange syntax
ISO 11788-1:1997	Electronic data interchange between information systems in agriculture—Agriculture data element dictionary—Part 1: General description
ISO/TC 31, TYRES, RIMS, AND VALVES	
AGRICULTURAL TYRES AND RIMS (SC 5)	
ISO 3877-1:1997	Tyres, valves and tubes—List of equivalent terms—Part 1: Tyres
ISO 3877-2:1997	Tyres, valves and tubes—List of equivalent terms—Part 2: Tyre valves
ISO 3877-3:1978	Tyres, valves and tubes—List of equivalent terms—Part 3: Tubes
ISO 3877-4:1984	Tyres, valves and tubes—List of equivalent terms—Part 4: Solid tyres
ISO 4250-1:1996	Earth-mover tyres and rims—Part 1: Tyre designation and dimensions
ISO 4250-2:1995	Earth-mover tyres and rims—Part 2: Loads and inflation pressures
ISO 4251-1:1992	Tyres (ply rating marked series) and rims for agricultural tractors and machines—Part 1: Tyre designation and dimensions
ISO 4251-5:1992	Tyres (ply rating marked series) and rims for agricultural tractors and machines—Part 5: Log skidder tyres
ISO 7867-1:1996	Tyres and rims (metric series) for agricultural tractors and machines—Part 1: Tyre designation, dimensions, and marking
ISO 7867-2:1996	Tyres and rims (metric series) for agricultural tractors and machines—Part 2: Service description and load ratings
ISO/TC 34, AGRICULTURAL FOOD PRODUCTS	
ISO 660:1996	Animal and vegetable fats and oils—Determination of acid value and acidity
ISO 711:1985	Cereals and cereal products—Determination of moisture content (Basic reference method)
ISO 712:1985	Cereals and cereal products—Determination of moisture content (Routine reference method)
ISO 1212:1995	Apples—Cold storage
ISO 1442:1973	Meat and meat products—Determination of moisture content
ISO 1444:1996	Meat and meat products—Determination of free fat content
ISO 1546:1981	Procedure for milk recording for cows

(Cont.)

Table 1.77. (Continued)

ISO 1673:1991	Onions—Guide to storage
ISO 1991-1:1981	Vegetables—Nomenclature—Part 1: First list
ISO 1991-2:1995	Vegetables—Nomenclature—Part 2: Second list
ISO 2167:1991	Round—headed cabbage—Guide to cold storage and refrigerated transport
ISO 2259:1972	Pots for propagation and transplantation made of peat and other plant material—Sampling
ISO 3513:1995	Chillies—Determination of Scoville index
ISO 3973:1996	Living animals for slaughter—Vocabulary—Bovines
ISO 4174:1980	Cereals and pulses—Measurement of unit pressure losses due to single-dimension air flow through a batch of grain
ISO 5223:1995	Test sieves for cereals
ISO 5524:1991	Tomatoes—Guide to cold storage and refrigerated transport
ISO 5527:1995	Cereals—Vocabulary
ISO 5559:1995	Dehydrated onion (Allium cepa Linnaeus)—Specification
ISO 5983:1997	Animal feeding stuffs—Determination of nitrogen content and calculation of crude protein content—Kjeldahl method
ISO 6320:1995	Animal and vegetable fats and oils—Determination of refractive index
ISO 6322-1:1996	Storage of cereals and pulses—Part 1: General recommendations for the keeping of cereals
ISO 6540:1980	Maize—Determination of moisture content (on milled grains and on whole grains)
ISO 6663:1995	Garlic—Cold storage
ISO 6883:1995	Animal and vegetable fats and oils—Determination of conventional mass per volume ("litre weight in air")
ISO 6886:1996	Animal and vegetable fats and oils—Determination of oxidation stability (Accelerated oxidation test)
ISO 7218:1996	Microbiology of food and animal feeding stuffs—General rules for microbiological examinations
ISO 7560:1995	Cucumbers—Storage and refrigerated transport
ISO 7702:1995	Dried pears—Specification and test methods
ISO 7703:1995	Dried peaches—Specification and test methods
ISO 7937:1997	Microbiology of food and animal feeding stuffs—Horizontal method for enumeration of *Clostridium perfringens*—Colony-count technique
ISO 7971:1986	Cereals—Determination of bulk density, called mass per hectolitre (Reference method)
ISO 8086:1986	Dairy plant—Hygiene conditions—General guidance on inspection and sampling procedures
ISO 9719:1995	Root vegetables—Cold storage and refrigerated transport
ISO 10272:1995	Microbiology of food and animal feeding stuffs—Horizontal method for detection of thermotolerant Campylobacter
ISO 10519:1997	Rapeseed—Determination of chlorophyll content—Spectrometric method
ISO 10621:1997	Dehydrated green pepper (*Piper nigrum L.*)—Specification
ISO 11290-1:1996	Microbiology of food and animal feeding stuffs—Horizontal method for the detection and enumeration of *Listeria monocytogenes*—Part 1: Detection method
ISO 704:1987	Principles and methods of terminology
ISO/TC 41, PULLEYS AND BELTS (INCLUDING VEEBELTS)	
ISO 433:1991	Conveyor belts—Marking
ISO 3410:1989	Agricultural machinery—Endless variable-speed V-belts and groove sections of corresponding pulleys
ISO 5289:1992	Agricultural machinery—Endless hexagonal belts and groove sections of corresponding pulleys

(Cont.)

Table 1.77. (Continued)

ISO/TC 43, ACOUSTICS	
ISO 5131:1996	Acoustics—Tractors and machinery for agriculture and forestry—Measurement of noise at the operator's position—Survey method Amendment 1:1992
ISO 7182:1984	Acoustics—Measurement at the operator's position of airborne noise emitted by chain saws
ISO 7216:1992	Acoustics—Ag and forestry wheeled tractors and self-propelled machines—Measurement of noise emitted when in motion
ISO 7917:1987	Acoustics—Measurements at the operator's position of airborne noise emitted by brush saws
ISO/TC 45, RUBBER AND RUBBER PRODUCTS	
ISO 1401:1987	Rubber hoses for agricultural spraying
ISO/TC 98, BASES FOR DESIGN OF STRUCTURES	
ISO 4355:1981	Bases for design of structures—Determination of snow loads on roofs
ISO/TC 101, CONTINUOUS MECHANICAL HANDLING EQUIPMENT	
ISO 1050:1975	Continuous mechanical handling equipment for loose bulk materials—Screw conveyors
ISO/TR 9172:1987*	Continuous mechanical handling equipment—Safety code for crew conveyors—Examples of guards for trapping and shearing points
ISO/TC 113, HYDROMETRIC DETERMINATIONS	
ISO 1100-1:1996	Measurement of liquid flow in open changes—Part 1: Establishment and oepration of a gauging station
ISO 4373:1995	Measurement of liquid flow in open channels—Water-level measuring devices
ISO/TC 138, PLASTICS PIPES, FITTINGS AND VALVES FOR THE TRANSPORT OF FLUIDS	
ISO 8796:1989	Polyethylene (PE) 25 pipes for irrigation laterals—Susceptibility to environmental stress-cracking induced by insert-type fittings—Test method and specification
ISO 13480:1997	Polyethylene pipes—Resistance to slow crack growth—Cone test method
ISO/TC 145, GRAPHICAL SYMBOLS	
ISO 3864:1984	Safety colours and safety signs
ISO 7000:1989	Graphical symbols for use on equipment Index and synopsis Bilingual edition
ISO/TC 147, WATER QUALITY	
ISO 6107-1:1996	Water quality—Vocabulary—Part 1
ISO 6107-5:1996	Water quality—Vocabulary—Part 5
ISO 6107-6:1996	Water quality—Vocabulary—Part 6
ISO 6468:1996	Water quality—Determination of certain organochlorine insecticides, polychlorinated biphenyls and chlorobenzenes—Gas chromatographic method after liquid-liquid extraction
ISO 9963-2:1994	Water quality—Determination of alkalinity—Part 2: Determination of carbonate alkalinity

(Cont.)

Table 1.77. (Continued)

ISO 10229:1994	Water quality-Determination of the prolonged toxicity of substances to freshwater fish—Method for evaluating the effects of substances on the gowth rate of rainbow trout [Oncorhynchus mykiss Walbaum (Teleostei, Salmonidae)]
ISO 10253:1995	Water quality—Marine algal growth inhibition test with Skeletonema costatum and Phaeodactylum tricornutum
ISO 10304-2:1995	Water quality—Determination of dissolved anions by liquid chromatography of ions—Part 2: Determination of bromide, chloride, nitrate, nitrite, orthophosphate and sulfate in waste water
ISO 10304-3:1997	Water quality—Determination of dissolved anions by liquid chromatography of ions—Part 3: Determination of chromate, iodide, sulfite, thiocyanate and thiosulfate
ISO 11905-1:1997	Water quality—Determination of nitrogen—Part 1: Method using oxidative digestion with peroxodisulfate
ISO 11969:1996	Water quality—Determination of arsenic—Atomic absorption spectrometric method (hydride technique)
ISO 13395:1996	Water quality—Determination of nitrite nitrogen and nitrate nitrogen and the sum of both by flow analysis
ISO/TC 153, VALVES	
ISO 5208:1993	Industrial valves—Pressure testing of valves
ISO/TC 190, SOIL QUALITY	
ISO 10693:1995	Soil quality—Determination of carbonate content—Volumetric method
ISO 10694:1995	Soil quality—Determination of organic and total carbon after dry combustion (elementary analysis)
ISO 11074-1:1996	Soil quality—Vocabulary—Part 1: Terms and definitions relating to the protection and pollution of the soil
ISO 11261:1995	Soil quality—Determination of total nitrogen—Modified Kjeldahl method
ISO 11263:1994	Soil quality—Determination of phosphorus—Spectrometric determination of phosphorus soluble in sodium hydrogen carbonate solution
ISO 11269-2:1995	Soil quality—Determination of the effects of pollutants on soil flora—Part 2: Effects of chemicals on the emergency and growth of higher plants
ISO 11276:1995	Soil quality—Determination of pore water pressure—Tensiometer method
ISO 11466:1995	Soil quality—Extraction of trace elements soluble in *aqua regia*
ISO 13536:1995	Soil quality—Determination of the potential cation exchange capacity and exchangeable cations using barium chloride solution buffered at pH = 8, 1
ISO 14239:1997	Soil quality—Laboratory incubation systems for measuring the mineralization of organic chemicals in soil under aerobic conditions

*ISO/TR, technical reports, provide either interim progress reports or factual information/data different from that normally incorporated in an International Standard.

find it difficult to adopt ISO standards to replace existing national standards. However, the goal of one worldwide standard transcends these difficulties.

What organizations develop standards? Each country that is a voting member of ISO has one organization or agency identified as the ISO Member Body. Member Bodies with voting status on ISO TC 23 are listed in Table 1.78.

The principles of regional and national standard development around the world tend to be similar, but procedures vary from country to country. Standard development begins with identification of a need that can be met by conformance to a standard. Usually the

Table 1.78. ISO member bodies on ISO TC23

Australia:	Standards Australia (SAA)
Belgium:	Institut belge de normalisation (IBN)
Belarus:	Committee for Standardization, Metrology, and Certification (BELST)
Brazil:	Associacao Brasileira de Normas Tecnicas (ABNT)
Bulgaria:	Committee for Standardization and Metrology at the Council of Ministers (BDS)
Canada:	Standards Council of Canada (SCC)
China:	China State Bureau of Technical Supervision (CSBTS)
Czech Republic:	Czech Office for Standards, Metrology and Testing (COSMT)
Finland:	Finnish Standards Association (SFS)
France:	Association Francaise de Normalisation (AFNOR)
Germany:	Deutsches Institut fur Normung (DIN)
Hungary:	Magyar Szabvanyugy Testulet (MSZT)
Iran:	Institute of Standards and Industrial Research of Iran (ISIRI)
Israel:	Standards Institution of Israel (SII)
Italy:	Ente Nazionale Italiano di Unificazione (UNI)
Korea, Democratic Peoples Republic of:	Committee for Standardization of the Democratic People's Republic of Korea (CSK)
Mexico:	Direccion General de Normas (DGN)
Netherlands:	Nederlands Normalisatie-instituut (NNI)
Philippines:	Bureau of Product Standards (BPS)
Poland:	Polish Committee for Standardization (PKN)
Portugal:	Instituto Portugues da Qualidade (IPQ)
Romania:	Institutal Roman de Standardizare (IRS)
Russia:	Committee of the Russian Federation for Standardization Metrology and Certification (GOST R)
South Africa:	South African Bureau of Standards (SABS)
Sweden:	Standardiseringen I Sverige (SIS)
Turkey:	Turk Standardlari Enstitusu (TSE)
United Kingdom:	British Standards Institution (BSI)
United States:	American National Standards Institute (ANSI)

need has economic implications, that is, a standard will reduce manufacturing costs, a standard will reduce the farmer's cost of equipment, or safety will be enhanced, thereby reducing potential losses from injury. Once the need is identified, the project is assigned to a subcommittee or working group, whose members have the appropriate technical expertise to begin standard development. Literature searches, research, testing, and data collection may be necessary activities before actual writing can begin. Frequently several drafts are developed before consensus is achieved and a standard is formally adopted.

1.10.1. Standardization on Workplace Health and Safety in the European Union

Prior to the establishment of the European Committee for Standardization (CEN) in 1961, as a nonprofit technical and scientific association, manufacturers followed the work program of ISO and transferred the standards into their national standardization system (AFNOR in France, BSI in the United Kingdom, DIN in Germany, etc.).

In the mid-1970s when ISO TC23 become a technical committee separate from ISO TC22, Road Vehicles, a subcommittee was already devoted to "safety and comfort of the operator."

To better overcome existing trade barriers caused by different national legislative systems, regulations, and standards in Western Europe, the European Council established by mean of Directives the minimum requirements to harmonize and improve safety and health of workers (Article 118A).

Already the "old approach," built on article 100A of the Single Act of 1987, required that Directives achieve a high degree of protection. However, standards built upon this article had a limited field of application, were quickly obsolete, and were optional in their application.

The "new approach" has larger scopes. For example, the machinery directive 89/392/EEC of 14 June 1989, amended by council directives 91/368, 93/44, and 93/68, covers all machines (defined broadly as "mechanical products with at least one moving part subjected to mechanical risks"), interchangeable equipment modifying the function of a machine, and safety components. It contains the Essential Safety and Protection Requirements (ESR) the products must fulfill.

It is important here to note that agricultural and forestry tractors are outside the scope of this Directive. They are under Directives 74/150 and 88/297 (old approach).

The EU Member States had to adopt and publish the necessary regulations in order to comply with the Directive before 1 January 1992, and to apply them before 1 January 1993. Machines conforming to the previous national regulations were allowed to be placed on the market until 31 December 1994.

Conformance is established primarily through self-certification by the manufacturer or the authorized representative established in the community. They bear full responsibility for conformance, except for certain high-risk machines listed in annex IV, which requires the involvement of a notified body that performs EC-type examination on a representative machine. The notified bodies must be situated in the EC territory. All machines conforming to the Directive must bear the EC standardized label.

How can a manufacturer apply the ESR of the Directives? In fact, the task of defining the technical specifications for the design of the machinery was left to the standard-making organization, the CEN. Ad-hoc working groups that include experts from industry, public institutions, and user organizations develop projects. The resulting draft is reviewed for ESR conformance by a consultant, after which it is forwarded to be voted on by the national delegations for standardization, after being checked for ESR conformity by a consultant appointed by the Commission.

Three types of harmonized technical standards exist:

- Type A, standards on basic concepts applying to all machines
- Type B, common specific requirements for a group of machines
- Type C, specific requirements for a particular machine

For agricultural machinery, the technical committee of CEN is TC144. It has eight working groups: WG1 deals with general safety requirements, WG2 with tractors and self-propelled machines, WG3 with mobile machines and trailers, WG4 with portable and pedestrian-controlled machines, WG5 with stationary equipment, WG6 with portable forestry equipment, WG7 with powered lawn and garden equipment, and WG8 with forestry machinery. These agricultural machinery safety standards have been established since 1989:

ANNEX IV: (certification by notified bodies)
EN 608: Portable chain saws
EN 1152: Guards for power take-off drive shafts
SELF CERTIFICATION
EN 632: Combine harvesters and forage harvesters
EN 690: Manure spreaders
EN 703: Silage cutters
EN 706: Vine shoot tipping machines
EN 708: Soil working machines with powered tools
EN 774: Portable powered hedge trimmers
EN 786: Electrically powered walk-behind and hand-held lawn trimmers and lawn hedge trimmers

A type B standard on common safety requirements for all agricultural machines is under preparation (pr EN 1153).

The small number of safety standards is due in particular to the public inquiries (primary Questionnaire, CEN inquiry) and the voting procedures (formal vote) adopted, both of which are essential to assure consensus. During these procedures, member states have to control the level of safety prescribed by these standards, and manufacturers express what is economically and technically feasible (known also as the *actual state of the technique*).

These European standards have to be closely reviewed because they replace national standards, and because they cannot be modified. The whole study is undertaken in the three official languages of CEN: English, German, and French.

If a standard details the status of a European Harmonized standard (references published in the *Official Journal of the European Communities*), it offers the user of the document a presumption of conformance to the ESR of Directive 2.

If a Member State of the Commission considers that the harmonized standards do not entirely satisfy the essential requirements, the matter is brought before the Committee of Directive 83/189/EEC.

1.10.2. Relationships Between European and International Standardization

In order to promote worldwide harmonization between European and international standardization wherever feasible, and to avoid duplication of work, CEN established an agreement with ISO (the Vienna agreement) and CENELEC with IEC (the Lugano agreement). All IEC projects are automatically considered for adoption as CENELEC standards; however, it is not the same for CEN, which *can* adopt ISO projects and can pass them to ISO for inclusion on the program of the latter body, with such projects being then accepted as CEN standards.

1.10.3. E.U. Standards for Environmental Protection

The work of TC144 has been recently extended to fertilizer- and pesticide-spreading equipment to better control effects on the environment.

It also supports Directive 91/676/EEC, which aims to protect against nitrate pollution of water. WG3 is concerned with these machines:

- Sprayers for tractors
- Portable sprayers
- Solid fertilizer spreaders
- Slurry and manure spreaders

1.10.4. Other Directives for Machinery

Some machines that pose more than solely mechanical risks may be covered by other directives and standards; electrical risks by the Low Voltage Directive electromagnetic risk by the Electromagnetic Compatibility Directive, and so forth.

The EC marking normally reflects conformity to all these directives.

1.10.5. National Standards Development

In some countries there are multiple organizations responsible for standards development within areas or sectors of technology. For example, in the United States there are more than 180 standards-developing organizations coordinated and accredited by the American National Standards Institute, which itself does not develop standards. Most of these organizations are engineering societies or trade associations. The American Society of Agricultural Engineers is the principal standards-writing organization in the United States for agricultural tractors and machinery, irrigation equipment and systems, agricultural structures, drainage, electric power applications in agriculture, and agricultural processing. A subject index of ASAE Standards and Engineering Practices is available on the Internet at *http://asae.org/*.

1.10.6. Standards Searches

The American National Standards Institute supports a free search capability for global standards. Bibliographic information is provided on more than 225,000 approved standards. The internet address is *http://www.nssn.org/*.

References

1. Van Gheluwe. 1996. The objectives of the new approach directives: the freedom of movement of machinery and the high level of safety. The European Seminar on the market control of the machinery in the EEA pp. 8–17. Ministry of Labour. Finland.
2. Rowe, J. and Senneff., M. 1996. Technical requirements of the European Union's Machinery Directive 89/392/EEC. ASAE Technical Paper 96-1102.
3. American Society of Agricultural Engineers: 1998. ASAE Standards. St. Joseph, MI: Author.
4. International Organization for Standardization. 1996. ISO Memento, Geneve: Switzerland.

2 Mechanizations Systems

2.1. Systems Engineering, Operations Research, and Management Science

C. Parnell, B. Shaw, and B. Fritz

The Engineering Council for Professional Development defined engineering (Whinnery, 1965) as the profession in which the knowledge of science (biology, chemistry, physics, and mathematics) gained by study, experience, and practice is applied with judgment to develop ways to utilize, economically, the materials and forces of nature. Systems engineering incorporates a holistic view of the problem with mathematical tools to assist the engineer with decision-support information. These same tools are utilized in operations research and management science by consultants and managers. The Operations Research Society of America defined operations research as scientifically deciding how to best design and operate man–machine systems (Ravindran *et al.*, 1987), usually under conditions requiring the allocation of scarce resources. The terms *systems engineering, operations research*, and *management science* refer to the same process, which is the application of systems analysis to gain a better understanding of the problem and make decisions that increase profit. Agricultural managers and engineers are required to make many decisions that impact the economic viability of agribusinesses during the course of their careers. It is assumed that the quality of management and engineering decisions will be improved by the appropriate and proper application and understanding of results of systems analysis. The goal of the application of mathematical tools is to improve the quality of management and engineering decisions. The knowledge of systems tools is not a replacement for experience and judgment but can be used to augment the basis for the decision making process.

The power and low cost of personal computers and the availability of relatively inexpensive software has made it easier for many engineers and managers to utilize mathematical decision-support systems that would have been inconceivable just a few years ago. Care must be exercised in the selection of the mathematical method utilized. The engineer/manager should strive to have the problem dictate the systems tool rather than forcing the problem to fit the systems-analysis method.

The objective of this section is to introduce a limited number of mathematical tools used in systems engineering, operations research, and management science and provide example applications of their use. Although it is impossible to cover all of the techniques that can be used, an attempt is made to present the most common methods.

2.1.1. Optimization

Linear Programming

Linear programming (LP) is one of the most popular systems-analysis tools. It is relatively easy to learn and can be utilized on a wide variety of problems. An example of a typical goal of LP is to minimize cost or maximize profit. The most difficult task associated with LP is appropriate problem formulation. The general formulation of a linear program is as follows:

$$X_i = \text{decision variables} \tag{2.1}$$

$$\text{Max (or Min)} Z = \sum C_i^* X_i \tag{2.2}$$

subject to:

$$AX(\leq, =, \geq)B \tag{2.3}$$

$$X_i \geq 0 \tag{2.4}$$

where X_i = decision variables (these are the variables that are to be determined by the LP analysis; care must be taken to insure that the units used in defining the decision variables are appropriate), C_i = cost or profit coefficients (the cost coefficients have units such that for each incremental change of a decision variable, there is a corresponding increase or decrease in the net profit for a "maximize Z" or net cost for a "minimize Z" objective function), AX ($\leq, =, \geq$) B (these are the constraint equations associated with the problem formulation in matrix form. There are no limits on the number of constraint equations. The values associated with the matrix B are generally referred to as resources. The constraint equations can be in the form of less than or equal to ($\leq$), equal to (=) or greater than or equal to ($\geq$), and $X_i \geq 0$ (these equations are referred to as the nonnegative constraints; all of the variables are limited to values greater than 0).

Example

A feed-mill manager desires to minimize the cost of the weekly feed production. The available ingredients and costs per unit of ingredient are indicated in Table 2.1. The following limitations must be met in the feed-mixing process:

- At least 60% of the feed must be corn.
- At least 20% of the feed must be oats and milo.

Table 2.1.

Ingredients	Amount Available (tons)	Cost ($/ton)
Oats	98	98
Milo	98	82
Corn	140	105
Vitamin supplement	37.5	387
Molasses	37.8	180

- At least 5% of the feed but no more than 10% must be molasses.
- At least 5% of the feed must be vitamin supplement.
- No more than 12% of the feed can be milo.
- The mill must produce 175 tons of feed.

The agricultural manager wants to know the best mix of ingredients to minimize the cost per ton of feed.

Problem Formulation

Step #1 in the problem formulation is to define the decision variables.

$X1$ = tons of oats
$X2$ = tons of milo
$X3$ = tons of corn
$X4$ = tons of vitamin supplement
$X5$ = tons of molasses

(Note that it is important to include and keep track of units associated with each variable. In this case, the units are tons.)

Step #2 is to formulate the objective function. The goal of this problem is to determine the mass of each ingredient that will yield the minimum cost. The objective function is an equation that defines the cost as a function of the decision variables. The following is the appropriate objective function for this problem:

$$\text{MIN}Z = 98X1 + 82X2 + 105X3 + 387X4 + 180X5 \tag{2.5}$$

(Note that Z is in units of dollars. For example, the first term in the objective function ($98X1$) is the \$98 per ton of oats multiplied times $X1$ tons of oats. The result is the cost to the mill of the oat component of the feed.)

Step #3 is to formulate the constraint equations that place limits based upon resources available and limits defined by the problem. For example, there are only 98 tons of oats. One constraint equation would be as follows:

$$X1 \leq 98 \tag{2.6}$$

The remaining similar constraints for the ingredients are as follows:

$$X2 \leq 98 \tag{2.7}$$

$$X3 \leq 140 \tag{2.8}$$

$$X4 \leq 37.5 \tag{2.9}$$

$$X5 \leq 37.8 \tag{2.10}$$

For the constraint stating that at least 60% of feed mixture must be corn, we must first define the total mass of feed to be produced (Xt) as follows:

$$Xt = X1 + X2 + X3 + X4 + X5 \tag{2.11}$$

(The total mass of feed produced is the sum of the ingredient masses.)

In this simple example, the total feed produced per week (Xt) is 175 tons. The remaining constraint equations and explanations are as follows:

$$X3 \geq 105 \tag{2.12}$$

(at least 60% of Xt must be corn, $0.6 * 175 = 105$).

$$(X1 + X2) \geq 35 \tag{2.13}$$

(at least 20% of Xt must be milo and oats).

$$X5 \geq 8.75 \tag{2.14}$$

(at least 5% of Xt must be molasses).

$$X5 \leq 17.5 \tag{2.15}$$

(no more than 10% of Xt can be molasses).

$$X2 \leq 21 \tag{2.16}$$

(the mass of milo cannot exceed 12% of Xt).

$$X4 \geq 8.75 \tag{2.17}$$

(at least 5% of Xt must be vitamin supplement).

$$X1 + X2 + X3 + X4 + X5 = 175 \tag{2.18}$$

(175 tons of feed are needed.)
The final constraints are known as nonnegative constraints. This basically means that none of the decision variables can be less than 0.

$$X1, X2, X3, X4, X5 \geq 0 \tag{2.19}$$

Solution of the Example Problem

Once formulated, the LP problem can be solved using techniques such the simplex method, or the revised simplex method. Computer software packages are available for these solutions. The solution of the example problem is as follows: $X1 = 0$, $X2 = 21$, $X3 = 136.5$, $X4 = 8.75$, $X5 = 8.75$, $Z = \$6683.5$. The following results and observations can be made from the formulation and the final solution:

- The cost of ingredients per ton of feed is $38.19.
- Almost all of the corn was used. Additional corn will be needed for the following week.
- No oats were used and none will be used until the supplies of milo have been depleted. The reason for this is that milo is a less costly ingredient than oats. The manager may not want to purchase oats and incur the cost of storing this ingredient until the price of oats decreases.
- The minimum amounts of vitamin supplement and molasses are used. This is a consequence of the cost per ton.

- There are approximately 4-week supplies of molasses and vitamin supplement available.
- In approximately 4.5 weeks, the mill will deplete its supply of milo and will have to start using the available oats, which will slightly increase the feed costs.

2.1.2. Time and Project Management

One of the critical engineering/management decisions associated with a project that consists of numerous tasks is the selection of which course of action to take when a project must be completed in the shortest time possible. Decisions that result in an inefficient use of resources or a delay in the project completion can be costly. The two systems tools that are utilized for time and project management are the critical path method (CPM) and project evaluation and review technique (PERT).

CPM

The initial step in utilizing CPM is a detailed table with descriptions of the jobs or tasks that must be accomplished to complete the project, with the associated durations and predecessors for each job. The process of developing this project description must incorporate knowledge of the project and experience with the anticipated durations of each task, taking into account the resources available. The next step is the development of a graphical representation of the tabular project description. A network-flow model is a graphical representation of the initial tabular description, durations, and chronological order or sequence of jobs that must be completed for the project.

Example

A farmer desires a grain dryer and bin to store 10,000 bushels of grain. He needs this system prior to harvest time. He has contacted a systems engineer to perform the work. The engineer has developed a table (Table 2.2) of jobs associated with this project.

Once the project has been described with durations and predecessors, a graphical description (project network) of the project is formulated. This network consists of a series of nodes connected by directed lines. Each line connecting a node represents a

Table 2.2.

Activity (Jobs)	Description	Predecessors	Duration (wk)
a	Excavate and prepare sites for concrete slabs	—	2
b	Order and get delivery of bin and dryer	—	7
c	Pour and finish concrete slabs	a	4
d	Erect grain bin	b, c	3
e	Install grain dryer	b, c	5
f	Provide electrical requirements	b	3
g	Install conveying equipment	d, e, f	5
h	Test the system	g	2
i	Clean the site	h	1

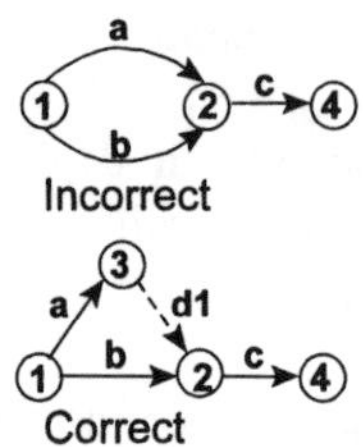

Figure 2.1.

specific job and time duration. When constructing this network there are several rules that must be followed. The first is that when analyzing concurrent activities, two lines cannot be connected to the same node. To overcome this, a new node and a dummy line should be used, as indicated by Fig. 2.1.

In Fig. 2.1, the new line, d1, has a duration of 0. The second rule is that no job or jobs leaving a node may start until all jobs entering that node are completed. For example, referring to Fig. 2.1, job c, which leaves from node 2, cannot start until both jobs a and b are completed. So, if job a has a duration of 7 weeks, and b, 5 weeks, job c cannot start until after 7 weeks. When looking at an overall project network, the longest path from the start of the entire project to its completion is referred to as the *critical path*. Continuing with the previous example, the critical path would be 1–3–2–4 with a duration of 7 weeks plus the duration of job c. Another important term is *activity slack*. Activity slack is the amount of time a specific job can be delayed without affecting the overall project duration. Again, using the above example, because we know that activity 2–4, or job c, cannot start until 7 weeks, and activity 1–2, or job b, has a duration of 5 weeks, job b can be delayed, starting 2 weeks after the start of job a, and still be completed by 7 weeks so that job c can start at 7 weeks. This means that activity 1–2 has a slack time of 2 weeks.

Following the steps outlined above, a nodal network for the example project has been developed, as shown in Fig. 2.2.

The next step, once the project nodal network has been constructed, is to determine the critical path of the overall project. This allows a manager to determine what key jobs must be targeted in order to insure that the project is completed in the least time. There are several software packages on the market that allow the user to input the different jobs, their predecessors, and the job durations and output project duration, critical path, and slack times associated with the different jobs. For smaller projects, this analysis can be done by hand. This is accomplished by the use of a CPM table. This table contains

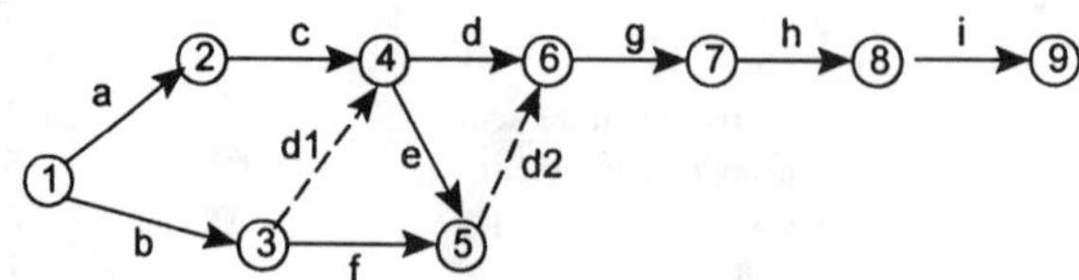

Figure 2.2.

Table 2.3.

Activity/Job	Duration	ES	EF	LS	LF	Float
1–2/a	2	0	2	1	3	1
1–3/b	7	0	7	0	7	0
2–4/c	4	2	6	3	7	1
3–4/d1	0	7	7	7	7	0
3–5/f	3	7	10	9	12	2
4–5/e	5	7	12	7	12	0
4–6/d	3	7	10	9	12	2
5–6/d2	0	12	12	12	12	0
6–7/g	5	12	17	12	17	0
7–8/h	2	17	19	17	19	0
8–9/i	1	19	20	19	20	0

the following column headings: activity, duration, early start (ES), early finish (EF), late start (LS), late finish (LF), and float or slack (F or S). ES and EF are the earliest times that a job can be started and completed. LS and LF are the latest times that a job can be started and finished. The float or slack is the activity slack associated with a specific job. It is equal to LF minus EF, or LS minus ES. (If the analysis is performed correctly LF − EF = LS − ES.) Activities that have a float of zero are on the critical path. Table 2.3 is the result of analysis for the example described previously.

Thus the critical path is 1–3–4–5–6–7–8–9, with a duration of 20 weeks.

PERT

PERT follows the same procedure as CPM analysis with the exception of the addition of probabilistic times for the job durations. For each job, three different times are given, a, m, and b, with a an optimistic time, m the most likely time, and b the pessimistic time. In order to determine the expected time for the job, the following relationship is used:

$$\text{time}_{\text{expected}} = t = \frac{a + 4m + b}{6} \tag{2.20}$$

In addition, the variance associated with the duration of each job can be determined using the following relationship:

$$\text{variance} = \sigma^2 = \left(\frac{b - a}{6}\right)^2 \tag{2.21}$$

Using the same table as for CPM analysis, and using the expected time, t, calculated from a, m, and b, the critical path and the expected duration of the critical path can be determined. In order to calculate the variance of the critical path, the variances of the jobs in the critical path are summed. The probability of the project being completed in a certain time can now be determined using the normal distribution. This is accomplished by determining the number of standard deviations away from the mean (μ, duration of critical path) of the Z value.

$$Z = \frac{x - \mu}{\sigma} \tag{2.22}$$

Using this Z value, a normal distribution table, critical path duration (μ), and standard deviation σ (σ = sqrt σ^2), one can determine the probability that a certain project will be completed in duration X.

2.1.3. Queuing Theory

Queuing theory describes the operation of waiting lines and servers. There are numerous instances in agricultural operations in which waiting lines are integral to the efficiency of an operation. In the United States, long lines can be seen at country grain elevators and cotton gins during the harvesting season. If a systems tool can be used to reduce waiting time, a significant number of customers will not be lost to a competitor. The understanding of queuing theory is necessary for many simulation models. Two queuing systems with a Poisson arrival rate and an exponential service rate are discussed in this section: single-server (M/M/1) and multiserver models (M/M/c).

Single-server M/M/1

Figure 2.3. is a graphical description of a single-server single waiting line with four customers. One of the customers is being served and three are waiting in line. The typical questions asked are as follows:

What will be the average waiting time in the queue (waiting line)?
What will be the average waiting time in the queue system (time in the queue plus time required for service)?
What is the average number of customers in the queue?
What is the average number of customers in the queue system?
What is the probability that 0,1,2, etc. customers are in the queue system?

The assumptions for the queuing-theory analysis of the single-server model example given here are that the calling population is infinite and that the queue discipline is first-come first-serve. The information needed to work with the M/M/1 model is the arrival rate (λ) and the service rate (μ). Both the arrival and service rates are expressed in units of average number of customers per time period. For example, typical units for λ and μ would be customers per hour. Classical queuing-theory analysis requires that the service rate exceed the arrival rate, $\mu > \lambda$. There are several relationships that have been developed to describe the characteristics of a single-server queuing system. The

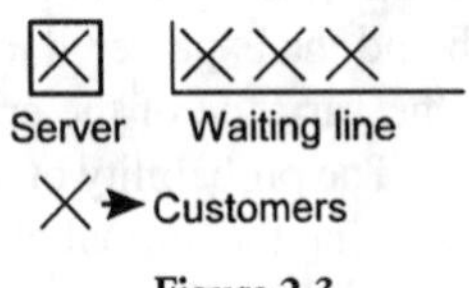

Figure 2.3.

probability that no customers are in the queue system is

$$P_0 = \left(1 - \frac{\lambda}{\mu}\right). \tag{2.23}$$

The probability that n customers are in the queue system is

$$P_n = \left(\frac{\lambda}{\mu}\right)^n \cdot P_0 = \left(\frac{\lambda}{\mu}\right)^n \left(1 - \frac{\lambda}{\mu}\right). \tag{2.24}$$

The average number of customers in the queue system (in the server and the waiting line) is

$$L = \frac{\lambda}{\mu - \lambda}. \tag{2.25}$$

The average number of customers in the waiting line is

$$L_q = \frac{\lambda^2}{\mu(\mu - \lambda)}. \tag{2.26}$$

The average time a customer spends in the total queue system (waiting line and server) is

$$W = \frac{1}{\mu - \lambda} = \frac{L}{\lambda}. \tag{2.27}$$

The average time a customer spends in the queue waiting to be served is

$$W_q = \frac{\lambda}{\mu(\mu - \lambda)}. \tag{2.28}$$

The probability that the server is idle is

$$I = 1 - U = 1 - \frac{\lambda}{\mu}. \tag{2.29}$$

U is also known as the *utilization factor*, which is the probability that the server is busy.

An alternative queue system is one referred to as “jockeying” (Fig. 2.4). An example of jockeying is the queue systems in a supermarket. In reality, this is multiple M/M/1 queues.

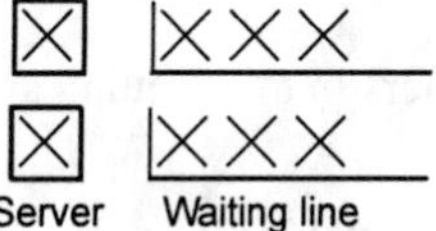

Figure 2.4.

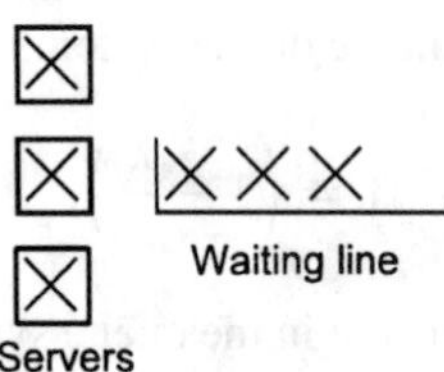

Figure 2.5.

Customers in a jockeying queue system choose a waiting line and stay in that line. There is no moving back and forth between lines. In order to perform a queuing theory analysis, it is assumed that the customers divide themselves equally between the lines, and the new arrival rate is one half the original rate, and the equations associated with the M/M/1 queuing theory are used. The service rate remains the same.

Multiple-Server Systems

In a multiserver queue system, a single waiting line is being served by multiple servers (Fig. 2.5). An example of a multiserver queue system can be observed at the check-in desk at an airport.

Information needed to analyze multiserver queue systems includes λ, arrival rate; μ, service rate; and c, number of servers. In contrast to the M/M/1 queue system, where the service rate was required to be greater than the arrival rate ($\mu > \lambda$), for the multiserver queue the product of number of servers and the service rate must exceed the arrival rate ($c\mu > \lambda$).

The probability that there are no customers in the system (all servers are idle, no one in waiting line) is

$$P_o = \frac{1}{\left[\sum_{n=0}^{n=c-1} \frac{1}{n!}\left(\frac{\lambda}{\mu}\right)^n\right] + \frac{1}{c!}\left(\frac{\lambda}{\mu}\right)^c \left(\frac{c\mu}{c\mu - \lambda}\right)} \tag{2.30}$$

The probability of n customers in the queuing system is

$$P_n = \frac{1}{c!\, c^{n-c}}\left(\frac{\lambda}{\mu}\right)^n \cdot P_o; n > c$$

$$P_n = \frac{1}{n!}\left(\frac{\lambda}{\mu}\right)^n \cdot P_o; n \le c. \tag{2.31}$$

The average number of customers in the queuing system is

$$L = \frac{\lambda\mu(\lambda/\mu)^c}{(c-1)!\,(c\mu - \lambda)^2} \cdot P_o + \frac{\lambda}{\mu}. \tag{2.32}$$

The average time a customer spends in the queuing system (waiting in line and being served) is

$$W = \frac{L}{\lambda}. \tag{2.33}$$

The average number of customers in the queue is

$$L_q = L - \frac{\lambda}{\mu}. \tag{2.34}$$

The average time a customer spends in the queue waiting to be served is

$$W_q = W - \frac{1}{\mu} = \frac{L_q}{\lambda}. \tag{2.35}$$

The probability that a customer arriving in the system must wait for service (all servers busy) is,

$$P_w = \frac{1}{c!}\left(\frac{\lambda}{\mu}\right)^c \frac{c\mu}{c\mu - \lambda} \cdot P_o. \tag{2.36}$$

Queuing-Theory Example

To demonstrate the concepts of single- and multiple-server queuing systems, the following example is provided. This example problem is from Ravindran *et al.* (1987).

A small town has one grain elevator that services all of the trucks bringing in the harvested wheat crop. Before the trucks can dump the load, they must go through a check-in process that involves being weighed, grain samples being taken, and a few other details. As it is the farmers' goal to get their crops in as quickly as possible, and because the trucks cannot return and pick up another load until the previous one is dumped, the farmers have a vested interest in the efficiency of the grain elevator. Suppose the average arrival time for trucks is 6.67 minutes, and the average service time is 6 minutes. Using the notation described previously, the service rate, μ, in units of customers per hour is 10 trucks per hour. The arrival rate, λ, is 9 trucks per hour. It has been observed that the trucks spend, on average, 1 hour at the elevator. This delay is considered intolerable by the farmers. The following three options have been suggested to increase the rate of service.

- Adding sideboards to the trucks (permitting slightly larger loads) will lengthen the interarrival time of the trucks to 10 minutes. At the same time, an extra worker at the check-in station would shorten the average service time to 4 minutes. This option has a cost of $30,000.
- The addition of another complete check-in station, including scale, test equipment, and personnel, would essentially double the service capacity. Arriving trucks would be required to form a single line and the truck at the front of the line would move to the first available check-in station. This option has a cost of $400,000.
- The addition of an entire new grain-elevator facility on the other side of town would not only double service capacity but would split the arrivals between each facility. This option would cost $1,000,000.

Solution. From the analysis of the problem description, the existing and first option are M/M/1 queue systems. The changes from the existing system to the first option are that the arrival rate is shortened and the service rate increased. The second option, with its single line to two servers, is a multiple-server model. The service rate for each server stays the same, and the arrival rate to the single line stays the same, but the analysis must be done using the multiple-server model equations. Option three creates a new server but each server has its own line. Option three is a jockeying queue system. The service rate stays the same as the initial problem, but the arrival rate is halved. The analysis can be done using the single-server model equations.

Table 2.4 contains the results of the queuing-theory analysis for the existing and the three options.

Table 2.4.

Proposal Option	Model	Arrival Rate, λ (trucks/h)	Service Rate, μ (trucks/h)	Waiting Time, W (minutes)	Utility, U (% time)
Existing	M/M/1	9	10	60	90
1	M/M/1	6	15	6.67	40
2	M/M/2	9	10	7.52	38
3	M/M/1	4.5	10	10.91	45

Note: Utility is the percent of time that the server is busy. For the M/M/2 system, it represents the percent of time one of the servers is busy $(1 - P_o)$.

2.1.4. Simulation

Simulation is the most widely used systems tool. It is a different tool in that the use of simulation does not result in analytical solutions that are the values of decision variables that result in maximum profit or minimum cost as with LP. It does not define the appropriate sequence of jobs that will result in the shortest duration of a project consisting of many interrelated jobs, as do CPM and PERT. It is, however, a very powerful tool.

Simulation is the mathematical representation of a real-world system. A good simulation model can be used for testing purposes without the expense of building and testing the real thing. The U.S. space program could not have achieved its present level of success without simulation. Simulation is used to estimate insect populations, concentrations of pollutants downwind from sources, operation of grain dryers and cotton gins, and the impact of yields of agricultural products on market demands, among many other applications. Monte Carlo simulation consists of estimating the results of statistical variations of independent variables on dependent variables. Once developed, the simulation model can be used to predict means and standard deviations of dependent variables with random variations of independent variables. In general, the information gained from the simulation process would be too expensive and would take too long for an engineer/manager to attempt to obtain similar information by conventional building and testing methods. Simulation is also the most often misused systems tool. Simulation modeling and testing should be used to gain insight and understanding of a problem, but the results should always be viewed with skepticism. Simulation results should never be used in a manner implying that the results are exact.

An added benefit of simulation is the knowledge gained from formulating the simulation model. The knowledge gained from the trial-and-error process of formulating a mathematical model that represents the results of a "real-world" system including statistical variations of independent variables can provide insight for the decision making process. It is essential that the engineer/manager have or be able to obtain detailed knowledge of the system that will be represented by the simulation model. Many of the decisions made in the trial-and-error process of formulating a model are based upon common sense and practical knowledge of the system being modeled.

In this section, three continuous probability distributions are introduced that can be used to represent the variation of independent variables. These are the uniform distribution, the exponential distribution, and the triangular distribution. With these three distributions, an engineer/manager can formulate a simulation model for any problem or system. The quality of the results is dependent (in part) upon how well the statistical variation of the selected distributions fit the real-world variations. The equations that can be used to determine the value of an independent variable as a function of a uniform random number are presented.

It is not essential that the engineer/manager have access to a simulation software package to utilize this tool. Sophisticated simulation modeling can be performed using a calculator, paper, pencil, and the three statistical distributions presented here. However, simulation-modeling packages can greatly improve the power of the modeler.

Random Numbers and Random-Number Generation

The random-number generator in most calculators and software packages has a uniform distribution of random numbers between 0 and 1. An ideal uniform distribution between 0 and 1 can be described by the following: If one hundred random numbers were chosen, and the number of occurrences between 0 and 0.1, 0.1 and 0.2, 0.2 and 0.3, ..., 0.9 and 1.0 were tallied, an ideal uniform distribution (0,1) would have 10 numbers in each range. The use of a uniform random-number generator (0,1) is essential to the utilization of the simulation tool presented in this section.

Uniform Distribution

If the value (X) of an independent variable is deemed to be uniformly distributed between A and B, the probability that X will equal any number from A to B is the same. This is represented in Fig. 2.6.

If the random-number generator were used to give a value for R between 0 and 1, the equation for independent variable X that is deemed to be uniformly distributed ranging

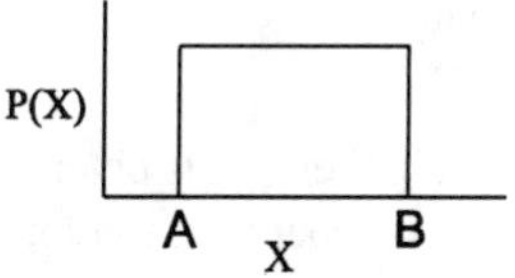

Figure 2.6.

from A to B would be as follows:

$$X = A + (B - A)R \tag{2.37}$$

Example

Suppose the arrival rate of grain trucks were uniformly distributed between 10 and 20 minutes; values for the random variable X could be determined using Eq. 2.33. The expectation of the modeler would be that if one were to measure the intervals between truck arrivals for several days or weeks and plot the frequency as a function of the arrival intervals, the plot would be similar to Fig. 2.6.

$$X = 10 + 10R \tag{2.38}$$

Exponential Distribution

A plot of probability versus values for an independent variable that is exponentially distributed is shown in Fig. 2.7. There are a number of operations that are more accurately represented by an exponential distribution than a uniform distribution. One of the classical queuing assumptions is that the service times are exponentially distributed. The repair time of breakdowns in a cotton gin or grain elevator may be better estimated by an exponential distribution. A simple description of Fig. 2.7 is that there are more frequent values for random variable X near the origin than there are further away from the origin. A practical example for arrival intervals of grain trucks at a grain elevator is as follows: Data taken on arrival intervals for grain trucks at an elevator indicate that an arrival interval of less than 10 minutes occurred 400 times; 10–20 minutes, 150 times; 20–30 minutes, 20 times; over 30 minutes, fewer than 5 times.

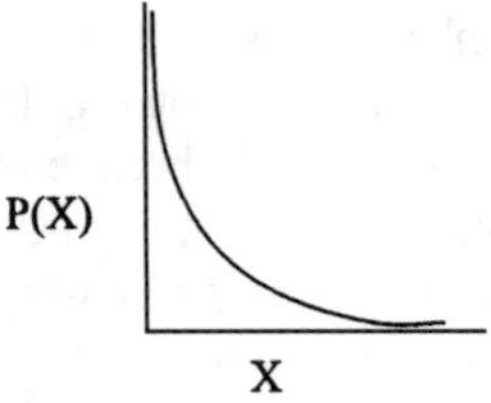

Figure 2.7.

Values for the random variable X that is deemed to be exponentially distributed can be determined using the following equation:

$$X = -\frac{1}{\lambda}\ln R \tag{2.39}$$

where λ is the mean of the distribution.

Example

The average service rate of a grain elevator is determined to be 10 minutes per truck or 6 trucks per hour Values for the independent random variable X can be determined using Eq. (2.39).

$$X = -1/6 * \ln R(\text{hours per truck}) = -10 * \ln R(\text{minutes per truck}) \tag{2.40}$$

Triangular Distribution

A plot of probability versus values for an independent random variable X that is characterized by a triangular distribution is shown in Fig. 2.8. A triangular distribution can be described by a frequency of occurrence that is small for $X = A$ and $X = B$ but high for $X = D$ where D is a value for X between A and B. An example of a triangular distribution describing a random variable such as the ginning rate is as follows: A ginner may say, "Most of the time we average 10 bales per hour. There are times when I can gin 15 bales per hour on good cotton but on rough cotton I can only get 6 bales per hour." This is an ideal description of a triangular distribution. The modeler would not want to use a uniform or exponential distribution to describe the ginning rate in the model.

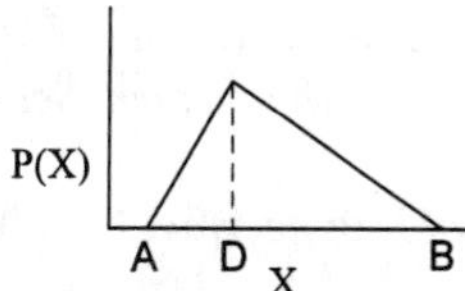

Figure 2.8.

The triangular distribution can be represented mathematically in terms of A, D, B, and random number R using the following equations:

$$X = A + \sqrt{(D - A)(B - A)R} \quad \text{for} \quad 0 \leq R \leq \frac{D - A}{B - A} \tag{2.41}$$

$$X = B - \sqrt{(B - D)(B - A)(1 - R)} \quad \text{for} \quad \frac{D - A}{B - A} < R \leq 1 \tag{2.42}$$

Example

The time between the arrival of cars at a body shop is exponentially distributed, with an interarrival time of 6 weeks. The time required for the repair work at the body shop is triangularly distributed with a low end of 4 weeks, a high end of 9 weeks, and a mode of 6 weeks. Table 2.5 represents the simulation of 6 cars in this system.

Table 2.5.

Car	R_1	Arrival Interval (wk)	Arrival Clock (wk)	Enter Facility Clock (wk)	Wait Time (wk)	Queue Length	R_2	Repair Time (wk)	Depart Time (wk)	Time in System (wk)
1	.19	9.9	9.9	9.9	0	0	.48	6.2	16.1	6.2
2	.32	6.8	16.7	16.7	0	0	.82	7.3	24	7.3
3	.23	8.8	25.5	25.5	0	0	.09	4.9	30.4	4.9
4	.19	9.9	35.4	35.4	0	0	.35	5.9	41.3	5.9
5	.67	2.4	37.8	41.3	3.5	1	.77	7.1	48.4	10.6
6	.36	6.1	43.9	48.4	4.5	1	.91	7.8	56.2	12.3

The following equations, which are based on Eqs. (2.39), (2.41), and (2.42) were used to calculate the arrival interval (AI) and the repair time (RT), based on the given information.

$$AI = -6\ln(R) \tag{2.43}$$

$$RT = 4 + \sqrt{10R} \qquad \text{for\_}0 \leq R \leq 0.4 \tag{2.44}$$

$$RT = 9 - \sqrt{15(1-R)} \quad \text{for\_}0.4 \leq R \leq 1 \tag{2.45}$$

References

Hillier, F. S. and Lieberman, G. J. 1995. *Introductions to Operations Research.* New York: McGraw-Hill.

Ravindran, A., Phillips, D. T., and Solberg, J. J. 1987. *Operations Research: Principles and Practice.* Needham Heights, MA: John Wiley & Sons.

Taylor, B. W. III. 1993. *Introduction to Management Science*, 4th ed. Dubuque, IA: Wm. C. Brown Publishers.

Whinnery, J. R. 1965. *The World of Engineering.* New York: McGraw-Hill.

2.2. Agricultural Mechanization Strategy

A. G. Rijk

2.2.1. Definitions

Agricultural mechanization embraces the use of tools, implements, and machines for agricultural land development, crop production, harvesting, preparation for storage, storage, and on-farm processing. It includes three main power sources: human, animal, and mechanical. The manufacture, distribution, repair, maintenance, management, and utilization of agricultural tools, implements, and machines is covered under this discipline with regard to how to supply mechanization inputs to the farmer in an efficient and effective manner.

Hand-tool technology is the simplest and most basic level of agricultural mechanization: the use of tools and simple implements using human muscle as the main power source.

Draft-animal technology refers to implements and machines utilizing animal muscle as the main power source.

Mechanical-power technology is the highest technology level in agricultural mechanization. It embraces all agricultural machinery that obtains its main power from sources other than muscular power.

Within each of these three levels of mechanization technology, degrees of sophistication must be distinguished. For example, a simple, locally made single-axle tractor without differential gears and gear box, a single-axle tractor with gearbox and power take-off, and a 70-kW tractor are all examples of mechanical-power technology, but with large differences in sophistication and capability.

Terms such as "intermediate technology" and "selective mechanization" are either inappropriate or have no practical meaning. The term *appropriate mechanization* may be used, referring to the level of mechanization and how it is used for a specific situation.

"Appropriateness" can be determined only after carefully considering the technical, economic, and social characteristics of each situation. No generalizations can be made concerning the appropriateness of a particular type of mechanization or particular agricultural tool, implement or machine for rural development. For the purpose of simplicity, the term *mechanization* in this section will be used to cover all levels of mechanization technology and their degrees of sophistication.

The purpose of an *agricultural-mechanization strategy* (AMS) is to create a policy, institutional, and market environment in which farmers and other end users have the choice of farm power and equipment suited to their needs within a sustainable delivery and support system. "Farmers and others" refers to all end users of farm power, tools, and equipment, such as small family-operated farms, commercial farm businesses, farmers' organizations, irrigation groups, contractors, government operators, and primary agricultural produce processors [1]. An AMS deals with manual, draft-animal, and mechanical power; the utilization of tools, implements, and machinery; and their supply and maintenance. The strategy may cover importation and domestic manufacture of tools, equipment, and machinery; their repair and maintenance; relevant training and extension programs; improvement of draft-animal health services and breeding programs; and promotion of financing systems for the purchase of draft animals and machinery.

Policies are established by governments to achieve specified objectives. *Strategies* define the ways in which policies are to be implemented. With the emphasis towards market liberalization and the recognition that the private sector is the most important actor to develop an economy, *AMS formulation* emphasizes the creation of conditions conducive to the adoption of appropriate farm tools, implements, and machinery in a most effective and efficient way. The output of an AMS consists largely of policy and institutional recommendations and reforms but also may include specific programs and projects. In a dynamic environment, conditions change over time, and therefore an AMS will need to be regularly refined, revised, and adjusted; an AMS should be dynamic.

2.2.2. Introduction

Agricultural mechanization in developing countries has been much criticized because it often has failed to be effective and has been blamed for exacerbating rural unemployment and causing other adverse social effects. This was largely the result of experiences from the 1960s until the early 1980s, when large quantities of tractors were supplied to developing countries either as gifts from donors or on very advantageous loan terms. In particular, projects that were designed to provide tractor services through government agencies had a miserable record. These projects proved unsustainable because of the intrinsic inefficiencies of government-run businesses. An overvalued foreign-exchange rate and low real interest rates made agricultural machinery artificially cheap as compared with labor and draft animals. These experiences, often combined with a very narrow perception and lack of knowledge about mechanization, namely the one-sided promotion of tractors and other capital-intensive mechanical-power technology, caused the aid community to largely turn its back on mechanization. At the same time there are many examples in which mechanization has been very successful, contributing to increased food production, productivity, and advancement of rural economies: privately owned, shallow-tube wells for irrigation in South Asia; axial-flow threshers in Southeast

Asia; single-axle tractors in Thailand; and various forms of farm mechanization in many parts of China.

The introduction of agricultural technology, including mechanization, is a complex process. The formulation of an AMS requires comprehensive knowledge of many aspects of agriculture in its broadest sense. An AMS very much depends on country-specific characteristics of the economy, its level of development, and the agriculture sector. This means that the formulation process for an AMS cannot be prescribed in a simple set of guidelines. Therefore, the purpose of this section is to aim at a better understanding of the process of mechanization, to provide broad guidelines for strategy formulation, and to address the major issues involved. Those who will be involved in the formulation of an AMS also should read the Food and Agriculture Organization's guide for AMS preparation [1].

The requirements for an AMS must be kept in the proper context. AMS formulation should not become an overbearing, time-consuming exercise, absorbing excess manpower and other resources, as compared with the actual constraints that need to be resolved by agricultural engineering. Preferably, mechanization technology should be considered in the context of an overall (agriculture) technology strategy.

2.2.3. Need for Mechanization and Productivity-enhancing Technology

Farm power—consisting of manual labor, agricultural tools, draft animals, tractors, implements, equipment, and machinery—is an essential farm input. In almost any agricultural production system, the annual expenditure on farm power, whether on labor, draft animals, or fuel and depreciation of machines, greatly exceeds the costs of other inputs such as agrochemicals and seeds. In many developing countries, agricultural production and food security are adversely affected because of insufficient use of farm power, low labor productivity, or labor scarcity. The need to improve agricultural labor productivity is increasingly recognized. In a case such as pump sets for irrigation, the need for machinery is undisputed. Rather than agricultural mechanization, it would be preferable to use the term *farm-power* or *labor productivity–enhancing technology* to recognize not only the importance of manual labor and hand tools, draft animals, and mechanical power, but also other issues related to labor scarcity, such as cropping and farming systems.

Finding solutions to environmental problems in agriculture requires improved agricultural tools and machinery, including tools for soil tillage and pesticide application. Similarly, machines are required to assist with postharvest loss reduction and on-farm processing. Thus, it is now (again) recognized that agricultural mechanization is crucial in the fight against hunger and poverty, and at the same time it is crucial to address environmental and health concerns. To avoid recurrence of past mistakes such as those described in the introduction, efficient mechanization strategies are required.

The term *mechanization* unfortunately is often very narrowly perceived, while its real purpose, namely, enhancing the productivity of land and labor, often is not well understood. In fact, an AMS ought to be part of an agricultural-technology strategy, which is to be part of an overall agricultural-development strategy. In this context, three principal purposes of mechanization may be summarized as follows:

- *To increase labor productivity:* The introduction of machinery to substitute for labor ("labor-saving") is a common phenomenon associated with the release of labor for employment in other sectors of the economy or to facilitate cultivation of a larger area with the same labor force.
- *To increase land productivity:* The purpose of mechanization is here to produce more from the existing land. Machinery is a complementary input, required to achieve higher land productivity, for example, through the introduction of pump sets, or through faster turnaround times to achieve higher cropping intensity. In labor surplus economies, net labor displacement or replacement should be avoided.
- *To decrease cost of production:* Introduction of a machine may lower production costs or offset increased costs of draft animals or labor.

Usually, in various degrees, a combination of the three objectives will be achieved. Additional benefits to the user may be associated with a reduction in the drudgery of farm work, greater leisure, or reduction of risk. These are subjective benefits and difficult to translate into cash. Frequently, mechanization increases an individual's workload, can be hazardous to health, and may reduce the social interactions associated with farm work.

2.2.4. Adoption Process for Mechanization and Labor Productivity–Enhancing Technology

When reviewing the process of applying labor-saving (or labor productivity–enhancing) innovations in agriculture, it is a serious but frequent mistake to assume that this can be achieved only through applying mechanical-engineering technology. In this context, nine stages in the process of enhancing labor productivity may be distinguished [2]:

- *Stage I: Application of improved hand-tool technology:* This process started in prehistoric times when early civilizations developed stick and stone tools that were the only means to enhance labor productivity. In many parts of the world, hand tools are the only technology used in agriculture, and even in highly mechanized agricultural systems, improved hand tools are still important.
- *Stage II: Draft animal power application:* At this stage, animal muscle power is substituted for human power—a process that began in ancient civilizations. A large variety of implements and machines have been developed that use animals as the principle power source.
- *Stage III: Stationary power substitution:* Mechanical power is substituted for human and animal power and is used in stationary operations. Stationary operations are mechanized first because motive power sources required to move across the field are technically more complex and therefore require higher investment. Typically, operations mechanized at this stage are paddy dehusking, grain milling, pumping water, and threshing.
- *Stage IV: Motive power substitution:* At this stage, substitution of mechanical power for muscle power takes place for field operations. It focuses on power-intensive field operations (for example, plowing), and machinery is of relatively simple design and easy to operate. Mechanization is still straightforward, and crop-production practices are usually unchanged. At stages III and IV, mechanization takes advantage of the lower costs of new power sources as compared with traditional ones.

- *Stage V: Human-control substitution:* At this stage the emphasis is on substitution of the human-control functions. Depending on the complexity of the control function and the degree of its mechanization, machinery becomes increasingly complicated and costly. A potato lifter is simple in design, but fruit- and cotton-harvesting machinery is complex and expensive .
- *Stage VI: Adaptation of cropping practices:* This stage features the adaptation of the cropping system to the machine. For example, removing weeds in broadcast crops cannot be done with machines, but row seeding and seed drills may be introduced to facilitate mechanization of weeding. Other examples include the increase in row distance to accommodate heavier and larger machinery to speed up field operations.
- *Stage VII: Farming-system adaptation:* The farming system and production environment is changed to facilitate further increase in labor productivity and to benefit from economies of scale, necessary to make the investment in expensive machinery financially feasible. An example of this is the rapid decline of mixed farming systems in Europe since the late 1960s, when farmers specialized in either dairy, poultry, hog, or crop production. Some crops that are difficult to mechanize may disappear if acceptable substitutes become available, or if they can be produced in countries with low labor costs. At this stage, investments in land development, land consolidation, and rural infrastructure are often needed to facilitate advanced degrees of mechanization.
- *Stage VIII: Plant adaptation:* This stage features the adaptation of the plant and animal to the mechanization system. Mechanization has advanced to a stage at which engineering alone can no longer provide further gains in labor productivity. Breeders increasingly take into account the suitability of new varieties for mechanized production.
- *Stage IX: Automation of agricultural production:* This stage is progressing in countries with high labor costs and sophisticated demands on production and quality. Examples are automated rationing of concentrate feeding for individual dairy cows based on their milk production, and sprinkler irrigation systems activated by soil moisture.

This sequence of mechanization is generally identifiable at an individual farm, although when considering the agriculture sector as a whole in a particular country, the stages are usually less pronounced because of the diversity of the sector, and several stages may occur simultaneously. However, when formulating an AMS, the different options for enhancing land and labor productivity, as well as their economic and financial implications, must be well understood. Sometimes, rather than advocating mechanization of certain operations, alternative options may be more attractive. For example, a frequently made mistake is to propose the introduction of mechanical rice transplanters to offset increased labor cost, although changing to broadcast rice is in most cases technically and financially a better solution. Rural development programs must take into account the future needs for agricultural mechanization. Thus, the design of irrigation and drainage systems and the field size and layout must take into account the access of machines to fields and the width and strength of bridges. Commercial tree-crop plantations must take into account the possibility of future labor scarcity, and thus the tree variety and planting pattern must be able to accommodate future mechanized operations. These examples explain why a holistic approach and multidisciplinary input in strategy formulation are very important.

2.2.5. Basic Guidelines and Principles for Strategy Formulation

The appropriate choice and subsequent proper use of mechanized inputs into agriculture have a significant effect on agricultural production and productivity, on the profitability of farming, and on the environment. In most cases, the mere application of advanced tools, draft animals, or machines does not by itself lead to increased yields but usually reduces the cost of production and counteracts peak periods of labor shortage. However, the benefits achievable by using advanced and improved inputs such as irrigation, better seed, fertilizers, and pesticides cannot be fully realized without an increased application of farm power. In situations in which land is not a constraint and area expansion is feasible, increases in farm-power input have led to direct increases in production by simply increasing the land area or the number of animals that one person can handle.

In agricultural production, farm power is only one input along with land, irrigation, fertilizer, seed, crop chemicals, and so forth. The level of its use is one of a mix of management decisions a farmer has to make in order to maximize production and income. Therefore, in a free-market situation, it is inappropriate for governments to have a stand-alone policy on mechanization, except as a component contributing towards the realization of a broader agricultural and socioeconomic policy. To have a policy to mechanize or, for example, to replace draft animals with tractors, would imply that mechanization is an end in itself, whereas it should only be a strategy to maintain or increase agricultural production or labor productivity.

Some basic principles must be kept in mind in the discussion on AMS:

- History has shown that the process of mechanization is driven by changes in relative prices, in particular cost of labor versus cost of capital.
- The reasons for mechanization are economic. The driving force behind mechanization is the farmer's effort to increase or maintain net income.
- Mechanization is demand-driven: Ultimately it is the farmer who will decide what machine to buy, from whom to buy it, and how to use it.
- Experience has shown that mechanization must be left to the private sector as much as possible. Government should not actively get involved in the manufacture, import, distribution, or repair of agricultural machinery or its operation but rather should provide the incentives for private-sector response.

Broad guidelines for the formulation of an agricultural technology strategy may be summarized as follows:

- Where land is abundant but labor is a limiting production factor, mechanization can increase production per worker and the area under cultivation.
- Where land is scarce but labor is in surplus, biological and chemical technology, such as high-yielding varieties and intensive cropping systems, should be emphasized to increase land productivity. In such cases certain types of mechanization technology (for example, pumpsets) may be required as supporting complementary inputs to biological and chemical technology or to reduce the cost of production.

- Where both land and labor are underutilized because of distinct seasonality, mechanization technology is required to eliminate labor-supply bottlenecks (usually for land preparation).
- Where there is a shortage of both labor and land, a combination of labor-saving mechanization and biological and chemical ("land-saving") technology should be applied to achieve high productivity of both labor and land.
- Where the cost of traditional power sources such as human labor and draft animals has become high, mechanization is required to reduce the costs of agricultural production.

A key indicator of the need for labor-enhancing technology (mechanization) is the cost of rural labor (the wage rate) for specific agricultural operations in peak and slack seasons. A low wage rate indicates that there is no significant labor scarcity and the promotion of labor-replacing technology would be unjustified, except for technology that increases employment (for example, pumpsets). In this case a preliminary evaluation would indicate that there is no need for a comprehensive AMS. Only when there is a significant need for labor productivity–enhancing technology and low labor productivity is a constraint that deserves priority consideration is a comprehensive AMS warranted.

Unlike in a centrally planned command economy, in a free-market economy the supply and type of technology, including farm power, should be demand-driven. Governments must refrain from making decisions that stipulate by which means and to which extent agriculture will be mechanized. This should be decided by the farmer to best suit the farming enterprise, taking into consideration many factors, including the farm-household composition. Choice of farm-power input is just be one of a number of choices the farmer has to make. Decisions are based on a mix of complicated issues involved and possible options, with economic rationale paramount. Thus, a farming-systems approach is needed when formulating an AMS. The government's responsibility is to create an enabling environment that ensures that the farmers' needs for inputs, including for farm power, can be met. If national considerations require that farmers use more efficient or productive inputs (for example, using pumpsets to increase food production and food security, using more environmentally friendly pesticide application technology), then the government should provide the incentives for the farmers to make the investments and for the suppliers to provide and service this technology.

2.2.6. The AMS Formulation Process

Within a general agricultural policy and development plan, a government develops strategies to achieve policy objectives and targets. A strategy on mechanization should be just one of a number of such strategies. A strategy should not be confused with a plan, which stipulates specific government actions. If a government decides that agricultural mechanization should be left completely to the private sector, then that policy establishes the parameters for a strategy without the government's operational involvement. In this case, the role of the government is limited to implementing policies that create an enabling environment and encourage the private sector to do what is desirable for the nation.

The dominating philosophy for development nowadays is that governments should provide the basic conditions that encourage private individuals to take appropriate initiatives and make sound investments that contribute to national-development objectives, with minimal government intervention. Thus, the development of a sustainable agricultural mechanization subsector should take place with minimum direct government intervention. However, that does not mean that agricultural mechanization can be neglected in the formulation of national agricultural policy. On the contrary, very special attention must be paid to the role of the public sector versus the private sector and the side effects of other policies on the use of engineering inputs in agriculture. In many developing countries, the government has a role to play in extension, education, training, and awareness creation, but the private sector may even be able to take over these functions to some extent. If the political will does not exist to change an adverse policy and institutional environment, any development strategy will be inefficient and ineffective. In that environment, the formulation of a comprehensive AMS should not be pursued. In that case the work should be limited to creating awareness of the effect of the adverse policy and institutional environment on the overall economy and on the agriculture sector. This requires an emphasis on macro- and sector-economic analysis rather than on farming systems and farm power.

Who Should Formulate the Strategy?

The first issue to decide is who will be in charge of formulating the AMS. Usually there is within the ministry of agriculture a sector-planning unit, or within the national planning authority a unit dealing with agriculture. These are usually the most appropriate units to formulate the strategy because of their analytical capabilities and responsibility to address policy and institutional reform matters and because of their continuity. Experience has shown that technical universities, engineering departments, and institutes usually are not suitable because of their largely technical orientation and their vested interests in matters related to technology.

How to Proceed with Formulation?

When formulating the strategy, the first step to be carried out is a review of the agriculture sector and the demand for and supply of farm power. For typical farming/cropping systems, farm-power supply and demand profiles should be presented. Collection of farm-management data, prices, costs of inputs (in particular of labor, draft animals, and rental rates for mechanical-power technology) and their projection into the future is crucial to assess the viability of different types and levels of mechanization technology. This should lead to a descriptive projection of the medium- and long-term supply and demand for farm power assuming different scenarios. This initial work is the preliminary evaluation, which has to conclude whether there is a need for a detailed, comprehensive AMS. The latter includes analysis of national inventories (draft animals, tractors, etc.), domestic manufacture and assembly (tools, implements, tractors, etc.), importation and distribution of farm tools and machinery, description and analysis of major farming systems in relation to the use of farm power, and bottlenecks in the agricultural production system that may be solved by agricultural engineering.

It is important that the strategy is objective and unbiased, that its ownership is with the government, and that it has the support of high-level decision makers. This requires that the stakeholders (government, farmers, private sector) understand the purpose of the AMS and be actively involved from an early stage. This may be achieved through formal and informal workshops and seminars, with active participation of the stakeholders, aiming at a broad consensus on the problems identified and the solutions proposed.

The purpose of strategy formulation is not to collect detailed statistical information. The purpose of the fieldwork is to identify technical and financial constraints and policy issues that impinge on farm mechanization, to diagnose problem areas and bottlenecks, and to make comprehensive recommendations that address these problems and constraints. Care must be taken that scarce resources are not wasted on baseline studies or extensive data collection in particular, because these data often remain largely unused because of time constraints or the limited value they add to the analysis. The need for and collection of data must be proportionate to the problem identified during the initial analysis and the complexity of the possible solutions to solve the problem. An experienced person should be able to obtain reliable and relevant information through a rapid appraisal methodology and meetings with well-informed persons.

Composition of the Strategy Formulation Team

For the analysis and formulation of the strategy, an experienced multidisciplinary team is required that has a thorough understanding of the issues involved. However, rather than having a large number of staff directly involved, preferably no more than three experts should be assigned full-time. These staff must have multidisciplinary expertise, something that is difficult to find. The core group of the team usually consists of a farm management/agricultural economist, an agricultural-mechanization engineer, and an agronomist. The team must consult a large and diverse group of persons and institutions and in particular farmers and private entrepreneurs, because these two groups essentially will make the decisions whether or not to mechanize and commit the necessary investments. The strategy formulators must have a broader role than simply determining the way to implement a set of government policies. They must be able to analyze implications of existing government policies, their consequences for agricultural production and productivity, and economic and social concerns, and to suggest possible alternatives. At regular intervals, the team must organize informal seminars where findings and proposals are presented to a wide audience, issues and options are discussed, and feedback is obtained. The draft strategy document should be discussed at a high-level meeting with policy formulators, decision makers, private dealers, distributors, and representatives from universities, farmers' organizations, the manufacturing industry, and financing institutions.

The team must fully understand the process of agricultural development and mechanization in the highly industrialized countries. Striking similarities can be observed. Answers to questions must be critically evaluated because they may be biased or represent only a personal view. Simple questions can put into the proper perspective the perceived problems in mechanization, such as lack of private initiative, the problem of spare parts supply and training of operators. Why is the trucking and bus transport sector in most African countries completely in private hands, sustainable, and highly competitive? Why are soft drinks and basic medicines available in almost any rural village in Southeast Asia? Why is it that privately owned motorcycles and cars are not junked after a few years of operation? More general, why are the problems and constraints that appear so persistent with agricultural mechanization almost nonexistent with other technologies? The answers from different sources and objective analyses will establish the real problems and issues to be addressed in the AMS.

Outline of the Strategy Document

The broad outline of the strategy document should reflect that there are in principle three key groups participating in the process of mechanization, namely:

- The *demand side:* This is the end user (usually the farmer), who is concerned about getting the needed farm power on a timely basis and at the lowest possible cost. In principle, end users do not care whether a machine is rented or owned, whether it is locally manufactured or imported, and so forth, as long as their requirements and concerns are met satisfactorily. Machinery-hire services (contractors) are included in the demand side. In fact, contractors often have their own farming businesses.
- The *supply side:* This involves importers, distributors, dealers, local manufacturers, and repair services. They are in the business of providing a good or service, principally to make a profit. They could equally sell or service something else, and many often do. In the past, the supply side often has not developed because of an adverse institutional and policy environment. This often has been the reason for a government to take over the role of the suppliers, rather than analyzing why the suppliers were not developing and how they could become activated by instilling an enabling policy environment.
- The *government:* In its broadest sense, the government must be considered in the first place as a facilitator, to eliminate market failure and to ensure that supply meets the demand in an efficient and satisfactory manner. For this purpose, the government can provide institutional support (for example, extension, training, and credit) and incentives. It can stimulate mechanization by implementing a favorable policy environment, for example, related to import duties, taxes, subsidies, financing terms and conditions, and so forth.

Mechanization will not proceed in a sustainable manner if any one of these three groups does not fully meet its role and responsibility. Thus, the formulation of a strategy must deal comprehensively with each of the them.

Four distinct stages can be distinguished during the AMS formulation and its implementation (adapted from [1]):

- Analysis of present situation
  - Economy and policy environment
  - Agriculture sector
  - Farming systems and use of farm power
  - Agricultural tools and machinery supply
  - Relevant institutions
- Future scenarios
  - Developments in the national economy
  - Implications for agriculture
  - Developments in farming systems
  - Farm-power and equipment requirements
  - Development of the agricultural-machinery industry

- The strategy
  - Roles of the government and the private sector
  - Policy and institutional recommendations
  - Programs and projects
  - Strategy advocacy and proposals for legislative action
- Strategy implementation
  - Strategy advocacy
  - Formulation of proposals for administrative and legislative actions
  - Approval of these proposed actions

The final AMS document should be structured along the first three distinct phases of AMS formulation. The need for the last phase (strategy implementation) and how to proceed with it should be addressed in the final document, but actual implementation may be time-consuming and may require additional expertise. To attract the attention of key high government policy and decision makers, it is essential that an AMS executive summary report is produced as a separate document, summarizing the key issues, consequences, and proposed solutions. Proposed policy and institutional reforms should be presented in a matrix.

Implementation of the Strategy

A well-prepared strategy defeats its purpose if it is not going to be implemented. Therefore, the strategy team should propose the course of actions to be undertaken to ensure that the strategy will be implemented. During formulation, it is therefore important to ensure that all parties involved participate in workshops and seminars where the findings are presented and recommendations discussed. Representatives of farmer's organizations, dealers, manufacturers, and other participants should be invited to review and comment on proposals and recommendations. However, pressure groups may become active to safeguard their interests. For example, if imported agricultural machinery is subject to high duties to protect the domestic machinery industry, it can be expected that a drastic reduction in these duties will meet severe resistance from the manufacturers and the ministry of industry. If warranted, the strategy should then come up with alternative assistance to the industry without the farmers carrying the burden for this support. Ownership of the strategy should rest with the ministry that is most concerned with agricultural development. This ministry must actively pursue implementation of the recommendations. After the strategy is completed, a strategy-advocacy team should be made responsible for publicity, ensuring that the recommendations are endorsed by the legislative bodies and are being implemented. The strategy formulation becomes meaningful only after its recommendations have been implemented.

2.2.7. Some Frequently Raised Issues

The strategy-formulation team often will be confronted with comments, statements, or popular views regarding agricultural mechanization. Some of the most common ones are addressed here:

- *Mechanization replaces labor.* If there are no distortions in the cost of capital versus the cost of labor (which has become the case in most developing countries after voluntary or imposed structural adjustment), mechanization will be introduced only in response to labor scarcity (i.e., when labor is being drawn away from the agriculture sector).

- *Mechanical-power technology requires high capital investment and therefore is not available to small farmers.* Using farm machinery does not mean the farmer has to own it. Small farmers have successfully mechanized in many countries by hiring machinery services (contractors). Some countries have successfully adopted a policy for small farmers to acquire machinery with the purpose of earning additional income as contractors.
- *Farms are small and holdings are fragmented; therefore machinery cannot be used.* Suitable machinery is available to operate on small and fragmented lots, as can be witnessed in Asian agriculture. However, the expensive and complex agricultural machinery developed in the newly industrialized countries of East Asia is usually not suitable for low-income developing countries.
- *Development of the local machinery-manufacturing industry is a prerequisite for successful mechanization.* In many highly industrialized countries, much of the machineries and many of the tools are imported. In many developing countries, imported tools and machinery are cheaper because domestic demand is too small to achieve economies of scale. This is particularly the case with the manufacture of engines used as the prime mover in agricultural machinery. In many developing countries, the agricultural-tool and -machinery industry has been highly protected from imports in order to pursue industrialization. In these cases, farmers have to pay for the protection of the industrial sector, something that should be strongly opposed by the ministry of agriculture. Instead of having farmers pay for the development of a domestic industry, other ways should be explored, such as tax breaks.
- *Agricultural mechanization is a male-dominated technology, and women are in need of special technology.* Technology is gender-neutral. The use of technology and the division of labor constitute a private decision that usually is determined culturally. In many cultures, women participate in the mechanization sector and operate the most sophisticated machines. Mechanization may be a means of freeing women and children from agricultural work to more rewarding occupations and education.
- *The agriculture sector should reduce fossil-energy consumption by putting more emphasis on draft-animal technology rather than on fuel-consuming mechanical-power technology.* The share of commercial energy consumption by the agriculture sector in developing countries is low, and the scope for fossil-fuel conservation is limited. The agriculture sector like other sectors of the economy, will optimize the use of this scarce resource, if it is not subsidized and equally priced compared with other sectors.

2.2.8. Key Policy Instruments for Formulation of an AMS

As is the case nowadays with economic development, the emphasis in an AMS will be on policy and institutional reform of direct and indirect relevance to agricultural mechanization. Many policies impinge directly or indirectly on agricultural mechanization. Examples are exchange-rate policy, policies influencing agricultural input and output prices, employment and wage-rate policies, land-ownership and tenure policies, and policies affecting agricultural financial markets. For strategy formulation for agricultural mechanization, it is important to have a good understanding of the effect of macro- and sector policies on the agricultural mechanization sector. Some key policies that directly influence agricultural mechanization are be discussed subsequently.

Subsidies

Subsidies exist in every country. A subsidy or financial incentives can be direct or indirect in the form of a pure income transfer, a negative tax, or off-budget assistance. Subsidies can occur in many forms, for example, as a targeted one-time, lump-sum cash payment, or as provision of (targeted) credit, foreign-exchange, or specific inputs (for example, fuel and fertilizer) below market prices. Subsidies are used to help redistribute income to the poor and improve economic efficiency and resource allocation in situations in which markets do not function well or to correct other distortions in the economy, or to bring the private market solution in line with the social optimum and to achieve economic benefits. Subsidies may be used to protect infant industries, aid the adoption of improved and new technologies, or offset the impact of temporary price shocks.

Poorly planned and managed subsidies result in economic distortions and misallocation of resources and may sometimes lead to political instability, while their financial burden can be too much for the government to bear. If the strategy-formulation team proposes specific subsidies (for example, to stimulate the use of more environment-friendly technology), the team must undertake comprehensive economic analyses to assess the need for, justification for, and effectiveness of the subsidy. The cost of this subsidy program to the government budget and the national economy also must be shown, as well as possible funding modalities, for example from a levy. The possible adverse implications must be thoroughly considered. If there is no economic justification but mere political and social considerations, most developing countries, unlike industrialized, wealthy countries, cannot afford or sustain subsidies for agriculture, because the nonagriculture sector is too small or weak to support the large agriculture sector.

The general approach worldwide is to discourage subsidies, except where they can clearly be justified on the grounds of efficiency, economic benefit, or equity. For most agricultural-mechanization technology, subsidies are not justified, have an adverse impact on efficient allocation of inputs, or cannot be implemented effectively to address the problems they were meant for but lead to undesirable and unanticipated consequences for income distribution. Whenever subsidies are nevertheless deemed necessary, they must be made transparent along with their explicit justification. Transparency refers to the extent to which the costs of the subsidy are apparent to taxpayers and to which prices are perceived to be lower through the subsidy. Where subsidies are justified as a temporary measure, a program for phased elimination is to be provided.

Credit for Agricultural Machinery

The cost of capital versus the cost of labor, reflecting their relative scarcities, is the most important factor in determining the rate of adoption of agricultural mechanization. In the past, agricultural credit often contained an element of subsidy, thereby making the investment and use of machinery artificially cheap. This caused an array of problems such as unemployment and providing socially and politically influential farmers with access to cheap capital, that was not available to small and poor farmers. Often the credit obtained for agricultural machinery was ultimately used for nonagricultural purposes. For many farmers, access to credit is a constraint rather than the interest rate. Therefore, credit should be available at market prices with adequate margins for risk and transaction costs. It should be up to the investor (the farmer) to decide for what investment the credit is wanted. The same applies to schemes that allocate scarce foreign exchange for certain investments or at a more favorable exchange rate. Equally important, the exchange rates should reflect the market rate and amount of foreign exchange freely tradable rather than targeted at certain investments. Let the investor decide what technology to acquire.

Targeted credit usually has been biased toward larger capital investments. For example, credit for new tractors may be available, while there is no credit available for draft animals or an engine overhaul for a tractor. Targeted credit usually has been made available at lower than market rates through government-owned financial institutions, such as agricultural development banks or development finance institutions. These credit programs not only have distorted the financial markets but also have had an adverse impact on the financial performance of many development finance institutions. In some cases, the institution has been actively involved in the importation and distribution of the mechanization technology, thereby undermining the development of a sustainable distribution and service network. In particular, international financing institutions and bilateral aid programs have been contributing to this problem by importation through international tender or from the donor country. This has been a major cause of the bad experience with mechanical-power technology in many developing countries. This problem can be resolved by assisting private entrepreneurs in establishing a distribution and servicing network and giving them access to credit and foreign exchange through commercial banks to import equipment and to establish and stock service and maintenance facilities. An additional advantage of this approach is that, after the project has been terminated, a system is in place to continue the importation, distribution, and repair of the equipment.

In a few developing countries, agricultural-equipment distributors have successfully adopted financing modalities, usually with the involvement of a commercial bank, such as hire purchase or finance lease, similar to the practice in industrialized countries, or with cars and consumer goods in some developing countries. These schemes do not require government involvement, and their terms and conditions reflect the true cost of credit operations, including the risk involved in the financing of agricultural machinery.

Taxes and Duties

Taxes and duties can be considered as negative subsidies, and their purposes may be similar, but in addition they generate revenue for the government. Frequently, there are inconsistencies with taxes and duties that need to be addressed during strategy formulation. For example, often there are low or no import duties on agricultural machinery, but high duties on replacement parts. This has an adverse effect on the maintenance and repair of machinery and on the utilization of the capital investment. A principal argument against taxes and duties on imported tools and machinery is that, if food imports are free of taxes and duties, the inputs to produce that food within the country also should be free of taxes and duties. If governments want to increase food production or produce food more efficiently, then the use of modern inputs should be encouraged rather than punished. This argument will have the support of the ministry of agriculture but not of the ministry of industry, which will have to be convinced that the high import duties to protect an infant agricultural-machinery industry are not justified. There are other ways of promoting an industry rather than having the farmers pay for it.

Private, Cooperative, or Government Ownership

There is overwhelming evidence that private ownership is the only way to ensure efficient mechanization. In most countries, cooperative ownership of agricultural machinery has failed because of management problems. The record on government-operated machinery-hire services is very poor, so this mode can never be recommended. Private contractors are most appropriate to provide mechanization technology to small farmers. Government-operated machinery-hire services often are directly or indirectly subsidized by the taxpayer and therefore prevent the emergence of a cost-effective private-sector contracting business. The same applies for the equipment to develop and maintain rural infrastructure, such as irrigation and drainage systems. Government agencies should not operate their own equipment but award contracts for infrastructure development and maintenance through competitive bidding to private contractors. If these contractors do not exist, then the government should stimulate the emergence of this private business.

Input and Output Prices

Costs of inputs (fertilizer, seed, labor, fuel, machinery, etc.) and prices for agricultural products will be the most important factor for farmers in deciding what to produce and how much and what technology to apply. During the strategy-formulation process, it is therefore important to look at prices and how they affect the farmer's decision. For this, it is necessary to make some farm budgets under different assumptions. This is also important to understand why farmers do what they are doing and to assess whether the proposed technology is financially attractive to the farmer. In addition, it is important to compare farm-gate prices with border prices of imported food and feed. If for political reasons domestic farm prices are kept lower then the economic costs of imports (in most cases a wrong policy), than there is a justification for higher farm-gate prices (which will increase output) or for a subsidy on the agricultural input. The latter solution may cause inefficiencies, so it would be a better policy to increase farm-gate prices.

Public Investments

Typical public investments may include the following.

Supporting Institutions and Services

These include research and development (R&D); education, training, and extension; support services such as veterinary services; infrastructure development; and land consolidation. It is frequently assumed that, in developing countries, the government must undertake these activities because the private sector is not yet developed enough. However, government institutions are usually also very weak and largely ineffective or inefficient. This often is caused by very low government salaries and low budgets for operational programs, making it impossible to attract and retain competent staff and to implement substantive programs.

Other solutions must therefore be considered. In the case of important commodities that go through distinct market channels, for example export or processing of industrial crops, a levy may be imposed to fund an autonomous body for R&D, extension, training, and so forth for that commodity sector. The development program and management of this autonomous body may be supervised by representatives from all parties involved in the commodity's production, marketing, trade, and processing. The autonomous body would be able to pay attractive salaries to staff and consultants and to be independent from government bureaucracy. Another alternative is subcontracting research and training to autonomous and financially independent institutions or to the private sector. This may seem expensive, but in the long run it will be much more cost-effective, in particular if the recurrent costs of government institutions are taken into account.

In industrialized countries, the R&D of agricultural-engineering inputs is dominated by the private sector, and this is also the case to a limited extent in many developing countries. Nevertheless, it is often argued that the dominantly small-scale agricultural machinery industry is financially too weak to undertake R&D. This is then used as a justification to establish public-sector R&D institutions. Other modalities should be explored, such as financial support and incentives to private-sector or autonomous institutions to undertake R&D to resolve identified problems in a specific time frame.

In many developing countries, impressive infrastructure has been established to undertake R&D on agricultural engineering, often with the help of external assistance. A common phenomenon is that these institutions plan their research programs on academically perceived problems, rather than identifying the actual constraints faced by the farming sector. Often, when the perceived problem is technically "resolved," the new gadget is added to the display area of the R&D complex "awaiting commercialization," which never takes place, usually because of lack of interest from the farmer. R&D institutions' performance and justification must be based on the principle that unless the new machine is commercially produced and used, their research budget and effort have not yielded a return.

A mechanization strategy provides a sound basis and priorities for a comprehensive agricultural-engineering R&D program. In addition, the R&D institution has to go through the following sequence during the implementation of its R&D program:

- Identify, in collaboration with potential end users, the constraints that can be resolved by engineering technology, and establish priorities for R&D topics for which there is a commercial demand.
- Make a preliminary cost estimate of the tool or machine to be developed and an analysis of the potential benefit to the end user; almost any problem can be technically resolved, but the cost may be prohibitive and lower-cost alternatives may exist.
- Involve the end users and manufacturers in all stages of the R&D program. Preferably the workshop facilities and technicians of a manufacturer should be used in developing the prototype as a joint effort.

- Once a satisfactory functioning prototype has been developed, incorporate design modifications that will reduce the manufacturing, operation, and maintenance costs.
- Assist the private entrepreneur in the manufacture and promotion (e.g., through training, extension, and demonstrations in farmers' fields) until the stage at which commercial demand is generated.

With the intention of protecting the farmer, governments have established testing and evaluation centers and minimum quality standards. Often their efforts have been counterproductive, by prescribing standards that are difficult to achieve given the level of manufacturing technology or raw material domestically available or lead to unnecessary expenses, and that are not required from the perspective of the end user. In the case of testing and evaluation, the requirements of the various clients are very different and therefore, when a testing and evaluation program is being proposed, it is crucial to establish its purpose and to understand the requirements of the different clients. These clients are:

- *Suppliers:* These include manufacturers, importers, agents and dealers. They see testing and evaluation as part of the marketing effort for their product.
- *Regulators:* They include the policy makers who restrict the free supply of machinery through legislation on importation, standards, health and safety, and so forth.
- *Finance institutions:* They have an interest in the machinery being of adequate quality to ensure that the investment generates the expected cash flow to repay the loan. There are examples where the testing and evaluation requirements of a targeted credit program significantly improved domestic machinery quality.
- *Advisers:* These are extension officers, consultants, and journalists of technical magazines. They need to be informed about the quality and performance of the machinery they recommend.
- *Users:* These are farmers, contractors, and managers of agricultural enterprises, who have to make investment decisions. They have little need for technical engineering parameters. For the user, it is more pragmatic to evaluate the performance of machinery under realistic field conditions and to have its performance, suitability, and financial benefits judged by the user himself or herself.

In the case of education, the courses at academic institutions often are patterned after those of industrialized countries, with the content of little relevance to the problems of the local farmer. In addition, agricultural engineering is frequently under the department of mechanical engineering, leading to graduates trained in design, when they should be trained in how to adapt or make existing technology cater to the farmer's needs.

Supply System

This includes importation, local manufacture, distribution, repair, and maintenance. These activities, including machinery-hire services, should always be undertaken by the private sector. If the private sector is nonexistent or weak, the government should provide assistance and incentives for it to enter into this type of business. The government must make foreign technical assistance and loans, for the supply of specific machinery conditional on the fact that the manufacturer is well represented in the country, or that the manufacturer must set up domestic support facilities as part of the contract. In particular, in the case of the previously centrally planned command economies, the supply system is not developed, and an agricultural mechanization strategy needs to substantiate how to establish a demand-driven agricultural mechanization supply and services sector [3].

References

1. FAO. 1997. *Agricultural Mechanization Strategy Preparation: A Guide.* Rome: Agricultural Engineering Service.

   The Agricultural Engineering Service (AGSE) of the Food and Agriculture Organization (FAO) of the United Nations has extensive experience with the topic and has undertaken several strategy studies in developing countries as well as in Eastern Europe. AGSE is the principle and unique source for information and assistance on agricultural mechanization strategy. Earlier publications from AGSE on this topic followed a central-planning approach to mechanization and focused on the collection of detailed information, with the policy environment considered exogenous and to be taken into account, rather than emphasizing the need for policy reform. This approach is now considered obsolete and not to be pursued. This new AGSE bulletin reflects the latest thinking of the Agricultural Engineering Service of FAO on the formulation of an agricultural mechanization strategy. It is a manual prescribing the various steps and requirements to assist those who are involved in actual strategy formulation work.

2. Rijk, A. G. 1989. *Agricultural Mechanization Policy and Strategy.* Tokyo: Asian Productivity Organization.

   This book deals with definitions, terminology, and concepts concerning the formulation of an agricultural mechanization strategy. It provides a comprehensive review of the agricultural mechanization process and addresses important developmental issues of mechanization in relation to crop production, farm-family income, employment, social change, transfer of mechanization technology, and fossil energy consumption. It discusses the need for mechanization policy and strategy, and its place in the agricultural planning process, and provides for general policy and strategy guidelines. As a case study the book addresses agricultural mechanization in Thailand and describes a computer model that was used for analyses of mechanization policy and strategy in that country. The book contains an extensive list of relevant literature.

3. FAO. 1997. *Farm Mechanization in Former Centrally Planned Economies.* Rome: Agricultural Engineering Service.

   Following the demise of the central-planning regimes, the gradual privatization and transition to a market economy have profound effects on the farm mechanization subsector. AGSE has significant experience with the formulation of agricultural mechanization strategy in the formerly centrally planned economies of Eastern Europe. The mechanization requirements in these countries are large, while the problems and issues related to agricultural mechanization are different from those in developing countries. The report analyzes the problems with mechanization of the former centrally planned economies, provides useful information about the issues involved in restructuring the mechanization subsector, and provides proposals for the establishment of a private sector demand-driven mechanization system.

2.3. Transfer of Technology

P. S. Lammers

2.3.1. Introduction

The development of new techniques by research facilities and of new machinery by the industries is the driving force for the transfer of technology (TT). TT is a current problem for both industrialized and nonindustrialized countries worldwide. In the first case, the rate of TT is vital for sustainable development of economics and for progress in economic competition. As for nonindustrialized countries, TT is necessary for beneficial use of existing techniques, their application in agriculture, and their adaption to socio-economic conditions.

The term *transfer of technology* is most frequently used for the import of agricultural machinery into the less industrialized countries. Technology transfer in this sense is based on the idea of supporting less industrialized countries with advanced technology from industrialized countries, thus promoting their agriculture. All persons involved in this process are aware of the fact that there are numerous preconditions that have to be met in order to be successful with this kind of strategy.

In industrialized countries, TT also takes place by transferring industrial technology and research innovations into agricultural technologies. In these countries, progress in technology is related to progress in industrial innovation and research.

An estimated two thirds of the worldwide expenses for research and development (R&D) are invested by the European countries, the United States, and Japan, while developing countries spend less than 10% of the entire R&D sum [1]. Low expenses for R&D are expected to cause low development, especially a slow pace of progress in technology, with such progress being synonymous with economic welfare. In this context, TT generally stands as a benefit in itself without any exceptions. It is, however, important to define the targets of TT. With respect to agriculture and concentrating on plant production, definitions of targets include quality improvement, the avoidance of drudgery, reduction of losses in harvesting processes, and an increase in yields in combination with other measures (e.g., plant protection, fertilization, and resource saving). With the same emphasis, the responsibility of the persons involved in promoting TT has to be mentioned. Engineers have a special responsibility for the design and development of machinery. Technology developed by engineers should be in accordance with care for human welfare (e.g., safety) and with the saving of natural resources such as energy.

2.3.2. TT in Industrialized Countries

TT in industrialized countries is no less a challenge than in other countries. But TT in these countries stands for the transfer of innovations from research and development facilities to the industries that disseminate and launch it in the market. According to Blöser and Glaser [2] forms of TT in this respect are:

- Research on request
- Cooperative research of institutions of the industrial sector and research institutions

- External consultation
- Transfer of personnel
- Training and further education

Innovations presuppose changes in socio-economic conditions even in industrialized countries. In these countries, TT can be defined as having control of a timely and economic utilization of new technologies, which has to be accompanied by the preparation of an infrastructure and market structure for new products, new processes, and advanced technical know-how.

The capabilities of TT are dependent on the following key factors reported by Theis [3]:

- Information systems
- Transfer of personnel
- Setting up of business ventures
- Dissemination activities and extension services

A basis for an information system includes catalogs presenting the experiences and technical knowledge offered by research institutions and the information demanded by industries. At any rate, offers and demands must find common ground to make technology transfer feasible. The transfer of personnel is the most efficient way of spreading knowledge and technical know-how. But frequently this transfer occurs only in big companies that employ academic staff. Medium- and small-sized industries are disconnected from this kind of TT because the they do not have the personnel structure and cannot afford to employ academic personnel. Taking into account that the capability of innovation is strongly dependent on personnel transfer, other forms of this method also must be considered, such as job rotation with universities and cooperation within projects. Another approach to a realization of research results of universities is to set up new commercial ventures. In this respect, it can be mentioned that in almost all industrialized countries, industrial parks have been established during the past few years to encourage university graduates to set up new commercial enterprises and to make the first steps easier to take. Supporting access to the technological know-how of the universities additionally is a matter of disseminating their knowledge, which can be organized by means of workshops using print and audiovisual media and extension-service facilities.

Obstacles that hamper the transfer of technology between universities and industries lie on both sides. Frequently medium-sized industries have no interest in TT; new projects appear too risky or they are connected with very specific requirements. Research facilities are handicapped by bureaucracy and restricted by regulations and laws focusing on theoretical approaches, or they are interested only in long-term investigations and research. Furthermore, they suffer from a lack of TT facilities.

There are various governmental tasks in promoting TT, including the support of transfer facilities and the establishment of industrial parks, as well as launching programs for the promotion of TT. Governmental promotion can take the form of direct or indirect project promotion or of indirect specific measures. In the case of direct project promotion, the government defines the activities to be carried out. Indirect promotion measures imply grants intended as financial incentives without definition of the activities they are to be spent for (e.g., the promotion of consultation and advisory services). Indirect promotion takes place if the companies use project funds but make their own decisions how to accelerate TT. However, the experience of the past decade is that governmental programs have little impact on accelerating TT.

In agriculture a typical technical innovation has an unsteady course, which can be illustrated by the adoption of new technologies and new machinery by young farmers. These farmers plan and invest for their whole professional life but make their decisions within a relatively short period, which determines the occasions for access to innovation in agriculture. This is especially valid as far as investments in farm buildings and other long-term investments are concerned. TT, in consequence, should primarily be directed to young farmers, but at the same time it must be considered that not all farmers always have the same readiness for the acceptance of innovations.

In agriculture, innovations and new technologies characteristically aim at increasing yields per hectare and production per worker [4]. Now in a period of surpluses of agricultural products, improving the efficiency of agricultural production is the driving force for the implementation of advanced technology. In addition, there are other issues that become more and more important, such as environmental protection and the saving of resources.

In these industrialized countries, TT as an activity is postulated to proceed simultaneously with a reliable assessment of new technologies in order to take the responsibility for the future consequences and risks. This might lead to the assumption that the already approved, recognized, and reliable technology of the industrialized countries can be applied to less industrialized countries easily. Such theories are of a tempting simplicity, but they reduce the problem of TT to the question of how fast societies are able to learn to deal with new technologies.

2.3.3. TT in Less Industrialized Countries

In these countries there is a strong link between politics and progress in applying modern technology. TT therefore depends on policies. In fact, this relation is mainly due to the lack of investments in the private sector, including agriculture. But only on the basis of available funds is a transfer of technology (e.g., mechanization in agriculture) possible and, in consequence, the governments, as the main investors, control the diffusion of advanced technologies. In addition, governmental policies are aim at national welfare and benefit, which is not in every case in accordance with the farmers' interests. The production of cereals, for example, can be a national interest because of a higher demand for flour for human nutrition, but farmers can earn more money by growing cotton. Hence, the adaptation of mechanization in cereal production cannot be expected to be accepted by the farmers willingly.

TT can be structured as:

- Evaluating and upgrading traditional technology
- Development of technology by applying modern engineering knowledge through R&D
- Cooperative development of technology by joint ventures
- Evaluation and adoption of imported technology

Upgrading traditional technology is supposed to be the most self-reliant method for the progress in mechanization. Improvements of hand-held and animal-drawn implements can be developed by the farmers and local manufacturers. Khan [5] claims that design and production of simple low-cost machines is an important factor in the mechanization of very small agricultural farms in particular. But it is widely recognized that the adoption of bullock-drawn and hand-operated tools is slow in comparison with the rate of adoption of tractor-drawn implements and irrigation equipments in India [6].

During the past few years a remarkable increase in TT between less industrialized countries has occured. In this case, TT is practiced by exporting agricultural machinery from countries like Brazil, India, China, and Egypt to neighboring countries or overseas markets. These exporting countries are not considered industrialized, but they have succeeded in developing their own manufacturing industry for agricultural machinery. Low costs of producing agricultural machinery in combination with advanced technology and affordable investments in the low industrialized countries have contributed to this successful approach to TT. Apart from that, the imported technology is likely to be taken as more appropriate than the high-level technologies from the industrialized countries. But even between less industrialized countries basic prerequisites must be fulfilled for successful TT.

Prerequisites and Constraints Connected with TT in the Rural Areas

The transfer of technology is a procedure that consists of a source of advanced technology and an acceptor, which is synonymous with a demanding group of society. The source can be domestic or foreign manufacturing industries that offer the products, including technological progress, to the consumers who are, in this context, the farmers.

Key prerequisites include:

- Suitable infrastructure
- Education and skills
- Availability of funds

A suitable infrastructure includes information technology, appropriate market and land tenure structures, financial institutions, and communication and transport systems, as well as an energy supply. Education is directed toward the farmers to elevate their knowledge in agricultural practices and to make them familiar with advanced technologies and with the economics of new technologies. On the other hand, for a sustainable TT, it is an equally important precondition to develop an industry for the production of farm machinery. In consequence, there must be corresponding facilities in the rural areas for training both farmers and personnel working in the agroindustries. Long-term education, which is regarded as a governmental task, is necessary for the establishment of academic professionals in research institutions engaged in the field of agricultural engineering and in the agroindustries where engineering working methods are introduced.

Moreover, no form of TT can be successful without careful planning of the transfer activities and ideas. Planning basically includes the analysis of the existing rural situation: land tenure, local climate, socio-economic conditions, and cultivation methods.

Steps and Structure of TT

Technology transfer can only be planned if there is a satisfactory knowledge about the organization and the steps of the procedure. In general, the following items can be listed as a rough structure:

- Dissemination of information about new technologies
- Preparation of the prerequisites such as capital, education, and infrastructure
- Production on a price level that is appropriate to the economic benefit of the new technology

Concretely referring to agricultural machinery, steps of technology transfer are:

- Checking on economic feasibility and evaluation of the new technology by farmers
- Developing products
- Guiding the commercialization of prototypes
- Training manufacturers and artisans in the production and maintenance of the new agricultural machinery
- Engaging in marketing activities for the supply of advanced agricultural machinery in rural areas
- Disseminating using exhibitions, field days, and public demonstrations
- Instructing farmers in the application of the new technology and operation of new machinery
- Disseminating by directories, leaflets, and publications

The Role of Farmers

Sutton [7] expressed the efforts and willingness of farmers to take part in the process of TT as follows: Each farmer has a very clear idea of which type of design is best suited to his or her particular needs and will often go through considerable trouble and expense to obtain this.

In the matter of TT in the sense of developing machines for a limited sector (special crops) and on the condition of a full integration of farmers and engineers, Martinov *et al.* [8] proposed a particular concept. This proposal includes the identification of the need for machinery or implements as a first step (Fig. 2.9).

The growers' ideas are required from the beginning, and the engineers' ideas are linked to them to set up a prototype. Developing a machine that fulfills the technical requirements and that also can be manufactured economically is a challenge the engineers have to face, while the testing of prototypes is a common task for both partners in the developing process. The test results have to be evaluated and should be used for improvements. The final step is the dissemination of technical know-how, which in this case means skills needed for the application of the new machine, and not the technical knowledge of the engineer.

The Role of Farm Machinery Industries for Local Manufacturing

The introduction of suitable farm machinery and its local manufacturing in a country are two aspects of mechanization and should be tackled simultaneously. Khan *et al.* [9] found that a self-reliant approach to mechanization, based on indigenous manufacture with optimum use of local labor, production methods, and materials, seems to be the most appropriate strategy for many developing countries. Moreover, the advantages of

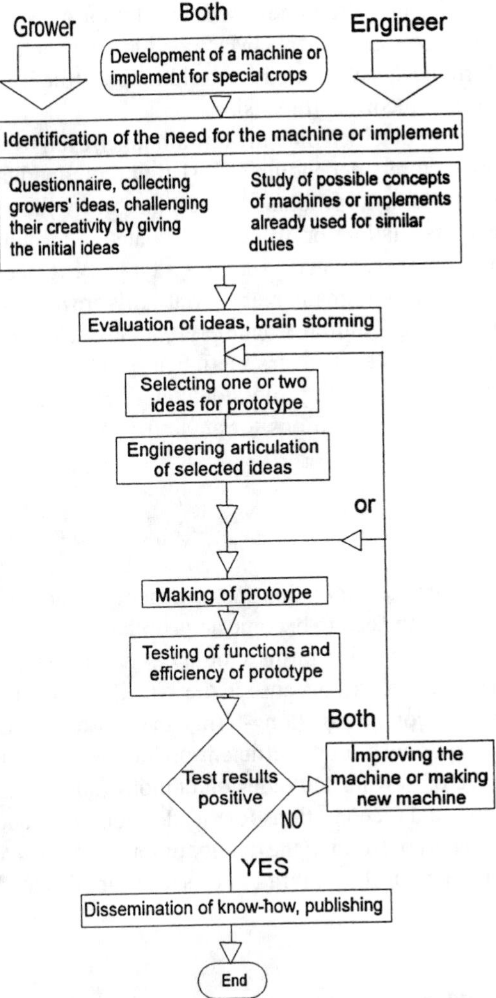

Figure 2.9. Flow chart for a concept of development and dissemination of agricultural machines for special crops.

locally produced machinery show in better maintenance and repairing possibilities. By contrast, Zaske [10] points out that Local manufacturers are not able to develop and produce advanced farm equipment by themselves.

The main weaknesses of local manufacturing are:

- Improper manufacturing facilities
- Raw material that often is not available and of poor quality
- Scarcity of managerial, engineering, and technical manpower

Manufacturing facilities are often one-sided and do not meet the requirements for manufacturing of agricultural implements. Machine tools for metal bending frequently are missing. Metal forming is substituted by welding constructions, which leads to a stretching of the shape and incorrect dimensions.

High quality steel can be obtained only in large amounts, even by industrialized countries. Small local manufacturers depending or insisting on individual procurement will not have success. Nevertheless, the problem can be managed by cooperative procurement through manufacturers' associations or by technical centers. Raw material of low quality requires an adapted design of machinery, which is a typical engineering assignment. In this context engineering and additionally managerial skills are necessary as a basis for an independent production of agricultural machinery that also is capable of development. The absence of accounting systems causes inefficient production, and thus there is no basis for cost-efficient methods. Both professional engineering and management make production flexible and are the conditions for an active participation of industry in the process of TT.

Facilities for TT

Demonstration Farms

Demonstration farms are dissemination instruments and should be involved in the training of young farmers who need to become acquainted with new and advanced technologies. Besides that, demonstration farms should be charged with the identification of suitable and appropriate technology, because in these facilities, the new farm machinery can be applied under local conditions. Long-term use of farm machinery on these testing sites will provide experience on the efficient operation, durability, and lifespan of the implements. The use of new technologies on demonstrations farms simultaneously points out the constraints and preconditions for an adoption of advanced technology. As examples, the requirements confronting the quality of repair services for machinery containing electronic equipment or the calibration of seeds for advanced precision seeders can be mentioned.

Technical Service Centers

These institutions aim at the promotion of the local manufacturing of agricultural machinery and spare parts as well as at the improvement of repair services. These centers preferably should be established in areas of intensive agriculture where blacksmiths and small manufacturers are already working.

Service centers can fulfill the following functions:

- Technical assistance and advisory service for the manufacturers
- Quality management
- Training in the operation of machine tools
- Workshop management
- Provision of alloy steel and other raw materials and shapes
- Provision of special production facilities as needed for heat treatment or gear and hydraulic-cylinder manufacturing

- Production of components to substitute for imported parts
- Interface between manufacturers and farm-machinery institutes

Examples of TT Successfully Carried out in Developing Countries

Pakistan[1]

To describe the process of technology transfer, an example is given for a less industrialized country. Pakistan is a country where the major crop is wheat. Traditionally, wheat has been harvested manually with sickles, tied in bundles, and then threshed with stationary wheat threshers. The mechanization of wheat harvesting was ushered in by the introduction of reapers of Chinese origin. Reapers cut and convey the wheat vertically. The stems are laid into continuous rows. The cut straw has to be bundled and then threshed as in the traditional procedure. A first attempt at mechanization using cutter binders of the European type, which cut and bundle the grain, was not successful due to the need for sisal rope for tying the bundles. Sisal rope caused additional costs and was not available in the local market. The implementation of reapers was not successful before changing the power tiller–driven version into a tractor-mounted version. In the main areas of wheat growing, tractors were already available at this time.

In pursuit of a successful implementation of reapers, attention must be paid to the following items:

- The manual cutting of grain is most inconvenient and labor consuming.
- A mechanization trying to introduce cutter binders results in an increase of losses.
- Tractors must be available.
- It must be possible to manufacture reapers locally.

To ensure a reliable machine, the adaptation was worked out by professional engineers, which helped to win the confidence of the farmers.The dissemination must be supported by training and leaflets.

The National Farm Machinery Institute (FMI) played a major part in the adoption and dissemination of the reaper technology. FMI tested and evaluated the Chinese reapers for 2 years. Defects were registered and prototypes of improved versions were set up. At the same time potential manufacturers were identified and thoroughly involved in testing and evaluation as well as in the improvement of prototypes. The FMI-Engineers studied the availability of raw material and various production methods to make the reaper cheaper in price and manufacturing. Moreover, information material in the form of blueprints and leaflets was prepared. Potential sites for demonstration were chosen considering the proposals of the manufacturers taking part in the project. At the beginning, the demonstration was carried out on large farms representing future areas of implementation, and also in the presence of provincial government officials and credit-giving banks. Later, the demonstration was expanded throughout the country. In order to ensure that farmers would not be confronted with any difficulties, they arranged for training of manufacturers and technicians in the use and repair of reapers. As an additional measure, just in the early years, FMI engineers stressed close cooperation with the manufacturers to achieve a successful operation of the reapers on the farmers' fields.

[1] Author: Dr. Abdul Shakoor Khan, Director of Farm Machinery Institute, Islamabad, Pakistan.

The Role of Organizations and Associations

A well-defined strategy of mechanization is the key to success for any implementation program. Agricultural mechanization organizations (AMOs) are policy makers and should therefore define the targets of innovation, select promising techniques, create guidelines, and organize training. AMOs act as linkages between industries and FMIs, and also cultivate contacts with foreign manufacturers of advanced technology. Their task is to set up standardization and test procedures and to give impulses to the establishment of test institutions. As policy makers representing the private sector, they are the counterparts of the governmental institutions and authorities in the development of strategies for TT. In so far as the farmers' associations are concerned, TT and mechanization are only part of their task, but an important part, as they express farmers' desires for new technology and support governmental institutions in analyzing the agricultural sector and developing strategies for TT.

The Role of Farm-Machinery Institutes

Institutions supporting the development of farm machinery by providing personnel with engineering skills are vital for the adaptation of existing technologies to local conditions, a reliable application of technologies, and progress in mechanization. Engineering skills are synonymous with technical know-how related to the development and design of machinery, transfer of knowledge about standards and measures in the field of testing, and quality control. FMIs can set up data banks registering suppliers of raw material and containing guidelines for an appropriate use of material. Furthermore, they are to support the manufacturers in production issues, such as the introduction of accounting systems to enable the manufacturers to produce with low costs. On a national basis, they can assist AMOs in standardization, and politicians as well as administration staff through information that can be gained through research.

FMIs should place a special emphasis on their efforts to promote engineering sections in enterprises manufacturing agricultural machinery. Engineering in manufacturing industries is not only development but also testing, documentation, and production, as well as after-sales service. Documentation, including bills of materials, drawings, and spare-part catalogs is missing in most manufacturing industries in nonindustrialized countries. Documentation, however, is necessary for quality production and for reliable service. Undocumented production of farm machinery is limited to simple machines that can be maintained only by universal tools and spares.

Because FMIs as research institutions in most developing countries receive public research funds, they must devote special attention to applied research and machinery design and development as well as to the dissemination of technological know-how among farmers and manufacturers. Moreover, the personnel of FMIs have an academic education that they frequently received in industrialized countries. Therefore, FMI personnel have access to advanced technologies and to know-how about machinery available in the industrialized countries. They can judge their functions and the conditions of their mission, the cost imposed by implementation, the requirements of maintenance and energy supply, and the skills needed to operate a newly introduced machine. Taking these capabilities of FMIs into account, it can be stated that they are a key factor in TT, for they also have the best knowledge and expertise about their own country's agriculture. Therefore, these institutions are highly qualified to guide TT. Because the benefit–cost ratio of farm mechanization is very low compared with many other inputs in agriculture, it is an extraordinarily challenging task that requires experts to implement advanced technology in the agricultural sector.

The Role of Education and Training

Using technology of a higher level is related to advanced skills and knowledge. The implementation of new technologies is a procedure linked with a kind of progress that is not restricted to technical issues but also involves progress in education and training. Education is seen as a long-term governmental task with the intention of elevating public education, which includes academic education and vocational training or the development of skills.

Training, in the case of promoting TT in agriculture, should be recommended to farmers and to personnel in the industries, administration, advisory service, and banks offering loans for agricultural machinery.

Special operational training preferably is aimed at young farmers, for they have a natural interest in new technologies. But successful implementation of new machines in agricultural processes demands more than just the ability to drive a tractor. Training therefore should be suggested not only to young farmers but to innovative farmers in general.

The instruction of farmers should include the following key items:

- Manpower training for efficient operation of the machinery
- Maintenance of machines and simple repair
- Cost of machines and expected benefit of machine use (cost-consciousness)
- Cooperative machine use
- Consideration of manufacturers and their staff as well as advisors for training

Training for manufacturers should embrace:

- Vocational training in operation of machine tools
- Quality control
- Accounting systems
- Workshop organization
- Raw material use
- Sales service including marketing and after sales service

TT has to be prepared by planning and keeping provisions of indispensable prerequisites, which is a task for highly educated personnel. Advisors or the personnel of an agricultural-mechanization organization should be trained for this kind of work.

The Role of the Government

As defined by Clarke, [11] the role of the government is in the following areas:

- Policies that affect the subsector
- Research and development
- Testing
- Education, training, and extension
- Supply of information

- Mechanization departments and a ministry of agriculture
- Consumer protection

The subsector includes the private sector related to agriculture, manufacturing entrepreneurs for agricultural machinery, and related institutions as well as associations dealing with TT. Policies affecting the subsector consist of numerous items such as policies influencing agrarian input prices as well as agricultural product prices, infrastructure, and energy and farm power, as well as land ownership and tenure policies. Additionally, governmental policies rule the financial markets for farmers and for manufacturers of agricultural machinery. The role of the government traditionally includes education and training, mechanization departments and ministries of agriculture, and research and development. All these items point out that there is a high importance to the governmental actions in the field of TT such as giving incentives, preparing conditions, and regulating the market of the subsector. But it must be emphasized that the other involved sectors are part of a chain for successful TT. A predominant role for the government bears the danger of neglecting interests in the process of TT that must not be neglected. On the other hand, the list given by Clark contains items and activities that can be delt with by the private sector or semigovernmental institutions. Testing and consumer protection are public assignments but this service, already well-proven in industrialized countries, can contribute to TT without being organized by state institutions.

The Protection of Intellectual Property and the Risk of a Possible Misuse of Know-how

As early as in the 1970s, a code was set up for TT, which includes conditions, prerequisites, and aids intended to protect the mutual interests of partners involved in TT. Part of the code is that each country acknowledging its contents is obliged to formulate, in the national law, a protection of intellectual property. In detail, the code proposes for the regulation of TT processes a restriction to research for the purpose of developing and improving the imported technology and to the adaptation of transferred technology to the conditions of the receiving country. The problem of how to deal with obligations after the expiration of property rights also is mentioned.

References

1. Gatz, W. 1980. Auswertung der Dokumentation der fünften Welthandelskonferenz. In *Wissenschaftliche Schriftenreihe des Bundesministeriums für Wirtschaftliche Zusammenarbeit*, vol. 37, 541–612. Baden-Baden: NOMOS.
2. Blöser, H., Glaser, E. A., 1988. *Der Technologietransfer in einer Strukturschwache Region*. Endbericht: Betriebswirtschaftliches Forschungszentrum an der Univeristät Bayreuth.
3. Theis, A. 1986. *Modellversuch Technologietransfer*. Abschlussbericht: Universität Tübingen.
4. Cheze, B. 1992. Coming technologies in agriculture. *Zemedelska Technika* 38:331–342.
5. Khan, A. U. 1992. Critical issues in transferring agricultural mechanization technologies in the developing world. *Agricultural Engineering Journal* 1(2):105–141.

6. Singh, G. 1990. Technology transfer: An experience at CIAE, Bhopal, India. *AMA* 21(3):66–72.
7. Sutton, D. H. 1990. Manufacturing in the third world. *Agricultural Engineering* (Autumn): 79–93.
8. Martinov, M., P. Schulze Lammers, and M. Tesic. 1997. Involving growers in development of mechanization for special crops. *AMA* 28(2):65–68.
9. Khan, A. U., D-M. Murray-Rust, and S. I. Bhuiyan. 1987. What developing countries need from today's engineers. *Agricultural Engineering* (1):14–16.
10. Zaske, J. 1988. Technologieentwicklung und Technologietransfer in der Agrartechnik der Enwicklungsländer. *Grundlagen der Landtechnik* 38(5):156–158.
11. Clarke, L. J. 1998. Agricultural mechanization strategy formulation concepts and methodology and the roles of the private sector and the government. Procceedings of the XII CIGR conference, vol. 3, pp. 1–15, Rabat/Morocco.

2.4. Field Machinery Management

D. Hunt

Good management of farm field machines is a substantial portion of a farm's economic success. Good machinery management includes understanding the mechanical principles and limitations of each machine, efficient operations in the field, appropriate machine maintenance, timely repair and replacement, and selection of a machine system.

Good machinery management requires an economic analysis of the actual or proposed management activities. This section presents simplified mathematical and economic concepts designed to be applied by on-farm managers to the problems in machinery management.

2.4.1. Field Operations

One aspect of machinery management is that of efficient operation of implements in the field. Tractor drivers need to understand that the operation of the implement attached to the tractor is their prime responsibility. Casual tractor steering and inappropriate tractor speeds can reduce an implement's efficiency.

Effective steering of the tractor–implement combination or self-propelled machine consists of several factors. Tillage implement paths often are overlapped to insure complete coverage. As shown in Fig. 2.10, this overlap is about 7.5% and reduces the effective width of the implement to 92.5% of its rated width. As a result, the field efficiency of the machine can never be above 92.5%. Steering to reduce excessive overlap improves operation efficiency. The problem is greatest for wide implements, as it is difficult for the centrally located tractor driver to judge the overlap occurring some distance away.

Overlap is wasteful in grain seeding machines, as is the unplanted space if successive paths are not adjacent. Guidance markers are used on many seed and chemical distributing implements in recognition of the difficulty for the tractor driver to use strictly adjacent paths. But such are effective only if the tractor driver pays constant attention to steering directly over the marker.

Figure 2.10.

Harvesting operations require guidance attention also. Cutterbars for both forage and grain harvesting are operated at something less than full width to insure that no area is missed from the previous path. But an excellent tractor driver will keep the effective implement width well above 95% to 97%. Row-crop harvesters may experience crop losses if the implement is not directed exactly over the row. If the row planter established misplaced rows, that inefficiency is repeated for all subsequent row-crop operations.

Steering assistance mechanisms are available for tractors and are helpful if their costs can be applied over many hectares. Global positioning technology may eventually be practical for farm field guidance, but until then careful human steering of machines is critical to field operations.

The speed of field operations is under the control of the tractor driver and is another important factor in efficient field operations. As the tractor driver may not be the machinery manager, the manager should stress to the driver that the objective of field operations is to minimize the time required while maximizing the effectiveness of the implement. Excessively slow field speeds waste time and in some cases may not produce good results for some tillage operations. Proper tillage speeds depend on the requirements of the seedbed and, of course, may be limited by the power available. Speeds should be limited to that maximum that does not throw the soil into an unacceptable position or produce an unacceptable fragmentation. Seeding operation speeds are limited by the ability of the implement to effectively meter and place the seed. Harvesting speeds are limited by excessive crop losses. Slower harvesting speeds may be necessary if wind-blown and overdry crop conditions are encountered.

As efficient engine fuel consumption depends on substantial load on the engine, increasing field speeds not only produces greater field capacities but also provides a more efficient loading on the power unit. Further increases in fuel efficiency without sacrificing field capacity can be realized by shifting the tractor to a higher gear and then reducing engine speed. Such is possible for light loads on tractors with excess power for the operation.

The tractor driver must not be uncomfortable with the faster speeds, and his degree of comfort may represent the upper limitation on such speeds. Some attention should be given to the production of smooth field surfaces so that the roughness does not jolt the driver or the machine excessively. The speed should not be so fast that stress is placed on the tractor driver.

Efficient field operations can be accomplished only with competent and alert tractor drivers. The need for alertness increases with the size and complexity of implements. In row culture for example, it is easy to observe the action of a two-row implement. An eight-row planter, cultivator, or harvester demands much more attention to discover malfunctions at one of the rows. Complex machines such as combine harvesters have many mechanical functions occurring simultaneously, and all require monitoring. Manufacturers are providing instrumentation that indicates the several operations, but such still requires gauge monitoring by the driver. The problem of driver alertness is recognized by manufacturers who provide visual and audio monitors to indicate malfunctions.

2.4.2. Field Patterns

Efficient field patterns are a responsibility of the machinery manager. Acceptable patterns depend on the field boundaries and the maneuverability of the self-propelled machine or the tractor–implement combination. Some common patterns (Fig. 2.11), are reported by Hunt [1], as is much of the following pattern analysis. Field patterns are identified by the type of implement turns required and do not depend on the shape of the field. Any decision that can produce a field pattern having a minimum number of unproductive turns is ideal. Long, rectangular fields can permit turning times as low as 2% to 3% of the total time, but such efficiency may not be achievable in small, irregular, or hilly fields. For such conditions the geometry of making a machine-engaged turn is important. Figure 2.12 describes an implement of width w and turning radius r proceeding through a 90-degree turn. Keeping the implement engaged means that no productive time is lost from turning, but the operation cannot be effective unless the implement is so maneuverable that the pattern does not leave significant unprocessed crescent areas in the field. For the shown implement geometry, these areas may be ignored for turns over 135-degrees, but sharper turns cause greater losses, as shown in Table 2.6. If the areas are not ignored, extra passes of the implement are required to process the missed crescents, and field efficiency is reduced.

Operation-restricted implements have special considerations for effective field patterns. A harvester with a cutterbar extending to the right of the tractor path is an example of an operation-restricted implement. In a large field with a turning headlands pattern, such a harvester initially must travel the full length of the headland before making the next turn into the crop (Fig. 2.13). Breaking through the crop (a breakthru) to establish a series of smaller areas to be processed produces a more time-efficient operation, as the travel times over the headlands are reduced. The cost of damage from the breakthru may be prohibitive for a high-value crop, but if a careful half-speed pick-up of the trampled crop is acceptable, an optimum pattern can be established. The optimum number of

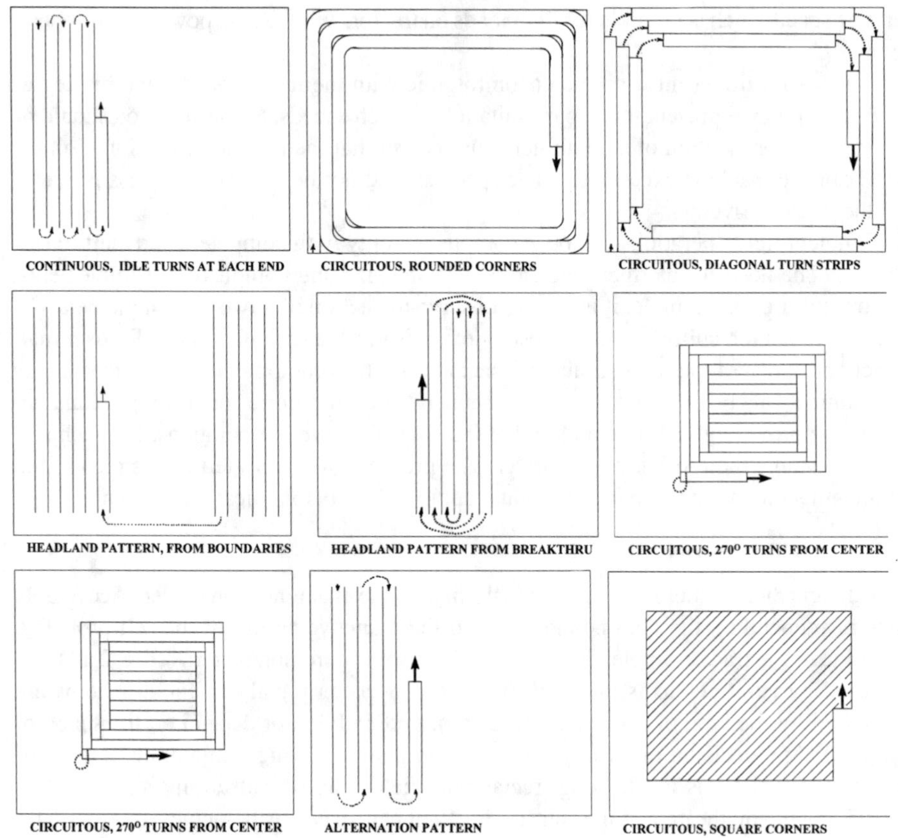

Figure 2.11.

breakthru passes is given as:

$$n = 0.5 + [(W^2 S_p)/(8 f w S_e)] \exp 0.5 \tag{2.46}$$

where n is the optimum number of breakthrus, W is the width of the field, (m), f is the length of the pass (m), w is the effective width of the cutterbar (m), S_p is the speed of the working harvester (km/h), and S_e is the speed across the headlands(km/h)

A mathematical solution of Eq. (2.46) seldom gives a whole number of breakthrus. In practice, one should choose the closest whole number. Realize that the number of breakthrus, n, is not the total number of areas processed. The total number of equal-sized harvest areas is $2n - 1$, which was used in the development of the pattern pictured in Fig. 2.13. Should the optimum number of breakthrus have been computed to be three, the initial travel of the implement for each area is indicated by the numbered arrows.

Self-propelled machines can perform a breakthru without any time or yield penalty. The appropriate turn on a headland is that which consumes the least time and may involve by-passing several path widths if the machine is not very maneuverable.

Table 2.6. Losses for a continuously engaged turn

Corner Angle (degrees)	Crescent Width (decimal of w)	Ratio of Extra Passes Required to Approaches
135	0.082	5:61
90	0.414	12:29
60	1.000	1:1
45	1.600	8:25
30	2.860	20:7

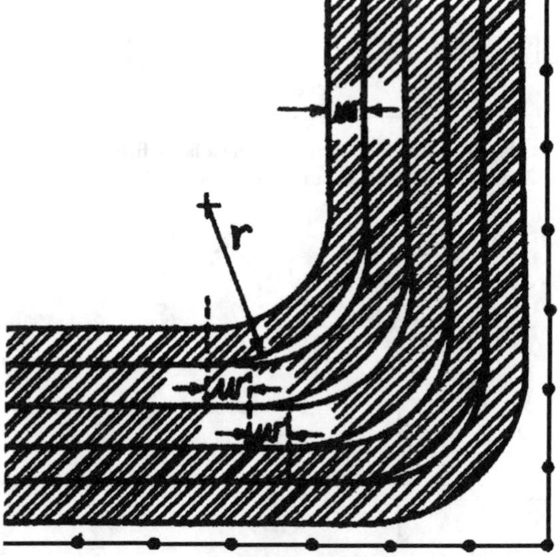

Figure 2.12. Engaged implement paths for a corner.

A more general turn problem occurs if an angled headland is encountered (Fig. 2.14). An implement having a width of w approaches the turning headland at an angle A. To completely cover the field area the implement must process the cross-hatched area in Fig. 2.14 as well. The length of the turn is increased by w/tan A over that of a 180-degree turn for a rectangular field. This extra travel causes losses in time, fuel, and distributed materials as the cross-hatched area is covered twice. This loss in time is extremely high if the angle is less than 30 degrees, as shown in Table 2.7, which was developed for a 4-m implement traveling at 4.8 km/h. If implement width is doubled, the losses are quadrupled.

Triangularly shaped areas arise frequently, and some conclusions about an efficient pattern for these fields can be developed. Figure 2.15 shows a triangular field with the implement paths parallel to the longer side of the field, because such minimizes the total number of turns. This pattern should be adopted, as only half of the headland turns are angled ones. The pattern in Fig. 2.16 is never more efficient because both headlands are angled. For irregular fields, alignment of the implement paths with the longest dimension of the field is most efficient, as the total number of turns is less.

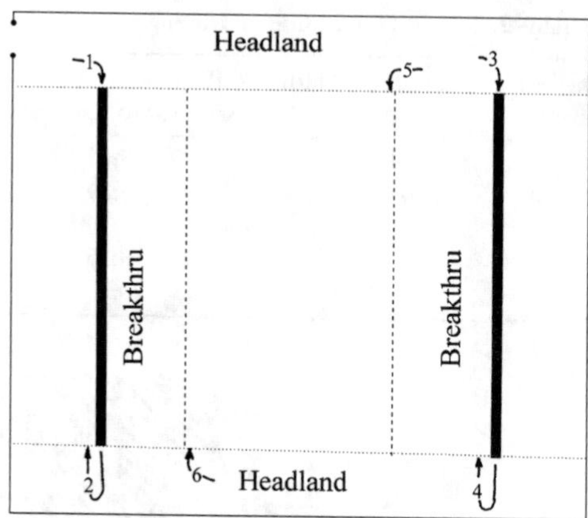

Figure 2.13. Breakthrus divide a large field into smaller areas, with less unproductive headland travel.

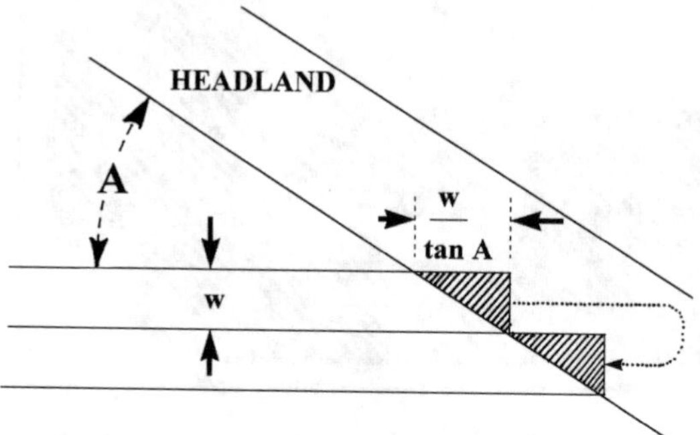

Figure 2.14. Angled headland loss.

Table 2.7. Turn-time losses for angled headlands

Angle (degrees)	Increased Turning Time (s)
90	0.00
75	0.80
60	1.73
45	3.00
30	5.20
20	8.24
10	17.01

2.4.3. Calibration

Good machinery management includes calibration tests of machines that distribute seed, fertilizer, and chemicals. Manufacturers provide dial settings to produce various application rates, but the variables under actual field operations may cause these settings

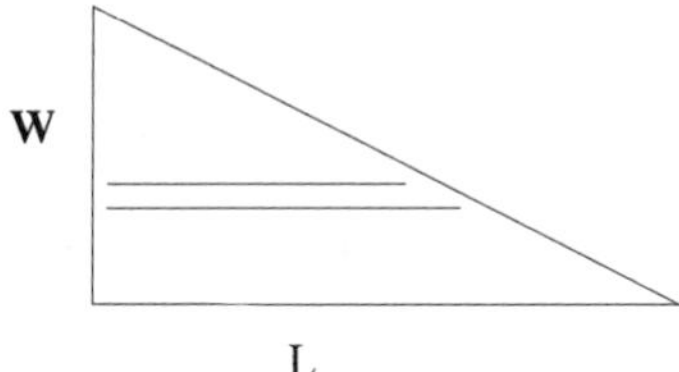

Figure 2.15. One-angle headland pattern.

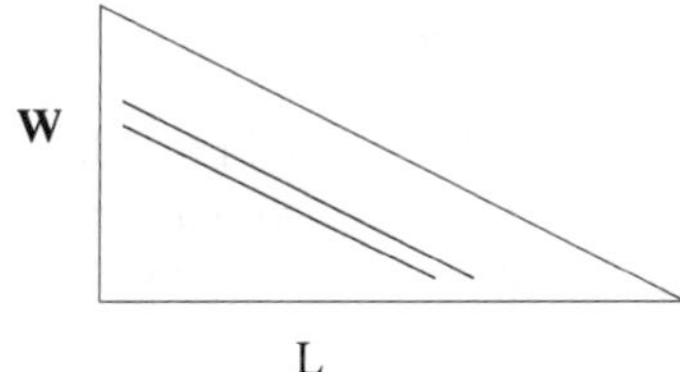

Figure 2.16. Two-angle headland pattern.

to be erroneous. Soil-surface conditions, physical characteristics of the materials being applied, wear in the metering mechanisms, and slackness in the control linkages contribute error. A calibration check conducted in the field with the material being applied should reduce error and provide confidence that the performance standards are being met. Uniformity of delivery from multiple metering mechanisms also is revealed from a calibration check.

Competence in simple arithmetic and being alert to the units that go with the numbers is a requirement for a machinery manager. In general, calibration is accomplished by catching the implement delivery over a measured distance. As an example, assume that the amount of wheat seed caught from a ground wheel–driven grain drill's eight delivery tubes spaced 16 cm apart is 0.5 kg during a field travel of 50 m. The desired answer is a seeding rate of kilograms per hectare. A unit factor statement assists in obtaining an answer by assigning proper dimensional units. The development of the appropriate unit factor statement starts by dividing the quantity of seed obtained by the distance traveled.

$$\_\,\text{kg/ha} = 0.5\,\text{kg}/50\,\text{m}$$

Because the rate per area is desired, the effective width of the metering mechanisms must be entered as a denominator.

$$\_\,\text{kg/ha} = 0.5\,\text{kg}/50\,\text{m} * 1/(8 * 16)\,\text{cm}$$

The remaining computation is one of producing the desired units with the use of unit factors. A unit factor of 1 m/100 cm is added to the statement in an inverted position so that the centimeter units cancel. Because 1 m is equal to 100 cm the numerical value of this factor is 1, or unity; thus the use of unit factors does not change the numerical values of the relationship:

$$\_\,\mathrm{kg/ha} = 0.5\,\mathrm{kg}/50\,\mathrm{m} * 1/(8 * 16)\,\cancel{\mathrm{cm}} * 100\,\cancel{\mathrm{cm}}/1\mathrm{m}$$

Adding another unit factor, 10000 m^2/1 ha, cancels the meter units, and the remaining units on the right side of the equality sign are equal to the desired ones on the left:

$$\_\,\mathrm{kg/ha} = 0.5\,\mathrm{kg}/50\,\cancel{\mathrm{m}} * 1/(8 * 16)\,\cancel{\mathrm{cm}} * 100\,\cancel{\mathrm{cm}}/1\,\cancel{\mathrm{m}} * 10000\,\cancel{\mathrm{m}}^2/1\,\mathrm{ha}$$

Only the numerical computation remains:

$$\underline{78.125}\,\mathrm{kg/ha} = 0.5/50 * 1/(8 * 16) * 100/1 * 10000/1$$

Essentially all calibrations follow this general procedure. A preliminary calibration plot can be accomplished in the shop by jacking up the implement and turning the drive wheel by hand. The number of revolutions of the drive wheel must be modified to include the effect of an effective rolling circumference, as this is less when measured in the field than when measured on concrete. Rubber-tired implements having filled seed containers have a reduced radius because the extra weight may cause the tire to flatten further and produce an even smaller rolling circumference. A half-filled container gives an average condition.

In a shop calibration test, 0.2 kg of seed are obtained from seven revolutions of the drive wheel having "a circumference of" 1.9 m. The unit factor statement follows:

$$\underline{117.5}\,\mathrm{kg/ha} = 0.2\,\mathrm{kg}/7\,\cancel{\mathrm{revs}}*1\,\cancel{\mathrm{revs}}/1.9\,\cancel{\mathrm{m}}*1/(8*16)\,\cancel{\mathrm{cm}}*100\,\cancel{\mathrm{cm}}/1\,\cancel{\mathrm{m}}*10000\,\cancel{\mathrm{m}}^2/1\,\mathrm{ha}$$

The ideal management practice is to produce calibration curves for each implement and for each material being distributed. A series of tests are run over the range of the metering mechanism's settings. The results are plotted and the resulting curve can indicate initial settings for intermediate rate applications. A final field check still may be required for best performance.

2.4.4. Loss Determination

A determination of the material losses from a harvesting implement allows the machinery manager to determine the harvesting efficiency of the operation and then suggest adjustments to reduce crop loss. The general procedure is one of operating the harvester over a known area under field conditions and then hand-gathering the losses from the soil surface. An example unit factor statement for a 3-m grain combine loss test illustrates the procedure. A 2 m by 1 m panel is carried under the combine and dropped during harvesting with the 2-m length parallel to forward motion. The loss under the panel consists of 60 g of grain dropped due to cutterbar action. The effluent dropped on the panel is 200 g of unthreshed heads and 30 g of clean threshed grain. Three unit factor statements reveal the loss in kilograms per hectare.

Cutterbar loss:

$$300\,\mathrm{kg/ha} = 60\,\cancel{\mathrm{gm}}/2\,\cancel{\mathrm{m}}^2 * 1\,\mathrm{kg}/1000\,\cancel{\mathrm{gm}} * 10000\,\cancel{\mathrm{m}}^2\,/\mathrm{ha}$$

Threshing cylinder loss:

$$333\,\text{kg/ha} = 200\,\cancel{\text{gm}}/(2 * 3)\,\cancel{\text{m}}^2 * 1\,\text{kg}/1000\,\cancel{\text{gm}} * 10000\,\cancel{\text{m}}^2/\text{ha}$$

(Notice that the width from which both the threshing and separating losses are obtained come from the full width of the 3-m cutterbar and not from the 1-m width of the panel.)

Separating loss:

$$50\,\text{kg/ha} = 30\,\cancel{\text{gm}}/(2 * 3)\,\cancel{\text{m}}^2 * 1\,\text{kg}/1000\,\cancel{\text{gm}} * 10000\,\cancel{\text{m}}^2\,/\text{ha}$$

Seldom can a harvester be adjusted to have zero losses, but a loss determination can indicate excessive loss. For this example, the cutterbar loss is excessive for ripe, upstanding grain. Slower forward speeds, less aggressive reel action, and lower cutterbar height are potential adjustments the machinery manager may suggest to the operator. The cylinder loss is excessive. More aggressive threshing action is indicated. The low amount of separating loss may indicate acceptable separating functions, but the condition of the grain in the grain tank should be examined for straw, chaff, weed seed, and other foreign matter. Should some be found, the fan speed may have to be increased or redirected.

Losses are not the only concern for harvesting. Damaged grain in the tank indicates overly aggressive threshing.

2.4.5. Field Adjustments

The machinery manager and the tractor driver should understand that only field operation experience will indicate the optimum implement adjustments. Variation in soil type, soil and crop moisture, crop residues, crop varieties, seed and chemical characteristics, weather, and so forth make operation of farm machinery a problem of constant adjustment. Conditions change, sometimes hourly or even every few minutes, and the operator must understand what elements of the machine system are adjustable and how adjustment will contribute to an optimum operation. Some adjustments the tractor driver may need to make are:

- Tillage: depth of working, desired placement of residues, degree of soil conditioning
- Seeding: seeding rate, seed depth, seed coverage, seed spacing
- Cultivation: depth, damage to crop, adequacy of weed kill, soil placement relative to the crop
- Chemical application: rate, placement, pollution control
- Harvesting: loss reduction, limit damage to harvested material, avoid gathering of foreign materials

2.4.6. Repair and Maintenance

Farm machines start deteriorating as soon as they are manufactured. Some of the deterioration is due to exposure to air, sunlight, and moisture. Iron rusts. Rubber oxidizes. Plastics and paints age. Such deterioration is retarded by storing machines in a dark shelter. But the most significant deterioration is from the use of the machine. The process of retarding or correcting deterioration is called *maintenance*. Repair costs are those expenditures for parts and labor to install replacement parts after a failure and to recondition renewable parts as a result of mechanical wear. Most data tables combine the estimated costs of maintenance and repair into one report as is that included in data developed some years ago for U.S. farms [2] .

The machinery manager is interested in performing those repair and maintenance activities that ensure reliability of the machine. The costs of downtime during critical field operations may appear to be so great that overmaintenance and repair is looked upon as prevention and worth the cost.

The tractor driver can undertake the maintenance activities when the machine is in use. Daily checks (sometimes even hourly) should be made for fuel, oil, and coolant levels, tension in belt and chain drives, loose bolted connections, needed grease applications, plugged radiator cores, and excessive build-up of dirt, dust, and crop materials around the engine and in the implement mechanisms. The machinery manager should emphasize to the tractor driver that many of these checks have personal-safety implications.

The term *preventative maintenance* is given to the periodic service of the machine as recommended by the manufacturer. Changing of lubricating and hydraulic oils; cleaning or replacing fuel, oil, and air filters; and replacing or refurbishing implement parts subject to rapid and predictable deterioration are performed in the belief that scheduled maintenance will prevent future failure.

Repair of farm machines may be done on the farm and is done in the event of an emergency breakdown. As farm machines and tractors become more complex, machinery managers are now referring these machines to the equipment dealer for annual or biannual comprehensive service, but remoteness from the dealership may inspire the machinery manager to maintain a well-equipped shop and experienced mechanics on the farm. Usually only the largest of farms should consider this investment, although a mechanically skilled tractor driver or machinery manager may economically undertake on-farm repair for the smallest of farms. If a farm shop is undertaken, substantial thought should go towards how large a supply of parts and maintenance items are economically worthwhile to avoid costly machine breakdown time.

2.5. Cost Analysis

D. Hunt

The most significant factor in good machinery management is economic. Unless the costs of machinery operations are contained, skillful operation and maintenance are for naught. This section analyzes the economic factors of machine operation, selection, and replacement.

2.5.1. Field Capacity of Machines

Machine operations can be measured by field capacity, a report of work accomplished in a unit of time. For field machines the common units are area per time, such as hectares per hour. The *theoretical capacity* is calculated from machine dimensions and operating forward velocity as:

$$\mathrm{TC} = SW/10 \tag{2.47}$$

where TC is the theoretical machine capacity (ha/h), S is the forward speed of the machine (km/h), and W is the operating width (m).

Several factors modify the expectations from the theoretical capacity equation. The designed operating width of the implement may not be achieved in actual field operations. Tillage implements often are operated overlapping the previous pass, which makes the effective width less than the design width. Cutterbar harvesters, even with skillful operators, are seldom operated at the designed width of cut. Row-crop seeding machines have effective widths that may be either greater or less than the design width, depending on the spacing of the present pass from the previous one. Inaccurate spacing during row-crop seeding is reflected in inaccuracies through subsequent cultivation and harvest.

Interruptions in machine operations reduce field capacity. Turning at the ends of a field, stopping to unload harvesters or to refill seeders, or any other times when the machine is disengaged from its work are time losses that reduce capacity. When only turning times, overlap, and area-related time stoppages are considered, an *effective field capacity* equation can be used to predict actual machine performance:

$$C = SWLe/(10L + DSWLe + 2.778St) \tag{2.48}$$

where

C is the effective field capacity (ha/h)
S is the average forward speed (km/h)
W is the implement design width (m)
L is the average length of the field (m)
e is the effective width of the machine (decimal of W)
D is the unproductive time (h/ha)
t is the time for a turn at row ends (seconds)

As an example, the computation of the effective capacity of a 2-m width-of-cut harvester traveling at 5 km/h follows: The harvester is stopped for 2 minutes to uncouple the trailing 4-ton yield-receiving wagon and to couple an empty wagon. The operator manages an average width of cut of 1.8 m. The average length of a pass across the field is 400 m and the average turning time is 25 seconds. The yield of the field is 8 tons/ha.

$$\frac{5 * 2 * 400 * (1.8/2.0)}{10 * 400 + 8/4 * 2/60 * 5 * 2 * 400 * 0.9 + 2.778 * 5 * 25}$$

$$C = 0.7848 \text{ ha/h}$$

The theoretical capacity is $2 * 5/10 = 1$ ha/h.

The ratio of effective field capacity to the theoretical capacity, $0.7848/1.0$ or 0.785, is called *field efficiency*. Field efficiency can be improved in this example. Reducing the number of turns by operating parallel to the long dimension of the field, making quicker turns, steering the harvester to make a more complete cut, and reducing the time spent in changing wagons (perhaps unloading the harvester into an accompanying wagon without stopping) will improve field efficiency and increase the effective capacity of the operation.

In many instances field capacity is not the most appropriate measure of machine performance. A machine's *material efficiency* describes the completeness of the harvest and the quality of the material harvested. Harvesting machines gather and process grain and forages. Other machines distribute seed and chemical materials. Tillage machines process soil materials. The effectiveness of a machine is not best evaluated by the rate of area coverage but by the effectiveness with which the material is processed.

For example, a harvester may recover 90% of a field's yield, but damaged grain and included foreign matter may reduce the value of the harvest. Damaged and contaminated yields are not necessarily valueless; therefore, the most realistic measure of material efficiency must be the comparison in value of the yield after harvest with that of its potential before harvest.

This economic concept of machine performance is difficult to compute. Tillage-operation efficiency must compare the actual performance with that of the ideal. Seeding-machine material efficiencies depend on the amount of the seed damaged and the losses associated with improper placement of viable seed. Cultivation efficiency evaluates such factors as weed kill and soil aeration. Chemical distributors are judged by the effectiveness of the quantity delivered and placement as compared with the ideal.

An example of the computation of material efficiencies is provided by a baled-hay harvesting system. The value of a bale perfectly dried and tightly bound is $4 per bale. The mower–conditioner misses the bottom 3 cm of each plant and drops some material at the cutterbar that is not placed in the windrow. Value of this loss is $0.20/bale. The baler pick-up mechanism misses some stems and shatters leaves onto the ground; value of the loss is $0.40/bale. Shattered leaves within the baler and lost to the ground are worth $0.30/bale. Additionally, if the field capacity of the system is too low, there may be losses of vitamin A due to sun-bleaching exposure in the windrow judged to be $0.35/bale. The value of the harvested bales is $4 – $1.25, or $2.75, and the material efficiency is thus 2.75/4 or 68.75%. Losses also may occur because of operations conducted at too-high or too-low moisture contents, but these losses are due to management error and should not be charged against the machine system.

Material efficiency is the most appropriate evaluation of machine performance and capacity. The difficulty of evaluating the factors, however, limits its use to the most sophisticated machinery managers. The field-capacity measure is a reasonable substitute. Machinery managers should always keep in mind that *maximizing the rate of area coverage by machines is not the ultimate goal.*

2.5.2. Costs of Operation

Costs of machinery operations always can be determined after the fact with good record keeping. Cost records are helpful for machinery managing purposes but the estimated costs of a proposed machine or practice is the more frequent management problem. These estimations help in making decisions as to whether a new machine is affordable, whether it should be used to replace an existing practice, and whether the capacity of a current machine is noncompetitive with a new machine.

Estimated costs are composed of expenses that occur independent of use, called *fixed costs*, and those that occur because of use, called *variable costs*. Fixed costs consist of the loss in value of the machine because of aging, wear, or obsolescence, which is described by the term *depreciation*. Other fixed costs are charges for interest on the investment in the machine, taxes, insurance, and a charge for shelter for the machine if any. Variable costs include the cost of labor, the cost of fuel energy, the cost of repair, and the maintenance costs associated with lubrication, sharpening, and consumables such as twine.

There are several ways to estimate depreciation. A realistic estimated depreciation is a straight-line reduction of the value over the normal life of the machine and is calculated as follows:

$$\mathrm{AD} = (P - S)/L \tag{2.49}$$

where AD is the estimated annual depreciation ($/y), P is the purchase price of the machine ($), S is the selling or salvage value of the machine ($), and L is the estimated ownership life of the machine (y).

An artificial rapid depreciation sometimes is allowed by taxing bodies. Completely depreciating a machine in the first few years of its life and then using the machine for years with no depreciation charge may be a financial strategy that encourages the purchase of new machines. It is not an accurate measure of the cost of operation unless the machine is actually replaced at the end of the shortened depreciation period.

Farm machines must compete against other enterprises on a farm for investment money. Thus an annual charge for interest (INT) must be included in the fixed costs. A simple estimate is an appropriate interest rate, i, multiplied by the average value of the machine over its lifetime:

$$\mathrm{INT} = i(P + S)/2 \tag{2.50}$$

If the interest rate is 7% per year, what would be the annual charge for the use of the money?

$$\mathrm{INT} = i(10000 + 1292)/2 = \$395.22/\mathrm{y}\ (\text{purchase price})$$

Charges for taxes, insurance, and shelter may be relatively small but should not be ignored if they exist. They can be estimated as a percentage of P.

For simplicity the fixed costs per year can be estimated to be a constant percentage of the purchase price of the machine. This constant percentage is called the *fixed-cost percentage*, FC%. Assuming a salvage value of 10% of the purchase price and a cost of 1.5% of P for taxes, insurance, and shelter, the fixed-cost percentage would be:

$$\mathrm{FC\%} = 100[0.9/L + (0.55i) + 0.015] \tag{2.51}$$

The *annual fixed costs* of a machine would be:

$$\mathrm{FC} = P(\mathrm{FC\%}/100) \tag{2.52}$$

For a 15-year life, an interest rate of 7%, and a purchase price of $10,000, the FC% would be 11.35% and the annual fixed costs would be $1135.

The variable costs for operating machines can be predicted with more confidence and are calculated on an hourly basis. The labor rates per hour and the consumption of fuel and oil and their prices are usually well known. Not so easily predicted are the repair and maintenance (R&M) costs. Past records in the United States have shown that average R&M costs can be expressed as a percentage of the purchase price of the new machine for the different machine classes. Sample annual R&M costs for several classes are shown here as percentages of the purchase price, P:

Tractors, 0.007%
Tillers, 0.035%
Seeders, 0.050%
Harvesters, 0.02–0.05%
Forage machines, 0.02–0.08%
Chemical applicators, 0.05%

An average R&M cost for a harvester priced at $10,000 with an estimated R&M% of 0.035% would be 10,000 * 0.00035, or $3.50 per hour of use.

An approximate *annual cost*, AC, can be developed as the sum of the annual fixed and variable costs and is useful in many machinery management planning decisions. Using previous notations:

$$\text{AC} = P(\text{FC\%}/100) + A/C[P(\text{R\&M}/100) + L + O + F + T] \quad (2.53)$$

where

AC is the annual cost ($/y)
$FC\%$ is the fixed cost percentage
C is the effective field capacity (ha/h)
R&M% is the estimated percentage charge for R&M
A is the annual area of machine operation (ha)
L is the operator's labor rate ($/h)
O is the engine oil consumption (changes) ($/h)
F is the fuel consumption ($/h)
T is the fixed costs of the tractor ($/h); $T = 0$ for self-propelled machines.

An example of calculation for the $10,000 harvester used previously is shown here for an annual use of 40 ha, an expected life of 15 years, a labor rate of $4/h, a fuel consumption of $5/h, and engine oil-change costs of $0.12/h. The tractor pulling the harvester has fixed costs of $8/h.

$$\begin{aligned}\text{AC} &= 10{,}000 * 0.1135 + 40/0.7848(3.50 + 4 + 0.12 + 5 + 8] \\ &= \$21{,}855.97 \text{ per year or } \$54.65 \text{ per hectare}\end{aligned}$$

2.5.3. Machinery Selection

Selection of machines to use on a farm is an important duty of a machinery manager. The decision includes judgment of the material efficiency of the machine, availability of dealer service, and the economic worth of the machine to the farm enterprise. In some instances the decision is one of whether to mechanize. Use of the previously developed annual cost equation allows the machinery manager to compare ownership of the machine with the alternatives of rented machines and custom operators as well as with alternative machine, animal, and even hand methods.

In comparing alternatives, the effect of timeliness must be considered in most areas of the world. Farm production is both season- and weather-dependent. Field operations must be accomplished in a rather limited time period. Machines with low field capacity may cause the tillage to be done at an improper moisture content; seeding may not be completed in season; chemical application may not be applied in time to avoid insect or fungus damage; and the harvest may be reduced in both quantity and quality.

Timeliness costs (the cost of untimely operations) are region-specific and difficult to estimate. The machinery manager can only predict such costs assuming average weather, soil conditions, and crop growth as established by history in that region. A simplifying assumption is that there is a single optimum week (or day!) in which a field operation should be accomplished. There is a timeliness cost for any operations occurring outside that optimum time. Simple linear reduction in the value of the operation seem acceptable for predicting timeliness costs.

As an example of timeliness-cost computation, consider the seeding of wheat. The manager determines that a particular week is optimum and that the historic probability of a working day in that week is 0.5. Ten hours of operation can be expected for a working day. Research indicates that the loss in yield and/or quality at harvest because of seeding before the optimum week is 20 kg/ha·d previous to and 30kg/ha·d after the optimum time. The value of the crop is $0.15/kg. The current animal-powered seeding operation has an effective field capacity of 0.2 ha/h (2 ha/d) for seeding 15 ha. The total seeding time required would thus be 75 hours or 7.5 days. At the 0.5 probability of a working day, only 3.5 working days could be expected during the optimum week and only 7 ha would be seeded. Optimistically, 4 ha of seeding would occur before the optimum and 4 ha afterward. Each 4 ha would require 2/0.55 or 4 calendar days to complete. The average loss for the early untimely planting is a triangular distribution whose area is one half the ordinate, 4 d ∗ 20 kg/ha·d (80 kg/ha), multiplied by the abscissa of 4 ha for a total loss of 160 kg (see Fig. 2.17). The timeliness cost for premature seeding would be 160 kg ∗ $0.15, or $24. Similarly, the loss for delayed planting would be 240 kg at a cost of $36. The total timeliness cost to be charged against the machine operation is $60.

Timeliness costs must be considered when comparing competing field operation strategies. An ultimate strategy to have zero timeliness costs would be to hire many custom operators so that all the seeding could be done in one optimum day. Whether

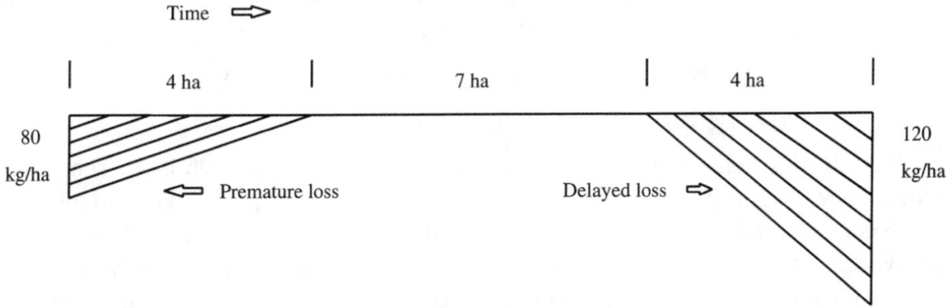

Figure 2.17. Example timeliness loss distribution.

such a strategy is acceptable depends on the availability of the custom operators and the price charged for the work. The animal-powered system described previously with its low field capacity has a low purchase price and low operating cost, but the timeliness costs have to be considered when comparing with a mechanized system. If a mechanized operation can seed the 15 ha in less than a week at an annual cost of \$100 with no timeliness costs, and the animal system has an annual cost of \$50 plus \$60 timeliness costs, the mechanized system is the more economical.

Selection of the optimum power for farm tractors and the corresponding optimum width and capacity of machines can be developed from the annual cost equation with an inclusion of timeliness costs. A machine that is too large for the operation has a high price and excessive total costs. A very small machine has a low purchase price, but the total costs of its use may be high because of increased labor and timeliness costs. An optimum-size machine having the least total costs exists between these two extremes.

An optimum-size machine can be determined if a purchase price per unit of capacity (usually related to the width of action of implements) can be developed. Assume that the price of a seeder is \$1500 plus \$600 for each meter of effective width. Such knowledge allows the derivation of an optimum width equation that is a rearrangement of the first derivative set to zero:

$$w = \{[1000 * A(L + T)]/(\mathrm{FC}\% * p * S * \mathrm{FE})\} \exp 0.5 \tag{2.54}$$

where

w is the least cost width of the machine (m)
A is the total area processed per year (ha)
L is the operator labor cost (\$/h)
T is the fixed costs of the tractor (\$/h); $T = 0$ for self-propelled machines
FC% is the fixed cost percentage
p is the price per additional unit of effective width (\$/m)
S is the forward working speed (km/h)
FE is the field efficiency (decimal)

This analysis assumes that repair and maintenance and fuel and oil consumption are directly related to the size of the machine. Therefore, these costs are independent of the size of the machine and are not a factor in selection of the least cost width.

An sample solution is presented for the wheat-seeding operations discussed under timeliness. The p for the seeder is \$600/m, labor rate is \$4/h, tractor fixed costs are \$3/h, S is 7 km/h, FE is 0.85, and the FC% is 11.35%:

$$w = [1000 * 15 * (4 + 3)/(11.35 * 600 * 7 * 0.85)] \exp 0.5$$

$$w = 2.59 \exp 0.5 = 1.61 \text{ m}$$

The effective field capacity would be 7 * 1.61 * 0.85/10, or 0.96 ha/h. The 15 ha would be seeded in 15.63 hours or 1.563 days. At the 0.5 probability of a working day, 3.125 actual working days would be anticipated. Because 3.5 days are available during the week, no timeliness penalties should be assessed this mechanized operation.

The three significant figures in the lowest-cost width indicate a precision that is not warranted. Inexpensive machines will have only a slightly increased annual cost for widths varying as much as 20% from the solution width. Expensive machines' costs increase sharply on either side of the lowest-cost width, but a variation of 5% should not be too costly.

A similar procedure can be used to estimate the lowest-cost tractor power for a farm. The computation is complicated, because all implements using the tractor have to be considered in this tractor-powered system. This system does not consist of the lowest-cost implements. The variable to be optimized is not the width of machines but the optimum rate of energy expenditure or power level. In equation form:

$$\mathrm{PWR} = \left\{\left[100/\mathrm{FC\%}\, t) * \sum (A_i E_i L_i / r_i)\right]\right\} \exp 0.5 \tag{2.55}$$

where

PWR is the least cost power level for the farm (kW)
A is the area of the individual field operations (ha)
E is the energy needed for the operation (kW·h/ha)
L is the operator's labor rate
r is the ratio of usable drawbar power to power take-off (PTO) power (decimal)
FC% is the annual fixed cost percentage for tractors
t is the price of incremental PTO power (\$/kW)
i is the subscript identifying the several areas of energies, labor costs, etc.
$\sum$ indicates a sum of the evaluations for each i operation

A sample solution follows for a three-operation farm having the values shown in Table 2.8. The price of tractor power, t, is \$300 per each additional kW, and the tractor FC% is 12%.

$$\mathrm{PWR} = \{100/(12 * 300) * [(30 * 12 * 3/0.8) + (15 * 5 * 4/0.8) + (15 * 15 * 6/0.7)]\} \exp 0.5$$

PWR = 101.4 exp 0.5 = 10.07 kW max. PTO power

This tractor might also be used for transportation, irrigation pumping, and so forth. These additional energy requirements add to the field energy requirements to produce a larger lowest cost for a tractor. Including timeliness costs leads to the larger, truly optimum sized tractor.

Now that the fixed costs of the tractor are known, the implement system selection may proceed. If an optimum implement's size is larger than the tractor has power to operate, the size of the implement must be reduced to fit the tractor. The total system cost can now be found, as all sizes of tractors and implements are known.

Table 2.8. Operations data for power optimization

Operation	Area (ha)	Energy (kW· h/ha)	r	Labor (\$/h)
Tillage	30 (twice)	12.0	0.8	3.00
Seeding	15	5.0	0.8	4.00
Harvest	15	15.0	0.7	6.00

The need to consider timeliness costs can be checked for the previous wheat-seeding example. The energy requirement for seeding and the optimum PTO power can be used to predict the field capacity of seeding:

$$10.07\,\text{kW} * 0.8 * 1\text{ha}/5.0\,\text{kW}\cdot\text{h} = 1.61 \text{ indicated ha/h}$$

At a field efficiency of 0.85, the actual capacity is only 1.369 ha/h. The calendar time for seeding 15 ha would be 10.96 h/0.5 or 2.19 days, well within the optimum week. No timeliness costs should be assessed. But suppose the seeding-time requirements indicate that 2 calendar days are required beyond the optimum time, and the penalty is 50 kg/ha·d. The total timeliness costs would be:

$$(50\,\text{kg/ha}\cdot\text{day} * 2)/2 * 27.38\,\text{ha} * \$0.15 = \$205.35$$

This cost would seem to be significant, and a larger seeder should be purchased for that operation. When there are timeliness costs, they must be considered in selecting optimum sizes of machinery.

2.5.4. Replacement Policies

Farm machines may need replacement because of accidents, because they are obsolete, because they no longer have the capacity needs of an expanded farm, or because a new design has a superior material efficiency. The more common replacement decision is to replace a worn existing machine with a new one to continue the same area usage and to gain a sense of increased reliability for the field operations.

The replacement decision depends on the analysis of a continuing record of actual (not estimated) R&M expenditures and a realistic evaluation of the current market value of the current machine. Fuel, labor cost, and oil and other consumables used are expected to be the same for the new machine as for the old and need not be considered in the replacement decision. Such costs are an important part of production expense and should be recorded and examined for other management decisions.

Typically, the actual costs of a new machine are high the first year because marketplace depreciation is very high. The repair and maintenance costs are usually low during the early life of a machine. Near the end of a machine's life the annual repair costs are likely to be high while the marketplace depreciation is low. Economic replacement occurs when the accumulated depreciation and R&M costs per unit of machine use fall to a minimum. Continued use of the machine after that point will cause increased costs. Table 2.9 shows an analysis for a machine purchased new. The data are obtained from R&M cost records and machinery marketplace values.

The machinery manager's replacement decision is based on the last line of the record sheet, the average accumulated cost. The high costs during the first few years are due to the high cost of actual depreciation. Only by keeping the machine for several years will these costs be reduced to reasonable values. If the accumulated costs continue to drop each year, the machine should *not* be replaced. Keeping the machine in the table through year 9 resulted in costs of $24.83 for all of the 230 ha of use since the machine was purchased. Replacement before year 10 may be indicated if a major repair is needed. Should this expense be as much as $412 with the same annual use and depreciation as

Table 2.9. Economic replacement worksheet example for new machine bought for $5000

	Year Just Ended								
	1	2	3	4	5	6	7	8	9
Value ($)	3000	2000	1700	1400	1200	1000	900	800	700
Depreciation ($)	2000	1000	300	300	200	200	100	100	100
Taxes, etc. ($)	80	72	64	60	58	45	40	30	25
R&M cost ($)	12	20	67	50	112	98	162	188	230
Annual cost ($)	2092	1092	431	410	370	343	302	318	355
Accumulated cost ($)	2092	3184	3615	4025	4395	4738	5040	5358	5713
Annual use (ha)	25	25	25	30	40	20	20	25	20
Accumulated use (ha)	25	50	75	105	145	165	185	210	230
Average accumulated cost ($/ha)	83.68	63.68	48.20	38.33	30.31	28.72	27.24	25.51	24.83

in year 9, the accumulated costs would then be $24.90 for every accumulated hectares, and replacement should have taken place at the end of year 9.

This replacement-policy analysis also is helpful in decisions about used machinery. Purchase of used machines has the potential for reduced production costs because of their reduced purchase prices. They do have, however, the disadvantages of potential larger repair costs, lack of the latest technology, and shorter useful lives. If a machinery manager could buy the machine in Table 2.9 at the market price at the end of year 4, the average accumulated costs would be as determined in Table 2.10. Even with the higher repair costs in the later years, the accumulated costs of the used machine are significantly lower than those for the new machine.

Purchase of used machinery can be an economic alternative to the purchase of new machines, but the farm operator should be skilled in machine operation and repair in

Table 2.10. Economic replacement worksheet example for used machine bought for $1400

	Year Just Ended								
	1	2	3	4	5	6	7	8	9
Value ($)	1200	1000	900	800	700	600	500	400	300
Depreciation ($)	200	200	100	100	100	100	100	100	100
Taxes, etc. ($)	58	45	40	30	25	20	15	15	15
R&M cost ($)	370	343	302	318	355	412	300	320	400
Annual cost ($)	628	588	342	448	480	532	415	435	515
Accumulated cost ($)	628	1216	1558	2006	2486	3018	3433	3868	4383
Annual use (ha)	40	20	20	25	20	20	20	20	20
Accumulated use (ha)	40	60	80	105	125	145	165	185	205
Average accumulated costs ($/ha)	15.70	20.22	19.48	19.10	19.88	20.81	20.80	20.91	21.41

order to avoid the likelihood of substantial downtime for the used machine. Should these stoppages be at a critical time, sizable timeliness costs can occur. In the preceding example, if the security of having more timely operations is worth $5/ha, the new machine should be purchased instead of the used one.

In summary, economic machinery management can be practiced skillfully with knowledge of the procedures described here and with the attention to record keeping to provide the needed data.

References

1. Hunt, D. R. 1995. *Farm Machinery Management*. Ames: Iowa State University Press.
2. *ASAE Standards*, 43rd ed. 1996. D487.2: Agricultural Machinery Management Data. St. Joseph, M. I. ASAE.

3 Trends for the Future

3.1. Sustainable and Environmental Engineering

R. Hegg

3.1.1. Definition and Background

In general, sustainable agriculture is "an agriculture that can evolve indefinitely toward greater human utility, greater efficiency of resource use, and a balance with the environment that is favorable both to humans and to most other species" [14]. Modern agriculture has gradually evolved, in the industrialized countries, to depend on high inputs of inorganic fertilizers and synthetic chemicals to produce monoculture cash crops. Mechanization was adopted rapidly in the mid-20th century due to scarcity and rising costs of labor. These practices in Europe and North America have led to increased soil erosion and contamination of surface and ground water. Many developing countries have adopted the same practices of the developed countries: high use of commercial fertilizers and pesticides and a few high-yield crops. This trend is based on the need to produce more food for increasing populations. The developing countries have experienced greater pest, disease, and weed problems, increased soil erosion, environmental hazards, and economic stress.

The recognition of these problems by concerned farmers has led to a strong worldwide trend, underway for years, to adopt a more sustainable form of agriculture. Sustainable agriculture is a form of agriculture that is both economically and environmentally compatible and socially acceptable. This means that the farmer must consider the long-term effects (20 or more years) of any agricultural practice adopted and not just look at the maximum profits on a yearly basis. Sustainable implies quasi–steady state conditions. There is no one definition of sustainable agriculture because appropriate practices can be used by large or small farms in developed or developing countries.

There was a growing movement of "humus farming" concepts from the 1930s through the 1960s, humus being the soil matrix that includes the nutrients, organic matter, organisms, and so forth. The focus was on maintaining or improving the soil fertility as the foundation for producing high-quality crops. Composting was encouraged as a practice that was compatible with humus farming. Humus farming was basically the same as methods called *organic*, *biological*, or *ecological agriculture*. The persons promoting the virtues of these alternative agriculture methods were a few innovative individuals who wanted to make sure that the world had a sustainable agricultural system.

Some of the common characteristics of sustainable agriculture are less inorganic fertilizer, more organic fertilizers (such as manure, municipal sludges, and crop residues incorporated into the soil), fewer pesticides, more use of biological pest-control practices, more use of combined animal and cropping systems, substitution of tillage for chemicals for weed control, more labor, smaller farms, less purchased energy, and better control of erosion.

Farmers will have to meet the challenges of using more advanced technology in ways that will increase productivity while protecting the environment [11]. The richer countries will have an easier time in accomplishing this. The developing countries will have more of a challenge, when about a billion people are malnourished and the social capital for environmental protection is not readily available.

Technology has been the force for moving agriculture toward higher productivity in the 20th century. Many of these technologies were a result of scientific research.

Wherever agriculture is practiced, insects, diseases, and weeds always are reducing productivity. Since 1950 chemicals have been widely used to control these constraints to farming. It appears that in the future there will be fewer of the potent chemicals because of the potential environmental damage they harbor. Chemical companies cannot justify the development or license-renewal costs for some pesticides. There are a wide array of alternative measures that can be used to replace some of the chemicals. These measures may include resistant plants, crop rotations, tillage practices, biological controls, and a minimum number of chemicals. There will be a need to build new equipment that is suitable to apply the new generation of chemicals, biological control agents, fertilizer, and weed-control agents. Biotechnology will be used to cross-breed plants so that the desirable traits can be utilized, all the while adhering to the principles of sustainable agriculture.

Farmers in developed countries with temperate climates may use the results of new research to reduce use of chemicals and lower production costs. They will be better able to coordinate the soil–plant nutrient cycles. In contrast, farmers in developing countries with tropical climates face much more acute problems in managing their natural resources. Large amounts of rainfall and high temperatures make it difficult to deal with intense disease and pest infestations. Poorer farmers are more limited in their opportunities to use technology, and they are more desperate to increase production even if the practices are environmentally damaging.

Many studies have tried to compare net incomes of farms with conventional systems (inorganic fertilizer, pesticides) and sustainable systems (crop rotations and tillage practices for soil fertility and moisture retention). The major shortcoming of these studies is the inability to put actual monetary values on the environmental damage to the natural resource base. Faeth [3] developed an equation that accounts for the long-term soil depreciation on a monetary basis.

$$\text{Soil-depreciation allowance} = [(y_o - y_n)/(n/RL)] * P_c * \{1 - [1/(1+i)^n]/i\}$$

where

y_o is the initial crop yield
y_n is the final crop yield
RL is the crop rotation length (years)
n is the period under consideration (years)
P_c is the crop price
i is the real interest rate (%)

Table 3.1. Net farm income under conventional and natural-resource accounting economic frameworks (in dollars per acre per year)

	Without Natural-Resource Accounting	With Natural-Resource Accounting
Gross operating margin	45	45
Soil depreciation	—	25
Net farm operating income	45	20
Government commodity subsidy	35	35
Net farm income	80	55

Source: [3].

Table 3.2. Net economic value under conventional and natural-resource accounting economic frameworks (in dollars per acre per year)

	Without Natural-Resource Accounting	With-Natural Resource Accounting
Gross operating margin	45	45
Soil depreciation	—	25
Net farm operating income	45	20
Off-site costs	—	46
Net economic value	—	(26)

Source: [3].

Tables 3.1 and 3.2 compare net farm income and net economic value per acre for the best conventional corn–soybean rotation over 5 years, with and without allowances for natural-resource depreciation, in Pennsylvania. The second column of Table 3.1 includes the soil-depreciation allowance ($25/acre/y) that was calculated for Pennsylvania. This is deducted from the gross operating margin and as a result decreases the net farm income, from $80/acre/y without natural-resource accounting to $55/acre/y with natural-resource accounting. Table 3.2 shows the net economic value ($/acre/y) if off-site costs (sedimentation, water quality degradation, etc.) are also included. From the $45/acre/y net farm income is deducted the $25/acre/y for soil depreciation and the $46/acre/y for the off-site costs, resulting in a negative $26/acre/y net economic value. Each region of each country in the world will have different soil-depreciation allowances and off-site costs, based on climate, soils, crop values, lakes, rivers, and so forth. The important point is that these values should be included in the overall calculations, but in most cases these values

Table 3.3. Rotation characteristics, Pennsylvania

Tillage/Rotation	Soil Erosion (tons/acre/y)	Off-Farm Erosion Cost[a] ($/acre/y)	Soil Depreciation ($/acre/y)
Conventional tillage			
Continuous corn	9.26	69	24.8
Corn–beans	6.07	47	24.6
Alternative cash grain[b]	4.25	32	(2.8)[c]
Alternative cash grain with fodder[d]	3.29	26	(8.4)
All hay	0.66	5	(4.8)
Reduced tillage			
Continuous corn	7.15	53	24.4
Corn-beans	5.29	41	23.8
Alternative cash grain[b]	3.49	27	(3.6)
Alternative cash grain with fodder[d]	2.49	20	(10.2)

Source: [3].

[a] Estimated using a damage cost of $8.16 per ton. Calculations weighted by crops and set-aside acreage.

[b] Alternative cash grain: organic corn-corn-beans-wheat/clover-barley.

[c] Parentheses indicate appreciation in soil asset values due to increased productivity.

[d] Alternative cash grain with fodder: organic corn-beans-wheat/clover-clover-corn silage.

Table 3.4. Rotation characteristics, Nebraska

Rotation	Soil Erosion (tons/acre/y)	Off-Farm Erosion Cost[a] ($/acre/y)	Soil Depreciation ($/acre/y)
Continuous corn	6.5	4.0	7.8
Corn–beans			
With inorganic inputs	3.7	2.3	3.0
With fertilizer	3.7	2.3	2.8
With organic treatment	3.1	2.0	(2.0)
Corns–beans–corn oats/clover			
With inorganic inputs	3.1	2.0	(1.3)
With fertilizer	3.1	2.0	(1.0)
With organic treatment	2.2	1.5	(4.0)

[a] Estimated using a damage cost of $8.16 per ton. Calculations weighted by crops and set-aside acreage.

are not readily available. This work of Faeth [3], is an example of how environmental affects can be included in evaluating the net economic value of a farming operation.

Tables 3.3 and 3.4 show some comparisons with different cropping practices in Pennsylvania (higher rainfall, eastern United States) and Nebraska (lower rainfall, midwest), respectively. The soil erosion is higher for Pennsylvania than Nebraska, which translates into higher off-farm erosion costs and also high soil-depreciation costs. Sustainable farming methods can cause increases in soil mass; therefore, there is a net gain in the soil-depreciation allowance. This analysis of Faeth [3] integrates information from four levels of sustainability: field, farm, region, and nation. In the densely populated northeast United States, the value of water is high ($8.16/ton of erosion) while in the less populated, dry northern plains the economic value of water is low, resulting in $0.66/ton of erosion. The USDA EPIC model was used to simulate the physical changes in the soil that would result from different agronomic practices.

3.1.2. Policy

The adoption of sustainable agricultural practices is somewhat dependent on existing policies in the particular country. If there are no incentives, or even disincentives, to the use of practices that conserve natural resources and energy, the farmers will not adopt them. Some of the policies include fiscal, land ownership, government support, energy, trade, environmental, and other regulations. Most of these policies are dependent on the governments having legislation in place and then enforcing that legislation, which encourages sustainable practices.

Fiscal policies that allow farmers to borrow funds, or provide subsidies, or give benefits for using energy-saving methods for implementing certain sustainable practices will be beneficial in the long term. If a policy is initiated that gives a tax break or subsidy to the farmer who implements certain environment-enhancing practices, such as erosion control or purchase of improved chemical applicators, then there will be a long-term benefit for the agriculture of that country. The difficulty is always the balance between the short-term economic benefits of nonsustainable agriculture (immediate returns on investment) and the long-term benefits of sustainable agriculture (maintaining the air, water, and soil resource base so that productivity can be maintained for centuries).

Policy towards agriculture should encourage practices that are best for sustainable agriculture. This has typically not been the case, but several countries are moving toward subsidy programs that are more supportive of sustainable agricultural practices. For example, the Sustainable Agricultural Adjustment Act of 1989 in the United States allows farmers using resource-conserving rotations, 100% planting flexibility, maintenance for crop-based acreage, waiver of crop set-aside acreage, and payments on forage crops.

To truly evaluate the sustainability of a particular farming practice extensive information must be known besides the traditional measures of annual profit per acre or crop yield. The environmental impacts must be calculated in monetary values so they can be factored into the overall economic analysis. The environmental impacts include groundwater depletion, soil productivity, off-site erosion damages, groundwater quality, wildlife effects, and human health.

3.1.3. Social, Economic, and Regional Differences

Three examples of evaluating the sustainability of certain agricultural practices are discussed here. Each country has its own unique situation of government policy, climate, soils, culture, technology, and farmers. The study tried to determine the effect of conventional agricultural practices and similarly resource-conserving practices.

India

Rice and wheat are the predominant crops in northwest India. With a favorable price structure (government policy), paddy rice production has grown rapidly. The high demand for water has dramatically dropped the groundwater table. Large inputs of fertilizer and pesticide are required for high yields. Estimating a 30-year production cycle showed that current practices are not sustainable. The recommended option for sustainability was to reduce tillage, reduce irrigation by 20%, and supplement inorganic fertilizer with manure. To implement these changes will require government-policy changes, education and a change in the farming practices.

Chile

Changes in trade and finance policies have helped revitalize agriculture in Chile, although this new expansion of agricultural production may be short-lived because soil erosion is degrading the soil. The land-tenure pattern is quite diverse, as it is divided between peasant and commercial farming. More than 70% of Chilean farms are peasant farms of less than 5 ha. The peasant farms only cover 10% of the arable land, although most of them (58%) are located on marginal lands where erosion and deforestation are common. The peasant and commercial farms will lose a great deal of their soil resource due to erosion, which is an unsustainable situation. Most of the peasant farmers do not benefit from technical advice from the government and university research. There is no national extension service to provide technical service to peasant farmers. Nongovernment organizations have developed some technical assistance programs for peasant farmers, but they only reach about 4% of these farmers. A linear-programming model showed that peasant farmers could maintain higher yields over a 30-year span if they adopted organic farming practices. The commercial farms would not see much benefit by switching to organic farming methods, because they do not have enough labor available.

Philippines

Rice production in the Philippines uses a great many pesticides. This disrupts the ecosystem's natural ability to cope with pests and leads to greater infestations later in the season. Some of the pesticides are very hazardous to human health if not applied correctly. Many farmers do not use safe application techniques, which results in high costs to human health that should be factored into the overall balance of sustainable agriculture. Between 1980 and 1987, 4031 cases of acute pesticide poisoning were reported, with 603 of them fatal. Underreporting is quite likely. The pesticide exposure can cause problems with heart, lungs, nerves, blood, and skin, so when the health costs are factored in, the benefits of natural control outweigh any other strategy.

The one constraint that almost everyone agrees with is how politically difficult it will be to change government subsidy and support programs to which farmers have become accustomed. For sustainable agriculture to become prominent, policies must be adopted to encourage appropriate agricultural practices.

3.1.4. Components

There are numerous "inputs" into a national or world agricultural system that produce the necessary food and fiber "outputs." These inputs include the basics of soil, water, air, seed, animals, nutrients, labor, technology, and capital (money). These are combined with government policies that sustain an agricultural system that allows the farmers to survive and prosper economically. This section defines and describes the importance of each of these input components to an overall sustainable agricultural system. There is a tremendous diversity in agricultural production throughout the world, ranging from the subsidence peasant who barely provides enough food and fiber for his or her family to the highly mechanized, intensive farms of Europe and North America where each farmer may produce food for 50 to 100 other people.

Soil

Since 1945, 1.2 billion ha, an area as large as China and India, have been seriously degraded as productive soils for agriculture [4]. Erosion has destroyed an estimated 430 million ha of arable land, which is about 30% of the land being cultivated in 1994. The symptoms of unsustainable agriculture are the prevalence of salinization, erosion, soil compaction, soil waterlogging, freshwater pollution, and depletion and desertification throughout the world.

The soil is the most critical component of sustainable agriculture. It must be maintained and protected, because without it food production will drop dramatically. If excessive soil erosion occurs, it takes decades or centuries to replenish or restore soils to their former condition if soil erosion is allowed. The primary reason for avoiding soil loss is the need for water retention. Second, there is concern with loss of nutrients. There is competition for agricultural land due to urban development. This is of particular concern in countries that already have high populations and are unable to provide for their own food and fiber needs. The rural land owner cannot retain ownership if there is a competing use for the land so that the price is driven up beyond its agricultural value. This land is then lost permanently from the agricultural base. It takes local or national governmental laws to prevent this conversion of valuable agricultural land to nonagricultural uses.

Table 3.5 lists some values for arable land per capita in 20 selected countries [9]. The lower the number, the higher the population density: Egypt has 0.05; The Netherlands, 0.06; and China, 0.08 arable ha per capita. Canada and Australia have dispersed populations over a large land area, with 1.75 and 2.90 arable ha per capita, respectively. If agriculture is more highly mechanized the arable land per farmer will be substantially higher than for nonmechanized countries: France, 12.81; United Kingdom, 11.22; versus China, 0.20; Egypt, 0.40; and India, 0.78 ha arable land per farmer.

Water

An adequate and dependable supply of water is necessary for sustainable agriculture. In most parts of the world the natural rainfall is the only water that is available for crop production. Agricultural production can be enhanced with additional water (irrigation) during critical parts of the growing season, or irrigation as the sole source of water in arid climates. Large irrigation projects have been built throughout the world to collect, store, and transport water to individual farms. This has definitely increased worldwide food and fiber production.

In some cases, the water demands for irrigation have been so large that they have resulted in reduced amounts of water for industrial and public use. This has happened in the western United States, where the government has had to reduce the amount of water available for agriculture.

Groundwater is another major source of irrigation water. This requires some type of pumping system that lifts and transport the water to the designated location (field). Some

Table 3.5. Arable land and technological inputs in agriculture in 20 selected countries (1989–1990)

			Total Technological Inputs in Agriculture		
	Arable Land per Capita (ha)	Arable Land per Farmer (ha)	Percent of Arable Land Irrigated	Tractors and Harvesters (10^6 kg)	Nitrogen Fertilizer (10^3 million tons)
Burundi	0.20	0.44	6	0.8	2.00
Egypt	0.05	0.40	100	330	754.1
Ghana	0.08	0.42	1	26.7	5.78
Uganda	0.28	0.52	<1	26.5	0.25
Zimbabwe	0.29	1.07	8	126	82.2
Canada	1.75	99.52	2	13,898	1,197
Costa Rica	0.10	1.14	41	60.3	63.0
Mexico	0.27	2.50	22	1,492	1,293
United States	0.76	64.04	10	79,650	10,048
Argentina	0.81	21.52	6	3,870	99.4
Bangladesh	0.08	0.40	32	30.9	630
P.R. China	0.08	0.20	52	5,387	18,855
India	0.20	0.78	26	5,570	7,386
Japan	0.03	0.99	69	19,843	641
France	0.32	12.81	7	13,176	2,660
Germany FR	0.12	6.36	5	12,487	1,487
Italy	0.16	5.16	35	11,648	827
Netherlands	0.06	3.88	61	1,613	412
United Kingdom	0.12	11.22	2	4,568	1,582
Australia	1.67	194.92	1.4	164.5	1.2

Source: [9].

type of energy is necessary to move the water from the well to the field. The energy input required must be less than the energy gained from increased crop production, or it is not a sustainable system. Plant or animal production systems sometimes have to change to a system that has reduced water requirements when the cost for water is too high. When groundwater is pumped out faster than it is replenished the system is not sustainable.

Techniques can be improved to conserve and utilize water more efficiently. This may allow some irrigated agriculture to remain profitable and therefore sustainable. Irrigation water must be of such quality that it will not leave salt residues in the fields. An estimated 20% to 25% (about 10 million acres) of all irrigated land in the United States suffers from salt-caused yield reduction [2]. It is very difficult to restore land that has soil-salinity problems. In almost all cases, the land is abandoned. Soil-salinity problems are common throughout the world. There are numerous water-quality problems as a result of municipal, industrial, and agricultural practices where high population or intensive farming exists. Waste products enter the water supply, either surface water or groundwater, and the quality is reduced. If the water quality is such that plant- or animal-production systems are negatively affected it is not a sustainable situation. In typical agricultural regions the water is most directly affected by the sediment, fertilizers, or pesticides that run off into surface-water supplies or leach into groundwater supplies. This degradation of water, whether for drinking or irrigation, greatly affects the sustainability of agriculture.

Air

Atmospheric deposition of pollutants can have a negative effect on agricultural production and yields. Ozone is considered as the air pollutant affecting vegetation to the greatest extent in the United States [5]. The photochemical reactions involving nitrogen oxides and volatile organic compounds form ozone (O_3). The annual effects on crop production from ambient ozone are estimated to be comparable to losses from pests and diseases. Rising carbon dioxide levels raise temperatures, raise water levels by melting ice, and shift cropping areas nearer the poles. This increase in carbon dioxide is a result primarily of burning fossil fuels. There may be enhanced plant growth due to higher carbon dioxide levels. Other gaseous compounds such as methane, nitrogen oxides, and sulfur oxides can cause negative effects on plants if their concentrations become too high. If man's activities degrade air quality too much, this will reduce the agricultural production capacity regionally or even internationally. Air is not considered very limiting, in comparison with soil and water conditions, for agricultural production.

Nutrients

For plant and animal productivity to increase requires additional sources of nutrients, assuming no increase in land area. For plant systems this is fertilizer, primarily nitrogen, phosphorus, and potassium. These can be from an inorganic source (purchased fertilizer) or an organic source (manure and crop residues). There is a pretty direct relationship between the amount of fertilizer added and the increase in crop yield, within certain ranges. There has to be a profit for the farmer if he or she is going to justify regular purchases of nutrients to enhance yields and productivity, whether for plant production or animal production or a combination. Intensive production systems depend on high use of low-cost inputs (fertilizer and fuel) and short-term profitability. There must be a balance of inputs and outputs for sustainable agriculture.

Nitrogen fertilizer use is a measure of technological input, meaning that energy is required to produce the plant form of nitrogen available to plants (Table 3.5). There is a wide range in the amount of nitrogen fertilizer used for agriculture, depending on the size of the country and its level of technology.

The key to nutrient management is nutrient cycling [6]. Most of the nutrients are conserved in the cycle, but inputs from the atmosphere and losses due to erosion, leaching, denitrification (conversion of nitrate to nitrogen gas), and ammonia volatization must be considered. Agricultural systems differ from natural systems because nutrients are removed as the harvested product. If the system is to continue these nutrients must be replaced, and in conventional agriculture purchased commercial fertilizer is the way in which it is replaced.

Land application of manure is the common way of returning harvested nutrients back to the soil for cycling. Unfortunately, the nitrogen component of manure is not very efficiently returned to the soil system because of either gaseous loss or leaching below the root zone. Potassium can be lost from the cycle by leaching if the soils have low cation-exchange capacity. Phosphorous does not move easily in soil; it will remain in the root zone and, thus, not be lost from the cycle except by soil erosion or removal by run-off.

Nitrogen can be supplied by biological fixation to meet most or all of crops needs. If crop rotations (alternating various plant species after each harvest) are used correctly, a nitrogen-fixing crop can leave enough carry-over nitrogen to supply the nitrogen needs of the non–nitrogen fixing crop. An example of this is soybeans or forage legumes, which are nitrogen-fixing plants, followed by corn, which is not a nitrogen-fixing plant.

Plant Base

There must be an appropriate plant base (seeds) that will thrive in the particular condition in which it is grown. Throughout the world there are plants that have evolved over thousands of years in order to adapt to the soil and climate of a particular region.

In most cases there is a diversity of plants that grow synergistically. Man needs to recognize that the diversity is beneficial for nutrient utilization, pest control, moisture utilization, and soil protection. Scientific agriculture has developed methods to produce superior species of plants that have higher value and therefore more profitability. In many cases, monoculture cropping systems replaced the multicropping systems, potentially resulting in more pest infestations, plant disease, weed problems, and soil erosion.

Integrating Plant and Animal Systems

During the last half of the 20th century intensive agriculture has moved away from farming systems that combined plant production and animal production systems [12]. This has led to the extensive use of inorganic fertilizer on plant-production farms, rather than the cycling of manure as a nutrient source. On the animal production–only farms there is typically not enough land to agronomically utilize the manure. In this case, the manure is a waste that needs to be disposed of.

The essential component of animal-and-crop integrated agriculture is the cycling of manure through the plant–soil system. Long-term studies have shown that the crop yields obtained with manure can be comparable to or greater than those obtained using inorganic fertilizer [1]. Nutrients supplied from manure or other organic matter may be released more slowly to plants than nutrients derived from inorganic fertilizer. The combination of animal and plant production on the same farm can lower inorganic inputs and improve long-term soil fertility.

Conservation tillage practices are a vital part of a sustainable plant production system [7]. The particular practice depends on the crops grown, the size of farm units, and the level of technology available. The equipment can range from hand tools, in developing countries, to mechanized power units, in developing countries, equipped with electronic sensors. In developing countries, most farmers use low-input agriculture practices because those are all that is available. They have the constraint of minimal capital, so there are few purchased inputs (pesticides, fuel, and fertilizer), resulting in impoverished soils and low crop yields. The developed countries adopted cropping practices that require high amounts of purchased energy. Much of this energy input is to do tillage and seedbed preparation. In the United States [13] primary and secondary tillage consume 10 to 12 trillion kcal a year, while energy used for production of fertilizer is about 160 to 170 billion kcal. Thus, a reduction in tillage would reduce only a small portion of the energy input.

Table 3.6. Energy output and inputs in agriculture for 20 selected countries (1989–1990)

	Agricultural Output (10^6 kcal/ha arable land)	Agricultural Output/Farmer (kcal)	Total Energy Inputs (10^6 kcal/ha)	Inputs per Farmer (10^6 kcal)	Output/Input Energy Ratio
Burundi	4.43	1.95	0.2	0.09	22.0
Egypt	15.89	6.33	11.0	4.36	1.4
Ghana	11.67	4.90	0.3	0.13	36.7
Uganda	3.17	1.64	0.05	0.03	63.0
Zimbabwe	3.92	4.16	1.24	1.33	3.2
Canada	3.36	334.85	2.8	276.6	1.2
Costa Rica	10.35	11.75	8.4	9.6	1.2
Mexico	4.52	11.29	2.2	5.5	2.1
United States	6.03	386.12	4.4	284.4	1.4
Argentina	4.36	93.79	1.3	27.05	3.5
Bangladesh	7.50	2.98	2.1	0.84	3.6
P.R. China	13.02	2.65	5.7	1.17	2.3
India	3.81	2.98	1.4	1.09	2.7
Japan	15.02	14.80	46.7	46.0	0.3
France	10.54	135.08	9.9	127.2	1.1
Germany FR	13.37	84.97	18.5	117.4	0.7
Italy	8.69	44.86	14.0	72.2	0.6
Netherlands	14.93	58.00	36.5	141.8	0.4
United Kingdom	12.36	138.67	10.8	121.7	1.1
Australia	1.67	194.92	1.4	164.5	1.2

Source: [9].

Energy

The food supply of all developed countries is dependent on fossil energy [9]. The values in Table 3.6 are based on FAO statistics and include as inputs only fertilizers, irrigation, pesticides, and machinery and fuel for field operation. Other energy inputs, such as energy and machinery for drying crops, transportation of inputs and outputs to and from the farm, electricity, and construction and maintenance of buildings and infrastructures are not included. For the overall food system, the on-the-farm energy requirements are still only a fraction when food processing, packaging, distribution, shopping, and home preparation are included.

Agricultural output as an energy equivalent in the crops or animals produced on per unit arable land area (10^6 kcal/ha) ranges from 3 to 6 for countries such as Uganda, Canada, India, and the United States to 10 to 15 for countries that have more intensive agricultural systems (Egypt, China, Japan, Netherlands, United Kingdom). Countries with less intensive agriculture or with restrictive climatic conditions are not able to produce high outputs, although the output per farmer can be quite high (United States, 386; Canada, 335; and Australia 195 × 10^6 kcal/farmer). This high output is due to high inputs other than labor. Countries with low outputs (below 10 × 10^6 kcal/farmer) would be those like Egypt, Uganda, India, and China, which use more labor but less fertilizer and machinery.

Table 3.6 shows the comparison of the inputs per farmer from the 20 selected countries, covering developed and developing countries. These numbers range from 0.03×10^6 kcal/farmer for Uganda to the United States, which has 284×10^6 kcal/farmer. This tremendous range is again reflective of the purchased inputs in energy, water, machinery, and other production practices of developed countries.

The calculation of output divided by input is a final way to compare the energy balance among various countries, developed and developing. Uganda, with its very low input energy (0.05×10^6 kcal/ha) and low output (3.17×10^6 kcal/ha) yields an output/input ratio of 63.0. Ghana has 36.7 and Burundi 22.0, while all other countries in this comparison are less than 3.6. The country with the lowest value is Japan, at 0.3. Very high efficiency (high output/input ratios) means that those farmers will be in a state of poverty as a result of very poor economic performance.

Nitrogen fertilizers require a high energy investment. Most nitrogen fertilizer is produced by combining atmospheric nitrogen with hydrogen, usually from methane, at high temperature and pressure to produce ammonia. The energy consumed in production of fertilizer nitrogen is 18,000 kcal per kilogram of nitrogen [10]. This means that the availability and cost of nitrogen fertilizer depends mainly upon energy supply and cost.

The other major energy requirements for intensive agriculture besides nitrogen are fuel for the power units (tractors and combines), pumping for irrigation, and manufacture of pesticides. Table 3.5 provides information on the distribution of technological inputs (tractors plus harvesters and nitrogen fertilizer) for 20 selected countries. There is a very wide disparity among developed and developing countries. Highly mechanized countries such as France, Germany, and Italy have 11,000 to $13,000 \times 10^6$ kg of tractors and harvesters. The much larger populated countries of India and China have about 5000×10^6 kg of tractors and machinery. The country with the largest amount of tractors and machinery ($79,650 \times 10^6$ kg) is the United States, which has a population of approximately 260 million people.

Energy saving benefits of conservation tillage are less fuel consumption due to reduced field operations, less power required due to better soil structure, less time and labor required, the possibility of double cropping, and lower investment in farm machinery. In contrast, there are some activities with conservation tillage that require more energy: more herbicides for weed control, control of the high incidence of insects and pathogens, addition of fertilizers, and higher seeding rates. There are several forms of conservation tillage [8]. *Minimum* or *conservation tillage* is defined as a cropping system that leaves 30% or more of the crop residue on the soil surface. These systems have the advantage of lowering farm expenses, reducing run-off and soil erosion, increasing organic matter in soil and increasing soil moisture retention. Conservation tillage has been widely adopted in the United States partially due to government programs. Some examples follow.

(a) No-Till or Slot Planting. The soil is left undisturbed prior to planting. Planting is completed in a narrow seedbed about 2 to 8 cm wide. Weed control is accomplished primarily with herbicides.

(b) Ridge-Till (Includes No Till on Ridges). The soil is left undisturbed prior to planting. About one third of the soil surface is tilled at planting with sweeps or row cleaners.

Planting is completed on ridges usually 10 to 15 cm higher than the row middles. Weed control is accomplished with a combination of herbicides and cultivation. Cultivation is used to rebuild ridges.

(c) Strip-Till. The soil is left undisturbed prior to planting. About one third of the soil surface is tilled at planting time. Tillage in the row may be done by a rototiller, in-row chisel, row cleaners, and so on. Weed control is accomplished with a combination of herbicides and cultivation.

(d) Mulch-Till. The total surface is disturbed by tillage prior to planting. Tillage tools, such as chisels, field cultivators, disks, sweeps, or blades, are used. A combination of herbicides and cultivation is used to control weeds.

(e) Reduced-Till. This system consists of any other tillage and planting system not covered previously that meets the criterion for conservation tillage of 30% surface residue cover after planting.

Minimum tillage systems more nearly resemble natural ecosystems than conventional agricultural systems. Plant decomposition and nutrient release should result in more efficient, less environmentally degrading, and more sustainable ecosystems.

Multiple cropping is a system in which the same field is used to grow two or more crops in a year. It is most common in tropical regions but it has wide application in other regions also. The crops can be grown sequentially, one after the other, or simultaneously. The term for the simultaneous growing of plants is *intercropping*. There is some degree of overlap where one crop is planted or harvested before the other crop. This diversity of plant life gives several advantages: decreased pest and weed pressure, reduce erosion potential, decreased risk of total crop failure, and better utilization of the soil resources. The use of legumes for nitrogen fixation in the cropping system can be very beneficial in supplying nitrogen to the nonlegume crops (maize, wheat, barley, etc.).

3.1.5. Summary

With increasing world populations, limited fresh-water supplies, limited fossil fuels, and the fact that most nations in the world are food importers, the outlook for world food security is not optimistic. Many countries are exporters of grains, but they are not able to provide for the future demands of the large-population countries such as China and India. It will take the concentrated efforts of various governments to enact policies that will encourage sustainable agricultural practices. There are several critical components to establishing sustainable agricultural practices. The maintenance of the current land quality and quantity is probably the most important component. Following this are water supplies (quantity and quality), air, and nutrients (for plants and animals). An adequate or improved plant base (strains of cereal, fruit, or forages) will play a lesser role than those factors previously mentioned. A return to integrated plant and animal systems will allow for recycling of nutrients through the soil system, rather than importing feed for livestock and exporting plant nutrients from the farm when the systems are separated. All countries, developed and developing, are highly dependent on fossil-fuel supplies. The nonrenewable fossil-energy sources are being depleted and there is not a logical energy replacement at this time. Mechanization practices are very dependent on a portable fuel supply for tractors and other power units.

References

1. Baldock, J. O. and Musgrave, R. B. 1980. Manure and mineral fertilizer effects in continuous and rotational crop sequences. *Agronomy Journal* 72:511–518.
2. El-Ashry, M. T., Van Schifgaarde, J., and Shiffman, S. 1985. Salinity pollution from irrigated agriculture. *Journal of Soil and Water Conservation* 40:48–52.
3. Faeth, P. 1993. Evaluating agricultural policy and the sustainability of production systems: an economic framework. *Journal of Soil and Water Conservation* 48:94–99.
4. Faeth, P. 1994. Building the case for sustainable agriculture. *Environment* 36(1): 16–20, 34–39.
5. Heck, W. W., Taylor, O. C., Adams, R. M., Bingham, G., Miller, J., Preston, E., and Weinstein, L. 1982. Assessment of crop loss from ozone. *Journal of Air Pollution Control Association* 32:353–361.
6. King, L. D. 1990. Soil nutrient management in the US. In *Sustainable Agricultural Systems*, pp. 89–105. Ankeny, Iowa: *Soil and Water Conservation Society.*
7. Lal, R., Eckert, D. J., Fausey, N. R., and Edwards, W. M. 1990. Conservation tillage in sustainable agriculture. In *Sustainable Agricultural Systems*, pp. 203–209. Ankeny, Iowa: *Soil and Water Conservation Society.*
8. Mannering, J. V., Schertz, D. L., and Julian, B. A. 1987. Overview of conservation tillage. In *Effects of conservation tillage on groundwater quality*, pp. 3–17. Chelsea, MI: Lewis Publications.
9. Pimental, D. and M. Giampietro, 1994. *Food, Land, Population and the U.S. Economy.* Washington, DC: Carrying Capacity Network.
10. Pimental, D., Hurd, L. E., Bellotti, A. C., Forester, M. J., Oka, I. N., Sholes, O. D., and Whitman, R. J. 1973. Food production and the energy crisis. *Science* 182:443–449.
11. Pluckett, D. L. and Winkelman, D. L. 1995. Technology for sustainable agriculture. *Scientific American* 273(3):182(5).
12. Stinner, B. R. and Blair, J. M. 1990. Ecological and agronomic characteristics of innovative cropping systems. In *Sustainable Agricultural Systems*, pp. 123–140. Ankeny, Iowa: *Soil and Water Conservation Society*.
13. Stout, B. A. 1984. *Energy Use and Management in Agriculture.* North Scituate, MA: Breton Publishers.
14. Harwood, R. R. 1990. A history of sustainable agriculture. In *Sustainable Agricultural Systems*, pp. 3–19. Ankeny, Iowa: *Soil and Water Conservation Society*.

3.2. Precision Farming

H. Auernhammer and J. K. Schueller

3.2.1. Introduction

Crops and soils are not uniform but vary according to spatial location. Large-scale nonuniformities have long been countered with different cropping practices in different regions. But precision farming responds to spatial variability within individual fields or orchards. This leads to a more cost-effective and environmentally friendly agriculture by

- Increasing food production
- Optimizing the use of restricted resources of water and land
- Reducing environmental pollution
- Engaging the efficiency capabilities of intelligent farm machinery
- Improving the performance of farm management

Precision farming concepts include:

- More accurate farm work by better adjustments of settings and by improved monitoring and control mechanisms
- Localized fertilizing on demand in accordance with the variability of soils, nutrients, available water, and plant growth
- Weed and pest control by localized crop production needs
- Automated information acquisition and information management with well-structured databases, geographic information systems (GIS), highly sophisticated decision-support models, and expert-knowledge systems in integrated systems connected by standardized communication links (Fig. 3.1).

Precision farming is not a fixed system, but rather a set of general concepts that may have different physical realizations with

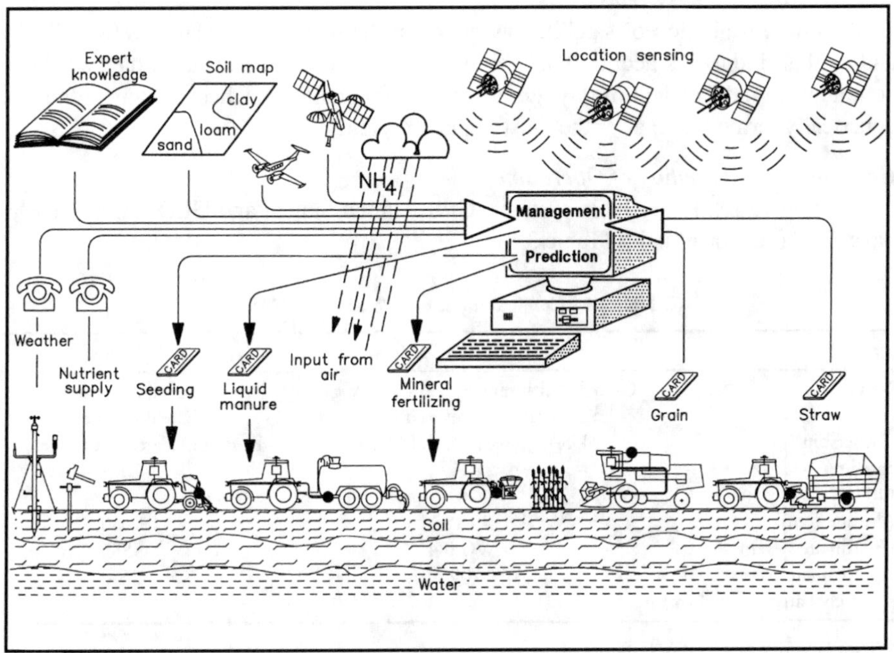

Figure 3.1. An integrated precision farming system.

- Different soil types under different climate conditions
- Different farm management systems and production levels
- Different mechanization solutions

3.2.2. Positioning in Precision Farming

Positioning is the key element in many precision farming systems. In most cases x–y (longitude and latitude) coordinates are sufficient. For some cases and for more sophisticated requirements, z (elevation) may be of interest. A categorization of accuracy requirements into four different classes can be seen in Table 3.7.

Table 3.7. Positioning accuracy requirements

Required Accuracy	Task	Example
±10 m	• Navigation	• Targeting of fields (machinery ring, contractor) • Targeting of storage area (forestry)
±1 m	• Job execution • Information • Documentation	• Local field operations such as yield monitoring, fertilizing, plant protection, soil sampling, action in protected areas • Automated data acquisition
±10 cm	• Vehicle guidance	• Gap and overlap control (fertilizing, spraying) • Harvesting without skips
±1 cm	• Implement (tool) guidance	• Mechanical weed control

Satellite Navigation Systems

With the installation of satellite navigation systems during the late 1980s, military and civilian industries acquired access to worldwide cost-free location data available continuously independent of daylight and weather conditions. These systems often are generically termed *GPS* after the most popular system.

Components and Method of Operation

The two currently available systems, GPS (United States) and GLONASS (Russia) have similar characteristics (Table 3.8).

Table 3.8. Configurations of GPS and GLONASS

	GPS-NAVSTAR	GLONASS
Name	Global Positioning System—NAVigation System by Time and Range	GLObal NAvigation Satellite System
Ownership	USA: Department of Defense	Russia: Department of Defense
Satellites	24	24
Orbits	6	3
Altitude (km)	20,183	19,100
Coordinate system	WGS84	SGS85
Approximate accuracy for civilians (2 drms) (m)	100[a]	35

[a] With Selective Availability intentional degradation.

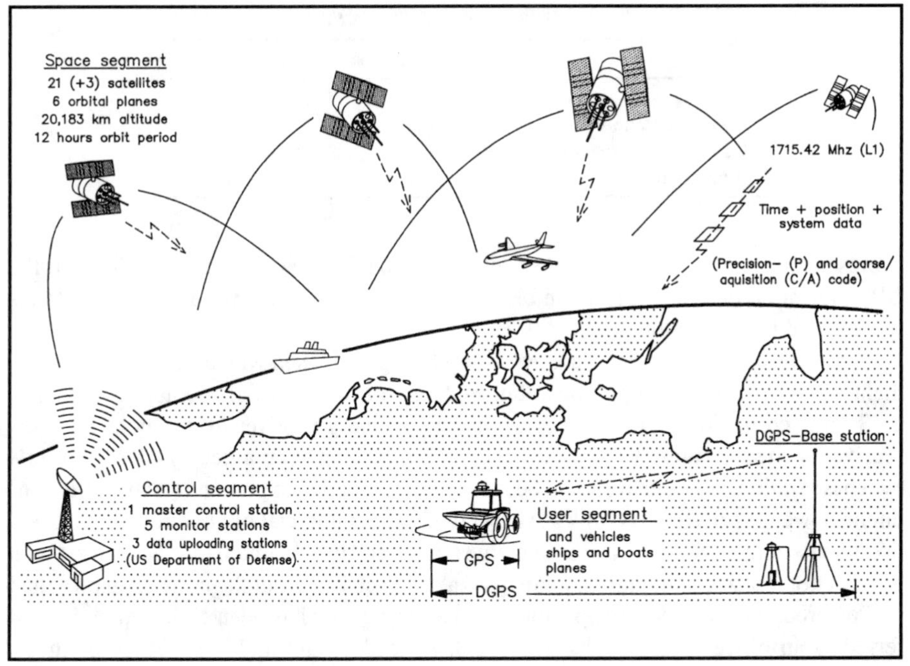

Figure 3.2. Global Positioning System (GPS).

GPS and GLONASS each consist of three segments (Fig. 3.2):

- The monitoring and control segment, which maintains the overall control of the system and is operated secretly by the system owner
- The space segment consisting of positioning satellites (currently 24) with a lifetime of up to 7 years in "blocks" of equal configuration and performance
- The user segment on land, at sea, and in the air with an unrestricted number of receivers.

Location is determined in each individual GPS or GLONASS receiver based upon at least three satellite signals (satellite position and time of signal generation). The location of the receiver is determined by the known satellite locations and the measurements of the ranges between the satellites and the receiver.

GPS and GLONASS have inaccuracies that depend upon atmospheric effects and the instantaneous geometric configurations of the satellites. In addition, the owner of GPS purposely degrades the civilian accuracy to 100 m, which is insufficient for agriculture. This inaccuracy must be corrected for GPS to be useful.

Differential Corrections

The accuracy of GPS location sensing can be improved by using an additional receiver at a fixed known position (e.g., the DGPS base station in Fig. 3.2). The additional receiver compares the GPS indication of its position to its known location to determine the instantaneous magnitude and direction of the GPS error. Assuming the same error

Table 3.9. Correction methods and agricultural operations suitability

Method	Mapping Operations	Control Operations
Real-time		
Own base station	Yes	Yes
Commercial correction	Yes	Yes
Post-processing	Yes	No

is present for the receiver at the unknown location of the moving farm equipment, the GPS indication of that receiver can be corrected easily for the error. Different methods can be used for correction (Table 3.9).

Real-time correction signals are necessary for operations in which agricultural equipment is controlled based upon position, such as fertilizer or pesticide application. The GPS error correction is transmitted from the stationary receiver to the mobile (rover) receiver on the agricultural equipment. The correction can be supplied by the rover user's own base station, although it should be close to the rover due to governmental restrictions on the power of private radio links. Alternatively, a commercial or governmental correction service may be used. Less financial investment and user maintenance is required with a correction service, although there may be a usage charge from the correction provider.

Postprocessing corrects the position data after the data has been collected and transferred to an office computer. No radio link is needed, and higher accuracy might be achievable. However, because accurate position information is not available until after the operation, control operations cannot use this technology. In addition, corrections must be obtained and used to process the uncorrected location data.

Other Location Systems

Other nonsatellite location-sensing systems may be of regional or farm-specific interest (Table 3.10).

Due to the large investment in sender installations, maintenance requirements, and the fulfillment of legal constraints, these positioning systems are concentrated in small areas or fulfill specific purposes in which they are competitive with satellite systems.

3.2.3. Concepts of Precision Farming Systems and Required System Elements

Precision farming is based on mapping systems, on real-time sensor–actuator systems, or on combinations of the two.

Table 3.10. Examples of nonsatellite location systems

	Infrared	Beacon	Network
Rover equipment	Sender/receiver	Receiver	Receiver
Fixed equipment	Reflectors	Senders	Commercial sender network
Range (km)	<5	<600	<400
Accuracy	10–20 cm	1–3 m	20–30 cm
Approximate 1997 price per mobile unit (US$)	5000	1500	4000

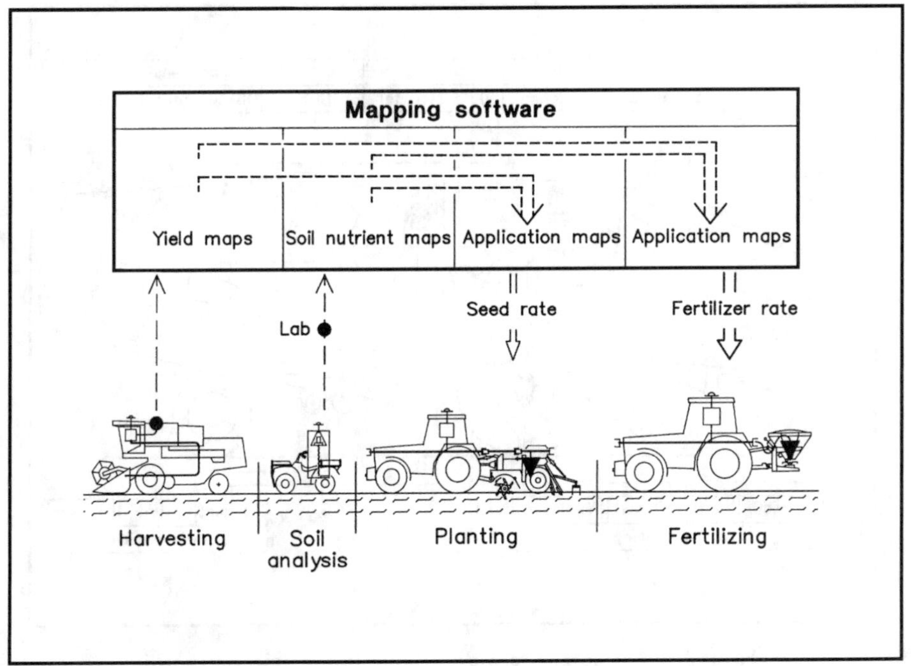

Figure 3.3. Systems and components for map-based precision farming.

Map-Based Systems

Map-based systems use GPS or other locator systems to establish a geographic basis for precision farming (Fig. 3.3). Components of map-based systems include:

- Locators to establish equipment position
- Sensors for yield and soil measurements
- Mapping software with color display and printing capabilities
- Controllers for map-based applications
- Actuators to perform the control

Map-based systems allow information to be gathered from various automatic and manual sources. Then a desired control map can be generated to guide such field operations as variable planting, irrigation, fertilization, or pesticide application.

Real-Time Systems

These systems do not require locators or mapping software and hardware. The relevant quantity is sensed and then an appropriate action is immediately taken. Examples include herbicide application based upon sensed organic matter and anhydrous ammonia application in growing maize (corn) based upon sensed soil nitrate level.

The primary limitation of real-time systems is that only the current sensor data can be used. For example, soil-type information is not available in the previous herbicide or anhydrous ammonia examples. Prior crop-yield data also are not available.

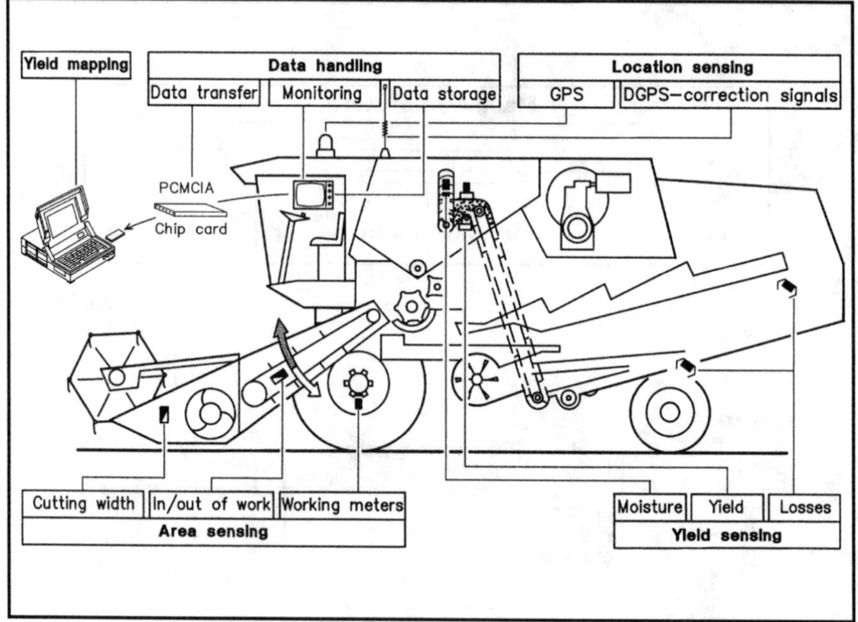

Figure 3.4. Components for yield mapping in a combine harvester.

Real-Time Systems with Maps

The most sophisticated systems combine the capabilities of map-based and real-time systems. Maps of yields, soil types, and nutrients can be used with real-time sensors of plant growth, soil moisture, and weed infestation to control field operations. Because such systems require all the components listed previously, they are complex and expensive. But they allow the optimization of the field operations. Intelligent decision software that decides the proper action based upon map and sensor data in real time is crucial.

3.2.4. Yield Mapping

The main target of agriculture is the production of crop yield. The farmer is therefore very interested in knowing yields. Yield measurement documents the result of the previous farm activities and can be used to plan for the coming crop. Components for most yield-mapping systems (e.g., Fig. 3.4) include:

- Yield sensor (and moisture sensor)
- Location-sensing system (usually GPS)
- Working-width sensor (or DGPS with decimeter accuracy)
- Monitoring and data-storage unit
- Data transfer to office computer
- Mapping software on the office computer

Grain Combine Harvester Systems

For grain measurement, on-the-go sensors are available to detect either the volume flow or the mass flow. The sensors usually are installed near the top of the clean grain

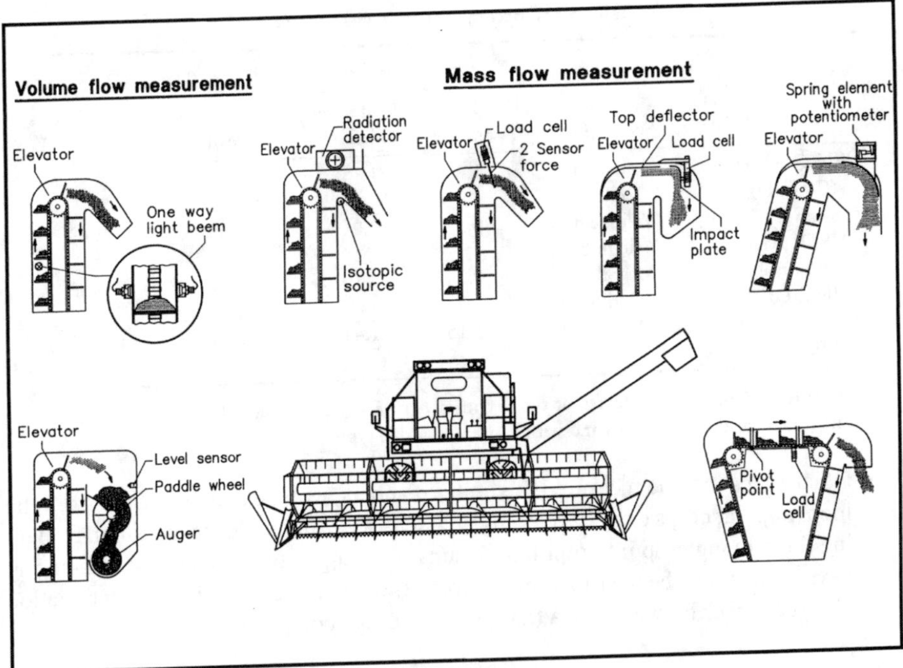

Figure 3.5. Grain flow sensors for yield mapping.

elevator and are approximated as having a 12- to 15-second time lag between when the grain is cut and the flow sensed. Several sensors are available (Fig. 3.5).

Depending on the measurement principle, the accuracy of yield sensors is influenced by factors such as moisture, density, throughput, elevator speed, and inclination. Each sensor therefore has its own accuracy under particular conditions (Table 3.11).

Calibration of the sensor is crucial and should be done frequently. Feedback calibration using field-transport wagons with weighing capabilities or scale measurements will improve the accuracy.

Other Continuous Crops

Yield sensors are in various stages of development and commercialization for many crops.

- Sugar-beet and potato harvesters may use belt weighing sensors (Fig. 3.6). Accuracy is influenced by heavy vibrations during transport and by dirt in these root crops.
- Sugar-cane harvesters similarly measure the mass flowing across a conveyor portion.
- Forage harvesters, mainly choppers, balers, and self-loading trailers, have a throughput of high-moisture material. High accuracy therefore also requires real-time moisture sensing. Available systems use various sensor systems (Fig. 3.7).
- Cotton yield may be measured by passing the picked bolls through a light beam, against an instrumented plate, or by accurately measuring the change in weight of the harvester's storage basket.

Table 3.11. Accuracy of grain yield sensors

	Sensor Type	Measurement Deviations in 1 s (%)			Field Tests (See Note)
		Flat Surface	15-degree Side Hill	15-degree Up/Down Hill	
Yield-O-Meter (Claas, D)	Volume flow cell wheel	1.80	1.15	1.68	3.86
CERES 2 (RDS, UK)	Volume flow light beams	0.94	3.04	9.49	3.43
Flowcontrol (MF, DK)	Mass flow radiometric	2.24	2.10	1.10	4.07
Yield Monitor (Ag-Leader, USA)	Mass flow force	3.15	1.41	1.64	4.06

Note: Results of tests conducted at the Technical University of Munich. Field tests from three years under various conditions and with various combine types.

- Other crops may be measured in similar manners, using existing sensors from the more popular crops or by developing new sensors. Volume can be measured by having the moving crop interrupt light beams or by a positive-displacement metering device. Mass can be measured by the force from a momentum change, by radiation absorbance, or by weight in a transport or storage component.

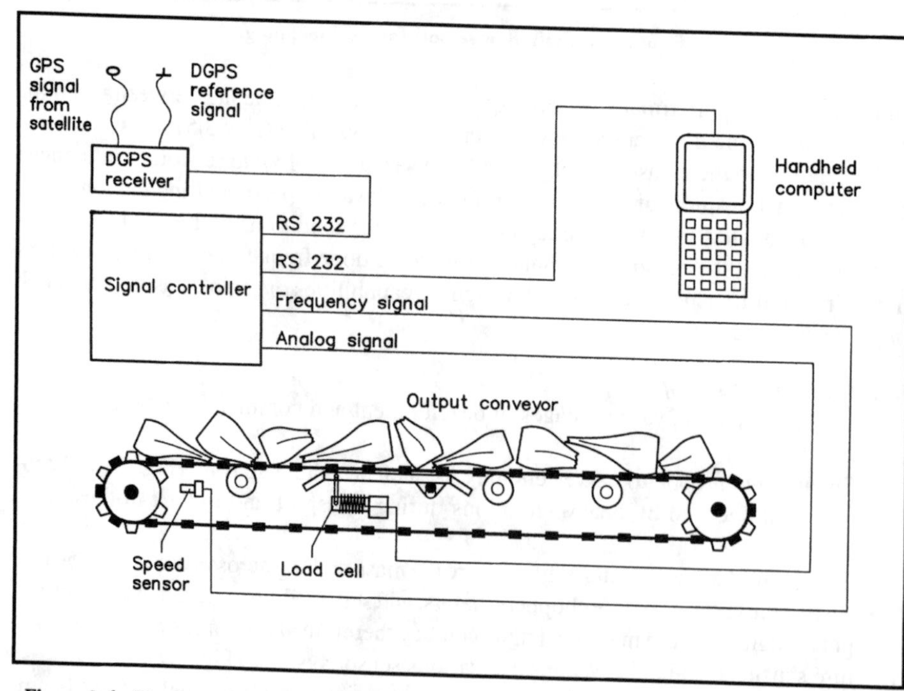

Figure 3.6. Fig. Sugar beet yield monitoring equipment setup and connection diagram (based on Walter, et al.).

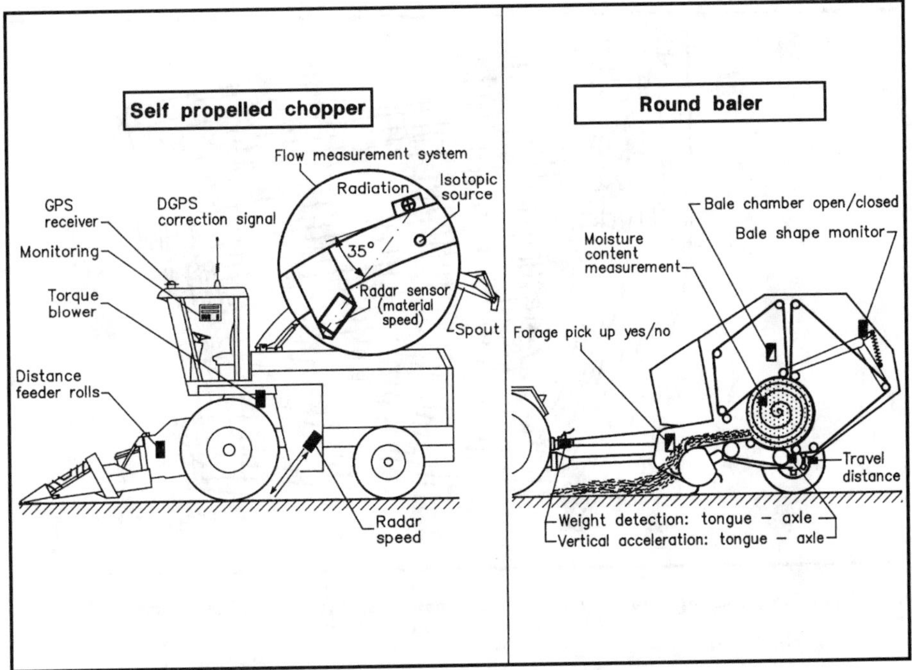

Figure 3.7. Forage sensing systems in a chopper (left) and a round baler (right).

Non-Continuous Yield

Some harvesting methods (for example, hay bales or fruit bins) will result in discrete units of yield at particular points rather than continuous crop output. If the bales or bins can be assumed to have an equal amount of crop, their positions are simply recorded. The density of position marks on a map indicates yield. If the assumption is invalid or better accuracy is required, the weight of each bale or bin is recorded through the use of load cells, strain gages, or hydraulic pressure on the loader or transport equipment.

Data Storage and Mapping

Signals from all sensors (continuous-yield, moisture, location, working-width) are processed to give yield per area in short-period cycles (usually once per second) and stored in an on-board controller. Signalling to the driver is necessary to monitor continued error-free functioning of the sensors and to show the actual situation of the work.

Data transfer to the on-farm office computer can use chip cards, PCMCIA (Personal Computer Memory Card International Association) cards, or radio links. Besides yield and positioning data, other information about the harvested plots may be added later for farm management.

Mapping programs in the office computer then can produce tracking maps and yield maps: Tracking maps show the work sequence and the accuracy of location sensing and can be further analyzed for task times (Fig. 3.8). Yield maps can be established using either grid or contour mapping. Both types present similar information, and the choice may depend upon the user.

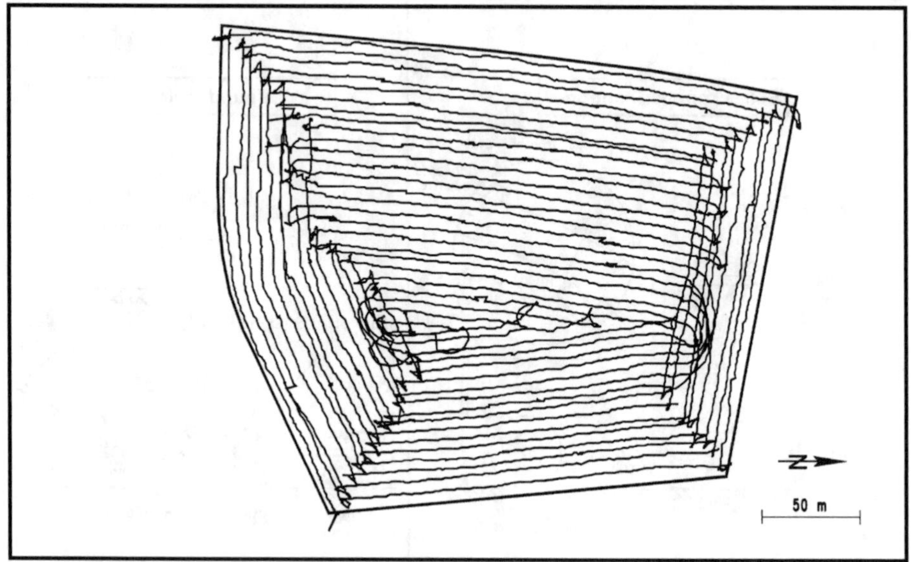

Figure 3.8. Tracking map of a maize (corn) chopping plot (DGPS, 4.5 m working width).

Yield maps show the large-scale variations in a field (Fig. 3.9). To understand the reality of a certain plot, several (perhaps at least three) consecutive yield mappings are necessary (Fig. 3.10). Strong correlations (perhaps 0.7–0.9) between different crop years confirm highly stable yield patterns.

3.2.5. Soil and Weed Mapping

Soil mapping for variable fertilizer application and weed mapping for pesticide application can be part of a complete precision-farming system.

Soil Sampling

Soil sampling may be done using different strategies, such as random, aligned, zoned, and repetitive.

- Random soil sampling involves choosing sampling points at random within a field or a portion of a field. Often, a field is divided into a grid and a random sample is taken within each grid square. Random sampling minimizes some of the systematic errors possible in sampling.
- Aligned sampling takes samples on a systematic grid in the field. Aligned sampling minimizes the maximum distance to sample locations within a field and has simplifying advantages during some interpolations and other analyses. But it is susceptible to systematic errors, such as being confounded with natural or cultural patterns.
- Zoned sampling can be done after stable yield and soil patterns are established. Within the zones of uniform yields and soils, sampling is done randomly at a lower

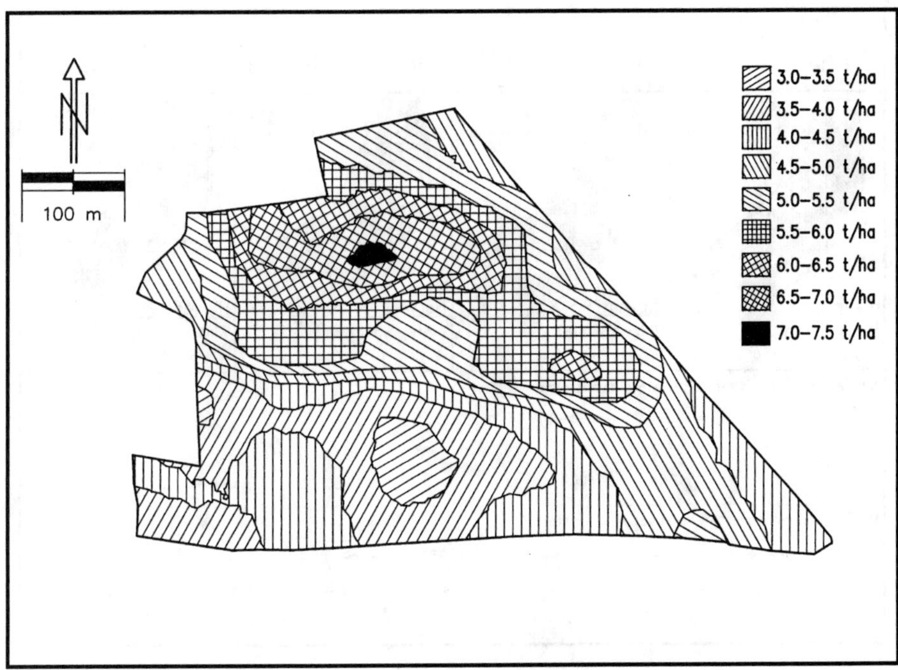

Figure 3.9. Contour yield map example (16.6 ha wheat field in southern Germany, 5.2 t/ha average yield).

density. This reduces the number of needed samples and the total costs of soil sampling by assuming that the soil properties follow the zone boundaries.

- Repetitive soil sampling tries to monitor the changes of soil nutrients over time. The sample location is accurately determined with DGPS and revisited during subsequent samplings in other years. The original locations of sampling may have been determined by one of the other methods.

Weed Mapping

Weed mapping is in an earlier state of development than crop yield or soil mapping. The following types of systems (Fig. 3.11) can be used:

- Manual mapping uses the detection of weed infestations by people combined with precise DGPS location to generate maps. Such mapping only locates areas of heavy infestation for later pesticide application using spot sprayers.
- Plant-coverage systems detect the overall plant coverage of the soil and distinguish between plant coverage without weeds and plant coverage with additional weeds or between fallow fields without weeds and fallow fields with weeds. Because the sensor measures only the amount of green plant, increased crop may be falsely sensed as increased weeds.
- Image processing analyzes images on the basis of size, shape, color, or location and transmits the location and type of weed or crop. Components of a sample

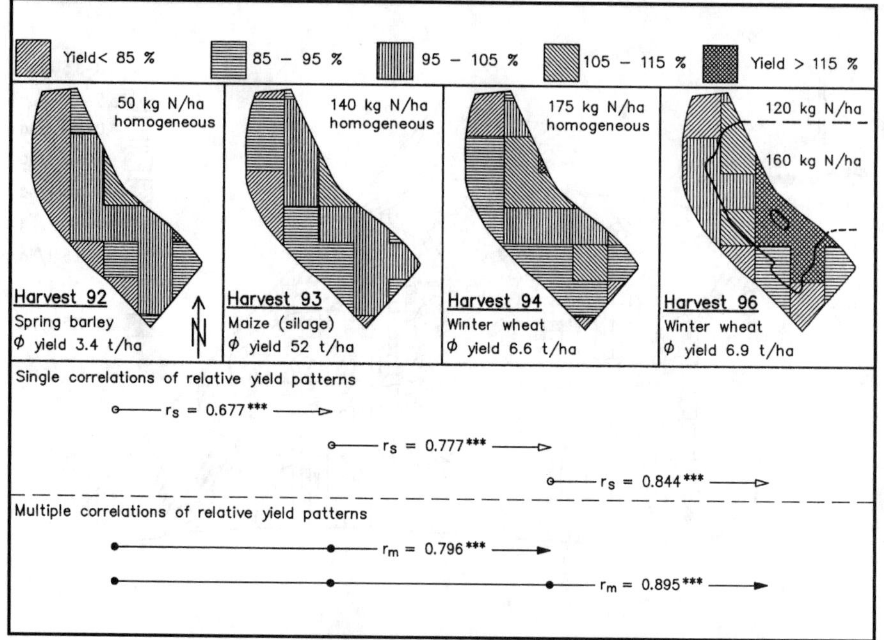

Figure 3.10. Consecutive relative yield maps and their pattern correlations (3.7 ha, 50 m × 50 m grid size).

system include a CCD (charge-coupled-device) camera, high-speed image analyzer, database with image information, controller, and sprayer actuator. Such a system is more complex and costly but will distinguish between crop and weeds if performing properly.

Remote Sensing

Remote sensing uses overhead images of a field or farm to indicate weed infestations or crop health. The images usually are acquired from satellites, airplanes, or remotely piloted vehicles. The presence of weeds between crop plants or rows of plants may be detected, or changes in the crop health, such as size, shape, maturity stage, or color, may be determined to indicate variations in insects, fungal infections, salinity, drainage, or other problems. Remote sensing can use single or multiple visible, infrared, or radar frequencies. Analyses often take the relative reflectance of solar wavelengths for differentiation (Fig. 3.12). Spot sprayers that turn on and off, or vary application rate, then may be programmed to spray according to the remote-sensing maps.

3.2.6. Control of Field Operations

Yield-, crop-, soil-, or pest-mapping indicates nonuniformities. The farmer may use those maps to develop understanding or make decisions. Often those decisions will be to perform field operations in a manner varied to correspond to those nonuniformities. Controlling fertilization, pesticide application, tillage, planting, or irrigation is a common part of precision farming. Fertilizer and pesticide application have seen the most

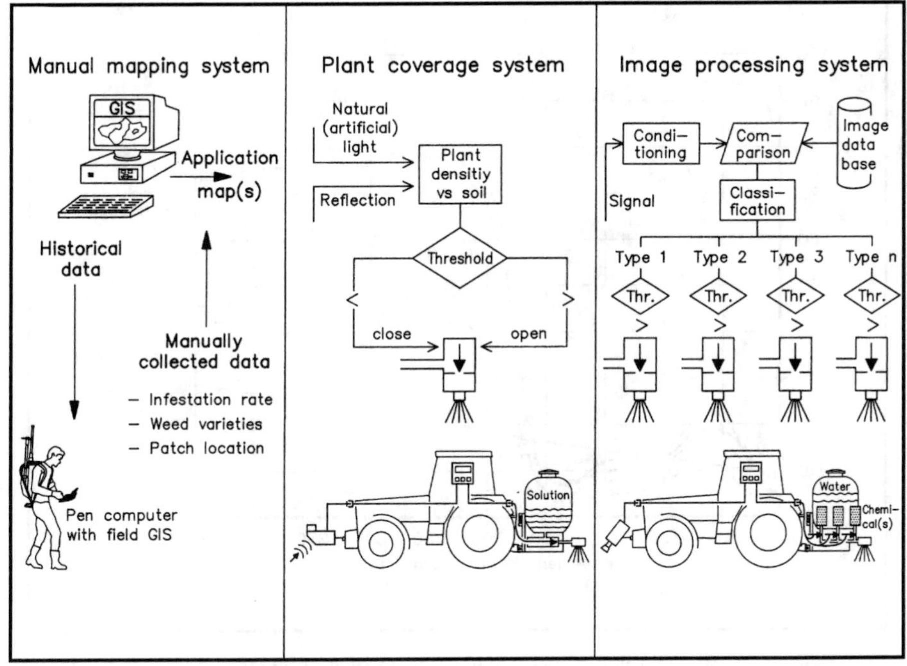

Figure 3.11. Weed mapping systems.

common implementations of precision control, because suitable equipment is available and nutrients and chemicals should only be applied as needed for both economic and environmental reasons.

Requirements and System Components

Field operation control requires an accurate real-time locator to indicate the equipment's position in the field, a map of the designed operation setpoints in a control computer, and an actuator on the equipment that can implement the controller's commands. The location must be accurate and the dynamics of the equipment must be compensated for so that the exact action occurs where it is desired.

Fertilization

The application rate of granular fertilizers (and similar materials such as lime) usually is varied by a microprocessor controlling hydraulics that actuate a variable-speed metering wheel, a variable-position gate, or a variable-speed chain conveyor. If the mixture of the nutrients is to be varied as well as the rate, multiple product bins and delivery systems are needed (Fig. 3.13) on the applicator, and the material-handling system must be able to efficiently refill the multiple bins.

Liquid fertilizer may be applied in a similarly variable manner. Variable rates can be achieved by varying either pump speeds, recirculation valve flow, or flow at individual nozzles through a variable pressure drop before the nozzle or turning the nozzle flow on and off with pulse-width modulation.

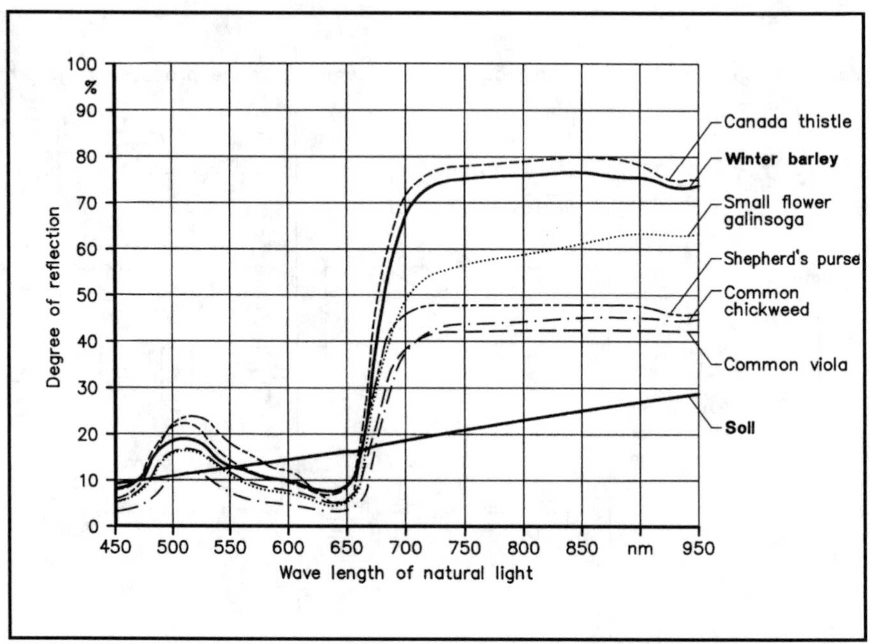

Figure 3.12. Wavelength reflectance of weeds in a barley culture (based upon Biller).

Both granular and liquid applicators must be designed for adequate dynamic response and to make sure the spread pattern is satisfactory at varying application rates. Variable mixture applicators can be designed to mix near nutrient storage, although then there will be significant time delays for changed mixtures to be transmitted to the soil, or mix near the nozzles or other distribution device, although then care must be taken to get complete mixture and additional hoses or conveyors are required.

Pesticide Application

Liquid pesticides may be applied either by varying the amount of premixed pesticide-carrier liquid or by injecting a variable amount of pesticide into a relatively constant flow of carrier (Fig. 3.14). Many of the same design concerns for fertilizer applicators apply to pesticide applicators, whether liquid or granular.

Other Controller Operations

Planting or seeding also may be controlled variably to achieve precision farming goals by varying variety, often in response to soil type or topographic position; population, in response to varying productivity potential; or depth, to find sufficient moisture.

Depth would be determined by real-time sensing while variety and population control would be more likely to be map-based.

Irrigation may be variably controlled by either maintaining full flow but changing the length of time of application, or maintaining length of time of application while reducing water flowrate.

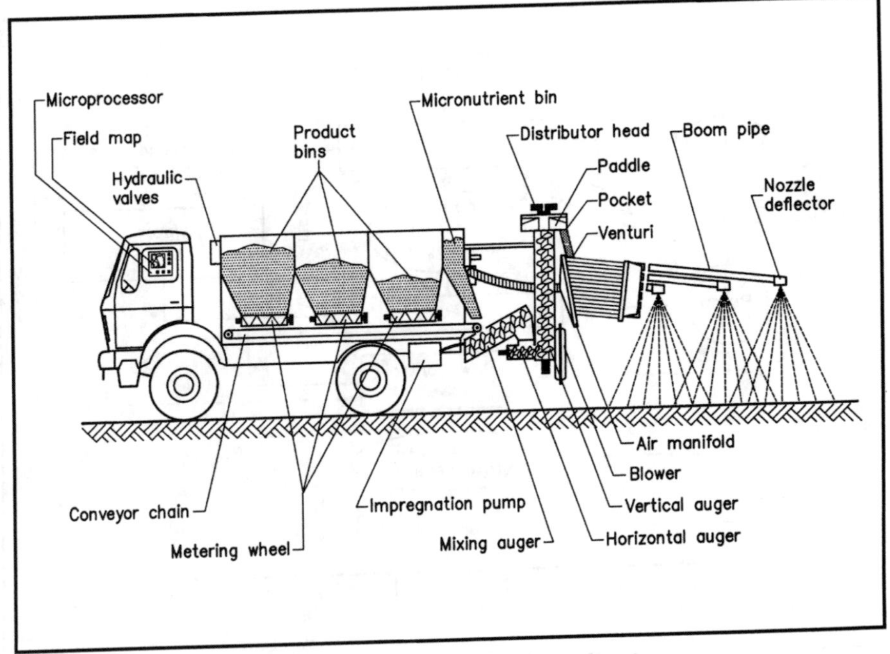

Figure 3.13. Granular fertilizer application system.

Tillage and landforming operations also may be controlled in either real-time or map-based manners.

3.2.7. Information Management

Sophisticated precision farming integrates a variety of computerized tools. Safe and reliable information transfer among all these tools needs standardized communication lines, standardized interfaces, and powerful software tools.

BUS Systems on Mobile Equipment

The advancements in agricultural electronics have led to a wide variety of controllers and electronic components. For example, fast and reliable communications are required between a tractor and the various implements attached to it (Fig. 3.15).

Compatibility is insured by communications standards. Two such standards, both using Controller Area Network (CAN), are as follows:

- The German LBS (Landwirtschaftliches BUS-System [Agricultural BUS-System]), codified as DIN 9684/2-5, is based on the 11-bit identifier of CAN V2.0A. It connects a maximum of 16 controllers, including the user terminal.
- The ISO 11783 standard works with the extended identifier of CAN V2.0B and is able to connect a maximum of 32 controllers. Its detailed structure using the ISO/OSI layer model and an additional six parts of special definitions tries to cover all requests of agricultural tractor–implement combinations.

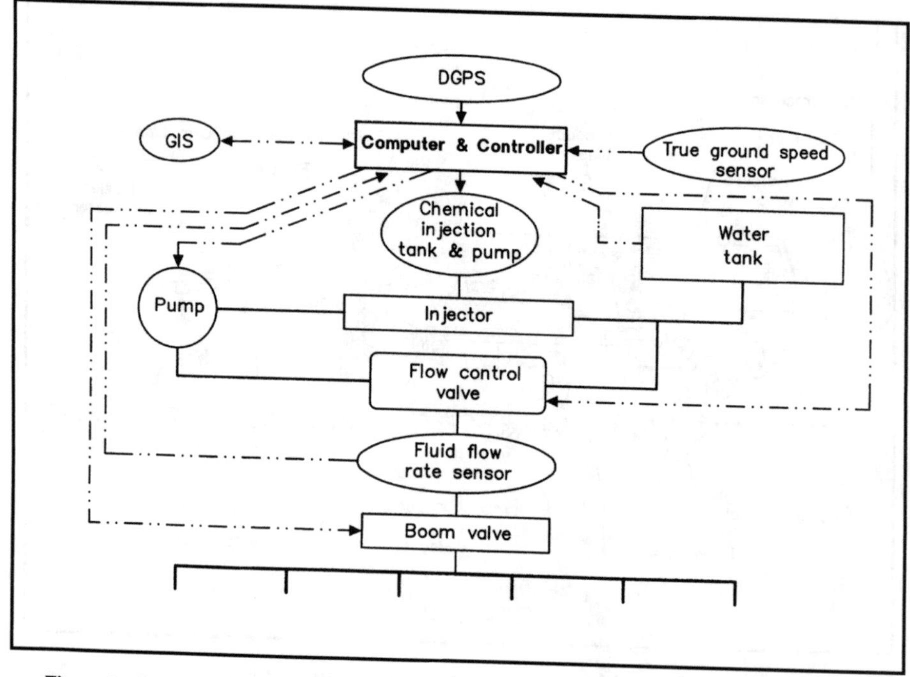

Figure 3.14. A variable-rate liquid pesticide applicator (based upon Clarke and McGuckin).

Compatibility between both standards is achieved in the overall function and in the physical layer.

Data Transfer to and from Farm Management

Complete precision farming systems are centered in the office computer. All information goes to and from this main unit. Yield and soil maps are displayed on it and desired control maps are generated on it. Data transfer can be done using human media transfer with chipcards or PCMCIA cards, or by bringing a portable computer to the office computer. Alternatively, radio links can be used. GPS differential corrections, remote sensing data, soil test laboratory results, and so forth can be obtained from e-mail or the Internet.

Data Management and Geographic Information Systems

The capabilities of computerized data gathering generate large volumes of data, which must be handled efficiently. Because precision farming data has position attributes, it usually is manipulated by geographic information systems. Such a system's representation of a field may contain layers of

- Soil type and topography
- pH and cation exchange capacity (CEC)
- Crop yields
- Weed maps
- Fertilizer and pesticide application maps

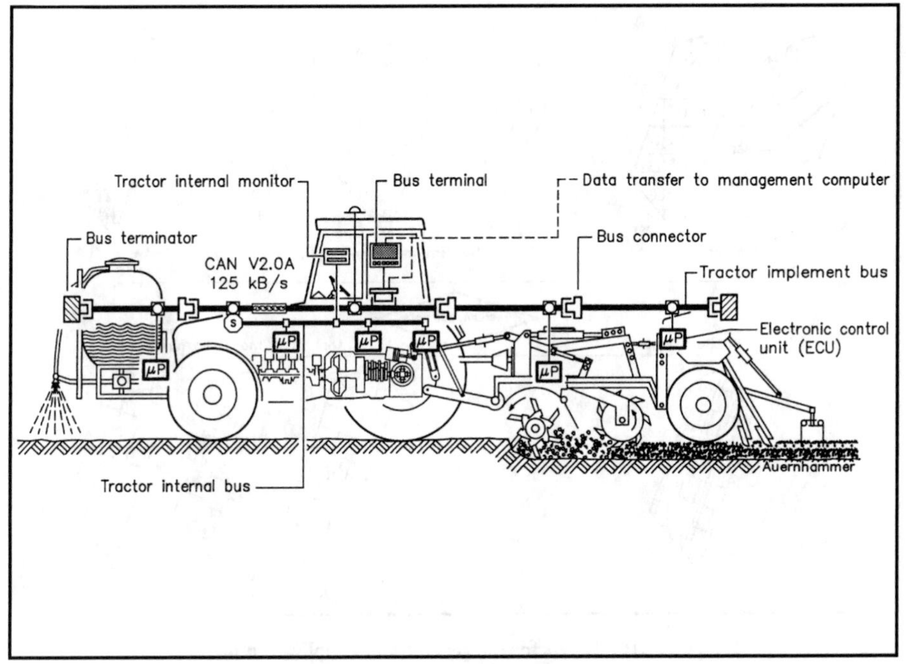

Figure 3.15. Communication between tractor and implement using DIN 86924/2-4.

These various layers can be analyzed or combined manually or automatically to generate a control map for field operations (Fig. 3.16).

Decision-support Systems

The control maps for map-based field operations must be generated according some sort of decision system. Even if the decisions are made manually, the volumes of data and the complexities of crop production favor a decision-support system. For example, the phosphorous application rates in Fig. 3.16 were calculated by a computer program for each area to remedy deficiencies and to provide sufficient nutrients for a crop of wheat. The input data included soil type, soil-test data, and yield potential based upon past yields.

The decision-making computer program can be deterministic, based upon rules or formulas. The computer determines the correct control action for each small part of the field or orchard based upon the geographic information system's data layers and the guidelines written into the decisionmaking program. It also can be stochastic, based upon computer simulations. Validated crop-growth models are run with different field-operation control strategies for representative weather scenarios in each field portion. The strategy with the maximum economic return and acceptable risk is used to establish the field operation control map.

Real-time systems must have control algorithms that immediately vary the actuator to the appropriate output based upon the sensor data.

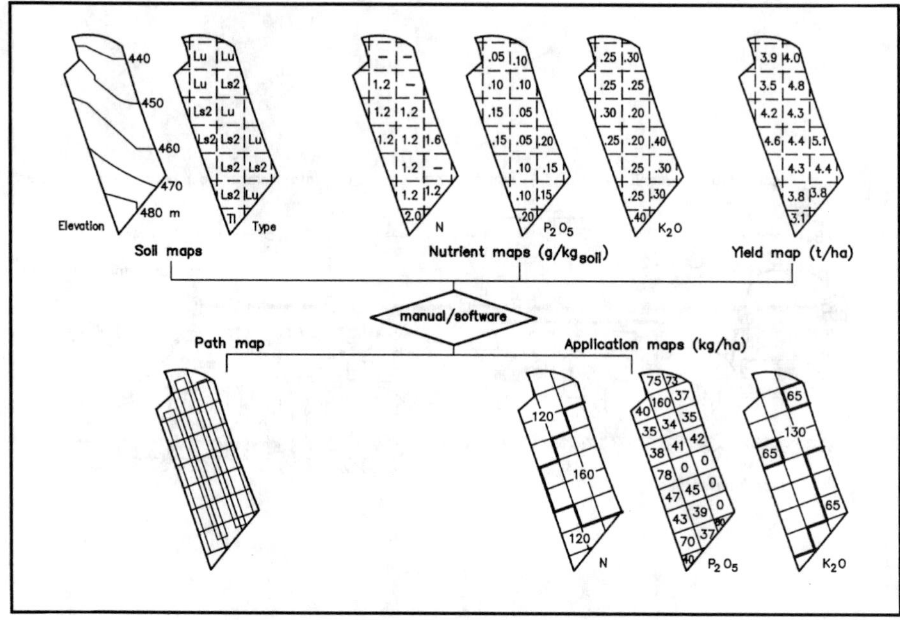

Figure 3.16. Layers for the generation of application maps.

References

1. Auernhammer, H., ed. 1994. GPS in agriculture [special issue]. *Computers and Electronics in Agriculture* 11(1).
2. Auernhammer, H. and Frisch, J., ed. 1993: *Landwirtschaftliches BUS-System LBS* [Agricultural BUS-System LBS]. Arbeitspapier: Muenster-Hiltrup.
3. Robert, P. C., Rust, R. H., and Larson, W. E., ed. 1993: *Proceedings of the First Workshop on Soil Specific Crop Management.* Madison, WI: ASA/CSSA/SSSA.
4. Schueller, J. K. 1992. A review and integrating analysis of spatially-variable control of crop production. *Fertilizer Research* 33:1–34.
5. Searcy, S. W., ed. 1991. *Automated Agriculture for the 21st Century.* ASAE Publication 11–91. St. Joseph, MI: ASAE.
6. Stafford, J. V., ed. 1996. Spatially variable field operations [special issue]. *Computers and Electronics in Agriculture* 14(2/3).

Index

acceleration (vibr.), 160
acces/exit (cab), 157, 159
"Ackerman" steering, 135
active ingredients, 274, 307
actuator, 602–603
adaptation, 540
adjuvants, 274
aerial application, 285–6
aerial logging systems, 497
aerohydroponics, 475–6
aeroponics, 475
aerosol and mists, 275, 278
aggregates, *see* soil aggregates
agitation, 290–2
  air sparging, 292
  hydraulic jet, 292
  mechanical, 292
  venturi-siphon, 292
Agricultural Engineering Service (AGSE), 553
agricultural mechanization, 1, 536
  levels, 1
Agricultural Mechanization Organizations, 562
agricultural tractor, 115
air, 593
air cleaner, 53
air conditioning, 155
air cushion vehicles, 496
air delivery ratio, 46
air handling capacity, 46
air seeders, 226
air-to-air intercooler, 51
air-to-water intercooler, 51
American National Standards Institute, 520
American Society of Agricultural Engineers, 520
AMS formulation, 537
angled headland, 569
anhydrous ammonia, 256
animal performance, 39
animal power, 22
animal skidding, 489
animal traction, 37–9
animal's draft capability, 24
animal's energy potential, 24–30
animal-drawn carts, 29
animal-drawn equipment for weeding, 37
animal-drawn seeders, 35
animal-drawn sowing equipment, 37
annual cost, 578
annual fixed costs, 577
appropriate mechanization, 536
ard ploughs, *see* ards
ards, 34
articulated (tractor), 117, 134
artificial lighting, 476–7
auger fertilizer spreader, 254–5
automation, 540
axe, 482
axle performance diagram, 142
axle suspension, 160

back strap, 26
balloon logging, 497
band seeders, 228
basic speeds (transm.), 145
basic tire loads (BTL), 129
beater, 450, 451
beet topping, 381, 390–2
belly strap, 26
belt conveyor, 469
belts, 441, 446, 448
benches, 472
berries, 452, 453
best forest management practices, 502
bias ply, 128
biological pest control, 272–3
blade, 433, 442, 445–7, 449
  chopping blades, 438
  lifting blades, 442, 443
block chassis design, 129
boom
  dry, 283
  wet, 283
boost, 50
bore, 46
boundary fertilizer spreading, 266
bovine crosses, 25
brake power, 48
brake thermal efficiency, 48
brake torque, 48
brakes, 131–4, 158
breakthru, 567
breast band, 26
breeching strap, 26
broadcast seeding, 228
broadcasting, 35
brush-cutter, 447, 449
BSFC, 49
bucking, 486

buffaloes, 22, 25
bulk seeding, 226–36
bundles, 434, 436, 437

cab, 117, 155, 160, 163
cab suspension, 116, 160
cab width, 157
cable crane, 496
cable logging systems, 494–6
cable-drawn implements, 214
calibration, sprayer, 304–306
  flow check, 305–306
  ground speed, 304
  timed flow method, 304–305
caliper brake, 133
camels, 25, 29
cane stumps, 435
cane whole, *see* whole cane
capsules, 438–40
carbon dioxide (CO_2), 47
carbon monoxide (CO), 47
cart transport, 29, 38
carting, 30
caster angle, 135
categories, 95
categories (hitch), 166
cattle, 22, 38
CEEMAT, 33
cell trays, 471
CEN, 517, 519
center of gravity (tractor), 123
central shaft concept (FWD), 131
cerametallic (clutch), 151
cetane rating, 47
chain harrow, 203
chain variator, 149
chainsaw, 483
characteristics, 95
chassis, 129–31
chemical
  closed-transfer, 303–304
  mixing and disposal, 303
chemical control, 269, 273–80
chisel plow, 210
chisel-type tools, 192
chisels, 209
choice of animals, 25–6
chopped cane, 437–8
chopper, 333
circular displacement frequency, 157
clay soils, 186
cleaning drum, 450
cleaning equipment, 303
cleaning shoe, *see* cleaning unit
cleaning systems, 439
cleaning unit, 311, 319–21
  airflow, 320, 321
  air velocity, 320, 321
  bulk phase, 321
  fan types, 321
  flight number, 320
  flight phase, 321
  fluidization, 321
  forces, 319, 320
  law of convection, 322
  loss, 323
  main parts, 320
  slope sensitivity, 323
  tailings, 312, 319
  theory, 322–3
  trash, 325
  winnowing steps, 319, 320
cleaning of young stands, 498
clearance dimensions (cab), 159
clearance zone (cab test), 163
climate, 155
climate computers, 478
clod breaking, 189
closed center (hydr.), 174
closed loop control circuit, 154, 155
coefficient of variation, 263
coffee, 452–3
collar, 26, 27
comb harrow, 36
combine harvesters, 311, 446, 604–605
  area capacity, 325
  attachments, 331–3
  automatic steering, 330, 331
  capacity, 325, 326
  cleaning unit, 311, 319–21
  compact combines, 340
  control systems, 330
  conventional, 311, 312, 326
  crop properties, 325
  functional components, 311–12
  harvest conditions, 325
  header, 332–3
  hillside, 312, 322
  loss, 323
  main elements, 311
  nonconventional, *see* rotary
  performance, 323–8
  performance characteristic, 318
  power requirement, 314, 317, 318, 327

productivity, 311
pull-type combines, 312
rotary combines, 315, 316, 327
rotary separator, 315, 316
self propelled, 312
separator, 311
straw chopper, 333
straw walker, 314
testing, 327
threshing, 311, 312, 317, 318
throughput, 317, 321, 327, 328, 331
workrate, 325, 327
combing, 413–14
comfort, 153
comfort boundary, 159
communications standards, 613
compression ratio, 46, 47
compression-ignited (CI), 41
computer, in precision agriculture, 604, 607–10, 613–15
computerized instruments, 39
concave, 313, 445, 450, 451
cone penetrometer, 119
conservation tillage, 594, 596
conservation tillage equipment, 210
conservation tillage systems, 208–10
constant brake power, 52
constant flow system, 174
constant pressure system, 174
contact action, 273, 274
contact pressure index (tires), 122
container benches, 472
continuous effort, 23
continuously variable transmission (CVT), 116, 148–50
contour line (field), 125
contractors, 547, 550
control forces (driver), 157, 158
control of seeders, 235
depth control, 224
control strategies, 477
control strategies (hitch), 169
controller area network (CAN), 613–15
convergence distance (hitch), 168, 169
conveyor, 433, 434, 436, 437, 440, 446
cooperative, 550
cost, 110
costs variable, 576
costs, annual, 578
cotton, 438–41
cotton picker, 438, 439–40
cotton stripper, 439, 440–1
coulter attached to share, 198
coutrier, 31
cows, 26
crank dead center (CDC), 41
crawler-type carrier, 458–9
riding type, 458
walking, 458
walking and riding, 458
credit, 548–9
creeper range (transm.), 147
critical path method (CPM), 525–7
crop rotations, 594
crop spraying, 301–302
crusher, 437, 438
crusting soils, 33
cultivation tines, 32
cultivator, 33, 34
cultivator shares, 34
cultivator, rotary, *see* rotary cultivator
cut into sections, *see* chopped cane
cutters, 432–4
CVT, 148
cylinder, 445, 446, 450, 451

daily feed ration, 28
dam forming, 207–208
damping (seat), 161
danish trolleys, 472
debarking, 486
decision rules, 478
decision support systems, 615
deep loosening, 214
deep plow, 212
defoliation, 438, 444
delimbing, 485–6
demand, 102
demonstration farms, 560
depreciation, 577
description, 95–101
desiccation, 438–9
design for safety, 153, 163
detachment, 409, 411, 415
deutz five-point-cycle (eng.), 139
developing countries, 22
development facilities, 554
diagram of tractive forces, 23
diaphragm check valve, 293
dibbers, 5, 6
dibbling, 35, 36
diesel engines, 116, 137–9
differential lock, 147
diffusion-limited burning, 46

digestible energy, 28
diggers, 1, 11, 13
digging, 411
direct seeding, 33, 34, 37
direct-injection (DI), 42
disc brakes, 132
disk, 191
disk coulter, 198
disk harrows, 34, 203
disk plow, 199, 203, *see also* plow disk
displacement, 45, 46
displacement (engines), 139
displacement (hydr. pumps), 174
ditch plow, 212
dividers, 433
donkeys, 22, 25, 29, 30, 38
draft, 106
draft (plow), 170
draft animal power, 539
draft animal power unit, 22–4
draft animal technology, 536
draft animals, 22, 34, 38, 451
draft control, 165
draft force, 22, 33
draft resistance, 22
draught-animal technology, 1
drawbar power, 122–3
drawn implements, 190–3
drift
 droplet, 302
 reduction, 302–303
 vapor, 302
drilling, 35, 36, 228
drip emitters, 474
drip tape, 474
drip-and-trickle irrigation, 473–4
driver, operator, 154
drivers, 433
driver-seat system, 160
driving fittings, 26
droplet size, 275
dry sowing, 35
drying, 441, 444, 446
dual tires, 117
duckfoot shares, 38
dust, 155
duties, 549–50
Dyna drive, 204

ears, 450, 451
earthing-up, 38
earthing-up ridger, 32
education, 552
effective field capacity, 575
efficiency, 315, 325
efficiency field, 106, 300
efficiency tractive, 106
encrustation, 187
energy, 592, 595–7
 human consumption rates, 1, 2
energy expenditure, 23, 24
energy potential of draft animals, 22
energy requirements, 23
engine (tractor), 117, 137
engine performance map, 51
engine power, 326–7
engine selection, 53–4
engine-power technology, 1
engines, 41
engines characteristics, 101
environmental conditions, 24
environmental engineering, 585–98
equipment, choice, 300–302
equipment for tillage, 31–5
equipment for transport, 29–31
equivalence ratio, 44, 45
ergonomics, 485
ergonomics and safety, 304
erosion, 585, 591
essential safety and protection requirements, 518
European Committee for Standardization, 517, 519
European Union, 517
evaluation, 552
evaporation, droplet, 276
exhaust emissions, 47
exhaust smoke, 47
expansion coefficient, 278
exponential function, 231
exposure limit (vibr.), 159

factory, 431, 432, 437, 438
family of engines, 51
fans, capacity, 301
farm cultivation, 38
farm equipment, 29–38
farm power, 538
farm-machinery institute, 562–3
farming-systems, 540, 542
fatigue (driver), 159, 160
fatigue (transm.), 142
fatigue-decreased proficiency boundary, 159

feed table, 437, 450, 451
feedback-feedforward control, 478
felling, 481–5, 487
felling tools, 482–4
Ferguson system, 116, 166, 170
fertilization, 611–13
fertilizer, 585, 586, 589, 590, 593, 595
fertilizer application
  boundary spreading, 266
  dusty, 254
  granular, 254
  liquid, 256–8
  powder, *see* fertilizer application, dusty
  site specific, 267
fertilizer distributors
  aerial, 255–6
  auger, 254–5
  boom, 252–5
  electronics, 267–8
  liquid, 256–8
  multiple hoppers, 266–7
  oscillating spout, 248–52
  particle trajectories, 260–1
  pneumatic, 253
  qualifications, 241
  spinning disc, 241–8
  spread pattern, 261–6
fertilizer properties
  aerodynamic resistance, 258–60
  coefficient of friction, 244, 248, 252
  density, 260
  diameter coefficient, 260
fibreglass fingers, 453
field
  capacity, 326
  efficiency, 325
field operations, 105–10
field speeds, 566
finger machines, 441
fire, 432
fire protection, 498
fixed costs, 576
fixed displacement pumps, 174
fixed-cost percentage, 577
flame weeding, 271
flat attachment (wheels), 136
flexible teeth, 445
floating, 160
floating implement (hitch), 168, 169
floats, 202
flow controller (hydr.), 174
flow on demand (hydr.), 174
fluid coupling, 116, 145, 151, 152
fluid power, 172–4
foliar applications, 277
folling tools, 482–4
Food and Agriculture Organization (FAO), 553
forage, 605
forage cereals, 374–80
  the methods, 374–6
  machines and equipment, 376–80
    forage harvesters, 376–80
forage harvesting, 109
force control (hitch), 170
force control by upper link, 170
force sensor (hitch), 171, 172
Fordson tractor, 115, 116
forehead yokes, 26
foreign material, 439, 441
forest fertilization, 498
forest regulations, 502
forest roads, 501–502
forest site preparation, 500
forest soil restoration, 499
forest stand establishment, 499–501
forest stand protection, 498–9
foreword, 348–9
formulations, 273–5
forwarding (full carying), 492
four-cycle engines, 42
four-wheel carrier, 457
  articulated steering, 457
  both walk-along and ride-on, 457
  ride-only, 457
  walk-along, 457
four-wheel carts, *see* trailers
four-wheel drive, 117, 124–5, 130–1
four-wheeled trailers, 29
four-wheeler, 117
frame chassis, 116, 130
frequency (vibr.), 160
friction mean effective pressure, 48
friction power, 48
froehlich tractor, 116
front end loader, 116
fruit, 607
fruit harvesting, 408, 422, 429
  damage, 410, 424
fuel consumption, 139, 566
fuel economy, 41, 54
fuel equivalent power, 43
fuel injection, 43
fuel tank, 117, 139

fuel tank capacity, 52
fungicides, 274
  contact, 274
  systemic, 274
furrow-opener, 211
future trends, 308–10
  direct injection metering, 309
  drift control, 309
  electrostatic charging, 308
  global positioning, 308–309
  precision spraying, 308
  tunnel sprayer, 309–10

gauge wheels, 224
gearbox, 140
gearbox input load, 142
geographic information systems (GIS), 599, 614–15
ginnery, 439
gleaning machines, 441
global positioning system (GPS), 600–602
GLONASS, 600–601
good grain, 451, 452
governor regulation, 52
governor's maximum, 51
governor's maximum point, 52
grab, 435, 437
grain, 464–5, 604–605
grain
  damage, 313, 314, 317, 318, 325
  harvesting, 311
  header, 332–3
  loss, 318, 323
  loss sensor, 328
  separation, 313, 317, 318, 322
  throughput, 324, 330
grain damage, 314, 317, 318, 325
granular applicators, 306–308
  application rate, 306, 307–308
  calibrating, 307
  metering, 306–307
grape, 417–19
grapple skidding, 491–2
graters, 17, 19
greenhouse climate control, 477–9
gross traction, 118
groundnut lifter, 32
groundnut oil, 446
groundnuts, 441–6
guidance implement, 565
gynophore, 445
  gynophore saws, 446

half track, 117
hand force (steering), 135
hand picking, 439, 441
hand sowing, 35
hand tool technology, 536
hand-tool technology, 1, 3
hand-tools, 5–6
handwork, *see* labor
hardpan, 185
harness, 26
harnessing, 489
harnessing systems, 26, 27
harrow disk, 191
harrow element, 203
harrow rotary for ridging, *see* rotary harrow for ridging
harrow, chain, *see* chain harrow
harrow, conventional, 203
harrow, disk, *see* disk harrow
harrow, power, *see* power harrow
harrow, reciprocating, *see* reciprocating harrow
harrow, rotary, *see* rotary harrow
harrows, 34, 202–204
harvest, 431, 432, 437–9, 441–4, 446, 449, 452, 453
harvester operation, 487
harvesters
  chopper-harvesters, 437–8
  combine harvester, 446
  full stick harvesters, 433–4
  power-driven harvesters, 431
  shaking harvesters, 449
  small machines, 432
  stackers, 434
  windrowers, 433–4, 443–4
harvesters chopper-harvesters, 437–8
harvesting, 311
  burning, 432
  cotton harvesting, 438
  crushing, 431
  cutting, 431–3, 438, 442, 444, 449
  groundnut harvesting, 441
  hand harvesting, 431, 448–9
  harvesting cost, 440
  harvesting heads, 440
  harvesting methods, 431
  mechanised harvesting, 431, 432, 438
  whole stick harvesting, 432–7
harvesting aids, 426–8, 429
harvesting hooks, 12, 13
hay, 607

HC (unburned hydrocarbons), 47
head dead center (HDC), 41
header, 311, 332
  corn header, 332
  grain header, 332
  loss, 323
  pickup header, 332
  stripper header, 311, 332
head yokes, 26
headland, 567
health and safety, 153
heaps, 431–3, 435, 441
heel force (plow), 168
helicopter logging, 497
herbicides, 274
  contact, 274
heterogeneous combustion, 46
high clearance, 98–100
high idle point, 52
high-pressure sodium lamps, 476
hillside operation, 125–6
hire company, 452
hitch, 117, 166
hitch categories, 166
hitch points, 166
hitching systems, 26
hoeing, 38
hoes, 1, 3, 5–8, 10, 11, 13, 14, 16, 34
hopper, 434, 439, 440, 446, 451
horizontal resistance to draft, 23
horses, 22, 25, 29, 30, 38
hullers, 19, 20
human cutters, *see* labor
human fatigue, 153
human work
  compensating attributes, 3–4
  manual rates, 1, 2
  output, 3
human-control, 540
humid zones, 436
hump cattle, 28
humpless cattle, 26
hydraulics, 165, 176, 607, 611
hydrostatic CVTs, 149
hydrostatic power split CVT, 150

image processing, 609–10
implement, 1, 5
implement control, 168–9
implement turns, 567
implements, *see* drawn implements and pto-driven implements
impurities, 439–40, 453
in-field measurements on working animals, 39
incentives, 548
inclined moving apron, *see* shacking aproon
independent PTO, 144, 145
indicated mean effective pressure, 48
indicated power, 47
indicated thermal efficiency, 48
indirect injection (IDI), 42
inertial shakers, 421–24
inflation pressure, 122
information management, 613–15
information systems, 328
insecticides, 274
instantaneous effort, 23
institutions, 550–2
integrated pest management (IPM), 269
intellectual property, 564
intercooling, 49–51
interest charge, 577
International Electrotechnical Commission, 504
International Organization for Standardization, 504
interrow cultivators, 38
interrow weeding, 37
inversion of sucrose, 432
inversion, temperature, 280
irrigation, 590–2, 595, 612
irrigation systems, 472–6
ISFC, 49
ISO, 504
  member bodies, 516–17
  technical committee 23, 504–505

king pin (steering), 135
knives, 11, 15–16

labor, 431–3, 435, 439, 447, 449, 453
labor productivity, 539
labor-saving, 539
land consolidation, 540
land productivity, 539
land reclamation, 212–13
Landwirtschaftliches BUS-system (LBS), 613–15
large agricultural ventures, 449
large plantations, 453
lateral seed distribution, 231
leaf area index (LAI), 277, 279
leaf-stripping system, 434
lifters, 11, 12, 14, 440–2, 449

lifting, 381, 436, 441, 443, 444, 446, 447, 449
lifting force (hitch), 168
line of draft, 23
line of draft (hitch), 168
linear programming, 522–5
liquid fertilizer
  advantages, 256
  anhydrous ammonia, 256
  application, 256–8
  disadvantages, 256
  solutions, 257
  suspensions, 257
livestock management, 28–9
load capacity of cart, 30
load collective, 142
load control, 52
load index tires, 129
load-sensing (LS) (hydr.), 116, 175
loader, 434–7, 449
  continuous loaders, 436–7
  discontinuous loaders, 435–6
  dumping clearance, 465
  dumping reach, 465
  forklift, 466
  front-end loader, 435
  hoist and truck, 465
  pushloader, 436
  self-propelled, 466
  swivel loaders, 435–6
  tractor-mounted, 465
  universal elevator, 465
loading, 431–7, 447–9
loamy soils, *see* silty soils
locomotion (off road), 492–4
log extraction, 488
longitudinal seed distribution, 231
losses, 311, 323, 451, 452
  behavior, 324
  characteristic, 324
  curves, 322, 326
  processing, 323
  saparating, 323
low correct inflation, 122
low section height tires, 129
lugging range, 52

machetes, 1, 3, 6, 7, 11, 13, 14, 16
machine, 1, 5
machine safety, 164
machinery directive (EU), 164
machinery hire, 552
machines and environment, 502
maintenance, 559
management, 110–13
management science, 521–36
manufacturers' associations, 560
manure, 586, 590, 594
manure burying coulter, 198
mapping, 602–10
mass, 606
mast hight (hitch), 166, 167
master clutch, 143, 151
material efficiency, 576
material other than grain (MOG), 311
  separation, 317, 318
  throughput, 318, 330
matrix, *see* soil matrix
meadow-type forages, 349–74
  harvest, treatment and storage methods, 350–7
    barn drying, 354
    dehydration, 355–6
    harvest and use of green forage, 350
    silage, 356–7
    traditional haymaking, 351–4
  machines and equipment, 357–4
    big balers, 371
    conditioners, 359–62
    conventional balers, 365
    handling and storage, 373–4
    hay cubers, 372–3
    mowers, 357–8
    rakes windrowers, 362–3
    round balers, 365–70
    self-loading wagons, 363–4
    stack wagons, 371–2
    tedders, 362
    wrapping and ensilage, 373
mechanical, 292
mechanical efficiency, 48
mechanical precision drilling, 218–22
mechanical sowing, 35
mechanical transmission, 100
mechanical-power technology, 536
mechanics, 103–105
mechanization, 537
mechanization strategy, 536–53
metal wheels, 31
metering of seeds, 226–8
metering rollers for seed, 226
methane, 593
millet, 450–2
minimum tillage, 596, 597
mixed mode control (hitch), 170, 172

mobile blowers, 453
model reference adaptive system, 479
moisture, 605
moldboard, 190
moldboard plow, 34, 35, 190, 194–9
moldboard plow bottom, 190
moldboard plow bottom, main parameters, *see* moldboard plow bottom
moldboard, cylindrical, 191
moldboard, forms, 191
moldboard, semidigger type, 191
moldboard, standard type, *see* moldboard, universal type
moldboard, universal type, 191
mole drain plow, 212
monitor, 295
  nozzle, 295
  system, 295
monitoring of seeders, 235
monorail, 468
monowheeler, 455
mortars and pestles, 16, 17
motive force, 23
motive power, 539
motor truck, 460–3
  farm truck, 461–3
  subcompact truck, 460–1
motorization concept (tractor), 138–9
mounted forks, 436
muffler, 53
mulch tillage, 208
mules, 22, 38
multipass effect, 124, 136
multiple-row seeders, 36
multipurpose toolbars, 32, 34, 38
multiwheel carrier, 457–8
  riding, 457–8
  super-pivot turn, 458
  walking, 457
Munich research tractor, 129, 162

narrow track tractor, 117
natural circular frequency (vibr.), 160
navigation, 600
nebraska tests, 116, 120, 125
neck yokes, 26, 27
net traction, 118
nitrate, 603
nitrogen, 593, 596
no-till, 187
no-till machine, 211
no-tillage planting, 209
no-tillage system, 210–11
noise, 155, 162
noise control, 161–2
noise load spectra, 162
noise pressure level, 162
noise equivalent level, 162
non-cane material, 434
non-chemical control, 269
nozzle, 611
nozzle bar, 470
nozzle materials, wear life, 299–300
nozzle types and patterns, 296–9
  controlled droplet, 298
  disc and core, 298
  even flat fan, 296
  flat fan, 296
  flooding fan (deflector), 296
  full cone, 298
  hollow cone, 298
  pneumatic, 298
  rotary, 298–9
nozzle, functions, 295–300
  fan overlap, 296
  flow rate, 295
  low drift, 296
  pattern, 295–6
  size of droplet, 295
number mediam diameter (NMD), 275
Nutrient-film technique, 475
nutrients, 593–4

obsolete, 582
OECD Test, 120, 125, 164
on demand flow, 175
one-row seeders, 36
open center (hydr.), 174
openers for seeders, 228
operations research, 521–36
optimization, 522–5
optimum power, 110, 580
optimum-size machine, 580
ordering forest machines, 502
organic framing, 590
organic matter, 603
oscillating spout spreader
  description, 248
  equation of motion, 251
  kinematics, 249–50, 252
  particle motion, 250–1
  propulsion, *see* oscillating spout spreader, kinematics
  spout design, 249–50, 252

overhead irrigation, 473
overlap, 565
overturning, 125, 126–7
overturning fatalities, 164
ownership, 550
oxen, 26, 33, 38
oxides of nitrogen (NOx), 46
ozone, 593, 595

pack saddles, 29
pack transport, 26, 29, 38
pannier baskets, 29
parameter-optimized controller, 479
park, concentration, 103
payload (tractor), 127
PCMCIA card, 607, 614
peat block, 471
pedal force (brake), 132
pedal position, 156, 157
pendulum test (ROPS), 116, 164
performance (tractor), 143
pest, 269
pesticide application, 612, 614
pesticides, 269, 586, 590, 594, 595
  contact, 269, 273
  systemic, 269, 273
phosphorus, 593
physical properties
  fertilizer, *see* Fertilizer properties
picker, 440, 441, 452
plant tops, 446, 447, 449
plant-growth models, 477
planters, 217–40
  transplanters, 238
planting, 500–501, 612
planting method, 35
plastic deformation, 188
plows, 32, 194, 447
plow disk, 191
plow frame, 33
plow, chisel, *see* chisel plow
plow, deep, *see* deep plow
plow, disk, *see* disk plow
plow, ditch, *see* ditch plow
plow, lining system, 197
plow, moldboard, *see* moldboard plow
plow, mole drain, *see* mole drain plow
plow, overload protection systems, 198
plow, rotary, *see* rotary plow
plow, semimounted, 197
plow, special tools, 198
plow, swing, *see* swing plow
plow, tractor mounted, one way, 196
plow, tractor mounted, reversible, 196
plow, trailed, 196
plowing, 31–4
plows of Asian design, 35, 36
plumbing and controls, 293–5
  control valves, 294–5
  pressure gauge, 293–4
  pump connections, 293
  relief valve, 294
  throttling valve, 295
  unloader valve, 294
ply rating (tires), 129
pneumatic, 298
pneumatic fertilizer spreader, 253
pneumatic tires, 116, 127
pods, 441, 444–6
  pod losses, 442
policies, 537, 547–8
policy, 589
porosity (soil), 120
porous hose, 474
position sensor (hitch), 172
positioning, 600–602
potassium, 593
potato, 605
potato harvest, 397–408
  harvest conditions, 398
  tuber damage, 407–408
  two-stage harvesting, 404–406
potato harvester
  discharge elevator, 402–403
  elevator digger, 406
  haulm separation, 400–402
  hopper, 402–403
  lifting, 397–8
  multi-row, 403–405
  self-propelled, 404
  separation rollers, 401
  separation units, 401–402
  separation units, electronic, 401
  sieving, 400
  single-row, 403–405
  windrower, 406
power, 22
power consumption by humans, 1–2
  production by human, 1–2
power harrow, 206–207
power index (tires), 128
power shift (transm.), 116, 147
power split CVTs, 116, 150
power steering, 116, 134

power take off (PTO), 115, 116, 143–4
power unit, 22
power-driven equipment, 431
power-intensive, 539
precision drilling, 218
  air-assisted, 222–4
  depth-control, 224
  inside-feeding, 219
  mechanical, 218
  outside feeding, 219
  punch seeders, 221
precision farming, 598–616
precision seeding, 36
premixed burning, 46
pressure bulbs (soil), 121
pressure guage, 293
preventative maintenance, 574
prices, 112
primary tillage implements, driven, 200–202
primary tillage machinery, 194–202
production, 102
productivity-enhancing technology, 538–9
project management, 525–8
proportional flow control value, 175
pruning, 498
public investments, 550–3
puddling, 35
pull output, 24
pulling, 411–12
pulpers, 453
pump
  capacity, 290
  centrifugal, 287–8
  characteristics, 290, 291
  diaphragm, 289–90
  nonpositive displacement, 286
  piston, 288
  positive displacement, 286
  roller, 287
pumps (hydr.), 149, 171, 173

quality standards, 552
queuing theory, 528–32

R&D, 550
rachis, 450
radial ply tires, 128
raft system, 476
railway, 437
raked blade, 188
random load fatigue theory, 142
range principle (transm.), 145
raw material, 559
real-time, 602–604
reciprocating harrow, 207
reduced tillage, 208, 209
remote sensing, 610, 612
repair and maintenance, 578
residue cutting, 211
resistance force, 23
resitance effort, 22
rest period, 2, *see also* human work rates
rice harvesting, 333–41
ridge, 34
ridge cultivation, 31, 35
ridge till, 187
ridger, 31, 34, 208
ridging, 34, 35
ridging plow, 34, 35
riding comfort, 157
rigid tines, 34
rim, 136
ripper, 33
ripping, 33
RMS, 157, 161
robotic harvesting, 428–9
rockwool, 476
roller dibbler, 471–9
rollers, 208, 439, 440, 443, 444
rolling benches, 472
rolling coefficient, 30, 31
rolling implements, 34
rolling resistance, 118
rollover protective structures (ROPS), 163, 164
root-mean-square values (RMS), 157
roots, 444, 446–8
rotary cultivator, 200
rotary harrow for ridging, 208
rotary harrows, 35, 36, 206
rotary plow, 200–201
rotating drums, 439, 453
rotating spindles, 439
rotovators, 206
row seeding, 35
rural infrastructure, 540
rural labor, 542

sacking device, 445, 450, 451
safety, 154, 163–5
safety frame, 164
safety in forest work, 487–8
sampling, 437, 438
sand, *see* sandy soils

sandy soils, 186
satellite, 600
saw, 482
scarifiers, 34
scythes, 11–13, 14
seat, 117, 159, 160
seat belt, 164
seat index point (SIP), 155
seat reference point (SRP), 155
seating position, 155
secondary cultivations, 38
secondary tillage, 37–8, 188
secondary tillage machinery, 202–208
secondary tillage tools, drawn, 202–205
secondary tillage tools, rotating, 204
security in forest work, 487–8
seed
  distances in the row, 232
  exponential distance function, 231
  incorporation into soil, 227–8
  lateral distribution, 231–3
  longitudinal ditribution, 231–3
seed cotton, 439, 440
seed-spacing and yield, 232, 233
seedbed, 185
seedbed combinations, 204–205
seeders, 34, 36, 469–71
seeding depth, 228–31
  variation of, 229, 233
seeding lines, 469–71
seedling trays, 471
seedrate, sensing, 234
self-lining effect, 194
self-propelled, 435, 440, 441, 446, 466
self-tuning regulator, 479
semiaerial logging, 494–6
sensing of seedrate, 234
sensors, 602–606
separating losses, 323
separator, 311
  conventional, *see* straw walker
  efficiency, 315
  rotary, 315
  theory, 317–19
services, 550–2
shakers, 442, 446
  shaker-windrowers, 444
shaking, 444, 447, 452
  shaking apron, 444
  shaking lifters, 442, 448
  shaking system, 443
shaking, shakers, 415–16
  amplitude, 417, 419, 420, 422
  frequency, 415, 417, 419, 420, 422
share, 190
share, chisel, 192
share, diamond pointed, 192
share, duckfoot, 192
share, straight, 192
share, stubble, 192
share, sweep, 192
share, twisted, 192
share, winged, 192
shares, 192
shellers, 17–20
shift elements (transm.), 151
shift pattern (transm.), 146
shift up throttle down, 154
shovels, 7, 11, 14
shredding, 449
sickles, 1, 11, 12, 14
sideways shaft concept (FWD), 131
sieves, 440, 445, 446, 451
signal mix (hitch), 172
silty soils, 186
simulation, 532–6
single-row, 403–406
site specific sowing, 234
skidding, 488–90
skimmer, 198
slatted moldboard, 35, 36
sledges, 29
slip (tires), 118
slippage, 107
slope limits, 127
slopes, 125, 433, 435, 453
small disk cutter, 432
small estates, 433
small farms, 435
small fruits, 416–17
small planters, 432
soil aggregates, 187
soil applications, 276
soil classification, 186
soil compaction, 120–2
soil conditions, 185–7
soil cutting, 211
soil engaging components, 188–91
soil failure, 188
soil improvement, 185
soil mapping, 608–609
soil resistance of plow heel, 23
soil resistance to forward movement, 23

soil sealing, 33
soil texture, 185
soil, failure pattern, 189
soil-less culture, 475–6
soil-salinity, 592
soils matrix, 187
solarization, 271
sowing, 33, 35–7
sowing densities, 35
sowing implements, 35
spades, 6, 7, 14
spading machine, 201–202
spark ignited (SI), 41
speeds, tractors, 565
speaking plant approach, 479
special conditions, tillage systems for, 211–15
specific fuel consumption (SFC), 49
specific torque, 48
spectral uneveness density (roads), 157
speed concepts (transm.), 141–2
speed symbol (tires), 129
speeds (tractor), 141, 142
spikelets, 450
spinning disc spreader
  description, 241–2
  disc design, 242, 245–8
  equation of motion, 242–5
  particle motion, 245–8
sprayers equipment, 281–6
  air blast, 284–2
  air-assisted boom, 285
  boom, 283–4
  boomless, 284
  motorized, 283
  tower, air curtain, 285
sprayers hand, 281–2
  battery operated, 282
  compressed air, 281
  controlled droplet, 282
  knapsack, 282
spraying
  low volume (LV), 279
  orchard, 302
  ultra low volume (ULV), 279
spread pattern
  classification, 263–4
  coefficient of variation, 263
  determination, 263
  evaluation, 263
  overlap, 265
spring tine harrows, 37
spring tines, 38
standard tractor, 117
standardization, 503–20
standards
  for agriculture, 506–16
  definition of, 503
  reasons for development, 504
static benches, 472
stationary power, 539
steam traction engines, 115
steering, 134–6
steering angle, 134
steering assistance, 566
steering force, 158
steering, tractor 565
step ratio (transm.), 142
stock watering, 28
straight rigid tines, 38
strainers and screens, 292–3
  basket strainer, 292
  line strainer, 292
  nozzles strainer, 292
strategies, 537
straw walker, 311, 312, 314
  auxiliaries, 315
  efficiency, 315
  loss, 323
strip tillage, 209
stripper, 441
stripper fronts, 336–7
stripper-gatherer, 342
stripping, 414–15
stroke, 46
subirrigation, 474–5
subsidies, 548
subsoilers, 447
subsoiling, 31
subsoiling share, *see* subsoiling tine
subsoiling tine, 33
substrate computers, 478
sugar, 431, 605–606
sugar-beet harvesting, 381
  beet topping, 381, 390–2
  cleaning, 394–5
  complete harvesters, 383–9
  leaf stripping, 381, 390, 396
  loader, 383
  loading, 381, 395
  multistage harvesters, 381–3
  quality harvesting, 395–7
  self-loading trucks, 383
  self-propelled six-row, 386
  self-propelled three-row, 385

sugar-beet harvesting (*cont.*)
self-propelled two-row, 385
soil tare, 397
transmissions, load sensing, 397
supply system, 552–3
surface tillage, 31, 34
surface tillage implements, 34
sustainable agriculture, 585, 586, 589
sustained effort, 23
swaths, 381, 431, 432
sweep, *see* share, sweep
swing plow, 199
symbols (fluid power), 172, 173
symbols (transm.), 145
synchronizer (transm.), 151
systemic action, 273
system
cropping systems, 431
farming system, 431, 441
systems engineering, 521–36

tanks
capacity, 290
materials, 290
sprayer, 290
taurines, 28
taxes, 549–50
technical service center, 560–1
temperature intergration, 479
testing, 552
texture, *see* soil texture
theoretical capacity, 574
thinning of young stands, 498
three-point hitch, 165
threshers
cereal thresher, 450
stationary thresher, 441, 444
thresher-picker, 444, 445–6
uprooter-threshers, 446
threshing, 311, 312, 317, 318, 441, 443, 444–6, 450, 451
axial, *see* rotary threshing unit
clearance, 314
concave, 312, 313
cylinders, 312–14
device, 312
diameter, 313
hold-on threshers, 312
loss, 313, 314, 317, 323
main parts, 313
multiple cylinder, 313, 314
peripheral speeds, 312–14, 318
power threshers, 341
respbars, 312
rotary threshing unit, 312, 315–18
spike-tooth threshing, 312
tangential threshing unit, 312–14, 317, 318
wire-loop, 312
threshing device, 445
throughput, 321, 327, 328, 330, 331
tillage, 24, 31, 184
tillage implements, 32
tillage machinery, 184–215
tillage system, 193–215
tillage system, conventional, 193–208
tillage systems for special conditions, *see* special conditions, tillage systems for
tillage tools, driven, 205–208
tillage tools, operations performed, 195
tillage, *see also* conservation tillage systems, mulch tillage, no tillage planting, reduced tillage, strip tillage, tillage systems
tillage, energy saving, 210
tillage, general importance, 184–8
tillage, socio-economic aspects, 187–8
tillage, specific tractor power, 194
tilling on flooded soils, 36
time management, 525–8
timeliness, 579
tine, harrow, 192
tine, rigid, 192
tine, spring, 192
tine tillage, 31
tined implements, 31, 34
tines, cutting edges of, 193
tip-carts, 31
tires, 127, 492
toolbar, 33
tools, *see* hand-tools
toothed cylinder, 36
topper, 433
topping, 444, 486
topping device, *see* topper
torque amplifier, 116
torque backup, 138
torque reserve, 52, 54, 138
tossing, 444
track laying tractor, 117
track width, 126, 136–7, 157
track-in-track driving, 134
tracks, 494
traction, 118
traction efficiency, 120

traction limit, 142
traction tires, 127–9
tractive force, 22–4
tractor (two axles), 115–76
tractor BUS systems, 613–15
tractor mounted, 441
trailers, 31, 436–8, 463–5
combine harvester and heavy equipment, 465
four wheel tractor, 463–4
grain, 464–5
hydraulic tipper, 464
walking tractor, 463
training, 555
training cattle, 28
tramlines, 234–5
transaxle, 139
transfer function (vibr.), 160
transfer of technology (TT), 554
transmission life, 141
transmission loads, 142–3
transmission maps, 144
transmissions, 139–151
transplanters, 238, 471–2
transplanting of rice, 238
transport, 110, 431, 432, 436–9, 447, 449
transportation, 431
transportation, 472
transportation system, 472
trashing, 432, 433
travel reduction, 118
tree fellers, 484–5
tri-wheeler, 456–7
walking, 456
walking and riding, 456–7
triangular share, 35, 36
tubers, 446–9
turbocharging, 49–51
two-cycle CI engines, 41
two-cycle SI engines, 41
two-wheel drive, 122–4
two-wheel tractors, 95–101
two-wheeled carts, 29, 31
types, 95–101

uneveness (roads), 157
unit factor, 571
upper link sensing (hitch), 170
uprooters, 443
uprooting, 441–4, 446, 452
use of animals, 26–9, 38
use of animals power, 38
use of draft animals, 38–9
used machinery, 583

vacuum, insect control by, 272
valve timing, 42
variable displacement pumps, 149, 174
vegetables, 408, 427
venturi spreader, 265
vermiculite dispenser, 471
vertical load, 23
vibrating picking device, 452
vibrations, 155, 157–61
virtual hitch points, 168
volume, 606
volume mediam diameter (VMD), 275

walking tractor, 115
waste, 440, 446, 451, 452
water transportation, 498
water-raising, 39
wavelength, 157
wear, 190
wedge, 188
weed mapping, 609–11
weeders, 34
weeding, 34, 38
weeding tines, 38
weeding, mechanical, 269–70
blade cultivators, 269
brushing machine, 270
chain harrow, 270
disk weeder, 270
hoeing machine, 269
weighing, 437
weight of implement, 23
wet disc brakes, 132
wet sowing, 35
wet transplanting, 35
wettability, 278
wheels, 31
whole cane, 431, 435, 437, 438
winch skidding, 490–1
windrows, 431, 433–6, 441, 443–5
wing-tined cultivator, 210
wings, *see* share, winged
winnower, 452, 453
winnowing system, 445
withers yokes, 26, 27
wood comminution, 487

wooden spokes, 31
work efficiency of animals, 40
work load (operator), 154
work rate, 300–301
working conditions, 23, 24, 36
working range (transm.), 141
working width, 435, 442, 443, 449
workrate, 452
worldwide, 504

yield mapping, 604–608
yokes, 26

zebus, 28
zero slip (tires), 118, 144